CONCORDIA UNIVERSITY CHICAGO
3 4211 00188 4454
WITHDRAWN
Y0-BEB-366

ANNUAL REVIEW OF EARTH AND PLANETARY SCIENCES

EDITORIAL COMMITTEE (1980)

A. L. ALBEE
J. G. CHARNEY
F. A. DONATH
T. H. JORDAN
A. R. PALMER
G. SCHUBERT
F. G. STEHLI
G. W. WETHERILL

Responsible for the organization of Volume 8
(Editorial Committee, 1978)

J. G. CHARNEY
F. A. DONATH
T. H. JORDAN
A. R. PALMER
G. SCHUBERT
F. G. STEHLI
J. B. THOMPSON JR.
G. W. WETHERILL
A. L. ALBEE (Guest)

Production Editor E. P. BROWER
Indexing Coordinator M. A. GLASS

ANNUAL REVIEW OF EARTH AND PLANETARY SCIENCES

FRED A. DONATH, *Editor*
University of Illinois—Urbana

FRANCIS G. STEHLI, *Associate Editor*
Case Western Reserve University

GEORGE W. WETHERILL, *Associate Editor*
Carnegie Institution of Washington

VOLUME 8

1980

ANNUAL REVIEWS INC. 4139 EL CAMINO WAY PALO ALTO, CALIFORNIA 94306

KLINCK MEMORIAL LIBRARY
Concordia Teachers College
River Forest, Illinois 60305

ANNUAL REVIEWS INC.
Palo Alto, California, USA

COPYRIGHT © 1980 BY ANNUAL REVIEWS INC., PALO ALTO, CALIFORNIA, USA. ALL RIGHTS RESERVED. The appearance of the code at the bottom of the first page of an article in this serial indicates the copyright owner's consent that copies of the article may be made for personal or internal use, or for the personal or internal use of specific clients. This consent is given on the condition, however, that the copier pay the stated per-copy fee of $1.00 per article through the Copyright Clearance Center, Inc. (P.O. Box 765, Schenectady, NY 12301) for copying beyond that permitted by Sections 107 or 108 of the US Copyright Law. The per-copy fee of $1.00 per article also applies to the copying, under the stated conditions, of articles published in any Annual Review serial before January 1, 1978. Individual readers, and nonprofit libraries acting for them, are permitted to make a single copy of an article without charge for use in research or teaching. This consent does not extend to other kinds of copying, such as copying for general distribution, for advertising or promotional purposes, for creating new collective works, or for resale.

REPRINTS The conspicuous number aligned in the margin with the title of each article in this volume is a key for use in ordering reprints. Available reprints are priced at the uniform rate of $1.00 each postpaid. The minimum acceptable reprint order is 5 reprints and/or $5.00 prepaid. A quantity discount is available.

International Standard Serial Number: 0084-6597
International Standard Book Number: 0-8243-2008-5
Library of Congress Catalog Card Number: 72-82137

Annual Reviews Inc. and the Editors of its publications assume no responsibility for the statements expressed by the contributors to this Review.

FILMSET BY TYPESETTING SERVICES LTD, GLASGOW, SCOTLAND
PRINTED AND BOUND IN THE UNITED STATES OF AMERICA

Siebert Grant 128078 G

Annual Review of Earth and Planetary Sciences
Volume 8, 1980

CONTENTS

SOME RELATED ARTICLES APPEARING IN OTHER *ANNUAL REVIEWS*

From the *Annual Review of Astronomy and Astrophysics*, Volume 17 (1979)

Computer Image Processing, Ronald N. Bracewell
Digital Imaging Techniques, W. Kent Ford, Jr.
Martian Meteorology, Conway B. Leovy
Stellar Occultation Studies of the Solar System, James L. Elliot

From the *Annual Review of Ecology and Systematics*, Volume 10 (1979)

The Regulation of Chemical Budgets over the Course of Terrestrial Ecosystem Succession, Eville Gorham, Peter M. Vitousek, and William Reiners
Late Cenozoic Fossil Coleoptera: Evolution, Biogeography, and Ecology, G. R. Coope
The Burgess Shale (Middle Cambrian) Fauna, Simon Conway Morris
Biogeographical Aspects of Speciation in the Southwest Australian Flora, Stephen D. Hopper

From the *Annual Review of Fluid Mechanics*, Volume 12 (1980)

Coastal Circulation and Wind-Induced Currents, Clinton D. Winant
Models of Wind-Driven Currents on the Continental Shelf, J. S. Allen

From the *Annual Review of Materials Science*, Volume 9 (1979)

Acoustic Microscopy with Microwave Frequencies, A. Atalar, V. Jipson, R. Koch, and C. F. Quate

From the *Annual Review of Microbiology*, Volume 33 (1979)

Population Ecology of Nitrifying Bacteria, Lawrence W. Belser
Initiation of Plant Root–Microbe Interactions, Edwin L. Schmidt
Biology of Oligotrophic Bacteria, S. I. Kuznetsov, G. A. Dubinina, and N. A. Lapteva

From the *Annual Review of Physical Chemistry*, Volume 30 (1979)

Kinetics of Thermal Gas Reactions with Application to Stratospheric Chemistry, Frederick Kaufman
Modeling Chemical Processes in the Stratosphere, Julius S. Chang and William H. Duewer

ANNUAL REVIEWS INC. is a nonprofit corporation established to promote the advancement of the sciences. Beginning in 1932 with the *Annual Review of Biochemistry*, the Company has pursued as its principal function the publication of high quality, reasonably priced Annual Review volumes. The volumes are organized by Editors and Editorial Committees who invite qualified authors to contribute critical articles reviewing significant developments within each major discipline.

Annual Reviews Inc. is administered by a Board of Directors whose members serve without compensation. The Board for 1980 is constituted as follows:

Dr. J. Murray Luck, Founder and Director Emeritus of Annual Reviews Inc.
Professor Emeritus of Chemistry, Stanford University

Dr. Joshua Lederberg, President of Annual Reviews Inc.
President, The Rockefeller University

Dr. James E. Howell, Vice President of Annual Reviews Inc.
Professor of Economics, Stanford University

Dr. William O. Baker, *President, Bell Telephone Laboratories*

Dr. Robert W. Berliner, *Dean, Yale University School of Medicine*

Dr. Sidney D. Drell, *Deputy Director, Stanford Linear Accelerator Center*

Dr. Eugene Garfield, *President, Institute for Scientific Information*

Dr. William D. McElroy, *Chancellor, University of California, San Diego*

Dr. William F. Miller, *President, SRI International*

Dr. Colin S. Pittendrigh, *Director, Hopkins Marine Station*

Dr. Esmond E. Snell, *Professor of Microbiology and Chemistry, University of Texas, Austin*

Dr. Harriet A. Zuckerman, *Professor of Sociology, Columbia University*

The management of Annual Reviews Inc. is constituted as follows:

John S. McNeil, Chief Executive Officer and Secretary-Treasurer

William Kaufmann, Editor-in-Chief

Charles S. Stewart, Business Manager

Sharon E. Hawkes, Production Manager

Ruth E. Severance, Promotion Manager

Annual Reviews are published in the following sciences: Anthropology, Astronomy and Astrophysics, Biochemistry, Biophysics and Bioengineering, Earth and Planetary Sciences, Ecology and Systematics, Energy, Entomology, Fluid Mechanics, Genetics, Materials Science, Medicine, Microbiology, Neuroscience, Nuclear and Particle Science, Pharmacology and Toxicology, Physical Chemistry, Physiology, Phytopathology, Plant Physiology, Psychology, Public Health, and Sociology. In addition, five special volumes have been published by Annual Reviews Inc.: *History of Entomology* (1973), *The Excitement and Fascination of Science* (1965), *The Excitement and Fascination of Science, Volume Two* (1978), *Annual Reviews Reprints: Cell Membranes, 1975–1977* (published 1978), and *Annual Reviews Reprints: Immunology, 1977–1979* (published 1980). For the convenience of readers, a detachable order form/envelope is bound into the back of this volume.

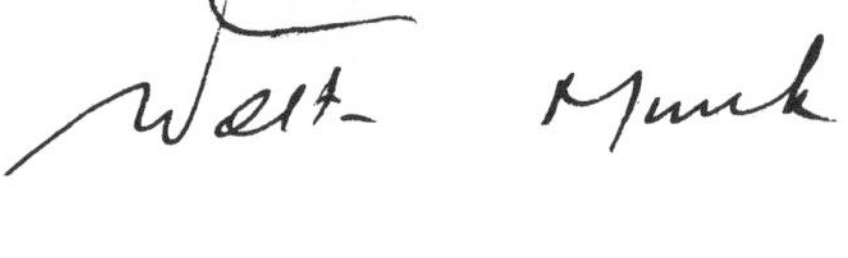

Ann. Rev. Earth Planet. Sci. 1980. 8: 1–16
Copyright © 1980 by Annual Reviews Inc. All rights reserved

AFFAIRS OF THE SEA

×10122

Walter H. Munk
Scripps Institution of Oceanography, University of California–San Diego, La Jolla, California 92093

PROLOGUE

I have been asked by the Editors to describe my experience as a student, teacher, and investigator, and to relate this experience to current trends in our field.

I was an unlikely person for a career in oceanography. When I was born in Austria in 1917 the country had already lost its tenuous hold on a last piece of coastline. My maternal grandfather was a private banker in Vienna, and left enough to provide adequately for his five children, but not for his grandchildren. In 1932 at the age of fourteen I was sent to a boys' preparatory school in upper New York state to finish high school, and to be subsequently apprenticed to a financial firm my grandfather had helped to found. I barely managed the preparatory school; it went bankrupt the following year. I then did a three-year stint at the financial firm; the firm folded the year after.

Somehow I talked my way into Cal Tech and graduated in 1939 in "Applied Physics." At the time I was in love with a Texas girl who vacationed in La Jolla, so I managed to get a job at the Scripps Institution of Oceanography in the summer of 1939. The Texas romance has been outlasted for some forty years by my romance with oceanography. In 1950, when I declined an offer by C.-G. Rossby to join his Department of Meteorology at the University of Chicago, Rossby told me that "anybody with any imagination changes jobs once a decade." But I am still here!

I loved Scripps from the time I spent the first night at the Community House, on the site now occupied by the Institute of Geophysics and Planetary Physics. Harald Sverdrup, the famed Norwegian Arctic explorer, was Director. The staff, including one secretary and a gardener, totalled about 20. After the first summer, I went back to Cal Tech for a Master's Degree in Geophysics under Beno Gutenberg. But next summer

0084-6597/80/0515-0001$01.00

I was back at La Jolla, requesting to be admitted for study towards the PhD Degree in Oceanography. Harald Sverdrup gave my request his silent attention for an interminable minute, and then said that he could not think of any one job in oceanography that would become available in the next ten years. I promptly enrolled, and for some time I constituted the Scripps student body.

This was the time of the German occupation of Austria, and a general war seemed imminent. I enlisted in the US Army, and spent 18 months in the Field Artillery and the Ski Troops. Peacetime service became dull, and I was glad when the opportunity came to join Harald Sverdrup, Roger Revelle, and Richard Fleming in a small oceanographic group at the US Navy Radio and Sound Laboratory at Point Loma (now NUSC). A week later Japan attacked Pearl Harbor. For the next six years I worked on problems of amphibious warfare. I did not get back to the PhD dissertation until 1947, and then only under threat of dismissal. My thesis was written *de novo* in three weeks and is the shortest Scripps dissertation on record. As it turned out, its principal conclusion is wrong (1,2). Last year when UCLA (where Scripps degrees were awarded in the forties) called to offer me the Distinguished Alumnus Award, I thought for a moment they were going to cancel my degree.

During my career I have worked on rather too many topics to have done a thorough job on any one of them; most of my papers have been superseded by subsequent work. But "definitive papers" are usually written when a subject is no longer interesting. If one wishes to have a maximum impact on the *rate* of learning, then one needs to stick out one's neck at an earlier time. Surely those who first pose a pertinent problem should be given some credit, and not just criticized for having failed to provide the final answer.

The underlying thread to my work consists of a theme and some habits. The theme is a kind of Earth spectroscopy: to collect long data series and then to perform spectral analysis of high resolution and reliability. We developed a system of computer programs called BOMM (for Bullard, Oglebay, Munk, and Miller), which at one time was widely used. The procedure has been rewarding in the studies of ocean waves and tsunamis, tides, Earth wobble and spin, variations in gravity, and the scattering of sound and radio waves. Oceanographers have long been familiar with discrete spectra, but when I entered the field the corresponding analysis of continuous (or noisy) processes was not familiar to oceanographers (although it had been applied in optics and acoustics for generations). The reason, I believe, has to do with the difficulty of measuring very low frequencies with resonant analog devices. The change did finally occur, but not until corresponding numerical methods had become accessible through the development of fast-speed computers.

Now to return to the habits:

1. I do not spend much time in polishing lectures. The excuse is that, in a small class, students learn more if they participate in halting derivations and have the joy of pointing out blunders, than if they are handed the subject on a silver platter. As a student, I listened to a series of lectures by a famed Scandinavian geophysicist who had selected each word in advance, including the layout of the blackboard. I found the lectures uninspiring.
2. I become intrigued with new techniques (spectral analysis, array processing, sensitive pressure transducers, radio backscatter, acoustic sensing) before knowing what purpose they might serve. It is a case of a solution looking for a problem. Here my excuse is that if you can apply a significant technical innovation to a field of general interest, then you cannot help but learn new things. I do not propose this procedure to everyone, but for me it has worked well.
3. I do not like to read other people's papers. The outcome has been that I have entered fields with little or no modern literature, to have left them some ten years later in a state of lively participation and an increasing flux of publications.

WAVES AND THE WAR, 1942–1950

In 1942, Harald Sverdrup and I were told of the Allied preparations for an amphibious winter landing on the northwest coast of Africa. The coast is subject to a heavy northwesterly swell, with breakers exceeding six feet on two out of three days during winter. Yet practice landings in the Carolinas were suspended whenever the breakers exceeded six feet because of broaching of the LCVP landing craft. The problem, simply put, was to do the landing during the one-third time when the waves are lowest.

We started work to predict waves on the basis of weather maps. The prediction consisted of three steps. 1. The height H and period T of the storm-generated sea were related to the wind speed U, the storm fetch F, and duration D through four dimensionless relations:

$$\frac{gH}{U^2}, \frac{gT}{U} \quad \text{as functions of} \quad \frac{gF}{U^2}, \frac{gD}{U}.$$

(Waves are either fetch-limited or duration-limited.) We scrounged together observations from oceans, lakes, and wave tanks and came out with rather pleasing scatter plots (3) extending over 6 octaves in gF/U^2. 2. Subsequent attenuation beyond the storm area was estimated from wave dispersion and geometric spreading. 3. The transformation in shallow water was computed from principles of conservation of energy flux (4). This method for predicting sea, swell, and surf was taught to classes of Navy and Air Force weather officers, and was applied widely to amphibious landings in the Pacific and Atlantic theaters of war. For the Normandy invasion, the waves were correctly predicted to be high but manageable.

The empirical relations have held with minor modifications to the present day. The principal shortcoming is that a complex wave spectrum is poorly described by just two parameters, height and period. (Pierson and Neuman subsequently extended the prediction to the entire wave spectrum.) We tried to calibrate the predicted heights and periods in terms of estimates by coxswains during landing exercises, and to compare these estimates to wave records taken at the same time and place. This led to the definitions of *significant* heights and periods as being appropriate to the averages of the highest third of the waves present.

These were exciting and rewarding times. In retrospect, we should not have sanctified our work by calling it a *theory* of wave prediction; it was empiricism, pure and simple, with a few dispersion laws thrown in. There is still no theory giving the observed wave dimensions, notwithstanding the important contributions by Phillips and Miles, though Hasselmann's work on nonlinear wave coupling has come close to providing the right magnitudes.

At the end of this period, we applied what we had learned to geologic processes in shallow water (5), and took a first stab at calculating longshore currents from obliquely incident waves (6). I earned some consulting money by calculating wave forces on offshore structures (7) according to $au^2 + b \cdot \mathrm{d}u/\mathrm{d}t$, with u designating a horizontal component of orbital velocity. The wave climate was established from hindcasts based on historical weather maps. One drilling rig for which I had calculated the wave forces collapsed in a storm.

EARTH WOBBLE AND SPIN, 1950–1960

One way to study the planet Earth is to observe irregularities in its rotation. In this manner one can learn about the growth of the core, the variable distribution of glaciation, air and water mass, global winds, bulk viscosity, and so forth. In each case the information is related to certain integral quantities (moments) taken over the entire globe. This is the weakness of the method—and its strength.

Astronomers were the first to attempt to exploit, for geological purposes, the irregularities they had discovered. They did this in the naive faith that the simplicity of celestial mechanics could be carried over to messy objects like the Earth. To account for inconsistencies in the latitude measurements they spoke of the "proper motion" of observatories; and to explain a rather sudden decrease in the Earth spin around 1920, they raised the Himalayan complex by one foot.

My interest in this subject was first aroused by a statement in 1948 by Victor Starr that a seasonal fluctuation in the net angular momentum of the atmosphere must be accompanied by "... undetectable inequalities in the rate of the Earth's rotation," the net angular momentum of the planet Earth being conserved. How big is undetectable? I started reading about clocks, and learned that M. and Mme. Stoyko of the Bureau de l'Heure in Paris had in fact discovered in 1936 that the length of day in January exceeds that in July by 2 ms. This was based on precision pendulum clocks, and subsequently confirmed with crystal clocks. A very simple calculation showed that this measured variation agreed in amplitude and phase with that inferred from the seasonal wind variation (8). Within a year we learned that the meteorologists had overestimated the strength of the westerly jets, and the inferred variation in the length of day came down by a factor of three (9). Three years later the astronomers found periodic errors in the right ascensions of the FK3 catalogue, and this led to a similar reduction in their estimates. The conclusion was the same as it had been in the first place, that a seasonal variation in the length of day is largely the result of a seasonal variation in the westerly winds (10). And so it stands today.

There was a curious dichotomy between those who measured latitude and so inferred the *wobble* of the Earth relative to a rotation axis fixed in space, and those who compared sidereal time to ephemeral time (later to atomic time) and thus inferred the variable *spin*. But wobble and spin are the three components of one vector, and for geophysical processes the observations of latitude and time had better be discussed together. For example, by considering the relative magnitude of the three components, Revelle and I put some upper bounds on the melting of the Greenland ice cap (11), and we suggested that the surprisingly large decade variation in the Earth's spin (without discernible wobble) could be related to variations in the angular momentum of the fluid core. After five years of scattered publications touching many diverse aspects of geophysics, I joined forces with Gordon MacDonald to prepare an encompassing geophysical discussion of the rotation of the Earth (12). This is now an active field, and I am pleased that a book by Kurt Lambeck is about to come out, following ours by just 20 years. One of the most interesting topics involving the Earth's rotation is the problem of tidal friction. This topic is intimately connected with the evolution of the Earth-Moon system. We estimated the tidal dissipation at 3×10^{12} watts from the modern astronomic observations, about the same from Babylonian eclipses, and 2×10^{12} watts from oceanographic measurements. By now the Babylonian observations have been reworked, additional ancient observations have been uncovered, and the global ocean tidal models have given much tighter estimates. Artificial satellites have provided independent evidence. Prehistoric data on tidal friction are now obtained by counting the number of daily striations per annulation in corals, going back to Devonian times when the year had 400 days. As I understand it, all the evidence is now consistent with a dissipation rate of 3.6×10^{12} watts.

SUN GLITTER AND RADAR CLUTTER

In 1953, Charles Cox and I spent the better part of a year measuring the statistics of surface slopes from photographs of sun glitter. The principle is simple: if the sea were glassy calm, the reflected sun would appear only at the horizontal specular point. In fact, there is a myriad of sun images wherever a surface facet is appropriately tilted to reflect the sun into the camera. The horizontal specular point at the center of the glitter is brightest because the probability of zero slope is highest. The outlying glitter is from the steepest facets that are the least probable. On a windy day when the slopes are relatively steep, the glitter area is large. The slope statistics are readily derived from the distribution of intensity (film density) about the specular point (13).

We managed to get an Air Force B17 (Flying Fortress) and started to work off Monterey. There was plenty of wind, but no sun. We then transferred to Maui, Hawaii, where we had plenty of sun but no wind. Finally, on the last available flying day, we recorded over the Alenuihaha Channel with winds up to 30 knots.

It was found that up/down wind components in mean-square slope exceeded the crosswind components by a factor of two, and that both components increased roughly linearly with wind speed. These results have been found useful in a variety of different problems. Just this year, 25 years since our experiment, Blyth Hughes repeated the measurements by a quite different method, but with substantially the same numerical results.

What was missing in the glitter experiment is a measure of the relative contribution to the mean-square slope from waves of different lengths (e.g. the slope spectrum). It is remarkable that the very concept of continuous (or noisy) spectra was unknown to oceanographers in 1946 (when we worked on wave prediction), at a time when these spectral methods were being routinely applied in optics and acoustics. The reason is, I believe, that spectral intensities are most readily measured with resonant filters, and these are simple in optics and acoustics, but difficult to come by at low oceanographic frequencies. It was not until computer algorithms provided numerical filters, equivalent to the analog filters, that the concepts of continuous spectra were really accepted by the oceanographic community.

The strength of the glitter experiment is that it provides a useful statistics directly, in contrast to the usual procedure for taking long data series and subsequently performing statistical analyses. I was on the lookout then for an equivalent method that could give *spectral* distributions directly. In 1952, Crombie and Barber reported such an experiment. They backscattered a radio wave from the sea surface. For a resonant interaction both wavenumbers and frequencies must add up:

$$\mathbf{k}_1+\mathbf{k}_2=\mathbf{k}_3, \qquad \omega_1+\omega_2=\omega_3.$$

Subscripts 1 and 3 refer to the outgoing and backscattered radio waves, and 2 to the resonant ocean wave. For radio backscatter, $k_1 = -k_3$ and so $k_2 = 2k_3$. The associated ocean wave frequency is $\omega_2 = \sqrt{g/k_2}$. But for resonance $\omega_2 = \omega_3 - \omega_1$ is the radio Doppler shift. Thus for a given radio wavelength $2\pi/k_3$ the radio Doppler shift is predicted at $\sqrt{g/2k_3}$, and this was beautifully confirmed by Doppler measurements.

With modern equipment the Doppler line is 50 dB above background, and there is not much that one cannot do with 50-dB signal-to-noise ratio. Robert Stewart found a slight departure of the measured Doppler from the predicted Doppler, and demonstrated that this slight departure was due to surface current. For a radio wavelength λ and resonant ocean wavelength $\frac{1}{2}\lambda$, the current is measured to an effective depth $\lambda/4\pi$. By conducting the measurements at a series of radio wavelengths, Stewart could estimate the current *shear*. By placing the radio receiver on a moving jeep one can form a synthetic aperture and measure the directional distribution of ocean waves (14). Very roughly this falls off with angle θ measured relative to the wind link $\cos^2(\frac{1}{2}\theta)$ at relatively high winds and high frequencies. We had guessed at something of this sort in our original wave-prediction scheme.

W. Nierenberg and I got a joint Stanford-Scripps effort underway, and participated in some of the analyses. We showed (15) that the backscatter cross section $\sigma = k^4F(k)$ is a direct measure of the saturation constant B for the Phillips surface wave spectrum $F(k) = Bk^{-4}$. This connects two independently measured empirical constants: the typical

"−23 dB" backscatter cross section of the sea surface, and the saturation constant 0.005. I believe that the opportunities for gathering ocean wave statistics by radio backscatter have by no means been exhausted.

SOUTHERN SWELL, 1956–1966

To the delight of the surfing community in California and Hawaii, occasional trains of tremendously long and regular swell roll into shore in the summer months. As part of our wartime wave activity, John Isaacs had arranged for aerial photographs to be taken over California beaches. One of these pictures provided a textbook example of regular undulations being transformed over the shelf. Working backwards and allowing for wave refraction in shallow water, I estimated a deep-water direction of SSW, and a deep-water length of 2000 feet! The inference was that we were seeing the effect of storms in the southern hemisphere winter, some 5000 n. miles away. There had been evidence in the Atlantic of ocean swell coming from very far away, particularly as a result of the work of Norman Barber and Fritz Ursell in the U.K. Could ocean swell provide useful information about distant storms (16)?

Frank Snodgrass came to La Jolla in 1953 in what was to be a wonderful partnership for the two of us for over 23 years. He adapted a "Vibrotron" transducer to measuring pressure fluctuations on the shallow seafloor, the purpose being to explore oscillations with frequencies even lower than those of the swell. We anchored off the Mexican island of Guadalupe for a week (17). Frequency analysis of the records (using an IBM 650 at Convair) showed a series of "events" starting with 1-mm waves of 50-mHz frequency, and ending a few days later with 10-cm amplitudes at 80 mHz. We computed the source distance and time as follows: the group velocity for waves of frequency f is $g/(4\pi f)$. This means that a wave disturbance travels at group velocity v over a distance x in a time $t-t_0$:

$$v = \frac{x}{t-t_0} = \frac{g}{4\pi f}, \quad \text{or} \quad f = \frac{g}{4\pi}\,\frac{t-t_0}{x}.$$

The slope of the line $f(t)$ gives the source distance x, and the zero intercept gives the source time t_0. And here it was, a distant southern swell (even though we had anchored on the eastern side of Guadalupe Island). But there was one problem: one of the events gave a source distance of half the Earth's circumference, and though the Pacific is big, it is not that big. Could it be that the waves had originated in the Indian Ocean and had traveled half way around the world along a great circle route, entering the Pacific between Antarctica and New Zealand?

I had by then become fascinated with array theory developed in radio astronomy, and was anxious to try it out in the oceans. Snodgrass installed a triangular array off-shore from San Clemente Island (18). The results were spectacular. The very far wave sources all were within a beam from 210° and 220° true, which is the angle subtended by the window between Antarctica and New Zealand, as seen from San Clemente.

This led to the final and most ambitious of these undertakings (19). We established six wave stations along a great circle route from New Zealand to Alaska to track wave packets originating in the great southern storm belt. Frank Petersen took the station at Cape Palliser, New Zealand. I monitored from Samoa, where my daughters Edie and Kendall (ages 4 and 2), my wife, and I lived in a palm *fale* in the village of Vailoa Tai at the southwestern coast of Tutuila. Gordon Groves and a radio operator went to the uninhabited equatorial island of Palmyra. Klaus Hasselmann recorded at Hawaii. The floating instrument laboratory FLIP was stationed north of Hawaii, to make up for the lack of coral islands in the cold northern Pacific. Our one graduate student at the time, Gaylord Miller, volunteered for Yakutat, Alaska. Snodgrass established the stations, using seafloor pressure recorders connected by cable to shorebased digital paper punches. We recorded for three months with 98% data return. Gordon Groves got into a battle with his radio operator, whom we had to take off the island by plane. FLIP ran out of cigarettes and I had some problem keeping them on station for another two weeks (via amateur radio).

The stations were spaced with the preconceived view that the principal loss in wave energy would be by wave-wave interaction as the southern swell crossed the tradewind sea. As it turned out, the energy loss was virtually complete within one diameter of the southern storms. Beyond that, for the next 10,000 km, the loss was less than 2 dB and not measurable beyond the effects of geometric spreading and dispersion.[1]

SURFBEATS, EDGE WAVES, AND TSUNAMIS, 1958–1965

Between the ocean swell and the tides there were 10 octaves of unexplored frequency space, only occasionally excited by storm "tides" and by earthquake-generated tsunamis. We found sea level oscillations of 1 to 2

[1] The National Science Foundation funded an educational film, "Waves Across the Pacific," which was widely distributed. I am pleased with the film but not the script; for once I would like to see a film about oceanography which shows it as it is. Things do not turn out as planned; improvisation is a way of life. The final lesson is hardly ever a response to the pre-expedition question.

minute periods at the foot of Scripps pier; these were clearly related to *groups* of incoming swell (20), e.g. the frequency of this "surfbeat" equals the bandwidth Δf of the incoming waves. This experience gave Klaus Hasselmann and me a first opportunity to practice a generalization of power-spectral analysis to nonlinear processes (21), following an important suggestion of John Tukey.

Frank Snodgrass was anxious to apply his experience in measuring bottom pressure fluctuations of low amplitude and low frequency to this part of the ocean wave spectrum. He installed a longshore array of transducers to determine the dispersion relation $\omega(k)$ in the frequency range $\frac{1}{2}$ to 60 cycles per hour (cph). The empirical $\omega(k)$ was then compared to a theoretical $\omega(k)$ for gravitationally trapped edge waves. The agreement is so good (22) that I suspect readers have simply assumed that the plotted curves were fitted to the empirical points, rather than having been derived independently. (This might explain why the paper has not been noticed.) It is my only experience of an oceanographic experiment that gave unequivocal confirmation to a previously derived theory.

This work gave us the impetus to explore the low-frequency wave background on the California continental borderland, away from the coastal edge. The result was a dull, featureless, and quite reproducible spectrum (23) which forms the background to the tsunami studies subsequently conducted by Gaylord Miller (24). The source of this background is not known.

We pushed the measurements to lower and lower frequencies, down to the tides and eventually through and beyond the tidal line spectrum (25).

TIDES, 1965–1975

The incentive to go seriously into tides came from a number of directions. The study of Earth rotation had provided the initial fascination with the global dimensions of bodily and fluid tides. Further, the ultimate limit to the prediction of the tidal line spectrum is set by the low-frequency continuum, and this limit had been ignored by the tidal community, who had been spoiled by a favorable signal-to-noise ratio. David Cartwright and I made a caustic remark (26) that "noise-free processes do not occur except in the literature on tidal phenomena...."

It always pays to know the ultimate limits set by instruments or by nature. Some of the weaker tidal lines routinely included in the harmonic method turned out to be hopelessly contaminated by the noise continuum and might as well be omitted. From these considerations, Cartwright and I proposed a "Response Method of Tidal Prediction," which consists of using station records to compute the transfer function between the tide-producing forces and the station response. This differs somewhat from the classical harmonic method, which independently evaluates the amplitudes and phases of the principal tidal constituents. In some tests conducted by Zetler et al (27) the response method comes out slightly ahead of the harmonic method, but here we have improved one of the few geophysical predictions that already works well.

The third and predominant consideration for working on tides came from an instrumental development. Frank Snodgrass had found that a newly developed quartz crystal pressure transducer was superior to the Vibrotron pressure transducer, our mainstay for some years. Starting in 1965, the quartz transducers were incorporated into capsules freely dropped to the seafloor and subsequently recalled by acoustic command from a surface vessel. (The free-fall technique became commonplace in the early 1970s.) Working with Jim Irish and Mark Wimbush we first did some deep-sea drops off California and located the M_2 amphidrome (the point where the tidal component has zero amplitude) in the Northeast Pacific (28, 29). This was followed by three drops between Australia and Antarctica, spanning the latitudes where the sublunar point travels around the southern oceans at the speed $\sqrt{gh}$ of free waves (30). A very naïve theory predicts a resonant amplification at such latitudes. We didn't believe the theory, but made the measurements anyhow. The result was a rather dull transition from south Australian to Antarctic tides.

We had organized an international SCOR working group on deep-sea tides, and numerous measurements were being made, particularly by Cartwright in the U. K., and by Mofjeld of NOAA, Miami. Snodgrass participated in an international calibration experiment in the Bay of Biscay. The latest IAPSO publication shows 108 pelagic tide stations have by now been occupied by a number of different investigators. The results have been useful as a check on the numerical modeling of tides.

Our last drops were made in 1974 south of Bermuda in $5\frac{1}{2}$ km of water, as part of the MODE bottom experiment. We discovered unexpected and still unexplained pressure fluctuations at subtidal frequencies that are coherent over 1000 km (31)! With regard to tides, an analysis led by B. Zetler was in splendid agreement with the traditional Atlantic cotidal charts (32). Two independent drops in the same area gave the following M_2 amplitudes and phases:

32.067 cm and 2.5° Greenwich epoch,
32.074 cm and 2.6° Greenwich epoch.

When it comes to four-figure accuracies, it is no longer oceanography. Further, satellite altimetry looked increasingly promising for future measurements of deep-sea tides. It was time to move on.

INTERNAL WAVES, 1971–1978

In 1958, Owen Phillips proposed from simple dimensional considerations that the distribution of surface elevation variance with wavenumber k varies as k^{-4} (L^2 per unit k_x per unit k_y). This "saturation spectrum" has turned out to be a most useful representation of high-frequency surface waves. Could something as simple and as useful be done about internal waves?

Christopher Garrett and I looked at existing evidence and found it consistent with a spectrum that falls off with horizontal wavenumber as k^{-2} and with vertical wavenumber as m^{-2}. The original model proposed in 1972 (33) has gone through a series of revisions, which have been referred to as GM75, GM79 ... to make explicit the built-in obsolescence. The surprising thing has been the degree of universality of the model spectrum. This indicates a saturation as in the case of the Phillips surface wave spectrum. But whereas the Phillips spectrum is white in curvature (and vertical acceleration), the internal wave spectrum is white in shear and isopycnal slope, suggesting a different saturation process. The entire ocean column is never very far from instability, and occasional internal breakers may play an important role in turbulence and finescale mixing. The essential work remains to be done.

OCEAN ACOUSTICS, 1975–

Regardless of the role played by internal waves in ocean microprocesses, there can be no doubt that they are a dominant source of fluctuation in sound speed. Clark and others have recorded time series of acoustic phase and intensity over a 1000-km transmission path between Eleuthera and Bermuda. From these observations one can infer the mean-square phase rate along any one of the multiple paths that connect source and receiver. The result is $\langle \dot{\phi}^2 \rangle = 1.6 \times 10^{-5}$ sec^{-2}. Fred Zachariasen and I have derived the theory for computing this parameter, given only the mean sound speed structure and an internal wave spectrum (34). For GM75, the result is $\langle \dot{\phi}^2 \rangle = 2.5 \times 10^{-5}$ sec^{-2}. This was the beginning of a major effort led by Roger Dashen and Stan Flatté to derive sound transmission statistics, given the spectrum of the variability in sound speed in ocean space and time (35). Since WWII the acoustic and oceanographic communities have gone their separate ways; I think that we have made a contribution towards bridging this gap.

In 1976, Peter Worcester and Frank Snodgrass set two deep moorings, with an acoustic source and receiver on each mooring. Oppositely directed acoustic transmissions gave information about the 25 km of intervening ocean. Variations in the *average* of the two travel times told something of the fluctuations in the temperature structure; *differences* in the travel times (with and against the current component) gave information about the water movements. This is a powerful technique for measuring ocean features on a scale of tens of kilometers, and it wetted our appetite for acoustic monitoring of the intense mesoscale features, with typical dimensions of 100 km.

At a range of 1000 km an ideal acoustic pulse is received as a complex series of subpulses, one along each of a series of multipaths. Our work predicts the effective spread of the subpulses, and hence the time *resolution* between separate paths. It also predicts the decorrelation time of the pulse structure, and hence the interval at which independent samples are taken. Typical values are 50 ms and 5 minutes. On this basis, Carl Wunsch and I have estimated that we can measure week-to-week fluctuations in acoustic travel time along a fixed path to an accuracy of 20 ms. But the expected variations from mesoscale ocean eddies are many times this large. Accordingly, we have proposed to measure the variable travel times between a series of moored acoustic sources each transmitting to a series of acoustic receivers, and then to construct three-dimensional charts of sound speed (essentially temperature) by an appropriate inverse theory (36). The idea is very simple; in the case of a warm eddy (say) all those rays that pass through the eddy will come in early by something like a quarter second, whereas the other transmissions are not affected. If the eddy is shallow, then only the early steep paths are affected; if it is deep, then the late flat paths (near the sound axis) are affected as well.

We have formed a joint venture involving Robert Spindel of Woods Hole, Carl Wunsch of MIT, Birdsall at Michigan, and our group at Scripps. In November, 1977, Spindel put out a 2000-m deep mooring south of Bermuda, which our graduate student, John Spiesberger, monitored at a coastal station 1000 km distant. There are about a dozen distinct arrivals, and those can be clearly traced over the two-month transmission period. The identification of each of these arrivals with a distinct ray path remains to be done; this will be my principal task in 1979. If all goes well, we shall put out 4 sources and 6 receivers in late 1980 and attempt to monitor a MODE-sized area by acoustic means over a period of four months.

So much for the main topics that have kept me busy. Nearly all the work has been done in collaboration with others; the bibliography at the end of the chapter is a way to make explicit my indebtedness to so many people. My principal collaborators have been Roger Revelle, Charles

Cox, George Carrier, Klaus Hasselmann, Gordon MacDonald, David Cartwright, Bernard Zetler, Fred Zachariasen, and Chris Garrett. My partnership with Frank Snodgrass lasted through 23 happy and constructive years. He retired in 1976 to become a farmer in Oregon, and I have never quite recovered from this loss.

I have two or three graduate students at a time (the most I can manage), and work closely with them. Among them have been Gordon Groves, Earl Gossard, Charles Cox, June Pattullo, Mohammed Hassan, Gaylord Miller, Mark Wimbush, Jim Irish, Jim Cairns, Gordon Williams, Peter Worcester, and now John Spiesberger and Mike Brown. I have learned more from them than they have learned from me.

Among my teachers are threee men in particular: Harald Sverdrup taught me how to write, and how to treat each observation with great care and respect (so much of this is lost in computer analyses and plots). Roger Revelle introduced me to the romance of work at sea, and showed me his style of broad-range inquiry. Carl Eckart taught me some classical physics. I have always regretted that I did not learn more physics before becoming absorbed in oceanography. I also regret that I am so poor at building and repairing gear (I was sheltered from this as a boy).

In 1958 we started a branch of the University-wide Institute of Geophysics (later Geophysics and Planetary Physics) on the Scripps campus. Roger Revelle and Louis Slichter made this possible. My wife, Judith, chose the laboratory site and the multi-level, one-story redwood construction. At the time, I was working on solid-earth geophysics, and this is reflected by the early appointments. I became rather lonely when my interest returned to the sea. We are now fairly evenly divided, with Freeman Gilbert looking after solid-earth geophysics. The birth and coming-of-age of IGPP has been one of my most rewarding experiences.

This biographical sketch would be unbalanced without some comments on an association with the United States Navy which spans my entire career (except for an interlude in World War II when my security clearance was suddenly withdrawn). The Office of Naval Research has given our work generous and effective support ever since ONR was formed, not only with money but in other ways as well. I owe a deep gratitude to this remarkable organization. At the same time I have been able to serve the Navy in different ways. In 1946, Bill Von Arx and I surveyed the circulation of Bikini lagoon and assessed its flushing rate prior to an underwater nuclear explosion. In 1951, working with Roger Revelle, John Isaacs, Willard Bascom, and Norman Holter, we monitored at close range the oceanographic effects of a very large thermonuclear explosion. And in recent years, largely through my association with JASON, I have been involved in a diverse set of Navy problems.

The 1951 nuclear test IVY-MIKE almost brought my scientific career to an end. Revelle, Isaacs, and I had expressed to high authority our fear that the thermonuclear shock to which Eniwetok Atoll was to be subjected might trigger a submarine landslide.[2] This, in turn, could generate a tsunami of destructive dimensions over much of the Pacific. Accordingly quiet plans were made for a possible evacuation of many low-lying areas all over the Pacific. Scripps moored two small rafts to a nearby seamount 36 miles from ground zero, with wave instruments attached to each mooring. I was aboard one of the rafts. The Scripps vessel HORIZON stood within sight of both rafts. Observers on the rafts were to signal any suspicious event to the HORIZON, which maintained open contact to the Flag Ship MT. McKINLEY, so that signals could flow instantly to Navy personnel standing by at the evacuation sites.

I should stress that the probability for a destructive wave was very, very small, and in fact nothing happened. After witnessing the explosion at this close range, and seeing no wave signal for 11 minutes thereafter (the computed time was 6 minutes following the landslide), we transferred to the HORIZON and steamed north at full speed to avoid radioactive fallout (unsuccessfully as it turned out). We returned in two days to pick up the rafts and instrumentation. I unspooled the records, checking the time marks made prior to my leaving the raft. Within 90 seconds following the final time mark, my wave recorder had malfunctioned, giving a signature equivalent to a huge tidal wave. It is true the "event" occurred too late to be consistent with computations, but I rather think that under the existing stress (and having in mind the possibility of a delayed landslide), had I seen this record signature I would have given the signal, and thus set into motion the evacuation of thousands of people from hundreds of sites. Under the circumstances, I would have left the expedition at the first landfall, and not come home.

S. Vigilio di Marebbe
January, 1979

[2] There are very few earthquakes in the area of the Pacific atolls.

Literature Cited

1. Munk, W. H. 1947. Increase in the period of waves traveling over large distances; with application to tsunamis, swell, and seismic surface waves. *Trans. Am. Geophys. Union* 28:198–217
2. Munk, W. H. 1949. Note on period increase of waves. *Bull. Seismol. Soc. Am.* 39:41–45
3. Sverdrup, H. U., Munk, W. H. 1946a. Empirical and theoretical relations between wind, sea and swell. *Trans. Am. Geophys. Union* 27:823–27
4. Sverdrup, H. U., Munk, W. H. 1946b. Theoretical and empirical relations in forecasting breakers and surf. *Trans. Am. Geophys. Union* 27:828–36
5. Munk, W. H., Traylor, M. A. 1947. Refraction of ocean waves: a process linking underwater topography to beach erosion. *J. Geol.* LV:1–26

6. Putnam, J. A., Munk, W. H., Traylor, M. A. 1949. The prediction of longshore currents. *Trans. Am. Geophys. Union* 30: 337–45
7. Munk, W. H. 1948. Wave action on structures. *Petrol. Tech.* 11: 1–18
8. Munk, W. H., Miller, R. L. 1950. Variations in the Earth's angular velocity resulting from fluctuations in atmospheric and oceanic circulation. *Tellus* 2: 93–101
9. Mintz, Y., Munk, W. H. 1951. The effect of winds and tides on the length of day. *Tellus* 3: 117–21
10. Mintz, Y., Munk, W. 1954. The effect of winds and bodily tides on the annual variation in the length of day. *Mon. Not. R. Astron. Soc., Geophys. Suppl.* 6: 566–78
11. Munk, W., Revelle, R. 1952. On the geophysical interpretation of irregularities in the rotation of the Earth. *Mon. Not. R. Astron. Soc., Geophys. Suppl.* 6: 331–47
12. Munk, W., MacDonald, G. J. F. 1960. *The Rotation of the Earth: A Geophysical Discussion.* Cambridge Univ. Press. 323 pp.
13. Cox, C., Munk, W. 1954. Statistics of the sea surface derived from sun glitter. *J. Mar. Res.* 13: 198–227
14. Tyler, G. L., Teague, C. C., Stewart, R. H., Peterson, A. M., Munk, W. H., Joy, J. W. 1974. Wave directional spectra from synthetic aperture observations of radio scatter. *Deep-Sea Res.* 21: 989–1016
15. Munk, W., Nierenberg, W. A. 1969. HF radar sea return and the Phillips saturation constant. *Nature* 224: 1285
16. Munk, W. 1951. Ocean waves as a meteorological tool. *Compendium Meteorol.* 1090–1100
17. Munk, W. H., Snodgrass, F. E. 1957. Measurements of southern swell at Guadalupe Island. *Deep-Sea Res.* 4: 272–86
18. Munk, W. H., Miller, G. R., Snodgrass, F. E., Barber, N. F. 1963. Directional recording of swell from distant storms. *Philos. Trans. R. Soc. London Ser. A* 255: 505–84
19. Snodgrass, F. E., Groves, G. W., Hasselmann, K. F., Miller, G. R., Munk, W. H., Powers, W. H. 1966. Propagation of ocean swell across the Pacific. *Philos. Trans. R. Soc. London Ser. A* 259: 431–97
20. Munk, W. H. 1949. Surf beats. *Trans. Am. Geophys. Union* 30: 849–54
21. Hasselmann, K., Munk, W. H., MacDonald, G. J. F. 1963. Chapter 8. In *Time Series Analysis*, ed. M. Rosenblatt, pp. 125–139. New York: Wiley
22. Munk, W. H., Snodgrass, F. E., Gilbert, F. 1964. Long waves on the continental shelf: an experiment to separate trapped and leaky modes. *J. Fluid Mech.* 20: 529–54
23. Snodgrass, F. E., Munk, W. H., Miller, G. R. 1962. Long-period waves over California's continental borderland. Part I. Background spectra. *J. Mar. Res.* 20: 3–30
24. Miller, G. R., Munk, W. H., Snodgrass, F. E. 1962. Long-period waves over California's continental borderland. Part II. Tsunamis. *J. Mar. Res.* 20: 31–41
25. Munk, W. H., Bullard, E. C. 1963. Patching the long-wave spectrum across the tides. *J. Geophys. Res.* 68: 3627–34
26. Munk, W. H., Cartwright, D. 1966. Tidal spectroscopy and prediction. *Philos. Trans. R. Soc. London Ser A* 259: 533–81
27. Zetler, B., Cartwright, D., Berkman, S. 1979. Some comparisons of response and harmonic tide predictions. *Int. Hydrogr. Rev.* In press
28. Munk, W. H., Snodgrass, F. E., Wimbush, M. 1970. Tides off shore–transition from California coastal to deep-sea waters. *Geophys. Fluid Dynam.* 1: 161–235
29. Irish, J. D., Munk, W. H., Snodgrass, F. E. 1971. M_2 amphidrome in the northeast Pacific. *Geophys. Fluid Dynam.* 2: 355–60
30. Irish, J. D., Snodgrass, F. E. 1972. Australian-Antarctic tides. *Am. Geophys. Union Antarctic Res. Ser.* 19: 101–16
31. Brown, W., Munk, W. H., Snodgrass, F., Mofjeld, H., Zetler, B. 1975. MODE bottom experiment. *J. Phys. Oceanogr.* 5: 75–85
32. Zetler, B., Munk, W. H., Mofjeld, H., Brown, W., Dormer, F. 1975. MODE tides. *J. Phys. Oceanogr.* 5: 430–41
33. Garrett, C. J. R., Munk, W. H. 1972. Space-time scales of internal waves. *Geophys. Fluid Dynam.* 2: 225–64
34. Munk, W. H., Zachariasen, F. 1976. Sound propagation through a fluctuating stratified ocean: theory and observation. *J. Acoust. Soc. Am.* 59: 818–38
35. Flatté, S. M., ed., Dashen, R., Munk, W. H., Watson, K. M., Zachariasen, F. 1979. *Sound Transmission Through a Fluctuating Ocean.* Cambridge Univ. Press
36. Munk, W. H., Wunsch, C. 1979. Ocean acoustic tomography: a scheme for large scale monitoring. *Deep-Sea Res.* 26A: 123–61

Ann. Rev. Earth Planet. Sci. 1980. 8:17–34
Copyright © 1980 by Annual Reviews Inc. All rights reserved

PLATFORM BASINS[1]

×10123

Norman H. Sleep

Department of Geophysics and Department of Geology, Stanford University, Stanford, California 94305

Jeffrey A. Nunn and Lei Chou

Department of Geological Sciences, Northwestern University, Evanston, Illinois 60201

INTRODUCTION

Thick sedimentary deposits have formed on continental crust throughout much of geological time. These basins can be conveniently classified into (*a*) Atlantic marginal basins, such as the continental shelf of the Eastern United States, (*b*) interior platform basins such as the Michigan basin (Figures 1 and 2), and (*c*) fault-controlled or rift basins, such as the Triassic grabens of the Eastern United States and the Boston basin.

Fault-controlled basins are clearly related to horizontal deviatoric stresses in the lithosphere. They are often directly associated with failed branches of rift systems which become mid-oceanic ridges (Burke & Dewey 1973). The mechanics of rift basins have been reviewed by Bott (1979) and will not be further discussed in this paper.

The similarity in the subsidence history of Atlantic-type margins and interior platform basins has been recognized for some time. These basins are a few hundred kilometers wide and contain a few kilometers of flat-lying sedimentary rocks. Sedimentary units generally thicken toward the center of the basins. This thickening, general concordance of facies and isopachs, and the presence of deeper water facies near the basin centers indicate that these basins subsided relative to their surroundings concomitant with deposition. Faulting is often present, but does not dominate the structure of the basin. The region of greatest deposition is generally consistent for successive units for periods of over a hundred million years.

Any physical theory of platform basin subsidence must explain continual subsidence for periods of greater than 100 million years. Inter-

[1] This work has been supported by NSF Grant EAR 78-13644.

0084-6597/80/0515-0017$01.00

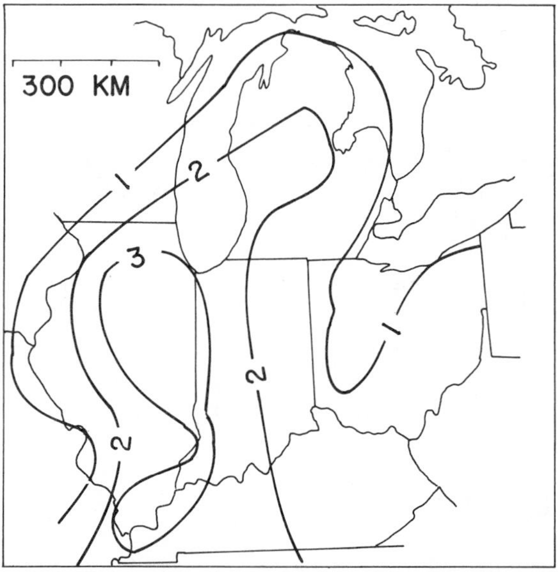

Figure 1 Index map showing isopachs of Cambrian sediments in kilofeet (mainly after Cook 1975). Note that the Michigan basin is not well defined in this time interval.

ruptions in sedimentation and changes in basin geometry with time must either be explained by the physical mechanism or attributed to sedimentological causes. Mechanisms proposed for Atlantic-type margins which depend on the adjacent oceanic lithosphere cannot explain the subsidence of interior basins. A plate tectonic explanation of interior basins is not obvious.

MECHANISMS OF SUBSIDENCE

In older literature the origin of mountain belts and the differences between oceans and continents are attributed to deformations caused by the cooling of the earth. Platform tectonics was considered insignificant and thus was not usually the subject of speculation.

Sediment Loading

The accumulation of sediments in an initially small depression loads the lithosphere and causes further subsidence. The amount of accumulation in a continental interior region that is initially near sea level is quite limited if the region remains in isostatic balance. For a process occurring beneath water the total accumulation of sediment is $(\rho_m - \rho_w)/$

$(\rho_m - \rho_s)$ times the initial depth of the depression, where ρ is density and subscripts m, w, and s indicate mantle, water, and sediment. For a mantle density of 3.3 gm/cm^3 and a crustal density of 2.6 gm/cm^3, a factor of 3.3 results. An improbably deep initial depression of one kilometer would thus be needed to explain the three to four kilometers of sediment in the Michigan basin.

Sediment loading does greatly amplify subsidence caused by other processes. By assuming appropriate mechanical properties of the lithosphere, the subsidence due to sediment loading can be removed to obtain the "driving load" causing subsidence (Watts & Ryan 1976). The driving load on the Atlantic shelf of the United States underlies the shelf. This precludes a major contribution to subsidence of the shelf by sediment loads on adjacent oceanic crust as proposed by Dietz (1963) and Walcott (1972).

Thermal Contraction

Oceanic crust subsides as it spreads away from mid-oceanic ridges due to thermal contraction of the lithosphere. Because Atlantic-type margins

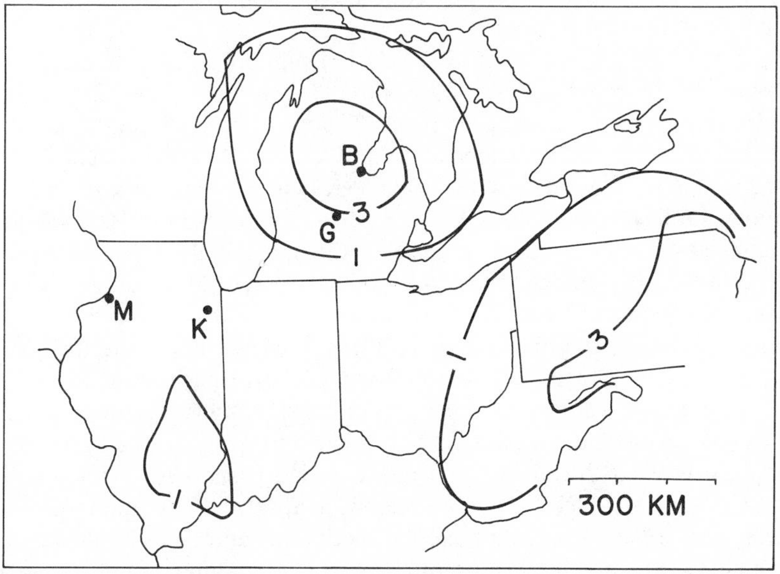

Figure 2 Index map showing isopachs of Silurian sediments in the Michigan basin and in the Appalachian basin and an isopach of Pennsylvanian sediments in the Illinois basin (mainly after Cook 1975). Regions of Middle Ordovician through Devonian subsidence are indicated by the former isopachs. Carboniferous subsidence occurred in southern Illinois. Locations of Pennsylvanian coal are indicated by letters (B, Bay City; G, Grand Ledge; K, Kankakee; and M, Moline).

form when a new mid-oceanic ridge begins spreading beneath a pre-existing continent, the subsidence of Atlantic shelf basins may be reasonably attributed to thermal contraction of lithosphere following the break-up (Vogt & Ostenso 1967, Schneider 1969, Sleep 1971). Subsidence of interior platform basins has been attributed to thermal contraction of the lithosphere (Sleep 1971, Sleep & Snell 1976, Haxby et al 1976) even though a heating event is not obvious.

Replacement of the mantle lithosphere with hot material from depth is necessary to produce enough expansion that subsequent contraction may cause a basin to form (Sleep & Snell 1976). Uplifts and subsequent subsidence have also been attributed to thermal expansion in oceanic areas such as the Hawaiian swell (Crough 1978).

Thermal contraction is easily modeled mathematically. If the basin always remains filled to sea level and no eustatic sea-level changes, sediment compaction, or variations in sediment density occur, the depth to a bed in a well would be

$$\text{Depth} = U_0[\exp(\text{age}/t_0) - 1] \tag{1}$$

where U_0 is a slowly varying function of position and t_0 is the thermal time constant of the lithosphere.

Sleep (1971) showed that this relationship provided a good fit to observed data from the East Coast of the United States, Gulf Coast, and Florida and the deviations from the fit could be attributed to known variations in the depth of deposition and known eustatic changes in sea level. He also showed that the subsidence of interior platform basins follows similar curves. Watts & Ryan (1976) and Steckler & Watts (1978) examined subsidence data from the East Coast in detail, made corrections for compaction, the rigidity of the lithosphere, eustatic changes of sea level, and depth of deposition using newly available sedimentological data, and showed that the driving load was probably thermal contraction.

For thermal contraction to produce a platform basin it is necessary that the freeboard of the crust be reduced. Otherwise, the uplifted material would contract back to its original level prior to the expansion. Surficial erosion during thermal uplift is one process for thinning the crust (Hsu 1965, Sleep 1971, LePichon et al 1973). For uplifts of one kilometer, erosion may thin the crust so that around three kilometers of sediments accumulate. The amount of crustal thinning is sensitive to the rate at which erosion occurs. Mathematical models to determine the duration and amount of erosion are easily constructed (LePichon et al 1973).

Surficial erosion is insufficient to explain the eventual amount of subsidence in many basins. There is no evidence of extensive erosion immediately before post-middle Ordovician subsidence in the Michigan basin (Sleep & Snell 1976), Carboniferous subsidence in the Illinois basin and in the basins in Kansas (Sleep 1971), and Mesozoic subsidence in the North Sea (McKenzie 1978). The amount of subsidence of Atlantic shelf basins is too large and the break-up unconformity too brief for surficial erosion to have thinned the crust (Steckler & Watts 1978, McKenzie 1978, Bott 1979). Subsurface processes, therefore, must have thinned or loaded the crust of these basins.

Phase Changes

Bowie (1927) proposed that phase changes within the earth's crust or upper mantle amplify the subsidence due to sediment loading. The process begins when minor initial declivity accumulates sediments which then load the crust and increase lithostatic pressure at depth. The increased pressure causes phase changes at depth which increase the density of the crust and cause further subsidence. The amplification continues until a great thickness of sediments is deposited. After a period of time, the geothermal gradient at depth comes into equilibrium with the sediment pile near the surface and leads to temperature increases at depth. The increase in temperature reverses the phase change and expands the crust. Erosion of the uplift related to the expansion thereby lowers the pressure causing additional reversal of the phase change and more uplift. The initial accumulation of sediment is thus uplifted and the cycle may repeat itself when the geothermal gradient comes into equilibrium with the eroded uplift. Numerical modeling of this hypothesis is possible but quite difficult (Joyner 1967, van de Lindt 1967, O'Connell & Wasserburg 1967, Mareschal & Gangi 1977).

The hypothesis has the advantage that the lithosphere is in isostatic equilibrium throughout the subsidence and uplift processes. Uplift (and presumably orogeny) is inherent to this mechanism. The main difficulty with the hypothesis is that uplift does not always follow subsidence. Platform basins such as the Belt series and the Michigan basin remained undeformed for periods of hundreds of million years. Orogenic deformation appears to be associated with plate boundaries and not directly dependent on sedimentary deposits. Plateau uplift affects areas without regard to the thickness of sedimentary cover. The uppermost mantle and lower crust are also not likely to be made of materials, such as basalt or eclogite, that can undergo volumetrically significant phase changes (see, for example, Ringwood 1975).

Phase changes (or metamorphism) may also densify the lower crust concomitant with a heating event (Falvey 1974). At least some sections of a new Atlantic margin, however, would be expected to have lower crustal rock types unsuitable for this mechanism. In particular, older terrains would have been heated and metamorphosed sometime in their pasts. The amount of subsidence would be expected to be strongly dependent on basement geological province if this mechanism were important. Haxby et al (1976) avoid this difficulty by proposing that the basaltic material intruded during the heating event subsequently transforms to eclogite. Hinze et al (1972) favor slowly reacting phase changes to produce gradual subsidence.

Crustal Stretching and Loading

During continental break-up the edge of the continental crust is stretched and loaded with igneous intrusions (Artenjev & Artyushkov 1971, McKenzie 1978, Bott 1979). Thinned and faulted crustal blocks have been observed seismically off the Bay of Biscay margin of France (de Charpal et al 1978). The tendency for light continental crust to flow out over the more dense oceanic crust like oil over water is expected to further thin the crust at the margin (Bott 1971, 1979, Bott & Dean 1972). These processes should become inactive once the lithosphere at the margin has cooled to subcrustal depths.

Crustal stretching is geometrically unlikely for an interior platform basin where no free edge of the continental crust exists. Haxby et al (1976) suggest that loading concomitant with heating is caused by basaltic intrusions which crystallize to eclogite. Removal of the lowermost crust into the mantle during the heating event is another possibility.

A rift valley or graben underlies certain platform basins such as the North Sea. Beaumont (1978) has suggested that regional isostatic compensation of the thinned crust, along with sediment load within the graben, produced the subsidence in the North Sea. Continued crustal thinning beneath the graben (or thermal contraction) is necessary to explain the observed duration of subsidence.

Sublithospheric Processes

Change in the density of the asthenosphere might cause subsidence. Low angle subduction may replace the asthenosphere with the slab and cause rapid subsidence (Cross & Pilger 1978). This mechanism might be applicable to the more gradual subsidence of other basins.

Melt migration in the asthenosphere is another possible cause of basin subsidence that is difficult to model physically (Scheidegger & O'Keefe 1967, Sloss & Speed 1974).

MICHIGAN AND ILLINOIS BASINS

Of the mechanisms discussed above only thermal contraction (along with redistribution of the load by the flexural rigidity of the lithosphere and amplification of the load by sediment accumulation) is well enough understood to be compared with observed data. Paleozoic subsidence in the Michigan and Illinois basins is discussed below to appraise the relevance of thermal contraction to subsidence and to illustrate the difficulties of interpreting actual data (Figures 1 and 2).

Subsidence in the Michigan and Illinois basins began in Late Cambrian time. The bulk of subsidence in the Michigan basin occurred from middle Ordovician through Devonian time (Figure 3). In Illinois rapid subsidence occurred from upper Cambrian through Ordovician time and again in Carboniferous time (Figure 3). Cretaceous and Tertiary sediments of the Mississippi embayment extend into Southern Illinois. The basins have not been highly tectonized subsequent to deposition. The Southern part of the Illinois basin, however, is faulted and the rocks south of the basin have been uplifted after deposition. The Michigan basin thus is considered in more detail.

Episodes of Subsidence

Determination of the duration of independent subsidence episodes is central to determining the mechanism of subsidence. Thermal contraction continues after the heating event and the relative subsidence of the basin continues whether or not the basin is exposed by a eustatic lowering of sea level. Sloss & Speed (1974) consider that the sequences of sediments between cratonwide unconformities also reflect separate mechanical episodes of subsidence. McGinnis (1970) and Ervin & McGinnis (1975) consider these episodes to be caused by dense intrusions inferred from gravity surveys and by variations in the stress transmitted from nearby plate margins.

Sleep (1976) showed that reasonable eustatic changes in sea level (300 m peak-to-peak measured on land) could cause the major unconformities observed in rapidly subsiding platform basins. The Illinois basin continued to subside relative to its flanks during the sub-Pennsylvanian unconformity. Sedimentary evidence on whether subsidence continued during other major unconformities is not available.

Different sequences of heating events are needed for the Illinois and Michigan basins. The subsidence in Illinois between upper Cambrian and lower Mississippian time can be fit with a single thermal subsidence

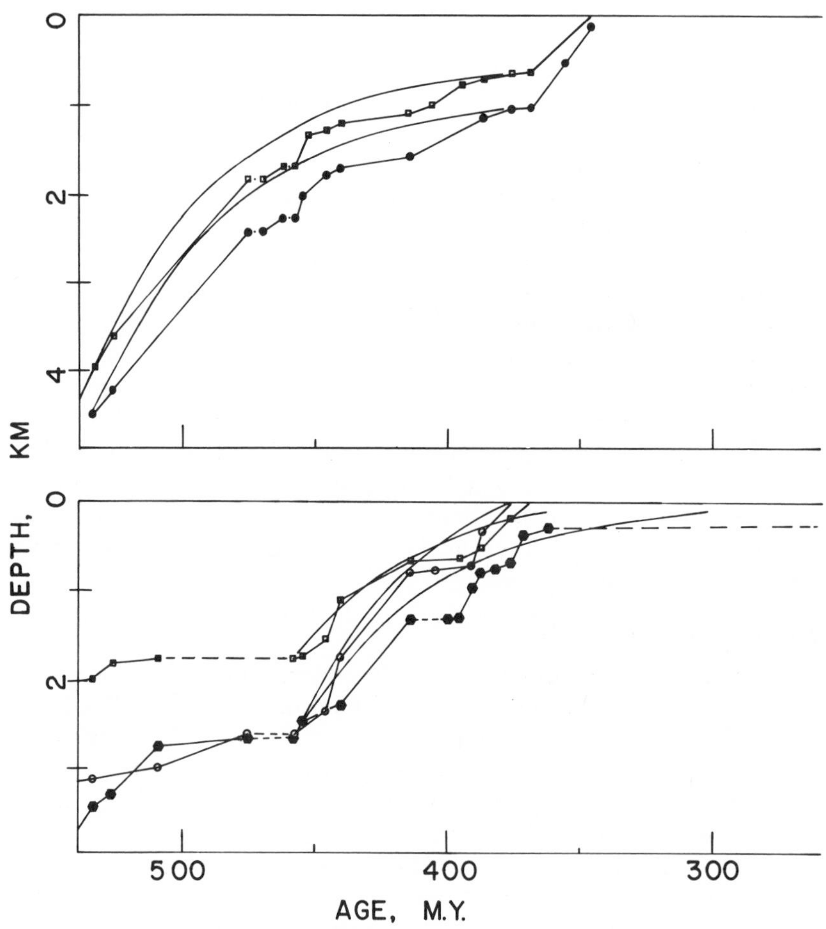

Figure 3 Depth to horizons in wells is plotted as a function of absolute age. Fifty-million-year exponentials expected if thermal contraction causes subsidence are shown for reference (*smooth curves*). In Illinois pre–Middle Ordovician (before 458 m.y.) subsidence predominates and renewed subsidence occurs in the Carboniferous period (*above*). In Michigan (*solid hexagons, below*) and the Appalachian basin (*open symbols, below*) pre–Middle Ordovician subsidence is less than later subsidence. *Horizontal dashed lines* indicate major unconformities. Data sources: *open squares above*, Johnson County Illinois, B. Farley #1, tops by E. Atherton, Illinois Geological Survey (unpublished); *solid circles above*, Pope County Illinois, M. L. Streich Comm. #1, tops by E. Atherton, Illinois Geological Survey (unpublished); *solid hexagons below*, Michigan (Hinze et al 1978); *open squares below*, Ohio, well 6-6 (Adkinson 1966); *open circles below*, New York, well 7-6 (Adkinson 1966).

curve (Figure 3), but a second event is needed for rapid Carboniferous subsidence. The gradual subsidence in Michigan before Middle Ordovician cannot be associated with the same heating event as later subsidence. The need for multiple heating events decreases the attractiveness of the thermal contraction hypothesis. The simplest sequence of events compatible with thermal contraction can be envisioned as follows:

1. The principal heating event occurred in middle to upper Cambrian time before the deposition of the widespread basal sandstones. The crust was thinned or loaded enough for a few kilometers of subsequent sedimentation to occur. The area of eventual subsidence related to this event extended throughout Michigan, Illinois, Indiana, and southward to the continental margin (Cook 1975). Igneous activity conceivably related to this event occurred on the northeast periphery of the Michigan basin (Doig 1970) and rift valleys with igneous rocks occurred in the Arbuckle region of Oklahoma (Cook 1975). The Reelfoot rift in southern Illinois as well as the Rome trough to the east may be associated with this event (Ervin & McGinnis 1975, Buschbach & Atherton 1979), but hard data is not available. Paleomagnetic poles obtained from Keweenawan(?) rocks at four kilometers depth in the Michigan basin are compatible with remagnetization during a middle or late Cambrian heating event, although the heating event could have been somewhat earlier (van der Voo & Watts 1978). The Illinois basin continued to subside from thermal contraction until it was affected by Carboniferous events.
2. A second heating event occurred in the Michigan basin and the Appalachian basin in Ohio and Western Pennsylvania and Western New York. The subsidence in the Appalachian basin resembles the subsidence in Michigan in that the Cambrian and Lower Ordovician units are thin compared with the Middle Ordovician units. The Appalachian basin became overfilled with sediments following the Acadian orogeny (see Schepp 1963) and ceased to accumulate sediments after that time. Except for minor Carboniferous faulting, the Michigan basin subsided due to thermal contraction without further interruption. There is no direct igneous evidence for this heating event.
3. Carboniferous subsidence was related to a heating event in Southern Illinois and to the south where the basin deepened in the direction of preserved drainage patterns (Howard 1979). The Carboniferous region of subsidence is more spatially restricted than either of the two earlier events.
4. The heating event related to break-up of the Gulf of Mexico produced the Mississippi embayment, which extends into Southern Illinois.

Some igneous activity is associated with time event but most of the igneous bodies in Southern Illinois are lower Paleozoic (see Ervin & McGinnis 1975 for a review).

The absence of direct evidence of all but the pre–upper Cambrian and the Mississippi embayment heating events is an objection to the thermal construction hypothesis. It is unclear whether the amount of drilling in the deep parts of the basins is sufficient to establish an absence of igneous activity.

Depth of Crustal Loading

Gravity anomalies associated with platform basins constrain the depths and positions of the loads that drive basin subsidence. The Michigan basin is best suited for this analysis because of the lack of later tectonic events and the availability of data. The edge of the continent at Atlantic margins produces large gravity anomalies which cannot be separated from those due to loads driving subsidence.

The reflexural rigidity of the lithosphere tends to spread out the subsidence due to driving loads. The accumulating sediments, which are less dense than mantle, displace the crust mantle interface downward causing a negative free air anomaly. If the load driving subsidence is shallow, the negative anomaly due to subsidence would be broader than the positive anomaly due to the load. A net positive anomaly would thus occur near the center of the basin. If the load was quite deep the negative anomaly due to subsidence would be narrower than the anomaly due to the load. This would cause a negative anomaly at the center of the basin (Walcott 1972, Sleep 1976, Haxby et al 1976).

Quantification of these inferences is not completely straightforward because the gravity anomaly prior to heating and subsidence is unknown. For example, a positive gravity anomaly through the center of the Michigan basin appears to be associated with a Keweenawan rift similar to Lake Superior (see Sleep & Sloss 1978 for summary). In analogy to Lake Superior a thicker but higher density crust may exist beneath the gravity anomaly. Even though this feature may be isostatically compensated a positive anomaly flanked by negative anomalies would be expected.

The stress within the lithosphere following the heating event is unknown. If the lithosphere is viscoelastic and can creep, relaxation of any initial stress would produce subsidence (or uplift) unrelated to the driving load. The effect is probably not important, as a heating event would be expected to weaken the lithosphere.

Glacial erosion and deposition in the Michigan basin is strongly correlated with the outcrop pattern around the basin. A positive anomaly of up to several milligals on the Lower Peninsula of Michigan may be related to deposition on the Peninsula and erosion from the surrounding Great Lakes (Sleep & Snell 1976).

An example of a theoretical gravity calculation using the actual geometry of the Michigan basin is illustrated in Figure 4. In agreement with

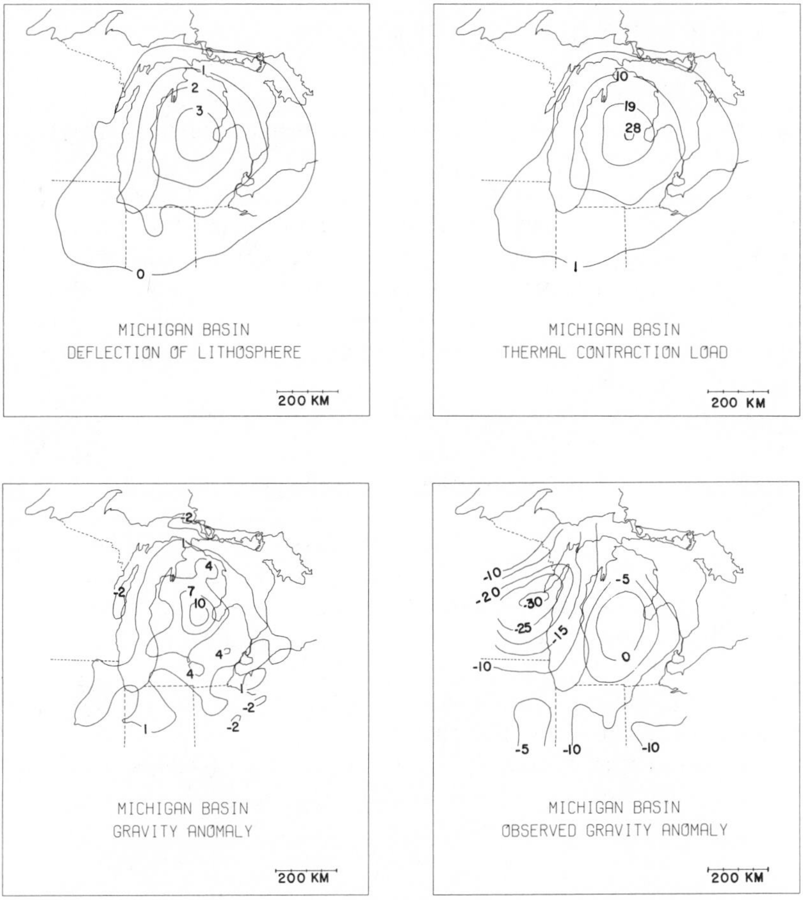

Figure 4 Subsidence model of the Michigan basin was computed with a flexural rigidity of 1×10^{31} dyne-cm, a viscoelastic decay time of 1 m.y., and a sediment-mantle density of 0.69 cm/gm^3. The observed sediment thickness in kilometers (*top left*) is artificially thinned to zero on the south. The driving load in 10^7 dyne/cm^2 is concentrated near the center of the basin (*top left*). The observed (after McGinnis et al 1979) and computed gravity anomalies in mgal show positive closures at the center of the basin (*below*).

observation a small positive closure is obtained. Increasing the depth of the crustal loading to subcrustal depths would cause a negative gravity anomaly. Increasing the flexural rigidity or the viscoelastic time constant of the lithosphere would cause the computed load to be more centered on the basin and thus increase the positive gravity anomaly. Asthenospheric driving loads can be excluded as a large negative gravity anomaly would occur at the basin center.

LATER VERTICAL MOVEMENTS

The thermal contraction hypothesis for basin subsidence implies that basins become quiescent after deposition. Direct determination of relative vertical movements after deposition is relevant to this implication and to removing eustatic effects from the subsidence data. The lack of large relative vertical movements since deposition over much of the interior and the lack of large uplifts on arches relative to adjacent shields indicates that basins are the active elements of the platform. The Pennsylvanian and Cretaceous of mid-western United States are discussed below to illustrate the methods and difficulties in determining the amount of post-depositional uplift.

To determine tectonic elevation changes or eustatic sea levels from outliers of sediment on the craton it is necessary to correct for the effects of isostasy. The loads of sediments subsequent to the time of deposit must be removed and the inferred elevation of deposition thus increased above the current elevation. The load of rocks present at the time of deposition which have been subsequently eroded must be added thus decreasing the inferred elevation of deposition. For isolated outliers the inferred elevation of deposition must be decreased by ρ_s/ρ_m times the elevation of the outlier above the surrounding plains. For $\rho_s = 2.6$ gm/cm^3 and $\rho_m = 3.3$ gm/cm^3 erosion of 100 m of sediment would produce 80 m of uplift and 20 m of elevation change.

The net amount of erosion and deposition must be averaged several hundred kilometers around the studied location to accurately remove the effect of isostasy. Even though the calculations to do this are not very difficult, it is seldom possible to obtain accurate enough data to warrant detailed calculations.

Pennsylvanian of Illinois, Iowa, and Michigan

Pennsylvanian rocks are preserved on the northern flank of the Illinois basin, across the Mississippi River in Iowa, and in the interior of the Michigan basin. The Pennsylvanian structural relief in these areas can be attributed to differential subsidence following and during deposition.

128078
G

The depth of burial of these rocks is relevant to determining the net amount of late stage subsidence as well as the amount of relative vertical movement subsequent to basin formation.

The depth of burial can be constrained by the degree of organic metamorphism or rank of coals in the Pennsylvanian section. The reflectance of vitrinite, which increases during organic metamorphism, increases from 0.47% to 0.60% in central Illinois to more than 1.0% in southeastern Illinois (Harvey et al 1979). In southeastern Illinois, the contours of equal rank dip more steeply than the structural contours of individual beds. This implies that either the heat flow was higher in the past in the southeastern part of the basin or that the southeastern part of the basin has been extensively uplifted (Damberger 1971). Uplift is likely because the basin was open to the south during deposition (Howard 1979) and because the structural uplift on the Pascola arch south of the basin is about 4 km (Buschbach & Atherton 1979). The 1.5–2.0 km of erosion estimated in southeastern Illinois by Damberger (1971), who preferred the igneous intrusion explanation, is thus compatible with gross structures. A minimum uplift of 2.5 km can be obtained by comparison of the vitrinite data from Pennsylvanian rocks in a well in Oklahoma (Hood et al 1975).

For Pennsylvanian rocks of central Illinois, the percentage of inherent seam moisture, which decreases during organic metamorphism, and the calorific value of the coal, which increases during organic metamorphism, are more sensitive indicators of organic metamorphism than vitrinite reflectance. The percentage of moisture in the Herrin coal varies from 20% in the northwest part of the outcrop belt (near Moline) and also across the Mississippi River into Iowa to 13% to the south and east to where the formation is about 200 m deep (Damberger 1971, 1974). This variation is largely correlative with the elevation of the coal seam and about the amount expected from variations within a single vertical section if little differential vertical movement has occurred since subsidence. Across the northern part of the outcrop belt over 160 km between Moline and Kankakee the percent of moisture varies from 20% to 17% and the structure controls deepen by 60 m (Figure 3). From Damberger (1971, Figure 5) the expected change in the structural contour is 100 m. Thus it can be concluded that less than 100 m of differential vertical movement has occurred in the central Illinois and adjacent Iowa, a region several hundred kilometers in extent, since the end of Pennsylvanian subsidence.

Pennsylvanian rocks in the Michigan basin dip toward the center of the basin and extend between about 300-m elevation and sea level. Moisture analyses (adjusted to mineral matter free) range from 14.5% to 10.4%

with an average of 12.7% for Bay City area coals (Andrews & Huddle 1948). These analyses are probably reliable, as the calorific value–moisture correlation of Damberger (1971, Figure 3) conforms to the data. The depth at which the samples were obtained is unknown but is probably between 100-m and 150-m elevation. Samples of coal collected at 240-m elevation at Grand Ledge, Michigan (Figure 2), were analyzed for vitrinite reflectance by John Castaño of Shell Oil. The most reliable and least weathered sample had a reflectance of 0.54%. This is equivalent to 13% moisture (Hood et al 1975, Damberger 1971) and in agreement with the older data.

Taking a reference elevation of 150 m and 19% moisture and 13% moisture for northern Illinois and Michigan, respectively, about 280 more meters of erosion have occurred in Michigan (from Damberger 1971, Figure 5). A relative uplift of 60 m of Michigan amplified by isostasy would produce this difference in depth of erosion. It is probable that both northern Illinois and Michigan continued to subside after the Pennsylvanian (Jurassic sediments are preserved in the Michigan basin) and that the northern part of the Illinois basin subsided more and thus was not as deeply eroded.

Determination of absolute rather than relative depths of erosion for Illinois requires extrapolation of published relationships to lower temperatures and longer times of burial. A temperature of 20–30°C over 300 m.y. is needed to produce the lowest grade coals in Illinois (21% moisture, from Hood et al 1975, Figure 3, and Teichmüller & Teichmüller 1968, Figures 18 and 19). This temperature range is similar to the mean temperature of tropical and subtropical areas. It is not unlikely that Illinois was in hot latitude belts during much of the last 300 m.y. Little burial and subsequent erosion in northern Illinois is thus required.

Lower Carboniferous coals in the Moscow, USSR, coal field show less organic metamorphism than northern Illinois coals. The moisture ranges between 33% and 40% (Yablokov 1962, p. 7). The elevation of these rocks is 200 m, similar to northern Illinois. The difference in coal rank between Moscow and northern Illinois could be attributed to either colder average climates in Moscow since Carboniferous time or a greater depth of burial and subsequent erosion in Illinois. Without detailed paleoclimatic reconstructions, coal-rank data are most useful for comparing nearby regions.

Basin Flanks

The depth of erosion can be determined from the relief of monadnocks and from the elevations of outliers surrounding the basin, such as deposits formed during the upper Cretaceous transgression.

Sleep (1976) obtained eustatic elevations corrected for isostasy of 325 m for Late Turonian–Early Coniacian and 250 m for Late Cenimanian for Cretaceous deposits in Minnesota. Hancock & Kauffman (1979) obtain values from 660 m in the Middle Maastrichtian to 260 m in the upper Turonian for England and adjacent Europe. These values are not corrected for isostasy and the estimated water depths are large (100–200 m) and hence less reliable. More eustatic sea-level estimates from North America are obtained below.

The Dakota Formation has been interpreted as deposits of a near-shore fluvial-deltaic environment, grading into the overlying marine Graneros Formation (Tester 1929, Bowe 1972). Because eustatic sea-level changes can be estimated more accurately using data from littoral deposits, the elevation of the top of the Dakota Formation, of Middle Cenomanian age (Cobban & Reeside 1952), is used to estimate the eustatic amplitude of the Upper Cretaceous transgression in Iowa. This contact is difficult to define because of its gradational nature. Using well logs from 90 irregularly distributed locations (Iowa Geological Survey, unpublished), we find the average elevation of the top of the Dakota to be 314.54 ± 20.63 m. The elevation of the Cretaceous of Iowa may have been decreased by 50 m owing to loading by Pleistocene deposits, which have an average thickness of 70 m. Thus, the sea during Middle Cenomanian time in Iowa was about 364 m above present sea level.

The elevation of the Ashville Formation of Early Cenomanian age in Manitoba is about 305 m without correction for isostatic compensation (Simpson 1975). The depth of deposition of the Ozan Formation of Early Cenomanian age in McCurtain County, Oklahoma, was less than 600 feet (Krancer 1978); the present elevation of the Ozan Formation is 350 feet. Therefore, the amplitude of the eustatic change during Early Cenomanian time in Oklahoma is about 950 feet (290 m).

The present level up to 375 m of Cretaceous rock on the shield requires around 135 m of erosion to reduce the Michigan basin to its present level of 240 m near Grand Ledge, Michigan, with no allowance for isostasy, which would increase the amount of erosion up to five times. The minimum erosion of 300 m for the interior of Michigan obtained from the coal data is thus compatible with erosion of material deposited during eustatic highstands without any tectonic uplift of the area. The relief of monadnocks in the unglaciated parts of Wisconsin is also around 300 m.

The lower grade coal regions of Illinois require tectonic subsidence as the average elevation is slightly lower in Illinois than Michigan. In a purely erosional model deeper erosion would be required to reduce Illinois to its present level. The amount of additional subsidence is probably less than 100 m. This subsidence probably occurred during Tertiary and

Cretaceous time because earlier subsidence would have caused accumulation of sediments and thus excessive organic metamorphism. The downwarp in Illinois may be a continuation of the Mississippi embayment to the south.

CONCLUDING REMARKS

The causes of vertical movements in interior platform basins are still poorly understood. The hypothesis that thermal contraction causes subsidence is preferred, not because of strong positive evidence, but by analogy to subsidence along Atlantic margins and mid-oceanic ridges. Other mechanisms have not yet been formulated sufficiently to allow quantitative comparison with observations.

The data establishing vertical movements have largely been collected for other purposes and when available are often scattered in state publications. Organic metamorphism studies appear to be able to supply information on depths of erosion that is not available from classical stratigraphy. Continued analysis is necessary to reduce geological data from sedimentary basins to a form useful to geophysicists.

Literature Cited

Adkinson, W., ed. 1966. Stratigraphic cross section of Paleozoic rocks, Colorado to New York. *Am. Assoc. Petrol. Geol. Cross Sect. Publ. 4*. 58 pp.

Andrews, D. A., Huddle, J. W. 1948. Analyses of Michigan, North Dakota, South Dakota, and Texas coals. *US Bur. Mines Tech. Pap. 700*. 106 pp.

Artemjev, M. E., Artyushkov, E. V. 1971. Structure and isostasy of the Baikal rift and the mechanism of rifting. *J. Geophys. Res.* 76:1197–1211

Beaumont, C. 1978. The evolution of sedimentary basins on a visco-elastic lithosphere: theory and examples. *Geophys. J. R. Astron. Soc.* 55:471–98

Bott, M. H. P. 1971. Evolution of young continental margins and formation of shelf basins. *Tectonophysics* 11:319–27

Bott, M. H. P. 1979. Subsidence mechanism at passive continental margins. *Am. Assoc. Petrol. Geol. Mem.* 29:3–9

Bott, M. H. P., Dean, D. S. 1972. Stress systems at young continental margins. *Nature Phys. Sci.* 235:23–25

Bowe, R. J. 1972. *Depositional history of the Dakota Formation in eastern Nebraska.* MS thesis. Univ. Nebraska, Lincoln

Bowie, W. 1927. *Isostasy*. NY: Dutton. 275 pp.

Burke, K., Dewey, J. F. 1973. Plume-generated triple junctions: key indicators in applying plate tectonics to old rocks. *J. Geol.* 81:406–33

Buschbach, T. C., Atherton, E. 1979. History of the structural uplift of the southern margin of the Illinois basin. In *Deposition and Structural History of the Pennsylvanian System of the Illinois Basin*, ed. J. E. Palmer, R. R. Dutcher, pp. 112–15. Urbana: Ill. State Geol. Surv.

Cobban, W. A., Reeside, J. B. 1952. Correlation of the Cretaceous formations of the Western interior of the United States. *Bull. Geol. Soc. Am.* 63:1011–44

Cook, T. D., ed. 1975. *Stratigraphic Atlas North-Central America.* Houston: Shell Oil Co.

Cross, T. A., Pilger, R. H. Jr. 1978. Tectonic controls of late Cretaceous sedimentation, western interior, USA. *Nature* 274: 653–57

Crough, S. T. 1978. Thermal origin of mid-plate hot-spot swell. *Geophys. J. R. Astron. Soc.* 55:451–70

Damberger, H. H. 1971. Coalification pattern of the Illinois basin. *Econ. Geol.* 66:488–94

Damberger, H. H. 1974. Coalification patterns of Pennsylvanian coal basins of the

eastern United States. *Geol. Soc. Am. Spec. Pap.* 153: 53–74
de Charpal, O., Guennoc, P., Montadert, L., Roberts, D. G. 1978. Rifting, crustal attenuation and subsidence in the Bay of Biscay. *Nature* 275: 706–11
Dietz, R. 1963. Collapsing continental rises: an actualistic concept of geosynclines mountain building. *J. Geol.* 71: 314–33
Doig, R. 1970. An alkaline province linking Europe and North America. *Can. J. Earth Sci.* 7: 22–28
Ervin, C. P., McGinnis, L. D. 1975. Reelfoot rift, reactivated precursor to the Mississippi embayment. *Geol. Soc. Am. Bull.* 86: 1287–95
Falvey, D. A. 1974. The development of continental margins in plate tectonic theory. *J. Aust. Petrol. Explor. Assoc.* 14: 95–106
Hancock, J. M., Kauffman, E. G. 1979. The great transgressions of the Late Cretaceous. *J. Geol. Soc. London* 136: 175–86
Harvey, R. D., Crelling, J. C., Dutcher, R. R., Schleicher, J. A. 1979. Petrology and related chemistry of coals in the Illinois basin. See Buschbach & Atherton 1979, pp. 127–42
Haxby, W. F., Turcotte, D. L., Bird, J. M. 1976. Thermal and mechanical evolution of the Michigan basin. *Tectonophysics* 36: 57–75
Hinze, W. J., Bradley, J. W., Brown, A. R. 1978. Gravimeter survey in the Michigan basin deep borehole. *J. Geophys. Res.* 83: 5864–68
Hinze, W. J., Roy, R. L., Davidson, D. M. 1972. The origin of late Precambrian rifts. *Geol. Soc. Am. Abstr. Prog.* 4: 723
Hood, A., Gutjahr, C. C. M., Heacock, R. L. 1975. Organic metamorphism and the generation of petroleum. *Am. Assoc. Petrol. Geol. Bull.* 59: 986–96
Howard, R. H. 1979. The Mississippian-Pennsylvanian unconformity in the Illinois basin—old and new thinking. See Buschbach & Atherton 1979, pp. 34–43
Hsu, K. J. 1965. Isostasy, crustal thinning, mantle changes, and the disappearance of ancient land masses. *Am. J. Sci.* 263: 97–109
Joyner, W. 1967. Basalt-eclogite transition as a cause for subsidence and uplift. *J. Geophys. Res.* 72: 4977–98
Krancer, A. E. 1978. *Nannofossils of the Ozan Formation (Cretaceous), McCurtain County, Oklahoma.* MS thesis, Univ. Oklahoma (Abstr. in *Okla. Geol. Notes* 38: 113–14)
Le Pichon, X., Francheteau, J., Bonnin, J. 1973. *Plate Tectonics* (Vol. 6 of *Developments in Geodynamics*). NY: Elsevier. 300 pp.
Mareschal, J. C., Gangi, A. F. 1977. Equilibrium position of a phase boundary under horizontally varying surface loads. *Geophys. J. R. Astron. Soc.* 49: 757–72
McGinnis, L. D. 1970. Tectonics and the gravity field in the continental interior. *J. Geophys. Res.* 75: 317–31
McGinnis, L. D., Wolf, M. G., Kohsmann, J. J., Ervin, C. P. 1979. Regional free air gravity anomalies and tectonic observations in the United States. *J. Geophys. Res.* 84: 591–602
McKenzie, D. 1978. Some remarks on the development of sedimentary basins. *Earth Planet. Sci. Lett.* 40: 25–32
O'Connell, R., Wasserburg, G. 1967. Dynamics of a phase change boundary to changes in pressure. *Rev. Geophys.* 5: 329–410
Ringwood, A. E. 1975. *Composition and Petrology of the Earth's Mantle.* NY: McGraw-Hill. 618 pp.
Scheidegger, A., O'Keefe, J. 1967. On the possibility of the origin of geosynclines by deposition. *J. Geophys. Res.* 72: 6275–78
Schepp, V., ed. 1963. Symposium on Middle and Upper Devonian Stratigraphy of Pennsylvania and Adjacent States. *General Geol. Rep. G39.* Penn. Geol. Survey. 301 pp.
Schneider, E. D. 1969. The deep-sea—a habitat for petroleum. *Undersea Technol.* 10(Oct): 32–57
Simpson, F. 1975. Marine lithofacies and biofacies of the Colorado Group (middle Albian to Santonian) in Saskatchewan. In *Cretaceous System in the Western Interior of North America. Geol. Assoc. Can. Spec. Pap.* 13: 553–87
Sleep, N. H. 1971. Thermal effects of the formation of Atlantic continental margins by continental break-up. *Geophys. J. R. Astron. Soc.* 24: 325–50
Sleep, N. H. 1976. Platform subsidence mechanisms and "eustatic" sea-level changes. *Tectonophysics* 36: 45–56
Sleep, N. H., Sloss, L. L. 1978. A deep borehole in the Michigan basin. *J. Geophys. Res.* 83: 5815–19
Sleep, N. H., Snell, N. S. 1976. Thermal contraction and flexure of mid-continent and Atlantic marginal basins. *Geophys. J. R. Astron. Soc.* 45: 125–54
Sloss, L. L., Speed, R. C. 1974. Relationships of cratonic and continental-margin tectonic episodes. In *Tectonics and Sedimentation*, ed. W. R. Dickinson, pp. 98–119. *Soc. Econ. Paleontol. Mineral. Spec. Publ. 22*
Steckler, M. S., Watts, A. B. 1978. Subsidence of the Atlantic type margin off

New York. *Earth Planet. Sci. Lett.* 41: 1–13

Teichmüller, M., Teichmüller, R. 1968. Geological aspects of coal metamorphism. In *Coal and Coal-Bearing Strata*, ed. D. Murchison, T. S. Westall, pp. 233–67. NY: Elsevier

Tester, A. C. 1929. The Dakota stage of the type locality. *Iowa Geol. Surv. Ann. Rep.* 35: 199–332

van de Lindt, W. 1967. Movement of the Mohorovicic discontinuity under isostatic conditions. *J. Geophys. Res.* 72: 1289–97

van der Voo, R., Watts, D. R. 1978. Paleomagnetic results from igneous and sedimentary rocks from the Michigan basin borehole. *J. Geophys. Res.* 83: 5844–48

Vogt, P. R., Ostenso, N. A. 1967. Steady state crustal spreading. *Nature* 215: 810–17

Walcott, R. I. 1972. Gravity, flexure and the growth of sedimentary basins at a continental edge. *Geol. Soc. Am. Bull.* 83: 1845–48

Watts, A. B., Ryan, W. B. F. 1976. Flexure of the lithosphere and continental margin basins. *Tectonophysics* 36: 25–44

Yablokov, V. S., ed. 1962. *Atlas of Coals of Moscow Coal Basin, Vol. I.* Tula, USSR: Cent. Bur. Tech. Inf. (In Russian)

Ann. Rev. Earth Planet. Sci. 1980. 8 : 35–63
Copyright © 1980 by Annual Reviews Inc. All rights reserved

GEOCHEMISTRY OF EVAPORITIC LACUSTRINE DEPOSITS

×10124

Hans P. Eugster
Department of Earth and Planetary Sciences, Johns Hopkins University, Baltimore, Maryland 21218

INTRODUCTION

To most of us, evaporites are the products of sea water evaporation. Sea salts have long been important to man and they have played an increasingly significant role since the industrial revolution (see Multhauf 1978). To the geologist, the thick accumulations of marine salts found throughout the geologic column are testimony to unusual conditions of climate, hydrology, sedimentation, and tectonics. And yet years of study have left unresolved some of the basic controversies with respect to depositional processes and environments. In part this is because the giant deposits of the past appear to lack contemporary analogues.

In contrast, scant attention has been paid to continental evaporites, even though thousands of saline lakes exist on all continents except Europe. Saline lake deposits, however, are usually restricted and ephemeral. Very few nonmarine evaporites can rival their marine counterparts on economic terms, the Green River Formation of Wyoming and Searles Lake of California being the exceptions.

This neglect is unfortunate, because evaporitic lacustrine deposits abound in problems of significance to climate, hydrology, tectonics, sedimentation, geochemistry, and mineralogy. Their compositional range is much wider than that of marine evaporites and the formational processes can be studied right now in a variety of settings. Quantitatively insignificant compared to marine rocks, chemical sediments of closed basins can teach us much about important processes that have modified the earth's surface throughout geologic time. Those of us who have spent time and effort studying such deposits have not been disappointed.

For this review, which focuses mainly on geochemical and related

0084-6597/80/0515-0035$01.00

aspects, I rely heavily on the experience of my colleagues L. A. Hardie, B. F. Jones, and J. P. Smoot and the earlier summaries of Hardie et al (1978), Smoot (1978), and Eugster & Hardie (1978), which were concerned with other aspects of the same topic.

CONTINENTAL EVAPORITIC BASINS: PHYSICAL ASPECTS

Hydrologic Conditions

Evaporitic basins are hydrologically closed, that is, evaporation must exceed inflow. This condition presupposes an arid to semiarid climate. However, evaporitic deposits can form only when a sufficient supply of solutes is available from various inflow sources. These two seemingly contradictory requirements are reconciled in local orographic deserts, where high mountains provide a rain shadow for the basin floors and also act as a precipitation catchment. For this reason, many closed basins are found in the lee of large mountain chains. Good examples are the salt pans and playas of the Western Great Basin (US), the salars of Bolivia, Argentina, and Chile, the salt lakes of Afghanistan, India, and China, and the Rift Valley lakes of E. Africa. Many of these basins were initiated tectonically, and we note the interplay between tectonics, climate, and sedimentation. Geographic latitude has little effect and some of the most arid conditions are found in arctic climates.

The hydrologic setting of a closed basin is defined by climate and tectonics and in turn has a pronounced effect on sedimentation and the depositional environment in which chemical sediments accumulate. Typically, a basin has a low, flat floor, which is surrounded by mountains, providing considerable topographic gradients. Dilute inflow may be by perennial rivers, by ephemeral streams, by storm runoff, and by perennial springs. Examples of salt lakes with perennial rivers are the Great Salt Lake of Utah, Lake Chad of Central Africa, the Dead Sea, the Caspian Sea, and many others. If inflow is primarily ephemeral, the basin is most likely occupied by a salt pan which is seasonally wet or dry and is fed largely by springs and storm runoff. Such salt pans are often small, but they may occasionally reach giant size; the Uyuni salar of Bolivia, for example, covers an area of some 15,000 km^2.

Hydrologic parameters of closed basins have been discussed by Langbein (1961), while the sedimentological aspects were summarized by Hardie et al (1978). Although this review emphasizes geochemical aspects of evaporitic lacustrine deposits, it must be kept in mind that for closed basins hydrology, sedimentation, water chemistry, and mineralogy are intimately intertwined.

The Sedimentological Framework

Sedimentation in arid environments has been discussed by a number of authors such as Reeves (1968), Glennie (1970), Picard & High (1973), and McKee (1979). We follow Hardie et al (1978) in identifying the sedimentological framework of closed basins by depositional subenvironments: 1. alluvial fan; 2. sand flat; 3. mud flat; 4. saline mud flat; 5. perennial salt lake; 6. ephemeral salt pan; 7. perennial stream flood plain; 8. ephemeral stream flood plain; 9. spring. Each subenvironment is defined by a specific set of characteristic depositional processes. Because of the arid setting, *alluvial fans* are important for all closed basins. They form by catastrophic storm events which collect large amounts of sediment from the steep mountain slopes and deposit them in cone-shaped wedges of very coarse sediment which build out towards the valley floor. Surface and capillary evaporation lead to the formation of pore and surface cements, mostly in the form of alkaline earth carbonates. All of the coarse sediment is dropped on the fan itself, but sand-sized material may accumulate in *sand flats* which fringe the toe of the alluvial fan. *Mud flats* are typical of the valley floor, where gradients are extremely gentle. Dry mud flats are normally exposed to the air and hence are heavily mud cracked. If the water table is shallow, efflorescent crusts form on the surface by evaporative pumping (Hsü & Siegenthaler 1969). *Saline mud flats*, on the other hand, are usually soaked with brine. They are characterized by saline minerals, such as gaylussite ($Na_2CO_3 \cdot CaCO_3 \cdot 5H_2O$), gypsum ($CaSO_4 \cdot 2H_2O$), and mirabilite ($Na_2SO_4 \cdot 10H_2O$), growing interstitially in the mud.

The topographic low of the basin is occupied either by a perennial salt lake or by an ephemeral salt pan. A perennial river is essential for maintaining a *perennial salt lake* and hence perennial lakes are much less common than salt pans (see Langbein 1961). Their lake level and salinity are subject not only to seasonal but also to long-term fluctuations in response to climatic variations (see Figure 1). Because of flooding with dilute water, such lakes are often chemically stratified, with a dense bottom brine overlain by fresher surface water. The bottom waters may be anoxic and consequently free of burrowing benthos.

Where perennial streams cannot form for lack of inflow, the topographic low is occupied by an ephemeral *salt pan*. Such pans are examples of playa lakes or "dry lakes." Some of the morphologic, hydrologic, and sedimentologic aspects of playas have been discussed by Cooke & Warren (1973), Glennie (1970), Mabbutt (1977), Neal (1975), and Reeves (1968, 1972). Salt pans are special cases of playas, because they contain a permanent body of saline water, most of which is present within the

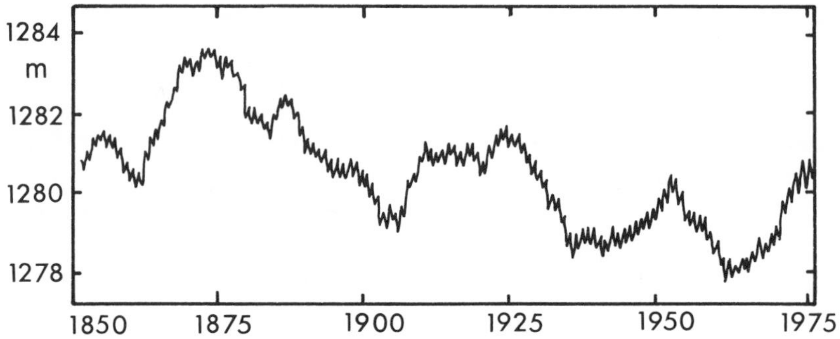

Figure 1 Elevation above sea level, in m, of Great Salt Lake, Utah, illustrating seasonal and long-term variations of closed-basin lakes.

sediment, especially in the interstices of the saline minerals which characterize the central part of the playa. Saline springs often issue on the fringes of such salt pans and represent a major solute source. Seasonal flood waters, which dissolve efflorescent crusts, are another important source of solutes.

Stream flood plains are important subenvironments of closed basins and their depositional elements have been summarized in books such as those of Allen (1970), Reineck & Singh (1975), Glennie (1970), and Picard & High (1973).

Springs are common in all closed basins, because under arid conditions much of the water must be transported underground to escape evaporation. Dilute inflow issues in the form of perennial springs at the foot of the mountains or the toes of the alluvial fans. In the absence of a shallow ground water table such waters flow a short distance and then seep back into the porous sediment. They will eventually act as recharge for a saline ground water body which accumulates near the center of the basin. It is this saline ground water which feeds the springs at the perimeter of the salt lake or salt pan.

CHEMISTRY AND MINERALOGY OF CONTINENTAL BRINES AND EVAPORITES

The chemistry of marine evaporites is confined by the constancy in composition of sea water. No such constraint exists for continental evaporites, and the variability of continental brines and evaporites is a characteristic attribute. Neighboring basins, separated only by a watershed, may have widely differing chemistry and mineralogy. The cause of this divergence is one of the basic questions we must address.

Saline lakes, their waters and deposits, have been discussed in many publications. Good summaries can be found in Grabau (1920), Clarke (1924), Lotze (1957), and Livingston (1963). The vast majority of continental brines can be represented by four cations and three anions: Na, K, Ca, Mg, Cl, SO_4, and HCO_3–CO_3. In addition, some deposits are rich in borates and nitrates. Equilibria must be studied in the standard quinary sea water system, Na–K–Mg–Cl–SO_4 (see, for instance, Braitsch 1971), with the addition of Ca and HCO_3–CO_3. A detailed classification scheme based on combining Na and K has been given by Eugster & Hardie (1978).

For broad comparisons, five major brine types can be distinguished: (*a*) Na–CO_3–Cl, (*b*) Na–CO_3–SO_4–Cl, (*c*) Na–SO_4–Cl, (*d*) Na–Mg–SO_4–Cl, and (*e*) Ca–Mg–Na–Cl. For detailed discussions of brine evolution, Na and K must be treated separately, SiO_2 must be added, as well as B, Br, NO_3, PO_4, Sr, Li, and any other constituent that may be of concern.

It will come as no surprise that the mineralogy of nonmarine evaporites is very complex. A list of the more common saline constituents is given in Table 1. While some of these salts may occur as bedded units, others obviously form as authigenic products from interstitial brines. Bedded deposits are commonly formed, for instance, by gypsum, halite, mirabilite, thenardite, trona, epsomite, and borax, and such beds presumably accumulated in an ephemeral or perennial salt lake. Characteristic authigenic phases are gaylussite, pirssonite, glauberite, bloedite, northupite, and dawsonite. Such minerals may form interstitially in the sediments associated with an active salt lake or salt pan, such as those of the saline mud flats, or they may have formed subsequently, after burial, in response to changing conditions. A good example of the latter kind are the extensive deposits of disseminated shortite in the Green River formation (see Milton 1971).

Associated with many evaporite deposits is a suite of minerals not by themselves saline, but indicative of saline conditions. The best-known examples are the zeolites of alkaline saline lakes (see Surdam & Sheppard 1978 for a recent summary), such as clinoptilolite, erionite, analcime, the sodium silicates such as magadiite, and the borosilicates such as searlesite. The most extraordinary list of authigenic minerals of a saline formation has been assembled for the Green River formation (Milton 1971).

On the basis of mineralogy alone it is not always possible to differentiate between marine and nonmarine evaporites. Halite is the most common constituent in both cases and gypsum is widespread also among continental environments. Minerals such as glauberite, polyhalite, epsomite, bloedite, sylvite, and tachhydrite are not diagnostic. On the other hand, because sea water is dominated by Cl and SO_4, saline carbonate

Table 1 Saline minerals of nonmarine evaporites

Brine type	Primary minerals		Authigenic minerals	
(a) $Na–CO_3–Cl$	halite	$NaCl$	gaylussite	$Na_2CO_3 . CaCO_3 . 5H_2O$
	nahcolite	$NaHCO_3$	pirssonite	$Na_2CO_3 . CaCO_3 . 2H_2O$
	natron	$NaCO_3 . 10H_2O$	shortite	$Na_2CO_3 . 2CaCO_3$
	thermonatrite	$Na_2CO_3 . H_2O$	northupite	$Na_2CO_3 . MgCO_3 . NaCl$
	trona	$NaHCO_3 . Na_2CO_3 . 2H_2O$	hanksite	$9Na_2SO_4 . 2Na_2CO_3 . KCl$
			aphthitalite	$K_3Na(SO_4)_2$
			tychite	$Na_2CO_3 . MgCO_3 . Na_2SO_4$
			dawsonite	$NaAlCO_3(OH)_2$
(b) $Na–CO_3–SO_4–Cl$	burkeite	$Na_2CO_3 . 2Na_2SO_4$		
	halite	$NaCl$		
	mirabilite	$Na_2SO_4 . 10H_2O$		
	naholite	$NaHCO_3$		
	natron	$Na_2CO_3 . 10H_2O$		
	thenardite	Na_2SO_4		
	thermonatrite	$Na_2CO_3 . H_2O$		
(c) $Na–SO_4–Cl$	gypsum	$CaSO_4 . 2H_2O$	glauberite	$CaSO_4 . Na_2SO_4$
	glauberite	$CaSO_4 . Na_2SO_4$		
	halite	$NaCl$		
	mirabilite	$Na_2SO_4 . 10H_2O$		
	thenardite	Na_2SO_4		

(d) Mg–Na–SO_4–Cl	bischofite	$MgCl_2 . 6H_2O$	bloedite	$Na_2SO_4 . MgSO_4 . 4H_2O$
	bloedite	$Na_2SO_4 . MgSO_4 . 4H_2O$	glauberite	$CaSO_4 . Na_2SO_4$
	epsomite	$MgSO_4 . 7H_2O$		
	glauberite	$CaSO_4 . Na_2SO_4$		
	gypsum	$CaSO_4 . 2H_2O$		
	halite	NaCl		
	hexahydrite	$MgSO_4 . 6H_2O$		
	kieserite	$MgSO_4 . H_2O$		
	mirabilite	$Na_2SO_4 . 10H_2O$		
	thenardite	Na_2SO_4		
(e) Ca–Mg–Na–Cl	antarcticite	$CaCl_2 . 6H_2O$		
	bischofite	$MgCl_2 . 6H_2O$		
	carnallite	$KCl . MgCl_2 . 6H_2O$		
	halite	NaCl		
	sylvite	KCl		
	tachyhydrite	$CaCl_2 . 2MgCl_2 . 12H_2O$		

minerals do not form in marine evaporites. Hence trona, gaylussite, burkeite, northupite, hanksite, dawsonite, and similar phases are exclusive to continental evaporites. Some deposits have formed under the influence of both marine and continental processes. This is true particularly for near-shore lagoons that are accessible also to continental waters. In such cases assemblages of saline minerals may not be sufficient to separate the diverse inputs.

In order to account for the chemistry and mineralogy of a particular closed basin and for the variability from basin to basin, we next focus on two aspects of mineral–water interaction: solute acquisition and brine evolution. Solute acquisition is the evolutionary branch that will define the composition of the dilute inflow waters and hence provide us with the starting material for brine formation. Evaporative concentration is the sum of those processes that lead to brine evolution. Mineral precipitation is an important aspect of brine evolution and it leads to the accumulation of chemical sediments of which the evaporites are an integral part.

CONTROLS ON INFLOW COMPOSITIONS: SOLUTE ACQUISITION

Using a computer model built on a suggestion by Garrels & MacKenzie (1967), Hardie & Eugster (1970) were able to demonstrate that the key to the great variability of brine compositions from basin to basin can be found in the composition of the dilute inflow waters. In a sense, the final outcome of evaporative concentration in a particular basin is sealed at the time the dilute waters have acquired their solutes. As we shall see, some subsequent modification of paths is possible, but inflow composition remains the single most effective parameter.

Rain water and weathering reactions are the chief solute sources. Rain water can contribute to all of the seven principal solutes of dilute waters, excepting silica, but the most important contributions will be to Na, Cl, SO_4, and HCO_3. The amounts vary greatly with distance from the sources, particularly for sea spray and pollution.

Many kinds of weathering reactions contribute to the solute load of inflow waters. One of the most effective is the congruent dissolution of soluble minerals such as gypsum or halite, which can lead to very high concentrations, and is an important mechanism for recycling evaporites. Calcite also dissolves congruently in CO_2-charged waters by a reaction such as

$$CaCO_3 + H_2CO_3 \rightarrow Ca^{2+} + 2HCO_3^-.$$

Quartz dissolution is possible, but it is not the principle mechanism for charging water with H_4SiO_4.

Silicates usually dissolve incongruently, but rain water charged with atmospheric CO_2 is not very aggressive. Soil waters, on the other hand, can pick up much more CO_2 and are more effective because of their lower pH. Silicate weathering has been likened to a vast acid-base titration. The best known example is the weathering of Sierra Nevada quartz-monzonites (Garrels & Mackenzie 1967), which can be formulated as follows:

$$\underset{\text{(albite)}}{NaAlSi_3O_8} + CO_2 + \tfrac{11}{2}H_2O \rightarrow \tfrac{1}{2}\underset{\text{(kaolinite)}}{Al_2Si_2O_5(OH)_4} + Na^+ + HCO_3^- + 2H_4SiO_4.$$

A feldspar is altered to a clay mineral and in the process the waters are charged with Na^+, HCO_3^-, and silica. Other silicates provide the additional cations Ca^{2+}, Mg^{2+}, and K^+, but such waters are always dominated by bicarbonate, derived from the atmosphere or from soil processes.

Redox reactions are important in weathering, both in terms of iron silicate dissolution and sulfide oxidation. In oxidizing environments, iron behaves as a lateritic element. Sulfide oxidation represents a significant source of acidity as can be seen by the simplified reaction,

$$\underset{\text{(pyrite)}}{FeS_2} + 15O_2 + 8H_2O \rightarrow 2Fe_2O_3 + 8SO_4^{2-} + 16H^+,$$

which proceeds most readily by bacterial pathways.

In summary, the composition of the dilute inflow waters depends largely on the minerals present in the watershed. Igneous and metamorphic rocks yield silica-rich, soft Ca–Na–HCO_3 waters, unless there are sulfides present to contribute SO_4^{2-}. Limestones produce hard Ca–HCO_3 waters poor in silica. Basic and ultrabasic rocks are likely to give alkaline Mg–HCO_3 waters. Cl is contributed from many sources and it can accumulate as a residual element in any setting.

Ideally, a detailed knowledge of the rocks that are being affected in the headlands should make it possible to predict inflow composition. But usually the reverse is more powerful: using inflow water compositions as tracers for the weathering reactions. It is of course not necessary that all solutes be acquired from near-surface interactions. Some hot springs may be associated with very extensive and deep hydrothermal systems and they may be the sources of important constituents such as B or Li.

Some closed basins have inflow waters that are homogeneous, because the bedrocks are uniform. In such cases, brine evolution can be readily documented. Other basins receive several contrasting inflow types and the complexities then are compounded by mixing of waters at various stages.

BRINE EVOLUTION

In a closed basin, all surface or near-surface waters are subject to evaporative concentration. Capillary evaporation is important whenever ground water is close enough to the surface and it is responsible for the efflorescent crusts that are so characteristic of closed basins. Evapotranspiration may also contribute to this process, but because of the sparse vegetation, this effect is often secondary. Evaporative concentration increases the solute load and eventually must lead to mineral precipitation. In fact, Hardie & Eugster (1970) found that many dilute spring waters of closed basins are already supersaturated with respect to calcite as they emerge, because their P_{CO_2} is higher than that of the atmosphere.

Fractionation by Mineral Precipitation

Precipitation of one or more mineral phases has a profound effect on the subsequent development of the water composition. This is due to the combined constraints of equilibrium and mass balance. For instance, precipitation of calcite is governed by the relation

$$Ca^{2+} + CO_3^{2-} \rightarrow CaCO_3 \qquad K = (a_{Ca^{2+}})(a_{SO_4^{2-}})(a_{H_2O})^2.$$

Consequently, as long as the water is in equilibrium with calcite, the activities of Ca^{2+} and CO_3^{2-} must vary inversely. On the other hand, the mass balance condition is that 1 mole of calcite forms by removing 1 mole of Ca^{2+} and 1 mole of CO_3^{2-} from solution. This condition implies that the Ca : CO_3 ratio must change, unless it was precisely one at the outset. Hence, evaporative concentration combined with calcite precipitation will enrich the residual solution in the more abundant species and deplete it in the other. The greater the initial disparity between the two species, the faster the enrichment-depletion process takes place. In this manner, calcite precipitation acts as a branching point or chemical divide in the sense that waters that are initially carbonate-rich will experience a further relative enrichment of carbonate and depletion of calcium and vice versa.

As calcite precipitates, the Mg/Ca ratio in the solution increases, first producing low-Mg calcite and eventually high-Mg calcite and protodolomite.

After calcite, the next mineral to reach saturation is often gypsum, which precipitates according to

$$Ca^{2+} + SO_4^{2-} + 2H_2O \rightarrow CaSO_4 \cdot 2H_2O \qquad K = (a_{Ca^{2+}})(a_{SO_4^{2-}})(a_{H_2O})^2.$$

It provides a chemical divide with respect to calcium and sulfate in the

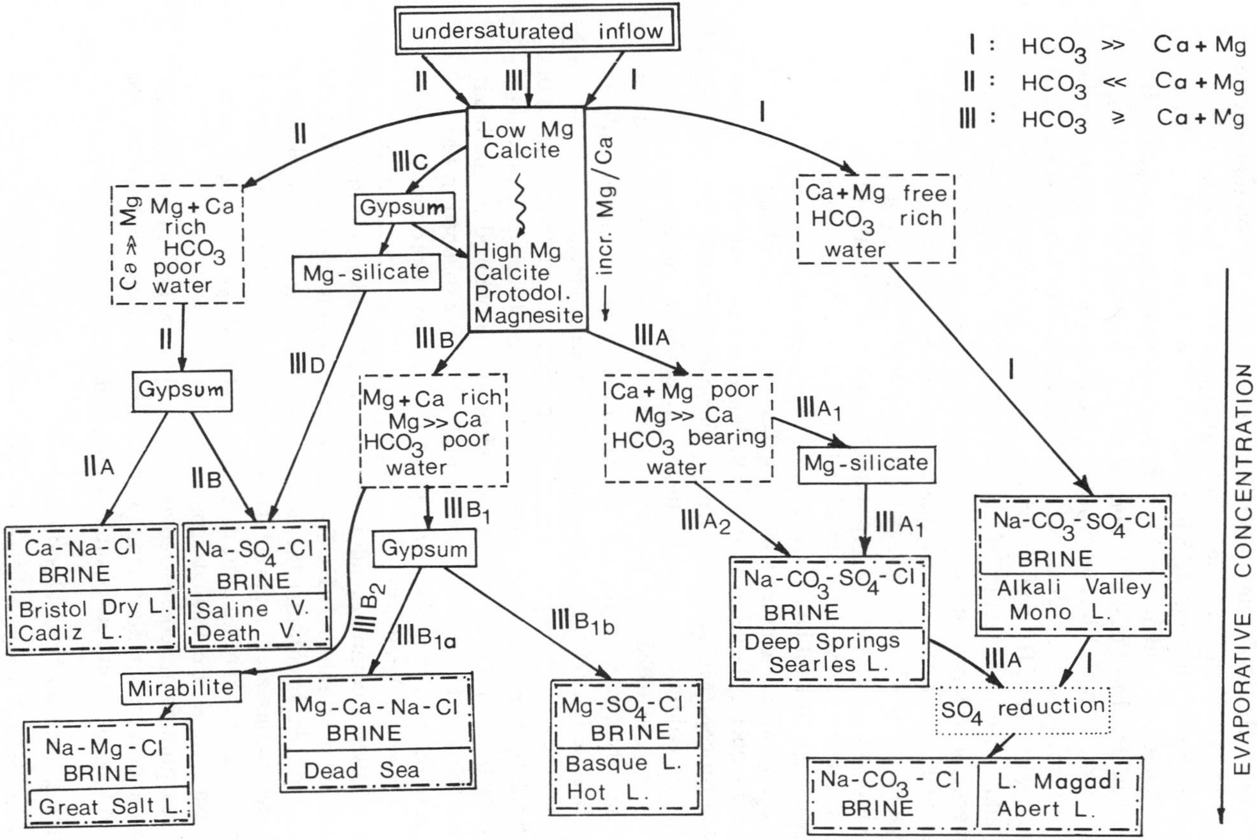

Figure 2 Brine evolution flow diagram (from Eugster & Hardie 1978), showing critical precipitates (*solid rectangles*) and resulting brines, together with examples of salt lakes.

same manner as calcite provided a chemical divide between calcium and carbonates. Hardie & Eugster (1970) formulated a brine evolution scheme based on such chemical divides. Following Garrels & Mackenzie (1967), Mg was removed in sepiolite. The revised version, shown in Figure 2, uses Mg-calcites and dolomite as Mg precipitates. This scheme, which is discussed in detail in Eugster & Hardie (1978), illustrates the fate of three inflow waters: path I has an excess of bicarbonate over alkaline earths (soft water), path II an excess of alkaline earths over bicarbonate (hard water), while path III is intermediate. Path I is the typical parent water for alkaline brines devoid of Ca and Mg. The normal precipitation sequence is calcite → trona; gypsum never appears along this path. Examples are brines such as those of Alkali Valley, Oregon, or Lake Magadi, Kenya. In contrast, path II produces brines poor in HCO_3, which normally precipitate gypsum after calcite. The path then divides into a sulfate-rich (Saline Valley, Death Valley, California) and a sulfate-poor branch (Bristol Dry Lake, Cadiz Lake, California).

In many waters, neither HCO_3 nor Ca + Mg are dominant (path III) and consequently large amounts of alkaline earth carbonates are formed. In the process, Mg becomes strongly enriched over Ca, and high-Mg calcite or dolomite become abundant. Eventually, either Ca + Mg or HCO_3 are exhausted, and the brines then move to the alkaline side or become saturated with gypsum. The latter divide again separates sulfate-rich (Basque Lake) from sulfate-poor (Dead Sea) products.

A scheme such as that of Figure 2 is capable of accounting for most of the brine types encountered on the earth's surface. It demonstrates forcefully the fact that the nature of the end product is defined by the input. Hence the variability of the continental brines is explained in terms of the variability of the inflow waters. In the process, a pronounced fractionation and differentiation has taken place. This is best illustrated in Figure 3, which compares inflow waters with closed basin brines. Inflow anions are predominantly HCO_3, whereas the cations are mixtures of Na, K, Ca, and Mg. In contrast, the cations of the brines are dominated by Na and the anions are mixtures. To a considerable extent, this shift in emphasis is caused by alkaline earth carbonate and gypsum precipitation.

If Na and K are treated separately, if silica is included, if sulfide-sulfate equilibria are considered as well as additional minor constituents, then mineral precipitation alone is not sufficient to account for the observed fractionation between inflow and brines.

Other Fractionation Mechanisms

A comparison of brine evolution in a number of closed basins led Eugster & Jones (1979) to propose the following five major fractionation mechanisms:

1. Mineral precipitation
2. Dissolution of efflorescent crusts and sediment coatings
3. Sorption on active surfaces
4. Degassing
5. Redox reactions.

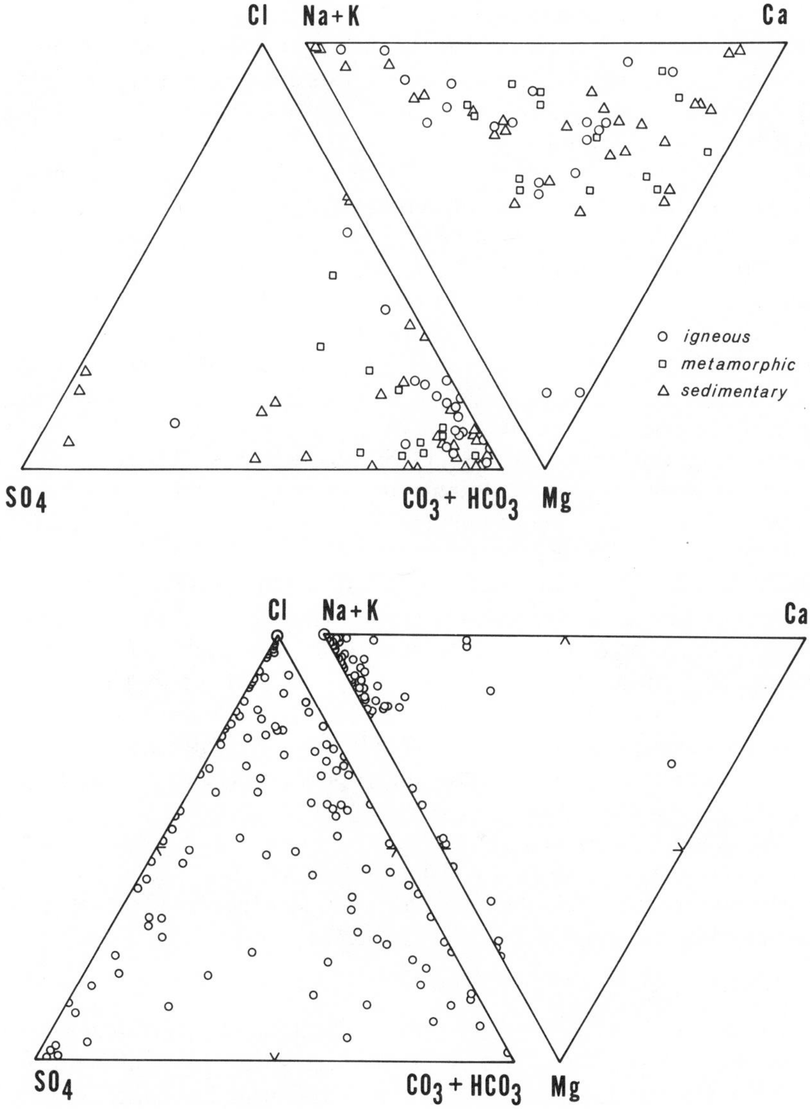

Figure 3 Comparison of dilute inflow waters with closed basin brines. From Hardie & Eugster (1970).

Efflorescent crusts and soil coatings represent another powerful process for fractionation and for increasing the solute load. Such crusts are common in all closed basins with sufficient inflow. They form either from ephemeral runoff by surface evaporation or by capillary evaporation from shallow groundwater. When rain or dilute runoff comes in contact with these crusts, only the most soluble constituents are taken back into solution, such as the sodium chlorides, carbonates, and sulfates, while the less soluble phases, such as the alkaline earth carbonates and silica, remain in the crusts. Since efflorescent crusts and soil coatings are ubiquitous in closed basins and generally short-lived, the importance of this process is obvious (Smith & Drever 1976, Jones et al 1977, Eugster & Hardie 1978). Drever & Smith (1978) used a laboratory demonstration to underscore the importance of wetting-drying cycles for the formation and fractional dissolution of intrasediment coatings. It should be emphasized that partial dissolution of efflorescent crusts does not lead to fractionation between components of the same soluble phases, such as Na and Cl, Na and CO_3, or Na and SO_4, but the differences in solubility between halite, trona, and thenardite are sufficient to produce fractionation between Cl, (HCO_3+CO_3), and SO_4. This, more than anything else, is responsible for the lateral zonation of saline minerals so frequently observed in salt pans (Hunt et al 1966, Jones 1965). This zonation is mirrored in the salinity of the interstitial brines, with the most concentrated brines at the center of the pan, near the topographic low, and the more dilute waters at the fringes.

Differential uptake and exchange of cations on active surfaces, such as provided by clays, has been blamed for a variety of solute losses. For instance, Eugster & Jones (1979) compared the extent of Na over K enrichment during evaporative concentration in a number of closed basins. They found the most extensive fractionation to have occurred at Lake Magadi which contains essentially no clay minerals. Volcanic glasses and silicate gels were thought to provide the surfaces on which exchange could take place.

Degassing of CO_2 in response to equilibration with respect to the atmosphere or to temperature changes leads to a decrease in the total carbonate species, while the carbonate alkalinity remains constant, as is indicated by the reaction

$$2HCO_3 \rightarrow CO_3^{2-} + CO_2 + H_2O.$$

It is one of the important mechanisms by which the pH of alkaline waters rises during evaporative concentration. Sulfate can be removed from brines by bacterial reduction to H_2S and subsequent degassing.

Sulfate is the most important solute responsive to redox reactions.

Many instances of sulfate loss from solution have been blamed on bacterial reduction (Kuznetsov et al 1963, Goldhaber et al 1977, Jones et al 1977) and associated sulfide precipitation. An interesting case of sulfur oxidation acting as a source for SO_4^{2-} has been described for some Lake Chad waters by Eugster & Maglione (1979).

Behavior of the Major Solutes

The number of closed basins for which sufficient water chemistry data are available to document solute behavior during evaporative concentration is no more than ten. In some of these, such as Saline Valley, California (Hardie 1968), inflow waters are so varied that the final products are a complex mixture that is difficult to interpret. Through an intrabasin comparison, Eugster & Jones (1979) have attempted to delineate the basic behavior of the major solutes.

The first task consists of finding an indicator of the degree of evaporative concentration. This should be a solute that does not participate in any of the fractionation mechanisms and hence is conserved in the solution. In many instances, no single solute satisfies this condition over the whole concentration range and it is then necessary to use bridges. Among the major constituents, we have found Cl to be the most useful, except for concentrated brines saturated with respect to chloride minerals. Minor solutes such as Br, B, and Li may also be useful.

Figure 4 illustrates the fractionation between Na and Cl over a concentration factor of 30,000 for the Lake Magadi, Kenya, basin. Among

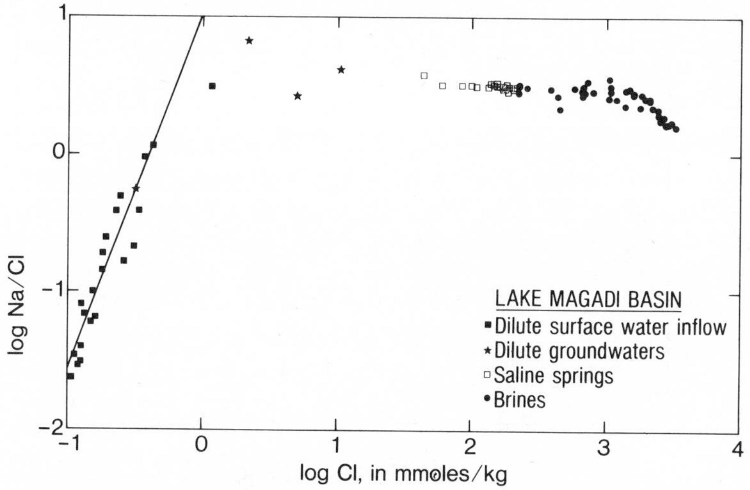

Figure 4 Fractionation between Na and Cl during evaporative concentration for Lake Magadi. From Eugster & Jones (1979).

the dilute inflow (*squares*), Na is acquired more quickly than Cl, which points to feldspar weathering as the principal source for Na. From ground waters (*stars*) to spring waters (*open squares*) to all but the most concentrated brines (*dots*), Na/Cl remains constant, because no fractionation occurs. The most concentrated brines are saturated with respect to trona, $NaHCO_3 \cdot Na_2CO_3 \cdot 2H_2O$, and precipitation clearly leads to a decrease in Na/Cl.

Carbonate species are subject to removal from solution by a variety of mechanisms, including precipitation of alkaline earth carbonates, degassing, and crystallization of saline carbonates. Figure 5 shows an example, again for the Lake Magadi basin and using Cl as monitor. Carbonate species are removed gradually and over the whole concentration range as befits the multiple removal mechanisms available.

Ca and Mg can be enriched or depleted during evaporative concentration, depending upon the initial $HCO_3/Ca+Mg$ ratio. At Magadi, they

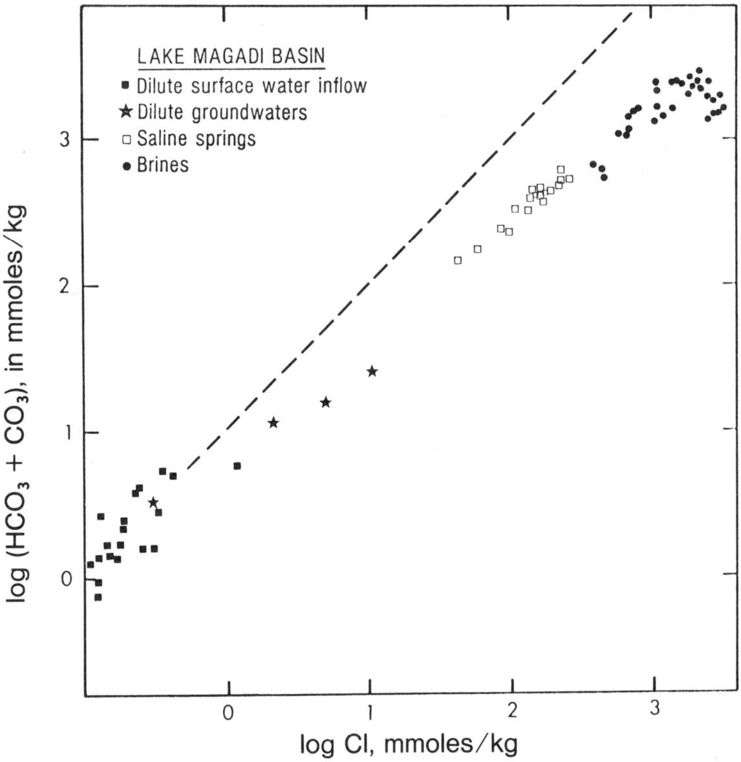

Figure 5 Fractionation between $(HCO_3 + CO_3)$ and Cl during evaporative concentration for Lake Magadi. From Eugster & Jones (1979).

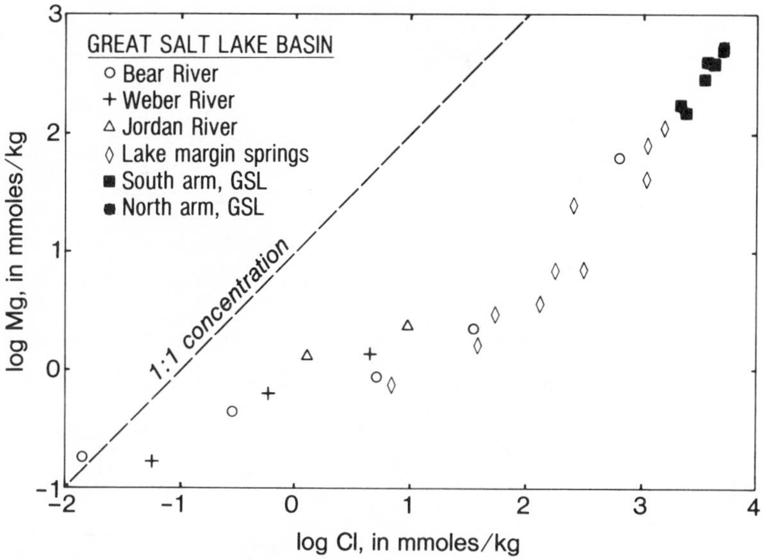

Figure 6 Fractionation between Mg and Cl during evaporative concentration for Great Salt Lake. From Eugster & Jones (1979).

are rapidly removed by precipitation as carbonate cements. In contrast, Figure 6 illustrates a case of Mg-enrichment as found at Great Salt Lake, Utah. The river waters show a distinct loss of Mg, compared to Cl, presumably due to carbonate precipitation. Once the waters have become impoverished in HCO_3, however, Mg is conserved along with Cl during evaporative concentration. This is the case for the lake margin springs and the lake brines.

In the brine evolution model of Figure 4, K and Na were combined, but in fact these solutes exhibit pronounced fractionation, as illustrated in Figure 7 for Lake Magadi. From dilute waters to hot springs, Na is enriched over K by a factor of 100, presumably by adsorption on active surfaces. The lake brines show no further enrichment of Na over K; in fact, the most concentrated brines are enriched in K due to trona precipitation.

Sulfate can behave in a complex fashion, since it is subject to mineral precipitation (e.g. in gypsum) and adsorption, as well as redox reactions.

Silica is perhaps the least understood of the major solutes. In dilute waters it is most readily removed by diatoms and similar organisms and some lake waters are strongly undersaturated even with respect to quartz (<1 ppm SiO_2). Evaporative concentration does not appreciably raise the silica level above that of the inflow. It has been suggested that during

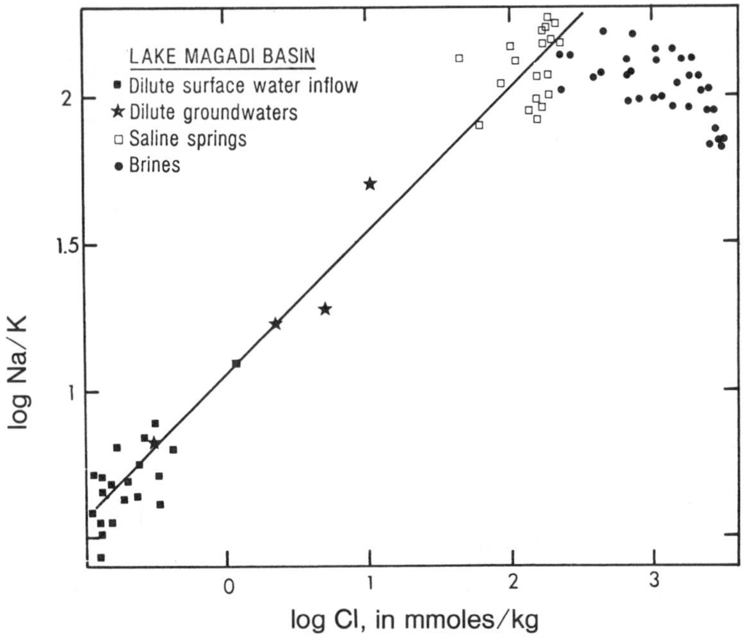

Figure 7 Fractionation between Na/K and Cl during evaporative concentration for Lake Magadi. From Eugster & Jones (1979).

this period, silica is lost from solution mainly in the form of opaline and similar cements. In alkaline brines, silica concentrations eventually increase due to the polymerization of H_4SiO_4 at high pH.

Behavior of the Minor Solutes

Data on minor solutes in closed-basin waters is very limited. Jones et al (1977) discussed Br, B, F, and PO_4 at Magadi and Vine (1976) has assembled information on Li. Boron contents of brines from Teels Marsh, Nevada, were reported by Surdam & Sheppard (1978). Smith (1979) believes that a single hot spring was responsible for the high boron content of the Searles Lake brines.

In principle, minor constituents can give as much information on evaporative concentration processes in closed basins as major solutes, but because of their low abundance in the dilute inflow, they are more difficult to trace. Nevertheless, many economically important deposits, such as those of nitrates, phosphates, borates, and lithium, are based on unusual enrichment of minor constituents. Most of these are connected with unusual input sources rather than unusual processes.

ACCUMULATION OF CHEMICAL SEDIMENTS

Mineral precipitation is, as we have seen, one of the main fractionation mechanisms affecting brine evolution. We must now focus on the fate of the precipitates.

Carbonate Production and Deposition

Alkaline earth carbonates are by far the most abundant products in all but the most unusual closed basins. A good example is the Green River formation which consists predominantly of dolomite and calcite. Similarly, at Searles Lake, California, the muds of the Mixed Layer, Bottom Mud, and Parting Mud are mainly dolomite, aragonite, and calcite (Smith 1979).

PERENNIAL SALT LAKES Alkaline earth carbonates are known to accumulate in active perennial salt lakes such as Great Salt Lake (Utah), Dead Sea, Lake Natron (Tanzania), Lake Balaton (Hungary), and others. Carbonates can be carried into the lake as clastics, they can be precipitated in the water column and settle to the bottom, or they can precipitate near-shore, for instance, in connection with bioherms (algal mounds).

Clastic input, contributed by rivers or by sheet floods, will probably consist mainly of reworked material derived from surrounding mud flats. Eugster & Surdam (1973) and Eugster & Hardie (1975) have suggested that much of the dolomite found in the oil shales of the Green River formation has this provenance. Peloidal fabric, the presence of intraclasts, and current structures are used as indicators.

The lower member of the Lisan formation, deposited from the precursor of the Dead Sea, is finely laminated and consists of dark-light couplets (Begin et al 1974). Calcite (reworked microfossils), dolomite, quartz, and clay minerals make up the dark couplets, which are considered to be of detrital origin, while the light couplets contain mainly precipitated aragonite.

The most detailed study of precipitation mechanisms of carbonates in a standing body of water has been carried out by Kelts & Hsü (1978) on Lake Zürich. Although referring to a freshwater setting, the conclusions are applicable to saline lakes. An earlier study by Nipkow (1920) has provided the basis for the "Lake Zürich model," used by Bradley (1929) to explain the "varves" of the Green River formation. According to Kelts & Hsü (1978), warming of surface waters in spring leads to supersaturation with respect to $CaCO_3$. Supersaturation is enhanced by algal blooms which remove CO_2 from the water and increase pH. In consequence, a fine rain of calcite and aragonite particles is initiated, which continues throughout the summer and results in the accumulation of a mm-thick carbonate lamina. Other material such as diatoms is included. During the fall, dead algal matter settles to the bottom, forming a dark lamina, the upper part of a couplet. Bioturbation is absent because of eutrophication. Couplets are continuous and are shown to be annual.

The Dead Sea also has carbonate laminae as part of couplets, but the other part is made up of clastic material (Neev & Emery 1967). Ca and HCO_3 are brought in by the fresher waters of the Jordan River and supersaturation with respect to aragonite takes place in response to surface evaporation. This produces a more-or-less continuous rain of carbonate particles which is interrupted by aperiodic floods bringing in detrital mud. The floods are not annual and in the last 70 years only 15–20 couplets have formed.

Great Salt Lake illustrates a further modification of the simple settle-out model. A considerable amount of the bottom sediment consists of brine shrimp fecal pellets (Eardley 1938). We have recently found that these pellets consist of a mixture of aragonite and detrital matter, mainly quartz and clays. Since *Artemia gracilis* is a filter feeder, the fecal pellets represent a sample of the mineral matter suspended in the water column. Aragonite probably formed in response to photosynthesis by algal blooms as indicated by a rise in pH of the surface waters.

Carbonate deposition near shorelines can take a variety of forms. Great Salt Lake is well known for its oolite beds (Eardley 1938) but the origin of the oolites is not yet clear (Sandberg 1975). Algal mounds or bioherms are common in many saline lakes and are important contributors to the carbonate rock record. At Great Salt Lake, Halley (1976) found the algal mounds to belong to an earlier, less saline period. Much has been written generally about the interaction between algal and carbonate sediment (for recent summaries see Walter 1976, Flügel 1977), but much more detailed information is needed from closed-basin situations. Smoot (1978) has discussed the significance of bioherms for part of the Green River formation.

Mixing of dilute, Ca–HCO_3 waters with alkaline lake brines may also lead to carbonate deposition near shore. At Lakes Magadi and Natron, this process leads to the formation of a pisolite pavement (Eugster 1979). The chemical aspects of mixing effects were delineated by Wigley & Plummer (1976).

EPHEMERAL SALT PANS Salt pans are not themselves environments where substantial Ca–Mg–CO_3 deposition takes place, though they may be surrounded by carbonate-rich mud flats. This is because salt pan brines are either poor in Ca + Mg or in CO_3. Alkaline earth carbonates present on the salt pan therefore are of clastic derivation or have formed by some unusual process, such as mixing of brines with dilute waters.

An interesting case of calcite pisolite growth in an active salt pan has been described by Risacher & Eugster (1979) from the Bolivian Andes. Hot springs, highly supersaturated with respect to atmospheric CO_2, bring in enough Ca to cause large supersaturation with respect to calcite after cooling and degassing. The result is a remarkable surface deposit reminiscent of cave formations.

MUD FLATS Mud flats associated with salt lakes can be very important for production, reworking, and deposition of Ca–Mg carbonates. The outer fringes of the flats may have waters dilute enough to precipitate substantial amounts of Ca–Mg carbonates in response to capillary evaporation. In saline, water-soaked mud flats a soft, micritic carbonate mud may accumulate, as it does at Deep Springs, California (Jones 1965, Clayton et al 1968), whereas intrasediment coatings and hard surface crusts are more typical for dry mud flats. The nature of calcrete surface crusts has been described by Goudie (1973), Read (1976), and Smoot (1978).

Pore waters of mud flats surrounding ephemeral salt pans are often zoned laterally with the most concentrated brines near the center. Because of calcite precipitation near the fringes, the Mg/Ca ratio commonly increases towards the center. This may lead to dolomitization of already-precipitated calcite, a process invoked by Eugster & Hardie (1975) to account for the high dolomite-to-calcite ratio of the Wilkins Peak Member of the Green River formation.

Mud flats are important locations not only for producing, but also for storing, reworking, and depositing carbonate sediment. Eugster & Hardie (1975) attributed the peloidal fabric of the carbonate-rich mud stones of the Wilkins Peak member to ripping up and transporting of mud-crack polygons by sheet floods associated with storm events. In general, laminations formed by such ephemeral, high-energy events can be distinguished from those of a quiet, standing body of water (Smoot 1978).

SPRING DEPOSITS Smoot (1978) has emphasized the importance of spring deposits for carbonate production in closed basins. Such deposits may be in the form of tufa mounds, travertine terraces, or surface crusts. They are particularly abundant near the toes of the alluvial fans, where dilute waters are abundant, and in the floor of ephemeral stream channels (Barnes 1965, Golubic 1969, Slack 1967). Dunn (1953) has associated tufa mounds with water mixing.

Deposition of Evaporites

PERENNIAL SALT LAKE Few details are available on the deposition of evaporite minerals in perennial salt lakes. We know that the Dead Sea is presently precipitating halite in the South Basin, which is in the process of drying up, whereas the lower water mass of the North Basin, which is just saturated with respect to halite (Lerman 1967, Lerman & Shatkay 1968), did deposit a massive halite unit 1500 years ago (Neev & Emery 1967). Great Salt Lake deposited halite during two periods in historic time, 1933–1934 and 1959–1963, and halite deposition appears to continue at the present time in the N-arm, where by 1972 a 1.5-m-thick halite crust had accumulated (Whelan 1973).

Precipitation of gypsum has been observed in the surface waters of the Dead Sea (Neev & Emery 1967), but gypsum apparently does not accumulate at the bottom, presumably because of bacterial reduction during settling. If it did, we might expect to find a laminated deposit consisting of aragonite-gypsum couplets. Aragonite and gypsum laminae have been described from the Lisan formation (Begin et al 1974).

From a perennial, nonalkaline salt lake in a regressive phase we might initially expect laminations consisting of detritus + a $CaCO_3$ phase precipitated in the water column. Increasing evaporation may cut out the detrital material and produce $CaCO_3$ + gypsum couplets. Eventually halite saturation will occur, represented by gypsum-halite couplets and eventually perhaps by halite beds. Finally, whatever residual salts are present, such as Mg or K, will deposit. No real salt lake follows this simple evolution and repressive phases will be interrupted by transgressions, reversing the depositional cycles. Furthermore, perennial saline lakes are prone to density stratification, with heavy brines at the bottom overlain by a more dilute water mass. This has been observed for the Dead Sea (Neev & Emery 1967) and Great Salt Lake (Whelan 1973). Begin et al (1974) have reported freshwater diatom laminae in the White Cliff member of the Lisan formation, which represents a saline stage of Lake Lisan. The diatoms must have been derived from a bloom in a dilute epilimnion.

EPHEMERAL SALT PAN The deposition of evaporites in ephemeral salt pans has been studied in a variety of modern settings (see Bonython 1955, Jones 1965, Hunt et al 1966, Hardie 1968, Eugster 1970, 1979, and others). Most of these pans are strongly zoned with respect to their interstitial brines and hence also their evaporite minerals which form by capillary evaporation. As mentioned earlier, the solute load is highest near the center and at the surface, presumably because of partial dissolution of efflorescent crusts. Less concentrated brines occur near the fringes of the pan and also below the center, producing a lateral as well as vertical zonation (see, for instance, Jones 1965). In steep-sided troughs, such as Lake Magadi, only the vertical zonation is obvious, as revealed by drill cores (Surdam & Eugster 1976, Eugster 1979). High-density brines are underlain by more dilute brines and there seems to be no tendency for overturn, because the brines are interstitial to the sediment.

In ephemeral salt pans, evaporite minerals can accumulate in one of two modes: as bedded surface deposits or as intrasediment precipitates. Efflorescent crusts, important as they are for brine evolution, are recycled too quickly to be able to accumulate in sizable deposits.

Bedded surface deposits may be very local and thin or they may be very large, depending upon the size of the pan and time of accumulation. At Lake Magadi, a trona deposit 7–40-m thick extends over 75 km^2 and is thought to have formed in less than 10,000 years (Eugster 1970). The salar of Uyuni in Bolivia has a bedded halite deposit of unknown thickness, covering 15,000 km^2. According to our model (Eugster & Hardie 1975), the many extensive trona beds of the Wilkins Peak member of the Green River formation have also formed in an ephemeral salt lake setting.

It is characteristic of bedded salt pan deposits that they are largely monomineralic. This is related to the fact that only the most soluble constituents of the efflorescent crusts will be dissolved and transported to the center of the pan, producing a chemically simple brine. Upon evaporation of this brine, a single, saline bed, usually in the mm–cm range, will be deposited. Flooding of the central pan is normally associated with storm runoff or spring thaws and such waters may bring in detrital material. This material will settle out in a thin mud layer before the saline mineral layer is deposited, producing a dark-light couplet. Couplets of 1–3-cm thickness have been described from the trona deposit at Lake Magadi (Eugster 1969).

Saline minerals accumulating in salt pans initially produce a sparry crystal meshwork with a porosity of up to 50%. Pores are occupied by interstitial brine. Considerable diagenetic changes are necessary to produce the massive deposits of the fossil record.

Surrounding the central salt pan we often find porous sediment containing interstitial brines. Capillary evaporation will initiate the growth of euhedral crystals or crystal aggregates of saline minerals within the porous sediment. The crystals can grow either by displacing the surrounding sediment or by including it. This is a common mode of growth for gypsum nodules, magadiite, gaylussite, glauberite, trona, and many other saline minerals. If there is a water table with fairly stable elevation, sizable beds of minerals can form in this manner. Jones (1965) has described the mineral zonation at Deep Springs playa (see Figure 8), which consists, from the fringes to the center, of the sequence: calcite → dolomite → gaylussite → thenardite → burkeite. At Saline Valley, Hardie (1968) noted the following mineral zones: alkaline earth carbonates → gypsum → gypsum + glauberite → glauberite → glauberite + halite → halite.

Mineral zonation appears to respond rapidly to changes in playa hydrology. If a playa grows, interstitial saline minerals will adjust to changes in brine composition and the zone boundaries will move outward, underscoring the dynamic nature of the playa system.

POSTDEPOSITIONAL PROCESSES

Evaporites are highly susceptible to postdiagenetic changes because of their solubility, their initially high porosity, and the presence of interstitial brines. Many of these brines are capable of interacting also with less soluble minerals, such as alkaline earth carbonates and silicates, and the result can be a varied and often characteristic suite of authigenic minerals.

Saline Minerals

Because their solubilities are strongly dependent upon P_{CO_2}, trona and nahcolite commonly form from interstitial brines, with the addition of CO_2 provided, for instance, by bacterial processes. In the Green River formation, nahcolite occurs as spherical concretions within the alkaline earth carbonate muds (Dyni 1974), while trona forms rosettes of bladed crystals which disrupt the bedding (Bradley & Eugster 1969, Eugster & Hardie 1975).

If the interstitial brines are Na_2CO_3-rich, minerals such as gaylussite ($Na_2CO_3 \cdot CaCO_3 \cdot 5H_2O$), pirssonite ($Na_2CO_3 \cdot CaCO_3 \cdot 2H_2O$), or shortite ($2CaCO_3 \cdot Na_2CO_3$) are likely to form by reaction with $CaCO_3$-minerals. Gaylussite is a very common constituent of Recent muds of saline alkaline lakes, while pirssonite forms at a somewhat higher salinity or temperature. In the Searles Lake muds, Eugster & Smith (1965) have used such mineral relations to document variations in the activity of H_2O of the interstitial brines. Shortite has only been encountered in the Green River formation and it probably formed at somewhat elevated temperatures (Bradley & Eugster 1969).

Gypsum and anhydrite can form within the sediment by authigenic processes, such as the reaction of $CaCO_3$ with sulfate-rich waters, although primary growth by mixing of waters is also possible. The latter process is thought to be responsible for the gypsum nodules of the Trucial Coast sabkhas where gypsum forms near the water table with continental runoff

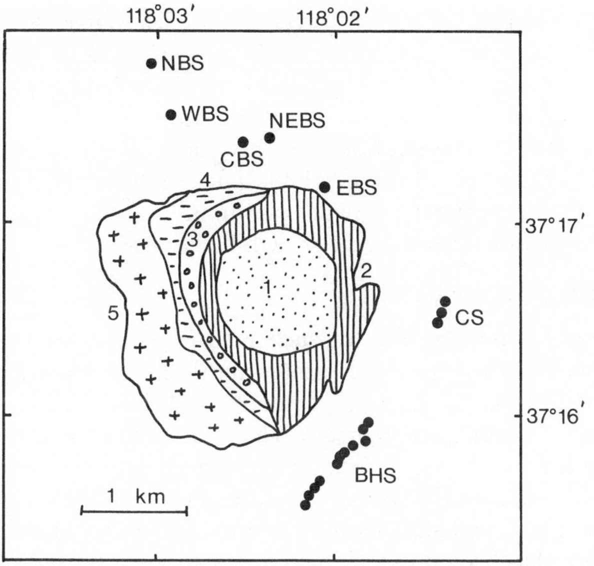

Figure 8 Mineral zonation of the Deep Springs playa. Modified from Jones (1965). *1*: burkeite, *2*: thenardite, *3*: gaylussite, *4*: dolomite, *5*: calcite. Black circles are mineral springs. From Eugster & Hardie (1978).

providing most of the Ca and sea water the necessary sulfate (Butler 1969). Oxidation of iron sulfides or organic sulfur is a mechanism that should not be overlooked for the formation of gypsum or anhydrite in areas where waters appear to be too low in sulfate. A *Thiobacillus* species is usually involved in the oxidation, which can produce acid waters that may dissolve $CaCO_3$ and locally lead to the formation of gypsum. Examples from the Chad basin, Africa, have been discussed by Eugster & Maglione (1979).

Double and triple salts, such as glauberite ($CaSO_4 \cdot Na_2SO_4$), burkeite ($Na_2CO_3 \cdot 2Na_2SO_4$), hanksite ($9NaSO_4 \cdot 2Na_2CO_3 \cdot KCl$), northupite ($Na_2CO_3 \cdot MgCO_3 \cdot NaCl$), tychite ($Na_2CO_3 \cdot MgCO_3 \cdot Na_2SO_4$), aphthitalite ($K_3Na(SO_4)_2$), polyhalite ($K_2SO_4 \cdot MgSO_4 \cdot 2CaSO_4 \cdot 2H_2O$), bloedite ($Na_2SO_4 \cdot MgSO_4 \cdot 4H_2O$) commonly form by authigenic reactions where the requisite starting materials are present. Hardie (1968) accounted for the presence of glauberite at Saline Valley, California, by reaction of previously formed gypsum with Na-rich brines. At Searles Lake, California, similar reactions were invoked for the presence of burkeite, tychite, northupite, hanksite, and aphthitalite (Eugster & Smith 1965). Some of these complex salts also have primary stability fields which can be reached by evaporation of natural waters. Glauberite and polyhalite, for instance, can form directly from sea water (Braitsch 1971), while burkeite forms a bedded surface deposit at Deep Springs (Jones 1965). In salt lakes of Turkey, polyhalite seems to form both by primary precipitation and by authigenic reactions (Irion 1973).

Careful textural observations may help document particular mechanisms, but in the case of these soluble minerals there may be no sharp division between primary growth by evaporative concentration and diagenetic formation or authigenic growth by brine-sediment interaction.

Associated Minerals

Zeolites and associated silicates, such as searlesite ($NaBSi_2O_6 \cdot H_2O$) and K-feldspar ($KAlSi_3O_8$), are common authigenic products of lake sediments. They form by reaction of interstitial brines with volcanic glass, silicate gels, or silicate precursors. They are typical for alkaline waters, such as sodium carbonates, because the high pH promotes silica solubility and hence reactivity. They take the place of clay minerals, such as montmorillonite, of nonalkaline settings. A recent summary of lacustrine zeolite occurrences has been provided by Surdam & Sheppard (1978).

At Searles Lake, California, and Teels Marsh, Nevada, searlesite ($NaBSi_2O_6 \cdot H_2O$) grows in response to high boron concentrations. In the Green River formation, several authigenic boro-silicates occur, namely searlesite, reedmergnerite, garrelsite, and leucosphenite (Milton 1971), along with other sodium-rich silicates, such as acmite, riebeckite, natrolite, and elpidite, all of which document the high activities of Na and SiO_2 in the interstitial brines. The unusual Zr-silicate elpidite ($Na_2ZrSi_6O_{12}(OH)_6$) is a case in point. Presumably it formed by interaction of a detrital zirconium mineral, such as zircon, with alkaline brines.

Among the many authigenic minerals of saline lacustrine deposits there are also some nonsilicates that are quite insoluble, such as fluorite (CaF_2), celestite ($SrSO_4$), and dawsonite ($NaAlCO_3(OH)_2$). At Magadi, fluorite is thought to form by interaction of fluoride-rich alkaline brines with $CaCO_3$ minerals (Surdam & Eugster 1976), a process also held responsible for some Oregon occurrences (Sheppard & Gude 1974). Celestite and dawsonite both occur in the Green River formation (Milton 1971) and presumably formed in a similar manner. Celestite has been reported to form in Turkish salt lakes by Irion (1973).

CONCLUSIONS

In the salt lakes of arid regions nature has provided us with fascinating laboratories in which processes of weathering, mineral formation, and deposition can be studied, processes that have significance well beyond evaporative settings. To understand and interpret their deposits fully, we must be cognizant of climatic, hydrologic, geochemical, biologic, and sedimentologic as well as tectonic aspects. At this time we probably know most about the geochemical controls, something about the sedimentological framework, and least about the interplay between climate, hydrology, and tectonic events. Lacustrine deposits represent records brief in geologic terms, but complete and unusually sensitive as indicators of environmental changes. We have a long way to go before we can adequately read this record.

ACKNOWLEDGMENTS

My understanding of lacustrine deposits owes much to my colleagues L. A. Hardie, B. F. Jones, and J. P. Smoot. Work supported by grants from the Petroleum Research Fund and the National Science Foundation.

Literature Cited

Allen, J. R. L. 1970. *Physical Processes of Sedimentation.* London: Allen & Unwin. 248 pp.

Barnes, I. 1965. Geochemistry of Birch Creek, Inyo County, California, a travertine depositing creek in an arid climate. *Geochim. Cosmochim. Acta* 29:85–112

Begin, Z. B., Ehrlich, A., Nathan, Y. 1974. Lake Lisan, the Pleistocene precursor of the Dead Sea. *Geol. Surv. Israel Bull. 63.* 30 pp.

Bonython, C. W. 1955. The area, volume, and salt content in Lake Eyre, South Australia. *Rep. Lake Eyre Comm., R. Geog. Soc. Aust.*

Bradley, W. H. 1929. The varves and climate of the Green River Epoch. *US Geol. Surv. Prof. Pap. 158-E*, pp. 87–110

Bradley, W. H., Eugster, H. P. 1969. Geochemistry and paleolimnology of the trona deposits and associated authigenic minerals of the Green River Formation of Wyoming. *US Geol. Surv. Prof. Pap. 496-B.* 71 pp.

Braitsch, O. 1971. *Salt Deposits: Their Origin and Composition.* New York: Springer. 297 pp.

Butler, G. P. 1969. Modern evaporite deposition and geochemistry of coexisting brines, the Sabhka, Trucial Coast, Arabian Gulf. *J. Sediment. Petrol.* 39:70–89

Clarke, F. W. 1924. The data of geochemistry. *US Geol. Surv. Bull. 770.* 841 pp.

Clayton, R. N., Jones, R. F., Berner, R. A. 1968. Isotope studies of dolomite formation under sedimentary conditions. *Geochim. Cosmochim. Acta* 32:415–32

Cooke, R. V., Warren, A. 1973. *Geomorphology in Deserts.* Berkeley: Univ. Calif. Press. 374 pp.

Drever, J. I., Smith, C. L. 1978. Cyclic wetting and drying of the soil zone as an influence on the chemistry of ground water in arid terrains. *Am. J. Sci.* 278: 1448–54

Dunn, J. R. 1953. The origin of deposits of tufa in Mono Lake. *J. Sediment. Petrol.* 23:18–23

Dyni, J. R. 1974. The stratigraphy and nahcolite resources of the saline facies of the Green River Formation in north-west Colorado. *Guidebook Rocky Mountain Assoc. Geol.* 74:111–21

Eardley, A. J. 1938. Sediments of the Great Salt Lake, Utah. *Am. Assoc. Petrol. Geol. Bull.* 22:1305–1411

Eugster, H. P. 1969. Inorganic bedded cherts from the Magadi area, Kenya. *Contrib. Mineral. Petrol.* 22:1–31

Eugster, H. P. 1970. Chemistry and origin of the brines of Lake Magadi, Kenya. *Mineral. Soc. Am. Spec. Pap.* 3:215–35

Eugster, H. P. 1979. Lake Magadi, Kenya, and its percursors. *Bat-Sheva Seminar III.* In press

Eugster, H. P., Hardie, L. A. 1975. Sedimentation in an ancient playa-lake complex: The Wilkins Peak member of the Green River formation of Wyoming. *Geol. Soc. Am. Bull.* 86:319–34

Eugster, H. P., Hardie, L. A. 1978. Saline lakes. In *Chemistry, Geology, and Physics of Lakes*, ed. A. Lerman, pp. 237–93. New York: Springer.

Eugster, H. P., Jones, B. F. 1979. Behavior

of major solutes during closed-basin brine evolution. *Am. J. Sci.* 279:609–31

Eugster, H. P., Maglione, G. 1979. Brines and evaporites of the Lake Chad basin, Africa. *Geochim. Cosmochim. Acta* 43:973–81

Eugster, H. P., Smith, G. I. 1965. Mineral equilibria in the Searles Lake evaporites, California. *J. Petrol.* 6:473–522

Eugster, H. P., Surdam, R. C. 1973. Depositional environment of the Green River formation of Wyoming: A preliminary report. *Geol. Soc. Am. Bull.* 84:1115–20

Flügel, E., ed. 1977. *Fossil Algae.* New York: Springer. 372 pp.

Garrels, R. M., Mackenzie, F. T. 1967. Origin of the chemical composition of some springs and lakes. In *Equilibrium Concepts in Natural Water Systems. Am. Chem. Soc. Adv. Chem.* 67:222–42

Glennie, K. W. 1970. *Desert Sedimentary Environments.* New York: Elsevier. 222 pp.

Goldhaber, M. B., Aller, R. C., Cochran, J. K., Rosenfeld, J. K., Martens, C. S., Berner, R. A. 1977. Sulfate reduction, diffusion, and bioturbation in Long Island Sound Sediments: Report of the FOAM Group. *Science* 277:193–237

Golubic, S. 1969. Cyclic and noncyclic mechanisms in the formation of travertine. *Verh. Int. Verein Limnol.* 17:956–61

Goudie, A. 1973. *Duricrusts in Tropical and Subtropical Landscapes.* Oxford: Clarendon. 169 pp.

Grabau, A. W. 1920. *Geology of the Non-Metallic Mineral Deposits.* New York: McGraw-Hill. 435 pp.

Halley, R, B. 1976. Textural variation within Great Salt Lake algal mounds. In *Stromatolites*, ed. M. R. Walter, pp. 435–45. New York: Elsevier

Hardie, L. A. 1968. The origin of the Recent non-marine evaporite deposit of Saline Valley, Inyo County, California. *Geochim. Cosmochim Acta* 32:1279–1301

Hardie, L. A., Eugster, H. P. 1970. The evolution of closed-basin brines. *Mineral. Soc. Am. Spec. Publ.* 3:273–90

Hardie, L. A., Smoot, J. P., Eugster, H. P. 1978. Saline lakes and their deposits: A sedimentological approach. *Int. Sediment. Assoc. Spec. Publ.* 2:7–41

Hsü, K. J., Siegenthaler, C. 1969. Preliminary experiments on hydrodynamic movement induced by evaporation and their bearing on the dolomite problem. *Sedimentology* 12:11–25

Hunt, C. B., Robinson, T. W., Bowles, W. A., Washburn, A. L. 1966. Hydrologic basin, Death Valley, California. *US Geol. Surv. Prof. Pap. 494-B.* 138 pp.

Irion, G. 1973. Die anatolischen Salzseen, ihr Chemismus und die Entstehung ihrer chemischen Sedimente. *Arch. Hydrobiol.* 71:517–57

Jones, B. F. 1965. The hydrology and mineralogy of Deep Springs Lake, Inyo County, California. *US Geol. Surv. Prof. Pap. 502-A.* 56 pp.

Jones, B. F., Eugster, H. P., Rettig, S. L. 1977. Hydrochemistry of the Lake Magadi Basin, Kenya. *Geochim. Cosmochim. Acta* 41:53–72

Kelts, K., Hsü, K. J. 1978. Freshwater carbonate sedimentation. In *Lakes: Chemistry, Geology, Physics*, ed. A. Lerman, pp. 295–323. New York: Springer

Kuznetsov, S. I., Ivanov, M. V., Lyalikova, N. N. 1963. *Introduction to Geological Microbiology.* New York: McGraw-Hill. 252 pp.

Langbein, W. B. 1961. Salinity and hydrology of closed lakes. *US Geol. Surv. Prof. Pap. 412.* 20 pp.

Lerman, A. 1967. Model for the chemical evolution of a chloride lake—The Dead Sea. *Geochim. Cosmochim. Acta* 31:2309–30

Lerman, A., Shatkay, A. 1968. Dead Sea brines: degree of halite saturation by electrode measurements. *Earth Planet. Sci. Lett.* 5:63–66

Livingston, D. A. 1963. Chemical composition of rivers and lakes. *US Geol. Surv. Prof. Pap. 440-G.* 64 pp.

Lotze, F. 1957. *Steinsalz und Kalisalze.* Berlin: Borntraeger. 465 pp.

Mabbutt, J. A. 1977. *Desert Landforms.* Cambridge: MIT Press. 340 pp.

McKee, E. D., ed. 1979. Continental arid climate lithogenesis. *Sediment. Geol.* 22: 1–120

Milton, C. 1971. Authigenic minerals of the Green River formation. *Wyoming Univ. Contrib. Geol.* 10:57–63

Multhauf, R. P. 1978. *Neptune's Gift: A History of Common Salt.* Baltimore: Johns Hopkins Univ. Press. 325 pp.

Neal, J. T., ed. 1975. Playas and dried lakes. *Benchmark Papers in Geology 20.* Stroudsburg, Penn.: Dowden, Hutchinson & Ross. 411 pp.

Neev, D., Emery, K. O. 1967. The Dead Sea: Depositional processes and environments of evaporites. *Israel Geol. Surv. Bull. 41.* 147 pp.

Nipkow, F. 1920. Vorläufige Mitteilungen über Untersuchungen des Schlammabsatzes im Zürichsee. *Zeitschr. Hydrologie* 1:100–22

Picard, M. D., High, L. R. 1973. *Sedimentary Structures of Ephemeral Streams.* New York: Elsevier. 223 pp.

Read, J. F. 1976. Calcretes and their distinction from stromatolites. In *Stromatolites*, ed. M. R. Walter, pp. 55–71. New York: Elsevier. 790 pp.

Reeves, C. C. 1968. *Introduction to Paleolimnology*. New York: Elsevier. 228 pp.

Reeves, C. C. ed. 1972. Playa lake symposium. *Int. Center Arid Land Studies Publ. 4*. Lubbock, Tex. 334 pp.

Reineck, H. E., Singh, I. B. 1975. Depositional sedimentary environments. New York: Springer. 439 pp.

Risacher, F., Eugster, H. P. 1979. Holocene pisoliths and encrustations associated with spring-fed surface pools, Pastos Grandes, Bolivia. *Sedimentology* 26:253–70

Sandberg, P. 1975. New interpretations of Great Salt Lake ooids and of ancient non-skeletal carbonate sedimentology. *Sedimentology* 22:497–537

Sheppard, R. A., Gude, A. J. 3rd. 1974. Chert derived from magadiite in a lacustrine deposit near Rome, Malheur County, Oregon. *J. Res. US Geol. Surv.* 2:625–30

Slack, K. V. 1967. Physical and chemical description of Birch Creek, a travertine depositing stream, Inyo County, California. *US Geol. Surv. Prof. Pap. 549-A*. 19 pp.

Smith, C. L., Drever, J. I. 1976. Controls on the chemistry of springs at Teels Marsh, Mineral County, Nevada. *Geochim. Cosmochim. Acta* 40:1081–93

Smith, G. I. 1979. Subsurface stratigraphy and geochemistry of late Quaternary evaporites, Searles Lake, California. *US Geol. Surv. Prof. Pap. 1043*. 130 pp.

Smoot, J. P. 1978. Origin of the carbonate sediments in the Wilkins Peak Member of the lacustrine Green River formation (Eocene), Wyoming, U.S.A. *Int. Assoc. Sedimentol. Spec. Publ. 2*, pp. 109–27

Surdam, R. C., Eugster, H. P. 1976. Mineral reactions in the sedimentary deposits of the Lake Magadi region, Kenya. *Geol. Soc. Am. Bull.* 87:1739–52

Surdam, R. C., Sheppard, R. A. 1978. Zeolites in saline, alkaline-lake deposits. In *Natural Zeolites; Occurrence, Properties, Use*, ed. L. B. Sand, F. A. Mumpton, pp. 145–73. New York: Pergamon.

Vine, J. D. 1976. Lithium resources and requirements by the year 2000. *US Geol. Surv. Prof. Pap. 1005*. 162 pp.

Walter, M. R., ed. 1976. *Stromatolites*. New York: Elsevier. 790 pp.

Wigley, T. M. L., Plummer, L. N. 1976. Mixing of carbonate waters. *Geochim. Cosmochim. Acta* 40:989–95

Whelan, J. A. 1973. Great Salt Lake, Utah: Chemical and physical variations of the brine, 1966–1972. *Utah Geol. Mineral. Surv. Water Resources Bull. 17*. 24 pp.

Ann. Rev. Earth Planet. Sci. 1980. 8: 65–93
Copyright © 1980 by Annual Reviews Inc. All rights reserved

CRATERING MECHANICS—OBSERVATIONAL, EXPERIMENTAL, AND THEORETICAL

×10125

H. J. Melosh

Division of Geological and Planetary Science, California Institute of Technology, Pasadena, California 91125

INTRODUCTION

Craters are the dominant landform on the Moon, Mercury, Mars, and many other solar system bodies. Old planetary surfaces, such as the lunar highlands, are so heavily cratered that new craters only obliterate old craters: no remnant of an original uncratered surface exists.

The word crater (from Greek κρατερ, meaning cup or bowl) is a non-generic term used to describe the approximately circular rimmed depressions observed in abundance on the surfaces of airless solar system bodies and occasionally even found on the earth. Crater diameters range from 0.1μm to more than a thousand kilometers. The origin of these features was hotly debated until about twenty years ago (see Shoemaker 1962 for a review of this controversy). It is now generally accepted that most craters are produced by the impact of meteorites of different sizes, although a small minority of craters are probably due to volcanism.

Impact cratering is thus an important geologic process. This review concentrates on the mechanics of this process. The role of impact cratering in the growth and evolution of planetary regoliths (Lindsay 1976, Gault et al 1974, Shoemaker et al 1970), erosional and depositional processes on airless bodies (Ross 1968, Soderblom 1970, Gault 1970), and the use of crater statistics to establish a geological timescale for other planets (Chapman & Jones 1977, McGill 1977, Shoemaker et al 1963) are not treated here.

The impact of a meteorite on a planetary surface (the "target") at perhaps 30 km sec^{-1} initiates an orderly (but rapid) sequence of events

0084-6597/80/0515-0065$01.00

which produces the observed crater. The formation of an impact crater can be divided into three major stages (Gault et al 1968). The first stage begins when the meteorite initially contacts the target surface. During this stage intense shock waves form in the meteorite and in the target. Much of the meteorite's kinetic energy is transferred to the target. During the second, excavation, stage the shock wave propagates into the target and establishes a flow field that opens the crater and ejects material from the crater cavity. When excavation finally ceases, the cavity partially collapses and is altered by various gravitational and elastic rebound processes during the modification stage.

These last two stages are the same for meteorite impact craters as for craters produced by explosives (either chemical or nuclear) detonated at shallow depths below the surface. Since there is far more experimental data available for large explosions than for large impacts, the similarity between the two processes has been widely exploited to learn more about impact craters, in spite of uncertainties about the validity of this comparison.

Although small impact craters (less than about 1 m) can be produced and studied in the laboratory, most of what is known about the mechanics of large cratering events is deduced from the final impact crater itself. Large impacts are too rare and occur too rapidly to be observed directly. It is thus appropriate to begin this review with a description of three "type" craters whose features must be explained by a correct theory of cratering mechanics.

OBSERVED CRATER MORPHOLOGY

Features whose gross morphology can be described as "roughly circular rimmed depressions" occur over a wide range of sizes on planetary bodies. Lunar rocks are pitted by tiny craters ranging from 0.1 μm (Clanton & Morrison 1979) up to several millimeters (Hörz et al 1971) in diameter. There is no break between these microcrater populations and populations of macroscopic craters with diameters from centimeters to nearly a thousand kilometers. The detailed morphology of craters does, however, depend upon their size. This review is focused on craters larger than a few meters in diameter. Within this size range, there are three principal morphologic types (Howard 1974). Although these types grade into one another, and departures from the type forms are common, it is useful to make this threefold distinction. Thus, craters can usually be classified as simple, bowl-shaped, rimmed depressions (small craters, up to $\sim$15 km diameter on the moon), complex craters with terraced walls, flat floors, and central peaks (intermediate size, roughly 15 to 150 km diameter on the moon), or, finally, as basins with two or more concentric mountainous rings (largest size, 150–1200 km diameter on the moon). Although the diameters at which these morphologic transitions occur are apparently functions of material properties and surface gravity, the same general progression from simple craters to complex craters to basins with increasing crater size is seen on all planets. These three major morphologic types are well illustrated by the Arizona Meteor Crater, the lunar crater Copernicus, and the Orientale basin on the moon.

Arizona Meteor Crater

The Arizona Meteor Crater is a simple, bowl-shaped crater located on the southern Colorado plateau in north-central Arizona. The crater's rim-to-rim diameter is approximately 1100 m, its somewhat eroded raised rim stands 47 m (originally about 67 m) above the surrounding plain, and its original depth below the level of the plain was 150 m (Roddy 1978). Meteor Crater thus has a depth/diameter ratio of nearly 0.2, similar to the depth/diameter ratio of small fresh lunar craters (Pike 1974). The crater was formed when a large iron meteorite struck a flat-lying Paleozoic rock section consisting of a thin surface veneer of Moenkopi sandstone, underlain by the Kaibab limestone, which is in turn underlain by the Coconino sandstone. The crater is of Pleistocene age (Shoemaker 1963). In plan form the crater is more square than circular. A concentration of small vertical tear faults occurs at the corners of the square. The diagonals of the square parallel regional joint directions, suggesting structural control as the reason for the noncircular planform. Although the velocity and mass of the impacting meteorite will probably never be known precisely, one of the current best estimates (Bryan et al 1978) puts the total energy of the meteorite at about 4.5 Megatons (1.88×10^{16}J), its velocity at about 15 km sec^{-1}, and its mass near 1.7×10^8 kg, thus requiring a body of iron roughly 30 m across.

A cross section of the crater (Figure 1) shows that its floor is underlain by a lens of mixed breccia nearly 150 m thick. This breccia is mainly composed of shattered blocks of Coconino sandstone, finely crushed and fused sandstone, and meteoritic material in the form of fine spherules dispersed in glass. Large blocks of Kaibab limestone occur near the top of the lens. These blocks were evidently displaced downward before coming to rest. This lens is overlain by a 10-m thick layer of breccia composed of rock fragments derived from all formations intersected by the crater, along with oxidized meteoritic material. This breccia layer is graded upward from coarse to fine and is interpreted by Shoemaker (1963)

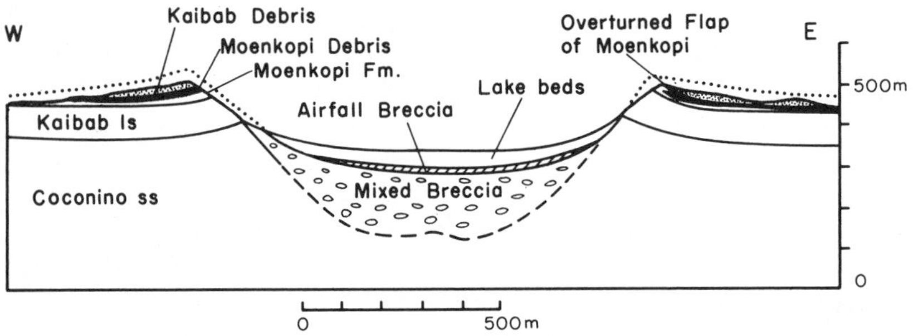

Figure 1 Cross section of the Arizona Meteor Crater, after Shoemaker (1960). The dotted line is approximately the original profile of the crater. Note the plastically uplifted rim, the overturned flap of Moenkopi on the east rim, and the inverted stratigraphy of the ejecta on the west rim.

as an airfall deposit from debris thrown to great height. Approximately 30 m of Pleistocene lake beds now overlie the airfall breccia. Gravity surveys (Regan & Hinze 1975) show a negative Bouguer anomaly centered on the crater, presumably associated with the low density breccia and fractured rock beneath the crater. Other terrestrial craters show similar negative gravity anomalies (Innes 1961).

Figure 1 shows that the upraised rim of Meteor Crater is due to a combination of two effects. Roughly one quarter to one half of the rim's height is due to local uplift of the original ground surface around the crater, thus causing originally horizontal bedding in the walls of the crater to dip outward. This permanent deformation of the rock adjacent to the crater is also seen in nuclear explosion craters, and is evidently due to plastic deformation and flow of the rock surrounding the crater (Nordyke 1961, Carlson & Jones 1965). The remainder of the rim height is due to material ejected from the crater. Shoemaker (1963) and Roberts (1964, 1965) noted that near the rim this ejected material retains the stratigraphy of the rock inside the crater, but is inverted. The "overturned flap" of Moenkopi overlain by Kaibab debris, in turn overlain by Coconino debris (an inverted stratigraphy), can be seen in Figure 1.

Much of Meteor Crater's original ejecta blanket has been removed by erosion. However, even now it is characterized by large blocks of rock on the rim and hummocky topography for several hundred meters beyond it. Fresh lunar craters comparable in size to Meteor Crater have similar blocky rims with hummocky topography grading into dune-forms beyond 0.1 to 0.3 crater radius. This topography extends out to the edge of the continuous ejecta blanket at about one crater diameter away from the rim (Schultz 1976, p. 186). Secondary craters occur beyond this range.

Copernicus Crater

The lunar crater Copernicus, a complex crater with terraced walls, flat floor, and central peaks, is located in Mare Procellarum, south of the Carpathian Mountains. Its bright ray system makes it a conspicuous feature, especially at full moon, and is consistent with the fresh appearance and young age of the crater (possibly 900 m.y.; Eberhart et al 1973: this age is based on the interpretation of Apollo 12 KREEP material as Copernicus ejecta). The flat floor of Copernicus lies 2.7 km below the level of the surrounding plain, and its rim rises 1.1 km above the plain (Pike 1976). The crater's rim-to-rim diameter is 93 km. In planform the crater rim is scalloped, consisting of a wreath of coalescing alcoves, each about 10 km across. A zone of terraces begins at the base of a scarp inside the rim, descending stepwise across a zone 15 or 20 km wide to the flat inner floor of the crater. These terraces appear to be slump blocks, each 2 to 5 km wide and up to 10 km long. Their tops are rotated so that most of them now dip outward, away from the crater (Howard 1975). The 54-km diameter flat floor of the crater consists of a unit of hummocky material, interpreted as fractured bedrock (breccia?), and a textured unit, interpreted as solidified impact melt. Frozen ponds of (presumably) impact melt occur on the rim and among the terraces. Leveed channels and flows of once molten material appear on the crater walls and floor. It is significant that none of these features share the structural deformation of the crater walls, indicating that the terraces and flat floor formed shortly after the impact, before the impact melt had time to solidify.

Copernicus has a striking group of low central peaks which occupy the inner 15 km of the crater floor. Although the highest of these peaks rises nearly a kilometer above the surrounding floor, its summit is still about 1.7 km below the level of the surrounding plain. Little is known of the subsurface structure of Copernicus. Eroded terrestrial impact craters such as Sierra Madera (Wilshire & Howard 1968) and Gosses Bluff (Milton et al 1972) show structural central uplifts which may be analogous to the central peaks of Copernicus. Copernicus is associated with a strong negative Bouguer gravity anomaly, suggesting that it, like Meteor Crater, is underlain by a large lens of low density breccia and fractured bedrock (Dvorak & Phillips 1977).

Copernicus has an extensive ejecta blanket. Continuous ejecta, which obliterate pre-impact topography, lie within about one crater diameter of the rim. The deposit is divided into a hummocky facies, which is blocky near the rim and extends 15 to 20 km away from the rim, and a radially lineated facies consisting of low subradial ridges. This facies continues to the edge of the continuous ejecta (Schmitt et al 1967). Discontinuous subradial ridges extend somewhat beyond this edge, along with loops, chains, and clusters of secondary impact craters. Secondary craters are common from the edge of the continuous ejecta out to several hundred kilometers from the crater rim. Bright rays, which are observed only as albedo features and hence contain little material, extend more than 500 km from the crater.

The Orientale Basin

The young multiringed basin Orientale is located on the western limb of the moon at about 20° S. The inner relatively flat floor lies 2 to 3 km below the distant surrounding plains. The basin is surrounded by a well-defined inward facing scarp 2 to 7 km high and about 900 km in diameter. This scarp sharply defines the inner edge of the Cordillera Mountain ring. Within the Cordillera escarpment, a second mountainous ring, which forms the Rook Mountains, has a diameter of about 600 km and is also bounded on the inward side by a 2 to 7 km high scarp (Howard et al 1974). The Rook escarpment is not as well defined as the Cordillera. The terrain between the Rook and Cordillera escarpments slopes gently outward from the crest of the Rooks to the foot of the Cordillera. Two other rings of irregular mountain peaks occur within the Rooks, having diameters of about 480 and 320 km (Moore et al 1974). The ratios of adjacent ring diameters for Orientale and other basins are said to be close to 1/2 or $\sqrt{2}$ (Baldwin 1963, p. 332; Hartmann & Wood 1971), although there is considerable scatter in this relation (Howard 1974).

The inner basin of Orientale is 4 to 8 km below the crests of the Rook and Cordillera escarpments. The inner basin is partially flooded with mare basalt, but otherwise has a fissured, hummocky floor which is interpreted as impact melt and fallback (Scott et al 1977). Although the Orientale basin is generally believed to be of impact origin, no one has yet succeeded in demonstrating which, if any, of the rings corresponds to the rim of the original impact crater. Modification of the crater after its excavation was rapid, since ejecta draping the Cordillera indicate that the escarpment was present when the impact ejecta arrived (McCauley 1977). Like other fresh craters, Orientale is associated with a negative Bouguer gravity anomaly (a small positive anomaly near its center is correlated with the mare fill). The anomaly is not as large as might be expected by extrapolation from smaller craters, however, implying at least partial isostatic compensation.

The overlapping ejecta blankets of large basins are the basis of lunar stratigraphy (Howard et al 1974). Orientale's ejecta blanket is recognized as the Hevelius formation, and is divided into several facies. An enigmatic knobby deposit lies between the Rook and Cordillera escarpments. This deposit (called the Knobby facies of the Montes Rook formation by McCauley 1977) is generally bounded by the Cordillera escarpment, although lobes of this material locally transgress the scarp. Beyond the Cordillera, the Inner facies of the Hevelius formation is characterized by radial elongate ridges and troughs. Occasional flow lobes, leveed channels, and transverse ridges within craters (deceleration dunes?) suggest that the deposit originated as some kind of ground-hugging flow. This facies extends 300 to 600 km from the Cordillera scarp where it grades into the Outer facies, a unit that is characterized by weakly lineated terrain. Secondary crater chains and clusters occur beyond about 1000 km from the Cordillera scarp.

Multiringed basins similar to Orientale and its lunar counterparts occur on Mercury (Gault et al 1975, McCauley 1977) and Mars (Wilhelms 1973), indicating that such basins are produced by some aspect of the cratering process for very large impacts.

CRATERING MECHANICS

The formation of an impact crater is a continuous process that begins when the meteorite (projectile) first contacts a planetary surface (target) and ends with the final motions of debris around the crater. However, it is convenient to divide this process into several stages, each of which is dominated by a different set of physical phenomena. Three major stages are recognized.

Compression or Penetration Stage

During the compression stage the meteorite first contacts the surface of a planetary body, generating strong shock waves in both the meteorite and surface rocks. Compression ceases when the shock finally engulfs the entire meteorite, by which time the meteorite has penetrated some distance into the surface and transferred much of its energy to the underlying rocks. The compression stage lasts for as long as it takes the shock to traverse the meteorite, which may be only 10^{-6} sec for laboratory experiments on millimeter size projectiles, but ranges from 10^{-3} to 10^{-1} sec for 10-m to 1-km meteorites.

When the rapidly moving meteorite strikes the surface, a high pressure region bounded by strong shock waves develops at the interface.[1] One

[1] It is not strictly necessary that the "shock waves" move faster than the speed of sound in either target or projectile. In the case of a subsonic "shock wave" (more accurately termed a plastic wave) the elastic precursor to the slowly moving front of strong compression is not large enough to significantly affect the target. The discontinuous "shock front" description is thus a good approximation. For more on this topic see Zel'dovich & Raizer (1967, Chapter 11) and Kolsky (1953).

shock wave propagates from the meteorite-target interface downward into the target, while the other propagates upward into the body of the meteorite (Figure 2). Target material is at rest until the downward propagating shock front reaches it, compressing and accelerating it downward behind the shock. Similarly, projectile material moves downward at its full impact velocity until the shock front propagating into the projectile reaches it, compressing it and slowing its initial velocity. Boundary conditions at the interface between the projectile and target require that the pressure and particle velocity behind both shocks are the same.

The pressure, shock velocity, particle velocity, and internal energy of the shocked material can be roughly estimated by assuming that the impact approximates the impact of two infinite plane plates that have the same composition as the projectile and target. At this stage the material properties of the projectile and target are adequately specified by the relation between the velocity of a shock front moving into unshocked material, U_p for the projectile and U_t for the target, and the particle velocity u_p, u_t in each material behind the shock front (in the reference frame where the unshocked material is at rest). Thus, for the projectile and target, respectively,

$$U_p = f_p(u_p) = A_p + B_p u_p + C_p u_p^2 + \dots \quad (1)$$

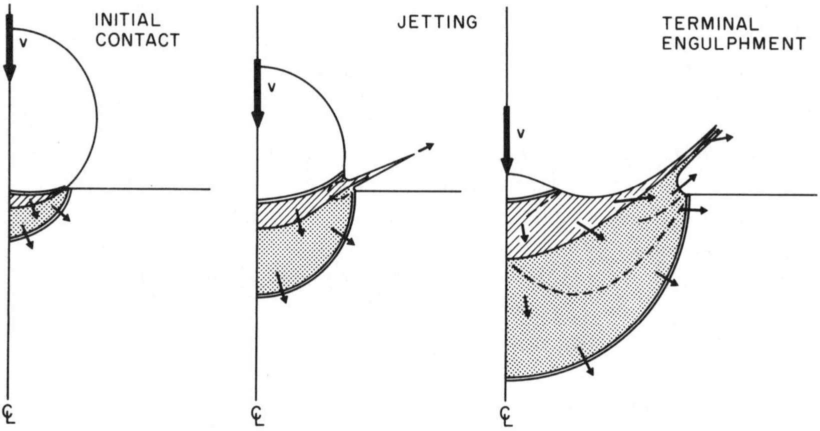

Figure 2 The penetration stage. When the meteorite, moving at velocity *v*, first strikes the target, shock waves propagate into the meteorite and into the target. Shocked meteoritic material is diagonally ruled; shocked target material is stippled. The arrows represent particle velocity. After the initial contact, rarefaction waves (dashed) propagate from free surfaces into the shocked material, resulting initially in jetting, then in the ejecta curtain as terminal engulfment occurs and brings the penetration stage to a close. After Gault et al (1968).

$$U_t = f_t(u_t) = A_t + B_t u_t + C_t u_t^2 + \dots \qquad (2)$$

These relations are material properties and must be measured experimentally. Fortunately, however, the first two terms in the Taylor expansions (1) and (2) frequently give an adequate representation of the shock velocity, even up to several tens of km sec^{-1}. More sophisticated numerical computations use a Tillotson (Mie-Gruneisen type) equation of state (Ahrens & O'Keefe 1977, Bryan et al 1978). The pressure P behind the shock, the density ρ of shocked material, and its specific internal energy E (per unit mass) are related to the initial pressure P_0, density ρ_0, and specific internal energy E_0 by the Hugoniot equations

$$\rho_0 U = \rho(U - u), \qquad (3)$$

$$P - P_0 = \rho_0 U u, \qquad (4)$$

$$E - E_0 = \frac{P + P_0}{2}\left(\frac{1}{\rho_0} - \frac{1}{\rho}\right) = \tfrac{1}{2}u^2. \qquad (5)$$

These equations are derived from the conservation of mass, momentum, and energy (respectively) across the shock front. The equations are valid in the frame of reference where unshocked material is at rest: they must be transformed to the moving coordinate frame of the projectile before being used there. Application of the Hugoniot equations (3), (4), and (5) to both the projectile and target, along with the material properties (1) and (2) and boundary conditions at the interface yields a unique solution for the pressure, particle velocity, shock velocities, and internal energies in both the projectile and target (Shoemaker 1963, Gault & Heitowit 1963). The pressures derived from the planar impact approximation are the highest attained during the impact process. These pressures range from 1 to about 10 Mbar for impacts at velocities of 10 to 30 km sec^{-1}, depending upon the projectile and target compositions (Gault 1974).

The phenomenon of "jetting," which occurs shortly after the contact of the projectile with the target, is due to these high initial pressures. The first material ejected from the site of an impact is in the form of a low angle hydrodynamic "jet" that has a velocity several times that of the impacting projectile (Gault et al 1963). This material is the most strongly shocked, hence the hottest, ejecta and is observed in the form of an incandescent liquid or superheated vapor spray. Since the amount of material involved in the jet is a small fraction of the projectile mass, the phenomenon is probably not quantitatively important in the mass or energy balances of the impact process, although it may contribute to the production of impact melt, especially for low velocity oblique impacts (Kieffer 1977). Jetting occurs as the shock waves spread away from the initial contact point into the target and projectile, leaving a high pressure region of strongly shocked material behind. This high pressure region is immediately adjacent to the zero pressure free surface at the sides of the projectile. A strong pressure gradient thus accelerates shocked material radially outward toward the free surface precisely as a shaped charge accelerates its liner (Birkhoff et al 1948). At the same time, rarefaction waves propagate inward toward the axis of the projectile, relieving the initial high shock pressure.

The compression stage ceases when the shock front propagating into the projectile, weakened by rarefactions moving inward from its sides, finally reaches the back of the projectile. As before, when the high pressure region behind the shock confronts a free surface, the pressure drops to zero, shocked material is accelerated toward the free surface (upward in this case), and rarefaction waves propagate back into the high pressure shocked region. These rarefaction waves move at the velocity of sound in the *shocked* material, which is larger than the original shock velocity, and unload the high pressure shocked material to zero pressure along an isentrope. Some of the internal energy of the shocked material is thus converted to kinetic energy (by acceleration down the pressure gradient toward the free surface). The rest appears as heat (Gault & Heitowit 1963). Depending upon the original shock pressure and the thermodynamic equation of state of the projectile, the unloaded material may be in the form of a hot solid, liquid, or a gas (Ahrens & O'Keefe 1972).

At the end of the compression stage, the entire mass of the projectile has been engulfed by the shock wave and compressed to high pressures. Jetting has ceased, although the ejecta curtain is beginning to form. The interface between the projectile and the target has penetrated about half a projectile diameter below the original target surface and continues to move downward at a speed given roughly by the particle velocity in the planar impact approximation. The projectile has become badly distorted, flowing almost hydrodynamically to form a kind of liner to the cavity that is developing [see O'Keefe & Ahrens (1977a) for illustrations of this phenomenon]. A strong, roughly hemispherical, shock front spreads into the target, engulfing more material as time goes on. The pressure behind this shock front is lower than that in the shock itself due to the encroachment of rarefaction waves from behind. A "detached shock" thus forms (Bjork et al 1967) in which the high pressure region is isolated on an expanding hemispherical surface. The main result of the compression stage is the transfer of energy from the projectile to the target. Roughly half (Gault & Heitowit 1963) or perhaps more (O'Keefe & Ahrens 1977a, Bryan et al 1978) of the initial kinetic energy of the projectile is deposited in the target either as internal or kinetic energy. Rarefaction waves starting from the back of the projectile propagate rapidly into the target, unloading the shocked material. Eventually the rarefactions catch up with the shock front spreading into the target, rapidly reducing its strength. This rapid reduction of the shock pressure begins several projectile diameters away from the impact point (Ahrens & O'Keefe 1977).

Since a large fraction of the projectile's kinetic energy is deposited in a small volume of the target, many authors have compared a meteorite impact to an explosion. Some of the earliest writers on meteorite impact (e.g. Boon & Albritton 1937, and other references in Shoemaker 1962), suppose that the impact event is best described as the penetration of the projectile, deposition of its energy, and a final explosion identical in most respects to an ordinary chemical or nuclear explosion. Although impact and explosion processes are not strictly analogous (Shoemaker 1962), especially since the projectile deposits much more momentum in the target than an explosion does, the comparison is often useful. Oberbeck (1971) clarified the relation between impact and explosion craters experimentally and showed the importance of the effective depth of burst.

Both the form and diameter of an explosion crater are strong functions of the depth of burst of the charge. Equal size charges which are either very near the surface or deeply buried produce small craters. There is an intermediate depth (the optimum depth of burst) that produces the largest craters (Nordyke 1961). The central problem in relating impact to explosion craters is to determine an effective depth of burst for the meteorite's energy.

Most efforts to compute the effective depth of burst center around the depth of penetration of a hydrodynamic jet. Birkhoff et al (1948) showed that a high velocity jet of material with density ρ_j and length L penetrates a distance D into a target of density ρ_t where

$$D = L\sqrt{\rho_j/\rho_t}. \tag{6}$$

This relation was verified for the lengthwise impact of rods into metals by Christman & Gehring (1966). It may, however, be a poor approximation for spherical projectiles, although a similar dependence of the depth of energy deposition on the projectile/target density ratio is obtained from the planar impact approximation. Shoemaker (1960, 1963) used Equation (6) to estimate the effective depth of burst for the meteorite that formed Meteor Crater. He obtained a depth of 120 to 150 m below the original ground surface (4 to 5 projectile diameters) by assuming that the jet forms when the rarefaction from the back of the meteorite just reaches the meteorite-target interface. He supposed further that the jet consists of the unloaded meteorite plus the shock-compressed target. Shoemaker chose this prescription because Meteor Crater is structurally similar to the crater produced by the deeply buried Teapot Ess nuclear explosion. More recent work by Bryan et al (1978) suggests, however, that a depth of burst of about 85 m (2 to 3 projectile diameters) gives a better fit. Oberbeck (1971) compared small explosion craters to an impact crater produced in sand by a 2 km sec^{-1} aluminum projectile, finding that the explosions best simulate the impact for a depth of burst nearly equal to the projectile length.

These results indicate that the effective depth of burst is roughly equal to D in Equation (6), where L is the projectile diameter. Equation (6) correctly predicts that a low density projectile impacting a high density target produces a crater that has a smaller effective depth of burst than a crater produced by a high density projectile impacting a low density target (Bjork et al 1967). This qualitative difference suggests, for example, that the impact of an icy cometary object on a silicate planetary surface produces a crater like that of a shallow near-surface explosion, whereas an iron meteorite impacting an icy surface produces a crater like that of a deeply buried explosion. Such relations, however, have not yet received detailed theoretical or experimental study. It is not known for certain whether the effective depth of burst depends upon velocity or how it is altered for oblique impacts. Thus, although the impact/explosion crater analogy is potentially very useful, it has not yet been formulated precisely.

Crater Excavation Stage

The crater cavity expands rapidly during the excavation stage. Target material is either ejected ballistically or pushed out of the crater by plastic deformation of the surrounding rock. A roughly hemispherical shock wave spreads away from the point of impact, engulfing increasing volumes of the target and becoming weaker in the process. Rarefaction waves propagating from the free surface reach the roughly hemispherical shock front, weakening it further. Eventually, the shock front becomes an ordinary elastic wave.

Target material entering the spreading shock wave is suddenly compressed and accelerated downward and outward in a direction normal to the shock front. Later, when rarefaction waves arrive from the horizontal free surface, shocked target material is unloaded isentropically and given an upward velocity increment (i.e. it is accelerated toward lower pressures). Depending upon the strength of the shock, unloading may leave the target in the form of a hot solid, liquid, or gas. The impact melt that is observed in terrestrial and lunar craters is produced during the early stages of crater excavation (Dence 1971) when shock pressures are high. The upward velocity increment, when added to the initial radial outward-and-downward velocity due to the shock, leads to a net upward velocity for those portions of the target that are already close to the surface (Gault et al 1968). This material is ejected ballistically from the growing crater. Deeper portions of the target merely have their radial velocities slowed and deflected. The depth of the transition between ejected material and radially displaced material is found to be about 1/3 of the final crater's depth in small scale impact experiments (Stöffler et al 1975).

As the shock wave weakens, the pressure jump across the shock, the shock velocity, and the particle velocity behind the shock decline. This is a result of the Hugoniot Equations (3) to (5) along with conservation of the system's total energy, which requires that the energy per unit mass E must decline as the initial energy is spread over more of the target. Gault & Heitowit (1963) used a semi-analytic approach to derive the rate at which the shock pressure declines. They assumed a linear relation between shock velocity and particle velocity (1) and (2), a hemispherical expanding shock wave, and that 1/2 of the projectile energy is deposited in the target. To simplify the calculation they also assumed that the pressure and energy density are uniform between the expanding hemispherical cavity and the shock front. Their results for an iron meteorite impacting a basalt target show an initial rapid pressure decline ($P \sim r^{-2.6}$, where r is the radial distance from the impact point). At still lower pressures, $P \sim r^{-1.5}$. These results do not agree with the more recent numerical computations of O'Keefe & Ahrens (1975) and Ahrens & O'Keefe (1977) who used a Tillotson equation of state. The numerical results indicate that the peak shock pressure is lower than the approach of Gault & Heitowit predicts and that the pressure decays less rapidly. Thus, more of the target sees lower pressures, yielding less vaporization and more melting of the target. Dence (1968) and Robertson & Grieve (1977) attempted to observe the decay of the shock pressure by studying shock effects in the rocks around terrestrial impact structures. These attempts are complicated by the displacement of the shocked material during excavation of the crater.

The course of crater excavation is determined by the transport of mass, pressure, and energy. Even in the case of axial symmetry the material equation of state and complex, changing geometry of the problem make numerical methods the only practical means of making theoretical predictions. The most comprehensive numerical models solve finite difference approximations to the full Eulerian equations of motion from before the impact until the last motions in the crater (or, more often, until the investigator runs out of computer funds). Various versions of these models incorporate material strength and gravity (Dienes & Walsh 1970, Kreyenhagen & Schuster 1977). No practical models, so far, are three dimensional, so that oblique impacts cannot be studied. The most severe problem faced by numerical computations is the large expansion of the crater. A finite difference grid fine enough to resolve the strong pressure gradients in the beginning of crater excavation is too fine to be practical for the later stages. Most numerical codes thus resort to rezoning (imposing the old solution on a coarser grid) at various times during the computation. Eulerian codes, moreover, do not resolve the sharp shock front, which becomes spread over several grid spacings.

Although the Eulerian code approach can correctly model all of the physical processes occurring during the impact, it is very costly and few models can be computed. Maxwell (1977) devised a simple model of the excavation stage alone whose results compare well with the results of Eulerian codes but which is much cheaper to run, although the relation of this model to physical processes is obscure. In Maxwell's "Z-model" the cratering flow is assumed to be incompressible and the radial velocity u_r is a power function of r, the distance from the impact point. Thus, $u_r = \alpha(t) r^{-Z}$ where $\alpha(t)$ is an undetermined function of time and Z is a constant. Energy conservation must be imposed on the flow and can be used to determine $\alpha(t)$. Gravity can also be incorporated into the model. The Z-model produces plausible-looking flow fields and may become a useful method of extending Eulerian code calculations to later times.

An important result of the work with Eulerian codes is the concept of "Late Stage Equivalence" (Dienes & Walsh 1970). The results of a number of numerical computations show that the late stage flow field is unchanged if the projectile diameter L (or, for nonspherical projectiles, the diameter of an equal mass sphere) and velocity v are varied in such a way that Lv^n is constant. The parameter n is found to be 0.58 ± 0.01. If the flow depended upon projectile energy W alone, n would be 2/3 (cube root scaling: distances are scaled by $W^{1/3}$; see Glasstone 1957), whereas if the flow depended upon momentum alone, n would be 1/3 (Öpik 1969). A value $n = 0.58$ is consistent with the $W^{1/3.4}$ scaling determined empirically for explosion craters (Nordyke 1961). Equivalence of the flow fields for two impacts ensures that the later, strength-dependent part of the flow is equivalent also, leading to the relation

$$H/L = K(\rho_p/\rho_t)^{1/3}(v/c)^n \tag{7}$$

where H is the crater depth, diameter, or some other characteristic dimension, ρ_p and ρ_t are projectile and target densities, and c is the speed of sound in the target. The quantity K is a function of target strength. Although K is constant for a given target material, it must be determined by experiment. Late Stage Equivalence is probably violated when gravity is important in the flow.

As the shock wave expands away from the impact point, it initiates radially outward flow of the material it engulfs. At first, where shock pressures are high, this flow is essentially hydrodynamic. Later on, as the shock weakens and particle velocities u_p decrease, gravity and the strength of the target play increasing roles. The importance of these two effects can be estimated by a dimensionless parameter analogous to the Reynolds number. If the yield stress of the target is Y, then material strength becomes important when the stagnation pressure of the flow, $\rho_t u_p^2$, becomes comparable to Y, i.e.

$$R_S = \rho_t u_p^2 / Y \lesssim 1. \tag{8}$$

The importance of gravity can be gauged by the Froude number F,

$$F = u_p^2 / gl \tag{9}$$

where l is a characteristic length scale and g is the gravitational acceleration. Gravity is important when F is smaller than one. Gravity becomes relatively more important as the crater size scale l increases.

Material strength resists expansion of the crater, resulting in smaller craters. Gravity has the same effect, since it makes it more difficult to eject material from the crater cavity. In strong materials the shock wave also expends much of its energy crushing the target. This "energy of comminution" consumes a significant fraction (10–25 %) of the total available energy in the impact (Gault & Heitowit 1963). Unlike gravity, comminution is independent of the crater size scale: its relative importance is the same in large and small craters. Continued plastic deformation of target material crushed by the shock is responsible for part of the raised rim observed around craters (Nordyke 1961).

Although a target of uniform strength is a reasonable first approximation for theoretical models, planetary surfaces usually contain materials of varying strength. Layers, or strata, of differing strengths are one of the most common variations. Such layering can arise either from a difference of rock units or from the presence of a water table. Quaide & Oberbeck (1968) studied the form of impact craters produced in a two-layer target that had a weak layer (noncohesive sand) over a strong substrate. They found that the crater morphology depends upon the ratio between the crater's rim-to-rim diameter D and the thickness h of the weak layer. In thick layers, $D/h < 4$, the crater is a normal, simple, bowl-shaped crater with a depth/diameter ratio of about 0.2. A small central mound appears on the floor of craters in layers with $4 < D/h < 8$ (this is one of at least three ways in which central peaks are produced). The base of the mound occurs nearly at the top of the strong layer. A central pit surrounded by a bench occurs for $D/h > 9$. This bench forms on top of the strong layer and produces apparently concentric craters for $D/h \gg 10$ when the bench occurs high on the crater wall. This sequence of morphologic types is observed on the lunar maria for small ($D \lesssim 200$ m) craters and has been successfully used to map the regolith's thickness.

The crater cavity grows at a rate given roughly by the average particle velocity behind the shock. Explosion experiments by Peikutowski (1977) show that the initially hemispherical cavity expands until it reaches its maximum depth (the depth of the "transient crater"). The diameter of the crater continues to increase, however, until the cavity has a depth/diameter ratio of about 0.2. The crater cavity is usually much larger than the original projectile. For a 15-km-sec^{-1} impact, the crater may be 10 times deeper and 50 times larger in diameter than the meteorite that produced it. The time required to form the crater is approximately proportional to crater size: about one second for a 10-km diameter crater, ten seconds for a 100-km diameter crater. The excavation stage thus takes several orders of magnitude longer than the compression stage.

The ballistic ejecta from the crater are expelled during the excavation stage. Target material, which is initially accelerated radially outward from the impact point, receives an upward momentum impulse from the rarefactions propagating down from the horizontal free surface. A "transition region" in the flow develops along the side of the expanding cavity. Target material deeper than the transition region continues to flow downward, whereas shallower material is ejected from the crater (Gault et al 1968). Experiments (Stöffler et al 1975) show that target material deeper than about 1/3 the depth of the final (simple) crater is

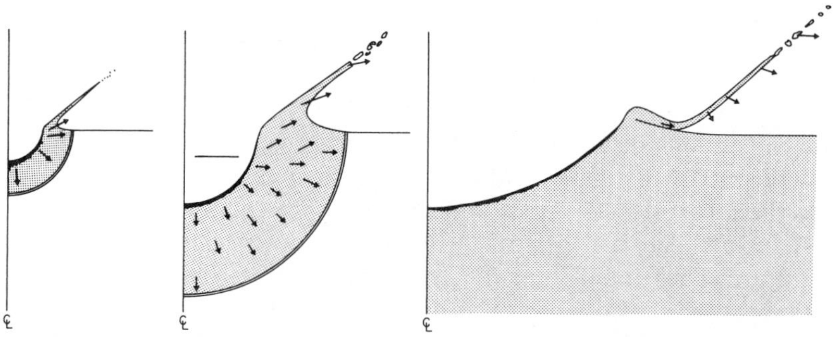

Figure 3 The excavation stage. The initially small hemispherical cavity grows at a rate given roughly by the particle velocity behind the rapidly expanding shock front (double lines). Shocked material is stippled. The cavity eventually stops increasing in depth. Ejecta is derived from a level above that indicated by the horizontal line in the middle frame. Finally, the crater diameter stops increasing, the overturned flap and raised rim are formed, and ejecta begins falling back to the surface. After Maxwell (1977).

not ejected at all. This fact is important for the interpretation of the provenance of material found in impact crater ejecta blankets.

The ejection velocity at the crater rim is barely adequate to heave material out of the crater. The crater stops growing at this point. The lowest velocity ejecta drop onto the crater rim, forming the overturned flap (Figure 3). Most of the ejected material leaves the crater at angles near $45° \pm 10°$ to the horizontal. The ballistic ejecta form a curtain whose walls dip inward at about 45°. This curtain sweeps across the target surface from the crater rim to the maximum range of ejection (Oberbeck 1975, Oberbeck et al 1975). The range of each ejected fragment is determined by its ejection angle and velocity. Since its velocity depends upon the magnitude of the shock pressure the fragment was subjected to, there is a crude relation between range of ejection and shock pressure: the more strongly shocked ejecta travel farthest, although this relation is somewhat modified by the depth from which the material originated (material from the surface is less strongly shocked than that from deeper down at the same radius; Ahrens & O'Keefe 1978). Some material is ejected at such high velocity that it can escape the planet entirely. The amount of material that thus escapes is important for planetary accretion: if the average impact ejects more mass than it brings in, then net erosion is occurring, whereas if, on the average, less mass is ejected, the planet is accreting. The crossover point between erosion and accretion is mainly a function of the impact velocity and surface gravity (Gault et al 1963, O'Keefe & Ahrens 1977b).

The thickness of the ejecta blanket is a function of distance from the center of the crater, R. The ejecta blanket is thickest on the crater rim (located at R_c) and thins outward. The total volume of the ejecta approximates the volume of the crater (with appropriate corrections for the plastically upraised rim and bulking of the ejecta). The thickness of the ejecta blanket also varies azimuthally, giving rise to the ray structure. Carlson & Jones (1965) documented this azimuthal variation for nuclear explosion craters and showed that the radial variation of ejecta blanket thickness δ (measured in mass/unit area) can be fit by an equation of form

$$\delta = k(R_c)(R/R_c)^{-m} \quad \text{for} \quad R \geqq R_c \tag{10}$$

where m is roughly 3 ± 1. Further study (McGetchin et al 1973) supported this value of $m \approx 3$ around impact craters, although there are some uncertainties (Settle et al 1974), especially in the dependence of k on R_c. Numerical computations have been done for shallow nuclear explosions (Trulio 1977), nuclear surface bursts (Seebaugh 1977), and impacts (Ahrens & O'Keefe 1978). These results show departures from $m = 3$, although experimental work on small, half-buried explosive charges indicates that $m = 3$ works well (Andrews 1977). Stöffler et al (1975) find that m varies between 2.8 and 3.5 in small-scale impact experiments. Thus, at the present time, $m \approx 3$ seems to be a good first estimate.

The thickness of the ejecta blanket around large lunar basins is more controversial. Estimates by Short & Forman (1972) and McGetchen et al (1973) differ by a factor of five in some places. Moreover, Oberbeck (1975) and Oberbeck et al (1975) argue that the high velocity basin ejecta scour the target surface, impart their outward momentum to it, and incorporate large quantities of pre-impact surface material into the "ejecta" blanket deposit. This mobilization of the target surface (which Oberbeck terms "ballistic sedimentation") may also be responsible for the ground-hugging flow characteristics observed in the ejecta of Orientale.

The ejecta distribution of impact craters is modified by a number of circumstances, such as oblique impact, nearby impacts, target properties, and target topography (Gifford et al 1979). Gault & Wedekind (1978) showed experimentally that the ejecta distribution becomes noncircular for low angle impacts, with high velocity rays extending downrange of the projectile and a wedge-shaped region of sparse ejecta lying uprange, producing a bilaterally symmetrical ejecta pattern. Such patterns are seen around lunar craters, such as Proclus, and missile impacts (Moore 1976).

When two or more impacts occur close together and nearly simultaneously (as commonly happens for secondary impact craters, which are produced in chains and clusters), straight or V-shaped dunes occur between them. This pattern is known as the "herringbone pattern" when it occurs along crater chains. Oberbeck & Morrison (1974) investigated these dunes experimentally and found that they are due to the interference of the ejecta blankets of the secondary craters.

Some Martian craters have peculiar, lobate ejecta blankets whose origin is presently uncertain. Carr et al (1977) termed these craters "rampart craters." Gault & Greeley (1978) showed that similar ejecta blankets could be produced by impacts into fluid mud. It is widely believed that the presence of permafrost in the martian soil is responsible for this type of ejecta blanket, but the mechanics of this process are not presently understood.

Modification Stage

Modification of the bowl-shaped transient crater begins after excavation is complete and the ejecta are launched (although some of the ejecta may not land until many minutes after the impact). Modification is not extensive for simple, bowl-shaped craters: some of the target and projectile material which failed to make it over the crater rim drains back down the walls and pools in a breccia lens on the crater floor. Local collapse of the inner crater wall occurs where it is steeper than the angle of repose.

Larger impact craters, however, are drastically modified; slumping partially fills the crater, central peaks rise, and multiple rings may form. There is currently no fundamental understanding of these phenomena. Most workers agree that slumping and central peaks are due to gravitational collapse (Quaide et al 1965, Settle & Head 1979). The evidence for this interpretation is largely from the appearance of the terraces (which resemble terrestrial slumps in undrained clay slopes), the general association of central peaks with slump terraces (Howard 1974), and the break in the depth/diameter curve (Figure 4; Pike 1974), which correlates with the appearance of terraces in the craters. The depth/diameter ratio for small lunar craters is constant at about 0.2. This ratio is similar to the depth/diameter ratio computed theoretically for impact and explosion craters just after excavation. However, craters larger than about 15 km in diameter on the moon have depths between 3 and 5 km, nearly independent of diameter (Figure 4). This sudden change in character suggests that some kind of strength threshold is exceeded for craters larger than about 15 km. If this is correct, then craters on Mercury and Mars should collapse at smaller diameters than lunar craters (Mercury and Mars have roughly the same surface gravity, so that the crater diameter at collapse should be the same on both if the target strengths are the same). Although some evidence supports gravitative collapse on Mercury (Gault et al

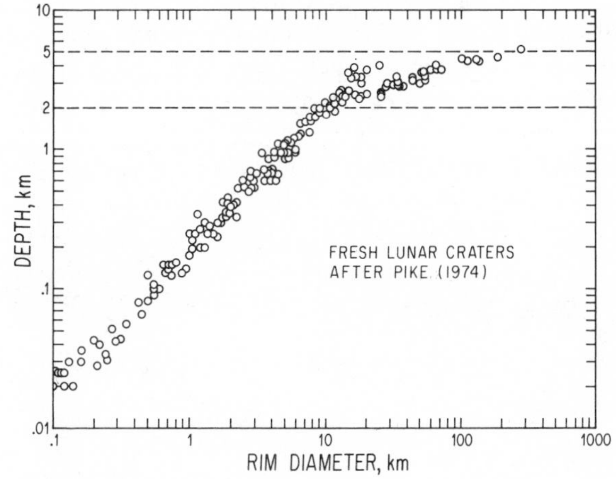

Figure 4 The relation between depth and diameter of fresh lunar craters. Note the break in slope near a diameter of 15 km and the approximately constant depth of 3–5 km for larger craters. After Pike (1974).

1975), Mars (Carr et al 1977), and the Earth (Hartmann 1972), discrepancies suggest that other factors may be important (Cintala et al 1976).

A surprising aspect of the idea that the transition from simple to complex craters is due to mechanical failure is the low strength required: for slumping to begin at 15-km diameter on the moon, the surface rock must fail under differential stresses of only 20 to 30 bars. Furthermore, this strength cannot increase as the overburden pressure increases. That is, failure must occur as if there were less than about 2° of internal friction (McKinnon 1978). Such a failure criterion is more typical of water-saturated clays than of dry rock, whose strength depends markedly on overburden pressure, with angles of internal friction ranging from 30° to 45°. Melosh (1977) showed that a perfectly plastic failure criterion can explain the flat floors, nearly constant crater depth, and slump terrace widths (which are roughly independent of crater size) of complex craters on the moon. The difficulty comes in understanding how such a peculiar failure law is realized, because lunar rocks do not contain water nor are they shock-heated to high enough temperatures to produce much impact melt beneath the crater. Since slumping is fast, as evidenced by undeformed impact melt overlying crater floors, this failure law must be due to dynamic effects in the rapidly flowing, shock-processed debris around the crater (Melosh 1979). Explosion cratering calculations also require a strength reduction after passage of the shock wave in order to fit the observations (Swift 1977). If this result is correct and the necessary failure characteristics are realized dynamically, plastic slip-line calculations (Melosh 1977) show that the collapse of large craters takes place along deep-seated faults. Such faults have been mapped around the deeply eroded Sierra Madera structure by Wilshire & Howard (1968). While the walls of the crater collapse downward and inward along steeply dipping faults to form terraces, the central region rises in plug-like fashion until the crater depth reaches 3 to 5 km (on the moon), at

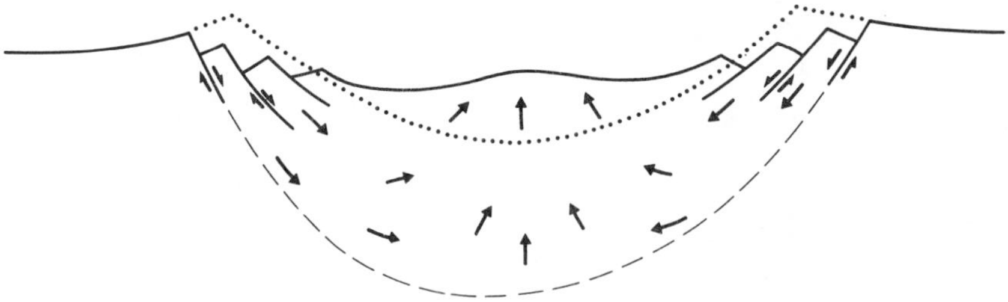

Figure 5 Motions inferred to occur during the slumping of an initially bowl-shaped transient crater (dotted line) to form a complex crater. The dashed line shows the limit of the deformation. After Dence (1971).

which point the crater becomes stable (Figure 5). This model does not predict central peaks, although it does produce a flat floor.

The origin of central peaks is highly controversial. There are at least three ways in which central peaks can arise, and all of them have been suggested to explain complex craters. The occurrence of a central mound over the strong layer in a layered target has already been mentioned. The mound occurs only for a limited range of crater diameters, being replaced by a pit at larger diameters. Many authors have proposed a hydrodynamic explanation (one of the first was Gilbert 1893) in which the central peak is likened to the central jet that rises from the center of a collapsing crater in water. Scott (1967) showed that this mechanism could operate only if the lunar surface rocks around the collapsing cavity had an effective viscosity between 3×10^7 and 3×10^{10} poise. Although this seems too low, perhaps it could also be realized by a dynamical flow mechanism. Finally, central peaks are commonly observed in surface explosion craters, although they do not occur for buried charges (Milton & Roddy 1972, Piekutowski 1977). In this case the central peak appears to be mainly due to rebound of the shock-compressed material beneath the crater. During unloading the downward velocities of material beneath the crater reverse, buckling up the central area. Ullrich et al (1977) studied this phenomenon in great detail for a 20-ton TNT explosion, concluding that some combination of "elastic" rebound and gravitative collapse is involved. Rebound central peaks are probably not the cause of the simple to complex crater transition observed on the moon because the formation of rebound peaks in surface explosions does not depend on crater size. The effective depth of burst for most impact craters is presumably larger than that necessary for the production of a rebound peak. However, in special cases where the depth of penetration is unusually low, rebound central peaks may arise. This may have occurred in the 40-m diameter crater produced by the Apollo 14 SIVB stage on the moon. This crater has a central peak, unlike nearby craters of similar diameter (Masursky et al 1978, p. 126), perhaps because of the low depth of penetration of the empty booster, although the peculiar structure of the projectile (a large hollow cylinder of low *average* density) may have had an important effect. Low density cometary impacts might also produce central peaks by this mechanism: too little is known of the appropriate conditions to be certain at present.

On a still larger size scale, suggestions for the origin of multiple ring basins also run through a gamut of possibilities. These ideas include various collapse models (Hartmann & Wood 1971, Howard et al 1974, Melosh & McKinnon 1978), which infer that the rings are fault scarps produced during the rapid collapse of the transient crater by gravity, the "frozen tsunami" model (Van Dorn 1968), which likens the rings to gravity waves produced by impact into an inviscid fluid, and the "nested crater" model (Hodges & Wilhelms 1978), which relates the rings to layers of different material properties in the moon. None of these models has achieved a consensus, and all have mechanical difficulties: perhaps the ultimate truth lies in some combination of them. Multiple ring-like structures have also been produced by surface explosions (Roddy 1976). Even in this relatively well-controlled case their origin is not understood.

The previously discussed collapse processes are rapid, occurring within minutes to hours after the impact. On the much longer geologic time scale, the crater is subject to slow erosion by other impacts, large and small, which may eventually batter the crater beyond recognition, as well as to the more usual sorts of aeolian and fluvial erosion, when they are possible. Many craters presently seen on planetary surfaces have been partially filled by lava flows, which sometimes bury craters completely. Isostatic adjustments may also occur for the larger craters, deforming the crater and reducing its relief by the viscous flow of material within the planet's interior. The relaxation of craters in a uniform viscosity half space was studied by Scott (1967) and Daneš (1962). Since small craters on the Moon, Mercury, and Mars have not flowed in the fashion predicted by these authors, the viscosity of the upper 50 to 100 km in these planets must exceed about 10^{26} poise. The recently observed craters on Ganymede, however, may fit the viscous relaxation model better (Smith et al 1979). The actual rheology of the terrestrial planets, however, is better approximated by an elastic lithosphere overlying a viscous or fluid interior. The first effects of isostatic adjustments in this rheology are regional doming beneath the crater and radial normal faulting extending several crater radii from the rim (Melosh 1976). Such post-impact radial normal faults occur around the Imbrium basin on the moon, where they cut mare basalts that are about 600 m.y. younger than the basin excavation event.

SCALING

Many attempts have been made to relate important impact crater dimensions such as diameter, depth, or volume to the total energy and other simple parameters of the impacting meteorite. These "scaling relations," although of great utility, usually lack a firm theoretical foundation and great caution must be used in extrapolating outside the diameter range in which they have been established (White 1971). The earliest scaling laws were derived from explosion experiments (Glasstone 1957) and related parameters such as diameter D and depth H to the total energy

W raised to the one-third power. Thus, if a crater of diameter D_0, depth H_0 is produced by a charge that released an energy W_0, then a charge of energy W should have diameter $D = D_0(W/W_0)^{1/3}$ and depth $H = H_0(W/W_0)^{1/3}$. Since all linear dimensions scale as $(W/W_0)^{1/3}$, the depth of burst must be similarly scaled for the explosions to be truly equivalent. These results were refined by Nordyke (1962) who included nuclear explosion data to show that a power 1/3.4 fits better than 1/3. Nordyke also examined the effects of the scaled depth of burst.

None of these empirical scaling laws contain the effect of gravity in a fundamental way. Theoretical arguments based on energy partition (Gault & Moore 1965, O'Keefe & Ahrens 1978) suggest that, whereas most of the impact energy is consumed by heating and fracturing of the target in small craters, progressively more energy is expended in overcoming gravity for larger craters. Thus, the volume V of small craters should scale as W directly (which, if H/D is constant, implies that H and D scale as $W^{1/3}$ with no dependence on gravity), while for large craters V should scale as $W^{5/4}g^{-1}$ [again assuming H/D is constant, the gravitational work in excavating a crater of volume V is of order $\rho g V H/2$, so that if this is a major fraction of the impact energy W, dimensions scale as $(W/\rho g)^{1/4}$]. Small-scale impact experiments (Gault & Wedekind 1977) show an intermediate $g^{-0.165}$ dependence of diameter on gravity, while explosion tests in a centrifuge (Schmidt 1977) indicate that diameter scales as $g^{-1/12}$. This disagreement may reflect the fundamental differences between impact and explosion events: the effective depth of burst in impact experiments may be a function of velocity, making comparison with explosion experiments meaningless until this variation is defined. Indeed, if Equation (6) is correct and the effective depth of burst depends only on projectile size and density, then higher velocity impacts of the same size object have shallower *scaled* depths of burst and may produce a different sort of crater. The scaling of impact craters by energy alone may thus be incorrect if the velocity plays a major independent role. O'Keefe & Ahrens (1977a) demonstrated the importance of impact velocity for energy partition and showed, in particular, that the volume of impact melt is a strong function of velocity.

The only rigorous kind of scaling (known as "dynamical scaling"; Dienes & Walsh 1970) can be derived from the Euler equations of motion including strength. These equations are unchanged when all length and time scales are changed by the same amount, so that velocity is unaltered. Thus, shock pressure, velocities, and the equation of state are the same before and after scaling. In this way, for example, the impact of a 1-cm diameter iron sphere impacting a basalt target at 15 km sec^{-1} can be related to the impact of a 1-km diameter iron sphere on basalt at 15 km

sec^{-1} simply by multiplying all length and time scales by 10^5. Unfortunately, gravity violates this simple scaling rule.

The size of an impact crater also depends strongly upon the angle of impact (Gault 1973). Although oblique hypervelocity impacts produce circular craters (except at grazing angles; Gault & Wedekind 1978), the shallower the angle of impact, the smaller is the resulting crater.

The best scaling relation for impact craters at the present time is probably that of Gault (1974), who relates crater diameter D (in m) to projectile density ρ_p(gm cm^{-3}), target density ρ_t and kinetic energy W (joules) of the meteorite by

$$D = 0.027\, \rho_p^{1/6}\, \rho_t^{-1/2}\, (\sin\theta)^{1/3}\, W^{0.28}. \tag{11}$$

This equation is supposed to be valid for craters larger than 100 m in diameter on the moon. The nature of the target must be considered in more detail for smaller craters.

The ejecta blanket of an impact crater scales very differently from the crater itself. Since the range of an ejected fragment depends primarily upon ejection velocity and gravity, it is largely independent of crater size. The ejection velocity is a function of the shock pressure, hence the impact velocity, not the size of the projectile. Thus a 1-m diameter crater and a 1-km diameter crater produced by meteorites with the same velocity send similarly shocked ejecta to a range of 10 km. The larger impact, of course, sends more material to this distance, but the ejecta velocity upon landing is roughly the same (ignoring the slight difference in range caused by the finite radius of the crater). This scaling effect may explain part of the facies changes observed in the ejecta blankets of progressively larger craters. Ejecta landing, say, one-half crater diameter from the rim of two different sized craters has a proportionately larger velocity for the larger crater. Thus, the ejecta within the continuous blanket should show more signs of target surface scour and high energy deposition as crater size increases (Oberbeck 1975). Gravity (and atmospheric drag, if present) also plays a role in regulating the range of crater ejecta, since ballistic ejecta of a given velocity travel farther in lower gravity. Gault et al (1975) noted that the secondary crater pattern around fresh Mercurian craters is proportionately smaller than that around lunar craters, supporting this idea.

CONCLUSION

Present understanding of the mechanics of impact cratering decreases as the size of the crater increases. The early stages of compression and excavation are reasonably well understood, although many details, especially the role of gravity and oblique impact, need further investigation. The major area for future research is probably the late stage modification of large craters. The mode of formation of complex craters and the origin of multiple rings are highly controversial. Unfortunately, these phenomena cannot be directly investigated experimentally. A combination of geologic work on terrestrial craters, photogeologic studies of planetary craters, theoretical computations, and well-designed scale model experiments must be relied upon to resolve these questions. Although the study of impact and explosion craters already has a long history, it is by no means complete.

ACKNOWLEDGMENT

I thank Don Gault, Tom Ahrens, and Dave Roddy for many useful discussions on impact cratering, as well as a host of others too numerous to name individually. This work was partially supported by NASA grant NSG-7316.

Literature Cited

Ahrens, T. J., O'Keefe, J. D. 1972. Shock melting and vaporization of lunar rocks and minerals. *The Moon* 4:214–49

Ahrens, T. J., O'Keefe, J. D. 1977. Equations of state and impact-induced shock-wave attenuation on the moon. In *Impact and Explosion Cratering*, ed. D. J. Roddy, R. O. Pepin, R. B. Merrill, pp. 639–56. Oxford: Pergamon. 1301 pp.

Ahrens, T. J., O'Keefe, J. D. 1978. Energy and mass distributions of impact ejecta blankets on the Moon and Mercury. *Proc. Lunar Planet. Sci. Conf. 9th*, pp. 3787–3802

Andrews, R. J. 1977. Characteristics of debris from small-scale cratering experiments. In *Impact and Explosion Cratering*, ed. D. J. Roddy, R. O. Pepin, R. B. Merrill, pp. 1089–1100. Oxford: Pergamon. 1301 pp.

Baldwin, R. B. 1963. *The Measure of the Moon*. Univ. Chicago Press. 488 pp.

Birkhoff, G., MacDougall, D. P., Pugh, E. M., Taylor, G. 1948. Explosives with lined cavities. *J. Appl. Phys.* 19:563–82

Bjork, R. L., Kreyenhagen, K. N., Wagner, M. H. 1967. Analytical study of impact effects as applied to the meteoroid hazard. *NASA Rep. CR-757*

Boon, J. D., Albritton, C. C. Jr. 1937. Meteoritic craters and structures. *Proc. Geol. Soc. Am.*, pp. 305–6

Bryan, J. B., Burton, D. E., Cunningham, M. E., Lettis, L. A. Jr. 1978. A two-dimensional computer simulation of hypervelocity impact cratering: Some preliminary results for Meteor Crater, Arizona. *Proc. Lunar Planet. Sci. Conf. 9th*, pp. 3931–64

Carlson, R. H., Jones, G. D. 1965. Distribution of ejecta from cratering explosions in soils. *J. Geophys. Res.* 70:1897–1910

Carr, M. H., Crumpler, L. S., Cutts, J. A., Greeley, R., Guest, J. E., Masursky, H. 1977. Martian impact craters and emplacement of ejecta by surface flow. *J. Geophys. Res.* 82:4055–65

Chapman, C. R., Jones, K. L. 1977. Cratering and obliteration history of Mars. *Ann. Rev. Earth Planet. Sci.* 5:515–40

Christman, D. R., Gehring, J. W. 1966. Analysis of high-velocity projectile penetration mechanics. *J. Appl. Phys.* 37:1579–87

Cintala, M. J., Head, J. W., Mutch, T. A. 1976. Characteristics of fresh Martian craters as a function of diameter: Comparison with the Moon and Mercury. *Geophys. Res. Lett.* 3:117–20

Clanton, U. S., Morrison, D. A. 1979. Hypervelocity impact craters less than 1000 Å diameter. *Lunar Sci. X*, pp. 212–14 (Abstr.)

Daneš, Z. F. 1962. Isostatic compensation

of lunar craters. *Res. Rep. RIR-GP-62-1.* Tacoma, Wash: Univ. Puget Sound Res. Inst.

Dence, M. R. 1968. Shock zoning at Canadian craters: Petrography and structural implications. In *Shock Metamorphism of Natural Materials*, ed. B. M. French, N. M. Short, pp. 169–84, Baltimore, Md.: Mono Book Co. 644 pp.

Dence, M. R. 1971. Impact Melts. *J. Geophys. Res.* 76: 5552–65

Dienes, J. K., Walsh, J. M. 1970. Theory of impact: Some general principles and the method of Eulerian codes. In *High-Velocity Impact Phenomena*, ed. R. Kinslow, pp. 45–104. New York: Academic. 579 pp.

Dvorak, J., Phillips, R. J. 1977. The nature of the gravity anomalies associated with large young lunar craters. *Geophys. Res. Lett.* 4: 380–82

Eberhardt, P., Geiss, J., Grogler, N., Stettler, A. 1973. How old is the crater Copernicus? *The Moon* 8: 104–14

Gault, D. E. 1970. Saturation and equilibrium conditions for impact cratering on the lunar surface: Criteria and implications. *Radio Science* 5: 273–91

Gault, D. E. 1973. Displaced mass, depth, diameter, and effects of oblique trajectories for impact craters formed in dense crystalline rocks. *The Moon* 6: 32–44

Gault, D. E. 1974. Impact cratering. In *A Primer in Lunar Geology*, ed. R. Greeley, P. Schultz, pp. 137–75. NASA Ames. 574 pp.

Gault, D. E., Greeley, R. 1978. Exploratory experiments of impact craters formed in viscous-liquid targets: Analogs for Martian Rampart craters? *Icarus* 34: 486–95

Gault, D. E., Guest, J. E., Murray, J. B., Dzurisin, D., Malin, M. C. 1975. Some comparisons of impact craters on Mercury and the Moon. *J. Geophys. Res.* 80: 2444–60

Gault, D. E., Heitowit, E. D. 1963. The partition of energy for hypervelocity impact craters formed in rock. *Proc. 6th Hypervelocity Impact Symp.* 2: 419–56

Gault, D. E., Hörz, F., Brownlee, D. E., Hartung, J. B. 1974. Mixing of the lunar regolith. *Proc. Lunar Sci. Conf. 5th*, pp. 2365–86

Gault, D. E., Moore, H. J. 1965. Scaling relationships for microscale to megascale impact craters. *Proc. 7th Hypervelocity Impact Symp.* 6: 341–51

Gault, D. E., Quaide, W. L., Oberbeck, V. R. 1968. Impact cratering mechanics and structures. In *Shock Metamorphism of Natural Materials*, ed. B. M. French, N. M. Short, pp. 87–99. Baltimore, Md.: Mono Book Co. 644 pp.

Gault, D. E., Shoemaker, E. M., Moore, H. J. 1963. Spray ejected from the lunar surface by meteoroid impact. *NASA Tech. Note D-1767*

Gault, D. E., Wedekind, J. A. 1977. Experimental hypervelocity impact into quartz sand—II, Effects of gravitational acceleration. In *Impact and Explosion Cratering*, ed. D. J. Roddy, R. O. Pepin, R. B. Merrill, pp. 1231–44. Oxford: Pergamon. 1301 pp.

Gault, D. E., Wedekind, J. A. 1978. Experimental studies of oblique impact. *Proc. Lunar Planet. Sci. Conf. 9th*, pp. 3843–75

Gifford, A. W., Maxwell, T. A., El-Baz, F. 1979. Geology of the Lunar Farside Crater Necho. *The Moon and the Planets* 21: 25–42

Gilbert, G. K. 1893. The moon's face; a study of the origin of its features. *Bull. Philos. Soc. Washington* 12: 241–92

Glasstone, S., ed. 1957. *The Effects of Nuclear Weapons.* Washington, DC: US At. Energy Comm. 597 pp.

Hartmann, W. K. 1972. Interplanet variations in scale of crater morphology—Earth, Mars, Moon. *Icarus* 17: 707–13

Hartmann, W. K., Wood, C. A. 1971. Moon: Origin and evolution of multiring basins. *The Moon* 3: 3–78

Hodges, C. A., Wilhelms, D. E. 1978. Formation of lunar basin rings. *Icarus* 34: 294–323

Hörz, F., Hartung, J. B., Gault, D. E. 1971. Micrometeorite craters on lunar rock surfaces. *J. Geophys. Res.* 76: 5770–98

Howard, K. A. 1974. Fresh lunar impact craters: Review of variation with size. *Proc. Lunar Sci. Conf., 5th.* pp. 67–79

Howard, K. A. 1975. Geologic map of the crater Copernicus. *US Geol. Surv. Misc. Geol. Inv. Map I-840*

Howard, K. A., Wilhelms, D. E., Scott, D. H. 1974. Lunar basin formation and highland stratigraphy. *Rev. Geophys. Space Phys.* 12: 309–27

Innes, M. J. S. 1961. Use of gravity methods to study the underground structure and impact energy of meteorite craters. *J. Geophys. Res.* 66: 2225–39

Kieffer, S. W. 1977. Impact conditions required for formation of melt by jetting in silicates. In *Impact and Explosion Cratering*, ed. D. J. Roddy, R. O. Pepin, R. B. Merrill, pp. 751–69. Oxford: Pergamon. 1301 pp.

Kolsky, H. 1953. *Stress Waves in Solids.* London: Oxford Univ. Press. 211 pp.

Kreyenhagen, K. N., Schuster, S. H. 1977. Review and comparison of hypervelocity impact and explosion cratering calculations. In *Impact and Explosion Cratering*, ed. D. J. Roddy, R. O. Pepin, R. B. Merrill, pp. 983–1002. Oxford: Pergamon. 1301 pp.

Lindsay, J. F. 1976. *Lunar Stratigraphy and Sedimentation.* Amsterdam: Elsevier. 302 pp.

Masursky, H., Colton, G. W., El-Baz, F., eds. 1978. *Apollo Over the Moon: A View from Orbit. NASA SP-362.* Washington, DC: GPO. 255 pp.

Maxwell, D. E. 1977. Simple Z model of cratering, ejection, and the overturned flap. In *Impact and Explosion Cratering,* ed. D. J. Roddy, R. O. Pepin, R. B. Merrill, pp. 1003–8. Oxford: Pergamon. 1301 pp.

McCauley, J. F. 1977. Orientale and Caloris. *Phys. Earth Planet. Inter.* 15:220–50

McGetchin, T. R., Settle, M., Head, J. W. 1973. Radial thickness variation in impact crater ejecta: Implications for lunar basin deposits. *Earth Planet. Sci. Lett.* 20:226–36

McGill, G. E. 1977. Craters as "fossils": The remote dating of planetary surface materials. *Geol. Soc. Am. Bull.* 88:1102–10

McKinnon, W. B. 1978. An investigation into the role of plastic failure in crater modification. *Proc. Lunar Planet. Sci. Conf., 9th,* pp. 3965–73

Melosh, H. J. 1976. On the origin of fractures radial to lunar basins. *Proc. Lunar Sci. Conf., 7th,* pp. 2967–82

Melosh, H. J. 1977. Crater modification by gravity: A mechanical analysis of slumping. In *Impact and Explosion Cratering,* ed. D. J. Roddy, R. O. Pepin, R. B. Merrill, pp. 1245–60. Oxford: Pergamon. 1301 pp.

Melosh, H. J. 1979. Acoustic fluidization: A new geologic process? *J. Geophys. Res.* In press

Melosh, H. J., McKinnon, W. B. 1978. The mechanics of ringed basin formation. *Geophys. Res. Lett.* 5:985–88

Milton, D. J., Barlow, B. C., Brett, R., Brown, A. R. Glikson, A. Y., Manwaring, E. A., Moss, F. J., Sedmik, E. C. E., Van Son, J., Young, G. A. 1972. Gosses bluff impact structure, Australia. *Science* 175: 1199–1207

Milton, D. J., Roddy, D. J. 1972. Displacements within impact craters. *24th Int. Geol. Cong., Montreal, Sec. 15,* pp. 119–24

Moore, H. J. 1976. Missile impact craters (White Sands Missile Range, New Mexico) and applications to lunar research. *US Geol. Surv. Prof. Pap. 812-B.* 47 pp.

Moore, H. J., Hodges, C. A., Scott, D. H. 1974. Multiringed basins—illustrated by Orientale and associated features. *Proc. Lunar Sci. Conf., 5th,* pp. 71–100

Nordyke, M. D. 1961. Nuclear craters and preliminary theory of the mechanics of explosive crater formation *J. Geophys. Res.* 66:3439–59

Nordyke, M. D. 1962. An analysis of cratering data from desert alluvium. *J. Geophys. Res.* 67:1965–74

Oberbeck, V. R. 1971. Laboratory simulation of impact cratering with high explosives. *J. Geophys. Res.* 76:5732–49

Oberbeck, V. R. 1975. The role of ballistic erosion and sedimentation in lunar stratigraphy. *Rev. Geophys. Space Phys.* 13: 337–62

Oberbeck, V. R., Morrison, R. H. 1974. Laboratory simulation of the Herringbone pattern associated with lunar secondary crater chains. *The Moon* 9:415–55

Oberbeck, V. R., Morrison, R. H., Hörz, F. 1975. Transport and emplacement of crater and basin deposits. *The Moon* 13: 9–26

O'Keefe, J. D., Ahrens, T. J. 1975. Shock effects from a large impact on the Moon. *Proc. Lunar Sci. Conf., 6th,* pp. 2831–44

O'Keefe, J. D., Ahrens, T. J. 1977a. Impact-induced energy partitioning, melting, and vaporization on terrestrial planets. *Proc. Lunar Sci. Conf., 8th,* pp. 3357–74

O'Keefe, J. D., Ahrens, T. J. 1977b. Meteorite impact ejecta: Dependence of mass and energy lost on planetary escape velocity. *Science* 198:1249–51

O'Keefe, J. D., Ahrens, T. J. 1978. Impact flows and crater scaling on the Moon. *Phys. Earth Planet. Inter.* 16:341–51

Öpik, E. J. 1969. The Moon's surface. *Ann. Rev. Astron. Astrophys.* 7:473–526

Piekutowski, A. J. 1977. Cratering mechanisms observed in laboratory-scale high-explosive experiments. In *Impact and Explosion Cratering,* ed. D. J. Roddy, R. O. Pepin, R. B. Merrill, pp. 67–102. Oxford: Pergamon. 1301 pp.

Pike, R. J. 1974. Depth/diameter relations of fresh lunar craters: Revision from spacecraft data. *Geophys. Res. Lett.* 1: 291–94

Pike, R. J. 1976. Crater dimensions from Apollo data and supplemental sources. *The Moon* 15:463–77

Quaide, W. L., Gault, D. E., Schmidt, R. A. 1965. Gravitative effects on lunar impact structures. *Ann. NY Acad. Sci.* 123:563–72

Quaide, W. L., Oberbeck, V. R. 1968. Thickness determinations of the lunar surface layer from lunar impact craters. *J. Geophys. Res.* 73:5247–70

Regan, R. D., Hinze, W. J. 1975. Gravity and magnetic investigations of Meteor

Crater, Arizona. *J. Geophys. Res.* 80:776–88
Roberts, W. A. 1964. Notes on the importance of shock crater lips to lunar exploration. *Icarus* 3:342–47
Roberts, W. A. 1965. Genetic stratigraphy of the Meteor Crater outer lip. *Icarus* 4:431–36
Robertson, P. B., Grieve, R. A. F. 1977. Shock attenuation at terrestrial impact structures. In *Impact and Explosion Cratering*, ed. D. J. Roddy, R. O. Pepin, R. B. Merrill, pp. 687–702. Oxford: Pergamon. 1301 pp.
Roddy, D. J. 1976. High-explosive cratering analogs for bowl-shaped, central uplift, and multiring impact craters. *Proc. Lunar Sci. Conf., 7th*, pp. 3027–56
Roddy, D. J. 1978. Pre-impact geologic conditions, physical properties, energy calculations, meteorite and initial crater dimensions and orientations of joints, faults and walls at Meteor Crater, Arizona. *Proc. Lunar Planet. Sci. Conf., 9th*, pp. 3891–3930
Ross, H. P. 1968. A simplified mathematical model for lunar crater erosion. *J. Geophys. Res.* 73:1343–54
Schmidt, R. H. 1977. A centrifuge cratering experiment: Development of a gravity-scaled yield parameter. In *Impact and Explosion Cratering*, ed. D. J. Roddy, R. O. Pepin, R. B. Merrill, pp. 1261–78. Oxford: Pergamon. 1301 pp.
Schmitt, H. H., Trask, N. J., Shoemaker, E. M. 1967. Geologic map of the Copernicus Quadrangle of the Moon. *US Geol. Surv. Misc. Geol. Inv. Map I-515*
Schultz, P. H. 1976. *Moon Morphology*. Austin/London: Univ. Texas Press. 626 pp.
Scott, D. H., McCauley, J. F., West, M. N. 1977. Geologic map of the west side of the Moon. *US Geol. Surv. Misc. Geol. Inv. Map I-1034*
Scott, R. F. 1967. Viscous flow of craters. *Icarus* 7:139–48
Seebaugh, W. R. 1977. A dynamic crater ejecta model. In *Impact and Explosion Cratering*, ed. D. J. Roddy, R. O. Pepin, R. B. Merrill, p. 1043–56. Oxford: Pergamon. 1301 pp.
Settle, M., Head, J. W. 1979. The role of rim slumping in the modification of lunar craters. *J. Geophys. Res.* 84:3081–96
Settle, M., Head, J. W., McGetchin, T. R. 1974. Ejecta from large craters on the Moon: Discussion. *Earth Planet. Sci. Lett.* 23:271–74
Shoemaker, E. M. 1960. Penetration mechanics of high velocity meteorites, illustrated by Meteor Crater, Arizona. In *Rep. Int. Geol. Congr., XXI Session, Norden, Copenhagen, Part XVIII*, pp. 418–34
Shoemaker, E. M. 1962. Interpretation of lunar craters. In *Physics and Astronomy of the Moon*, ed. Z. Kopal, pp. 283–357. New York & London: Academic. 538 pp.
Shoemaker, E. M. 1963. Impact mechanics at Meteor Crater, Arizona. In *The Moon, Meteorites, and Comets*, ed. B. M. Middlehurst, G. P. Kuiper, 4:301–36. Chicago & London: Univ. Chicago Press. 810 pp.
Shoemaker, E. M., Hait, M. H., Swann, G. A., Schleicher, D. L., Schaber, G. G., Sutton, R. L., Dahlem, D. H., Goddard, E. N., Waters, A. C. 1970. Origin of the lunar regolith at Tranquillity Base. *Proc. Apollo 11 Lunar Sci. Conf.*, pp. 2399–2412
Shoemaker, E. M., Hackman, R. J., Eggleton, R. E. 1963. Interplanetary correlation of geologic time. *Adv. Astronaut. Sci.* 8:70–89
Short, N. M., Forman, M. L. 1972. Thickness of impact crater ejecta on the lunar surface. *Modern Geol.* 3:69–91
Smith, B. A., Soderblom, L. A., Johnson, T. V., Ingersoll, A. P., Collins, S. A., Shoemaker, E. M., Hunt, G. E., Masursky, H., Carr, M. H., Davies, M. E., Cook, A. F., Boyce, J., Danielson, G. E., Owen, T., Sagan, C., Beebe, R. F., Veverka, J., Strom, R. G., McCauley, J. F., Morrison, D., Briggs, G. A., Suomi, V. E. 1979. The Jupiter system through the eyes of Voyager I. *Science* 204:951–72
Soderblom, L. A. 1970. A model for small-impact erosion applied to the lunar surface. *J. Geophys. Res.* 75:2655–61
Stöffler, D., Gault, D. E., Wedekind, J., Polkowski, G. 1975. Experimental hypervelocity impact into quartz sand: Distribution and shock metamorphism of ejecta. *J. Geophys. Res.* 80:4062–77
Swift, R. P. 1977. Material strength degradation effect on cratering dynamics. In *Impact and Explosion Cratering*, ed. D. J. Roddy, R. O. Pepin, R. B. Merrill, pp. 1025–42. Oxford: Pergamon. 1301 pp.
Trulio, J. G. 1977. Ejecta formation: Calculated motion from a shallow-buried nuclear burst, and its significance for high velocity impact cratering. In *Impact and Explosion Cratering*, ed. D. J. Roddy, R. O. Pepin, R. B. Merrill, pp. 919–57. Oxford: Pergamon. 1301 pp.
Ullrich, G. W., Roddy, D. J., Simmons, G. 1977. Numerical simulations of a 20-ton TNT detonation on the earth's surface and implications concerning the mechanics of central uplift formation. In *Impact and Explosion Cratering*, ed. D. J. Roddy,

R. O. Pepin, R. B. Merrill, pp. 959–82. Oxford: Pergamon. 1301 pp.
Van Dorn, W. G. 1968. Tsunamis on the Moon? *Nature* 220:1102–7
White J. W. 1971. Examination of cratering formulas and scaling methods. *J. Geophys. Res.* 76:8599–603
Wilhelms, D. E. 1973. Comparison of Martian and Lunar multiringed circular basins. *J. Geophys. Res.* 78:4084–95
Wilshire, H. G., Howard, K. A. 1968. Structural pattern in central uplifts of cryptoexplosion structures as typified by Sierra Madera. *Science* 162:258–61
Zel'dovich, Ya. B., Raizer, Yu. P. 1967. *The Physics of Shock Waves and High-Temperature Hydrodynamic Phenomena.* New York: Academic. 916 pp.

Ann. Rev. Earth Planet. Sci. 1980. 8:95–117
Copyright © 1980 by Annual Reviews Inc. All rights reserved

HEAT FLOW AND HYDROTHERMAL CIRCULATION

×10126

C. R. B. Lister

Departments of Geophysics and Oceanography, University of Washington, WB-10, Seattle, Washington 98195

The term "terrestrial heat flow" is usually applied to the loss of heat from the interior of the earth through the global surface. There are three principal mechanisms for this transport of heat through the surface: simple conduction under the influence of the "geothermal gradient," percolative flow of relatively low-temperature fluids, like water, steam, or brine, and the direct ejection of high-temperature rock materials such as lava or tephra. Historically, the mechanism of conduction has received the most attention because of the early observation that ambient temperature increased markedly in deep mines [e.g. Thomson (later Lord Kelvin) 1846].

The units of measurement for heat flow are either microcalories per square centimeter per second (HFU, heat flow unit) or, as used more recently by some authors, milliwatts per square meter. The new Systèrne Internationale unit has lost the association with entrained entropy that should be one function of a measure of heat, and I shall retain the older unit as used in most of the work cited. The bulk of the heat emitted by the earth moves through the upper layers of the crust by conduction. From the concept of a conductive flux comes the term "geothermal gradient," which is simply the rate of increase of temperature with depth. Often expressed in °C km^{-1}, it tends to be between 20 and 40°C km^{-1} in the upper crust, but must become much lower at greater depth because the melting point of the material is approached, and this cannot be exceeded without the occurrence of vigorous convection or other magmatic processes. The product of the geothermal gradient and the rock conductivity gives the heat flux.

The simple picture of a conducted flux breaks down when there is advective flow of material. Volcanoes have been observed since earliest times and are responsible for the traditional belief that the earth is hot

0084-6597/80/0515-0095$01.00

inside, although we still do not have any clear knowledge of how magma is generated and rises to the surface. The heat transported can be estimated fairly well for most types of volcano, inasmuch as it is just the product of the magma heat content and the volume of the effusion. The contribution to the total earth's heat loss from all currently active volcanoes is not negligible but is also not of major importance: it is of the order of 1%.

The interpretation of the third mechanism of heat transport, fluid percolation through pores and cracks in the rocks, is considerably more complex. Frequently, both the sources and sinks of the flow are buried underground, so that estimation of the transport is indirect. Substantial progress has been made in understanding one aspect of the problem, the mechanism by which fluid itself can cause rock cracking and thus permeability for its own further flow. Perhaps paradoxically, this has had the largest effect on our view of the ocean floor, which is the least accessible part of the earth's crust. A substantial proportion of the total oceanic heat flow is vented directly to the sea by open hydrothermal circulation, and much of the remainder is redistributed through the oceanic crust by fluid convection. I shall tend to emphasize the oceanic work in this review, but I shall also attempt to mention the most significant features of hydrothermal flow on the continents because they contain our only explored and exploited geothermal areas.

THE GLOBAL PICTURE

The early discussions of heat flow by Thomson not only had a bias toward the mechanism of conduction but focused on an explanation of the thermal gradient as caused by continued cooling of an initially hot and molten earth. An immediate discrepancy arose between the apparent time scale of the cooling and the time scale proposed by geologists for the long sequence of strata on the surface of the earth. The controversy was resolved by the discovery of radioactivity, and modern debate tends to be about where within the earth the emitted heat is generated. A substantial fraction of the heat conducted through the surface of the continents is provided by the decay of radioactive elements contained within the thick continental crust itself. This is not unexpected, inasmuch as the three most important radioactive elements, potassium, uranium, and thorium, separate preferentially into low-density acidic rocks. An accurate estimate of crustal heat cannot be made without knowing the nature of the continental lower crust. Some evidence from central North America suggests that a true compositional boundary may not exist at the depth where somewhat higher seismic velocities are usually found. There is good seismological evidence for a thick high-velocity, high-Q upper mantle under Precambrian cratons; and this, in turn, points to a low geothermal gradient and low mantle heat flow beneath the crust (Smithson & Decker 1974, Smithson 1978, Anderson & Archambeau 1964, Herrin 1969, Dorman 1969, Solomon 1972).

The oceanic regions have a relatively low concentration of radioactive elements in near-surface materials because the crust is thin and primarily basaltic in composition. The principal source of heat is the cooling of newly created lithosphere where material wells up from deeper within the mantle in response to the separation, or spreading, of tectonic plates (Bullard 1954, Langseth et al 1966, Oxburgh & Turcotte 1968). The upwelling is great enough on a global scale for the oceanic contribution to be the majority of the earth's heat loss, but the sources may be distributed in many possible ways between the bulk of the mantle and the core (e.g. Heirtzler et al 1968, Williams & Von Herzen 1974, Ringwood 1975).

The investigation of processes that transport heat from the earth's deep interior is fraught with many difficulties. We can estimate the present total heat loss (see the conclusions section of this article), but there is no guarantee that this is exactly in equilibrium with the present rate of heat generation in the interior. The thermal inertia is large, and there is some evidence of periodicity in continental geology that might suggest a periodicity in active sea-floor spreading. The oldest ocean floor is too young to offer any direct evidence of what went on during earlier geologic periods. Moreover, heat generated by the decay of radioactive elements is not constant through time and should have been two to three times as great in the earth's early history. The trend of current thinking is to assume (*a*) that approximate equilibrium exists, and (*b*) that a negligible fraction of the oceanic heat loss is generated near enough to the surface to be conducted up to it.

The amount of heat that can be brought to the surface purely by conduction is very clearly limited. Rock materials can remain solid only if the thermal gradient is less than the melting point gradient, so that, below the level at which the melting point is approached, the geothermal gradient must remain, on average, less than the gradient in the melting point due to increasing pressure. This latter gradient is known approximately, at least for the upper mantle (2°C km^{-1}), and a number of estimates have been made of the conductivity of rocks at elevated temperatures (Clark 1966, Fujisawa et al 1968). The influence of the radiative component of conductivity cannot be extrapolated well to upper mantle temperatures from existing data, but even the highest estimates would permit a conductive flux only a quarter of the observed mean oceanic heat flow. Hence, either there must be large-scale convective flow in the mantle or upward melt migration to carry the heat. This implies major chemical exchange between the lithosphere and the lower mantle, and possibly even between the mantle and core (Runcorn 1965, Fyfe 1978).

Some heat must be liberated from the earth's core because a source of energy is needed to run the dynamo that maintains the magnetic field. The amount depends on the electrical efficiency of the dynamo, which has been guessed at but is not known, and the thermodynamic efficiency of the convection or other motional force that provides the drive. In order of increasing efficiency, three main mechanisms have been proposed: simple thermal convection, precession, and chemical convection due to the gradual growth of the inner core (Bullard 1950, Malkus 1968, Loper 1978). Most authors agree that the energy required is within the current figure for the total earth's heat emission, but it is hard to see how they could conclude otherwise.

BASIC MEASUREMENT TECHNIQUES

The measurement of heat flow is beset by its own special problems and has given rise to an extensive literature. On the continents, the two principal sources of error are climatic fluctuations in the long-term surface temperature and the circulation of groundwater. The geothermal gradient is generally measured in boreholes drilled for mineral exploration or specifically for measurement of heat flow. Holes drilled in potential oilfield formations tend to be unsuitable because, if oil is found, fluid flows dominate the conductive gradient, whereas if oil is not found, the holes are plugged and abandoned before the drilling disturbance can subside (but see Carvalho & Vacquier 1977). Successful measurements require holes of substantial depth because the ratio of the geothermal temperature rise to plausible climatic change increases with depth, and the probability of disturbance by groundwater flow decreases (e.g. Diment et al 1965). The test of a good measurement is generally one of self-consistency: if the heat flow through a series of formations near the bottom of the hole is constant, it is presumed to represent the real heat flow of the region. Since the conductivity of continental rocks varies widely, separate measurements of thermal gradient and conductivity must be made for each rock formation down the hole. The tedium of the data acquisition, combined with the cost of the holes themselves, have severely limited the number and distribution of continental heat flow values. In some areas, the only values available have been obtained by marine-like techniques through the floors of lakes and fiords. These measurements are often subject to severe corrections and cannot be considered completely reliable (e.g. Hyndman 1976).

In some continental volcanic regions, a significant contribution to the overall heat flow is borne by the extrusive materials themselves. Surface flows cool rapidly without disturbing the local geothermal gradient noticeably, and steam and tephra eruptions dissipate the bulk of their energy directly into the atmosphere. A good way to demonstrate the potential volcanic heat flow is to estimate the thickness of basalt that would liberate the same amount of heat as a heat flow of 2 HFU persisting for 1 My (Megayear or 10^6 years). For a lava heat-content of 1350 cal/cc, a 440-m thickness would have to be extruded. A major conical volcano like Mt. Rainier, Washington, represents a substantial local heat output but not a large contribution to that of the whole mountain range in which it sits. On the other hand, major basalt-filled basins like the Columbia Basin of eastern Washington and the Deccan Traps of India can be equivalent to 2 HFU over a large area for intermittent periods of 3 My or more (Swanson et al 1979, Raymond & Tillson 1968). The world's largest oceanic hotspot, the Hawaiian/Emperor seamount chain, may put out as much as 10^9 cal/s, or 0.01% of the total global heat loss. Thus, extrusive activity forms a significant, but not major, component of that heat loss.

Measurement of heat flow through the ocean floor is in some ways more difficult than on land, yet easier once adequate equipment has been perfected. Much of the sea floor is covered by soft sediments, so that the thermal gradient can be measured merely by punching a probe or a coring tube a few meters into the bottom. The temperature rises at various depths are either recorded in the instrument or telemetered to the surface acoustically. The conductivity of the sediments can be measured on cores recovered, or directly *in situ* by heating a part of the apparatus. Such shallow penetration is sufficient in the deep ocean because the temperature of the bottom water is very stable. The thermal mass of the oceans is large enough to smooth out short-term fluctuations in the temperature, especially as the rate of replenishment of the deep water is relatively slow. The temperature of new bottom water, which is made only in the polar regions, is regulated by the freezing point of sea water. Present bottom temperatures in the ocean basins range from $-0.5°C$ in the South Pacific to $+4°C$ in the North Atlantic; hence, they are unlikely to have been much colder during the last ice age, and plausible changes would have a negligible effect on the present measured gradients (Bullard 1954, Ratcliffe 1960, Von Herzen & Maxwell 1959, Gerard et al 1962, Sclater et al 1969, Lister 1970a).

On the whole, the methods have worked well, and the results can be relied on to represent the heat flow through the sediments where the instrument has penetrated. Measurements attempted in water less than

1 km deep, or in small marine basins near a source of cold bottom water, are often seriously in error due to fluctuations in the temperature of the bottom water (e.g. Lachenbruch & Marshall 1968). Early intimations that variation might be intense on an extremely local scale (Bullard & Day 1961) were allayed by the consistency of values in other areas (Lister 1963). The effects of heat-flow refraction through topographic variations, and that of steady sedimentation or sudden slumping, were shown to be relatively small or rare on the deep ocean floor (Jeffreys 1938, Von Herzen & Uyeda 1963, Lachenbruch 1968). Consideration and rejection of possible sources of error due to sediment compaction, the oxidation of organic material, or a global oceanic warming convinced most workers that the ocean-floor measurements were a true measure of the earth's heat loss through the ocean floor (Bullard 1954, Bullard et al 1956, Von Herzen & Uyeda 1963).

INTERPRETATION OF THE MEASUREMENTS

The exploration phase of ocean-floor heat-flow measurement yielded values with a range from near zero to about 7 HFU and a mean a little above 1 HFU. It had already been established that a typical continental heat flow was about 1.4 HFU, so that the oceanic heat flow turned out to be remarkably similar. It was known that a substantial portion of the continental heat flux was generated within the continental crust by the decay of radioactive elements and that the basaltic rocks of the ocean floor contained little of these. Thus the result was something of a surprise and produced two schools of thought: one held that the radioactive elements beneath the oceans were present in the same amounts per unit area as in the continents, but had failed to differentiate upwards; the other held that the concentrations were too low for the heat generated in the much thicker layer to be conducted upward through material below the melting point, and that some convective process was bringing the heat to the surface from deep within the earth. The latter concept was instrumental in the development of the ideas of plate tectonics and continental drift, but the large scatter in the heat-flow values themselves precluded their use in refining the plate tectonic models.

The correlation of recognizable magnetic anomalies with the age of the ocean floor makes it possible to obtain mean values of the heat flow for provinces of equal age irrespective of the plate-tectonic spreading rate (e.g. Heirtzler et al 1968). An example of such means, and of the scatter associated with the measurements, is shown in Figure 1. Although some general decrease in the heat flow with age is observed, it is not convincing statistically; a decrease in the scatter with age is better defined. A representative theoretical curve is shown for comparison. Based on this kind of data, one would have to conclude that either the theoretical curve is wrong, or that there is something wrong with the measurements at the younger ages. The large variation among values obtained from a single province was considered cause for caution in the interpretation of the measurements.

The theoretical heat-flow versus age curves in the literature are based on calculations of heat conduction within flow regimes that model the process of sea-floor spreading. The analytic solutions cannot handle two-dimensional flow fields, and therefore these solutions are based on the evolution of the temperature distribution in a horizontally flowing half-space with an arbitrary temperature distribution along a vertical boundary where the spreading center should be. The complexities of achieving an adequate model by this method have occupied the literature for several years, but the details are not relevant here because all the reasonable models converge onto the same heat-flow curve within a million years (McKenzie 1967, Oxburgh & Turcotte 1968, Davis & Lister 1974, Sleep 1975). If new rock material upwells to the surface at magmatic tempera-

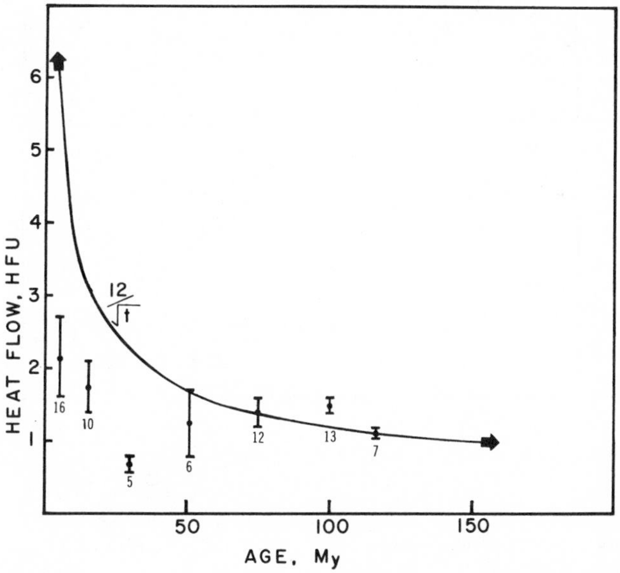

Figure 1 A plot of heat flow versus age for seven provinces in the South Atlantic Ocean. The error bars are for the standard errors of the means, and the numbers below the points give the sizes of the samples. Multiply the error bars by the square root of the sample counts to obtain the standard deviations of the raw measurements. A representative theoretical decay curve is shown for comparison (data after Sclater & Francheteau 1970).

tures, the heat flow in the young provinces should be much higher than observed. The one computer model of two-dimensional flow that appears to agree with the measurements suffers from too large a grid size and convergence problems in the critical region (Le Pichon & Langseth 1969).

Most of the doubts about the validity of the theoretical estimates were dispelled when the models were applied to ocean floor topography. Early gravity data showed that mid-ocean ridges were not associated with any significant free-air anomaly, and must therefore be close to isostatic equilibrium (Talwani et al 1965). Since the spreading centers form ridges and hot material is less dense than cooler material, it was a logical step to calculate the topography over the thermal models on the assumption of perfect isostasy, or floating on a semi-fluid upper mantle (equivalent to having a constant-pressure zone at depth). Comparison of the calculated topography with actual data, averaged from different ocean basins by the age-province method, showed remarkable agreement (Sleep 1969, Sclater & Francheteau 1970, Parker & Oldenburg 1973): see Figure 2.

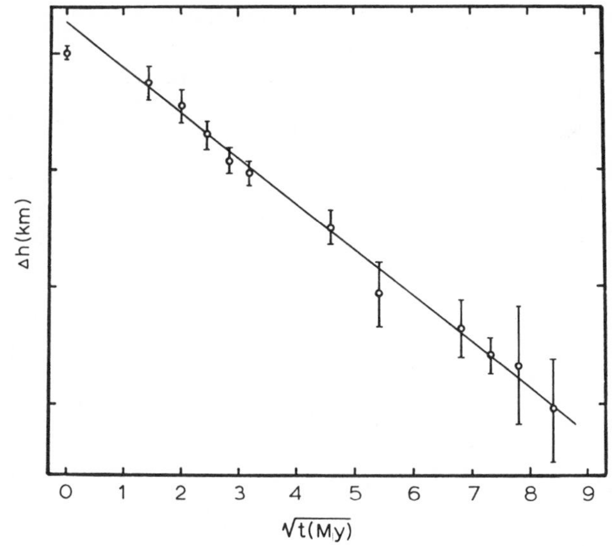

Figure 2 The decrease in topographic height from the ridge crest toward the flanks for the South Atlantic Ocean. Age is plotted as the square root of millions of years to produce a straight-line asymptote. The line has been fitted to the points because theory does not define the slope parameter. The poor fit of the crestal point has two main causes. There is often some difficulty in finding the true crest when there are relatively few bathymetric track lines across a ridge with many offsets (transform faults, or fracture zones). The true height is also hard to define when there is a rift valley at the crest: this is a dynamic feature associated with slow-spreading ridges (Lachenbruch 1976). Diagram from Davis & Lister (1974).

Thus it became clear that the discrepancy lay in the heat-flow measurements, not in the concept of sea-floor spreading or the analysis of its thermal consequences (e.g. Lister 1977b).

Iceland is a part of the mid-Atlantic Ridge and has long been famous for its hot springs and geysers. Thus it is not surprising that discussion linking hydrothermal circulation to cooling of normal ridge crests first appeared in print in an Icelandic publication (Palmason 1967). A similar suggestion was made by Talwani et al (1971) to account for some remarkably low heat-flow measurements near the crest of the Reykjanes Ridge—a sea-floor feature extending southward from Iceland itself. These authors rejected seawater circulation as an explanation of low heat flow on the flanks of the ridge even though the discrepancy is just as marked as that at the crest, and a careful analysis of alternative explanations was made. A similar analysis, proposing hydrothermal circulation at the ridge crest and endothermic metamorphic reactions beneath the flank, was made by Hyndman & Rankin (1972).

The elevation of a suggestion to a hypothesis requires the prediction of testable phenomena separate from the original discrepancy. This came with the publication of independent work on the Juan de Fuca Ridge region (Lister 1972). Heat-flow measurements near ridge crests have a feature even more dramatic than the disappointingly low means—a scatter often larger than the local mean itself. It was noticed quite early that surprisingly low measurements were made in flat-floored valleys in regions of otherwise rough topography, but although attribution to sediment slumping into the valleys was less than convincing, such explanations seemed to be accepted (Von Herzen & Uyeda 1963). The critical information needed to resolve this problem came from applying three separate newly available techniques to a region that fortunately contained a high heat-flow zone as well as the usual low ones. High-resolution, acoustic-reflection profiling provides information on the total thickness as well as the distribution of thin sediment cover on young oceanic crust, and precise bounds can be calculated for the heat-flow disturbances caused by possible slumping. Carbon-14 dating of the carbonate material in cores taken at actual heat-flow stations can, in critical instances, eliminate completely any possible disturbance to the gradient due to slumping (Lister 1970b). Finally, detailed two-dimensional mapping can provide a more complete picture of the sea-floor environment than widely spaced profiles (Lister 1971).

The thinness of the sediment cover in the Juan de Fuca area meant that slumping would have to be extraordinarily recent to depress the heat flow through the flat floors of the valleys—the thermal "time constant" of 50-m sediment is a mere 300 yr. The evidence from the cores was that the sediment was undisturbed for at least 14,000 yr since the last glaciation ended. Excellent constancy of heat flow over the depth of penetration ruled out water temperature fluctuations as the cause of the discrepancy, especially as the depth difference between hills and valleys was only about 300 m in many cases. The clinching evidence that thermal conduction could not be the dominant process came from a string of heat-flow measurements across the Explorer Troughs, off Vancouver Island. A parallel University of British Columbia reflection profile showed that the troughs were the least sedimented parts of the area and therefore the youngest and presumably hottest. Yet the heat flow in them was low (Figure 3). Neighboring, higher topography was sediment-covered and had a much higher heat flow. The only conceivable explanation for such a distribution is the circulation of water through the hard rocks of the oceanic crust.

It was possible to make some predictions with the circulation hypothesis that were testable with the original data. An area of unusually smooth basement west of the Juan de Fuca Ridge was draped by undisturbed hemipelagic sediments in a fine example of pure "snowfall" deposition. Having verified that pelagic sediment has a very low permeability, it was possible to predict that such an area may have local variability from convection beneath the sediment blanket but should have the correct mean heat flow for a cooling lithosphere of the appropriate age. The mean heat flow of this region agreed with the theoretical to within the experimental error. The variability on an abyssal plain, where sediment has filled in the basement topography, should be much less than in a "snowfall" deposition area. Heat flow from Cascadia Basin, east of the ridge crest, bore this out well. Finally, in areas where outcrops of basement are common, warm circulating water should be lost directly to the sea. The patches of sediment cover where measurements can be made should therefore show unusually low heat flow. There were two such areas in the original data set: immediately west of the ridge crest, and near some small seamounts further down the flank. In both, heat flow was found to be unusually low. The predictions and results could be summarized in a cartoon of the flow pattern in the three cases: open permeable medium, uniform impermeable thermal blanket, and abyssal plain infilling (Lister 1972).

HYDROTHERMAL CIRCULATION: THEORETICAL DEVELOPMENT

A parallel development to the above was a calculation by Bodvarsson & Lowell (1972) that demonstrated the ease with which a pipeline type of convective system could modify the heat flow of a region. They showed

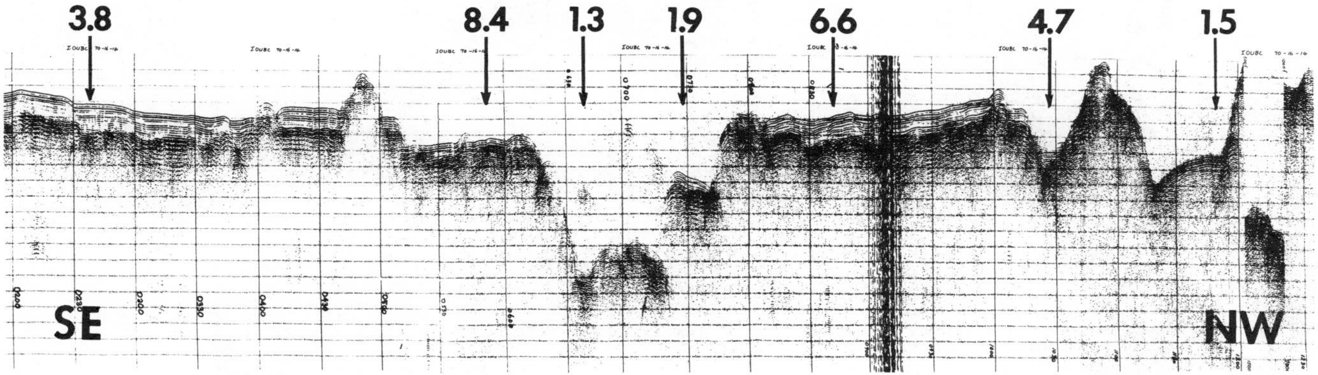

Figure 3 The original heat flow and seismic-reflection profile across the Explorer Ridge spreading zone that made the hypothesis of hydrothermal circulation inescapable. Reprinted from Lister (1972), airgun seismic profile courtesy of R. L. Chase, University of British Columbia. Vertical scale is 2.5 to 5 s of two-way travel time; vertical exaggeration: about 10:1. Numbers on arrows refer to HFU.

that cracks as little as 3 mm wide and a kilometer deep could convect away the heat, and that minute amounts of thermal contraction in the oceanic crust would be sufficient to generate these cracks. While the model might not be very realistic as it stands, the idea that thermal contraction cracks could so easily provide enough permeability for convective flow was to prove important. It was the key to the development of a complete, if highly preliminary, theory of water penetration into hot rock. The starting assumptions of the theory were simple enough: at time zero, a reservoir of cold fluid is poured onto an infinite horizontal surface of hot rock. The problem of rock cooling, cracking, and convection in the cracked region was to be treated one-dimensionally with no lateral variations permitted (Figure 4). In spite of the gross simplifications, the solution required consideration of interaction between thermal contraction strain, rock creep, crack propagation, convective and conductive boundary conditions, heat transport across a permeable layer, permeability of cracked solids, and variable fluid properties. Knowledge in many of these fields was still in a primitive state, particularly for the transient creep of rocks, propagation of cracks, and the fundamental heat-transport relation in porous medium convection. In addition, substantive errors were made in treatment of the critical boundary condition, which relates the speed of advance of the cracking front (separating the cracked and uncracked

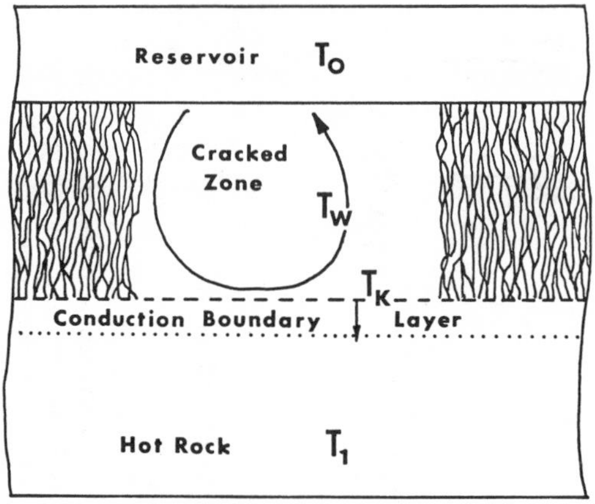

Figure 4 A cartoon of the main features of the one-dimensional model of water penetration into hot rock. The critical step is the concept of a cracking front separating a cold, permeable region from relatively undisturbed hot rock. Ahead of the cracking front is a thin conductive boundary layer where the hot rock cools, shrinks, and builds up the tension needed to cause the cracks to advance. The scales in a natural geothermal area may be several kilometers for the depth of the cracked zone, a few meters for the conductive boundary layer, and about one meter for the primary crack spacing. From Lister (1974).

regions) and the crack spacing; that spacing, in turn, controls the permeability of the cooled rock matrix (Lister 1974).

The primary qualitative result of the water penetration theory transcends all the simplifications and mistakes. The process is extraordinarily efficient; the cracks advance rapidly, cool a large volume of rock in a short time, and provide an intense localized source of geothermal output. A highly permeable matrix is produced for later groundwater flow, and the cracks are close enough together to allow major chemical interaction between the percolating fluids and the rock. Immediate predictions of the theory are that active geothermal systems should be short-lived transient phenomena even at fast-spreading mid-ocean ridges, but that relatively gentle circulation of cool water should persist in the residual permeable matrix, driven by heat conducted up from below. The latter type of circulation has been termed a "passive" geothermal system, to distinguish it from the intense high-temperature "active" phase of initial penetration (Lister 1980). This latest reference also includes corrections to the cracking boundary conditions and heat-transfer relation, and demonstrates conclusively that the major qualitative aspects of the results are insensitive to the details assumed.

The testing of such a theory is in itself a complex problem, especially in the presence of the usual geologic and geophysical variability. The assumption of one dimensionality is not as serious a limitation as might at first appear (e.g. Lister 1975), and the existence of a long string of closely spaced high-output geothermal areas in New Zealand suggests that "active" systems might really exist and that penetration diameters of 1–5 km may be preferred (Elder 1965, Bolton 1975). The implications for geothermal power production are profound, because the active-area resource is not so much a "reservoir" to be drawn down, as a convection plume to be tapped. Even here, distinctions between the rival concepts are hard to draw, since well-head pressures can be expected to drop solely because of mineral deposition in the active vents. The immediate reduction in natural surface venting following development at Wairakei would seem to support the plume-tapping hypothesis (W. A. J. Mahon, personal communication, 1974). The existence of a plume at Wairakei is fairly well established by the original pre-production borehole temperatures (Elder 1965).

HYDROTHERMAL CIRCULATION: FIELD TESTING OF THE THEORIES

Before considering the physical field evidence from the ocean floor, it is important to digress slightly and consider the geochemical implications and evidence. Long before marine geophysicists obtained a good grasp of the fundamentals of the heat-flow problem, geochemists had noticed that surface metalliferous sediments seemed to be concentrated near mid-ocean ridge crests. The correspondence of the heavy metals in the sediments, with those considered hydrothermally labile, led immediately to the suggestion that they were extracted from the oceanic crustal rocks by some form of hydrothermal circulation (Bostrom & Peterson 1969, Bender et al 1971, Corliss 1971, Dasch et al 1973, Piper 1973). Once the physical evidence for water circulation had been presented, there was almost an explosion of corroborative data obtained by analyzing the rocks in ophiolite suites, established as subaerially exposed sections of former oceanic crust (Gass 1968, Coleman 1971, Dewey & Bird 1971, Muehlenbachs & Clayton 1972, Hart 1973, Robertson & Hudson 1973, Spooner & Fyfe 1973, Hart et al 1974, Spooner et al 1974). There is good evidence that seawater is the fluid involved in the metamorphism of oceanic crust to several km of depth, and that the alteration takes place while the rocks are still part of the ocean floor. Minor occurrence of high-temperature alteration zones suggests that a good part of the metamorphism may take place while the crust is very young: in the terminology just discussed, in an "active" geothermal system.

The development of physical data to test the theory has proceeded more slowly because of the inherent experimental difficulties. Detailed mapping of heat flow requires both better instrument positioning than theretofore available and some means of making measurements more efficiently than by lowering an instrument from the surface for each one. The latter problem has been solved by the development of telemetering instruments capable of making several penetrations on each lowering (Corry, Dubois & Vacquier 1968, Von Herzen & Anderson 1972), and one approach to solving the positioning problem is the installation and use of a net of sea-floor acoustic transponders (McGehee & Boegemann 1966, Spiess et al 1966). A major areal investigation along these lines was conducted on the Galapagos Rift Zone spreading center (Klitgord & Mudie 1974, Williams et al 1974). Although the initial purpose of this was to disprove the existence of water circulation in oceanic crustal rocks (J. G. Sclater, personal communication, 1972), excellent and convincing evidence for hydrothermal circulation was obtained. A quasiperiodic fluctuation of heat flow was observed over subdued topography covered by an even layer of biogenic sediment. Some correlation existed between heat flow and topography, in the same sense as discussed by Lister (1972), and large areas covered by sediment mounds have since been shown to be through-sediment hydrothermal vents (Corliss et al 1979a, b). The investigation has been continued and has yielded the first three-dimensional map of heat flow on the ocean floor that has sufficient detail to delineate the heat-flow zones. Figure 5 reproduces this map and shows the remarkable large-scale zoning and the wide range of local heat-flow averages. Both are consistent only with large-scale hydrothermal circulation, and the recharge zones needed to balance the mounds and other hot springs.

The other successful physical approach has been to search for areas influenced by continentally derived sediments, where the sediment thickness is sufficient to even out the smallest scale of variability, and to develop an instrument of sufficient endurance to make long drift profiles with simultaneous acoustic-reflection profiling. The result of the first idea is a large grid-map of the northern end of the Juan de Fuca Ridge, an area of young ocean crust flooded by sediments of the Fraser River drainage in British Columbia. Originally undertaken as a purely exploratory exercise, the development of interpretative ideas paralleled the laborious process of data reduction and resulted in two new concepts (Davis & Lister 1977). The first is to analyze areas of variable sediment

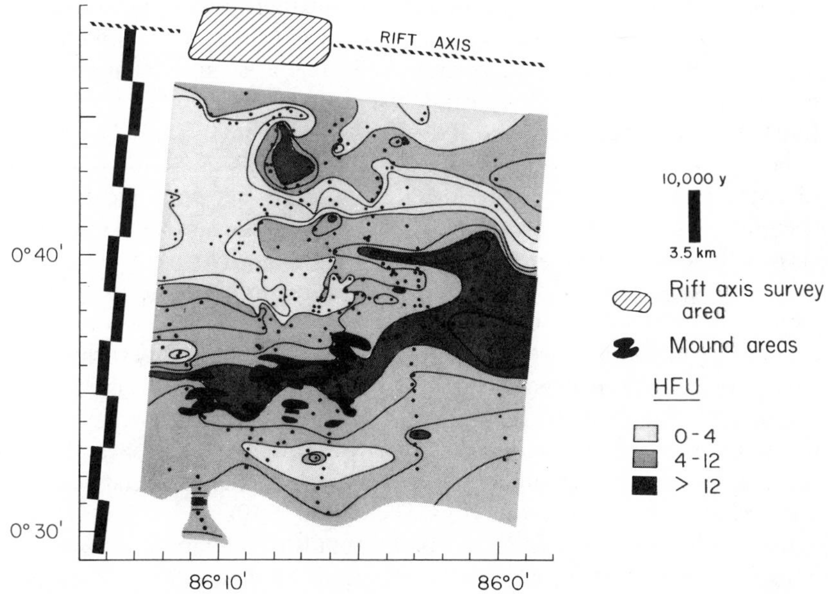

Figure 5 A map of the heat flow south of the Galapagos Rift Zone spreading center, contoured from the basis of more than 200 heat-flow measurements (small dots). Reproduced with the permission of R. P. Von Herzen and K. Green of the Woods Hole Oceanographic Institution. Detailed navigation was achieved by sonar transponder in addition to the usual satellite and radio methods.

cover in terms of basement temperature, making the well-supported assumption that the sediment permeability is normally too low to support thermally significant flow (e.g. Pearson & Lister 1973, Bryant et al 1974). The second is that heat flow can be redistributed over considerable horizontal distances beneath sediment thermal blankets, especially if there is venting at outcrops. Both concepts follow directly from the method of projecting heat flow onto acoustic-reflection profiles, pioneered by Lister (1972).

The logical extension of this method is to coordinate instrumentation to make repeated heat-flow penetrations, while the ship drifts slowly over the sea floor making an acoustic-reflection profile. With favorable weather and terrain, good relative positioning accuracy can be obtained without the complexity and expense of a full acoustic transponder net and a near-bottom survey. The critical step here was the invention of a mechanical configuration for heat-flow probe and temperature sensors that is rugged enough for an indefinite number of penetrations and yet has a short thermal response time. Due to Lister and reported in Hyndman et al (1978) the "violin-bow" concept employs a heavy strength member to resist bending on pullout, while the thermistors are contained in a tough alloy tube under tension, made thin to shorten the conductive thermal response time in the sediments. Simple as it may seem, it took nearly 20 years to make a real advance from the configurations of Bullard et al (1956) and Gerard et al (1962) (due to Ewing). The main difficulties have always been the lack of knowledge about what happens to instruments lowered by wire into the sea floor and, consequently, the need for a purely empirical approach to design. An example of the new class of data being generated is shown in Figure 6, where both variations on a very small scale (<2 km) and on a much larger scale (~ 10 km) seem to be present. These can be interpreted in the following way: the small-scale fluctuations are caused by near-surface permeability variations, where hydrothermal deposits tend to seal cracks near the surface but tectonic faulting tends to reopen paths for flow. The larger-scale variations are due to major convection patterns in a fairly thick (several km) permeable layer, in turn influenced by topography and localized venting and recharge at basement outcrops (Davis et al 1980).

One of the more interesting questions about hydrothermal circulation is its depth of penetration. In the development of the theory of water penetration by thermal contraction cracking, it was hard to limit the maximum depth of cracking by any plausible physical mechanism. This led to the suggestion that a compositional change, from water-resistant gabbroic minerals to easily serpentinized olivine, could be responsible (Lister 1974, 1977a). The physical evidence of the heat-flow distribution has turned out to be of little help. Experiments in Hele-Shaw cells have

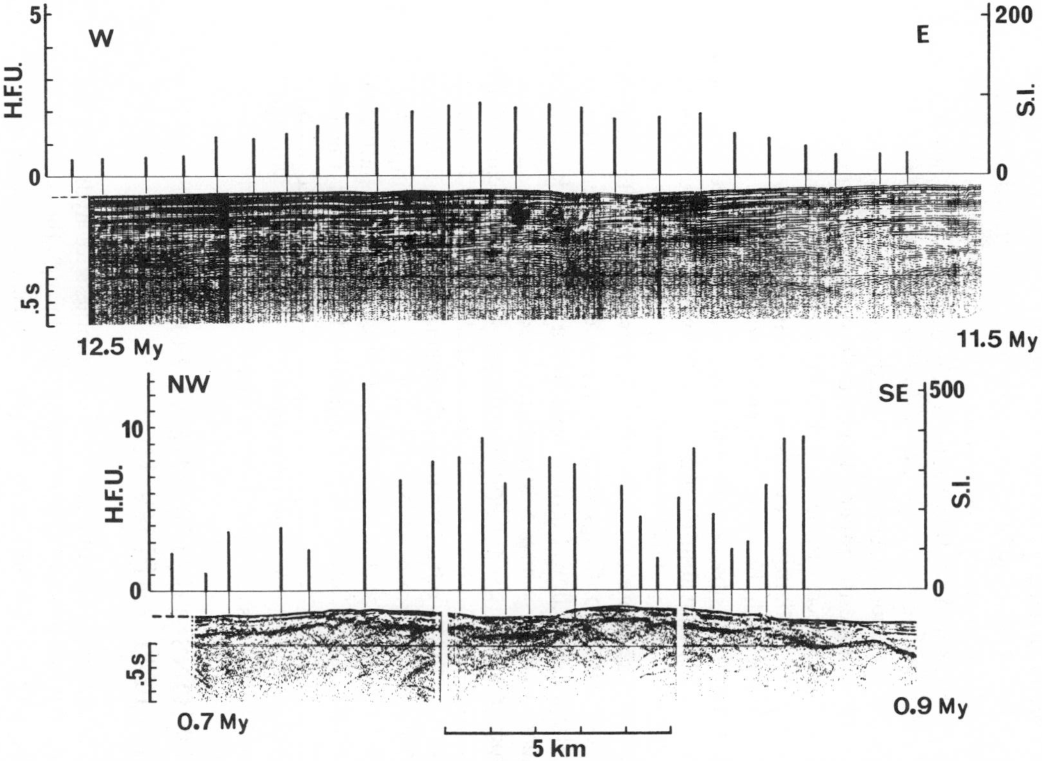

Figure 6 (*Upper*) A simultaneous heat-flow and airgun seismic-reflection profile taken on the sediment fan southwest of the Queen Charlotte Islands. Note the large scale of the lateral variation and the smoothing of the small-scale variations by the relatively thick sediment cover. Theoretical heat flow about 3.5 HFU. An intersecting NS drift profile (not shown) shows that the heat-flow variation is three-dimensional. (*Lower*) A simultaneous heat-flow and acoustic-reflection profile taken across the southeast flank of Explorer Ridge; the result of a single drift station. Vertical exaggeration of acoustic profile 3.4 : 1; the profile is about 50 km southwest along the ridge from that shown in Figure 3. Note the lack of correlation between the basement fault in the middle of the profile and the heat flow. Theoretical lithospheric heat flow is about 13 HFU. Both profiles reproduced by courtesy of Davis et al (1980).

shown that convection cells in a porous medium cannot be distorted readily from their natural shape by moderate forcing, such as by topography on the upper boundary of the permeable layer. However, if there are discharge and recharge zones where the upper boundary of the impermeable medium is underneath a thermal blanket of relatively impermeable sediment, large-scale regional flows can co-exist with cellular convection. A few of the Hele-Shaw experiments demonstrated the existence of stable cells in the presence of larger-scale bilateral flows (Hartline 1978, Davis et al 1980). What this means is that the large-scale heat flow variations under sediment-covered areas do not indicate the thickness of the permeable zone in any simple way, and new experiments must be conducted to determine the range of vented flows. Unfortunately, even the finding of what appears to be a thick boundary layer in the convective flow on the Mid-Atlantic Ridge does not indicate directly what the depth of circulation might be (Hyndman et al 1976).

The large scale of the observed heat-flow variations, coupled with the three-dimensionality shown in Figure 5 and elsewhere, is good evidence that rock permeability is fairly general and not confined to widely spaced fractures like those of Bodvarsson & Lowell (1972). The length of time that circulation continues is simply dependent on the permeability and the level of the heat flow. Stations less than a few tens of kilometers from basement outcrops show substantially lower heat flow than that expected from lithosphere of the appropriate age. When the means of all stations are compared with those of measurements selected to be far enough from outcrops to reflect the true lithospheric heat flow, differences persist out to ages as great as 80 My (Sclater et al 1976). Attempts have been made to categorize the cessation of circulation by individual spreading center. These are probably misguided because of the high degree of local geologic variability, the lack of precise knowledge about opportunities for venting near most stations, and statistical problems with the data themselves (e.g. Anderson et al 1977).

CONCLUSIONS: GLOBAL HEAT LOSS AND GEOCHEMICAL IMPLICATIONS

One of the interesting consequences of the hypothesis of hydrothermal circulation is that estimates of the total heat loss through the ocean floor can be made with a new confidence. Instead of trying to collate highly variable data points and make 1° or 5° square averages from the measurements, one can use the theoretical heat-flow decay curves with the best fits to the most reliable measurements (Lee 1970, Sclater et al 1976). Although most of the ocean floor has now been dated by the analysis of magnetic anomalies, a heat-flow estimate based on the relative area of ocean floor of various ages has not yet been made. Instead, it has been assumed that the ocean floor has all been generated by spreading at a ridge equal in length to the sum of all present-day ridges and with an intermediate spreading rate. Various methods have been used to account for the cooling of freshly extruded material, and different assumptions made about whether old ocean floor stabilizes at some arbitrary heat flow or continues to cool according to the square root of time. However, if all the numerical assumptions are brought into line with the most recent data, the estimates fall into a relatively narrow range (Davies 1979, Williams & Von Herzen 1974, Chapman & Pollack 1975, Langseth & Anderson 1979). The mean for the ocean floor is more like 2 HFU than the 1.4 estimated previously, or the 1.4 HFU still estimated for the continents. Not only is the estimate for the global heat loss raised significantly, but the former approximate equality between the continental and oceanic heat flows is removed. There is insufficient knowledge about general convection theory to apply this result to problems of convection within the earth.

Perhaps the most important implication of hydrothermal circulation is geochemical. On the one hand, a significant fraction of the water in the oceans may circulate through the rocks of the oceanic crust every million years. A major exchange of magnesium for calcium provides a large sink for the former and a source for the latter in the oceans, a possible explanation of why dolomite may no longer be formed in oceanic deposits. Access of water into hot rock cracked on a relatively fine scale permits the extraction of heavy metals in a hot, acidic, and sulphide-rich solution. These solutions either exude from the ocean floor, making heavy-metal precipitates available to ridge-crest sediments and manganese to sea-floor nodules; or, if trapped beneath a cap of relatively impermeable sediments, they can be the source of the massive sulphide deposits found in some parts of the earth (Sleep & Wolery 1978, Humphries & Thompson 1978, Bischoff & Dickson 1975, Corliss 1971, W. S. Fyfe, personal communication). Wherever thick impermeable sediments cover young ocean floor, trapped high-temperature hydrothermal systems may exist, and a few have been found already (Davis & Lister 1977). These systems may be of exploitable value, not so much for the thermal energy they contain, as for the increasingly scarce and valuable metals that should be present in their fluids.

ACKNOWLEDGMENTS

I wish to thank E. E. Davis and R. P. Von Herzen for contributing original diagrams for inclusion in this report. The National Science Foundation paid for the typist through grant OCE 77-17870 (A01).

Literature Cited

Anderson, D. L., Archambeau, C. B. 1964. The anelasticity of the earth. *J. Geophys. Res.* 69: 2071–84
Anderson, R. N., Langseth, M. G., Sclater, J. G. 1977. The mechanisms of heat transfer through the floor of the Indian Ocean. *J. Geophys. Res.* 82: 3391–409
Bender, M., Dymond, J. R., Heath, G. R. 1971. Isotopic analyses of metalliferous sediments from the East Pacific Rise. *Geol. Soc. Am. Abstr.* 3: 537
Bischoff, J. L., Dickson, F. W. 1975. Sea water-basalt interaction at 200°C and 500 bars; Implications for the origins of sea-floor heavy metal deposits and regulation of sea-water chemistry. *Earth Planet. Sci. Lett.* 25: 385–97
Bodvarsson, G., Lowell, R. P. 1972. Ocean floor heat flow and the circulation of interstitial waters. *J. Geophys. Res.* 77: 4472–75
Bolton, R. S. 1975. Recent developments and future prospects for geothermal energy in New Zealand. *Proc. 2nd UN Symp. on Development and Use of Geothermal Resources* 1: 37–42
Bostrom, K., Peterson, M. N. A. 1969. The origin of aluminium ferromanganoan sediments in areas of high heat-flow on the East Pacific Rise. *Mar. Geol.* 7: 427–47
Bryant, W. R., DeFlache, A. P., Trabant, P. K. 1974. Consolidation of marine clays and carbonates. In *Deep Sea Sediments: Physical and Mechanical Properties*, ed. A. L. Interlutzen. New York: Plenum
Bullard, E. C. 1950. The transfer of heat from the core of the earth. *Mon. Not. R. Astron. Soc., Geophys. Suppl.* 6: 36–41
Bullard, E. C. 1954. The flow of heat through the floor of the Atlantic Ocean. *Proc. R. Soc. London Ser. A* 222: 408–29
Bullard, E. C., Maxwell, A. E., Revelle, R. 1956. Heat flow through the deep sea floor. *Adv. Geophys.* 3: 153–81
Bullard, E. C., Day, A. 1961. The flow of heat through the floor of the Atlantic Ocean. *Geophys. J. R. Astron. Soc.* 4: 282–92
Carvalho, H. dS., Vacquier, V. 1977. Method for determining terrestrial heat flow in oil fields. *Geophysics* 42: 584–93
Chapman, D. S., Pollack, H. N. 1975. Global heat-flow: a new look. *Earth Planet. Sci. Lett.* 28: 23–32
Clark, S. P. Jr. 1966. High pressure phase equilibria. In *Handbook of Physical Constants*, ed. S. P. Clark Jr. *Geol. Soc. Am. Mem.* 97. 587 pp.
Coleman, R. G. 1971. Plate-tectonic emplacement of upper mantle peridotites along continental edges. *J. Geophys. Res.* 76: 1212–21
Corliss, J. B. 1971. The origin of metal-bearing submarine hydrothermal solutions. *J. Geophys. Res.* 76: 8128–38
Corliss, J. B., Dymond, J., Gordon, L. I., Edmond, J. M., Von Herzen, R. P., Ballard, R. D., Green, K., Williams, D. L., Bainbridge, A., Crane, K., Van Andel, Tj. H. 1979a. Submarine thermal springs on the Galapagos Rift. *Science* 203: 1073–83
Corliss, J. B., Edmond, J. M., Gordon, L. I. 1979b. Some implications of heat/mass ratios in Galapagos Rift hydrothermal fluids for models of seawater-rock interaction and the formation of oceanic crust. In *Origin of the Oceanic Crust, Proc. 2nd Ann. Ewing Symp.* Washington DC: Am. Geophys. Union
Corry, C., Dubois, C., Vacquier, V. 1968. Instrument for measuring terrestrial heat flow through the ocean floor. *J. Mar. Res.* 26: 165–77
Dasch, E. J., Hedge, C. E., Dymond, J. 1973. Effect of seawater interaction on the strontium isotope composition of deep-sea basalts. *Earth Planet. Sci. Lett.* 19: 177–83
Davies, G. F. 1979. Review and revisions of global heat flow estimates. *Geophys. Res. Lett.* In press
Davis, E. E., Lister, C. R. B. 1974. Fundamentals of ridge crest topography. *Earth Planet. Sci. Lett.* 21: 405–13
Davis, E. E., Lister, C. R. B. 1977. Heat flow measured over the Juan de Fuca Ridge: evidence for widespread hydrothermal circulation in a highly heat transportive crust. *J. Geophys. Res.* 82: 4845–60
Davis, E. E., Lister, C. R. B., Wade, U. S., Hyndman, R. D. 1980. Detailed heat flow measurements over the Juan de Fuca Ridge system and implications for the evolution of hydrothermal circulation in young oceanic crust. *J. Geophys. Res.* In press
Dewey, J. F., Bird, J. M. 1971. Origin and emplacement of the ophiolite suite: Appalachian ophiolites in Newfoundland. *J. Geophys. Res.* 76: 3179–3206
Diment, W. H., Marine, I. W., Neiheisel, J., Siple, G. E. 1965. Subsurface temperature, thermal conductivity, and heat flow near Aiken, South Carolina. *J. Geophys. Res.* 70: 5635–44
Dorman, J. 1969. Seismic surface-wave data on the upper mantle. In *The Earth's*

Crust and Upper Mantle. Am. Geophys. Union Monogr. 13:257–65

Elder, J. W. 1965. Physical processes in geothermal areas. *Am. Geophys. Union Monogr.* 8:211–39

Fujisawa, H., Fujii, N., Mizutani, H., Kanamori, H., Akimoto, S. I. 1968. Thermal diffusivity of Mg_2SiO_4, Fe_2SiO_4 and NaCl at high pressures and temperatures. *J. Geophys. Res.* 73:4727–33

Fyfe, W. S. 1978. The evolution of the earth's crust: modern plate tectonics to ancient hot spot tectonics? *Chem. Geol.* 23:89–114

Gass, I. G. 1968. Is the Troodos Massif of Cyprus a fragment of Mesozoic ocean floor? *Nature* 220:39–42

Gerard, R., Langseth, M. Jr., Ewing, M. 1962. Thermal gradient measurements in the water and bottom sediments of the western Atlantic. *J. Geophys. Res.* 67:785–803

Hart, R. A. 1973. A model for chemical exchange in the basalt-seawater system of oceanic layer II. *Can. J. Earth Sci.* 10: 799–816

Hart, S. R., Erlank, A. J., Kable, E. J. D. 1974. Sea-floor basalt alterations: some chemical and Sr isotopic effects. *Contrib. Mineral. Petrol.* 44:219–30

Hartline, B. K. 1978. Topographic forcing of thermal convection in a Hele-Shaw cell model of a porous medium. PhD thesis. Univ. Washington, Seattle

Heirtzler, J. R., Dickson, G. O., Herron, E. M., Pitman, W. C. III, LePichon, X. 1968. Marine magnetic anomalies, geomagnetic field reversals, and motions of the ocean floor and continents. *J. Geophys. Res.* 73:2119–36

Herrin, E. 1969. Regional variations of P-wave velocity in the upper mantle beneath N. America. In *The Earth's Crust and Upper Mantle. Am. Geophys. Union Monogr.* 13:242–46

Humphries, S. E., Thompson, J. 1978. Hydrothermal alteration of oceanic basalts by seawater. *Geochim. Cosmochim. Acta* 42:107–25

Hyndman, R. D. 1976. Heat flow measurements in the inlets of southwestern British Columbia. *J. Geophys. Res.* 81:337–49

Hyndman, R. D., Rankin, D. S. 1972. The Mid-Atlantic Ridge near 45°N. XVIII. Heat flow measurements. *Can. J. Earth Sci.* 9:664–70

Hyndman, R. D., Von Herzen, R. P., Erickson, A. J., Jolivet, J. 1976. Heat flow measurements in deep crustal holes on the Mid-Atlantic Ridge. *J. Geophys. Res.* 81:4053–60

Hyndman, R. D., Rogers, G. C., Bone, M. N., Lister, C. R. B., Wade, U. S., Barrett, D. L., Davis, E. E., Lewis, T., Lynch, S., Seemann, D. 1978. Geophysical measurements in the region of the Explorer Ridge off western Canada. *Can. J. Earth Sci.* 15:1508–25

Jeffreys, H. 1938. The disturbance of the temperature gradient in the Earth's crust by inequalities of height. *Mon. Not. R. Astron. Soc., Geophys. Suppl.* 4:309–12

Klitgord, K. D., Mudie, J. D. 1974. The Galapagos spreading centre: a near-bottom geophysical study. *Geophys. J. R. Astron. Soc.* 38:563–86

Lachenbruch, A. H. 1968. Rapid estimation of the topographic disturbance to superficial thermal gradients. *Rev. Geophys.* 6:365–400

Lachenbruch, A. H. 1976. Dynamics of a passive spreading center. *J. Geophys. Res.* 78:3395–3417

Lachenbruch, A. H., Marshall, B. V. 1968. Heat flow and water-temperature fluctuations in the Denmark strait. *J. Geophys. Res.* 73:5829–42

Langseth, M. G. Jr., LePichon, X., Ewing, M. 1966. Crustal structure of the mid-ocean ridges: heat flow through the Atlantic Ocean floor and convection currents. *J. Geophys. Res.* 71:5321–55

Langseth, M. G., Anderson, R. N. 1979. Correction. *J. Geophys. Res.* 84:1139–40

Lee, W. H. K. 1970. On the global variations of terrestrial heat flow. *Phys. Earth Planet. Inter.* 2:332–41

LePichon, X., Langseth, M. G. Jr. 1969. Heat flow from the mid-ocean ridges and sea-floor spreading. *Tectonophysics* 8: 319–44

Lister, C. R. B. 1963. A close group of heat-flow stations. *J. Geophys. Res.* 68: 5569–73

Lister, C. R. B. 1970a. Measurement of *in situ* sediment conductivity by means of a Bullard-type probe. *Geophys. J. R. Astron. Soc.* 19:521–32

Lister, C. R. B. 1970b. Heat flow of the Juan de Fuca Ridge. *J. Geophys. Res.* 75:2648–54

Lister, C. R. B. 1971. Crustal magnetisation and sedimentation near two small seamounts west of the Juan de Fuca Ridge, Northeast Pacific. *J. Geophys. Res.* 76: 4824–41

Lister, C. R. B. 1972. On the thermal balance of a mid-ocean ridge. *Geophys. J. R. Astron. Soc.* 26:515–35

Lister, C. R. B. 1974. On the penetration of water into hot rock. *Geophys. J. R. Astron. Soc.* 39:465–509

Lister, C. R. B. 1975. Qualitative theory on the deep end of geothermal systems.

Proc. 2nd UN Symp. on Development and Use of Geothermal Resources 1:459–63
Lister, C. R. B. 1977a. Qualitative models of spreading-center processes, including hydrothermal penetration. *Tectonophysics* 37:203–18
Lister, C. R. B. 1977b. Estimators for heat flow and deep rock properties based on boundary-layer theory. *Tectonophysics* 41:157–71
Lister, C. R. B. 1980. "Active" and "passive" hydrothermal systems in the oceanic crust: predicted physical conditions. In *The Dynamic Environment of the Ocean Floor.* Lexington, Mass: D. C. Heath. In press
Loper, D. E. 1978. The gravitationally powered dynamo. *Geophys. J. R. Astron. Soc.* 54:389–404
Malkus, W. V. R. 1968. Precession of the earth as the cause of geomagnetism. *Science* 160:259–64
McGehee, M. S., Boegemann, D. E. 1966. MPL acoustic transponder. *Rev. Sci. Instrum.* 37:1450–55
McKenzie, D. P. 1967. Some remarks on heat flow and gravity anomalies. *J. Geophys. Res.* 72:6261–73
Muehlenbachs, K., Clayton, R. N. 1972. Oxygen isotopes of fresh and weathered submarine basalts. *Can. J. Earth Sci.* 9:172–84
Oxburgh, E. R., Turcotte, D. L. 1968. Mid-ocean ridges and geotherm distribution during mantle convection. *J. Geophys. Res.* 73:2643–61
Palmason, G. 1967. On heat flow in Iceland in relation to the Mid-Atlantic Ridge. In *Iceland and Mid-Ocean Ridges*, ed. S. Bjornsom. *Soc. Sci. Islandica, Reykjavik, Publ.* 38, pp. 11–127
Parker, R. L., Oldenburg, D. W. 1973. Thermal model of ocean ridges. *Nature* 242:137–39
Pearson, W. C., Lister, C. R. B. 1973. Permeability measurements on a deep-sea core. *J. Geophys. Res.* 78:7786–87
Piper, D. Z. 1973. Origin of metalliferous sediments from the East Pacific Rise. *Earth Planet. Sci. Lett.* 19:75–82
Ratcliffe, E. H. 1960. The thermal conductivities of ocean sediments. *J. Geophys. Res.* 65:1535–41
Raymond, J. R., Tillson, D. D. 1968. Evaluation of a thick basalt sequence in south-central Washington. *Atomic Energy Comm. Res. Dev. Rep. BNWL-776.* 126 pp.
Ringwood, A. E. 1975. *Composition and Petrology of the Earth's Mantle.* New York: McGraw-Hill
Robertson, A. H. F., Hudson, J. D. 1973. Cyprus umbers: chemical precipitates on a Tethyan ocean ridge. *Earth Planet. Sci. Lett.* 18:93–101
Runcorn, S. K. 1965. Changes in convection pattern in the earth's mantle and continental drift, evidence for cold origin of the earth. *Philos. Trans. R. Soc. London Ser. A* 258:228–51
Sclater, J. G., Corry, C. E., Vacquier, V. 1969. *In situ* measurement of the thermal conductivity of ocean floor sediments. *J. Geophys. Res.* 74:1070–81
Sclater, J. G., Francheteau, J. 1970. The implications of terrestrial heat-flow observations on current tectonic and geochemical models of the crust and upper mantle of the earth. *Geophys. J. R. Astron. Soc.* 20:509–42
Sclater, J. G., Crowe, J., Anderson, R. N. 1976. On the reliability of oceanic heat flow averages. *J. Geophys. Res.* 81:2997–3006
Sleep, N. H. 1969. Sensitivity of heat flow and gravity to the mechanism of sea-floor spreading. *J. Geophys. Res.* 74:542–49
Sleep, N. H. 1975. Formation of oceanic crust: some thermal constraints. *J. Geophys. Res.* 81:4037–42
Sleep, N. H., Wolery, T. J. 1978. Egress of hot water from mid-ocean ridge hydrothermal systems: some thermal constraints. *J. Geophys. Res.* 83:5913–22
Smithson, S. B., Decker, E. R. 1974. A continental crustal model and its geothermal implications. *Earth Planet. Sci. Lett.* 22:215–25
Smithson, S. B. 1978. Modeling continental crust: structural and chemical constraints. *Geophys. Res. Lett.* 5:749–52
Solomon, S. C. 1972. Seismic-wave attenuation and partial melting in the upper mantle of North America. *J. Geophys. Res.* 77:1483–1502
Spiess, F. N., Loughridge, M. S., McGehee, M. S., Boegemann, D. E. 1966. An acoustic transponder system. *J. Hist. Navigation* 13:154–61
Spooner, E. T. C., Fyfe, W. S. 1973. Sub-sea-floor metamorphism, heat and mass transfer. *Contrib. Mineral. Petrol.* 42:287–304
Spooner, E. T. C., Beckinsale, R. D., Fyfe, W. S., Smewing, J. D. 1974. O^{18} enriched ophiolitic metabasic rocks from E. Liguria (Italy), Pindos (Greece), and Troodos (Cyprus). *Contrib. Mineral. Petrol.* 47:41–62
Swanson, D. A., Wright, T. L., Hooper, P. R., Bentley, R. D. 1979. Revisions in stratigraphic nomenclature of the Columbia River Basalt Group. *US Geol. Surv. Bull. 1457-G.* 59 pp.
Talwani, M., LePichon, X., Ewing, M. 1965. Crustal structure of the mid-ocean

ridges, 2. Computed model from gravity and seismic refraction data. *J. Geophys. Res.* 70:341–52

Talwani, M., Windish, C. C., Langseth, M. G. Jr. 1971. Reykjanes Ridge crest: a detailed geophysical study. *J. Geophys. Res.* 76:473–517

Thomson, W. (Lord Kelvin) 1846. Age of the earth and its limitations as determined from the distribution and movement of heat within it. PhD thesis. Univ. Glasgow, Scotland

Von Herzen, R. P., Maxwell, A. E. 1959. The measurement of thermal conductivity of deep sea sediments by a needle-probe method. *J. Geophys. Res.* 64:1557–63

Von Herzen, R. P., Uyeda, S. 1963. Heat flow through the eastern Pacific Ocean floor. *J. Geophys. Res.* 68:4219–50

Von Herzen, R. P., Anderson, R. N. 1972. Implications of heat flow and bottom water temperature in the eastern equatorial Pacific. *Geophys. J. R. Astron. Soc.* 26:427–58

Williams, D. L., Von Herzen, R. P. 1974. Heat loss from the earth: new estimate. *Geology* 2:327–28

Williams, D. L., Von Herzen, R. P., Sclater, J. G., Anderson, R. N. 1974. The Galapagos spreading center: lithospheric cooling and hydrothermal circulation. *Geophys. J. R. Astron. Soc.* 38:587–608

Ann. Rev. Earth Planet. Sci. 1980. 8: 119–44
Copyright © 1980 by Annual Reviews Inc. All rights reserved

DEFORMATION OF MANTLE ROCKS

×10127

Yves Gueguen and Adolphe Nicolas
Laboratoire de Tectonophysique, Université de Nantes, 44072 Nantes, France

I INTRODUCTION

Plate tectonics emerged as a concept in the sixties during the course of the "upper mantle" International Project. Although plate tectonics is largely a daughter of oceanographic investigations, its emergence during a period of special effort to understand the upper mantle is not accidental. The seventies International Project has been centered logically on geodynamics. Because the motor of plate geodynamics lies in the mantle, the effort spent on mantle investigations has understandably increased during the last decade. Surprisingly, the studies on mantle specimens are by and large directed more toward upper mantle mineralogy, petrology, and geochemistry than toward mantle plasticity. To our knowledge, fewer than ten institutes in the world devote activity to this aspect of the upper mantle. Similarly, there have only been two extensive reviews on plasticity in the upper mantle (Nicolas & Poirier 1976, Carter 1976). Other review papers (Moores 1974, Nicolas 1976) deal only partially with the subject.

The deepest mantle rocks, brought up by kimberlites, come from 200 km. Direct information on the upper mantle is therefore restricted to the shell extending from the Moho down to 200 km. Through the study of deformed peridotites that originated in this upper part of the mantle, two distinct bodies of information are obtained. A systematic investigation of specimens collected on a regional scale (massif, volcanic district) leads to models of the structure, kinematics, and dynamics of the upper mantle at that very scale. In contrast, a more general analysis of deformational microstructures, when compared with experimentally obtained ones, makes it possible to determine which flow mechanisms have been operating in the upper mantle and also introduces some constraints on the flow parameters.

Information on regional or general upper mantle geodynamics is obtained by combining experimental and theoretical investigations on

0084-6597/80/0515-0119$01.00

upper mantle minerals with observations and measurements on naturally deformed peridotites. Consequently, this paper will first review the recent developments in the domain of experiments and theories and will then show how, through the study of naturally deformed peridotites, our understanding of the upper mantle has progressed.

II EXPERIMENTAL AND THEORETICAL INVESTIGATIONS

II.1 *Deformation and Recrystallization Mechanisms in Olivine*

The problem of creep-controlling mechanisms is of major importance for extrapolation to mantle conditions. Various factors controlling dislocation creep are dislocation glide, cross slip, and climb. This last

Table 1 Slip systems in olivine

Slip direction	Slip plane	Temperature (°C)	Ref.[a]
[001]	(100)	<300	1
		<350	2
		800	4
	{110}	300–1000	1
		<150	2
		<1000	3
		800–1000	4
	(010)	800	4
		1600	5
[010]	(100)	"low temperature"	1
		800	4
[100]	(001)	1600	5
		1600	6
	(010)	>1000	3
		>1250	2
		800–1000	4
		1600	5
		1600	6
	(Ok1)	>1000	1
	(pencil)	>1000	2
		>1000	3

[a] References:
1. Raleigh (1968): polycrystals, high stress
2. Carter & Ave Lallemant (1970): polycrystals, high stress
3. Raleigh & Kirby (1970): polycrystals, high stress
4. Phakey et al (1972): single crystals, high stress
5. Durham et al (1977): single crystals, moderate stress
6. Jaoul et al (1979): single crystals, moderate stress

mechanism involves diffusion. Understanding plastic flow mechanisms requires 1. identification of the slip systems, creep laws, and dislocation microstructures in order to determine whether glide, cross slip, or climb is predominating and 2. data on diffusion to test the hypothesis of climb-controlled creep. The first aspect is now well documented from experiments in both single and polycrystals. Diffusion data are not yet complete. Other important problems are recovery, recrystallization, and paleo-stress determinations. They will be discussed below.

CREEP LAWS, DISLOCATION MICROSTRUCTURES, AND SLIP SYSTEMS

(*a*) *High stress experiments* (*1–15 kb*) The first experiments concerned with olivine plasticity were performed mainly with the Griggs apparatus (Griggs 1967). Raleigh (1968), Carter & Ave Lallemant (1970), and Raleigh & Kirby (1970) determined the slip planes using slip bands and subboundary external axis of rotation (Table 1). Phakey et al (1972) determined the slip systems between 600 and 1000°C using a transmission electron microscope (TEM) to study dislocations in olivine single crystals.

Carter & Ave Lallemant (1970) and Raleigh & Kirby (1970) have described creep laws of the type $\mathring{\varepsilon} = A\sigma^n \exp(-E/RT)$ (Table 2). Variations in the value of E were attributed to the H_2O content of olivine, and Blacic (1972) suggested that a water weakening mechanism similar to that observed in quartz existed in olivine. Zeuch & Green (1979) also

Table 2 Olivine creep laws, $\mathring{\varepsilon} = A\sigma^n \exp(-E/RT)$, ($\sigma$ in kb)

Sample[b]	A	n	E (kcal mole^{-1})	Ref.[a]
Dry dunite	1.2×10^{10}	4.8	120	1
Wet dunite	6.2×10^{6}	2.4	80	1
Dry dunite	10^{8}	5	106	2
Wet dunite	4.3×10^{8}	3	94	3
Dry dunite	4.3×10^{8}	3	126	3
Olivine $[110]_c$	1.3×10^{12}	3.6 ± 0.3	125 (estimate)	4
Olivine $[101]_c$	1.1×10^{12}	3.7 ± 0.2	125	4
Olivine $[011]_c$	6.3×10^{10}	3.5 ± 0.3	125 (estimate)	4
Forsterite $[101]_c$	7.1×10^{12}	3.5	135	5

[a] References:
1. Carter & Ave Lallemant (1970): polycrystals, high stress
2. Raleigh & Kirby (1970): polycrystals, high stress
3. Post (1973): polycrystals, high stress
4. Durham & Goetze (1977b): single crystals, moderate stress
5. Durham & Goetze (1977a): single crystals, moderate stress

[b] Subscript "c" refers to an imaginary cubic lattice.

reported differences in strengths between dry and wet olivine. Paterson (1979) concluded that no definite proof of a water weakening effect is yet available.

(b) Moderate stress experiments (100–1000 bars) More recently, many data have been collected at stresses more comparable to those inferred for the earth's mantle. They were obtained from experiments on single crystals in room-pressure creep machines (Kohlstedt & Goetze 1974, Durham & Goetze 1977a, b, O. Jaoul and C. Froidevaux, in preparation). Creep laws have the same form as those previously determined, but the parameters are different (Table 2).

Durham et al (1977) have described in detail the related dislocation microstructures, using a simple and powerful decoration technique introduced by Kohlstedt et al (1976) and making the dislocations visible in the optical microscope. Using this method, Durham et al (1977) found slip systems that compare well with previous results (Table 1). Extending the technique to forsterite, Jaoul et al (1979) obtained similar results (Table 1).

(c) Dislocation microstructures and creep-controlling mechanisms Gueguen (1979a) has classified into six types the dislocation microstructures observed in natural deformation (Figure 1). All are also observed in experimental deformation. On the basis of these various observations (Zeuch & Green 1979, Ricoult 1978, Durham et al 1977, Jaoul et al 1979), one can conclude that dislocation walls rarely develop in single crystal deformation. Such a difference between single and polycrystals may be important for extrapolating creep laws from single crystals to polycrystals.

Because the [100] Burgers vector is dominant at high temperature, we will not discuss in great detail the observations relating to the [001] Burgers vector, which dominates at low temperatures. When $b = [001]$, the free dislocations are mainly screws and these screws are very long and straight. This could be explained by a sessile dissociation. Vander Sande & Kohlstedt (1976) have reported evidence for such a dissociation. In that situation, cross slip of [001] screws could be the creep-controlling mechanism.

The same authors have also reported evidence of splitting of [100] screws. Poirier (1975) and Poirier & Vergobbi (1978) have suggested that cross slip of [100] screws could be the controlling step at high temperatures. This fits well with a microstructure of type 1 (Figure 1). On the other hand, Gueguen (1979b) has noticed that edges or mixed segments are dominant when (010) [100] and (001) [100] slip systems are active (Figure 2). These segments are very straight and they may be dissociated

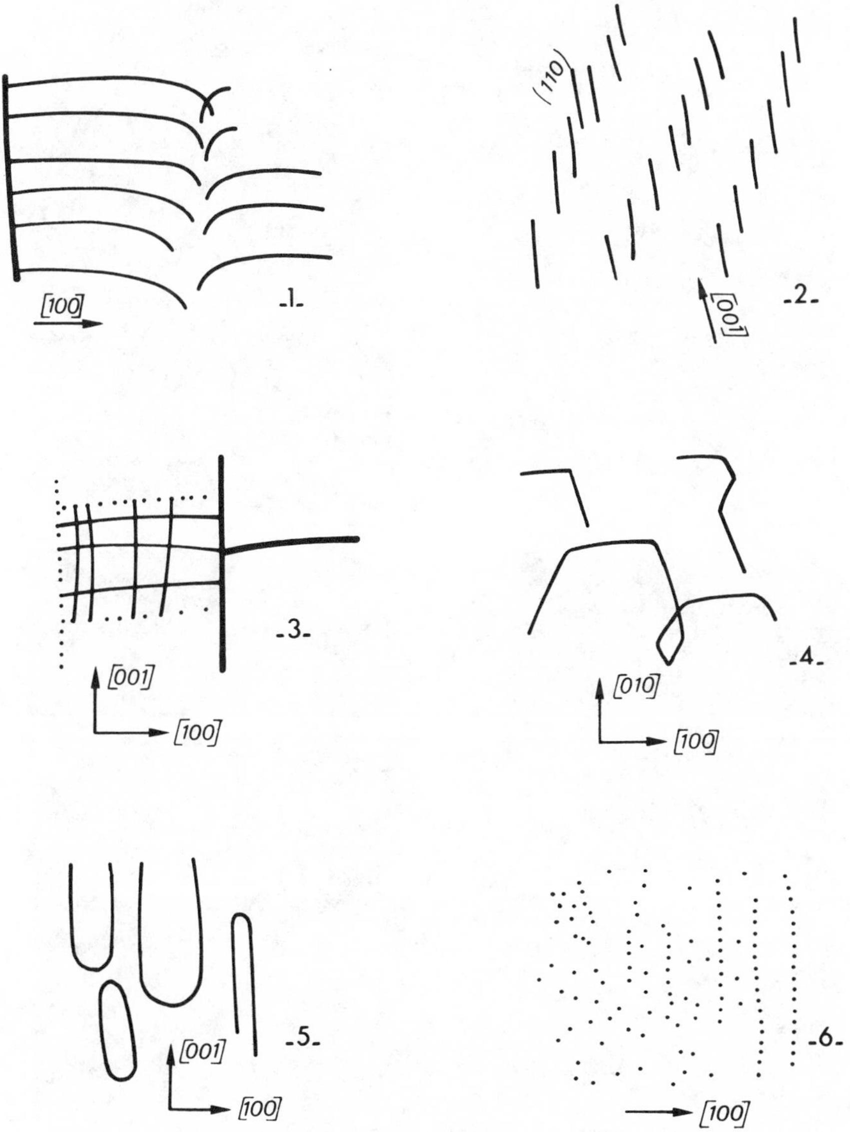

Figure 1 Six main types of dislocation microstructures (Gueguen 1979a): 1. (100) organization: [100] screws between (100) tilt walls, 2. (110) organization: [001] screws in {110} slip bands, 3. polygonized structure: (100) and (001) tilt walls, (010) twist walls, 4. (001) glide loops: [100] screws and ⟨110⟩ mixed segments, 5. (010) glide loops: [100] screws and [001] edges, 6. development of (100) tilt walls by climb of edges.

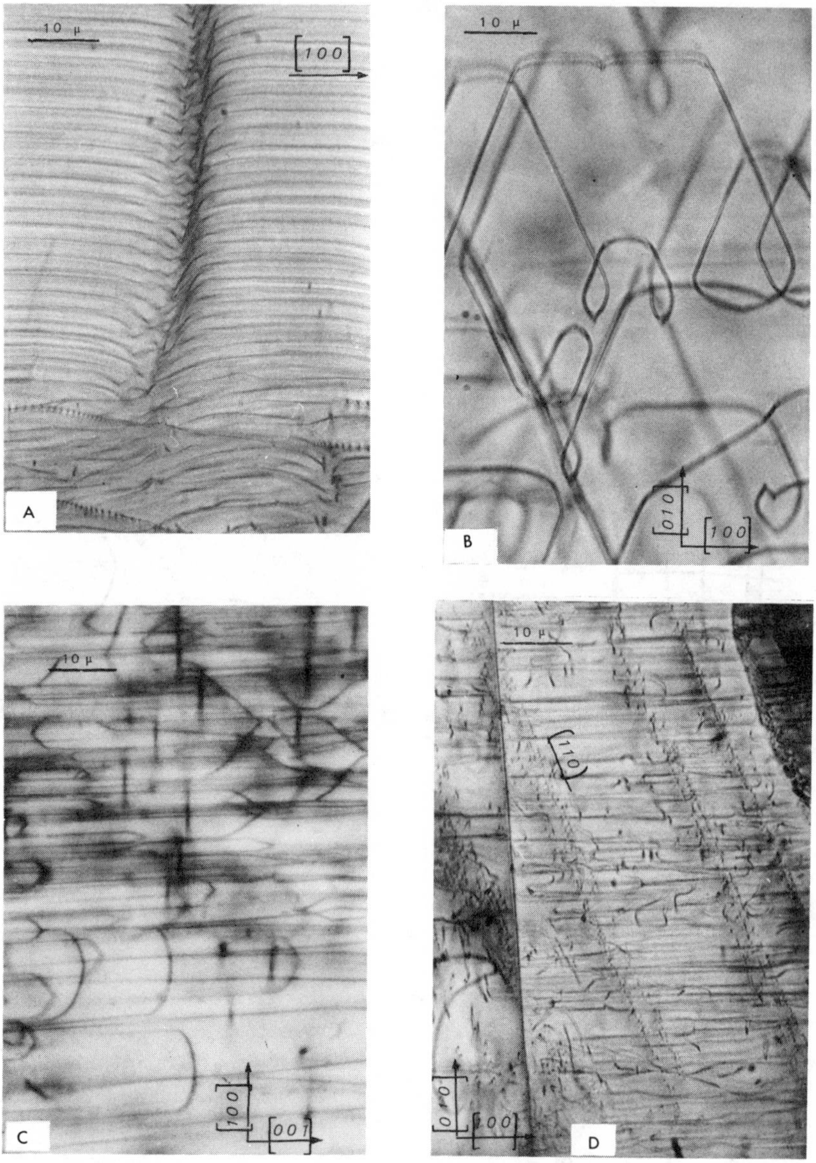

Figure 2 Microphotographs (optical microscope) of decorated dislocations corresponding to some of the 6 main types described in Figure 1. *A*. (100) organization: [100] screws crossing from one (100) tilt wall to another. This structure is type 1 in Figure 1. *B*. (001) glide loops: [100] screws and ⟨110⟩ mixed segments. This structure is type 4 in Figure 1. *C*. (010) glide loops: [100] short screws and [001] long edges. This structure is type 5 in Figure 1. *D*. (110) slip bands superimposed on a (100) organization. Slip bands are ascribed to a late low temperature deformation.

out of the glide planes. This could lead to high temperature creep being controlled either by double kink nucleation (glide-controlled creep) or by jog nucleation and diffusion (climb-controlled creep). It is likely that in fact there are several creep laws for olivine single crystals, depending on temperature and glide system.

Several authors have also considered the possibility of deformation mechanisms other than dislocation creep. Twiss (1976) has discussed the results of Post (1973) and concluded that superplastic creep could dominate in the earth's upper mantle. Schwenn & Goetze (1978) have investigated hot pressing of olivine and extracted from their data the parameters for Coble creep. Extrapolating these results. Goetze (1978) has suggested that Coble creep could dominate in the lithosphere. However, observations in natural peridotites do not support this contention as structures ascribed to superplastic creep have only been described in a single case (Boullier & Gueguen 1975).

POINT DEFECTS AND DIFFUSION

The above discussion stresses that there are three distinct mechanisms that could control creep in olivine: cross slip, kink nucleation (glide), or jog nucleation plus diffusion (climb). In such a situation, it is crucial to compare activation energy for creep and diffusion. However, diffusion data are still very incomplete so that no definitive conclusion can be drawn.

Interdiffusion data have been published for Fe-Mg in olivine between 1000 and 1200°C (Buening & Buseck 1973). The interdiffusion coefficient depends on temperature, oxygen partial pressure, and orientation. The activation energy is 57 kcal mole^{-1} for $T > 1125$°C and 29 kcal mole^{-1} for $T < 1125$°C. The P_{O_2} exponent is $n = 0.17$.

Reddy & Cooper (1978) and O. Jaoul and C. Froidevaux (in preparation) have measured oxygen self-diffusion in forsterite using radioactive O^{18} in ion microprobe and nuclear reaction techniques. Oxygen diffusion is much slower than Fe-Mg diffusion. However, it cannot be the step controlling creep since its activation energy has been found to be 85 kcal mole^{-1}, which is much lower than that for creep (120–140). No P_{O_2} dependence has been reported, but measurements have been done in pure forsterite only.

Silicon diffusion has not yet been measured, either in olivine or in forsterite. Windhager & Borchardt (1975) measured it in Co_2SiO_4 using Si^{30}. They found that, in this case, Si diffusion is faster than oxygen diffusion. They also observed that it depends on SiO_2 activity. Co_2SiO_4 is the only olivine-like crystal in which Si diffusion has been measured so far.

Diffusion coefficients in olivine depend not only on pressure and temperature but also on P_{O_2}, a_{SiO_2}, a_{FeO} where P_{O_2} is the oxygen partial pressure, a_{SiO_2} the SiO_2 activity, and a_{FeO} the FeO activity. If dislocation creep were diffusion-controlled, the creep law would be a function of these parameters. Results quoted above show examples of P_{O_2} dependence (Fe-Mg diffusion) and a_{SiO_2} dependence (Si diffusion). Schmalzried (1978) and Stocker (1978a, b) have considered the various possibilities for point-defect equilibrium concentrations and diffusion in olivine.

RECOVERY AND RECRYSTALLIZATION

Goetze & Kohlstedt (1973) have measured the kinetics of recovery in naturally deformed polycrystalline olivine. From the collapse of vacancy loops, they deduced a diffusion coefficient $D = 3.10^4 \exp(-135 \text{ kcal mole}^{-1}/RT)$, (in $cm^2\ s^{-1}$). However, the more complete set of data on recovery is that of Ricoult (1978, 1979) who has shown that several mechanisms were active during recovery so that the resulting kinetics is complex. If the kinetics follows a law of the type $d\rho/dt = k\rho^m$, where ρ is the free dislocation density, it can be shown that the order of the kinetics m is lower than 3. It is impossible to extract an activation energy from the temperature dependence of the constant k. Toriumi & Karato (1978) have concluded from similar studies on a more limited temperature-time range that $m = 2$ and $E = 110$ kcal mole^{-2}. However, Ricoult & Gueguen (1980) have shown that these authors have grossly underestimated the uncertainties in measuring the dislocation densities.

Theoretical considerations of the recrystallization processes have been recently discussed by several authors (Mercier 1976, Twiss 1977, Sellars 1978, Poirier & Guillope 1979). Experimental data on recrystallization in olivine have been obtained by Post (1973), Mercier (1976), and Zeuch & Green (1979). Recrystallization can occur during the process of deformation (dynamic recrystallization) or during annealing (static recrystallization). As far as natural deformation of olivine is concerned. Poirier & Nicolas (1975) have shown that two distinct mechanisms can operate during dynamic recrystallization. They are 1. nucleation of new grains and growth, and 2. progressive misorientation of subgrains. The former occurs at higher stresses and the latter at lower stresses.

PIEZOMETERS

Microstructural parameters are related to internal stresses. Then if one assumes that internal and external stresses are almost equal, it is theoretically possible to determine paleostresses from microstructural investigations. Nicolas (1978) has recently discussed the applicability of the different parameters to geological situations. From this discussion, it follows that only three "piezometers" have to be considered: dislocation densities, tilt wall spacings, and recrystallized grain sizes.

(*a*) *Dislocation densities* Kohlstedt et al (1976), Durham et al (1977), and Zeuch & Green (1979) have shown that the free dislocation density and the stress correlate well in experiments. Durham et al (1977) have found that $\sigma = 2.10^{-5}\,\rho^{0.6}$ (ρ in cm^{-2} and σ in kb). This is a particular example of the general relationship $\sigma = \kappa\mu b\rho^{1/2}$ observed for a number of materials (Nicolas & Poirier 1976). Limits for using this piezometer come from the fact that a given free dislocation population only represents the preceding few percent deformation or less. Annealing or late deformation may have taken place after the main deformation. This is indeed observed in many cases (Gueguen 1979a). If it can be shown that no late deformation has been superimposed, minimum stresses can be estimated. Using this method, Gueguen (1977) and Nicolas (1978) have estimated minimum stresses of 200–500 bars for peridotite xenoliths.

(*b*) *Dislocation walls* Bound dislocations certainly represent more than a few percent deformation. They are also less recovery sensitive (Ricoult 1979). Most of the bound dislocations in olivine are organized in (100) tilt walls so that only (100) wall spacing is considered as a possible candidate for paleostress estimates. Depending on the dislocation density within it, a wall can be optically visible in polarized light or not. Raleigh & Kirby (1970) have measured optically visible wall spacing and found that $\sigma = 190\,d^{-1}$ (σ in kb and d in micrometers). Goetze (1975a, b), Mercier (1976), and Durham et al (1977) have suggested different relationships which take into account the decorated walls. These are, respectively $\sigma = 17\,d^{-1}$, $\sigma = 10\,d^{-1}$, and $\sigma = 115\,d^{-1}$ (same units). The large disagreement among the data shows that no satisfactory experimental calibration is yet available.

(*c*) *Recrystallized grain size* Dynamic recrystallization of olivine is observed experimentally in high stress polycrystal deformation as well as in nature. Post (1973) first suggested that the recrystallized grain size is a function of stress: $\sigma = 19\,d^{-0.7}$ (same units as above). Goetze (1975b) and Mercier (1976) have obtained different correlations: $\sigma = 11\,d^{-0.5}$ and $\sigma = 40\,d^{-0.8}$, respectively. Twiss & Sellars (1978) have discussed the applicability of such relations. However, several important limits exist for the use of this piezometer. All the experimental data were obtained at high stresses in the Griggs machine. Since stress measurements in this apparatus are not very accurate (Post 1973), the experimental calibration is not satisfactory, especially if one considers that extrapolation over one order of magnitude or more is required for comparison with peridotite xenoliths from basalts. Moreover, Poirier & Guillope (1979) have suggested that the recrystallization mechanisms may be different at high stresses and low stresses. They have shown for halite that different mechanisms lead to different stress–grain size correlations. Consequently, use of this piezometer should wait for a better experimental calibration.

II.2 *Deformation and Recrystallization Mechanisms in Other Minerals*

Several minerals coexist with olivine in the upper mantle: pyroxenes, spinel, garnet. Recent data have been obtained on the first two and will be presented below. At depths greater than 400 km, the main mantle mineral is the spinel phase of olivine for which some qualitative data are also available.

CLINOPYROXENE DEFORMATION AND RECRYSTALLIZATION

Clinopyroxene is a minor constituent of the upper mantle. Lherzolites contain 5–15% of this mineral. Ave Lallemant (1978), Kirby & Etheridge (1980), and Etheridge & Kirby (1980) have deformed clinopyroxenite, websterite, and single crystals of diopside in the Griggs apparatus.

At low temperatures flow is dominated by slip parallel to $\langle 001 \rangle$ and by twinning on $\{100\}$ and $\{001\}$. At higher temperatures, flow is dominated by multiple slip, recovery, and recrystallization. The transition temperature between these mechanism fields decreases with decreasing strain rate. The second flow regime is dominant in naturally deformed clinopyroxene. The mechanical data fit the power law $\mathring{\varepsilon} = A\sigma^n \exp(-E/RT)$ where n seems to increase with stress (3.3–5.3) and E is higher for "wet" experiments (91 kcal mole^{-1}) than for "dry" ones (78 kcal mole^{-1}). Recrystallization takes place by the two mechanisms already reported for olivine. Subgrain rotation, however, is only observed after considerable strain and extensive subgrain development. Ave Lallemant (1978) has observed a recrystallized grain size–stress relationship $d = 60\ \sigma^{-0.9}$ (d in micrometers and σ in kb).

ORTHOPYROXENE DEFORMATION AND RECRYSTALLIZATION

Orthopyroxene is common in lherzolites as well as in harzburgites. Both contain up to 20–30% of this mineral. Ross & Nielsen (1978) have deformed polycrystalline enstatite in the Griggs apparatus. The principal mechanism of plastic deformation at low temperatures is the shear-induced ortho-enstatite to clino-enstatite transformation described previously (Coe & Kirby 1975). At high temperatures and low strain-rates, this mechanism is supplanted by slip, subgrain formation, and recrystallization. Kirby & Etheridge (1980) and Etheridge & Kirby (1980) have deformed polycrystalline enstatite and bronzite single crystals. The number of available slip systems is less than that for clinoenstatite, and recrystallization is in general more developed. The high temperature creep law for wet conditions is of the type $\mathring{\varepsilon} = A\sigma^n \exp(-E/RT)$ where the stress exponent n is about 3 and the activation energy E is 65 kcal mole^{-1} (Ross & Nielsen 1978). Etheridge & Kirby (1980) report the same two mechanisms for recrystallization as in clinopyroxenes. Recrystallization in single crystals of orthopyroxenes occurs more rapidly and at lower strains than in clinopyroxenes under the same conditions. Ross & Nielsen (1978) determined a recrystallized grain size–stress relationship: $d = 47\,\sigma^{-0.85}$ (d in micrometers, σ in kb).

DEFORMATION OF SPINELS

Spinel is a minor constituent present at shallow depths in the mantle. Spinel peridotite contains up to 5% spinel. The most frequent spinel in xenoliths is picotite although other spinels are also observed.

Most of the information available on spinel plasticity has been published in the metallurgical literature. Experiments have focused on high temperature deformation of spinel single crystals with variable stoichiometry, corresponding to the formula MgO $(Al_2O_3)_n$. At temperatures higher than 0.6 T_m, dislocation climb plays a major role (Veyssière et al 1978). The strength is observed to increase when stoichiometry is approached. These results have been recently discussed in terms of dissociation of dislocations (Veyssière et al 1978). Interaction between point defects and dislocations could strongly influence the dissociation. Point-defect concentrations are obviously high when spinel is nonstoichiometric. Doukhan et al (1979) have shown that in stoichiometric spinel $MgAl_2O_4$ the dislocations are dissociated out of their slip planes. Pure edges are long and straight. They are dissociated in the $\{110\}$ plane normal to the slip plane ("climb dissociation") with a dissociation width of the order of 100 Å.

DEFORMATION OF HIGH PRESSURE MANTLE MINERALS

Since the suggestion by Bernal (1936) that common olivine might transform to a spinel structure in the deep mantle, high pressure phase transformation of olivine has received much attention. However, investigation of plastic properties under pressures higher than 25 kb is yet out of the experimental possibilities. For that reason, studies of germanates (analogues of olivine) or metastable phases have been conducted in the recent years.

Vaughan & Coe (1978) have as a first step shown that Mg_2GeO_4 olivine has plastic properties very similar to those of forsterite. The same authors have determined creep laws for both the olivine and spinel structures of Mg_2GeO_4 of the type $\varepsilon^{\circ} = A\sigma^n \exp(-E/RT)$, where $n = 3.5$ and $E = 105$ kcal mole^{-1} for the olivine phase and $n = 2.2$ and $E = 73$ kcal mole^{-1} for the spinel phase. The extremely fine grain size of the spinel (3 μm) suggests that it deformed by a superplastic mechanism. Vaughan & Kohlstedt (1979) have analysed by TEM the dislocation structures of the spinel phase and concluded that stacking faults on {110} planes are dominant.

Poirier & Madon (1979) have reported similar observations on a natural $(Mg_{0.74}Fe_{0.26})_2$ SiO_4 spinel/ringwoodite from a shocked meteorite. The only defects observed by these authors are planar stacking faults in {110} planes of the type (110) a/4 [110].

It is well known that metals and some ceramics deform superplastically during polymorphic phase transitions (Nicolas & Poirier 1976). Sammis & Dein (1974) have suggested that this transformational superplasticity should be considered for transition zones in the upper mantle (400-km and 650-km depth). Phase transition limits in the mantle can move significant distances in response to changing surface loads.

III NATURAL DEFORMATION

III.1 *Petrofabric Analysis in Peridotites*

The recent development of experimental deformation has renewed the interest in petrofabric analysis of peridotites. With a better knowledge of slip systems in the rock-forming minerals and of deformation mechanisms, it has been possible to infer the plastic flow orientation in relation to the structures (foliation, lineation) visible in the field or in hand specimens. Considering that the subject has been reviewed recently (Nicolas & Poirier 1976), we will merely present two examples illustrating the general method and some recent developments.

The examples have been chosen in peridotites from ophiolite sections (F. Boudier and R. G. Coleman, in preparation and unpublished data). The first diagram (Figure 3*a*) represents the preferred orientation of olivine in a specimen of harzburgite derived from the top of the ultramafics beneath the cumulates. The preferred orientation is characterized by strong point maxima of the crystallographic axes [100], [010], and [001], which are respectively close to the mineral lineation, to the pole of the foliation, and to the third principal structural direction. Plastic flow operated dominantly with the (010) [100] system, which is the high temperature one (see Table 1). This is confirmed by the analysis of the dislocation substructure. The strength of the preferred orientation indicates that the strain was very large, as also deduced from the flattening and

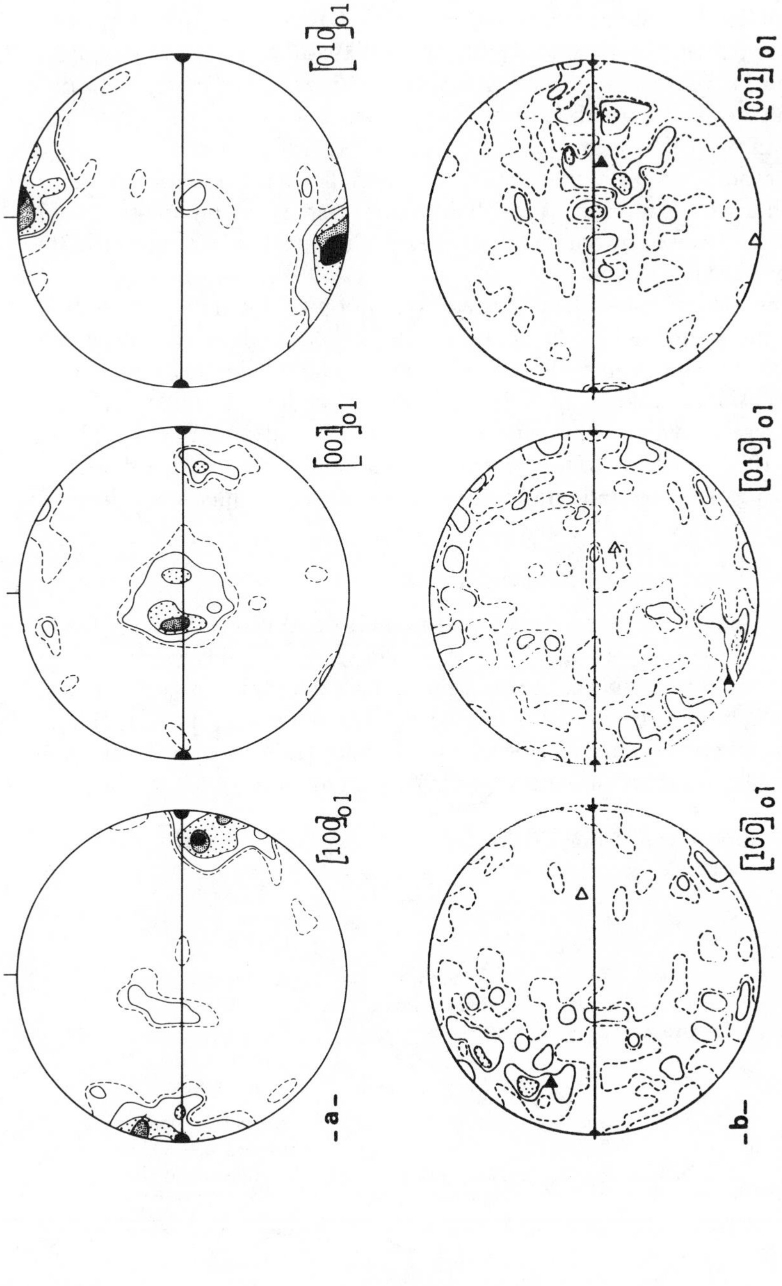

Figure 3 Olivine preferred orientation diagrams: (*a*) uppermost harzburgite from the Zambales ophiolite, (*b*) basal mylonitic from the Oman Ophiolite. Equal are projections, lower hemisphere; 100 olivine crystals; contours: 1, 2, 4, 12, and >14% for 0.45% area. Horizontal line: trace of foliation plane; E.W dots: trace of mineral lineation.

alignment of spinel grains. The slight obliquity (8°) between the crystallographic axes and the corresponding structural axes is accepted as evidence for a rotational (shear) flow regime, here with a dextral sense. Independently, the temperature during flow has been estimated at peridotite hypersolidus conditions (Nicolas et al 1980). At such temperatures, diffusion is very active, resulting in a typical mosaic structure. With diffusion providing enough degrees of freedom, it is speculated that plastic flow can operate with only one slip system being activated. Downwards, the harzburgites in the main mass of ultramafics have been deformed at subsolidus temperatures. Accordingly, they display an olivine-preferred orientation caused by dominant activation of the {0k1} [100] intermediate temperature slip system.

Figure 3*b* illustrates the preferred orientation of olivine in a mylonitic peridotite from the base of the peridotite sequence where flow occurred at ~800°C. The weak preferred orientation of [001] close to the stretching lineation is in keeping with activation of the low temperature [001] slip direction with various slip planes (Table 2) and possibly also along with [100]. Combination of various slip systems can explain the large scattering of the crystallographic axes in an otherwise intensely deformed rock.

III.2 *The Various Peridotites*

The distinction between crustal and mantle peridotites has been made clear by Thayer (1960) and Jackson & Thayer (1972). Three types were described, the stratiform, the concentric, and the alpine peridotites, the two former being generated in a crustal environment and the latter being tectonically emplaced from the mantle. Peridotites from stratiform complexes have been accumulated by gravity settling of minerals on the floor

Table 3 Various types of mantle peridotites

Form	Occurrence	Type of peridotite	Mantle location
Massifs	in Orogenic belts	Harzburgites within ophiolites	Oceanic mantle (~10–30 km)
		Lherzolites within granulites	Continental mantle (~30–100 km)
Xenoliths	in Alkali basalts	Lherzolites	Oceanic and continental mantle (~30–80 km)
	in Kimberlites	Lherzolites and harzburgites	Continental mantle (~50–200 km)

of a magma chamber located in the crust and filled with a basaltic magma. The structures at the various scales reflect their cumulate origin (Jackson 1961, Wager & Brown 1967). Deformation in these complexes is usually moderate and the cumulate structures are easily identified. In the concentric complexes, primary cumulate structures are often obliterated by subsequent plastic flow: the origin of these complexes is also more disputable.

Mantle rock occurrences at the earth's surface are either in ultramafic massifs outcropping in orogenic belts, or, as xenoliths, in alkali basalts and kimberlites (Table 3). Both sets are complementary: the xenoliths are small fragments of the contingently flowing mantle, whereas the massifs exhibit large-scale exposures of flow structures which are usually developed in the mantle and subsequently altered in crustal conditions. A common origin in the mantle can sometimes be inferred (Mercier & Nicolas 1975). From this very large and scattered sampling of the upper mantle, it is now clear that peridotites are by and large the dominant rocks (Maaloe & Aoki 1977). They are often layered with up to 10% of pyroxenites or dunites. If one considers only the alkali basalts, whatever their geological situation (intra-continental or intra-oceanic plate, accretion or subduction plate margin), the peridotite composition is restricted to that of a spinel lherzolite with 65–70% olivine, 20% enstatite, 5–10% diopside, 5% spinel.[1] The same composition is noted in the lherzolitic massifs. We believe that this reflects the mean composition of the upper mantle between 30 and 100 km, which is the depth range over which the mantle is sampled by these peridotites. However, the xenoliths from African kimberlites are more harzburgitic suggesting that the upper mantle beneath this shield may be more residual (Maaloe & Aoki 1977).

In the following sections, the various mantle rock occurrences (Table 3) will be examined for their flow structures. Special attention will be paid to the question of whether the flow represents a local lithospheric situation or a more general asthenospheric one. This is especially important for modeling large-scale convection.

III.3 *Peridotite Xenoliths from Basalts and Kimberlites*

It has been known for a long time (Ernst 1935, Collée 1963) that lherzolite xenoliths in alkali basalts have tectonic and metamorphic structures indicative of plastic flow. Systematic studies on lherzolite xenoliths in alkali basalts (Mercier & Nicolas 1975) and on lherzolite and harzburgite xenoliths in kimberlites (Boullier & Nicolas 1973, 1975) have led to the

[1] Exceptionally, the spinel coexists with another aluminous phase which at high pressure is garnet, at low pressure, feldspar, and at high P_{H_2O}, amphibole.

following results: 1. structural classifications can be proposed based on the fact that these mantle rocks constitute a suite with various degrees of strain, from nearly zero to very large; 2. the flow mechanism is dislocation slip and climb; 3. this plastic flow was achieved in the upper mantle (not in the magma during ascent); 4. commonly, the flow has been occurring up to the incorporation of the xenolith in the basaltic magma and therefore this mantle flow is connected with magma generation.

CLASSIFICATION

The nomenclature of deformed peridotite xenoliths coming from these studies has been blended by Harte (1977) with that proposed by other workers for kimberlites (Cox et al 1973, Harte et al 1975, Dawson et al 1975) and for basalts (Pike & Schwarzman 1977). It is sufficient here to recall the main terms in Harte's classification, being, however, conscious that xenoliths form a continuous suite. The *coarse* xenoliths (Figure 4*A*) correspond to grains 2 mm or greater which can be either equant or slightly tabular and with usually straight or smoothly curved grain boundaries. The strain is very small as deduced from the absence of grains elongated by plastic flow (porphyroclasts). The *porphyroclastic* xenoliths (Figure 4*B*) now have a well-defined foliation due to parallel orientation of plastically deformed grains. Recrystallization increases with increasing strain giving birth to *mosaic*-porphyroclastic structures. Once recrystallization is nearly complete the structure becomes granuloblastic with the possible addition of prefixes *equant* (Figure 4*C*) or *tabular* (Figure 4*D*) if one wants to specify the shape of the recrystallized grains (which are called *neoblasts*).

Several authors have investigated the dislocation microstructures of various xenoliths using transmission electron microscopy (TEM) (Green 1976, Green & Radcliffe 1972, Buiskool Toxopeus & Boland 1976). Gueguen (1977, 1979a, c) has used the decoration technique introduced by Kohlstedt et al (1976) in an extensive study of dislocation microstructures in xenoliths from both basalt and kimberlites. Such investigations are concerned with only the last 1% of strain, possibly modified by natural annealing. Therefore it is not surprising that the six main types of dislocation microstructures (Gueguen 1979a, c) (Figures 1 and 2) are not related to the structural types described above. These six types are frequently observed together in the same sample.

ORIGIN OF FLOW STRUCTURES IN XENOLITHS FROM KIMBERLITE

Great interest stemmed from the discovery of a possible relation between, on the one hand, composition, depth of origin, and temperature of equilibration estimated by pyroxene geothermometry and barometry

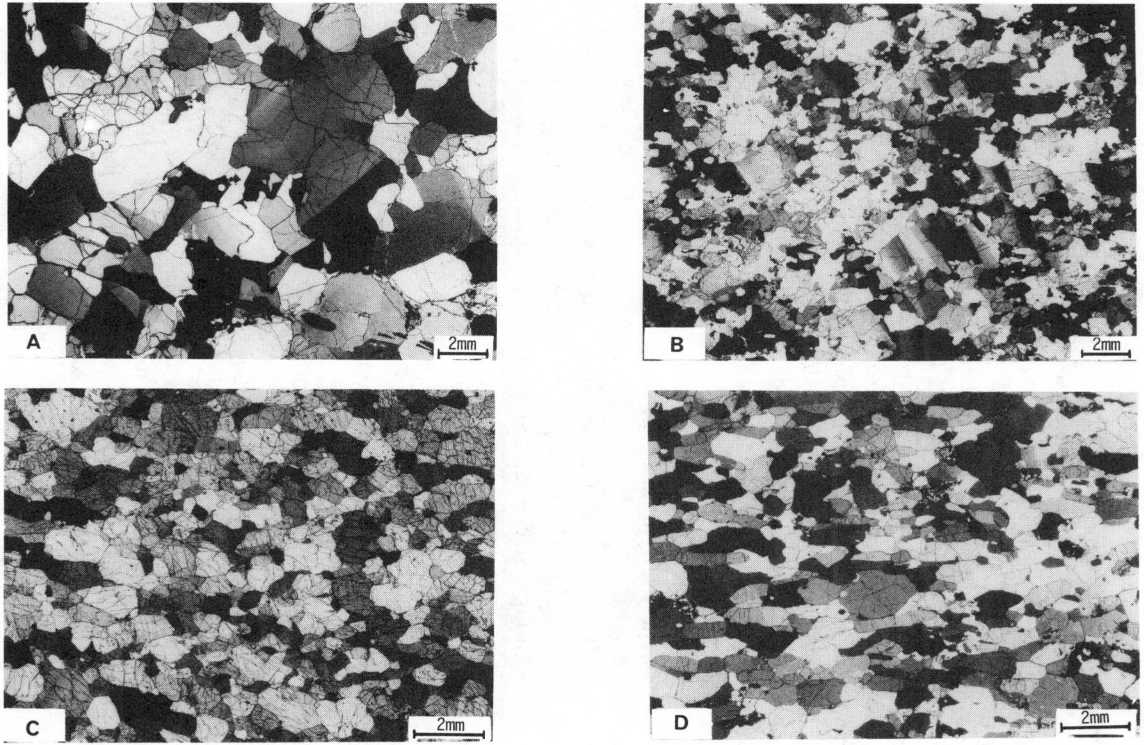

Figure 4 Microphotographs (crossed polarizers) of typical structures in peridotites from basalt xenoliths. Observation plane parallel to lineation and perpenducular to foliation. *A*. Coarse (protogranular) structure: no foliation, large dislocation substructure. *B*. Porphyroclastic structure: fairly tight substructure in olivine porphyroclasts, recrystallization in neoblasts. *C*. Equant granuloblastic (equant-equigranular) structure: complete recrystallization in olivine, poor foliation. *D*. Tabular granuloblastic (tabular equigranular) structure: complete recrystallization of olivine in tablets.

and, on the other hand, structures observed in xenoliths from kimberlite (Boyd & Nixon 1972). Nixon et al (1973) proposed that the lherzolite xenoliths equilibrated at high temperature ($\sim$1400°C) and great depth ($\leq$200 km) and, displaying structures indicative of large strains (the "sheared nodules"), had been extracted from the asthenosphere flowing beneath the South African shield. The shear flow would have raised the temperature by frictional heating and caused the generation of kimberlite magma. The less deformed, more superficial ($\sim$100 km), and cooler ($\sim$1000°C) xenoliths with a harzburgitic composition would have been picked up by the magma during its ascent through the overlying lithosphere. In an effort to explain extensive strain in some xenoliths and a kink observed on the geotherms deduced from pyroxene barometers and thermometers, Green & Gueguen (1974) presented a model of diapiric intrusion. The diapir, initiated at 300-km depth on a thermal anomaly, would adiabatically rise with a velocity due to thermal contrast with surrounding mantle until partial melts produced inside the diapir would be expelled. The melts of kimberlitic composition could incorporate fragments of the upper skin of the diapir (the "sheared lherzolites") along with fragments of the overlying lithosphere. The diameter of the diapir was estimated at 10–50 km.

The experimental calibration of piezometers based on dislocations and substructures (see Section II) has led Goetze (1975a) to question the diapir model. Although stress estimates are not accurate, minimum stresses of 500 bars have been experienced by the "sheared" nodules. This stands substantially above the 10–100 bars ascribed to the asthenosphere by geophysical methods (stress drops, gravity data, heat flux balance, and rheology). Consequently, strain rates were probably high and the shearing restricted to a narrow zone. This is also suggested by the observation of large strain gradients at the scale of a specimen. Recently the importance of syntectonic magmatic impregnation in the lherzolites has been realized (Harte et al 1975, Erlank & Rickard 1977). Combined with the necessity of comparatively fast strain rates, this has fostered the emergence of a new model (Mercier 1977). Possibly generated in a rising diapir as invoked above, a very mobile magma forces its way up along channels whose diameter is now at the meter rather than at the kilometer scale. The mantle at some distance above this intrusion is cracked with injection of magma and metasomatism; closer to the intrusion, the mantle suffers an intense plastic deformation. This forceful intrusion would become explosive once the gas pressure exceeds the load.

It is remarkable that the models proposed to explain the plastic flow associated in mantle rocks with kimberlite generation have progressively evolved from a plate-tectonic scale to the much smaller scale of local accident. There is now a consensus that the plastic flow in xenoliths from kimberlite in no way represents the asthenospheric flow. Let us now examine the situation for xenoliths from basalts.

ORIGIN OF FLOW STRUCTURES IN XENOLITHS FROM BASALTS

The deformation structures in xenoliths from basalt differ significantly. The strain is usually less intense without any evidence of strain gradients at the scale of the specimen. Superplastic flow, documented in xenoliths from kimberlite, is also unknown in basaltic occurrences. The stress as deduced from dislocation structure remains on the whole significantly lower in xenoliths from basalt than in those from kimberlite (Mercier 1976, Coisy & Nicolas 1978, Gueguen 1977), except in the case of a few porphyroclastic structures which have been ascribed to local shearing in the mantle (Mercier 1976, Coisy & Nicolas 1978). The temperature of equilibration is also lower (950–1200°C) and the depth of origin is shallower (30–80 km) as shown by the equilibration of xenoliths from basalt in the spinel lherzolite field (O'Hara 1967). Finally, their deformation occurred in nearly dry conditions, considering the virtual absence of hydrous phases (phlogopite and amphibole) or their late introduction into the spinel lherzolites (Francis 1978). This is also in sharp contrast with xenoliths from kimberlite where H_2O and CO_2 are well represented.

The question arises whether or not the basalt xenolith flow structures represent the large-scale horizontal flow in the asthenosphere. Nicolas (1978) approached this question from a flow stress point of view. The crude and disputable estimates based on dislocation structures and neoblast size in olivine (see Section II) yield for basalt xenoliths a minimum stress of 200 bars, which is high compared with the 10–100 bars ascribed to the asthenosphere by geophysical methods. The conclusion should be that the flow recorded in xenoliths from basalts does not reflect the asthenosphere flow but, considering the uncertainties in both estimates, it cannot be said that the conclusion is firmly based. Shaw (1973) has presented a model to explain the relation mentioned above between plastic flow and basalt generation, interpreting the plastic flow as representing the large-scale horizontal flow in the asthenosphere. In this model, basalts are generated in the asthenosphere by shear heating, assuming that the asthenosphere flow is intermittent, operating in a staccato fashion. This introduces the problem of the stability of the flow in a mantle with an olivine rheology. Schubert et al (1976), Melosh (1976), Yuen & Schubert (1977), and Schubert & Yuen (1978) have discussed this in great detail. Unstable solutions exist only if the shear stress is constant; with the accepted rheologies, the stress remains low (<50 bars).

Another possibility is that the flow registered in basalt xenoliths is produced during the rise of an asthenosphere diapir carrying heat and basaltic melts toward the surface (Basu 1975, Wilshire & Pike 1975, Francis 1978). A systematic study of basalt xenoliths from 70 volcanic vents scattered in the Massif Central (France) has shown that there was a tendency for the structural types to be geographically ordered (Coisy & Nicolas 1978). Vents with dominant granuloblastic-structured xenoliths tend to be surrounded by those with dominant porphyroclastic- and coarse-structured xenoliths. This somewhat concentric distribution is itself centered within a domain of thinner crust (24 km versus 30 km elsewhere) and abnormal mantle (Perrier & Ruegg 1973). It has been interpreted as due to the rising of a hot mantle diapir. The top and the margins of the diapir would accommodate a large plastic flow, now recorded in the deformed xenolith structures; the coarse structures would be derived from the outside areas where virtually no flow occurs. The geographical distribution makes it possible to estimate the size of the diapir at 50 × 100 km. The amount of uplift is at least 6 km.

III.4 *Peridotite Massifs*

The peridotite massifs derived from the upper mantle are emplaced in orogenic belts. They range in surface from tectonic inclusions extending over a few square meters to over 1000 km^2. Den Tex (1969) calls them *orogenic peridotites* and divides them into two subgroups depending upon whether the peridotites are parts of ophiolite complexes, i.e. *ophiolitic peridotites*, or tectonic bodies incorporated in the "root zone" of orogenic belts, constituted by high-grade polymetamorphic terrains. The latter *"root zone" peridotites* can be equated with Green's *"high temperature" peridotites* (1967). Jackson & Thayer (1972) divide the alpine type, using mineralogical and chemical criteria, into *lherzolite* and *harzburgite subtypes*. Nicolas & Jackson (1972) relate the two kinds of subdivisions by showing that the "root zone" peridotites are essentially lherzolites and the ophiolitic peridotites, harzburgites. The "root zone" lherzolite massifs are exceptionally numerous and well exposed in the alpine belt of the western Mediterranean; the ophiolitic harzburgite massifs occur much more extensively and are present, together with the other members of the ophiolitic complex, along most phanerozoic orogenic belts.

FLOW IN LHERZOLITE MASSIFS

For Nicolas & Jackson (1972) the lherzolite massifs represent the normal undepleted mantle as shown by their chemical similarity with basalt xenoliths. The harzburgites have originated inside the uppermost oceanic mantle; this mantle section and the overlying oceanic crust are then equated with an ophiolite section. If the lherzolites are considered as representative of a normal upper mantle, the difference from harzburgites in mineralogical and chemical composition is attributed to the 20–30% partial melting occurring at the oceanic ridges: through this process the lherzolites are depleted of their most fusible components and changed to harzburgites. One should therefore expect that at a 20–30-km depth beneath the oceanic crust the harzburgites grade into less depleted lherzolites. Indeed a few ophiolite massifs show in their basal part such a transition from harburgites to lherzolites. The same authors ascribed a subcontinental origin to the lherzolite massifs from the western Mediterranean in view of their dominant association with granulitic terrains. The massifs with a tectonic fabric like Lanzo would have been forcefully intruded from subcontinental upper mantle during alpine continent-continent collision (Boudier 1978). The massifs with a metamorphic structure and paragenesis equilibrated with the surrounding granulites (Lensch 1971) would represent either older intrusions or the normal transition between deep continental crust and the upper mantle, as exemplified by the Ivrea zone. Lherzolite massifs with a tectonic fabric may also be derived from the rifted edge of a continent (Cottin 1978), from oceanic transform faults where the crust may be thinner and the underlying mantle less depleted than usual, or even from deeper oceanic mantle subducted during a previous orogeny and plated beneath a young continental margin (Richard & Allègre 1980).

FLOW IN THE OCEANIC UPPER MANTLE

Information about flow in the uppermost oceanic mantle is obtained through 1. peridotite specimens dredged in the various oceanic environment, 2. peridotite sections in ophiolites, and 3. indirectly through the analysis of anisotropy in seismic velocity. This latter topic extends beyond the scope of this paper (for reviews see Nicolas & Poirier 1976 and Carter 1976). It is sufficient to recall that a seismic velocity anisotropy of a few percent has been measured in oceanic uppermost mantle with the fast velocity trending parallel to fracture zones; this phenomenon is best explained by a crystallographic preferred orientation in olivine such that the [100] mean orientation is parallel to the fast velocity direction. This crystallographic direction is also that of the dominant slip direction line at high temperature. Large plastic flow results in the orientation of [100] of olivine parallel to the flow line (Figure 3). It is concluded that the fast seismic velocity direction in oceanic uppermost mantle indicates what was the flow direction before the mantle became rigid due to temperature drop.

Structural studies on peridotite sections in ophiolites have been conducted by Nicolas et al (1980) and the various structures compared with those in peridotites from various environments. Typically the upper part of the peridotite section in ophiolites is a few kilometers thick, usually composed of coarse-granular to coarse-porphyroclastic textures indicative of moderate differential stress, strain, and solidus or hypersolidus temperature conditions for plastic flow. The petrofabrics in these peridotites are presented in Figure 3*a* and 3*b*. In many massifs, they grade downward into a basal zone of peridotites with fine-grained porphyroclastic textures, in contact with medium to high grade metamorphic rocks of oceanic origin. In the basal zone, both stress and strain greatly increase downward to peridotite mylonites (see Figure 3*b*) deformed at differential stresses above 1 kb near the contact with amphibolites. The temperature is probably around 800°C during this deformation.

Similar textures and mineral-preferred orientations are reported from a few suboceanic peridotites. In particular, coarse textures are found in peridotites from ridge environments and, typically, porphyroclastic textures in peridotites from trenches. Fracture zones have yielded both types. From this similarity and from considerations of the temperature, pressure, and stress conditions of plastic flow, it is concluded that the tectonic structures observed in the upper parts of ophiolitic peridotites were formed during asthenosphere flow at sea floor spreading sites. The tectonic structures from basal parts of the ophiolite massifs have been superimposed during upthrusting of an oceanic mantle and crust slice over oceanic crust in a trench environment (Nicolas & Le Pichon 1980).

IV SUMMARY

1. Most of the recent experimental data and theoretical studies are concerned with olivine single crystals. Although sophisticated experiments and detailed analysis of dislocations have led to a better insight into the problem of creep-controlling mechanisms, it is obvious now that the flow processes are more complex than previously thought.
   —Olivine creep is not controlled by oxygen diffusion. It is either controlled by Si diffusion or by another mechanism: cross slip, kink, and jog nucleations are possible candidates.
   —Several dislocation microstructures have been identified. The core structure of dislocations seems to play a fundamental role. It is likely that there are several creep laws for olivine single crystals, corresponding to particular conditions.
   —Extrapolation to polycrystals is not straightforward. In particular dislocations tilt walls are frequently observed in polycrystal deformation but not in single crystal deformation.
2. The problem of paleo-stress determination (piezometers) has also

received much attention. It is important, however, to note that better experimental calibrations and a better theoretical understanding of the various stress-microstructure relationships is necessary.

3. Deformation in peridotite xenoliths from kimberlites is probably due to flow along narrow channels filled with kimberlite magma, in response to the upward pressure exerted by the magma. In the case of xenoliths from basalts, the deformation could be due to the large-scale horizontal asthenospheric flow, although the interpretation of flow at the margins of mantle diapir some tenths of a kilometer in diameter is preferred.
4. Deformation in lherzolite massifs can result from forceful intrusions at continent-continent collision sites, the lherzolites being derived from mantle beneath or at the margin of continents. However, direct oceanic origin cannot be excluded because in transform faults the upper mantle seems to be more lherzolitic than elsewhere. Deformation in harzburgites from ophiolites (considered as oceanic crust and mantle) has been ascribed to ridge and to subduction zone environments.

ACKNOWLEDGMENTS

J. P. Poirier brought our attention to recent papers, M. S. Paterson reviewed the manuscript, and J. F. Violette provided the orientation data on specimens from Zambales ophiolite.

Literature Cited

Ave Lallemant, H. G. 1978. Experimental deformation of diopside and websterite. *Tectonophysics* 48: 1–27

Basu, A. R. 1975. Hot spots, mantle plumes and a model for the origin of ultramafic xenoliths in alkali basalts. *Earth Planet. Sci. Lett.* 33: 443–50

Bernal, J. D. 1936. Discussion. *Observatory* 59: 268

Blacic, J. D. 1972. Effect of water on the experimental deformation of olivine. In *Flow and Fracture of Rocks*, ed. H. C. Heard. *Geophys. Monogr. Ser.* 16: 109–15

Boudier, F. 1978. Structure and petrology of the Lanzo peridotite massif (Piedmont Alps). *Geol. Soc. Am. Bull.* 89: 1574–91

Boullier, A. M., Nicolas, A. 1973. Texture and fabric of peridotite modules from kimberlite at Mothae, Thaba Putsoa and Kimberley. In *Lesotho Kimberlites*, ed. P. H. Nixon, pp. 57–66. Lesotho Natl. Develop. Corp.

Boullier, A. M., Nicolas, A. 1975. Classification of textures and fabrics of peridotites xenoliths from South African kimberlites. In *Physics and Chemistry of the Earth*, ed. L. H. Ahrens, J. B. Dawson, A. R. Duncan, A. J. Erlank, 9: 97–105. Oxford: Pergamon

Boullier, A. M., Gueguen, Y. 1975. SP. Mylonites: Origin of some mylonites by superplastic flow. *Contrib. Mineral. Petrol.* 50: 93–104

Boyd, F. R., Nixon, P. H. 1972. Ultramafic nodules from the Thaba Putsoa kimberlite pipe. *Carnegie Inst. Washington Yearb.* 71: 362–73

Buening, D. K., Busek, P. R. 1973. Fe-Mg lattice diffusion in olivine. *J. Geophys. Res.* 78: 6852–62

Buiskool Toxopeus, J. M. A., Boland, J. N. 1976. Several types of natural deformation in olivine, an electron microscopy study. *Tectonophysics* 32: 209–33

Carter, N. L. 1976. Steady state flow of rocks. *Rev. Geophys. Space Phys.* 14: 301–60

Carter, N. L., Ave Lallemant, H. G. 1970. High temperature flow of dunite and peridotite. *Geol. Soc. Am. Bull.* 81: 2181–202

Coe, R. S., Kirby, S. H. 1975. The orthoenstatite to clinoenstatite transformation by shearing and revision by annealing: Mechanism and potential applications. *Contrib. Mineral. Petrol.* 52:29–55

Coisy, P., Nicolas, A. 1978. Regional structure and geodynamics of the upper mantle beneath the Massif Central. *Nature* 274: 429–32

Collée, A. L. G. 1963. A fabric study of lherzolites with special reference to ultrabasic nodular inclusions in the lavas of Auvergne (France). *Leidse Geol. Meded. 28*. 102 pp.

Cottin, J. Y. 1978. *L'association ultramafique-mafique de la région du Bracco (Apennin ligure, Italie)*. Thèse Doctorat 3ème cycle Paris VII. 211 pp.

Cox, K. G., Gurney, J. J., Harte, B. 1973. Xenoliths from the Matsoku pipe. In *Lesotho Kimberlites*, ed. P. H. Nixon, pp. 76–98. Lesotho Natl. Develop. Corp.

Dawson, J. B., Gurney, J. J., Lawless, P. J. 1975. Paleogeothermal gradients derived from xenoliths in kimberlite. *Nature* 257: 299–300

Den Tex, E. 1969. Origin of ultramafic rocks, their tectonic setting and history. *Tectonophysics* 7:457–88

Doukhan, N., Duclos, R., Escaig, B. 1979. Sessile dissociation in the stoichiometric spinel $MgAl_2O_4$. *J. Phys.* 40:381–87

Durham, W. B., Goetze, C. 1977a. A comparison of the creep properties of pure fosterite and iron-bearing olivine. *Tectonophysics* 40:T15–18

Durham, W. B., Goetze, C. 1977b. Plastic flow of oriented single crystals of olivine. I. Mechanical data. *J. Geophys. Res.* 82: 5737–53

Durham, W. B., Goetze, C., Blake, B. 1977. Plastic flow of oriented single crystals of olivine. 2. Observations and interpretations of the dislocation structures. *J. Geophys. Res.* 82:5755–70

Erlank, A. J., Rickard, R. S. 1977. Potassic richterite bearing peridotites from kimberlite and the evidence they provide for upper mantle metasomation. *2nd Int. Kimberlite Conf. Abstr.*

Ernst, T. 1935. Olivinknollen der Basalte als Bruchstücke alter Olivinfelse. *Nachr. Ges. Wiss. Goettingen Math. Phys. Kl. Fachgruppe 4* IB:147–54

Etheridge, M. A., Kirby, S. H. 1980. Experimental deformation of rock forming pyroxenes: recrystallization mechanism and preferred orientation development *Tectonophysics*. In press

Francis, D. M. 1978. The implication of the compositional dependence of texture in spinel lherzolite xenoliths. *J. Geol.* 86: 473–85

Goetze, C. 1975a. Sheared lherzolites: from the point of view of rock mechanics. *Geology* 3:172–73

Goetze, C. 1975b. Textural and microstructural systematics in olivine and quartz. *EOS, Trans. Am. Geophys. Union* 56:6

Goetze, C. 1978. Mechanisms of creep in olivine. *Philos. Trans. R. Soc. London Ser. A* 228:100–19

Goetze, C., Kohlstedt, D. L. 1973. Laboratory study of dislocation climb and diffusion in olivine. *J. Geophys. Res.* 78: 5961–71

Green, D. H. 1967. High temperature peridotite intrusions. In *Ultramafic and Related Rocks*, ed. P. J. Wyllie, pp. 212–22. New York: Wiley

Green, H. W. II, Radcliffe, S. V. 1972. Deformation processes in the upper mantle. In *Flow and Fracture of Rocks*, ed. H. C. Heard. *Geophys. Monogr. Ser.* 16:139–56

Green, H. W. II, Gueguen, Y. 1974. Origin of kimberlite pipes by diapiric upwelling in the upper mantle. *Nature* 249:617–20

Green, H. W. II. 1976. Plasticity of olivine in peridotites. In *Electron Microscopy in Mineralogy*, ed. H. R. Wenk, pp. 443–64. Berlin: Springer

Griggs, D. T. 1967. Hydrolitic weakening of quartz and other silicates. *Geophys. J. R. Astron. Soc.* 14:19–31

Gueguen, Y. 1977. Dislocation in mantle peridotite nodules. *Tectonophysics* 39: 231–54

Gueguen, Y. 1979a. Dislocations in naturally deformed terrestrial olivine: classification, interpretation, applications. *Bull. Mineral.* 102:178–84

Gueguen, Y. 1979b. High temperature olivine creep: evidence for control by edge dislocations. *Geophys. Res. Lett.* 6:357–60

Gueguen, Y. 1979c. *Les dislocations dans l'olivine des péridotites*. Thèse doct. Etat. Nantes Univ. 193 pp.

Harte, B., Cox, K. G., Gurney, J. J. 1975. Petrography and geological history of upper mantle xenoliths from the Matsoku kimberlite pipe. In *Physics and Chemistry of the Earth*, ed. L. H. Ahrens, J. B. Dawson, A. R. Duncan, A. J. Duncan, 9:477–506. Oxford: Pergamon

Harte, B. 1977. Rock nomenclature with particular reference to deformation and recrystallization textures in olivine bearing xenoliths, *J. Geol.* 89:279–88

Jackson, E. D. 1961. Primary textures and mineral associations in the ultramafic zone of the Stilwater complex, Montana. *US Geol. Surv. Prof. Pap. 358*. 106 pp.

Jackson, E. D., Thayer, T. P. 1972. Some criteria for distinguishing between stratiform, concentric and alpine peridotite-gabbro complexes. *24th Int. Geol. Cong.* 2:289–96

Jaoul, O., Gueguen, Y., Michaut, M., Ricoult, D. 1979. A technique for decorating dislocations in forsterite. *Phys. Chem. Miner.* 5:15–20

Kirby, S. H., Etheridge, M. A. 1980. Experimental deformation of rock forming pyroxenes: ductile strengths and mechanisms of flow. *Tectonophysics.* In press

Kohlstedt, D. L., Goetze, C. 1974. Low stress high temperature creep in olivine single crystals. *J. Geophys. Res.* 79:2045–51

Kohlstedt, D. L., Goetze, C., Durham, W. B., Vander Sande, J. 1976. A new technique for decorating dislocations in olivine. *Science* 191:1045–46

Lensch, G. 1971. Die Ultramafitite der zone von Ivrea. *Am. Univers. Saraviensis* 9: 5–146

Maaloe, S., Aoki, K. 1977. The major element composition of the upper mantle estimated from the composition of lherzolites. *Contrib. Mineral. Petrol.* 63: 161–73

Melosh, H. J. 1976. Plate motion and thermal instability in the asthenosphere. *Tectonophysics* 35:363–90

Mercier, J. C. 1976. *Natural peridotites: chemical and rheological heterogeneity of the upper mantle.* PhD thesis. State Univ. New York, Stony Brook. 710 pp.

Mercier, J. C. 1977. Peridotite xenoliths and the dynamics of kimberlite intrusion. *2nd Int. Kimberlite Conf. Abstr.*

Mercier, J. C., Nicolas, A. 1975. Textures and fabrics of upper mantle peridotites as illustrated by xenoliths from basalts. *J. Petrol.* 16:454–87

Moores, E. M. 1974. Geotectonic significance of ultramafic rocks. *Earth Sci. Rev.* 9: 241–58

Nicolas, A. 1976. Flow in upper mantle rocks: some geophysical and geodynamic consequences. *Tectonophysics* 32:93–106

Nicolas, A. 1978. Stress estimates from structural studies in some mantle peridotites. *Philos. Trans. R. Soc. London* 288:49–57

Nicolas, A., Boudier. F., Bouchez, J. L. 1980. Interpretation of peridotite structures from ophiolitic and oceanic environments. *Am. J. Sci.* In press

Nicolas, A., Jackson, E. D. 1972. Répartition en deux provinces des péridotites des chaînes alpines longeant la Méditerranée: implications géotectoniques. *Bull. Suisse Mineral. Petrogr.* 52:479–95

Nicolas, A., Poirier, J. P. 1976. *Crystalline Plasticity and Solid State Flow in Metamorphic Rocks.* London, New York: Wiley. 444 pp.

Nicolas, A., Le Pichon, X. 1980. Thrusting of young lithosphere in subduction zones with special reference to structures in ophiolitic peridotites. *Earth. Planet. Sci. Lett.* In press

Nixon, P. H., Boyd, F. R., Boullier, A. M. 1973. The evidence of kimberlite and its inclusions on the constitution of the outerpart of the earth. In *Lesotho Kimberlites,* ed. P. H. Nixon, pp. 312–18. Lesotho Natl. Develop. Corp.

O'Hara, M. J. 1967. Mineral facies in ultrabasic rocks. In *Ultramafic and Related Rocks,* ed. P. J. Wyllie, pp. 7–18. New York: Wiley

Paterson, M. S. 1979. The mechanical behaviour of rocks under crustal and mantle conditions. *The Earth, Its Origin, Structure and Evolution,* ed. M. W. McElwinny, pp. 469–89. London: Academic

Perrier, G., Ruegg, J. C. 1973. Structure profonde du Massif Central français. *Ann. Geophys.* 29:435–502

Phakey, P., Dollinger, G., Christie, J. M. 1972. Transmission electron microscopy of experimentally deformed olivine crystals. In *Flow and Fracture of Rocks,* ed. H. C. Heard. *Geophys. Monogr. Ser.* 16:117–38

Pike, J. E. N., Schwarzman, E. C. 1977. Classification of textures in ultramafic xenoliths. *J. Geol.* 85:49–61

Poirier, J. P. 1975. On the slip systems of olivine. *J. Geophys. Res.* 80:4053–61

Poirier, J. P., Nicolas, A. 1975. Deformation induced recrystallization due to progressive misorientation of subgrains, with special reference to mantle peridotites. *J. Geol.* 83:707–20

Poirier, J. P., Vergobbi, B. 1978. Splitting of dislocations in olivine, cross slip controlled creep and mantle rheology. *Phys. Earth Planet. Inter.* 16:370–78

Poirier, J. P., Guillope, M. 1979. Deformation induced recrystallization of minerals. *Bull. Mineral.* 102:67–74

Poirier, J. P., Madon, M. 1979. Transmission electron microscopy of natural $(Mg_{0.74}Fe_{0.26})_2SiO_4$ spinel. *EOS, Trans. Am. Geophys. Union* 60:370

Post, R. L. 1973. *The flow laws of Mt. Burnett dunite.* PhD thesis. Univ. Calif., Los Angeles. 272 pp.

Raleigh, C. B. 1968. Mechanisms of plastic deformation in olivine. *J. Geophys. Res.* 73:5311–5406

Raleigh, C. B., Kirby, S. H. 1970. Creep in the upper mantle. *Mineral. Soc. Am. Spec. Pap.* 3:113–21

Reddy, K. P. R., Cooper, A. R. 1978. Oxygen self diffusion in forsterite and $MgAl_2O_4$. *Am. Ceram. Soc. Bull.* 57:310

Richard, P., Allègre, C. J. 1980. Neodymium and strontium isotope study of ophiolite and orogenic lherzolite petrogenesis. *Earth Planet. Sci. Lett.* In press

Ricoult, D. 1978. *Recuit expérimental de l'olivine.* Thèse de 3ème cycle. Univ. Nantes. 75 pp.

Ricoult, D. 1979. Experimental annealing of a natural dunite. *Bull. Mineral.* 102:86–91

Ricoult, D., Gueguen, Y. 1980. Experimental studies on the recovery process of deformed olivines and the mechanical state of the upper mantle: a discussion. 1979. *Tectonophysics.* In press

Ross, J. V., Nielsen, K. C. 1978. High temperature flow of wet polycrystalline enstatite. *Tectonophysics* 44:233–61

Sammis, C. G., Dein, J. L. 1974. On the possibility of transformational superplasticity in the earth's mantle. *J. Geophys. Res.* 79:2961–65

Schmalzried, H. 1978. Reactivity and point defects of double oxides with emphasis on simple silicates. *Phys. Chem. Miner.* 2:279–94

Schubert, G., Froidevaux, C., Yuen, D. A. 1976. Oceanic lithosphere and asthenosphere: thermal and mechanical structure. *J. Geophys. Res.* 81:3525–40

Schubert, G., Yuen, D. A. 1978. Shear heating instability in the earth's upper mantle. *Tectonophysics* 50:197–205

Schwenn, M. B., Goetze, C. 1978. Creep of olivine during hot pressing. *Tectonophysics* 48:41–60

Sellars, C. M. 1978. Recrystallization of metals during hot deformation. *Philos. Trans. R. Soc. London Ser. A* 288:147–58

Shaw, H. R. 1973. Mantle convection and volcanic periodicity in the Pacific; evidence from Hawaii. *Geol. Soc. Am. Bull.* 84:1505–26

Stocker, R. L. 1978a. Point defect formation parameters in olivine. *Phys. Earth Planet. Inter.* 17:108–17

Stocker, R. L. 1978b. Influence of oxygen pressure on defect concentrations in olivine with a fixed cationic ratio. *Phys. Earth Planet. Inter.* 17:118–29

Thayer, T. P. 1960. Some critical differences between alpine type and stratiform peridotite gabbro complexes. *12th Int. Geol. Cong.* 13:247–59

Toriumi, M., Karato, S. I. 1978. Experimental studies on the recovery process of deformed olivines and the mechanical state of the upper mantle. *Tectonophysics* 49:79–95

Twiss, R. J. 1976. Structural superplastic creep and linear viscosity in the earth's mantle. *Earth Planet. Sci. Lett.* 33:86–100

Twiss, R. J. 1977. Theory and applicability of a recrystallized grain size paleopiezometer. *Pageoph* 115:228–44

Twiss, R. J., Sellars, C. M. 1978. Limits of applicability of the recrystallized grain-size piezometer. *Geophys. Res. Lett.* 5(3):337–40

Vander Sande, J. B., Kohlstedt, D. L. 1976. Observation of dissociated dislocations in deformed olivine. *Philos. Mag.* 34:653–58

Vaughan, P. J., Coe, R. S. 1978. Creep mechanisms in Mg_2GeO_4. *EOS, Trans. Am. Geophys. Union* 59:1185 (Abstr.)

Vaughan, P. J., Kohlstedt, D. L. 1979. Stacking faults in experimentally deformed Mg_2GeO_4 spinel. *EOS, Trans. Am. Geophys. Union* 60:370

Veyssière, P., Rabier, J., Garem, H., Grilhé, J. 1978. Influence of temperature on dissociation of dislocations and plastic deformation in spinel oxides. *Philos. Mag. A* 38:61–79

Wager, L. R., Brown, G. M. 1967. *Layered Igneous Rocks.* London: Oliver & Boyd. 588 pp.

Wilshire, H. G., Pike, J. E. 1975. Upper mantle diapirism: evidence from analogous features in alpine peridotite and ultramafic inclusions in basalts. *Geology* 3:467–70

Windhager, H. J., Borchardt, G. 1975. Tracerdiffusion und Fehlordnung in dem Orthosilikat CO_2SiO_4. *Ber. Bunsenges. Phys. Chem.* 79:1115–19

Yuen, D. A., Schubert, G. 1977. Asthenospheric shear flow: thermally stable or unstable? *Geophys. Res. Lett.* 4:503–6

Zeuch, D. H., Green, H. W. II. 1979. Experimental deformation of an "anhydrous synthetic dunite." *Bull. Mineral.* 102:185–87

Ann. Rev. Earth Planet. Sci. 1980. 8: 145–67
Copyright © 1980 by Annual Reviews Inc. All rights reserved

ARCHEAN SEDIMENTATION ×10128

Donald R. Lowe
Department of Geology, Louisiana State University, Baton Rouge, Louisiana 70803

INTRODUCTION

As a time unit, the Archean includes that interval between the formation of the earth, approximately 4.6 billion years (b.y.) ago, and the development of large areas of stable sialic crust about 2.6 to 2.4 b.y. ago. As a tectonic-stratigraphic unit, it is characterized by a distinctive bimodal association of granite-greenstone belts and high grade gneiss-granulite terranes. Although preserved terrestrial rocks represent only the last 1.3 to 1.2 b.y. of Archean time, they are of great importance as the major source of direct information on the origin and evolution of life, the development of the earth's crust and mantle, and the characteristics of the early ocean and atmosphere.

Relatively unaltered sedimentary rocks underlie large areas of many Archean shields. In a series of classic papers, Pettijohn (1934, 1943) demonstrated that these strata could be studied and interpreted in the same way as Phanerozoic sedimentary deposits. Subsequent stratigraphic and sedimentological work in the Canadian Shield has provided the backbone of information regarding its geologic and tectonic evolution (Goodwin 1973, 1977a, b, Goodwin & Ridler 1970, Turner & Walker 1973, Walker 1978, Walker & Pettijohn 1971).

The sedimentological histories of other shields have only recently been investigated. Although brief descriptions of sedimentary rocks have traditionally been part of areal investigations and have frequently accompanied volcanological, geochemical, and geochronological studies, the results of more systematic investigations in the South African (Eriksson 1978, 1979, Lowe & Knauth 1977, 1978, von Brunn & Hobday 1976), Western Australian (Barley et al 1979, Glover & Groves 1978), and Indian (Naqvi 1977, Naqvi et al 1978) cratons have appeared largely within the last five years. Although these areas represent one of the most vital and dynamic periods of earth history and possess an immense potential for

0084-6597/80/0515-0145$01.00

yielding fundamental information on the early earth through the application of modern stratigraphic and sedimentological techniques, they have until recently been bypassed in the course of sedimentary research. The recent spate of Archean studies, evidenced by the many new books on the subject (Glover & Groves 1978, McCall 1977, Salop 1977, Windley 1976, Windley & Naqvi 1978), includes some new sedimentological investigations. Hopefully these will provide a more comprehensive and systematic picture of Archean sedimentation, complementary to that emerging from geochemical, geochronological, and structural research.

The present review first attempts to synthesize a general picture of the lithology, stratigraphy, and sedimentology of Archean sedimentary rocks, principally in greenstone belts. The sedimentary evolution of these belts is then considered in terms of the constraints it places on interpretations of their tectonic development and the general evolution of the Archean crust.

DISTRIBUTION OF ARCHEAN SEDIMENTARY ROCKS

Most Archean sedimentary rocks are preserved within tectonic-stratigraphic units called greenstone belts. These belts are widely developed in Archean shield areas and are regarded by many as characteristic structural units of the Archean. They are best preserved and have been most completely studied in Canada, South Africa, Rhodesia, Western Australia, and India (Figure 1). Other greenstone belts in the Baltic Shield (Gaal et al 1978) or central and northern Africa (e.g. Williams 1978) are at least in part of Archean age but remain largely unknown. The general characteristics of Archean greenstone belts are discussed by Anhaeusser (1971a, b, 1973), Anhaeusser et al (1969), Engel (1968), Glikson (1970, 1976), Glikson & Lambert (1976), Goodwin (1973, 1977a), Goodwin & Ridler (1970), Turner & Walker (1973), Viljoen & Viljoen (1969a, b, c), and Walker (1978).

Greenstone belts typically occur as highly deformed volcanic and sedimentary keels, steeply infolded into surrounding, coeval to younger metamorphic and granitoid plutonic rocks. Adjacent to the plutons, the greenstone sequences commonly show narrow aureoles of higher grade metamorphism, but over large areas of most belts, the rocks are altered only to the lowest greenschist facies.

Anhaeusser et al (1969) have emphasized the remarkably similar rock association, stratigraphy, structure, metamorphism, and tectonic evolution of known Archean greenstone belts. The stratigraphic sections usually range between 10 and 25 kilometers thick and include distinct

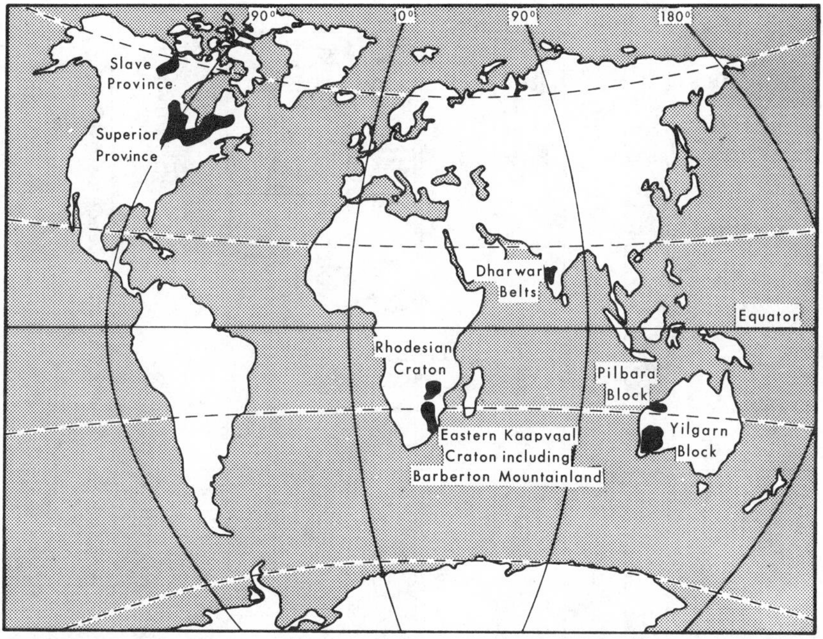

Figure 1 Generalized distribution of the principal Archean shields containing relatively unmetamorphosed greenstone belts discussed in this report.

volcanic and sedimentary sequences (Figure 2). In older belts, such as the Barberton Mountain Land (3.0 to 3.5 b.y.) and Pilbara Block (3.0 to 3.5 b.y.), there is usually a single thick volcanic sequence that is overlain by one main sedimentary succession (Hickman & Lipple 1975, Lipple 1975, Viljoen & Viljoen 1969a). In younger, generally post–3.0 b.y. old belts, such as those in the Canadian and Yilgarn shields, there are commonly several volcanic cycles, each capped by a thick sedimentary unit (Glikson 1976, Goodwin & Ridler 1970, Henderson 1972, McGlynn & Henderson 1970, Turner & Walker 1973).

Archean sedimentary rocks outside greenstone belts occur locally as platform deposits younger than greenstone belts or cratonization. The 3.0 b.y. old Pongola Supergroup (von Brunn & Hobday 1976) and slightly younger Dominion Reef strata of South Africa are indistinguishable from widespread Phanerozoic stable shelf carbonate-quartzite associations, and the auriferous 2.7 to 2.3 b.y. old Witwatersrand basin deposits (Pretorius 1975, Vos 1975) were deposited mainly by fluvial processes in an intracratonic basin. These types of stable shelf or cratonic basin deposits characterize many early Proterozoic sedimentary sequences and set them apart from the subjacent greenstone sedimentary assemblages.

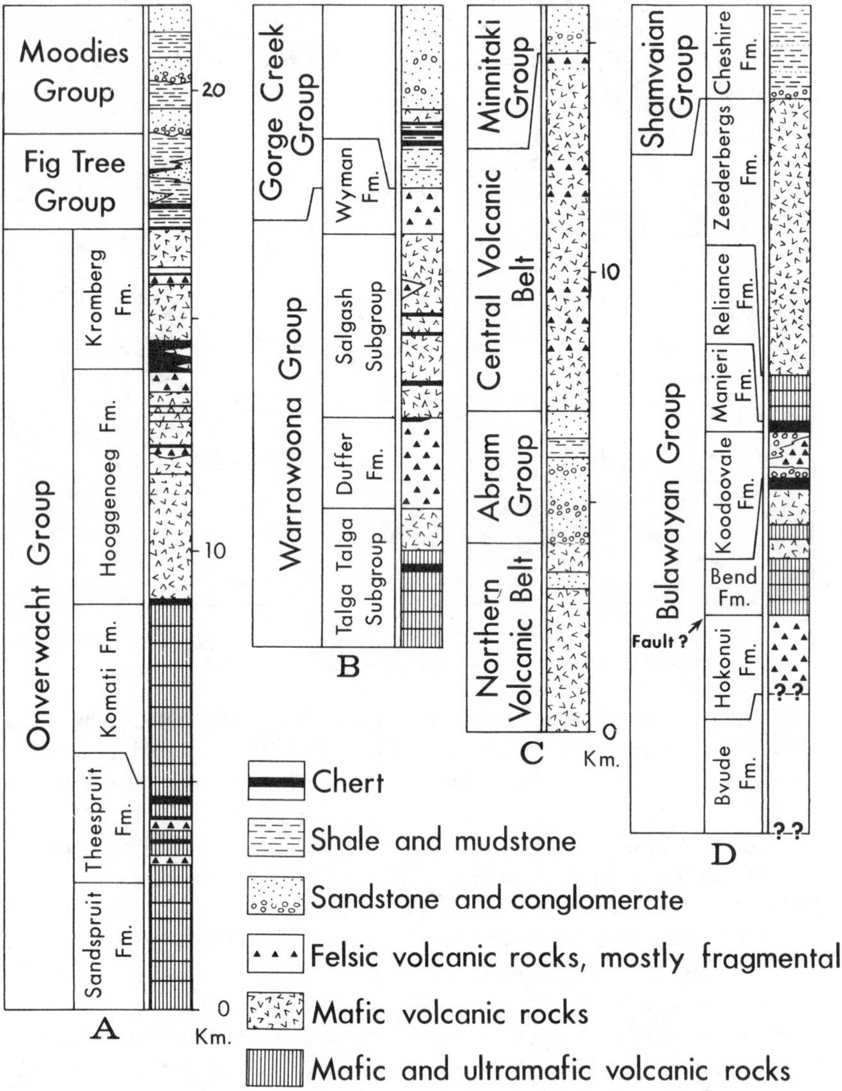

Figure 2 Representative greenstone belt stratigraphic sequences. All at same scale. *A*. Barberton Mountain Land, South Africa. Zwartkoppie Formation of Viljoen & Viljoen (1969a, c) is considered to be a facies of the upper Kromberg Formation and lower Fig Tree Group. *B*. Eastern Pilbara Block, Western Australia (Hickman & Lipple 1975). *C*. Sioux Lookout area, southern Superior Province, Canada (Turner & Walker 1973). *D*. Belingwe greenstone belt, Rhodesia. Bvude Formation is mainly hornblende schist, amphibolite, and calc-silicate rock with minor ultramafic schists and felsites (Orpen 1978).

GREENSTONE BELT VOLCANIC SEQUENCE

The characteristic rocks of greenstone belts are volcanic in origin. They include three principal varieties: mafic and ultramafic komatiites, tholeiitic basalt, and felsic lavas and fragmental rocks. The mafic and ultramafic lavas, both komatiites and tholeiites, tend to form massive sections, commonly several kilometers thick, made up of a large number of individual flow units. In general, individual greenstone belts tend to show a trend from komatiitic lavas at the base of the volcanic sequence to more tholeiitic and felsic units toward the top. The common development of pillows and, in the ultramafic units, crystalline quench textures suggests rapid subaqueous extrusion and crystallization (Viljoen & Viljoen 1969b).

Felsic volcanic rocks are best developed in the upper part of the volcanic sequences. They may occur in units up to several kilometers thick or as relatively thin zones of tuff. Most of the felsic material was extruded in a fragmental form and subsequently redistributed by sedimentary processes. Many workers have emphasized the presence in greenstone belts of volcanic-sedimentary cycles consisting from the base upward of mafic or ultramafic volcanic rocks, felsic volcanics, and a chert cap (Viljoen & Viljoen 1969a,c, Anhaeusser 1971b).

Felsic Volcanic and Sedimentary Rocks

Felsic volcanic rocks are a characteristic facies of the volcanic portions of Archean greenstone belts. Such units are present in the Theespruit and Hooggenoeg Formations of the Onverwacht Group in the Barberton Mountain Land of South Africa (Figure 2) and in the Gindalbie and Gundockerta Formations in the Eastern Goldfields Province of the Yilgarn Block (Williams 1973) and include the Hokonui and Koodoovale Formations in the Rhodesian Belingwe Belt (Orpen 1978; Figure 2), the Duffer and Wyman Formations of the Warrawoona Group in the Pilbara Block (Hickman & Lipple 1975), the Jackson Lake Formation and similar units in the Slave Province (Henderson 1972, Lambert 1978), and many formations in the Superior Province (Goodwin & Ridler 1970, Dimroth 1977, Dimroth & Demarcke 1978). The volcanic rocks in these units are typically dacitic, rhyodacitic, or rhyolitic in composition, and contrast strongly with the associated tholeiitic and komatiitic flows. In the Canadian Shield, felsic formations commonly include andesitic and other calc-alkaline volcanic rocks (Baragar & Goodwin 1969) which are largely lacking in other greenstone belts.

Although individual felsic units may persist over wide areas, many are markedly lenticular and most exhibit major thickness and facies changes

(Barley et al 1979, Goodwin & Ridler 1970). The thickest parts of felsic formations commonly reach several hundred to several thousand meters and are characteristically composed of coarse, poorly stratified fragmental material and flow rock. The principal rock types include coarse breccia, tuff breccia, tuff, and coarse-grained volcaniclastic sedimentary units which collectively formed cones of loose autoclastic, pyroclastic, and epiclastic debris constructed around active vents. The abundance of tuffs and the local development of shallow-water and terrestrial volcaniclastic deposits (Barley et al 1979, Williams 1973) indicate that the tops of these cones were commonly above sea level (Dimroth 1977). The building of large vent complexes probably produced broad subaerial volcanic edifices of considerable relief. Ash flow tuffs, lahars, and oligomict conglomerate, formed by penecontemporaneous erosion of the vent cones, are locally important constituents of terrestrial and shallow-water portions of these sequences (Dimroth & Demarcke 1978, Lambert 1978, Williams 1973).

Most of the vent cones are composed in large part of subaqueously redeposited fragmental volcanic debris. Although the vent cones locally developed narrow fringing aprons of shallow-water volcaniclastic sand and gravel, most of the debris apparently moved directly into deeper water within massive subaqueous sediment gravity flows (Barley 1978, Dimroth 1977, Lambert 1978, Orpen 1978, Tasse et al 1978, Viljoen & Viljeon 1969c). The deposits of these flows are massive to crudely graded and resemble those described from Phanerozoic sequences (Fiske 1963). The subaqueous slopes upon which the flows originated and across which they moved were probably relatively steep, perhaps locally approaching 10 to 25 degrees, the common range of the angle of repose of loose, water-saturated debris. They mark the first appearance of significant slopes within the greenstone belt basins.

Away from the vent complexes, felsic units tend to thin and fine (Lambert 1978). Massive debris flow deposits grade laterally away from the vents into sequences that reflect deposition from turbidity currents of progressively lower energy (Williams 1973). The most distal deposits are fine airfall tuffs. A similar, vertical sequence, formed as a result of waning vent activity, characterizes the deposits marginal to some vent complexes. In the Barberton Mountain Land, the upper felsic volcanic cycle in the Komati Gorge section of the Hooggenoeg Formation shows massive debris flow deposits (Viljoen & Viljoen 1969c) grading upward into turbidites dominated by Bouma A and B divisions which are in turn succeeded by gray cherts formed by the silicification of turbidites showing mainly B, C, and D Bouma divisions (Lowe & Knauth 1977).

The erosion of these volcanic complexes during the waning stages of felsic volcanism, accompanied by gradual subsidence, provided broad, low relief platforms upon which were deposited a variety of shallow-water sediments under volcanically quiescent conditions. Over 500 m of banded gray and white chert in the Barberton Mountain Land, originally considered to be part of the upper Hooggenoeg Formation (Viljoen & Viljoen 1969c, Lowe & Knauth 1977) but now known to mark the lowest Kromberg Formation (Figure 2), was deposited on top of an eroded felsic volcanic platform. The chert includes desiccation features and silicified gypsum indicating shallow-water to subaerial, low-energy deposition of a carbonate-silica-evaporite sequence (Lowe & Knauth 1977).

Away from the vent complexes, thin felsic tuff units are also capped by chert layers. Lowe & Knauth (1977, 1978) have shown that many of these chert units in the Barberton Mountain Land formed by silicification of fine-grained vitric pyroclastic material. In some cases, these were airfall units but others clearly represent current-deposited volcaniclastic sediment. Most are relatively thin, less than 10 m thick, and are underlain by or grade laterally into unsilicified felsic tuff or breccia. The chert ranges from light to dark gray or greenish in color and contains abundant microcrystalline chlorite and sericite impurities formed during devitrification and recrystallization of the volcanic ash. Hence, it is distinguished from chemical precipitates or silicified carbonate rock, which tend to lack intergrown phyllosilicate minerals.

In the Barberton Mountain Land this type of deposit is developed throughout the tholeiitic Hooggenoeg Formation and includes the well-known Middle Marker at the base of the Formation (Figure 2). Many of these chert units show brecciation and shrinkage cracks, intraformational conglomerates, cross-laminations, and units of accretionary lapilli suggesting deposition under shallow-water to subaerial conditions (Lowe & Knauth 1977, 1978).

Although some felsic volcanic units reach several thousand kilometers, most are less than 2 km in thickness. The frequency with which the vent cones became subaerial suggests that either the original water depths were not great or that regional epeirogenic uplift, possibly related to intrusive activity, accompanied felsic volcanism.

Non-Volcanogenic Sedimentary Rocks

Within most volcanic sequences, sedimentary rocks other than those obviously of volcanogenic origin make up a minor proportion of the stratigraphic column. The most common interflow units are cherts but thin layers of carbonate and terrigenous clastic rocks occur locally.

BANDED AND FERRUGINOUS CHERTS Perhaps the most widely distributed interflow sedimentary rocks are banded and/or ferruginous cherts, including banded gray and white chert, banded ferruginous chert, and a variety of iron-rich cherty rocks loosely referred to as iron-formation. Ferruginous rocks are largely absent in the Onverwacht Group, but banded gray and white cherts are well developed especially in the Hooggenoeg and Kromberg Formations (Figure 2). Thin ferruginous chert and iron-formation units, primarily of an oxide facies, are reported from many mafic and ultramafic volcanic sequences (Archibald et al 1978, Barley et al 1979, Beukes 1973, Condie & Harrison 1976, Horwitz & Smith 1978, Orpen 1978, Williams 1973, and many others).

These cherty layers are generally less than 20 m thick and commonly mark individual stratigraphic horizons over a wide region. In other cases they are lenticular and some single beds, when traced laterally, split into several units separated by volcanic flows (Beukes 1973). These minor interflow sedimentary units accumulated during breaks in the eruptive sequence. Some are accompanied by tuffaceous material indicating some subaerial volcanic activity, but many lack any associated detrital or pyroclastic material suggesting quiet sedimentation in areas far removed from subaerial vents or erodible source terranes. In some instances, these units show evidence of shallow-water to evaporitic depositional environments (Barley et al 1979, Dunlop 1978).

Banded chert and ironstone units up to several hundred meters thick occur in many greenstone belts (Goodwin 1973, Goodwin & Shklanka 1967, Orpen 1978, Lipple 1975). Most of these thicker units include sulfide and carbonate as well as oxide iron-formation facies. They tend to occur at stratigraphic breaks between contrasting volcanic units and represent major breaks in the extrusive sequence rather than simple interflow deposits.

Most of these deposits are probably chemical precipitates, although the primary sediments may have included a wide variety of ferruginous and non-ferruginous silicate and carbonate materials. They are typically finely laminated; lack scour, cross-stratification, clastic sediments, and other features indicative of current deposition; and were apparently deposited in quiet water. Some controversy exists as to the interpretation of the various iron-formation facies, however (compare Goodwin 1973 and Walker 1978), and it is not yet possible to estimate specific water depths for these deposits.

CARBONACEOUS CHERT Some interflow chert units in the Theespruit, Hooggenoeg, and Kromberg Formations of the Barberton Mountain Land and in the Talga Talga Subgroup in the Pilbara Block contain abundant carbonaceous particles that have yielded microstructures interpreted by many as ancient life forms (Engel et al 1968). These cherts contain a minor amount of sericite but are otherwise relatively free of non-quartzose silicate minerals. Those in the Kromberg Formation include granular and stromatolitic carbonaceous material and are interbedded with iron-rich dolomite showing evidence of shallow-water to supratidal deposition (Lowe & Knauth 1977). In the Pilbara Block, similar cherts are associated with barite deposits thought to represent replaced gypsum deposited in shallow-water to evaporite environments (Dunlop 1978).

CARBONATE ROCK Carbonate rocks are trace components of greenstone belt sequences, but are widely distributed, particularly in post–3.0 b.y. old belts. They generally occur at major breaks within the volcanic sequence, such as the Manjeri Formation on the Rhodesian craton (Wilson et al 1978), or at the contact between the volcanic and sedimentary greenstone belt sequences, such as in the Slave Province of Canada (Henderson 1975) and parts of the Rhodesian craton (Wilson et al 1978). Most of the carbonate is calcite, dolomite, ferruginous dolomite, or ankerite. These units are particularly important because they contain the oldest known stromatolites (Bickle et al 1975, Henderson 1975, Schopf et al 1971, Wilson et al 1978) and because they have provided important geochemical information on the composition of the Archean oceans and atmosphere (Schidlowski et al 1975, Veizer 1976).

Primary carbonate sediments may have been much more widely distributed than present occurrences would indicate. Many relatively pure chert layers include disseminated carbonate grains and may represent silicified carbonate or evaporite units (Dunlop 1978, Lowe & Knauth 1977). A large amount of very early diagenetic marine carbonate also occurs in carbonated tuff and volcaniclastic units (Lowe & Knauth 1979).

TERRIGENOUS CLASTIC ROCKS Terrigenous clastic sediments within the volcanic sequences show an irregular distribution. The most ancient preserved volcanic sequences, including the Onverwacht and Warrawoona Groups and the true greenstone belts of the Indian Shield (Naqvi et al 1978), virtually lack non-volcanogenic epiclastic rocks. The Onverwacht Group contains essentially no shale and the only sandstone is composed of locally derived felsic volcaniclastic material. No debris or other evidence has been found to indicate a continental, plutonic, metamorphic, or older sedimentary provenance (Lowe & Knauth 1977). The Warrawoona Group contains only a few thin beds of epiclastic sediment (Lipple 1975) and these appear to be of volcanic derivation.

Younger, post–3.0 b.y. old greenstone belts commonly include a small but important component of non-volcanogenic clastic detritus within the volcanic suite. Shale, argillite, mudstone, quartzose sandstone, and conglomerate containing clasts of older plutonic or metamorphic rocks have been described as interflow sediments from the Yilgarn Block (Williams 1973), the Bulawayan Group on the Rhodesian Craton (Harrison 1970, Orpen 1978, Wilson et al 1978), and from many formations in the Canadian Shield (Donaldson & Jackson 1965, Turner & Walker 1973).

BARITE Stratiform barite occurs in the Warrawoona Group in the Pilbara Block (Dunlop 1978) as interflow sediment associated with chert. It has been interpreted as including both primary barite sediment and diagenetic replacement after gypsum (Dunlop 1978, Barley 1978). Deposition apparently took place in shallow water. Barite in the Barberton Mountain Land is developed mainly in the Fig Tree Group and, locally, in the Onverwacht Group (Heinrichs & Reimer 1977).

GREENSTONE BELT SEDIMENTARY SEQUENCE

The upper formations of most greenstone belts consist predominantly of sedimentary rocks, commonly 1000 to 10,000 m thick. In contrast to sedimentary units in the subjacent volcanic sequences, these have been the object of numerous sedimentological studies, particularly in the Canadian Shield where their study dates back to the classic work of Pettijohn (1943).

These sedimentary sequences are made up largely of terrigenous detritus including shale, mudstone, siltstone, sandstone, and conglomerate. Many include a significant volcaniclastic component, and volcanic units, commonly of felsic, alkaline, or calc-alkaline type, are locally interbedded with the sedimentary rocks (Anhaeusser 1971b). The coarser detritus typically includes quartz, microcline, plagioclase, chert, and rock fragments indicating a provenance composed of uplifted parts of the subjacent volcanic and sedimentary sequences and a variety of plutonic and metamorphic rocks. The high Ni and Cr content of associated shales indicates the weathering of ultramafic source rocks (Naqvi 1976).

The sedimentary environments represented are generally like those in modern orogenic belts. Donaldson & Jackson (1965), Goodwin & Ridler (1970), Goodwin & Shklanka (1967), Henderson (1972), Ojakangas (1972), Pettijohn (1943), Turner & Walker (1973), Walker (1978), and Walker & Pettijohn (1971) have provided detailed discussions of the sedimentary portions of the Archean belts in the Canadian Shield. Their studies demonstrate the existence of a deep- to shallow-water clastic association dominated by thick turbiditic submarine fan or flysch sequences juxtaposed with subaerial alluvial-fan facies (Hyde & Walker 1977, Walker 1978). There seems to be a notable scarcity of shallow-water clastic shelf and shoreline sequences.

The Fig Tree Group in South Africa (Figure 2) is mainly a deep-water turbiditic graywacke-mudstone assemblage in northern outcrops (Anhaeusser et al 1969, Condie et al 1970, Reimer 1975) but, southward, includes oxide-facies banded iron-formation, volcaniclastic units, bedded barite, and shale-graywacke units (Reimer 1967, Heinrichs & Reimer 1977). This latter association reflects deposition in an area of volcanic and tectonic instability under mainly quiet subaqueous conditions but with local shallow-water deposits around tectonic highs or emergent volcanic cones. The clastic sediments include both locally derived volcaniclastic debris deposited around active vents and material derived by the erosion of larger tectonic uplifts (Condie et al 1970). The Moodies Group conformably overlies the Fig Tree Group in northern areas (Eriksson 1977, 1978, 1979), but steps unconformably across Fig Tree rocks to the south where it rests on units as old as the lower Kromberg Formation (Eriksson 1978). Detailed sedimentological studies of the Moodies Group (Eriksson 1977, 1978, 1979) have shown that it is composed largely of quartzose sandstone, orthoquartzite, and shale deposited in a variety of coastal environments including deltaic plain; tidal flats, channels, and shoals; barrier island; and shallow marine shelf.

The sedimentary sequences on other cratons are less well known than those in Canada and South Africa. The Shamvaian Group in Rhodesia includes several formations, apparently separated by unconformities (Stowe 1971), which show an alternation of argillaceous units and sandstone deposited in shallow water (Stowe 1971, Wilson et al 1978). Studies of thick sedimentary sequences at the top of the Dharwar Group in India, termed "geosynclinal" accumulations by Naqvi (1976), suggest deposition in alluvial, shallow-marine, and deep-sea environments (Naqvi 1977, Radhakrishna & Vasudev 1977). In the Pilbara Block of Western Australia, the pre–3.0 b.y. old Gorge Creek Group includes a succession of quartzite, feldspathic sandstone, and banded iron-formation showing some evidence of shallow-water sedimentation (Hickman & Lipple 1975, Lipple 1975). A younger post–3.0 b.y. Archean sedimentary group is apparently present as well, possibly including turbidites, but little is known of its stratigraphy and sedimentology (Fitton et al 1975). Studies by Dunbar & McCall (1971) indicate the occurrence of stratigraphically complex deep-sea sequences in parts of the Yilgarn Block of Western Australia, whereas other areas include shallow-water to terrestrial deposits (Marston 1978).

THE SEDIMENTARY EVOLUTION OF GREENSTONE BELTS

Greenstone belt sedimentary rocks display a number of well-defined secular trends reflecting the magmatic and tectonic development of the belts. In the lower part of the volcanic sequence, the character of sedimentation is fully controlled by the style and cyclicity of volcanism. The rapid submarine eruption of low-viscosity ultramafic and mafic flows constructed massive shield volcanoes of low relief, but produced little pyroclastic or autoclastic debris that could have fed local sedimentary systems. The paucity of interflow sediment must reflect the short time intervals between flows. The few sediments that did accumulate between eruptions are represented largely by banded and ferruginous cherts probably of chemical origin. There is little evidence for the existence of subaerial exposures subject to weathering and erosion. Sedimentation probably took place in quiet, extremely low-relief subaqueous to subaerial clastic-starved environments.

The first appearance of clastic detritus generally coincided with the initiation of felsic volcanism. The silicic magmas were probably of higher viscosity and contained more volatiles than the more mafic lavas, and explosive vesiculation and autobrecciation during eruption resulted in the construction of relatively steep vent cones composed largely of fragmental debris. These cones, which frequently became subaerial, were important sources of loose sediment and their tops, flanks, and the surrounding sea floor provided a range of sedimentary environments, including terrestrial, shallow-marine, slope, and submarine-fan, for its accumulation. The absence or extreme paucity of debris derived from mafic portions of the sequence and from non-volcanic sources indicates that, although magmatically active, most greenstone belts were essentially anorogenic during their volcanic stages.

There is no evidence that tectonic uplifts systematically accompanied the formation or volcanic development of greenstone belts. Although thin sedimentary quartzite and polymictic conglomerate layers occur within the volcanic sequences in some belts (Donaldson & Jackson 1965, Wilson 1964, Wilson et al 1978), radiometric dating of granitic and gneissic terranes around some belts indicates the existence of local areas of older sialic rocks (Wilson 1973, Wilson et al 1978), and local contacts occur where parts of the volcanic sequences rest unconformably on older sialic basement (Baragar & McGlynn 1976, Bickle et al 1975), there is generally no evidence within the greenstone belts themselves that associated older sialic blocks influenced their formation or early evolution, or served as major sediment sources. These older basement terranes were evidently of low relief and tectonically inactive until the later stages of greenstone belt development.

The transition into the sedimentary sequences marks a profound change in the tectonic and magmatic histories of greenstone belts. These rocks document the waning and ultimate cessation of volcanism, both the rapid effusion of mafic flows and the dominance of felsic pyroclastic volcanism as the main source of sedimentary debris. The decline in volcanic activity is clearly accompanied by major tectonic uplift of the marginal parts of the greenstone belts and of plutons that have intruded them. The abundance of quartz, microcline, and granitic detritus documents the first appearance of sialic basement as a major source of sedimentary debris. In greenstone belts that seem to have been associated with older basement blocks, such as those in the Canadian Shield, many of these orogenic uplifts appear to have been composed largely of older basement rock rather than subjacent portions of the greenstone belt itself.

The formation of high-standing source areas was accompanied by the development of adjacent subsiding sedimentary basins. The alternation of deep-sea flysch units and shallow-marine to terrestrial deposits within these basins indicates successive periods of rapid basin formation and filling. The common association of feldspathic and quartzitic sedimentary rocks indicates periods of tectonic uplift and orogeny followed by intervals of quiescence and maturation. No single tectonic or sedimentary trend characterizes all sedimentary sequences, but the overall direction seems to be toward the formation of substantial areas of subaerially exposed stable sialic crust. The termination of the greenstone belt cycle, which is not fully reflected in the sedimentary record, involves intrusion, deformation, and cratonization of the belts themselves.

THE TECTONIC EVOLUTION OF GREENSTONE BELTS

The development of the concept of plate tectonics has made possible an actualistic interpretation of sedimentary basins and basin deposits. Modern studies of sedimentary sequences have demonstrated that sedimentary deposits provide an important record of basin setting and the tectonic, magmatic, and metamorphic evolution of adjacent source terranes (Reading 1972, Dickinson 1974a, b).

Tectonic studies of Archean greenstone belts have focused, to a large degree, on the answers to two fundamental questions: (*a*) are greenstone belts primarily ensialic or ensimatic features, and (*b*) can the origin and evolution of greenstone belts be explained in terms of plate tectonics?

Although sedimentary studies cannot provide unique answers to these questions, they can place significant constraints on the answers that are ultimately offered and accepted. In the following discussion, we explore some of these constraints.

Tectonic Setting of Greenstone Belts

A relatively thick sialic crust exerts a profound influence on the tectonics of suprajacent basins and, hence, on basin sedimentation, particularly if there are local variations in its thickness or if it is involved in rifting or compression. A relatively thin sialic crust, however, might behave little differently than a thin layer of low-density sedimentary or volcanic rock within the greenstone sequence, may have little or no effect on local tectonics, could be radiometrically destroyed by metamorphism and plutonism, and, hence, may be essentially undetectable. Greenstone belts developed on such a crust would be effectively ensimatic.

It is evident that major enclaves of sial existed at least 3.5–3.8 billion years ago. These include the Amitsoq gneiss and Isua supracrustals in Greenland (McGregor 1973, Moorbath et al 1972), the Ancient Gneiss Complex of Swaziland (Hunter 1974), and the Rhodesian nucleus (Hawkesworth et al 1975, Moorbath et al 1977, Wilson 1973, Wilson et al 1978). Such thick masses of continental crust should have had clearly discernable effects on adjacent or suprajacent sedimentary basins. These effects seem to be visible in some greenstone belts but absent in others.

The oldest well-preserved belts, those active generally between 3.5 and 3.1 b.y. and cratonized by 3.0 b.y., including the Barberton Mountain Land (Anhaeusser 1973), belts in the Pilbara Block (Compston & Arriens 1968), and possibly the true greenstone belts in the Dharwar terrane (Naqvi 1976, 1977, Naqvi et al 1978) contrast significantly with younger greenstone sequences. The former tend to include major amounts of ultramafic komatiitic lavas, essentially lack calc-alkaline volcanic rocks (Hallberg 1972), and may show a more or less progressive evolution from the ultramafic through the tholeiitic to the sedimentary stages. The sedimentary rocks in the volcanic sequence are composed either of intrabasinal felsic volcanic debris or orthochemical sediments. There is an almost complete absence of non-volcanogenic detritus, including clay, quartz, feldspar, and identifiable plutonic, metamorphic, and sedimentary grains. The paucity of shale is particularly noteworthy inasmuch as many of the sedimentary units were deposited under low-energy subaqueous conditions. During their volcanic development, there could not have been any large land areas of either low or high relief in or adjacent to these greenstone basins.

Archean, Proterozoic, and Phanerozoic basins known to have developed on or adjacent to continental crust almost inevitably contain debris derived from it, especially quartz and clay. Stable shelf and associated offshelf deep-sea deposits are characterized throughout geologic time by significant quartz and clay components, and clastic units deposited in association with tectonically active areas of continental crust include microcline, plagioclase, and basement-derived rock fragments. If sialic crust existed locally around or beneath these pre–3.0 b.y. old greenstone belts, it must have been so thin as to be effectively absent. The sedimentary deposits indicate that the basins were effectively ensimatic and anorogenic.

The direct influence of older sialic blocks on these belts was negligible, but a somewhat different situation characterizes the pre–3.0 b.y. Sebakwian greenstone belts in Rhodesia. Although sparsely distributed and generally heavily altered, younger Sebakwian sequences contain small but significant amounts of basement-derived plutonic and metamorphic detritus within the volcanic sequence (Stowe 1971, Wilson 1973, Wilson et al 1978). The presence of quartzitic beds (Wilson 1964) and locally abundant interflow shaly units clearly indicates the exposure and weathering of older rocks, almost certainly the pre–3.5 b.y. old Rhodesian basement nucleus with which these belts are closely associated.

Post–3.0 b.y. old greenstone belts tend to show a pattern of development that also reflects the influence of older basement rocks. Those in the eastern Yilgarn Block and the Bulawayan belts of Rhodesia were generally intruded by plutonic rocks and stabilized about 2.7–2.6 b.y. ago with the main volcanism and sedimentation occurring between 2.9 and 2.6b.y. (Compston & Arriens 1968, Hawkesworth et al 1975, Moorbath et al 1977, Oversby 1975). These belts exhibit several well-developed volcanic-sedimentary cycles and include, in the volcanic sequences, minor but noticeable amounts of shale, quartzose sandstone, and, locally, conglomerate containing pebbles of granitoid and gneissic rocks (Harrison 1970) indicating presence of locally exposed older basement. The Rhodesian basement nucleus probably provided craton-derived debris to the Bulawayan belts, and a pre–3.0 b.y. old Pilbara-like basement may have existed in or adjacent to the Yilgarn Block (Oversby 1975, Horwitz & Smith 1978).

Canadian greenstone belts in the Slave and Superior Provinces were stabilized during the Kenoran Orogeny about 2.6–2.5 b.y. ago (Stockwell 1968). Greenstone belt volcanism and sedimentation generally occurred in the interval between 2.9 and 2.6 b.y. ago and clearly reflect the influence of surrounding and underlying (?) sialic basement (Goodwin 1977a,b, Goodwin & Ridler 1970, Goodwin & Shklanka 1967, Walker 1978). Major volcanic-sedimentary cycles are well developed and are commonly repeated several times within individual belts. Ultramafic lavas are apparently absent, and calc-alkaline volcanics are abundant (Goodwin

1977a, Goodwin & Ridler 1970). Thick units of shale and non-volcanogenic basement-derived sandstone and conglomerate occur, commonly at several levels throughout the volcanic sequence (Baragar & McGlynn 1976, Donaldson & Jackson 1965, Goodwin & Shklanka 1967, Salop 1977, Turner & Walker 1973). These sedimentary units document the existence of major, regionally extensive outcrops of older gneissic, granitic, and metamorphosed supracrustal sequences (Salop 1977) and provide an unambiguous model for greenstone belts developing on or adjacent to sialic crust.

The significance of the secular trend suggested by these observations is unclear. The essential termination of greenstone belt formation after the Archean suggests that some fundamental change in tectonic style occurred about 2.5 b.y. ago, perhaps related to the evolution of large areas of continental crust. If this crust developed progressively or episodically over a considerable part of the late Archean, then the probability of younger greenstone belts developing under the influence of sialic blocks would also increase with time. Such a trend may be reflected in the contrasts between older and younger preserved belts.

Although information is sparse in many areas, these comparisons suggest that there may be significant differences in greenstone belt sediments, volcanism, and general patterns of evolution and that these differences may reflect differences in the location of the greenstone belts relative to older sialic basement blocks. Greenstone belts developed on or adjacent to sialic basement contain unambiguous evidence of its presence. This evidence may include aspects of the stratigraphic and volcanic sequence, but will certainly include detritus derived from the basement by erosion.

Tectonic Evolution of Greenstone Belts

The formation and evolution of greenstone belts have been described in terms of virtually every known plate tectonic setting and many hypothetical tectonic styles. Early models generally depicted their development in non–plate-tectonic terms, commonly involving the vertical gravitational foundering of large mafic volcanic edifices constructed on an ensialic (Macgregor 1951) or ensimatic (Glikson 1971) crust. More recent hypotheses have tended to describe greenstone belts in terms of plate-tectonic regimes, including island arcs (Anhaeusser 1973, Engel 1968, Tarney et al 1976), rifts or spreading centers (Condie & Hunter 1976), and hot spots (Fyfe 1978). Once again, although information provided by sedimentary studies cannot yet resolve these differences in interpretation, it can place important constraints on proposed models.

The anorogenic character of greenstone belt volcanic development makes it difficult to interpret greenstone belt origins and early evolution in terms of rifting (e.g. Archibald et al 1978, Condie & Hunter 1976, Glikson & Lambert 1976). Modern rifts, whether ensimatic or ensialic, involve crustal block faulting and differential uplift, and there is little reason to suspect that substantially different conditions would have characterized Archean rifts. Phanerozoic sequences deposited within or adjacent to areas of rifting show basal detrital units composed of debris eroded from the uplifted rift blocks (Dickinson 1974b). The abundance in Archean greenstone belt volcanic sequences of rocks deposited in shallow-water environments indicates that block faulting or differential uplift would have formed subaerial exposures subject to weathering and erosion. The presence of interflow chemical deposits indicates that there were pauses in volcanism during which detrital sediments derived from such uplifts could have accumulated had any been produced. The paucity of non-volcanogenic detrital rocks within greenstone belt volcanic sequences, especially those outside of the Canadian Shield, indicates that greenstone belt formation did not involve major differential uplift and strongly suggests that these volcanic sequences did not fill fault-bounded rift depressions.

Propositions that greenstone belts developed in association with convergent plate junctions, particularly in a backarc or marginal sea setting (Tarney et al 1976), are somewhat more difficult to evaluate sedimentologically because both are dominated by volcanic and volcaniclastic rocks. Several characteristics of modern backarc and forearc basins are noteworthy when considering such models for Archean belts. Phanerozoic arc-related sequences almost always show a strong polarity, best characterized by classical eugeosyncline-miogeosyncline couples representing proximal- and distal-arc environments respectively. Although much work must be done before the facies of Archean greenstone belts are well understood, there is no evidence that their volcanic development was characterized by strong polarity. Felsic vent complexes seem to occur almost randomly, although careful paleocurrent, grain size, and isopach studies may suggest a more regular distribution. Felsic volcanic complexes also appear to be local intrabelt features rather than major zones defining greenstone belt margins. Other than felsic detritus, there is in the volcanic portions of greenstone belts an absence of volcaniclastic debris that could have been derived from associated volcanic arcs. Modern forearc and backarc deposits typically include abundant detritus derived by erosion of the arc itself. If volcanic arcs were associated with Archean greenstone belts, there is no evidence that they were reflected topographically or sedimentologically.

The overall sedimentological character of greenstone belt volcanic sequences does not coincide well with those of modern divergent or convergent plate junctions. Their anorogenic development is more like that of modern intraplate settings.

Although the sedimentary portions of greenstone belts include thick sections of detrital sediments, less is known of the sedimentology and tectonic implications of these units, in some respects, than of those in the volcanic sequence. It is apparent that these detrital formations reflect orogeny, but the nature of these orogenic movements is not clear. In the Barberton Mountain Land, which shows little evidence of an associated older basement, much of the debris was derived from penecontemporaneous volcanism, syngenetic plutons, and uplifted parts of the greenstone belt itself. Some Canadian sedimentary units, deposited in basins that reflect the influence of older basement throughout their volcanic development, are made up largely of basement-derived debris and contain little volcanic or greenstone belt detritus. In some belts, it has been suggested that deposition took place in linear, fault-bounded basins (Marston 1978, Rutland 1976), whereas in others the sedimentary sequences more closely resemble those along modern continental margins (Eriksson 1978, 1979). It may be that the tectonic environments represented by these orogens are as diverse as those represented by Phanerozoic systems.

Perhaps the most popular model for the orogenic phase of greenstone belt evolution is that of a backarc basin or marginal sea (Anhaeusser 1973, Condie & Harrison 1976, Engel 1968, Tarney et al 1976). This popularity lies in large part in the apparent insufficiency of other models and in part in the fact that volcanic arcs represent the principal modern sites where large-scale volcanism is accompanied and succeeded by plutonism and orogeny. Sedimentologically, this model has yet to be evaluated although information presently available suggests that there may be problems of polarity, volcanic versus tectonic sources, and an overall paucity of arc-derived volcanic detritus. A tectonic setting like that of modern volcanic arcs may have existed only during the volcanic-sedimentary transition in many greenstone belts.

SUMMARY

Although modern sedimentology and sedimentary petrography have proven to be powerful tools in the interpretation of Phanerozoic tectonics, their potential has not yet been realized in the study of Archean sequences. This review has attempted to summarize the sedimentary characteristics of Archean greenstone belts and to consider their implications regarding both the immediate environments and processes of deposition and the overall tectonic settings and development of greenstone belts. The paucity of suitable data severely hampers such an exercise at this time, but a number of studies beginning or already under way should add much factual data to our knowledge of greenstone belt sedimentary rocks.

The lower volcanic portion of greenstone belts is characterized by two main types of sedimentary deposits: (*a*) chemical sediments, dominated by a variety of cherts and iron-formation, and (*b*) units of felsic volcanic rocks and volcaniclastic sediments. Thick mafic and ultramafic flow sequences generally contain thin interflow units of chemical origin deposited in subaqueous, shallow-marine environments on broad shield volcanoes of exceedingly low relief. Felsic volcanic rocks are often developed in thick fragmental associations representing vent cones. These cones were commonly subaerial and provided abundant loose pyroclastic debris, slopes over which it was moved, and a variety of environments in which it was deposited. These environments include submarine-fan, slope, shallow-shelf, and terrestrial. Very little debris reached the basins from extrabasinal or tectonic sources and, in spite of intense volcanic activity, greenstone belts were probably anorogenic during this stage of their development.

The sedimentary sequence or sequences in greenstone belts show deposition of mixed volcanic and basement-derived detritus under orogenic conditions. Documented environments of deposition include deep-sea fan, shallow-shelf, deltaic, and alluvial, and generally point to the close association of topographically high source areas and adjacent deep basins of deposition. Uplifted sialic basement was subject to rapid erosion, but eventual tectonic stabilization and denudation often led to its yielding abundant quartzitic detritus.

Comparisons of preserved greenstone belts in southern Africa, Australia, India, and Canada suggest that those which lay near blocks of older sialic basement show slightly different sedimentary, volcanic, and stratigraphic histories than those developed in completely ensimatic settings.

The initial volcanic history of greenstone belts cannot be interpreted in terms of orogenic events. Rifting or compression involving the development of topographic relief was absent, even though sialic blocks lay adjacent to many belts. Sedimentary sequences deposited during the orogenic phase of greenstone belt development are grossly similar to those of many modern orogens, particularly to those deposited adjacent to modern island arcs. This similarity arises because island arcs represent the only modern terranes where volcanism is succeeded by large-scale plutonism, uplift, and exposure to yield a sedimentary sequence reflecting this progression of events. Archean greenstone belts may have evolved in settings like modern convergent plate junctions, they may have evolved in intraplate settings such as over intracrustal fractures or mantle plumes, or they may have evolved at uniquely Archean sites of volcanism and plutonism.

ACKNOWLEDGMENTS

The author's research on the sedimentology of Archean cherts in South Africa and Western Australia has been supported by the National Science Foundation under Grants EAR76-06095 and EAR78-01642, respectively.

Literature Cited

Anhaeusser, C. R. 1971a. The Barberton Mountain Land, South Africa—a guide to the understanding of the Archaean geology of Western Australia. *Spec. Publ. Geol. Soc. Aust.* 3:103–19

Anhaeusser, C. R. 1971b. Cyclic volcanicity and sedimentation in the evolutionary development of Archaean greenstone belts of shield areas. *Spec. Publ. Geol. Soc. Aust.* 3:57–70

Anhaeusser, C. R. 1973. The evolution of the early Precambrian crust of southern Africa. *Philos. Trans. R. Soc. London Ser. A* 273:359–88

Anhaeusser, C. R., Mason, R., Viljoen, M. J., Viljoen, R. P. 1969. A reappraisal of some aspects of Precambrian shield geology. *Bull. Geol. Soc. Am.* 80:2175–200

Archibald, N. J., Bettenay, L. F., Binns, R. A., Groves, D. I., Gunthorpe, R. J. 1978. The evolution of Archaean greenstone terrains, Eastern Goldfields Province, Western Australia. *Precambrian Res.* 6:103–31

Baragar, W. R. A., Goodwin, A. M. 1969. Andesites and Archean volcanism of the Canadian Shield. *Oregon Dept. Geol. Miner. Ind. Bull.* 65:121–42

Baragar, W. R. A., McGlynn, J. C. 1976. Early Archean basement in the Canadian Shield: a review of the evidence. *Geol. Surv. Can. Pap. 76-14.* 21 pp.

Barley, M. E. 1978. Shallow-water sedimentation during deposition of the Archaean Warrawoona Group, eastern Pilbara Block, Western Australia. *Publ. Geol. Dept. & Extension Serv. Univ. West. Aust.* 2:22–29

Barley, M. E., Dunlop, J. S. R., Glover, J. E., Groves, D. I. 1979. Sedimentary evidence for an Archaean shallow-water volcanic-sedimentary facies, eastern Pilbara Block, Western Australia. *Earth Planet. Sci. Lett.* 43:74–84

Beukes, N. J. 1973. Precambrian iron-formations of South Africa. *Econ. Geol.* 68:960–1004

Bickle, M. J., Martin, A., Nisbet, E. G. 1975. Basaltic and peridotitic komatiites and stromatolites above a basal unconformity in the Belingwe greenstone belt, Rhodesia. *Earth Planet. Sci. Lett.* 27:155–62

Compston, W., Arriens, P. A. 1968. The Precambrian geochronology of Australia. *Can. J. Earth Sci.* 5:561–83

Condie, K. C., Harrison, N. M. 1976. Geochemistry of the Archean Bulawayan Group, Midlands greenstone belt, Rhodesia. *Precambrian Res.* 3:253–71

Condie, K. C., Hunter, D. R. 1976. Trace element geochemistry of Archean granitic rocks from the Barberton region, South Africa. *Earth Planet. Sci. Lett.* 29:389–400

Condie, K. C., Macke, J. E., Reimer, T. O. 1970. Petrology and geochemistry of early Precambrian graywackes from the Fig Tree Group, South Africa. *Bull. Geol. Soc. Am.* 81:2759–76

Dickinson, W. R. 1974a. Sedimentation within and beside ancient and modern magmatic arcs. *Spec. Publ. Soc. Econ. Paleontol. Mineral.* 19:230–39

Dickinson, W. R. 1974b. Plate tectonics and sedimentation. *Spec. Publ. Soc. Econ. Paleontol. Mineral.* 22:1–27

Dimroth, E. 1977. Archean subaqueous autoclastic volcanic rocks, Rouyn-Noranda area, Quebec: classification, diagenesis and interpretation. *Geol. Surv. Can. Paper 77-1A*, pp. 513–22

Dimroth, E., Demarcke, J. 1978. Petrography and mechanism of eruption of the Archean Dalembert tuff, Rouyn-Noranda, Quebec, Canada. *Can. J. Earth Sci.* 15:1712–23

Donaldson, J. A., Jackson, G. D. 1965. Archaean sedimentary rocks of North Spirit Lake area, northwestern Ontario. *Can. J. Earth Sci.* 2:622–47

Dunbar, G. J., McCall, G. J. H. 1971. Archaean turbidites and banded ironstones of the Mt. Belches area (Western Australia). *Sediment. Geol.* 5:93–133

Dunlop, J. S. R. 1978. Shallow-water sedimentation at North Pole, Pilbara, Western Australia. *Publ. Geol. Dept. & Extension Serv. Univ. West. Aust.* 2:30–38

Engel, A. E. J. 1968. The Barberton Mountain Land: clues to the differentiation of the earth. *Trans. Geol. Soc. S. Afr.* 71 (annex): 255–70

Engel, A. E. J., Nagy, B., Nagy, L. A., Engel, C. G., Kremp, G. O. W., Drew, C. M. 1968. Alga-like forms in Onverwacht Series, South Africa: oldest recognized

lifelike forms on earth. *Science* 161:1005–8
Eriksson, K. A. 1977. Tidal deposits from the Archaean Moodies Group, Barberton Mountain Land, South Africa. *Sediment. Geol.* 18:257–81
Eriksson, K. A. 1978. Alluvial and destructive beach facies from the Archaean Moodies Group, Barberton Mountain Land, South Africa and Swaziland. *Can. Soc. Petrol. Geol. Mem.* 5:287–311
Eriksson, K. A. 1979. Marginal marine depositional processes from the Archaean Moodies Group, Barberton Mountain Land, South Africa: evidence and significance. *Precambrian Res.* 8:153–82
Fiske, R. S. 1963. Subaqueous pyroclastic flows in the Ohanopecosh Formation, Washington. *Bull. Geol. Soc. Am.* 74:391–406
Fitton, M. J., Horwitz, R. C., Sylvester, G. 1975. Stratigraphy of the early Precambrian in the West Pilbara, Western Australia. *Aust. CSIRO Miner. Res. Lab., FP-11.* 30 pp.
Fyfe, W. S. 1978. The evolution of the earth's crust: modern plate tectonics to ancient hot spot tectonics? *Chem. Geol.* 23:89–114
Gaal, G., Mikkola, A., Söderholm, B. 1978. Evolution of the Archean crust in Finland. *Precambrian Res.* 6:199–215
Glikson, A. Y. 1970. Geosynclinal evolution and geochemical affinities of early Precambrian systems. *Tectonophysics* 9:397–433
Glikson, A. Y. 1971. Primitive Archaean element distribution patterns: chemical evidence and geotectonic significance. *Earth Planet. Sci. Lett.* 12:309–320
Glikson, A. Y. 1976. Stratigraphy and evolution of primary and secondary greenstones: significance of data from shields of the southern hemisphere. See Windley 1976, pp. 257–77
Glikson, A. Y., Lambert, I. B. 1976. Vertical zonation and petrogenesis of the early Precambrian crust in Western Australia. *Tectonophysics* 30:55–89
Glover, J. E., Groves, D. I., eds. 1978. *Archaean Cherty Metasediments: Their Sedimentology, Micropalaeontology, Biogeochemistry, and Significance to Mineralization. Publ. Geol. Dept. Extension Serv. Univ. West. Aust. 2.* 88 pp.
Goodwin, A. M. 1973. Archaean iron-formations and tectonic basins of the Canadian Shield. *Econ. Geol.* 68:915–33
Goodwin, A. M. 1977a. Archean basin-craton complexes and the growth of Precambrian shields. *Can. J. Earth Sci.* 14:2737–59
Goodwin, A. M. 1977b. Archean volcanism in Superior Province, Canadian Shield. *Spec. Pap. Geol. Assoc. Can.* 16:205–41
Goodwin, A. M., Ridler, R. H. 1970. The Abitibi orogenic belt. In *Symposium on Basins and Geosynclines of the Canadian Shield*, ed. A. J. Baer, pp. 1–30. *Geol. Surv. Can. Pap. 70-40.* 265 pp.
Goodwin, A. M., Shklanka, R. 1967. Archean volcano-tectonic basins: form and pattern. *Can. J. Earth Sci.* 4:777–95
Hallberg, J. A. 1972. Geochemistry of Archean volcanic belts in the Eastern Goldfields region of Western Australia. *J. Petrol.* 13:45–56
Harrison, N. M. 1970. The geology of the country around Que Que. *Bull. Geol. Surv. Rhodesia* 67. 125 pp.
Hawkesworth, C. J., Moorbath, S., O'Nions, R. K., Wilson, J. F. 1975. Age relationships between greenstone belts and "granites" in the Rhodesian Archaean craton. *Earth Planet. Sci. Lett.* 25:251–62
Heinrichs, T. K., Reimer, T. O. 1977. A sedimentary barite deposit from the Archaean Fig Tree Group of the Barberton Mountain Land (South Africa). *Econ. Geol.* 72:1426–41
Henderson, J. B. 1972. Sedimentology of Archean turbidites at Yellowknife, Northwest Territories. *Can. J. Earth Sci.* 9:882–902
Henderson, J. B. 1975. Archean stromatolites in the northern Slave Province, Northwest Territories, Canada. *Can. J. Earth Sci.* 12:1619–30
Hickman, A. H., Lipple, S. L. 1975. Explanatory notes on the Marble Bar 1:250,000 Geological Sheet, Western Australia. *Geol. Surv. West. Aust.* (Microfiche)
Horwitz, R. C., Smith, R. E. 1978. Bridging the Yilgarn and Pilbara Blocks, Western Australia. *Precambrian Res.* 6:293–322
Hunter, D. R. 1974. Crustal development of the Kaapvaal craton. 1. The Archaean. *Precambrian Res.* 1:259–94
Hyde, R. S., Walker, R. G. 1977. Sedimentary environments and the evolution of the Archean greenstone belt in the Kirkland Lake area, Ontario. *Geol. Surv. Can. Pap. 77-1A*, pp. 185–90
Lambert, M. B. 1978. The Black River volcanic complex—a cauldron subsidence structure of Archean age. *Geol. Surv. Can. Pap. 78-1A*, pp. 153–58
Lipple, S. L. 1975. Definitions of new and revised stratigraphic units of the eastern Pilbara region. *Ann. Rep. Geol. Surv. West. Aust. 1974*, pp. 58–63
Lowe, D. R., Knauth, L. P. 1977. Sedimentology of the Onverwacht Group (3.4 billion years), Transvaal, South Africa,

and its bearing on the characteristics and evolution of the early earth. *J. Geol.* 85:699–723

Lowe, D. R., Knauth, L. P. 1978. The oldest marine carbonate öoids reinterpreted as volcanic accretionary lapilli, Onverwacht Group, South Africa. *J. Sediment. Petrol.* 48:709–22

Lowe, D. R., Knauth, L. P. 1979. Petrology of Archean carbonate rocks in the Onverwacht and Fig Tree Groups, South Africa. *Prog. Abstr. Ann. Meet. Geol. Assoc. Can.* 4:64

Macgregor, A. M. 1951. Some milestones in the Precambrian of Southern Rhodesia. *Proc. Geol. Soc. S. Afr.* 54:27–71

Marston, R. J. 1978. The geochemistry of Archaean clastic metasediments in relation to crustal evolution, northeastern Yilgarn Block, Western Australia. *Precambrian Res.* 6:157–75

McCall, G. J. H. 1977. *The Archean.* Stroudsburg, Pa.: Dowden, Hutchinson, & Ross. 505 pp.

McGlynn, J. C., Henderson, J. B. 1970. Archean volcanism and sedimentation in the Slave structural province. *Can. Geol. Surv. Pap. 70-40*, pp. 31–44

McGregor, V. R. 1973. The early Precambrian gneisses of the Godthab district, West Greenland. *Philos. Trans. R. Soc. London Ser. A* 273:343–58

Moorbath, S., O'Nions, R. K., Pankhurst, R. J., Gale, N. H., McGregor, V. R. 1972. Further rubidium-strontium age determinations on the very early Precambrian rocks of the Godthab district, West Greenland. *Nature* 240:78

Moorbath, S., Wilson, J. F., Goodwin, R., Humm, R. 1977. Further Rb-Sr age and isotope data on early and late Archaean rocks from the Rhodesian craton. *Precambrian Res.* 5:229–39

Naqvi, S. M. 1976. Physico chemical conditions during the Archaean as indicated by Dharwar geochemistry. See Windley 1976, pp. 289–98

Naqvi, S. M. 1977. Archaean sedimentation of Dharwars in the central part of the Chitradorga schist belt, Karnataka, India. *Geophys. Res. Bull.* 15:17–30

Naqvi, S. M., Rao, V. D., Narain, H. 1978. The primitive crust: evidence from the Indian Shield. *Precambrian Res.* 6:323–45

Ojakangas, R. W. 1972. Archean volcanogenic graywackes of the Vermilion District, northeastern Minnesota. *Bull. Geol. Soc. Am.* 83:429–42

Orpen, J. L. 1978. *The Geology of the Southwestern Part of the Belingwe Belt and Adjacent Country—The Belingwe Peak Area.* PhD thesis. Univ. Rhodesia, Salisbury. 242 pp.

Oversby, V. M. 1975. Lead isotopic systematics and ages of Archaean acid intrusives in the Kalgoorlie-Norseman area, Western Australia. *Geochim. Cosmochim. Acta* 39:1107–25

Pettijohn, F. J. 1934. Conglomerate of Abram Lake, Ontario, and its extensions. *Bull. Geol. Soc. Am.* 45:479–506

Pettijohn, F. J. 1943. Archean sedimentation. *Bull. Geol. Soc. Am.* 54:925–72

Pretorius, D. A. 1975. The depositional environment of the Witwatersrand goldfields: a chronological review of speculations and observations. *Inf. Circ. Econ. Geol. Res. Unit Univ. Witwatersrand 95.* 47 pp.

Radhakrishna, B. P., Vasudev, V. N. 1977. The early Precambrian of the southern Indian Shield. *J. Geol. Soc. India* 18:525–41

Reading, H. G. 1972. Global tectonics and the genesis of flysch successions. *Proc. Int. Geol. Cong., 24th, Montreal* 6:59–66

Reimer, T. O. 1967. *Die Geologie der Stolzburg Synclinale im Barberton Bergland (Transvaal-Südafrika).* PhD thesis. Goethe Univ., Frankfurt

Reimer, T. O. 1975. Untersuchungen über Abtragung, Sedimentation und Diagenese im frühen Präkambrium am Beispiel der Sheba-Formation (Südafrika). *Geol. Jahrb.* 17:3–108

Rutland, R. W. R. 1976. Orogenic evolution of Australia. *Earth Sci. Rev.* 12:161–96

Salop, L. J. 1977. *Precambrian of the Northern Hemisphere.* Amsterdam: Elsevier. 378 pp.

Schidlowski, M., Eichmann, R., Junge, C. E. 1975. Precambrian sedimentary carbonates: carbon and oxygen isotope geochemistry and implications for the terrestrial oxygen budget. *Precambrian Res.* 2:1–69

Schopf, J. W., Oehler, D. Z., Horodyski, R. J., Kvenvolden, K. A. 1971. Biogenicity and significance of the oldest known stromatolites. *J. Paleontol.* 45:477–85

Stockwell, C. H. 1968. Geochronology of stratified rocks of the Canadian Shield. *Can. J. Earth Sci.* 5:693–98

Stowe, C. W. 1971. Summary of the tectonic development of the Rhodesian Archaean craton. *Spec. Publ. Geol. Soc. Aust.* 3: 377–83

Tarney, J., Dalziel, I. W. D., De Wit, M. J. 1976. Marginal basin "Rocas Verdes" complex from southern Chile: a model for Archaean greenstone belt formation. See Windley 1976, pp. 131–45

Tasse, N., LaJoie, J., Dimroth, E. 1978. The anatomy and interpretation of an Archean volcaniclastic sequence, Noranda region,

Quebec. *Can. J. Earth Sci.* 15:874–88

Turner, C. C., Walker, R. G. 1973. Sedimentology, stratigraphy, and crustal evolution of the Archean greenstone belt near Sioux Lookout, Ontario. *Can. J. Earth Sci.* 10:817–45

Veizer, J. 1976. $^{87}Sr/^{86}Sr$ evolution of seawater during geologic history and its significance as an index of crustal evolution. See Windley 1976, pp. 569–78

Viljoen, M. J., Viljoen, R. P. 1969a. An introduction to the geology of the Barberton granite-greenstone terrain. *Spec. Publ. Geol. Soc. S. Afr.* 2:7–28

Viljoen, M. J., Viljoen, R. P. 1969b. The geology and geochemistry of the lower ultramafic unit of the Onverwacht Group and a proposed new class of igneous rocks. *Spec. Publ. Geol. Soc. S. Afr.* 2:55–86

Viljoen, R. P., Viljoen, M. J. 1969c. The geological and geochemical significance of the upper formations of the Onverwacht Group. *Spec. Publ. Geol. Soc. S. Afr.* 2:113–52

von Brunn, V., Hobday, D. K. 1976. Early Precambrian tidal sedimentation in the Pongola Supergroup of South Africa. *J. Sediment. Petrol.* 46:670–79

Vos, R. G. 1975. An alluvial plain and lacustrine model for the Precambrian Witwatersrand deposits of South Africa. *J. Sediment. Petrol.* 45:480–93

Walker, R. G. 1978. A critical appraisal of Archean basin-craton complexes. *Can. J. Earth Sci.* 15:1213–18

Walker, R. G., Pettijohn, F. J. 1971. Archean sedimentation: analysis of the Minnitaki basin, northwestern Ontario, Canada. *Bull. Geol. Soc. Am.* 82:2099–2130

Williams, H. R. 1978. The Archaean geology of Sierra Leone. *Precambrian Res.* 6:251–68

Williams, I. R. 1973. Explanatory notes on the Kurnalpi 1:250,000 Geological Sheet, Western Australia. *Geol. Surv. West. Aust.* 37 pp.

Wilson, J. F. 1964. The geology of the country around Fort Victoria. *Bull. Geol. Surv. S. Rhodesia 58.* 147 pp.

Wilson, J. F. 1973. The Rhodesian Archaean craton—an essay in cratonic evolution. *Philos. Trans. R. Soc. London Ser. A* 273:389–411

Wilson, J. F., Bickle, M. J., Hawkesworth, C. J., Martin, A., Nisbet, E. G., Orpen, J. L. 1978. Granite-greenstone terrains of the Rhodesian Archaean craton. *Nature* 271:23–27

Windley, B. F. 1973. Crustal development in the Precambrian. *Philos. Trans. R. Soc. London Ser. A* 273:321–41

Windley, B. F. ed. 1976. *The Early History of the Earth.* London: Wiley. 620 pp.

Windley, B. F., Naqvi, S. M. 1978. *Archaean Geochemistry.* Amsterdam: Elsevier. 406 pp.

Ann. Rev. Earth Planet. Sci. 1980. 8: 169–90
Copyright © 1980 by Annual Reviews Inc. All rights reserved

THE DYNAMICS OF SUDDEN STRATOSPHERIC WARMINGS[1]

✕10129

James R. Holton
Department of Atmospheric Sciences, University of Washington, Seattle, Washington 98195

INTRODUCTION

The normal global-scale circulation of the stratosphere is dominated by a zonally (longitudinally) symmetric mean flow that is westerly (*from* the west) in the winter hemisphere and easterly (from the east) in the summer hemisphere. Superposed on this zonal mean flow are longitudinally varying wave perturbations referred to as *planetary waves*. Both the zonal mean flow and the planetary waves are in approximate *geostrophic balance* (i.e. the horizontal winds blow parallel to the height contours on constant-pressure surfaces, with speed proportional to the height gradient, and with low heights on the left facing downstream in the northern hemisphere). Thus, the large-scale horizontal flow can be represented by maps of the topography of constant-pressure surfaces (see Figure 1). In the summer hemisphere the wave disturbances tend to be weak and transient so that the time mean circulation is to a good approximation an axially symmetric *anticyclonic* (clockwise rotating) vortex centered on the pole. The height contours on constant-pressure surfaces are nearly parallel to latitude circles with height increasing toward the pole. In the winter hemisphere, on the other hand, there are substantial quasi-stationary planetary waves of zonal wavenumbers 1 and 2 (1 and 2 wavelengths around a latitude circle). Thus, height contours in the cyclonic winter polar vortex depart significantly from axial symmetry due primarily to the "Aleutian high" formed by the planetary waves.

Occasionally, the amplitudes of the planetary wave disturbances in the winter hemisphere increase dramatically over the course of several days.

[1] Contribution No. 510, Department of Atmospheric Sciences, University of Washington, Seattle.

0084-6597/80/0515-0169$01.00

The normal cyclonic polar vortex then becomes highly distorted as shown, for example, in Figure 1, which illustrates the major stratospheric warming of January–February, 1979 (Quiroz 1979). Accompanying this wave amplification are strong poleward fluxes of heat due to the positive correlation between the perturbation temperatures and northward velocity components associated with the waves. The polar warming that occurs as a result of these anomalous fluxes is referred to as a *sudden stratospheric warming*. Because the atmosphere is in hydrostatic balance, the polar warming increases the *thickness* of the column of air between two constant-pressure surfaces. Thus, the warming raises the height of the upper-level pressure surfaces in the polar region. If sufficient warming occurs, this process may reverse the mean meridional pressure gradient so that the normal westerly (cyclonic) flow in the polar night is replaced by easterly (anticyclonic) flow as in Figure 1*c*. By internationally agreed convention a warming event in which the wind reversal penetrates to the 10-mb level (about 31 km) or below is referred to as a *major* stratospheric warming. In a typical major warming the temperature at the 10-mb level at the North Pole may increase by 40–60°C in less than a week, while locally over limited regions increases of more than 80°C are often recorded. Major warmings apparently occur only in the northern hemisphere, and only once every other year or so. Warmings of a less intense nature (*minor* warmings) occur fairly commonly throughout the winter in both hemispheres.

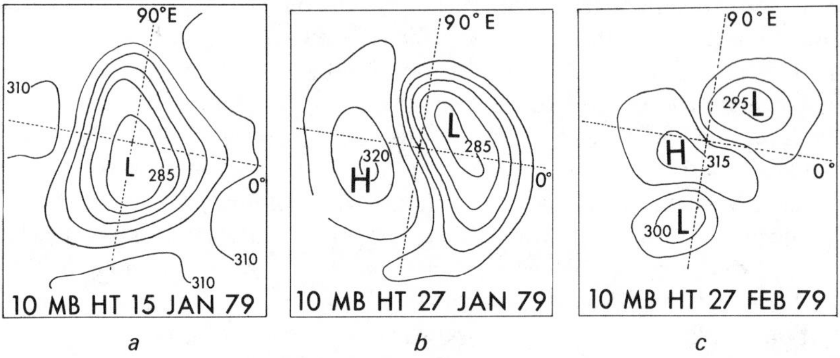

Figure 1 Evolution of the topography of the Northern Hemisphere 1-kPa (10-mb) height field during early 1979. Height contours are labeled in 100's of meters and are plotted at 500-m intervals. *L* and *H* denote height minima and maxima, respectively. The region poleward of approximately 20°N is depicted; 0° and 90°E longitude lines are labeled. Frame (*a*) depicts the normal winter polar vortex, (*b*) depicts a prewarming phase with a strong wavenumber one disturbance, (*c*) depicts a postwarming situation with a weak anticyclonic circulation in the polar region and a strong wavenumber two disturbance. Figure courtesy of R. S. Quiroz.

Sudden stratospheric warmings play an important role in the overall budgets of heat, momentum, energy, and trace constituents in the stratosphere. It has been known for many years (Reed et al 1963) that warmings occur in conjunction with enhanced wave energy fluxes from the troposphere to the stratosphere, which in turn cause amplification of the planetary waves in the stratosphere. The strong increases in poleward eddy heat fluxes due to the amplifying waves have traditionally been viewed as the primary cause of the sudden warmings. O'Neill & Taylor (1979) have recently shown, however, that meridional eddy momentum fluxes may also play a crucial role in some observed warmings.

The heat and momentum budgets of the polar stratosphere during sudden warmings cannot, however, be understood merely in terms of the anomalous eddy fluxes associated with the planetary waves. The zonally averaged meridional and vertical motions (which together compose the so-called *mean meridional circulation*) must also be considered. In fact, diagnostic studies (e.g. Mahlman 1969) have shown that during sudden warmings the heating due to the eddy heat flux convergence in the polar zone is partly balanced by cooling due to mean rising motion and consequent adiabatic expansion. Similarly, the zonal wind decelerations that accompany the polar warmings result from a small imbalance between the Coriolis force due to the mean meridional flow and the eddy momentum flux convergence.

Thus, a physical understanding of the sudden warming process in terms of conventional Eulerian dynamics requires that we account for the *signs* of the slight imbalances between forcing by the eddies and forcing by the mean meridional circulation. A primary object of this review is to show that, through using a hybrid Eulerian-Lagrangian theory, the apparent paradox of the mean meridional circulation nearly balancing the eddy forcing can be resolved, and the immediate cause of the observed warmings can be clearly elucidated.

Observational aspects of the sudden stratospheric warmings have recently been reviewed in depth in McInturff (1978). Briefer reviews and extensive bibliographies may also be found in Schoeberl (1978) and Quiroz et al (1975). Readers interested in climatological and synoptic aspects of the sudden warmings should consult these reviews. The present review focuses almost entirely on the dynamics of sudden warmings.

Sudden warmings provide a particularly dramatic illustration of the forcing of mean flows by wave motions. The topic of wave–mean flow interaction is an area in which much progress has been made within the last few years, primarily due to the work of Andrews & McIntyre (1976, 1978). (A good introduction to the subject is contained in McIntyre 1979a.) Andrews & McIntyre have shown that the dynamics of wave–mean flow interaction can be elucidated most effectively if the wave disturbances are described in terms of fluid particle displacements about the mean flow, rather than in terms of the traditional Eulerian perturbation velocity fields. This approach, which Andrews & McIntyre call the "generalized Lagrangian-mean theory" (it is actually a hybrid Eulerian-Lagrangian theory), provides clear physical descriptions of wave-induced mean flow acceleration and of wave transport processes in general, although there are limitations in the applicability of the theory to large amplitude waves. In this review we discuss the dynamics of the sudden warmings using both the Eulerian-mean and the Lagrangian-mean methodologies so that the reader may gain an appreciation of the relative merits of the two approaches.

DYNAMICAL BACKGROUND

In order to keep the mathematical development as simple as possible we will not attempt to model the stratosphere quantitatively, but will consider, rather, a prototype fluid system that qualitatively illustrates the important dynamical processes occurring in the stratosphere. The motion will be referred to a so-called midlatitude Cartesian β-plane with the (x, y, z) coordinate axes directed eastward, northward, and upward, respectively. The local vertical component of the planetary vorticity, expressed by the *Coriolis parameter* $f \equiv 2\Omega \sin \phi$ (where Ω is the angular speed of rotation of the earth and ϕ is latitude) will be treated as a constant everywhere except when it is differentiated with respect to y in the vorticity equation, in which case $\mathrm{d}f/\mathrm{d}y \equiv \beta$ will be considered constant. The β-plane approximation, which is used widely in dynamic meteorology, allows us to avoid the geometrical complexity of spherical coordinates while still retaining the dynamical effects due to the latitudinal gradient of the Coriolis parameter.

We will adopt the *Boussinesq approximation* in which the density is assumed to be a constant (ρ_0) everywhere except when coupled with gravity in the buoyancy force. The dynamical equations can then be written using as dependent variables the three Cartesian velocity components (u, v, w), pressure divided by the reference density (p), and the buoyancy acceleration (θ) defined as $\theta \equiv -\rho' g/\rho_0$, where ρ' is the deviation of density from the reference value ρ_0 and g is the acceleration of gravity. (The θ field in this prototype fluid is analogous to potential temperature in the atmosphere.)

Eulerian-Mean Equations

In the Eulerian approach the dependent fields, u,v,w,p,θ, are each split into two portions, a zonally averaged portion designated by an overbar, and the deviation from the zonal average, designated by a prime. Thus, for example, $u = \bar{u}(y, z, t) + u'(x, y, z, t)$. In the following analysis we assume that the perturbations (i.e. the planetary waves) are of sufficiently small amplitude so that we may linearize the perturbation equations. Following Andrews & McIntyre (1976) we designate the magnitude of the wave amplitude by the symbol a. Eulerian eddy fluxes, which are quadratic in the perturbation fields, will then have magnitudes $O(a^2)$.

Neglecting terms that are small for large-scale motions in midlatitudes, the perturbation equations may be written correct to $O(a)$ as follows:

$$\frac{\mathrm{D}u'}{\mathrm{D}t} - \left(f - \frac{\partial \bar{u}}{\partial y}\right)v' + \frac{\partial p'}{\partial x} = -X' \tag{1}$$

$$\frac{\mathrm{D}v'}{\mathrm{D}t} + fu' + \frac{\partial p'}{\partial y} = -Y' \tag{2}$$

$$\theta' = \partial p'/\partial z \tag{3}$$

$$\frac{\mathrm{D}\theta'}{\mathrm{D}t} + v'\frac{\partial \bar{\theta}}{\partial y} + w'N^2 = Q' \tag{4}$$

$$\frac{\partial u'}{\partial x} + \frac{\partial v'}{\partial y} + \frac{\partial w'}{\partial z} = 0 \tag{5}$$

where (1)–(5) are the x and y components of the momentum equation, the hydrostatic equation, the thermodynamic energy equation, and the continuity equation, respectively. Here, $\mathrm{D}/\mathrm{D}t \equiv \partial/\partial t + \bar{u}\partial/\partial x$ designates the rate of change following the mean zonal flow; X' and Y' designate mechanical dissipation of the perturbations, $N^2 \equiv \partial\bar{\theta}/\partial z$ is the square of the *buoyancy frequency*, which is assumed to be constant, and Q' is the buoyancy generation by diabatic processes.

The corresponding zonally averaged equations correct to $O(a^2)$ are

$$\frac{\partial \bar{u}}{\partial t} = f\bar{v} - \frac{\partial}{\partial y}(\overline{u'v'}) \tag{6}$$

$$+f\bar{u} = -\frac{\partial \bar{p}}{\partial y} \tag{7}$$

$$\bar{\theta} = \frac{\partial \bar{p}}{\partial z} \tag{8}$$

$$\frac{\partial \bar{\theta}}{\partial t} + \bar{w}N^2 = -\frac{\partial}{\partial y}(\overline{v'\theta'}) + \bar{Q} \tag{9}$$

$$\frac{\partial \bar{v}}{\partial y} + \frac{\partial \bar{w}}{\partial z} = 0. \tag{10}$$

The meridional momentum equation (7) expresses the fact that the mean zonal flow $\bar{u}$ is to a good approximation in geostrophic balance with the mean meridional pressure gradient. Combining (7) and (8) we obtain the well-known *thermal wind* relationship,

$$f\frac{\partial \bar{u}}{\partial z} = -\frac{\partial \bar{\theta}}{\partial y}, \tag{11}$$

which states that the vertical shear of the mean zonal wind is proportional to the mean meridional buoyancy gradient.

Quasi-Geostrophic Theory

The mean zonal wind is not the only portion of the flow that is in approximate geostrophic balance. It turns out that the perturbation horizontal velocity components are also nearly geostrophic so that $fu' \simeq -\partial p'/\partial y$ and $fv' \simeq \partial p'/\partial x$. A theoretical framework has been developed to utilize the geostrophic nature of the large-scale flow, while retaining those aspects of the departures from geostrophy that are dynamically important. This theory is referred to as the "quasi-geostrophic" theory (see Holton 1972, 1975, for detailed treatments).

The first step necessary for development of the quasi-geostrophic theory is to form the vertical component of the perturbation vorticity equation by taking $\partial(2)/\partial x - \partial(1)/\partial y$ to obtain

$$\frac{\mathrm{D}}{\mathrm{D}t}\left(\frac{\partial v'}{\partial x} - \frac{\partial u'}{\partial y}\right) + \left(f - \frac{\partial \bar{u}}{\partial y}\right)\left(\frac{\partial u'}{\partial x} + \frac{\partial v'}{\partial y}\right) + \left(\beta - \frac{\partial^2 \bar{u}}{\partial y^2}\right)v' = -\left(\frac{\partial Y'}{\partial x} - \frac{\partial X'}{\partial y}\right). \tag{12}$$

We next differentiate (4) with respect to z and substitute from (11) to obtain

$$\frac{\mathrm{D}}{\mathrm{D}t}\left(\frac{\partial^2 p'}{\partial z^2}\right) - f\frac{\partial^2 \bar{u}}{\partial z^2}v' + \frac{\partial w'}{\partial z}N^2 = \frac{\partial Q'}{\partial z}. \tag{13}$$

If we now approximate the relative vorticity in (12) geostrophically by letting

$$\frac{\partial v'}{\partial x} - \frac{\partial u'}{\partial y} = \frac{1}{f}\left(\frac{\partial^2 p'}{\partial x^2} + \frac{\partial^2 p'}{\partial y^2}\right)$$

and neglect $\partial \bar{u}/\partial y$ compared with f in the second term on the left, we may combine (12) and (13) with the aid of the continuity equation (5) to obtain the perturbation quasi-geostrophic potential vorticity equation,

$$\frac{\mathrm{D}q'}{\mathrm{D}t} + v'\frac{\partial \bar{q}}{\partial y} = -S', \tag{14}$$

where

$$q' \equiv \frac{1}{f}\left(\frac{\partial^2 p'}{\partial x^2} + \frac{\partial^2 p'}{\partial y^2}\right) + \frac{f}{N^2}\frac{\partial^2 p'}{\partial z^2},$$

$$\frac{\partial \bar{q}}{\partial y} \equiv \beta - \frac{\partial^2 \bar{u}}{\partial y^2} - \frac{f^2}{N^2}\frac{\partial^2 \bar{u}}{\partial z^2},$$

and

$$S' = \left(\frac{\partial Y'}{\partial x} - \frac{\partial X'}{\partial y}\right) - \frac{f}{N^2}\frac{\partial Q'}{\partial z}.$$

Combining the zonal mean equations (6)–(10) in an analogous fashion we obtain[2]

$$\frac{\partial \bar{q}}{\partial t} = -\frac{\partial}{\partial y}(\overline{q'v'}) - \overline{S} \tag{15a}$$

where

$$\bar{q} \equiv \frac{1}{f}\frac{\partial^2 \bar{p}}{\partial y^2} + \frac{f}{N^2}\frac{\partial^2 \bar{p}}{\partial z^2} + f, \tag{15b}$$

$$\overline{q'v'} = -\frac{\partial}{\partial y}(\overline{u'v'}) + \frac{f}{N^2}\frac{\partial}{\partial z}(\overline{v'\theta'}), \tag{15c}$$

and

$$\overline{S} = -\frac{f}{N^2}\frac{\partial \overline{Q}}{\partial z}.$$

Equations (14) and (15) are the perturbation and zonal mean quasi-geostrophic *potential vorticity* equations.

Vertical Wave Propagation

Before proceeding further it is useful for future reference to consider solutions of the wave equation (14) for zonally propagating disturbances of wave number k and phase speed c. For simplicity we assume that the mean wind shear is small enough so that $\partial\bar{q}/\partial y \simeq \beta$. Letting

$$p' = P_0(z) \exp\left[ik(x - ct)\right]$$

[2] The perturbation velocities u' and v' are evaluated geostrophically in (14) and (15).

we find by substitution into (14) that $P_0(z)$ must satisfy the vertical structure equation

$$d^2P_0/dz^2 + m^2P_0 = 0$$

where

$$m^2 \equiv [\beta/(\bar{u}-c) - k^2](N^2/f^2).$$

Vertical propagation requires $m^2 > 0$. Thus, we obtain the Charney-Drazin (1961) criterion which states that planetary waves forced from below will propagate vertically only if[3]

$$0 < \bar{u} - c < \beta/k^2. \tag{16}$$

Thus, planetary waves can propagate vertically only if the waves propagate *westward* relative to the mean wind, but at a speed less than β/k^2. Stationary waves ($c = 0$) can propagate vertically only when the mean wind is westerly and the wavelength long enough so that $\beta/k^2 > \bar{u}$. This propagation condition explains both why large amplitude stationary waves are found only in the winter hemisphere in the stratosphere and why only the longest planetary wave modes (zonal wavenumbers 1 and 2) have significant amplitude in the stratosphere. The easterly ($\bar{u} < 0$) winds of the summer hemisphere do not permit vertical propagation for any stationary planetary waves while short wavelength modes can not propagate even in westerly winds. As we shall see in Section 3 the blockage of wave propagation at the *critical level* where $\bar{u} - c = 0$ is important for the evolution of sudden stratospheric warnings.

The Charney-Drazin Theorem

Equations (14) and (15) constitute a special case of the more general law of conservation of potential vorticity, which states that for adiabatic, inviscid motions the ratio of the depth of a fluid column to the vertical component of absolute vorticity is conserved following the motion. For conditions under which (14) and (15) are valid (quasi-geostrophic flow and small amplitude waves) the wave–mean flow interaction problem is reduced to the analysis of a single scalar field, the potential vorticity q.

If the zonal mean source term $\overline{S}$ vanishes, (15) immediately shows that the zonal mean potential vorticity can only change if there is potential vorticity transport by the eddies ($\overline{q'v'} \neq 0$). If we multiply (14) by q and average the resulting equations zonally we obtain[4]

[3] In a compressible atmosphere the criterion becomes $0 < \bar{u} - c < \beta(k^2 + f^2/4H^2N^2)^{-1}$ where H is the scale height of the atmosphere.

[4] The potential vorticity flux was first expressed in the form (17) by Holton & Dunkerton (1978).

$$\overline{q'v'} = -\left[+\frac{1}{2}\frac{\partial}{\partial t}\overline{q'^2} + \overline{S'q'}\right]\Big/\left(\frac{\partial \bar{q}}{\partial y}\right). \tag{17}$$

Thus, provided $\partial\bar{q}/\partial y \neq 0$, there will be potential vorticity transport by the eddies only if the eddies are transient (changing in amplitude) or subject to dissipation. For steady, conservative waves $\overline{q'v'} = 0$ and there is no net forcing of the mean flow by the eddies. This is the famous Charney-Drazin nonacceleration theorem (Charney & Drazin 1961, Dickinson 1969).

In summary, it should be clear already from this simple analysis that the existence of nonzero eddy heat and momentum fluxes does not necessarily mean that the waves will drive changes in the mean flow. It is only through a nonzero potential vorticity flux due to wave transience and/or damping that changes in $\bar{q}$ will occur, and since both $\bar{q}$ and $\bar{u}$ are determined by the distribution of $\bar{p}(y,z,t)$, a steady $\bar{q}$ in general implies a steady mean flow. (Precise conditions for the vanishing of $\partial\bar{u}/\partial t$ are discussed following Equation (22) below.)

Wave–Mean Flow Interaction; Eulerian Model

In the above section we presented the Charney-Drazin nonacceleration theorem in its most compact form, in terms of the quasi-geostrophic potential vorticity. For a physical understanding of the contributions of eddy heat and momentum fluxes to wave–mean flow interaction it is necessary, however, to return to the approximate zonal mean equations (6)–(10).

Observational studies and numerical simulations indicate that in the zonal mean momentum equation (6) there is an approximate balance between the Coriolis acceleration $-f\bar{v}$ and the eddy momentum flux divergence $\partial(\overline{u'v'})/\partial y$. Similarly, in the zonal mean energy equation (9) there is an approximate balance between the adiabatic cooling term $\bar{w}N^2$ and the eddy heat flux divergence $\partial(\overline{v'\theta'})/\partial y$. Thus, both the momentum and heat budgets are maintained by approximate balances between driving by the eddy fluxes and by the mean meridional circulation. This near cancellation of the eddy forcing by the mean meridional circulation is not simply a remarkable coincidence; it is, rather, an inevitable consequence of the geostrophic and hydrostatic balance of the mean zonal flow expressed in (7) and (8). The twin constraints of geostrophic and hydrostatic balance imply that the mean zonal wind and buoyancy fields are not independent of each other, but are closely coupled through the thermal wind relationship (11). By changing the $\bar{u}$ and $\bar{\theta}$ fields, the eddy heat and momentum fluxes tend to destroy the thermal wind balance. In the process they generate unbalanced pressure gradient forces which immediately generate a mean meridional flow $(\bar{v}, \bar{w})$ which returns the zonal mean circulation to thermal wind balance.

Thus, the mean meridional circulation is to a large extent an *eddy-driven circulation* (see Holton 1972, pp. 228–34, for a more detailed discussion). It cannot be considered as an entity entirely separate from the waves. The splitting of the flow between eddy fluxes and the mean meridional circulation is simply an artifice of the Eulerian-mean approach to the dynamics. Although the Eulerian approach is simple to apply in observational and modeling studies, great care is required because, as mentioned earlier, even in sudden warming conditions the mean field changes $\partial\bar{u}/\partial t$ and $\partial\bar{\theta}/\partial t$ may be given by relatively small differences between two large terms.

An improvement on the traditional Eulerian scheme was suggested by Andrews & McIntyre 1976. They introduced a "residual" mean meridional circulation $(\bar{v}^*, \bar{w}^*)$ defined by

$$\bar{v}^* = \bar{v} - \frac{\partial}{\partial z}\left(\frac{\overline{v'\theta'}}{N^2}\right); \quad \bar{w}^* = \bar{w} + \frac{\partial}{\partial y}\left(\frac{\overline{v'\theta'}}{N^2}\right). \tag{18}$$

Substituting from (18) into (6) and (9) and using (15c) we obtain

$$\frac{\partial\bar{u}}{\partial t} = f\bar{v}^* - \frac{\partial}{\partial y}(\overline{u'v'}) + f\frac{\partial}{\partial z}\left(\frac{\overline{v'\theta'}}{N^2}\right) = f\bar{v}^* + \overline{q'v'} \tag{19}$$

and

$$\frac{\partial\bar{\theta}}{\partial t} = -\bar{w}^* N^2 + \bar{Q}. \tag{20}$$

We also note from (18) and (10) that

$$\frac{\partial v^*}{\partial y} + \frac{\partial w^*}{\partial z} = 0. \tag{21}$$

Using (11) to eliminate the time derivatives in (19) and (20) by taking $f\partial^2(19)/\partial z^2 + \partial^2(20)/\partial y\partial z$ we obtain with the aid of (21),

$$\frac{\partial^2\bar{v}^*}{\partial y^2} + \frac{f^2}{N^2}\frac{\partial^2\bar{v}^*}{\partial z^2} = -\frac{f}{N^2}\frac{\partial^2}{\partial z^2}(\overline{q'v'}) - \frac{1}{N^2}\frac{\partial^2\bar{Q}}{\partial y\partial z}. \tag{22}$$

Thus, if $\bar{Q}$ and $\overline{q'v'}$ vanish and $\bar{v}^*$ also vanishes on the boundaries, $\bar{v}^*$ must vanish everywhere and from (19) $\partial\bar{u}/\partial t = 0$, which is the complete statement of the Charney-Drazin nonacceleration theorem.

From (20) it is clear that the residual mean vertical velocity $\bar{w}^*$ is just that part of the $\bar{w}$ field which is not cancelled by the eddy heat fluxes. Although in theory it might be possible for the forcing by the potential vorticity flux to cancel the forcing due to diabatic heating, the two processes are independent so that such cancellation would be fortuitous. Thus, the existence of nonzero $\overline{q'v'}$ and/or nonzero $\overline{Q}$ in practice always drives mean flow accelerations. However, if the residual circulation is forced primarily by diabatic heating there will be a tendency for the $\bar{w}^*$ field to balance the diabatic forcing in order to keep the mean zonal wind and temperature fields in thermal wind balance. In such situations the residual circulation is referred to as the "diabatic circulation." Dunkerton (1978) has shown that this circulation is more relevant to the meridional and vertical transport of tracers in the stratosphere than is the Eulerian-mean meridional circulation.

In the case of a sudden warming, however, the forcing is primarily due to the nonzero potential vorticity flux caused by wave transience. The residual circulation is then itself a *wave-driven* circulation required to maintain the thermal wind balance in the presence of the mean flow acceleration driven by $\overline{q'v'}$.

Lagrangian-Mean Theory

As the previous section has indicated, the Eulerian-mean approach to wave–mean flow interaction problems leads to a rather artificial distinction between "wave" and "mean flow" processes. An alternative more fundamental approach is the so-called *generalized Lagrangian-mean* theory of Andrews & McIntyre (1978). By focusing on fluid particle displacements produced by the wave disturbances this theory partitions the flow between eddy fluxes and mean meridional motions in a manner that is more natural than in the Eulerian-mean theory.

In the Lagrangian-mean description, small amplitude wave motions are characterized by introducing a particle displacement vector (ξ', η', ζ') which specifies the location of a fluid particle relative to the position the same particle would have in the absence of the wave.[5] The components of the displacement vector $\xi'(x,y,z,t)$, $\eta'(x,y,z,t)$, $\zeta'(x,y,z,t)$ are themselves treated as Eulerian field variables, which distinguishes the Andrews & McIntyre theory from traditional Lagrangian approaches. For small amplitude waves the displacement field is related to the perturbation velocity field (u',v',w') as follows:

$$\frac{D\xi'}{Dt} = u' + \eta' \frac{\partial \bar{u}}{\partial y} + \zeta' \frac{\partial \bar{u}}{\partial z} \tag{23a}$$

[5] Andrews & McIntyre (1978) discuss how this definition is modified for waves of finite amplitude.

$$\frac{\mathrm{D}\eta'}{\mathrm{D}t} = v' \tag{23b}$$

$$\frac{\mathrm{D}\zeta'}{\mathrm{D}t} = w' \tag{23c}$$

where as before $\mathrm{D}/\mathrm{D}t = \partial/\partial t + \bar{u}\partial/\partial x$ is the rate of change following the mean zonal flow. The terms $\eta'\ \partial\bar{u}/\partial y$ and $\zeta'\ \partial\bar{u}/\partial z$ must be included in the first of these expressions since deviations are being measured relative to the mean at the undisplaced position, rather than the instantaneous position.

Corresponding to the displacement field is a generalized Lagrangian mean, which Andrews & McIntyre (1978) designate by $\overline{(\ )}^{\mathrm{L}}$, and which differs from the Eulerian mean $\overline{(\ )}$ in that $\overline{(\ )}^{\mathrm{L}}$ is an average along the wavy material line defined by the particle displacement field while $\overline{(\ )}$ is the average along a line parallel to the x-axis. Thus, the Lagrangian mean is an average that is taken over the same fluid parcels as the motion evolves in time, while the Eulerian mean is an average over those particles that happen to lie along a given line parallel to the x-axis at any instant. Therefore, as pointed out by Matsuno & Nakamura (1979), the Lagrangian-mean motion is the motion of the *center of mass* of a wavy material line which would be parallel to the x-axis in the absence of the wave motion. As defined here the Lagrangian mean must, therefore, be initialized using an undisturbed reference state (actual or hypothetical). Under sudden warming conditions the displacements become too large for the small amplitude theory to apply quantitatively, and there are a number of technical difficulties associated with applying the Lagrangian scheme in practice (McIntyre 1979b). However, the small amplitude theory is still adequate for a qualitative discussion of wave–mean flow interaction processes, which is the goal in this review.

Matsuno & Nakamura (1979) have shown that for small amplitude steady wave disturbances[6] the Lagrangian-mean versions of the zonal momentum, thermodynamic energy, and continuity equations are

$$\frac{\partial \bar{u}^{\mathrm{L}}}{\partial t} - f\bar{v}^{\mathrm{L}} = -\left(\overline{\frac{\partial p}{\partial x}}\right)^{\mathrm{L}} \tag{24}$$

$$\frac{\partial \bar{\theta}^{\mathrm{L}}}{\partial t} + \bar{w}^{\mathrm{L}} N^2 = \bar{Q}^{\mathrm{L}} \tag{25}$$

[6] McIntyre (personal communication) has pointed out that not only must the wave velocity and pressure fields be steady, but the amplitude of particle displacements must also be steady for (24)–(26) to be valid.

$$\frac{\partial \bar{v}^{\mathrm{L}}}{\partial y} + \frac{\partial \bar{w}^{\mathrm{L}}}{\partial z} = 0 \tag{26}$$

which should be compared to (6), (9), and (10). In this approximation, the Lagrangian-mean version of the thermodynamic energy equation has a form identical to that of the transformed Eulerian form (20), consistent with our earlier statement that $\bar{w}^*$ was approximately equal to the Lagrangian-mean vertical motion.

In the momentum equation the eddy flux term present in the Eulerian version (6) is replaced by the *radiation stress divergence* $\overline{(\partial p/\partial x)}^{\mathrm{L}}$, which unlike its Eulerian counterpart, $\overline{(\partial p/\partial x)}$, does not in general vanish. For small amplitude disturbances the Lagrangian mean can be related to the Eulerian mean by the Taylor series expansion

$$\begin{aligned}\overline{A}^{\mathrm{L}} = \overline{A} + \overline{\xi' \frac{\partial A'}{\partial x}} + \overline{\eta' \frac{\partial A'}{\partial y}} + \overline{\zeta' \frac{\partial A'}{\partial z}} + \frac{1}{2}\,\overline{\eta'^2}\,\frac{\partial^2 \overline{A}}{\partial y^2} \\ + \overline{\eta'\zeta'}\,\frac{\partial^2 \overline{A}}{\partial y \partial z} + \overline{\zeta'^2}\,\frac{\partial^2 \overline{A}}{\partial z^2} + O(a^3).\end{aligned} \tag{27}$$

Letting $A = \partial p/\partial x$ in (27) and noting from (23) and (5) that

$$\frac{\partial \xi'}{\partial x} + \frac{\partial \eta'}{\partial y} + \frac{\partial \zeta'}{\partial z} = 0$$

we obtain

$$\begin{aligned}\overline{\left(\frac{\partial p}{\partial x}\right)}^{\mathrm{L}} &= \frac{\partial}{\partial y}\overline{\left(\eta' \frac{\partial p'}{\partial x}\right)} + \frac{\partial}{\partial z}\overline{\left(\zeta' \frac{\partial p'}{\partial x}\right)} \\ &= -\frac{\partial}{\partial y}\overline{\left(p' \frac{\partial \eta'}{\partial x}\right)} - \frac{\partial}{\partial z}\overline{\left(p' \frac{\partial \zeta'}{\partial x}\right)}.\end{aligned} \tag{28}$$

Thus, the radiation stress divergence depends on the meridional and vertical gradients of the correlations between the perturbation pressure field and the slopes of the meridional and vertical displacements, respectively.

Comparing (24) with (19) we see that the radiation stress divergence is closely related to the potential vorticity flux of the Eulerian-mean theory. The momentum equation (24), unlike the thermodynamic energy equation, explicitly retains disturbance quantities in its Lagrangian mean form, namely the radiation stress divergence. As we shall see in the next section, the radiation stress divergence in fact provides the most fundamental measure of mean flow acceleration by the waves.

MATSUNO'S MODEL OF THE SUDDEN WARMINGS

In the previous section we have developed those aspects of wave–mean flow interaction theory required for a theoretical treatment of the sudden warmings. In particular, we have found that wave transience and/or dissipation are essential if planetary waves are to force changes in the mean flow. In this section we describe, using both the Eulerian and Lagrangian formalisms, how such processes can operate to produce the sudden warmings.

Matsuno's Eulerian Quasi-Geostrophic Model

Our current understanding of the dynamics of sudden warmings is based primarily on the classic work of Matsuno (1971), who used a quasi-geostrophic numerical model to simulate a sudden stratospheric warming. In addition to numerical calculations, Matsuno included a heuristic model which (although it ignores meridional wave propagation) provides an interesting conceptual guide to the warming process.

Matsuno considered the idealized case of a stationary planetary wave with positive buoyancy flux ($\overline{\theta' v'} > 0$) and vanishing momentum flux ($\overline{u'v'} = 0$) which is incident on a critical level ($\bar{u} = 0$) in the upper stratosphere. Now, according to the Charney-Drazin condition, a stationary planetary wave propagating vertically in a mean westerly wind ($\bar{u} > 0$) will be trapped at a critical level so that its amplitude must go to zero in the easterly ($\bar{u} < 0$) regime above the critical level. The buoyancy flux $\overline{\theta' v'}$ associated with the wave must, therefore, decrease from a finite positive value to zero at the critical level. Since the buoyancy flux divergence $\partial\overline{\theta' v'}/\partial y$ is nearly balanced by adiabatic cooling due to mean rising motion ($\bar{w} > 0$) the singularity in the vertical derivative of the buoyancy flux at the critical level will cause the wave-driven $\bar{w}$ to decrease

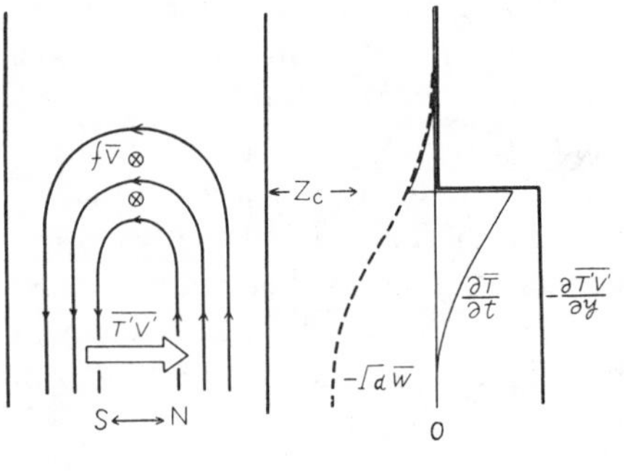

Figure 2 Schematic vertical profiles of the poleward eddy heat flux $\overline{V'T'}$, local temperature change $\partial\bar{T}/\partial t$, eddy heat flux divergence $-\partial(\overline{V'T'})/\partial y$, adiabatic heating $-\Gamma_d \bar{w}$ [Γ_d is analogous to N^2 in Equation (9)], and the Coriolis force due to the induced mean meridional motion $f\bar{V}$, for a planetary wave incident on a critical level Z_c. After Matsuno (1971).

rapidly with height and hence, through mass continuity, will drive a strong equatorward mean meridional flow ($\bar{v} < 0$) near the critical level (Figure 2). This mean meridional flow will in turn, through the action of the Coriolis force, induce a strong easterly acceleration ($\partial \bar{u}/\partial t < 0$) which will cause the critical level to descend in altitude. Maintenance of the thermal wind balance requires that there be warming below the critical level in the polar region as shown in Figure 3. Also shown in the figure is a pattern of compensating cooling in the equatorial region below the critical level and in the polar region above the critical level. Indeed observations do indicate that weak equatorial stratospheric cooling and polar mesospheric cooling accompany sudden stratospheric warmings.

One weakness of this heuristic description of the sudden warmings is that normally—until a warming is well under way—the winds are westerly throughout the polar winter stratosphere and mesosphere so that no critical level exists. Matsuno, however, pointed out that wave transience can provide an analogous mean flow deceleration mechanism. Numerous observational studies have shown that enhanced amplitudes of planetary wavenumbers 1 or 2 must precede the occurrence of a sudden warming. If the wave amplitude at the tropopause is increased at

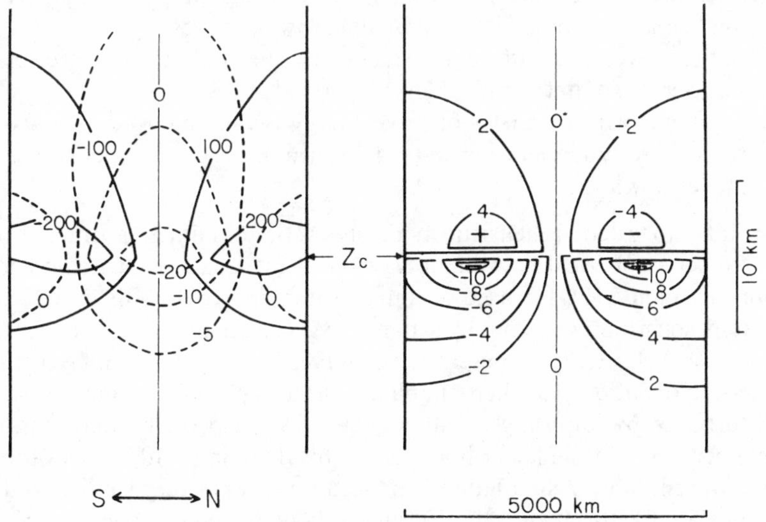

Figure 3 Schematic zonal mean field changes induced by the meridional circulation shown in Figure 2. The left-hand diagram shows changes in the isobaric heights (solid lines, m day^{-1}) and changes in the mean zonal wind (dashed lines, m s^{-1} day^{-1}). The right-hand diagram shows temperature changes (°C day^{-1}). The amplitude of the height perturbation of the wave incident on the critical level is assumed to be 500 m. After Matsuno (1971).

a time $t = 0$, wave energy will propagate upward at the vertical group velocity W_g so that at a time $t = t_0$ the increased amplitude, and associated increase in the buoyancy flux, will have penetrated to the level $z = W_g t_0$. Above that level the buoyancy flux must decrease with height, and an easterly mean flow acceleration will occur due to the same process that occurs at a critical level. In the transient wave situation, however, the mean flow deceleration and warming will be spread out over a deep layer rather than being concentrated in a thin layer near the critical level.

If the transient wave amplification is of sufficient amplitude and duration the mean wind deceleration which it drives may cause the mean wind to vanish at some level. At that point, according to Matsuno's argument, the critical level mechanism should apply and a rapid warming and descent of the $\bar{u} = 0$ line should occur.

In summary, Matsuno's heuristic model of the sudden warmings consists of the following sequence of events:

1. Quasi-stationary planetary waves of zonal wavenumbers 1 or 2 become anomalously large in the troposphere;
2. The growing waves propagate into the stratosphere;
3. The waves decelerate the mean winds causing the polar night jet to weaken and become distorted by the growing waves;
4. If the waves are sufficiently strong the mean flow may decelerate sufficiently so that a critical level is formed;
5. Further upward transfer of wave energy is then blocked and a very rapid easterly acceleration and polar warming occurs as the critical level moves downward.

Matsuno tested his heuristic model in a series of numerical simulations based on a spherical coordinate version of the quasi-geostrophic equations (14) and (15). He used an initial value approach with initial conditions consisting of a zonally symmetric geostrophically balanced mean flow $\bar{u}(y,z)$ chosen to represent the observed zonal mean winter stratospheric circulation. He then simulated the growth of a planetary wave disturbance by imposing an amplifying pressure perturbation of wavenumber 1 or 2 at the lower boundary of his domain ($\simeq 10$ km elevation). The wavenumber 2 simulation produced a warming very similar to that shown in Figure 1*c* for the February 1979 warming. After an initial growth period of about 10 days, the forced wavenumber 2 began to rapidly decelerate the mean flow, the polar vortex became distorted, and by day 20 had completely broken down to be replaced by a weak polar anticyclone. The temperature rise at the pole eventually exceeded 80°C in this simulation.

In many respects Matsuno's numerical calculations succeeded in simulating the observed sudden warmings, and at least superficially supported his heuristic model described above. However, a closer analysis of his numerical results, as well as recent observational studies, indicates that the actual dynamics of the sudden warming phenomenon is considerably more complex than the sequence of events postulated by Matsuno, primarily because meridional wave propagation as well as vertical propagation is involved.

Holton (1976) repeated Matsuno's simulations with a slightly different model. He found that even during the most rapid warming period there was a strong horizontal momentum flux divergence which nearly balanced the Coriolis acceleration so that the actual mean flow acceleration was a small residual as shown in Figure 4. The importance of horizontal eddy momentum fluxes has also been demonstrated observationally by Johnson (1977) and O'Neill & Taylor (1979).

When both momentum and heat fluxes are present it is no longer possible to deduce the sign of the local mean flow acceleration from the

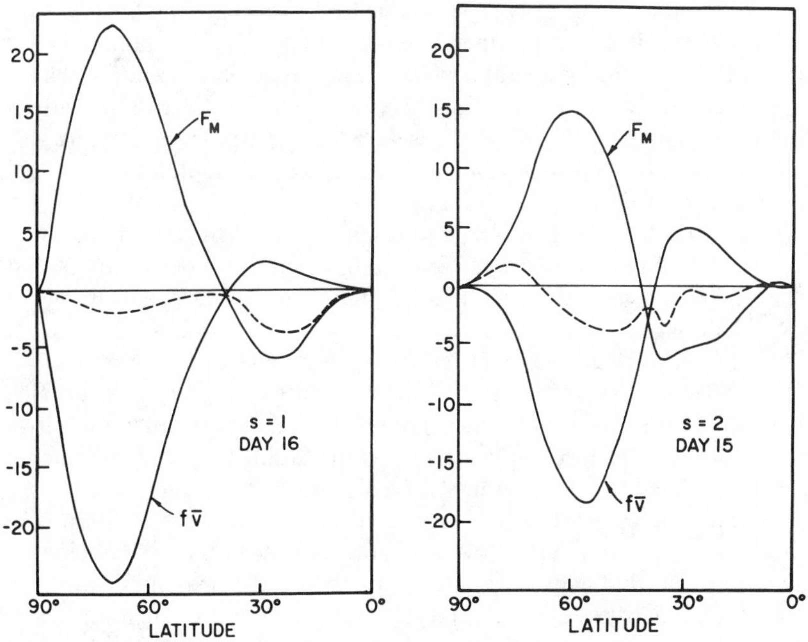

Figure 4 Horizontal eddy momentum flux convergence $F_M \equiv -\partial(\overline{u'v'})/\partial y$ and Coriolis force $f\bar{v}$ plotted as functions of latitude at the 40-km level in simulated sudden warmings forced by wavenumbers 1 (*left panel*) and 2 (*right panel*). Net accelerations $\partial\bar{u}/\partial t$ shown by dashed curves in units of m s^{-1} day^{-1}. After Holton (1976).

vertical gradient of the buoyancy flux alone. However, a modification of Matsuno's heuristic Eulerian model may still be applied if we consider the *potential vorticity* flux. The potential vorticity flux (15c) involves both the horizontal momentum flux convergence and the vertical gradient of the heat flux. According to (19) a negative (equatorward) potential vorticity flux will decelerate the mean zonal flow. This is of course consistent with Matsuno's model since near the critical level, or in the presence of transient growth, $\partial\overline{v'\theta'}/\partial z < 0$ so that $\overline{q'v'} < 0$. The form of the mean flow equation (19) demonstrates, however, that it is not necessary to consider the heat and momentum fluxes separately, but merely the potential vorticity flux! Now, according to (17) the potential vorticity flux will be negative for transient growing waves, or for damped waves. Thus, regardless of the relative contributions of the momentum and heat flux terms, if the waves are amplifying, $\overline{q'v'} < 0$ (provided $\partial\bar{q}/\partial y > 0$) and the zonal flow will decelerate. Johnson (1977) in his study of the January 1973 warming found that prior to the warming the momentum and heat flux contributions to $\overline{q'v'}$ nearly cancelled but that in the period of rapid warming there was a broad region of strong negative potential vorticity flux primarily due to the vertical heat flux gradient. O'Neill & Taylor, on the other hand, found that in the January 1977 warming the largest temperature changes occurred in association with a strong equatorward momentum flux in high latitudes, but with a relatively small heat flux. In that case too the potential vorticity flux was equatorward over a broad region of the stratosphere so that the modified version of Matsuno's model still applied.

In summary, if Matsuno's conceptual model is recast in terms of potential vorticity fluxes it is clear that mean flow deceleration and polar warming are an inevitable result of anomalous planetary wave amplification in the winter stratosphere. This model does not, however, elucidate the particular circumstances in which momentum fluxes rather than heat fluxes might dominate in the potential vorticity flux. Finally, it should be emphasized that from (20)–(23) it can be seen that a negative potential vorticity flux implies a poleward "residual" mean meridional flow ($\bar{v}^* > 0$) and compensating subsidence in the polar region below the level of maximum negative $\overline{q'v'}$. Thus, since the residual circulation is to a first approximation equal to the Lagrangian-mean meridional circulation, it is clear that actual air parcel motion is poleward and downward. Therefore, regardless of the structures of the heat and momentum fluxes for any particular warming, it remains true that for a warming to occur there must be a *Lagrangian-mean subsidence* and accompanying adiabatic compressional warming in the polar region. As a corollary, ozone and other passive tracers will also be transported poleward and downward during a sudden warming, so that warmings play a significant role in stratospheric trace species budgets.

A Lagrangian-Mean Model

Matsuno & Nakamura (1979) have presented an alternative conceptual model of the sudden warming, based on the Lagrangian-mean equations (24)–(26), but again allowing only vertical propagation. Now, for planetary waves p' and η' are approximately in phase so that we may neglect the first term on the right in (28).[7] The zonal mean momentum equation thus becomes approximately

$$\frac{\partial \bar{u}^{\mathrm{L}}}{\partial t} = f\bar{v}^{\mathrm{L}} + \frac{\partial}{\partial z}\left(\overline{p'\frac{\partial \zeta'}{\partial x}}\right). \tag{29}$$

Comparing (19) and (29) we see that for small amplitude disturbances when $\bar{u}^{\mathrm{L}} \simeq \bar{u}$ and $\bar{v}^{\mathrm{L}} \simeq \bar{v}^*$, the potential vorticity flux is approximately equal to the vertical gradient of the radiation stress $\overline{p'\partial\zeta'/\partial x}$.

The sign of the radiation stress can be evaluated by considering the longitude height profile for a vertically propagating amplifying planetary wave as shown in Figure 5. The heavy lines indicate wavy material lines at two levels in the fluid. Because the wave propagates westward relative to the mean flow, the maximum upward displacement of a material line occurs 1/4 cycle to the east of the maximum upward motion. From the figure it is clear that $p' > 0$ where $\partial\zeta'/\partial x > 0$ so that the radiation stress $\overline{p'\partial\zeta'/\partial x} > 0$. Thus, the fluid below a wavy material surface exerts a *westward* directed force on the fluid above. If we consider the block of fluid contained between the two wavy material lines in Figure 5, then since the wave amplitude decreases with height, the westward force exerted on the fluid due to the radiation stress across the lower material surface will be greater than the eastward force due to the radiation stress across the upper material surface, and the fluid between the two material surfaces will experience a net westward acceleration. If the wave motion is undamped and has settled down to a steady state below the lower material line in the figure, the radiation stress gradient will vanish in that region. If the wave is orographically forced by flow over undulating topography, then the radiation stress is essentially the pressure drag exerted by the undulating topography. Thus, we see that for westerly flow (which allows vertical wave propagation) the drag exerted on the

[7] This is a consequence of the approximate geostrophic balance and is *not* the same as assuming $\overline{u'v'} = 0$.

atmosphere by orographic forcing may cause the mean flow to decelerate many scale heights above the surface in the stratosphere. The waves serve to transfer the momentum of the earth vertically over a great depth (Uryu 1974, McIntyre 1977).

The mean flow deceleration caused by the radiation stress divergence tends to destroy the geostrophic wind balance so that a Lagrangian-mean meridional motion must arise which is directed poleward (down the pressure gradient) in the layer where the deceleration is occurring. By continuity there must then be a vertical circulation which consists of Lagrangian-mean sinking (rising) below (above) the decelerating layer in the polar regions, with a reversed circulation in low latitudes. Thus, the

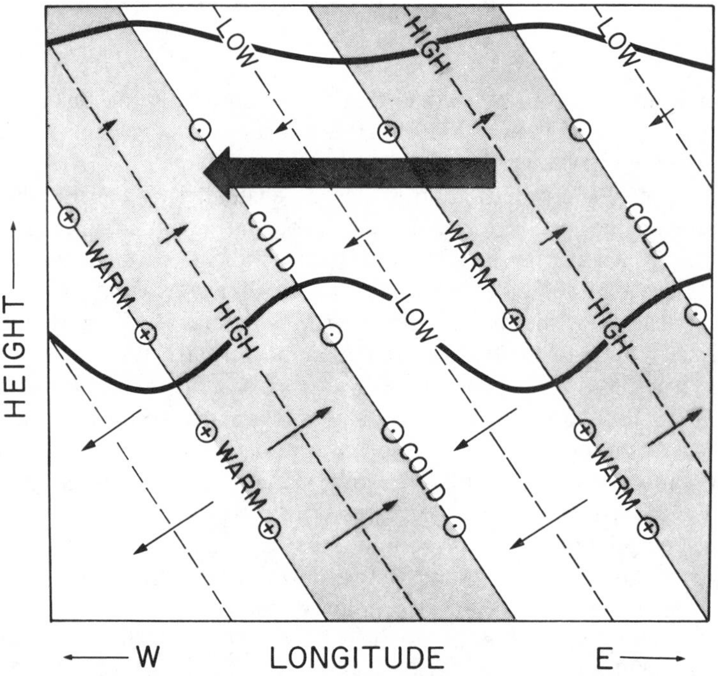

Figure 5 Longitude-height section showing pressure, temperature, and wind perturbations for an upward propagating stationary planetary wave in a region of mean westerly (eastward) zonal flow. Areas of high perturbation pressure are shaded. Small arrows indicate zonal and vertical wind perturbations with length proportional to wave amplitude. Meridional wind perturbations directed out of (into) page shown by circles with dots (crosses). Axes of temperature maxima and minima and pressure maxima and minima are labeled. Heavy wavy lines indicate material lines. The large shaded arrow indicates the net mean flow acceleration for the fluid contained in the region between the two wavy material lines due to the radiation stress divergence.

Lagrangian-mean model immediately explains the observed mesospheric high latitude and stratospheric low latitude coolings which are observed to occur in conjunction with the polar stratospheric warmings.

The Lagrangian-mean description of the sudden warming process given here is quite similar to the *modified* Eulerian description of the previous section. However, the Lagrangian-mean description relates the mean flow changes to the fundamental physical process—the radiation stress—rather than to a dynamical "tracer" (potential vorticity). Hence, the Lagrangian-mean description is the more fundamental description.

SUMMARY AND CONCLUSIONS

In this review we have concentrated on developing heuristic dynamical models of the sudden warmings. We have shown that the traditional Eulerian description, with its rather artificial division of the flow field into zonal mean and perturbation parts, does not yield a simple consistent physical model for the warmings.

The Lagrangian-mean approach, on the other hand, describes the wave–mean flow interaction process in terms of fundamental wave properties—and clearly shows how amplifying waves propagating vertically into the stratosphere can drive the sequence of events associated with a sudden warming.

There can be little doubt that the wave–mean flow interaction process described in this review is in fact the process responsible for the generation of stratospheric sudden warmings. Since this process depends on vertically propagating planetary waves, a complete model of the warmings must account for the tropospheric processes which lead to planetary wave amplification and subsequent propagation into the stratosphere. Observations indicate that the connection between tropospheric planetary wave behavior and the occurrence of stratospheric sudden warmings is quite subtle. No model has yet been able to successfully *predict* the occurrence of a sudden warming, and no simple set of precursor conditions has been established. In some cases planetary wavenumber 1 dominates in the warming process; in other cases (Figure 1, for example) wavenumber 2 dominates. Nonlinear interactions between wavenumbers 1 and 2, and the dependence of wave propagation characteristics on the meridional distribution of the mean flow (and its time evolution), are aspects not considered in this review which probably play important roles in the evolution of stratospheric warmings.

Progress in both observational and theoretical understanding of the sudden warmings has been rapid in recent years. Routine observations by satellite (Houghton 1978) and detailed dynamical simulation models should within the next decade provide answers to many of the remaining questions concerning the dynamics of the sudden warmings.

ACKNOWLEDGMENT

I wish to thank Mr. Roderick S. Quiroz of the National Meteorological Center, NOAA, for providing Figures 1*a–c*, and Dr. M. E. McIntyre for helpful comments on the manuscript. This work was supported by the Atmospheric Research Section, National Science Foundation, Grant ATM76-84633.

Literature Cited

Andrews, D. G., McIntyre, M. E. 1976. Planetary waves in horizontal and vertical shear: the generalized Eliassen-Palm relation and the zonal mean acceleration. *J. Atmos. Sci.* 33:2031–48

Andrews, D. G., McIntyre, M. E. 1978. An exact theory of nonlinear waves on a Lagrangian-mean flow. *J. Fluid Mech.* 89:609–46

Charney, J. G., Drazin, P. G. 1961. Propagation of planetary scale wave disturbances from the lower into the upper atmosphere. *J. Geophys. Res.* 66:83–109

Dickinson, R. E. 1969. Theory of planetary wave–zonal flow interaction. *J. Atmos. Sci.* 26:73–81

Dunkerton, T. 1978. On the mean meridional mass motions of the stratosphere. *J. Atmos. Sci.* 35:2325–33

Holton, J. R. 1972. *An Introduction to Dynamic Meteorology*. New York: Academic. 319 pp.

Holton, J. R. 1975. *The Dynamic Meteorology of the Stratosphere and Mesosphere*. Boston: Am. Meteorol. Soc. 218 pp.

Holton, J. R. 1976. A semi-spectral numerical model for wave–mean flow interactions in the stratosphere: applications to sudden stratospheric warmings. *J. Atmos. Sci.* 33:1639–49

Holton, J. R., Dunkerton, T. 1978. On the role of wave transience and dissipation in stratospheric mean flow vacillations. *J. Atmos. Sci.* 35:740–44

Houghton, J. T. 1978. The stratosphere and the mesosphere. *Q. J. R. Meteorol. Soc.* 104:1–29

Johnson, K. W. 1977. *Potential Vorticity Transport and Stratospheric Warmings*. PhD thesis. Univ. Md. 178 pp.

Mahlman, J. D. 1969. Heat balance and mean meridional circulation in polar stratosphere during warming of January 1958. *Mon. Weather Rev.* 97:534–40

Matsuno, T. 1971. A dynamical model of the stratospheric sudden warming. *J. Atmos. Sci.* 28:1479–94

Matsuno, T., Nakamura, K. 1979. The Eulerian- and Lagrangian-mean circulations in the stratosphere at the time of a sudden warming. *J. Atmos. Sci.* 36:640–54

McInturff, R. M., ed. 1978. Stratospheric warmings: synoptic, dynamic and general-circulation aspects. *NASA Ref. Publ. 1017*. 174 pp.

McIntyre, M. E. 1977. Wave transport in stratified rotating fluids. *Springer Lecture Notes in Physics* 71:290–314

McIntyre, M. E. 1979a. An introduction to the generalized Lagrangian-mean description of wave–mean-flow interaction. *Pure Appl. Geophys.* In press

McIntyre, M. E. 1979b. Towards a Lagrangian-mean description of stratospheric circulations and chemical transports. *Philos. Trans. R. Soc. London Ser. A*. In press

O'Neill, A., Taylor, B. F. 1979. A study of the major stratospheric warming of 1976/77. *Q. J. R. Meteorol. Soc.* 105:71–92

Quiroz, R. S. 1979. Tropospheric-stratospheric interaction in the major warming event of January–February 1979. *Geophys. Res. Lett.* 6:645–48

Quiroz, R. S., Miller, A. J., Nagatani, R. M. 1975. A comparison of observed and simulated properties of sudden stratospheric warmings. *J. Atmos. Sci.* 32:1723–36

Reed, R. J., Wolf, J. L., Nishimoto, H. 1963. A spectral analysis of the energetics of the stratospheric sudden warming of early 1957. *J. Atmos. Sci.* 20:265–75

Schoeberl, M. R. 1978. Stratospheric warmings: observation and theory. *Rev. Geophys. Space Phys.* 16:521–38

Uryu, M. 1974. Induction and transmission of mean zonal flow by quasigeostrophic disturbances. *J. Meteorol. Soc. Jpn.* 52:481–90

Ann. Rev. Earth Planet. Sci. 1980. 8: 191–204
Copyright © 1980 by Annual Reviews Inc. All rights reserved

THE EQUATORIAL UNDERCURRENT REVISITED

×10130

S. G. H. Philander

Geophysical Fluid Dynamics Laboratory/NOAA, Princeton University, Princeton, New Jersey 08540

1 *Introduction*

In September 1951 the Hugh M. Smith, a vessel of the Pacific Oceanic Fisheries Investigation of the US Fish and Wildlife Service, was fishing for tuna on the equator south of Hawaii. The winds and the surface currents were westward so that the ship drifted in that direction. The long line fishing gear, however, drifted to the east! Subsequent measurements revealed the existence of the eastward Equatorial Undercurrent, a major component of the current-systems in the tropical oceans. (The Equatorial Undercurrent is sometimes referred to as the Cromwell Current in the Pacific, and the Lomonosov Current in the Atlantic Ocean.) Figure 1 shows a section across this intense narrow eastward jet, which has a subsurface core symmetrical about the equator and in the thermocline. The Undercurrent is a permanent feature in the Pacific and Atlantic Oceans but is a transient phenomenon in the western part

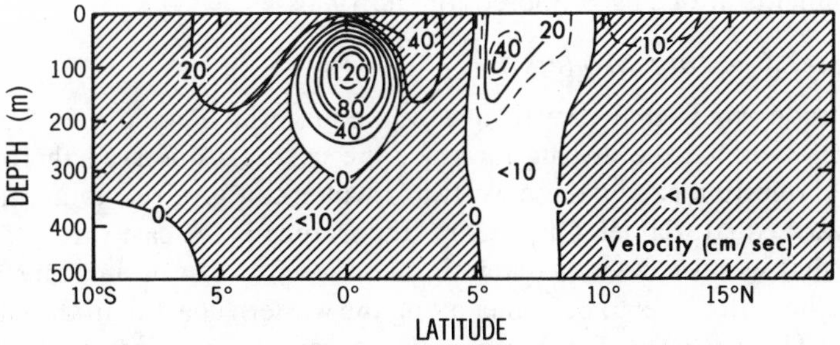

Figure 1 Meridional cross section of the zonal velocity component in the central Pacific. Shaded regions indicate westward flow, unshaded regions eastward flow. Velocities are geostrophic velocities except near the equator where the values are based on direct observations. The eastward Equatorial Undercurrent is centered on the equator; the North Equatorial Countercurrent is between 5°N and 9°N (after Knauss 1963).

0084-6597/80/0515-0191$01.00

of the equatorial Indian Ocean where it appears in the spring towards the end of the northeast monsoon season.

During the 1950s and 1960s, there were several oceanographic expeditions to the equator to define the structure of this current along different meridians, and there were numerous theoretical studies that attempted to explain the mean current as a response of the equatorial oceans to westward winds. [See Philander (1973) for a summary.] Over the past decade observational and theoretical studies have focused on the variability of this current. Because of large-scale experiments (such as GATE in the tropical Atlantic Ocean during the summer of 1974) we now know that the variability occurs over a spectrum of frequencies, and that meanders and fluctuations of the Undercurrent are particularly energetic at periods of a month (in the Atlantic and Pacific Oceans) and two weeks (in the Atlantic Ocean). On seasonal and interannual time scales the variability of the density field in the equatorial plane is known in outline but the few available measurements are inadequate for a description of the associated changes in the currents. (An intensification of the Undercurrent has, however, been implicated in the interannual occurrence of Los Niños, events associated with anomalously warm surface waters in the eastern equatorial Pacific Ocean.) Recent theoretical studies have concerned the generation and decay of equatorial currents (including the Undercurrent). These studies are of direct relevance to the Indian Ocean where the winds intensify and relax suddenly, and are of indirect relevance to variability in the Atlantic and Pacific Oceans where changes in the wind occur gradually (not abruptly). The studies suggest, for example, how phenomena such as El Niño can occur. We have now reached the stage where it is possible to design experiments that address specific questions concerning the variability of the equatorial oceans. Such experiments are being proposed for the 1980s.

2 *Steady-State Models*

The trade winds that prevail over the equatorial Pacific and Atlantic Oceans must be the driving force for the currents observed there. The central question is how these westward winds can give rise to an intense eastward equatorial jet. Of primary importance is the eastward pressure force that westward winds maintain in a closed ocean basin. (The winds cause the water level to be higher along the western coast than the eastern coast.) The existence of such a pressure force in the tropical Atlantic and Pacific Oceans can readily be inferred from the slope of the thermocline as shown in Figure 2. Isotherms slope upwards, towards the surface, as one proceeds from west to east. (Hence, the sea surface temperature in the western tropical Atlantic and Pacific Oceans is substantially higher

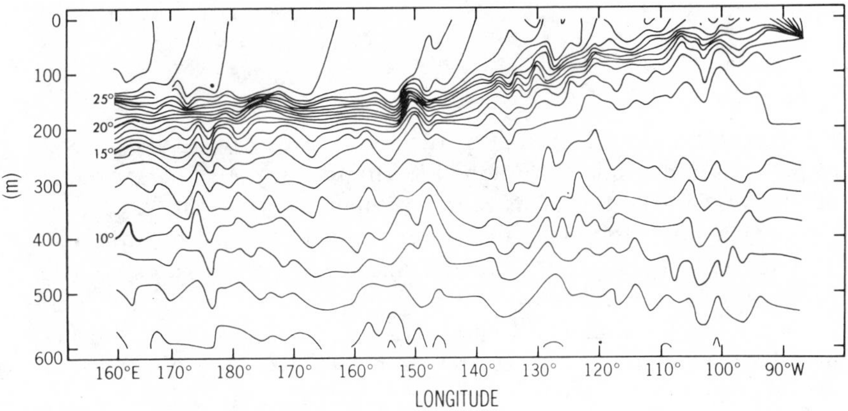

Figure 2 Isotherms in an equatorial plane in the Pacific Ocean (after Colin et al 1971).

than in the eastern side of these basins.) If the pressure forces are to be small near the ocean floor—there is no evidence to the contrary—then the sea level must be higher in the west (where the thermocline is deep and the water column has a low density) than in the east (where the thermocline is shallow and where the water column has a relatively high density). The observed slope of the thermocline therefore implies an eastward pressure force in the surface layers of the tropical oceans.

In a rotating fluid the Coriolis force causes motion to be along isobars (and not from a high to a low pressure region). Hence, an eastward pressure force is associated with equatorward geostrophic motion in both hemispheres. (We are concerned with the inviscid region below the mixed surface layer where the wind acts.) Near the equator, the Coriolis force is vanishingly small and cannot balance the pressure force. As a consequence, fluid particles acquire eastward momentum (from the pressure force) as they approach the equator. This eastward acceleration can approximately be estimated in the following manner. Assume that fluid particles remain on the same horizontal plane as they move equatorward, and assume that the motion is inviscid. Then the conservation of the vertical component of vorticity can be written

$$u_y - f = \text{constant}. \tag{1}$$

(Here f is the local vertical component of the rotation vector of the earth, β is its latitudinal derivative, and u_y is the relative vorticity of the fluid particle if zonal variations are neglected.) According to (1) a particle with a zero zonal velocity component at a distance Y_0 from the equator will, if it moves to the equator geostrophically, have a zonal speed u at the equator where

$$u = \frac{\beta Y_0^2}{2}.$$

If Y_0 corresponds to 3°N then the particle will move eastward with a speed of 100 cm/s at the equator! Fofonoff & Montgomery (1955) first pointed out that the conservation of the vertical component of vorticity of particles moving equatorward implies the existence of an eastward equatorial jet. This is a heuristic argument, however, because Equation (1) is approximate. It neglects the vertical movement of fluid particles into the divergent surface layers for example. Charney (1960), who constructed the first model of the Equatorial Undercurrent, studied the response of a constant-density equatorial ocean to westward winds. (His model-ocean is bounded below by a rigid surface which is taken to be the thermocline because its strong stratification inhibits the vertical transfer of momentum.) In this model there are two opposing processes that determine the structure of the currents at the equator. The westward winds give the surface particles a westward velocity component. This motion is confined to a shallow Ekman layer, except in the neighborhood of the equator where the westward momentum diffuses downwards to greater and greater depths. The westward winds also maintain an eastward pressure force which causes equatorward and eastward motion in subsurface layers, as explained above. Wind-induced equatorial upwelling advects this eastward momentum into the surface layers and thus counters the downward diffusion of westward momentum gained from the wind. The parameter $c = (H^5\tau\beta/v^3)^{1/3}$ measures the relative importance of the two opposing processes (the intensity of the wind-stress is τ, v is the coefficient for the vertical diffusion of momentum, H is the depth of the model-ocean).

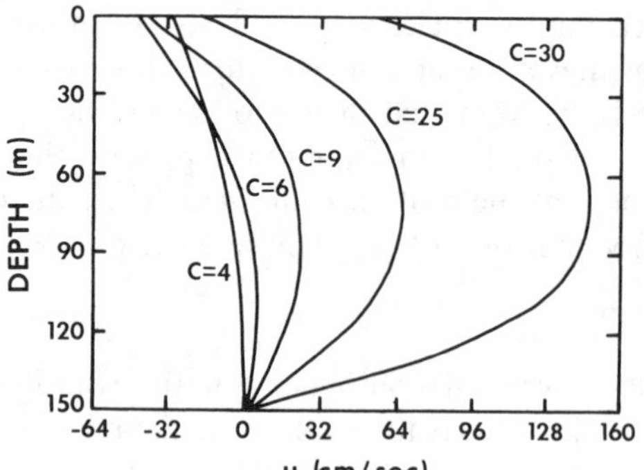

Figure 3 The zonal flow at the equator as a function of depth for different values of the parameter c defined in the text (after Charney 1960).

Figure 3 shows the zonal velocity at the equator for different values of c. For large values of c the eastward momentum gained nonlinearly from the pressure force is sufficiently large for the motion to be eastward at all depths. For small values of c the diffusion of westward momentum from the surface is sufficiently important for the equatorial motion to be westward at all depths. In the latter case the motion is linear and highly diffusive so that Equation (1) is inappropriate. This linear solution shows that the pressure force is necessary but not sufficient for the existence of the Equatorial Undercurrent. The solutions in Figure 3 are also counterexamples to the false analogy that is sometimes drawn between motion at the equator and motion in a nonrotating tank of water. In the tank there is an Undercurrent if variations in a direction perpendicular to the equatorial plane are suppressed. (In such a case westward motion in the surface layers must return eastward in subsurface layers to conserve mass.) In the oceans there can be a net zonal transport along the equator; the return flow is at a slightly higher latitude.

For reasonable values of the various parameters the Equatorial Undercurrent in Charney's (1960) model closely resembles the observed one. That nonlinear processes are indeed important can be verified by checking that the observed intensity U is related to the width L of the current by the expression

$$L = (U/\beta)^{1/2}.$$

[This follows from Equation (1) for the conservation of vorticity.]

Another feature of the model that can be checked is the equatorward motion at the depth of the core of the Undercurrent (in other words, in the thermocline). Katz et al (1979) and Weisberg et al (1979a) describe measurements that confirm such convergent motion. The success of the model is dependent on an appropriate choice for the value of the coefficient of viscosity ν. We return to this matter later.

Charney's model can be refined in a number of ways, most importantly by taking the stratification of the ocean into account. Stratification inhibits the vertical transfer of momentum so that the downward diffusion of westward momentum (in the case of westward winds) is insufficient to destroy the eastward momentum subsurface particles gain from the pressure force, even when the flow is linear. Thus, in Charney's (1960) constant-density model linear motion is westward at all depths, but in a stratified model a linear Undercurrent is possible (J. McCreary, manuscript in preparation). Linear models do not address the question of how the thermocline is maintained as a thermal boundary layer close to the ocean surface, and in effect do not explain how the depth scale of the Undercurrent is determined. (This current is in the equatorial thermocline.) In the nonlinear models of Semtner & Holland (1980) and Philander & Pacanowski (1980) the thermocline is maintained by a balance between the downward diffusion of heat and the upwelling of cold water. In an earlier study Philander (1973) argued that the zonal advection of cold water from the east plays an important role in the heat balance below the core of the Undercurrent, especially during periods of weak wind. Measurements below the thermocline show that the flow there is highly variable; a westward current is not consistently observed. It is possible that at those depths the transients have an important influence on the mean state.

Whereas western boundary currents such as the Gulf Stream depend on the wind-stress over the entire ocean basin, the Equatorial Undercurrent appears to be a response to the local equatorial winds. Its eastward transport is returned by the westward currents in which it is embedded; the recirculation takes place within 5° latitude of the equator. In the case of winds with no spatial structure the Undercurrent is most intense near the western coast and gradually loses fluid as it moves eastward.

The structure of the Equatorial Undercurrent in the models mentioned here is sensitive to the values assigned to the coefficients of eddy viscosity (ν) and diffusivity (k) which parameterize the vertical diffusion of momentum and heat. Oceanic measurements suggest values of the order of $\nu = 10, k = 1\ \text{cm}^2/\text{s}$—see Katz et al (1979)—but these values are approximate to within a factor of two or three at least. In the numerical models the intensity of the Undercurrent doubles when the value of ν is halved. A comparison between the simulated and observed Undercurrent should permit a determination of the appropriate value for ν. But the intensity of the Undercurrent is subject to such large fluctuations that it is necessary to understand the variability of this current before proceeding. We discuss aspects of this variability in the next sections.

Motion observed in the equatorial oceans corresponds to one of the many flow patterns possible between concentric rotating spheres. There are different flow patterns for different values of parameters such as the rate of rotation of the earth, the depth of the ocean, the intensity of the wind-stress, and the coefficients of viscosity and diffusion. From studies that explore flow patterns associated with non-oceanic parameter values, it emerges that the flow at the equator reverses with depth in a number of cases. Although these studies are of considerable interest, they are not of direct relevance to the observed Equatorial Undercurrent.

3 *Generation of the Equatorial Undercurrent*

An eastward pressure force appears to be a necessary condition for the existence of the Equatorial Undercurrent. In the Indian Ocean the northeast monsoons that maintain such a pressure force prevail between November and April only. Is that a sufficient period of time for the generation of an Equatorial Undercurrent?

Let us consider the response of a stratified ocean, initially at rest, to the sudden onset of uniform westward winds. (In the absence of any forcing, diffusion will modify the initial stratification. We assume that this happens on a time scale very long compared to the time it takes to generate the Undercurrent). Initially, the motion in the interior of the ocean basin is independent of longitude (as if there were no north-south coasts), is in the direction of the wind, and accelerates constantly. The flow is most intense near the equator where there is a jet with a half-width of about 200 km (Yoshida 1959). [The vicinity of the equator, which is distinct because of equatorial upwelling, is very similar to the neighborhood of a coast (Charney 1955, Gill 1975).] The flow continues to accelerate until the ocean floor starts to exert a drag, but long before this happens coastal effects become important. Waves, generated initially at coasts by the sudden onset of the wind, propagate across the ocean basin and establish a zonal pressure gradient which balances the wind-stress. It follows that, once the pressure gradients are established, the wind-stress no longer drives accelerating currents; a state of equilibrium would have been reached.

The vertical structure of the waves that effect the oceanic adjustment depends on the stratification of the ocean and is therefore different from the vertical structure of the wind-driven surface currents (which gradually diffuse downwards). In the case of a constant-density ocean the vertical scale of the waves is simply the total depth of the ocean. For a model with realistic stratification, the exceptionally sharp, shallow thermocline in the tropics causes certain wave modes to be trapped in and above the thermocline. Such modes are dominant in the adjustment of the upper ocean to a sudden change in the surface winds (Philander & Pacanowski 1980). The main point is that the directly wind-driven currents are confined to the shallow surface layers whereas the waves introduce a zonal pressure force over a greater depth, namely that of the thermocline. Once an eastward pressure force is established, in response to westward winds, an eastward Undercurrent will appear in the equatorial thermocline as explained in Section 2.

In the case of a nonrotating tank of fluid of depth h, gravity waves with speed $(gh)^{1/2}$ effect the adjustment to a new equilibrium state when the forcing function (which drives motion in the tank) suddenly changes. The situation is more complicated in the case of an ocean on a rotating sphere. The oceanic adjustment on time scales longer than a few days is accomplished by planetary wave-modes which are evanescent poleward of a certain turning latitude (Blandford 1966, Matsuno 1966, Lighthill

1969, Moore 1968, Cane & Sarachik 1979). The gravest modes, which have a significant amplitude within a few hundred kilometers of the equator only, propagate most rapidly. Modes that extend into higher latitudes travel more slowly. Hence, when there is a sudden change in surface winds, the oceanic adjustment is most rapid in the equatorial zone, and is progressively more gradual the higher the latitude. The waves most important in the adjustment of the equatorial oceans are the equatorially trapped Kelvin and Rossby waves which propagate eastward and westward respectively.

Figure 4*a*–*c* shows the evolution of motion at the equator in response to the sudden onset of westward winds in a nonlinear numerical model (Philander & Pacanowski 1980). Initially, the wind drives an accelerating westward surface jet while the temperature remains independent of longitude. The acceleration stops abruptly when zonal density gradients are established in the wake of an eastward-propagating wavefront indicated by the dashed line. This line corresponds to the equatorially trapped Kelvin wave. (A Rossby wave that emanates from the eastern coast is also evident in Figure 4*a*.) Once the zonal density gradient is established an Equatorial Undercurrent appears as expected. Because of upwelling the eastward momentum of this current is advected into the surface layers so that the westward jet there decelerates. Although small fluctuations persist for a long time because of repeated wave reflections, an equilibrium state is reached approximately 150 days after the onset of the winds. Cane's (1979) model gives a similar adjustment time, which depends on the speed of equatorial waves. [The corresponding adjustment time in midlatitudes is of the order of a decade (Veronis & Stommel 1956).] In Figure 4 the Equatorial Undercurrent in the equilibrium state extends from the surface to below the thermocline, and has its core at a depth of about 80 m.

According to the calculations described above, the Equatorial Undercurrent is generated in the wake of an equatorial Kelvin wave that propagates eastward from the western coast of a basin. The current exists within a matter of months after the onset of westward winds. Over the western equatorial Indian Ocean the northeast monsoons prevail from November onwards so that an Undercurrent should be present there by early March. Measurements confirm this and show that the Undercurrent usually persists into April and May by which time the winds have reversed direction. We next describe what happens to this current when the westward winds suddenly relax.

4 *Decay of the Equatorial Undercurrent*

In an equilibrium state a westward wind maintains an eastward pressure force and an eastward Equatorial Undercurrent in the thermocline. If the

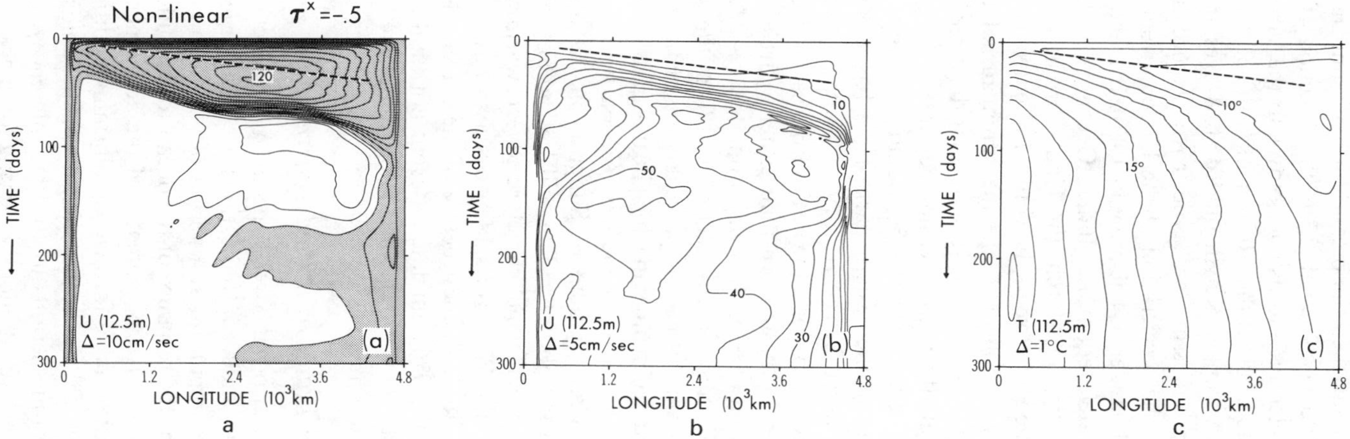

Figure 4 Evolution of the zonal velocity (at depths of 12.5 m and 112.5 m) and temperature (at a depth of 112.5 m) in the equatorial plane, in response to the sudden onset of westward winds. The dashed line corresponds to the initially excited Kelvin wavefront that is principally responsible for the adjustment of the upper ocean (after Philander & Pacanowski 1980).

winds should suddenly relax then the pressure force is unbalanced. The Undercurrent, which is flowing "downhill," in the direction of the pressure force, therefore accelerates and its core moves to the surface. Once again the waves most important in the adjustment of the ocean to a state of rest are the equatorially trapped Kelvin and Rossby waves. In the wake of these waves the eastward pressure force is destroyed. Hence the passage of wavefronts first stops the acceleration of the Undercurrent and then starts to decelerate it (Philander & Pacanowski 1980). It is therefore not surprising that in the western equatorial Indian Ocean where the northeast monsoons stop blowing in March, the Undercurrent persists into April and early May. The structure of the equatorial currents during its final decay is apparently complicated: the measurements of Luyten & Swallow (1976) in late May and June 1976 show that the direction of the zonal flow reverses repeatedly with increasing depth.

El Niño events in the equatorial Pacific Ocean—the appearance of anomalously warm water in the eastern part of the basin—apparently occur when the trade winds weaken (Wyrtki 1977). These westward winds usually cause the thermocline to slope downward from east to west so that heat storage is far greater in the western than eastern part of the basin (see Figure 2). A relaxation of the winds will ultimately result in a horizontal thermocline, which implies a transfer of heat from the western to the eastern side of the basin. In this way a weakening of the trades can result in El Niño. The detailed manner in which heat is transferred zonally depends on how the winds relax. In the next section we discuss what happens when the winds relax gradually. Suppose on the other hand that the winds stop blowing abruptly. This will lead to an acceleration of the Equatorial Undercurrent, which therefore advects warm water into the eastern equatorial Pacific. Waves (including a Kelvin wave) will finally stop this warming and initiate the destruction of the current. McCreary (1976) and Hurlburt et al (1976) simulate an event not unlike this one. There are at the moment no measurements to determine whether or not this sequence of events accompanies Los Niños.

5 *Variability of the Equatorial Undercurrent*

Variability near the equator occurs over a spectrum of frequencies and is primarily atmospherically induced. (Indications are that large-scale instabilities of currents are important at a period of about 1 month only. See Section 6.) Studies of the response of the ocean to the sudden onset and relaxation of the winds give us estimates of the adjustment time of the ocean. (This is the time it takes for new equilibrium conditions to be established after a sudden change in the intensity of the winds.) Knowledge of the adjustment time permits us to draw inferences concerning the response of the ocean to continuously varying forcing.

Suppose that the fluctuating winds vary only on a time scale long compared to the adjustment time of the ocean. In such a case the ocean is always in an adjusted state. In other words, the ocean is at each moment in equilibrium with the winds at that moment and has no "memory" of past winds. Thus, when the westward winds are weak the zonal slope of the thermocline is small and the currents are weak; when the winds are intense zonal density gradients are large and currents are strong. At these low frequencies there is essentially no phase lag between the forcing and response. The ocean in effect passes through a series of steady states.

Suppose that the fluctuating winds vary on a time scale short compared to the adjustment time of the ocean. In such a case the ocean is never in an adjusted state so that the winds will primarily excite waves and drive non-equilibrium currents.

According to the results of Section 3, the adjustment time for the equatorial zone of an ocean basin 5000-km wide is of the order of 150 days. (This time depends on how long it takes planetary waves to propagate across the ocean basin and hence varies with the size of the basin.) We therefore expect the seasonal cycle (and interannual variability) in the Atlantic Ocean to occur in phase with the seasonally (and interannually) varying surface winds. Katz and collaborators (1977) find that the zonal slope of the equatorial thermocline is indeed proportional to, and varies in phase with, the seasonally changing zonal wind-stress over the equatorial Atlantic Ocean. In the Pacific Ocean indications are that the zonal pressure gradient and wind-stress vary in phase on interannual time scales only (Barnett 1977). On shorter time scales, the seasonal cycle for example, there is zonal phase propagation associated with variations in zonal density gradients (Meyers 1979). It would appear that the seasonal time scale is longer than the adjustment time of the equatorial Atlantic but shorter than that of the Pacific Ocean. This difference between the two basins, which may be attributable to their different sizes, has an interesting consequence. Changes in the large-scale zonal slope of the thermocline occur on time scales longer than the adjustment time. Such changes are associated with a considerable zonal redistribution of heat, which gives rise to phenomena such as El Niño. It therefore seems plausible that whereas Los Niños can occur interannually in the Pacific Ocean, they are possible annually in the Atlantic Ocean. The seasonal suppression of upwelling along the coast of the Gulf of Guinea (in the eastern equatorial Atlantic) where the winds have little seasonal variability, has been interpreted as a seasonal El Niño (Hisard & Merle 1979).

Our understanding of the response of the equatorial ocean to forcing that varies over a spectrum of frequencies is clearly rudimentary, especially on time scales less than the adjustment time. There is a need for theoretical studies of the oceanic response to forcing that varies continuously (rather than abruptly) and there is a need for further measurements, particularly time series of the currents. In the Atlantic we do not know how the currents change when the zonal pressure gradient fluctuates seasonally. Measurements in the central Pacific show that the maximum speed of the Undercurrent can be as low as 75 cm/s, and as high as 160 cm/s. Because these measurements were made on isolated occasions, over an extended period of time, it is unclear whether the variability is interannual or seasonal. Instrumented moorings have now been deployed on the equator so that these questions will be resolved soon.

6 *Instabilities and Meanders*

The Equatorial Undercurrent is an intense jet with considerable latitudinal shear so that Raleigh instabilities, which cause the current to meander about the equator, would seem likely. A stability analysis of equatorial jets (Philander 1976) reveals that eastward jets are in general much more stable than westward jets. Equatorial divergence, and the latitudinal variation of the Coriolis parameter, are stabilizing for eastward, destabilizing for westward currents. For the Undercurrent to be unstable, its maximum speed would have to be of the order of 150 cm/s for several months. (The period of possible unstable waves is of the order of a few months.) This rules out instabilities in the Atlantic and Indian Oceans where the Undercurrent is too weak to be unstable. In the central Pacific Ocean the current can attain the necessary high speeds, but usually not for a sufficient length of time. We conclude that instabilities of the Undercurrent will occur very infrequently in the Pacific Ocean.

The surface flow at the equator is westward, the neighboring surface

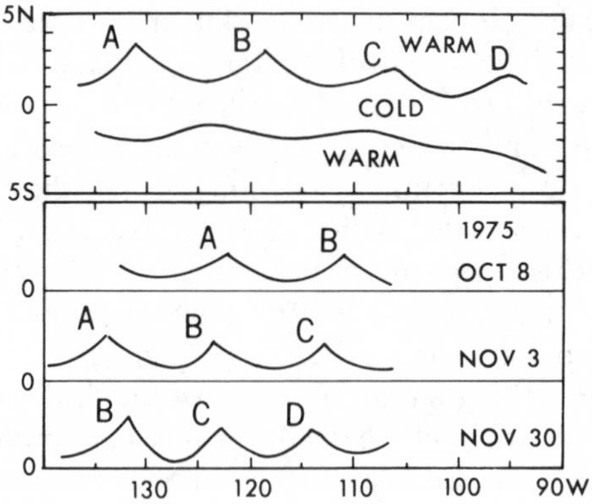

Figure 5 A schematic diagram of undulations of the equatorial Pacific front between the cold South Equatorial Current and warm North Equatorial Countercurrent (after satellite photographs of Legeckis 1977).

current to the north is eastward. In the Pacific both these currents are most intense in the late summer and autumn at which time the latitudinal shear near 3°N (the boundary between these currents) is sufficiently large to give rise to unstable waves with a wavelength of about 1000 km, a period of a month, and a westward phase speed of approximately 50 cm/s (Philander 1978). Satellite photographs of the sea surface temperature give striking evidence of such waves (Figure 5). These instabilities of the surface currents will cause the Undercurrent to meander—there is some evidence of this happening in the Atlantic Ocean—and the instabilities will radiate equatorially trapped waves into the deep ocean. (See Weisberg et al 1979b.) Because of this radiation of energy away from the unstable region the instabilities remain linear and contribute to variability at a period of one month only. They do not contribute to variability over a spectrum of frequencies. By contrast, instabilities of the Gulf Stream are highly nonlinear (because energy cannot be radiated away) and hence contribute to variability over a spectrum of frequencies.

Finally, we note that energetic westward-propagating meanders of the Equatorial Undercurrent in the Atlantic Ocean, with a period of about 16 days and a zonal wavelength in the neighborhood of 2000 km, are as yet unexplained (Duing et al 1975).

Summary

The Equatorial Undercurrent is a permanent feature of the current-systems in the tropical Atlantic and Pacific Oceans. Steady-state models successfully explain and simulate its main characteristics, but for a stringent comparison between models and measurements it is necessary to understand the variability of this current first. Theoretical studies of the oceanic response to sudden changes in the intensity of the winds explain the seasonal appearance of the Undercurrent in the Indian Ocean and permit inferences concerning the variability of this current in the other oceans. Observational and theoretical studies planned for the 1980s should enhance our understanding of this variability considerably.

Literature Cited

Barnett, T. P. 1977. An attempt to verify some theories of El Niño. *J. Phys. Oceanogr.* 7:633–47

Blandford, R. 1966. Mixed Rossby-gravity waves in the ocean. *Deep-Sea Res.* 13: 941–61

Cane, M. 1979. The response of an equatorial ocean to simple wind-stress patterns. *J. Mar. Res.* 37:233–52

Cane, M., Sarachik, E. 1979. Forced baroclinic ocean motions III. *J. Mar. Res.* 37:355–98

Charney, J. G. 1955. The generation of oceanic currents by winds. *J. Mar. Res.* 14:477–98

Charney, J. G. 1960. Non-linear theory of a wind-driven homogeneous layer near the equator. *Deep-Sea Res.* 6:303–10

Colin, C., Henin, C., Hisard, P., Oudot, C. 1971. Le Courant de Cromwell dans Le Pacifique central en fevrier. *Cah. ORSTOM Ser. Oceanogr.* 9:167–86

Duing, W., Hisard, P., Katz, E., Meincke, J., Miller, L., Moroshkin, K., Philander,

G., Rybnikov, A., Voigt, K., Weisberg, R. 1975. Meanders and long waves in the equatorial Atlantic. *Nature* 257:280–84

Fofonoff, N. P., Montgomery, R. B. 1955. The Equatorial Undercurrent in the light of the vorticity equation. *Tellus* 7:518–21

Gill, A. E. 1975. Models of Equatorial Currents. *Proc. Symp. Numer. Models Ocean Circ. Natl. Acad. Sci. Durham, Oct. 17–20, 1972*

Hisard, P., Merle, J. 1979. Onset of summer cooling in the Gulf of Guinea. *Deep-Sea Res.* 26 (Suppl. II):325–42

Hurlburt, H. E., Kindle, J. C., O'Brien, J. J. 1976. A numerical simulation of the onset of El Niño. *J. Phys. Oceanogr.* 6:621–31

Katz, E. and collaborators. 1977. Zonal pressure gradient along the equatorial Atlantic. *J. Mar. Res.* 35(2):293–307

Katz, E. J., Bruce, J. G., Petrie, B. D. 1979. Salt and mass flux in the Atlantic Equatorial Undercurrent. *Deep-Sea Res.* 26 (Suppl. II):137–60

Knauss, J. A. 1963. The Equatorial Current System. In *The Sea*, ed. M. N. Hill, Vol. 1, pp. 235–52. NY: Interscience

Legeckis, R. 1977. Long waves in the eastern equatorial Pacific; a view from a geostationary satellite. *Science* 197:1179–81

Lighthill, M. J. 1969. Dynamic response of the Indian Ocean to the onset of the southwest monsoon. *Philos. Trans. R. Soc. London Ser. A* 265:45–92

Luyten, J., Swallow, J. 1976. Equatorial Undercurrents. *Deep-Sea Res.* 23:1005–7

Matsuno, T. 1966. Quasi-geostrophic motions in equatorial areas. *J. Meteorol. Soc. Jpn.* 2:25–43

McCreary, J. 1976. Eastern tropical ocean response to changing wind systems: with application to El Niño. *J. Phys. Oceanogr.* 6:632–45

Meyers, G. 1979. Annual variation in the slope of the 14°C isotherm along the equator in the Pacific Ocean. *J. Phys. Oceanogr.* 9:885–91

Moore, D. W. 1968. Planetary-gravity waves in an equatorial ocean. PhD thesis. Harvard University, Cambridge, Mass.

Philander, S. G. H. 1973. Equatorial Undercurrent: Measurements and theories. *Rev. Geophys. Space Phys.* 2(3):513–70

Philander, S. G. H. 1976. Instabilities of zonal equatorial currents: I. *J. Geophys. Res.* 81(21):3725–35

Philander, S. G. H. 1978. Instabilities of zonal equatorial currents: II. *J. Geophys. Res.* 83:3679–82

Philander, S. G. H., Pacanowski, R. C. 1980. The generation and decay of equatorial currents. *J. Geophys. Res.* In press

Semtner, A. J., Holland, W. R. 1980. Numerical Simulation of Equatorial Ocean Circulation. *J. Phys. Oceanogr.* In press

Veronis, G., Stommel, H. 1956. The action of variable wind-stresses on a stratified ocean. *J. Mar. Res.* 15:43

Weisberg, R. H., Miller, L., Horigan, A., Knauss, J. A. 1979. Velocity observations in the equatorial thermocline during GATE. *Deep-Sea Res.* 26 (Suppl. II): 217–48

Weisberg, R. H., Horigan, A., Colin, C. 1979b. Equatorially trapped Rossby-gravity wave propagation in the Gulf of Guinea. *J. Mar. Res.* 37:67–86

Wyrtki, K. 1977. Sea level during the 1972 El Niño. *J. Phys. Oceanogr.* 7:779–87

Yoshida, K. 1959. A theory of the Cromwell Current and equatorial upwelling. *J. Oceanogr. Soc. Jpn.* 15:154–70

Ann. Rev. Earth Planet. Sci. 1980. 8:205–30
Copyright © 1980 by Annual Reviews Inc. All rights reserved

SEISMIC REFLECTION STUDIES OF DEEP CRUSTAL STRUCTURE

※10131

J. A. Brewer and J. E. Oliver
Department of Geological Sciences, Cornell University, Ithaca, New York 14853

Introduction

Somewhat like the imaginary inhabitants of Edwin A. Abbott's *Flatland*, geologists live in a world of two spatial dimensions, the surface of the earth. Yet in their task of understanding the three-dimensional earth, knowledge of the third dimension is vital. The means of exploring the depth dimension range from inference based on surface geology to sampling of deeper rocks by drill or studying igneous activity, through application of a variety of geophysical techniques. This review describes the rather recent application of the powerful seismic reflection profiling technique, which was developed by the petroleum industry in the search for hydrocarbons in sedimentary basins, to the study of the basement rocks of the continents. This method of study of the continental crust is in its infancy; only a few sites representing a tiny fraction of the continental basement have so far been explored. The results, however, are proving informative and critical for enhanced understanding of some of the fundamental questions of the geology of the continents.

Concentrated efforts to deepen our understanding of geology are timely. A society of four billion people gathered on, and largely dependent upon, the continents of the earth for their livelihood, must have a thorough understanding of its geological environment and its geological resources. The science itself is ready for a concerted effort of exploration of the continents. Plate tectonics stands as the basis for further advances. For all its current successes, this important concept will require modifications and development if it is to provide us with a thorough understanding of continental crust and its evolution. Possibly new concepts of comparable stature will be required.

The rate of demand by our society for increasingly rare minerals and

0084-6597/80/0515-0205$01.00

fuels shows no signs of slowing. If future needs are to be met, accurate means of locating and classifying such deposits must be developed. It is highly important to seek detailed knowledge of the structure and framework of continental crust in this stage of the development of the science of geology.

Various geophysical methods exist for studying crustal structure below the surface such as gravity, magnetics, electrical conductivity, heat flow, and seismic refraction and reflection methods. Studies of earthquake-generated waves are also informative. With the possible exception of seismic reflection methods, these techniques tend to average crustal structure over the dimensions of the experiment. Surface geological studies of rocks typical of the continental crust, such as those exposed in Precambrian shield areas and old eroded mountain belts, indicate a high degree of structural complexity and lithological heterogeneity, and a long and involved history of deformation covering, in some cases, at least 3800 m.y. (see, for example, Windley 1977).

Seismic reflection profiling at the present time has outstanding potential for study of the details of deep crustal structure. Seismic reflection studies of the earth's deep crust have been carried out in West Germany, Canada, the USSR, Australia, and elsewhere. At the moment, the most concerted effort is being made in the US under the auspices of the Consortium for Continental Reflection Profiling (COCORP). The COCORP program is accumulating large quantities of detailed data on the structure of the crust, and providing new insights into its history and mode of deformation.

Crustal Seismic Studies

Most controlled-source seismic studies of the crust use one of three methods: (*a*) seismic refraction, (*b*) wide angle seismic reflection, and (*c*) near-vertical-incidence seismic reflection. With these techniques acoustic signals are transmitted into the ground using explosives, weight drops, or vibratory sources. Reflections, refractions, or other characteristic modifications of these signals are caused by structural and compositional changes within the earth. The returning signals are received at the surface using geophones, devices that detect ground motion. The amplitude of the returning signals is a function of the geometry of the path, the velocity and density contrasts at structural or compositional changes, and the attenuation effects along the wave path. In crustal seismic refraction studies geophones are typically as much as 200 km or more away from the source point and the recorded signals traverse the whole thickness of crust, thus tending to average physical properties along their path (Braile & Smith 1975). In wide angle work, geophones are typically placed 50–150 km away from the source, and in near-vertical-incidence work, geophones are usually placed up to 5–10 km away from the source. Data collected with this latter method are most sensitive to lateral variations in crustal structure because signals propagate in a mainly vertical direction and each ray path samples only a limited horizontal segment of the crust. The different methods have their respective advantages; amplitudes of vertical ground motion are larger in the wide angle region than in the near-vertical region (Meissner 1967, McCamy et al 1962). Near-vertical-incidence data may be relatively insensitive to transition zones whereas refraction data are insensitive to low velocity layers. In some cases, velocities can be obtained with higher precision using wide angle and refraction techniques. These velocities, however, represent averages over the large lateral distances traveled by the seismic waves, so it is unclear exactly how they relate to deep crustal geology. Thus all three methods may give complementary and partially independent views of the earth's crust (Beloussov et al 1962, Dix 1965). To obtain as complete a picture as possible of the structure and velocity profile of the continental crust all three methods should be used. The major boundaries of the crust are in many areas defined by refraction work. This review will deal with recent studies of the continental crust using near-vertical-incidence seismic reflection techniques, which can reveal finer scale structures within this framework.

The term "seismic reflection profiling" refers to the field techniques in which by continually moving source and receivers a two-dimensional profile of the earth's crust is obtained. Early reflection studies of the crust often involved fixed source and receivers.

Past Seismic Reflection Studies of the Crust

The seismic reflection technique has been extensively developed over many years by the petroleum industry (Stommel & Graul 1978), and the highly sophisticated equipment and techniques now in use are the result of much research and capital investment. However, in hydrocarbon exploration the primary interest is in the sedimentary succession overlying the crystalline basement. Data are usually recorded to depths of up to 8–12 km (about 4–5 s two-way travel time of the seismic waves).

The first reported seismic reflections obtained from crystalline basement were found because seismic crews extended recording times in the course of standard exploration work (Junger 1951, Widess & Taylor 1959, Robertson 1963, Dix 1965, Dohr & Fuchs 1967). In these instances usually only small areas were covered, and sometimes it was not clear that the events observed on the seismic profiles represented basement structure or whether they were due to energy reverberating in the sedi-

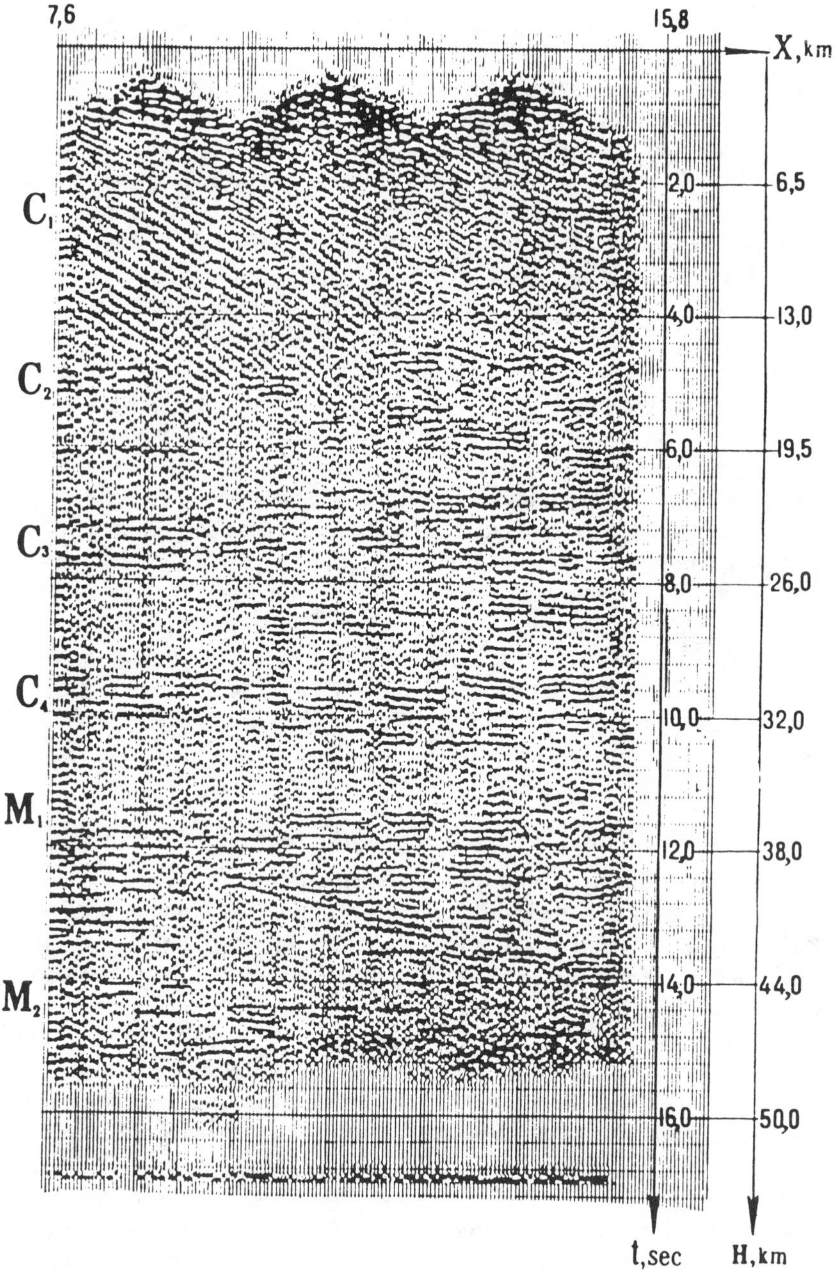
7,6
15,8
X, km
2,0
4,0
6,0
8,0
10,0
12,0
14,0
16,0
6,5
13,0
19,5
26,0
32,0
38,0
44,0
50,0
C1
C2
C3
C4
M1
M2
t, sec
H, km

mentary section (multiples), random noise, or some other artificial effect (Steinhart & Meyer 1961).

More recently, detailed and extensive seismic studies of the continental crust have been made in Europe and particularly in West Germany. Using refraction, wide angle, and near-vertical-incidence reflection techniques much information has accumulated on the nature of reflecting horizons in the deep crust, the nature of the Moho, and the variation of crustal structure (Meissner 1967, Dohr & Fuchs 1967, Fuchs 1969, Meissner 1973, Dohr & Meissner 1975, Mueller 1977). Seismic reflections recorded from the lower crust are generally characterized by short segments (2–3 km in length), frequent phase shifts, lower frequency cutoffs (10–12 Hz), and generally high amplitudes (Fuchs 1969). These last two facts suggest that the most probable cause of the reflected events are horizons of laminated zones of alternating high and low velocities rather than sharp (first order) discontinuities or transition zones (velocity gradients) (Fuchs 1969). Statistical analysis of crustal reflections in Europe suggests that three main discontinuities exist, at depths corresponding to the position (as defined by refraction work) of the top of the crystalline basement, the Conrad (a mid-crustal horizon thought to define the boundary between a mainly granitic upper crust and a mainly basaltic lower crust), and the Moho (Dohr & Fuchs 1967). As an example of how the different seismic techniques complement each other, wide angle reflection studies of the Moho suggest that it is a velocity gradient a few kilometers thick, whereas near-vertical-incidence studies suggest a superimposed detailed laminar or lenticular structure (Meissner 1973).

Deep seismic studies of the crust have been carried out over many years and on a regional scale in the USSR and Eastern Europe, through mostly wide angle reflection and refraction, but also some reflection profiling (Beloussov et al 1962, Sollogub et al 1968, 1975, Davydova 1975; Figure 1). These studies also show that lower crustal horizons produce complex waveforms suggesting transition zones of thin layers with alternating high and low velocities. The crust and upper mantle of the USSR has in some cases a complex block structure. These blocks are bounded by deep-seated faults frequently defined at depth by offsets in the Moho (Davydova 1975). The depth to the Moho varies with surface

---

← *Figure 1* Seismic reflection profile from area in Ukrainian shield. The profile is displayed in terms of the travel time of the seismic waves (unmigrated). Vertical scale is two-way travel time in seconds (with approximate depth conversion); horizontal scale is distance in kilometers. The profile is 8.2-km long. C_1–C_4 mark mid-crustal discontinuities; M_1–M_2 are apparent Moho reflections representing a rather complex structure in this area. The profile was recorded using explosive sources. From Sollogub et al (1975).

topography, and the Conrad can sometimes be traced with complicated relief.

Seismic reflection studies of the crust have also been carried out in Canada (Kanasewich & Cumming 1965, Clowes et al 1968, Clowes & Kanasewich 1972, Cumming & Chandra 1975, Mair & Lyons 1976, Berry & Mair 1977). These studies have served to delineate the Moho and Riel (the Canadian version of the Conrad discontinuity), again resolving considerable relief. Seismic reflection profiling has detected a buried Precambrian rift in Alberta (Kanasewich et al 1968). The crust in this area appears to have the same block-faulted character as is reported in the USSR (Clowes & Kanasewich 1972).

Limited seismic reflection surveys have been carried out in Australia (Mathur 1974, Branson et al 1976). Good Moho reflections have been obtained with indications of a multi-layered structure.

In the US, one of the first attempts to study the crust using seismic reflection techniques was carried out in Utah by Narans et al (1961). Their results suggested two reflecting horizons roughly correlative with those obtained from refraction profiles, but other events indicated a yet more complex structure. Hasbrouck (1964) tentatively identified the Moho using seismic reflections and also noted that his data indicated a complex crust and mantle structure. Dix (1965) documented many reflectors of low dip between 8- and 35-km depth in the Mohave desert. Deep crustal reflections have been recorded in Wyoming by Perkins & Phinney (1971). Apart from the present COCORP project, some of the most extensive studies of basement rocks using seismic reflection techniques are those of Smithson et al (1977a,b). In these studies numerous reflecting horizons were observed in different types of basement rock, showing that the reflection technique could image highly complex structures (Smithson 1978).

The well-documented and widespread occurrence of reflecting horizons in the continental crust, combined with the structures observed, suggests a complexity much greater than that expected from the classical, simple division of the crust into a "granitic layer" overlying a "basaltic" layer (Jeffreys 1926). In general, seismic horizons in the deep crust appear different from those characteristic of the sedimentary section, being complex transition zones, probably with layered structures, rather limited continuity, and a rather limited frequency spectrum (about 10–20 Hz). These structural constraints, combined with the observation that granulite facies rocks are abundant in stable Precambrian shield areas (Oliver 1969) and the fact that lower crustal seismic velocities are consistent with granulite facies mineralogy (Christensen & Fountain 1975), have led to highly complex crustal models of predominantly metamorphic rock of an overall intermediate composition, with interspersed igneous complexes (Smithson & Decker 1974, Smithson & Brown 1977). In the light of such a variable and complex crustal model some authors (e.g. Smithson 1978) believe that, in its simplest form, a pervasive mid-crustal discontinuity is an outmoded concept, or at least of minor importance.

The COCORP Project

The Consortium for Continental Reflection Profiling began operations in 1975. It represents the most concerted effort to date to apply the seismic reflection profiling technique to the study of the continental basement. The object is to map the basement rocks of the US and to focus on specific geological problems. In this context the work is part of the US Geodynamics Project. COCORP has built upon the success of other workers in obtaining seismic reflections from within the basement using more or less conventional exploration techniques (see references cited above). Seismic crews using state-of-the-art exploration technology are hired from the geophysical industry. Scientists from industrial, governmental, and academic laboratories advise the project as to the best areas to work, the equipment and techniques to use, and in the interpretation of the resulting data. Cornell University is the operating institution and prime contractor. A more detailed discussion of the field operations and data processing steps used to produce a typical COCORP seismic profile may be found in Schilt et al (1979).

The present practice of the COCORP project is to run long profiles over areas of geological interest (typically 50–200 km long). Data are usually recorded to 20 s two-way travel time (1 s of two-way travel time approximately equals 3-km depth in crystalline basement).

In general, it is important to have surface geological control. One of the most successful aspects of the method is that it is possible in some cases to follow to depth features on the seismic profiles that can be identified at or near the surface. Sometimes such features are of very great lateral extent and long lines are therefore necessary to trace them fully (e.g. see the discussion of the southern Appalachians data below). The VIBROSEIS[1] system is used by COCORP because it offers high quality data while maximizing efficiency and minimizing environmental impact. Previous use of the VIBROSEIS system in deep crustal reflection work has been described by Mateker & Ibrahim (1973), Fowler & Waters (1975), and Mair & Lyons (1976). COCORP uses five vibrating trucks which transmit into the ground linear sweep signals of about 8–40 Hz bandwidth. The returning echoes are collected by long linear arrays

[1] Registered Trade Mark of Continental Oil Company.

(typically 6–10 km in length) of geophones connected to a 96-channel truck-mounted recording system. About 24 geophones per channel are used. Data are collected and computer processed in such a way as to produce a 24-fold common-depth-point stacked seismic section, representing a cross section through the earth's crust in terms of travel time of the seismic waves. Important preliminary interpretations can be made on these time sections, but depth conversion and migration are important for detailed studies (Bloxsom 1978, Phinney & Jurdy 1979).

Geological Provinces Studied by COCORP

COCORP has operated successfully in a wide range of geological environments, with surface features ranging in age from the Precambrian to the present. Initial testing of the COCORP method was carried out in Hardeman County, Texas (Oliver et al 1976). Results were very encouraging, with many seismic events recorded from the basement. Figure 2 shows the location of areas in which COCORP has worked and sites under consideration for future work.

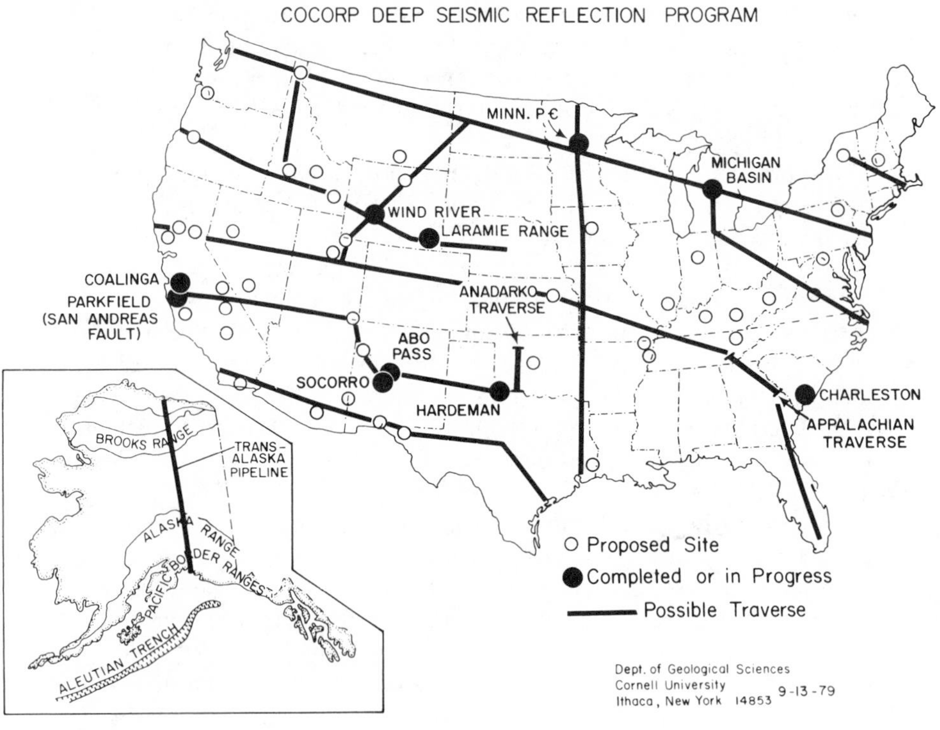

Figure 2 Status of the COCORP project as of September 1979.

The following is a brief description of some of the data collected by COCORP at the various sites, and their interpretation. Much of this work is published elsewhere in greater detail, and is referenced when this is the case. Examples of the seismic data are shown in order that the features they contain can be assessed by the reader. The vertical scale on the seismic profiles shown here is in terms of the two-way travel time of the seismic waves. For an approximate conversion to depth in kilometers, multiply two-way travel time in seconds by 3.

HARDEMAN BASIN, TEXAS The Hardeman basin is a late Paleozoic basin on the southern flank of the Wichita mountain system (Ham & Wilson 1967). Three seismic lines were recorded there in 1975. Line 1, a north-south line 17-km long is shown in Figure 3. A nearby well-penetrated Precambrian basement at a depth corresponding to 1.6 s (a date of 1265 ± 40 m.y. was obtained for basement cuttings from this hole). A zone of horizontal subparallel reflections, which arise from the sedimentary section of the basin, can be seen on Figure 3 down to about 1.5–1.6 s. A transparent zone underlies the base of this section. At times of about 2.9 and 3.7 s (about 9 and 11 km) are two bands of very strong reflections with pronounced lateral continuity and gentle dip. Below these, from about 4.0–15.0 s, the records show three main types of seismic character: (*a*) many short reflection segments of non-uniform distribution (in some places these may form subhorizontal groups but elsewhere they are discrete, isolated events); (*b*) dipping events of considerable lateral extent with a hyperbolic curvature, mainly between 5.0 and 12.0 s (these are almost certainly diffraction hyperbolae); (*c*) zones of no reflections (these must be interpreted with care because they could result from poor surface conditions or low signal/noise ratios, but if, for instance, they are underlain by reflectors then they could represent tightly folded or distorted structures, homogeneous bodies such as plutons, or areas with a gradual velocity gradient). The two strong bands at 2.9 and 3.7 s may be of sedimentary or volcanic origin. A high interval velocity between them (about 6 km/s) could indicate subsequent metamorphism if these units were originally sedimentary (Schilt et al 1979). Bloxsom (1978) has suggested that the layer between 2.8 and 3.7 s is a granitic, sill-like intrusion.

The deeper parts of the profile cannot be interpreted in detail at the present time because of the lack of other geophysical data or surface exposure in the area of the seismic lines. Considerable complexity is suggested by the different seismic characters discussed above (Oliver et al 1976). Schilt et al (1977) have made a three-dimensional study of the hyperbolic events and suggest that these may arise from numerous

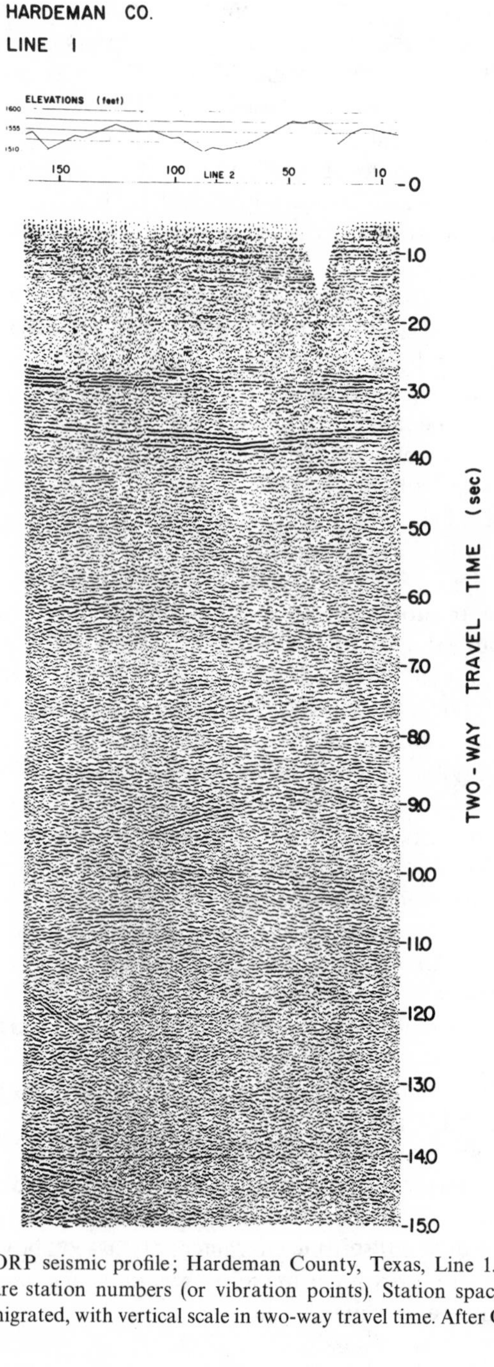

Figure 3 COCORP seismic profile; Hardeman County, Texas, Line 1. Numbers on the horizontal axis are station numbers (or vibration points). Station spacing is 134 m and the section is unmigrated, with vertical scale in two-way travel time. After Oliver et al (1976).

relatively small plutons possibly emplaced during rifting along the Wichita/Anardarko fault system.

The data from Hardeman County showed that COCORP methods could produce abundant and useful basement information. They also illustrated the importance of longer lines and good surface control.

THE WICHITA UPLIFT AND ANARDARKO BASIN, OKLAHOMA As a continuation of the Hardeman County lines, and in order to investigate the deep structure of the Wichita mountains and the flanking Anardarko basin [between which the structural relief is on the order of 12 km (Ham & Wilson 1967)], COCORP recorded a series of lines across the Wichita mountains in spring 1979. The survey is still incomplete and the data still under analysis, but initial results show that the strong and continuous 2.9 and 3.7 s events on the Hardeman County records are continuous to the southern margin of the Wichita mountains (60–80 km away) where they appear to be faulted out of the section. The deeper parts of the new profiles show rather more continuous coherent energy than the Hardeman County records with similar (but in general fewer) diffraction hyperbolae.

RIO GRANDE RIFT, NEW MEXICO The Rio Grande rift was the site of the first full-scale investigation of a specific geological problem by COCORP. Crustal extension along this rift began about 32–27 m.y. ago (Chapin 1979), and much diverse evidence indicates that rifting is continuing today (full literature citations are given in Brown et al 1979). COCORP recorded 155 km of reflection data transverse and parallel to the rift (Oliver & Kaufman 1976, Brown et al 1979). The objective of the COCORP work was to study the structure of an active rift, including the geometry of faulting, the nature of the crust-mantle boundary, and the possible existence of magma bodies at depth.

Figure 4 shows part of the seismic data from the traverse across the rift, and Figure 5 shows an interpreted line drawing of the main events seen on the complete traverse. These COCORP profiles show a crustal seismic response rather different from that of the Hardeman County profiles in that they lack the long-ranging events with a hyperbolic curvature. Instead hyperbolic events are rather localized and the deeper part of the crust, apart from the main features discussed below, is composed of short, discontinuous reflecting segments of variable concentration with generally low dip. Several zones of few reflectors exist, most noticeably just below the top of the basement, toward the base of the crust, and beneath the surface expression of the east-flanking fault of the rift boundary (Figure 5).

The east-flanking fault occurs at about vibration point (VP) 180 of

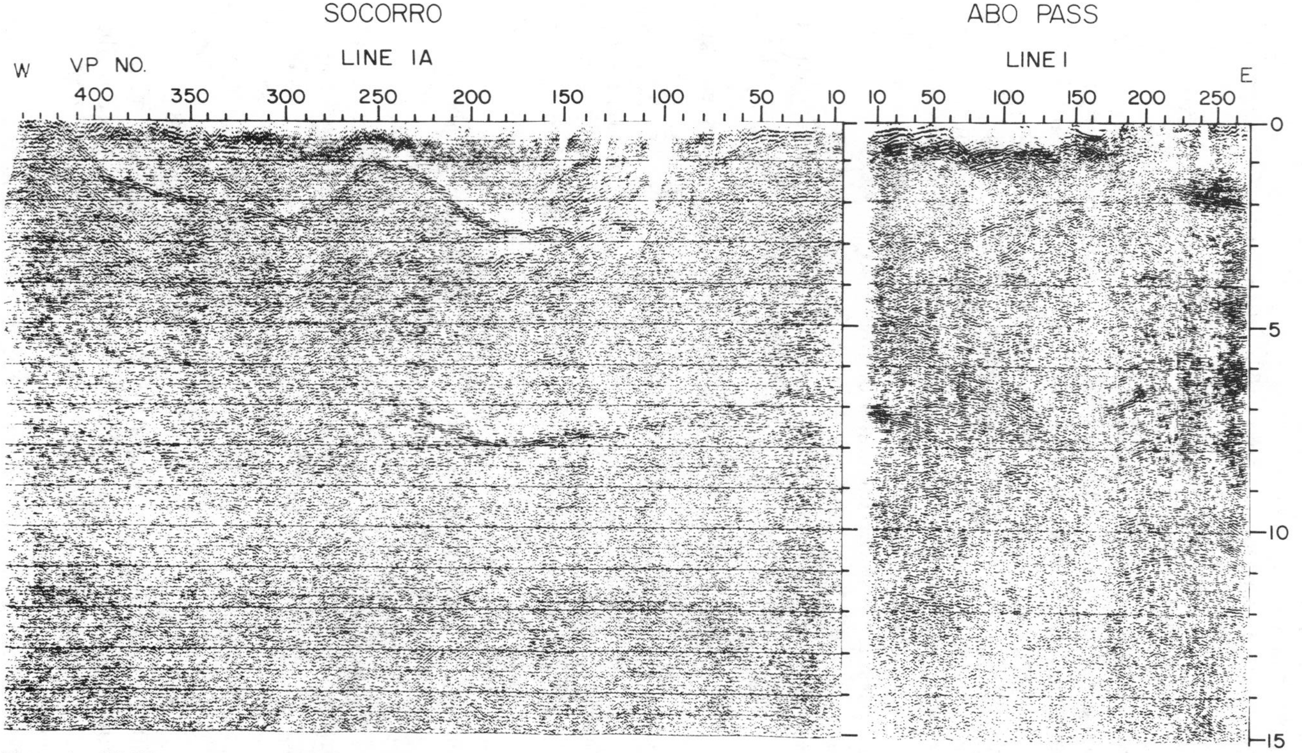

Figure 4 COCORP seismic profiles across the Rio Grande rift near Socorro, New Mexico. The entire width of the rift is traversed by these two profiles. Station spacing is 100 m on Line 1 and 134 m on Line 1A, and the section is unmigrated, with vertical scale in two-way travel time. After Oliver & Kaufman (1976) and Brown et al (1979).

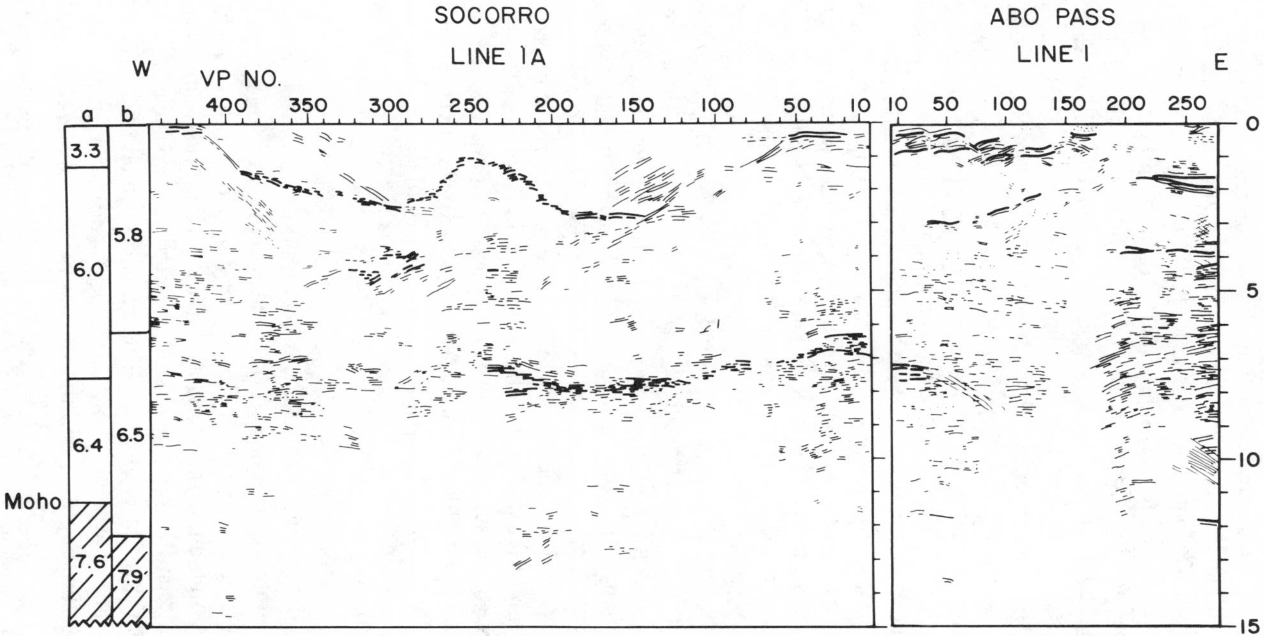

Figure 5 Line drawing interpretation of Figure 4. The proposed magma body can be seen at about 7 s in the middle of the profiles. Numbers in boxes at the left represent the velocity structure of the crust derived from refraction measurements. (*a*) After Olsen et al (1979), (*b*) after Toppazada & Sanford (1976). After Brown et al (1979).

Abo Pass Line 1 (Figure 5) and is defined at the surface by the transition from normal faulted strata in the middle of the rift (VP 10–180 of Abo Pass Line 1) to the rather event-free shallow Precambrian basement beneath the rift flank. A steeply dipping seismically transparent zone underlies the surface expression of the fault. If this steeply dipping zone does represent the eastern rift boundary, then it indicates that this boundary was not formed by listric faulting (Brown et al 1979). This character might be caused by a pervasively sheared fault zone, or a preexisting area of weakness along which faulting was localized. The character might also be caused by intrusions which developed during rifting. The western rift boundary (Figure 4, VP 410) shows no evidence of a similar transparent zone at depth. Surface mapping indicates that it is a high angle fault. On the seismic section the boundary is represented by a moderate angle (about 40°) event to about 2 s depth. Whether this represents a simple, shallowly dipping fault or a composite reflection from a closely spaced set of steeper step faults, or whether some anomalous waveform such as a reflected refraction is responsible for this event is not clear; however, it does appear that the western rift boundary is dissimilar to the eastern boundary.

The major feature on these COCORP profiles is a high amplitude complex group of events at about 7 s (Figures 4 and 5). These events are consistent in depth and dip with a magma body previously inferred from other geophysical data (Sanford et al 1977). Although this is not a unique interpretation of the event, geodetic evidence of recent uplift, high heat flow, and microearthquake data very strongly support the suggestion of a mid-crustal magma body in the middle of the rift. Model studies of this feature suggest that it is composed of numerous localized intrusive layers of variable thickness. Whether the magma represents in situ melting of the crust, or migration of magma from deeper levels and accumulation at some mid-crustal discontinuity, is not known at present (Brown et al 1979).

WIND RIVER MOUNTAINS, WYOMING In Wyoming there are a series of uplifts of Precambrian basement formed during the Laramide orogeny. The uplifts have diverse trends and are often flanked by faults which at the surface are high angle reverse faults to low angle thrust faults. Basement relief between the mountains and the intervening basins can be as much as 10 km (Prucha et al 1965). There has been much discussion in the literature about whether horizontal or vertical movements were responsible for these uplifts (see Smithson et al 1978 for citations). The Wind River mountains are one of the largest of the Laramide uplifts and are flanked on their southwest side by the Wind River thrust.

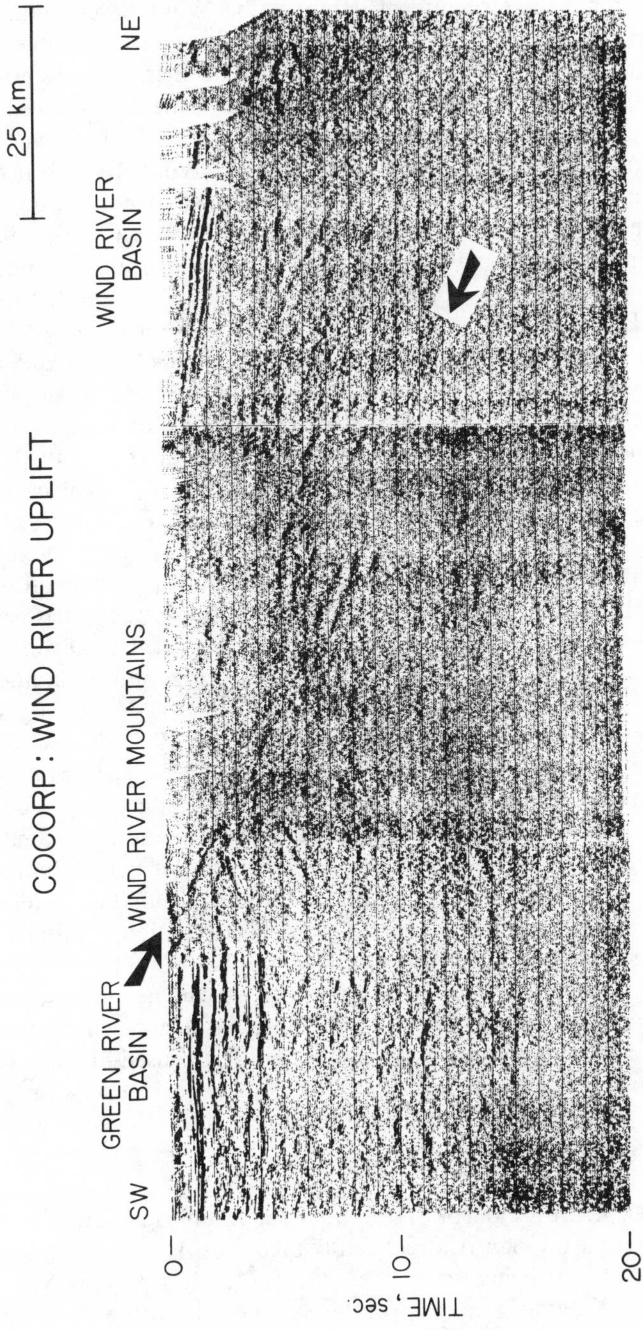

Figure 6 COCORP seismic profiles across the Wind River uplift, Wyoming. The profiles are unmigrated, with vertical scale in two-way travel time. The trace of the Wind River thrust is marked by arrows.

To determine the structure at depth of the Wind River mountains, COCORP collected 158 km of seismic profiles running from the Green River basin, across the Wind River mountains at their southeastern end, and into the Wind River basin. Figure 6 shows these seismic sections (unmigrated). A pronounced set of dipping events is observed on Figure 6 originating at the surface at the position of the Wind River thrust, and trending into the crust at a moderate angle. It can be traced to depths of approximately 24 km, and possibly as deep as 36 km. Changes in line direction indicate an average dip of 30°–45° of this feature, varying from easterly to northeasterly. This feature is interpreted as being the trace of the Wind River thrust, and thus the geometry indicates that crustal compression was the dominant factor in the formation of the Wind River mountains. At least 21 km of horizontal movements have occurred, and 13 km of vertical movements. This structure suggests that regional compression was the cause of the Laramide orogeny (Brewer et al 1980).

Apart from this obvious feature of the COCORP profiles, much interesting deep crustal structure can be observed. On Line 1A there are groups of events of marked curvature. Possibly they represent complex fold structures in the middle and lower crust (Smithson et al 1979). Distinct differences in the seismic response of the basement between the mountains and the basins (Figure 6) suggest larger scale lateral variations. Events under the Green River basin tend to be horizontal, and under the Wind River basin they tend to dip to the northeast. Under the Wind River Mountains, events above and below the Wind River thrust tend to have curvature. Because the surface geology changes considerably some of the apparent differences in basement structure might represent the effects of different surface conditions (multiples, attenuation, etc; Smithson et al 1979). However, it does appear that crustal structure under the Wind River mountains is considerably different from that under the basins on either side. Whether this reflects deformation resulting from movements along the Wind River thrust is not known at present.

It is interesting that the Wind River thrust can be traced into the crust by the fault zone reflections, indicating that a sufficient impedance contrast exists across the fault zone. This could be caused by thick bands of mylonitized rocks, several subparallel fault zones causing structural "tuning" of reflected signals, and/or juxtaposition of different rock types in a heterogeneous crust.

SAN ANDREAS FAULT, CALIFORNIA About 300 km of right lateral slip has accumulated on the San Andreas fault since the Miocene, with present rates of movement being about 5 cm/yr. The fault separates Mesozoic plutonic and metamorphic rocks of the Salinian block (the Pacific plate) on the southwest from highly disrupted late Jurassic–early Cretaceous oceanic crustal sequences of the Franciscan block (North American plate).

As a test of the COCORP method over this feature a 27-km-long seismic line was run near Parkfield, California in 1976, where the fault bifurcates to form a thin sliver. The line is short and crooked, and the data quality is not as good as has been observed elsewhere, but several interesting features may be observed (Long et al 1978).

Beneath the surface trace of the San Andreas fault there are several diffraction curves. This zone of diffractions extends to about 4 s (about 12 km). Below this lies a vertical blank zone about 4-km wide. Assuming that this zone is not a function of surface conditions (for instance, a near-surface region heavily disrupted by faulting might be acoustically opaque) then it is reasonable to suppose that this feature may be formed by similar mechanisms to the blank zone under the eastern boundary of the Rio Grande rift. Earthquakes on the San Andreas fault are generally limited to the upper 15-km depth, and a possible interpretation of the differing seismic character with depth of the fault zone is that the diffraction patterns correspond to a shallow zone of brittle fracturing, whereas the blank area represents a wide zone of disruption, perhaps created by ductile flow. If this interpretation is correct, it strongly constrains rheological models of the San Andreas fault. Most of the deeper crustal events are short and discontinuous. See Long et al (1978) for a discussion and interpretation of these data.

MICHIGAN BASIN, MICHIGAN This is a classic example of a cratonic basin. The Precambrian surface underlying it is an approximately circular depression (about 500 km in diameter) filled with Phanerozoic sediments that reach their maximum thickness of 4.5 km west of Saginaw Bay (Hinze et al 1975). The object of the profiling here was to investigate the deep underlying structure of the basin, and to try to determine what caused its formation.

The COCORP seismic profiles show strong subhorizontal reflections from the sedimentary rocks of the basin. Below these, and truncated by them, is a prominent trough-shaped event approximately 60-km wide, with about 3 km of relief and lying about 9 km below the surface at its deepest point (Jensen et al 1979). A conformable series of reflections underlie the trough-shaped event. The curvature of these events and discontinuities in the events suggest normal faulting of these and underlying horizons. Deeper reflections are sparse, consisting largely of discontinuous segments. These seismic data, together with gravity and magnetic data, strongly support the suggestion that the Michigan basin is underlain by a northwest-southeast–trending rift, probably closely related to the Keweenawan rifting event (about 1100 m.y. ago; Jensen et al. 1979). However, the relationship, if any, between this and the Phanerozoic basin is not clear. Not only did the basin start forming much later (late Cambrian–early Ordovician), but it is also circular, as opposed to the relatively linear Precambrian structure. It is difficult to see the connection between radially symmetric models of basin formation (Haxby et al 1976) and the rift structure.

THE SOUTHERN APPALACHIAN TRAVERSE COCORP studies in the southern Appalachians were initiated to investigate the Brevard Zone, an enigmatic fault zone in the Inner Piedmont, and the subject of a

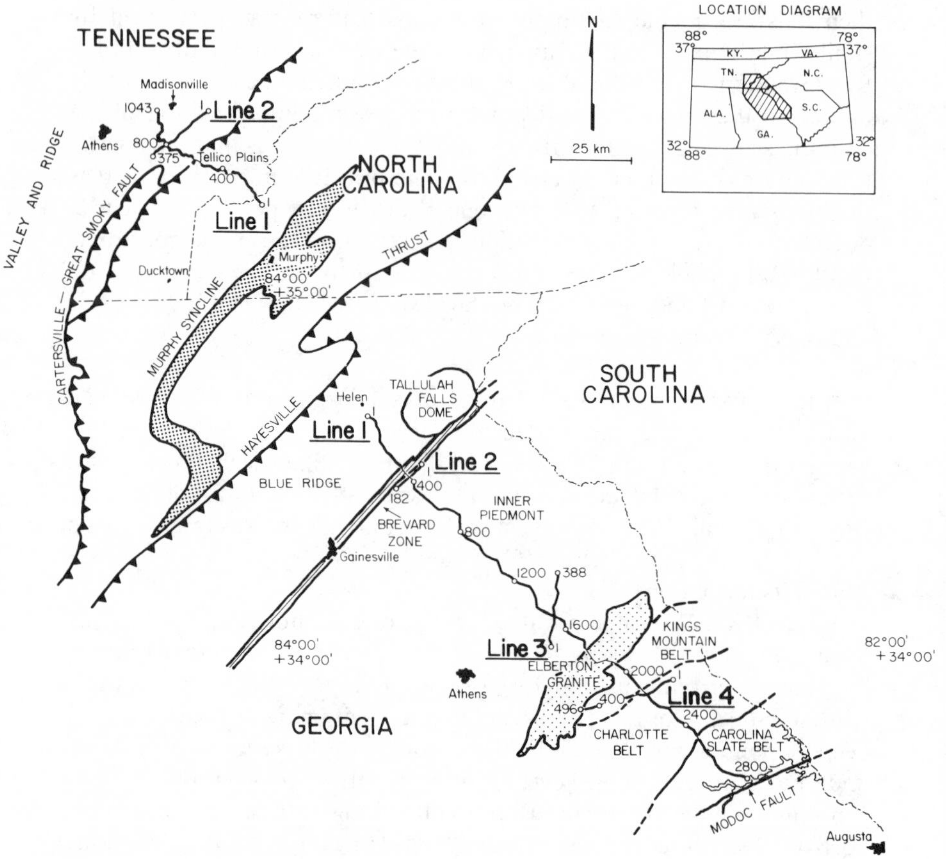

Figure 7 Line location map of the Southern Appalachian traverse showing the geological provinces traversed by the COCORP profiles. From Cook et al (1979).

whole suite of diverse interpretations (see Clark et al 1978 for examples). In the course of COCORP profiling, the Brevard Zone was found to be rooted in and underlain at 2.5–3.5 s by a band of subhorizontal laminated reflections, implying that a more profound structure existed at depth.

The COCORP profiles were extended in order to map out these layered events and determine their relationship to Appalachian geology (Figure 7). These data reveal that the major tectonic feature of the southern Appalachians is an allochthonous sheet of crystalline rocks which has overthrust sedimentary rocks along an extensive sole thrust. The seismic evidence for the existence of this overthrust sheet lies in the band of subhorizontal laminated reflections (Figure 8) which may be correlated

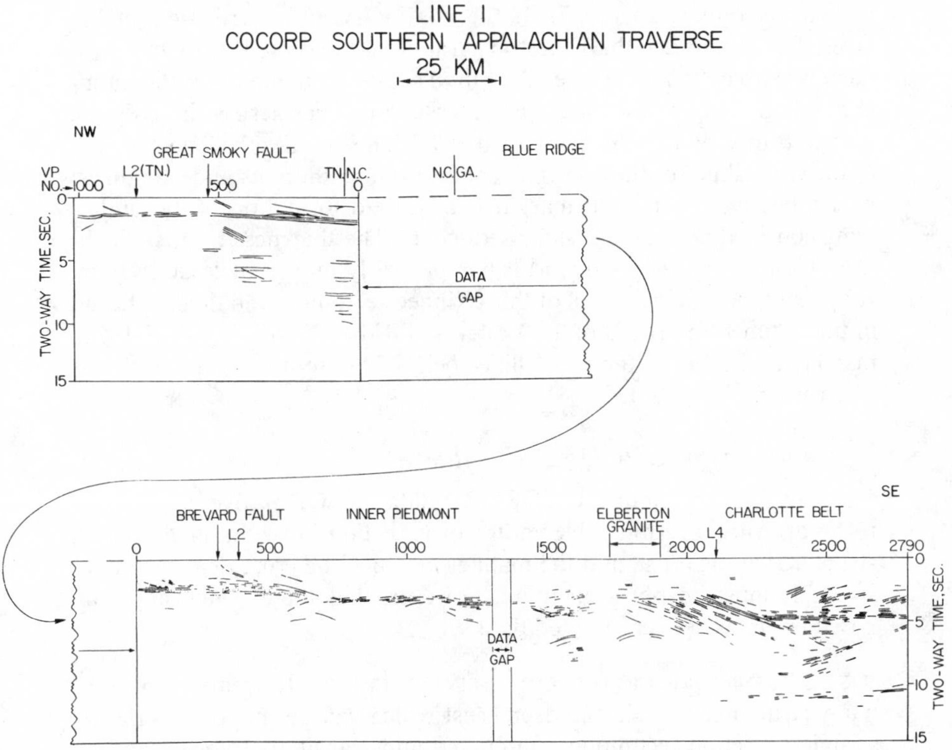

Figure 8 Interpreted line drawing (unmigrated) of the COCORP Southern Appalachian traverse. The Great Smoky fault marks the boundary in the northwest of the southern Appalachians allochthon against the Valley and Ridge Province. The base of the allochthon may be traced as a continuous band of events to the southeast, under the Inner Piedmont, Elberton Granite, and Charlotte Belt, and off the southeastern edge of the section. After Cook et al (1979).

with similar events in the Valley and Ridge Province. In the Valley and Ridge Province, these events are identified seismically and by well data as Cambrian and Ordovician sedimentary rocks (Harris & Milici 1977). From the Valley and Ridge these layered events may be followed to the southeast under the Blue Ridge and Inner Piedmont Provinces, with their thickness varying between 1 and 3 km (Cook et al 1979). The line crosses a large granite body, the Elberton Granite, which outcrops on the southeast edge of the Inner Piedmont. The granite appears to be seismically transparent with the layered events continuous underneath it, and these may be followed further to the southeast, increasing in thickness to about 6 km. Hatcher & Zietz (1978) on the basis of gravity, magnetic, and geological data suggested that the root zone of this allochthon may lie in the Kings Mountain belt (along strike from the Elberton Granite; Figure 7). If the continuity and correlation of the layered events under the Elberton Granite are accepted, then the sedimentary succession under the allochthon southeast of the Inner Piedmont is strikingly similar to that seen on seismic profiles across the Atlantic margin (Grow et al 1979, Cook et al 1979). This implies that most, or all, of the crystalline southern Appalachians is allochthonous and was thrust over a passive continental margin with little deformation of the underlying continental shelf and slope sediments. The thickness of this allochthon varies from 6–15 km, and it has probably moved at least 260 km. It appears that the concepts of thin-skinned tectonics, well demonstrated in the sedimentary rocks of the Valley and Ridge (Harris & Milici 1977), may be applicable to the crystalline rocks of the southern Appalachians as a whole.

Comparisons of COCORP Results

Comparative studies of the COCORP data allow tentative conclusions to be drawn concerning the nature of reflecting horizons in the crust, structures in the crust, and the manner in which the crust deforms under stress. It must be borne in mind, however, that these conclusions are based on an as yet very limited data set.

SEISMIC SIGNATURE OF THE CONTINENTAL CRUST It is apparent that many structures exist in the deep crust which can be mapped using the seismic reflection technique. Laminated and transitional zones may be the cause of many of these high impedance contrasts (see, for example, Fuchs 1969). Laminations may be discontinuous vertically and horizontally, with sporadic structural "tuning" of the reflected signals, perhaps causing the high amplitudes observed.

Allowing for such problems as low signal-to-noise ratio, poor surface conditions, lack of surface control, and possible near-surface reverberations, there are characteristics of the crust seen on COCORP profiles which may be interpreted geologically. Much of the data are composed of short, discontinuous (about 2–3 km long) reflecting segments. This type of crustal signature has been documented by other workers (e.g. Dohr & Meissner 1975), and can be interpreted in a variety of ways. This signature could be explained by a crust composed of metamorphic complexes. It could be caused by small or large scale folding of, or intrusion into, part of the crust that was originally layered, this layering being either sedimentary, volcanic, or gneissic banding. In a complexly deformed crust, some of these events may be originating away from the plane of the profile and, in general, migration of events is important for accurately locating them spatially (Smithson 1978, Bloxsom 1978). Examples of COCORP data where this type of signature is common include Hardeman County (Figure 3) and the Rio Grande rift (Figure 4).

Some COCORP profiles contain strong continuous events deep in the crust which tend to be fairly simple events with infrequent lateral phase changes. Areas where these are observed include Wyoming (about 11.0 and 15.0 s on line 1; Figure 6), Hardeman County (about 2.9 and 3.7 s; Figure 3), the continuation of these events on the Oklahoma data, and the Michigan basin data. These events may possibly be first-order discontinuities, and in some cases may reflect original layering of a sedimentary or volcanic character.

The Wyoming data so far are unusual in that they contain events showing marked curvature (Figure 6, Lines 1 and 1A) at middle and lower crustal levels that are not diffraction hyperbolae. These events might represent complexly folded layering, possibly caused by movements on the Wind River thrust.

Blank or transparent (within the signal bandwidth) zones are also common on the COCORP profiles. Assuming these are not artificial, they could represent highly deformed and folded rock, dips too steep to return reflected energy to the surface, gradational contacts associated with igneous activity (Smithson & Brown 1977), or homogeneous rocks such as uniform igneous intrusions.

With the exception of some features of the Wyoming data and Southern Appalachian data, and obviously hyperbolic events, seismic reflections from the lower crust are usually of fairly low dip (less than about 5°). This could partly reflect the field techniques used which by their nature preferentially enhance shallow dips. However, low dip could also reflect predominantly horizontal layering induced in the crust either by horizontal shearing motions (Phinney 1978) or by gravitational settling in an igneous or metamorphic event (Smithson 1978).

CHARACTER OF THE CRUST-MANTLE BOUNDARY Interestingly, this important boundary between the crust and mantle is not always observed on the COCORP data, and where it is observed its character is quite variable.

Meissner (1973), reviewing the appearance of the Moho under Germany on refraction and reflection records, concluded that it is a velocity gradient a few kilometers thick with a detailed laminar structure. Comparative studies of the COCORP data on the appearance of the Moho (Cook et al 1978) suggest considerable lateral inhomogeneity, with perhaps three basic seismic signatures: (*a*) simple fairly strong reflections (1–2 cycle wavelets), (*b*) wide bands of several reflecting segments, sometimes of more complex waveform, and (*c*) seismically transparent (within the signal bandwidth) zones.

The first type of signature implies a simple boundary, and examples are seen on Wyoming Line 1 at about 15 s (Figure 6), and possibly on the northeast side of the San Andreas fault at 9.0 s. The second type of boundary implies a complexly layered structure. This type could be produced by compositional layering (either an original structure, or created by horizontal shearing motions, or due to relict layering from earlier melting episodes) or this type could be caused by physical layering, such as interlayered solid and partial melts (Clowes & Kanasewich 1972, Meissner 1973, Berry & Mair 1977). Such boundary types appear on the Rio Grande rift data at 11–12 s (Figure 4). It is possible that the layered and podded structures described by Berckhemer (1969) and Mehnert (1975) from the Ivrea Zone could produce this type of seismic signature. The third type of seismic character, that of a transparent or blank zone, could represent a lack of sufficient impedance contrast, a highly deformed boundary, a velocity gradient, or simply a low signal-to-noise ratio.

It thus appears that the crust-mantle boundary may have a variable character, presumably in some way correlated with the geological history of the region under consideration, but in any case consistent with the ideas of a complex and laterally inhomogeneous crust developed in this review.

CRUSTAL FAULTING Other studies of the crust using seismic techniques, especially those of the Russians (Davydova 1975), emphasize a block-like nature, caused by penetrative faulting. These COCORP studies tend to support this view, at least for the shallower depths. The Wind River thrust, the eastern boundary of the Rio Grande rift, and the San Andreas fault (thrust, normal, and strike-slip faults respectively) can all be traced with a reasonable degree of confidence through much of the crust. The southern Appalachians data indicate an allochthon that has moved a considerable distance (at least 260 km), with apparently minimal deformation of the underlying foreland. These data imply that in some cases the crust fractures along fault zones which may remain discrete and localized throughout much of the crust and for considerable lateral extent. The seismic character of the Wind River thrust at depth, especially, may possibly be explained by discrete fault zones (Brewer et al 1980). The crust in this case appears to have fractured as a fairly rigid block. Structures produced by compressional tectonics have so far been very well resolved by COCORP methods. Possible reasons for this are that the resulting structures have a moderate to low dip, and that the mode of thrusting produces sufficient contrasts in the structure or composition of the rocks involved.

Conclusions

The continental crust is indeed worthy of detailed study in three dimensions, and seismic reflection profiling has the capability to do this. The data are as yet limited, and it would be premature to state their ultimate impact on problems such as the genesis and evolution of the crust, or whether they uphold the ideas of plate tectonics in continental crust or not. Structures such as the Wind River mountains and the southern Appalachians allochthon suggest that horizontally directed stresses are important, as might be generated by horizontally moving plates. It may well be that examples of horizontal movements, such as large allochthonous masses, are much more widespread than previously thought.

These first COCORP data demonstrate the utility of the method for providing information on possible hydrocarbon and mineral provinces. They strongly suggest the nature of the dominant tectonic regimes involved in the formation of the Rocky Mountains of Wyoming and in the southern Appalachians, in the former by indicating that basin sequences might extend under the flanks of some of the Laramide basement uplifts, and in the latter by suggesting that Lower Paleozoic sediments of the Valley and Ridge might extend for hundreds of kilometers under the southern Appalachians allochthon (Cook et al 1979).

The Rio Grande rift data, strongly supporting a mid-crustal magma body originally suggested by other workers, have enhanced the area as a prospect for future geothermal energy studies. Finally, data collected in Charleston, South Carolina, not discussed in this review for lack of space, bode well for the application of seismic reflection profiling to the evaluation of seismic hazards.

ACKNOWLEDGMENTS

The COCORP project is funded by the National Science Foundation as part of the US Geodynamics Project. COCORP is dependant upon the advice and help of numerous individuals from government, industry, and the scientific community. Much credit for the success of the operation must go to members of the Executive, Technical Advisory, and Site Selection committees. Members of the Executive committee are A. W. Bally of Shell Oil Co., M. B. Dobrin of the University of Houston, S. Kaufman of Cornell University, J. C. Maxwell of the University of Texas at Austin, R. P. Meyer of the University of Wisconsin at Madison, J. E. Oliver of Cornell University, and R. A. Phinney of Princeton University. We gratefully acknowledge the help and advice of the members of the COCORP project at Cornell in writing this review; they are S. Kaufman, L. D. Brown, D. S. Albaugh, F. A. Cook, L. W. Jensen, G. H. Long, and D. Steiner. Geosource Inc. has been the principal contractor of the COCORP data; the San Andreas fault data were collected by Compagnie Générale de Géophysique. This research was funded in part by NSF grants Nos. EAR 78-23672 and EAR 78-23673. Cornell Contribution to Geology No. 752.

Literature Cited

Beloussov, V. G., Vol'vovski, B. S., Vol'vovski, I. S., Ryaboi, V. A. 1962. Experimental investigation of the recording of deep reflected waves. *Bull. Acad. Sci. USSR, Geophys. Ser.* English transl. No. 8:662–69

Berckhemer, H. 1969. Direct evidence for the composition of the lower crust and Moho. *Tectonophysics* 8:97–105

Berry, M. J., Mair, J. A. 1977. The nature of the earth's crust in Canada. *Geophys. Monogr. Am. Geophys. Union* 20:319–48

Bloxsom, H. I. 1978. Migration and interpretation of COCORP's Hardeman County, Texas, data: Seismic expression of a batholith? *EOS, Trans. Am. Geophys. Union* 59:1137

Braile, L. W., Smith, R. B. 1975. Guide to the interpretation of crustal refraction profiles. *Geophys. J. R. Astron. Soc.* 40: 145–76

Branson, J. C., Moss, F. J., Taylor, F. J. 1976. Deep crustal reflection seismic test survey, Mildura, Victoria, and Broken Hill, New South Wales, 1968. *Dept. Nat. Res. Bur. Min. Res., Geol. Geophys. Rep. 83.* Canberra: Australian Govt. Publ. Service

Brewer, J. A., Smithson, S. B., Oliver, J. E., Kaufman, S., Brown, L. D. 1980. The Laramide orogeny: Evidence from COCORP seismic reflection profiles in the Wind River mountains, Wyoming. *Tectonophysics* 62:165–89

Brown, L. D., Krumhansl, P. K., Chapin, C. E., Sanford, A. R., Cook, F. A., Kaufman, S., Oliver, J. E., Schilt, F. S. 1979. COCORP seismic reflection studies of the Rio Grande rift. In *Rio Grande Rift: Tectonics and Magmatism*, ed. R. E. Riecker, pp. 169–84. Washington, DC: Am. Geophys. Union

Chapin, C. E. 1979. Evolution of the Rio Grande Rift: A summary. See Brown et al 1979, pp. 1–6

Christensen, N. I., Fountain, D. M. 1975. Constitution of the lower continental crust based on experimental studies of seismic velocities in granulite. *Geol. Soc. Am. Bull.* 86:227–36

Clark, H. B., Costain, J. K., Glover, L. 1978. Structural and seismic reflection studies of the Brevard ductile deformation zone near Rosman, North Carolina. *Am. J. Sci.* 278:419–41

Clowes, R. M., Kanasewich, E. R. 1972. Seismic attenuation and the nature of reflecting horizons within the crust. *J. Geophys. Res.* 75:6693–6705

Clowes, R. M., Kanasewich, E. R., Cumming, G. L. 1968. Deep crustal seismic reflections at near-vertical incidence. *Geophysics* 33:441–51

Cook, F. A., Brown, L. D., Kaufman, S. 1978. The nature of the Moho on COCORP reflection data. *EOS, Trans. Am. Geophys. Union* 59:389

Cook, F. A., Albaugh, D. S., Brown, L. D., Kaufman, S., Oliver, J. E., Hatcher, R. D. Jr. 1979. Thin-skinned tectonics in the crystalline southern Appalachians: COCORP seismic reflection profiling of the Blue Ridge and Piedmont. *Geology* 7:563–67

Cumming, G. L., Chandra, N. N. 1975. Further studies of reflections from the deep crust in Southern Alberta. *Can. J. Earth Sci.* 12:539–57

Davydova, N. I. 1975. DSS Studies of deep faults. In Davydova, N. I., ed. *Seismic Properties of the Moho Discontinuity. Akad. Nauk. SSSR*, pp. 66–75

Dix, C. H. 1965. Reflection seismic crustal studies. *Geophysics* 30:1068–84

Dohr, G., Fuchs, K. 1967. Statistical evaluation of deep crustal reflections in Germany. *Geophysics* 32:951–67

Dohr, G., Meissner, R. 1975. Deep crustal reflections in Europe. *Geophysics* 40:25–39

Fowler, J. C., Waters, K. H. 1975. Deep crustal reflection recordings using "VIBROSEIS" method—a feasibility study. *Geophysics* 40:399–410

Fuchs, K. 1969. On the properties of deep crustal reflectors. *Z. Geophys.* 35:133–49

Grow, J. A., Mattick, R. E., Schlee, J. S. 1979. Multichannel seismic depth sections and interval velocities over outer continental shelf and upper continental slope between Cape Hatteras and Cape Cod. In *Geological Investigations of Continental Margins*, ed. J. S. Watkins, L. Montadert, P. W. Dickersen. *Am. Assoc. Petrol. Geol. Mem.* 29:65–83

Ham, W. E., Wilson, J. L. 1967. Paleozoic epeirogeny and orogeny in the central United States. *Am. J. Sci.* 265:332–407

Harris, L. D., Milici, R. C. 1977. Characteristics of thin-skinned style of deformation in the southern Appalachians, and potential hydrocarbon traps. *US Geol. Surv. Prof. Pap. 1018.* 40 pp.

Hasbrouck, W. P. 1964. *A Seismic Reflection Crustal Study in Central Eastern Colorado.* PhD thesis. Colorado School of Mines. 133 pp.

Hatcher, R. D. Jr., Zietz, I. 1978. Thin crystalline sheets in the southern Appalachians, Inner Piedmont and Blue Ridge: Interpretation based upon regional aero-magnetic data. *Geol. Soc. Am. Abstr. Programs* 10:417

Haxby, W. F., Turcotte, D. L., Bird, J. M. 1976. Thermal and mechanical evolution of the Michigan basin. *Tectonophysics* 36:57–75

Hinze, W. J., Kellogg, R. L., O'Hara, N. W. 1975. Geophysical studies of basement geology of southern peninsula of Michigan. *Am. Assoc. Petrol. Geol. Bull.* 59:1562–84

Jeffreys, H. 1926. On near earthquakes. *Mon. Not. R. Astron. Soc. Geophys. Suppl.* 1:385–402

Jensen, L. W., Steiner, D., Brown, L. D., Kaufman, S., Oliver, J. E. 1979. COCORP deep crustal seismic reflection studies in the Michigan basin. *Mich. Basin Geol. Soc. Symp.* 16

Junger, A. 1951. Deep basement reflections in Big Horn County, Montana. *Geophysics* 16:499–505

Kanasewich, E. R., Cumming, G. L. 1965. Near vertical-incidence seismic reflections from the "Conrad" discontinuity. *J. Geophys. Res.* 70:3441–46

Kanasewich, E. R., Clowes, R. M., McCloughlan, C. H. 1968. A buried Precambrian rift in western Canada. *Tectonophysics* 8:513–27

Long, G. H., Brown, L. D., Kaufman, S. 1978. A deep seismic reflection survey across the San Andreas fault near Parkfield, California. *EOS, Trans. Am. Geophys. Union* 59:385

Mair, J. A., Lyons, J. A. 1976. Seismic reflection techniques for crustal structure studies. *Geophysics* 41:1272–90

Mateker, E. J., Ibrahim, A. K. 1973. Deep crustal reflections from a vibratory source. *43rd Ann. Meet. Soc. Expl. Geophys. Prog.*, pp. 21–22 (Abstr.)

Mathur, S. P. 1974. Crustal studies in southwestern Australia from seismic and gravity data. *Tectonophysics* 24:151–82

McCamy, K., Meyer, R. P., Smith, T. J. 1962. Generally applicable solutions of Zoeppritz' amplitude equations. *Seismol. Soc. Am. Bull.* 252:923–55

Mehnert, K. R. 1975. The Ivrea Zone: A model for the deep crust. *Neues Jahrb. Mineral. Abh.* 125:156–99

Meissner, R. 1967. Exploring deep interfaces by seismic wide angle measurements. *Geophys. Prospect.* 15:598–617

Meissner, R. 1973. The "Moho" as a transition zone. *Geophys. Surv.* 1:195–216

Mueller, S. 1977. A new model of the continental crust. *Am. Geophys. Union Monogr.* 20:289–317

Narans, H. D. Jr., Berg, J. W. Jr., Cook, K. L. 1961. Sub-basement seismic reflections in northern Utah. *J. Geophys. Res.* 66:599–603

Oliver, J., Kaufman, S. 1976. Profiling the Rio Grande Rift. *Geotimes* 21:20–23

Oliver, J., Dobrin, M., Kaufman, S., Meyer, R., Phinney, R. 1976. Continuous seismic reflection profiling of the deep basement, Hardeman County, Texas. *Geol. Soc. Am. Bull.* 87:1537–46

Oliver, R. L. 1969. Some observations on the distribution and nature of granulite facies terrains. *Geol. Soc. Aust. Spec. Publ.* 2:259–68

Olsen, K. H., Keller, G. R., Stewart, J. N. 1979. Crustal structure along the Rio Grande rift from seismic refraction profiles. See Brown et al 1979, pp. 127–43

Perkins, W. E., Phinney, R. A. 1971. A reflection study of the Wind River uplift, Wyoming. *Am. Geophys. Union Monogr.* 14:41–50

Phinney, R. A. 1978. Interpretation of reflection seismic images of lower continental crust. *EOS, Trans. Am. Geophys. Union* 59:389

Phinney, R. A., Jurdy, D. M. 1979. Seismic imaging of the deep crust. *Geophysics* 44:1637–60

Prucha, J. J., Graham, J. A., Nickelsen, R. P. 1965. Basement controlled deformation in Wyoming Province of the Rocky Mountain foreland. *Am. Assoc. Petrol. Geol. Bull.* 49:966–92

Robertson, G. 1963. Intrabasement reflections in southwest Alberta. *Geophysics* 28:910–15

Sanford, A. R., Mott, R. P. Jr., Shuleski, P. J., Rinehart, E. J., Caravella, F. J., Ward, R. M., Wallace, T. C. 1977. Geophysical evidence for a magma body in the crust in the vicinity of Socorro, N. M. *Am. Geophys. Union Monogr.* 20:385–404

Schilt, S., Kaufman, S., Long, G. H. 1977. Analysis of 3-d diffraction patterns from COCORP data. *EOS, Trans. Am. Geophys. Union* 58:436

Schilt, S., Oliver, J., Brown, L., Kaufman, S., Albaugh, D., Brewer, J., Cook, F., Jensen, L., Krumhansl, P., Long, G., Steiner, D. 1979. The heterogeneity of the continental crust: Results from deep crustal seismic reflection profiling using the VIBROSEIS technique. *Rev. Geophys. Space Phys.* 17:354–68

Smithson, S. B. 1978. Modelling continental crust: Structural and chemical constraints. *Geophys. Res. Lett.* 5:749–51

Smithson, S. B., Brown, S. K. 1977. A model for lower continental crust. *Earth Planet. Sci. Lett.* 35:134–44

Smithson, S. B., Brewer, J. A., Kaufman, S., Oliver, J. E., Hurich, C. 1978. Nature of the Wind River thrust, Wyoming, from COCORP deep reflection data and from gravity data. *Geology* 6:648–52

Smithson, S. B., Brewer, J. A., Kaufman, S., Oliver, J. E., Hurich, C. 1979. Structure of the Laramide Wind River uplift, Wyoming, from COCORP deep reflection data and from gravity data. *J. Geophys. Res.* 84:5955–72

Smithson, S. B., Decker, E. R. 1974. A continental crustal model and its geothermal implications. *Earth Planet. Sci. Lett.* 22:215–25

Smithson, S. B., Shive, P. N., Brown, S. K. 1977a. Seismic reflections from Precambrian crust. *Earth Planet. Sci. Lett.* 37:333–38

Smithson, S. B., Shive, P. N., Brown, S. K. 1977b. Seismic velocity, reflections, and structure of the crystalline crust. *Am. Geophys. Union Monogr.* 20:254–70

Sollogub, V. B., Pavlenkova, N. I., Ckekunov, A. V. 1968. Deep seismic research in the Ukraine. *Proc. 8th Assem. Eur. Seismol. Comm. 1954 Akakemiai Kiado, Budapest*, pp. 252–60

Sollogub, V. B., Grin, N. E., Gontovaya, L. I. 1975. On application of stacking method to deep crustal studies. *Boll. Geofis. Teor. Appl.* 17:79–84

Steinhart, J. S., Meyer, R. P. 1961. Explosion studies of continental structure. *Carnegie Inst. Washington Publ. 622.* 409 pp.

Stommel, H. E., Graul, J. M. 1978. Current trends in Geophysics. *Arab. J. Sci. Eng. Spec. Issue*, pp. 41–63

Toppozada, T. R., Sanford, A. R. 1976. Crustal structure in central New Mexico interpreted from the Gasbuggy explosion. *Bull. Seismol. Soc. Am.* 66:877–86

Widess, M. B., Taylor, G. L. 1959. Seismic reflections from layering within the Precambrian basement complex, Oklahoma. *Geophysics* 24:417–25

Windley, B. F. 1977. *The Evolving Continents.* New York: Wiley. 385 pp.

Ann. Rev. Earth Planet. Sci. 1980. 8:231–61
Copyright © 1980 by Annual Reviews Inc. All rights reserved

GEOMORPHOLOGICAL PROCESSES ON TERRESTRIAL PLANETARY SURFACES

×10132

Robert P. Sharp
Division of Geological and Planetary Sciences, California Institute of Technology, Pasadena, California 91125

INTRODUCTION

This review deals with features and processes on planetary surfaces, first by examining the impact of photographic explorations of Moon, Mars, and Mercury on studies of surface processes on our own planet, and second by treating matters related to current deformation of Earth's surface.

An unanticipated outfall from the space program has been a strong upsurge of interest in Earth-surface processes. This occurs because understanding and interpretation of features on other planetary surfaces is most effectively approached through analogues from Earth (Belcher et al 1971, Frey 1979, Hartmann 1974, Komar 1979, Malin 1974, 1977, Trevena & Picard 1978, Veverka & Liang 1975, as examples among others). Geologists and geomorphologists (students of landforms and surface processes) are able to make significant contributions to space exploration programs because of their background and experience with earthly forms and processes.

Planetary exploration has proved to be a two-way street. It not only created interest in Earth-surface processes and features as analogues, it also caused terrestrial geologists to look on Earth for features and relationships better displayed on other planetary surfaces. Impact cratering, so extensive on Moon, Mercury, and Mars, is a well-known example (Roddy et al 1978). Another is the huge size of features such as great landslides and widespread evidence of large-scale subsidence and collapse on Mars, which suggest that our thinking about features on Earth may have been too small scaled. One of the lessons from space is to "think big."

0084-6597/80/0515-0231$01.00

With some exceptions (Cotton 1944), volcanologists have been more concerned with mechanisms of eruption and with petrology (chemistry and physical characteristics) than with the morphological features created by volcanism. However, volcanic landforms are now receiving much greater attention because of exploration of Moon and Mars.

The space program has also generated a high level of interest in such phenomena as catastrophic floods, erosional and depositional work of the wind, mass movements (landslides, earthflows, creep), perennially frozen ground, ground ice, sapping processes, and large-scale crustal collapse. Planetologists, who perhaps once thought that permafrost was a type of commercial refrigerator, now use the word freely and confidently, although sometimes too loosely.

On Earth, the work of wind and of mass movements is so overshadowed by erosional and depositional products of water that the effects are hard to evaluate. Opportunity to observe the products of wind and mass movements on other planetary surfaces, particularly on Mars where they now dominate, brings a fuller appreciation of their relative effectiveness on Earth.

Surface features on other planets are largely fossil, affording a record of events and conditions extending back three to four billion years. Many earthly features are also fossil, but of more modest antiquity, a few million years at most. This is so because vigorous terrestrial weathering, erosion, and deposition quickly modify, erase, or bury antecedent forms and the records of past events. Exploration of other planetary surfaces provides opportunity to identify and evaluate features, processes, and conditions that may have been important on Earth at some earlier time but are unappreciated owing to the loss of record.

The problem of distinguishing impact craters from volcanic craters, the morphology of lava-flow complexes, the surface manifestations of ejecta blankets and tephra sheets (fragmental volcanic material), the products of eolian erosion, the creation of channels by processes other than fluvial erosion, the possibilities of ground-ice deterioration on a scale far surpassing anything seen on Earth, the creation of landscapes and landforms by sapping mechanisms, and the role of collapse in creating chaotic terrains, huge chasms, and complex networks of smaller chasms are all matters receiving attention owing to space exploration.

Recently, interest in surface features and processes on Earth has been further stimulated by desire for protection from natural hazards, such as earthquakes, floods, and landslides. Increased concern with soils, alluvium (stream deposits), colluvium (slope mantles), and other Holocene (the last 11,000 years) deposits mantling the earth is demonstrated by formation of the Society for Quaternary Geology (the geology of the last 2–3 million years), establishment of a *Journal of Quaternary*

Research, creation of Quaternary Research institutes, and an outflow of books and professional articles. New geological engineering companies dealing principally with surface problems have sprung up, and older firms have expanded rapidly in size, geographical distribution, and scope. Environmental impact statements for large projects, such as the Alaska pipeline or the liquid natural gas terminal on the Pacific Coast, demand a wide spectrum of talents and interests. Kirk Bryan used to say, with a twinkle in his eye, that a geomorphologist's (also read Quaternary geologist's) best assets were a strong back and a shovel. They have both now been replaced by the backhoe and bulldozer.

Concern with surface forms and processes on Earth affecting human activities directs attention to small-scale landforms (micro-morphology) in their relationship to fault activity and slope stability. Attention is also focused on micro-stratigraphy (history and layering) of surficial materials for the same purposes. Micro-stratigraphic studies are providing information on the recurrence interval of earthquakes of specified magnitude along faults of known historical activity. This is a method of earthquake forecasting, crude and approximate to be sure, but one that represents a solid approach to the subject.

Micro-stratigraphy and micro-morphology are both dependent upon micro-chronology, the determination of relatively short intervals of geological time. Radioactive, chemical, paleomagnetic, and tephra-chronological methods are all making contributions on this score. Determination of rates at which geological processes proceed and geological changes occur is a goal toward which progress is being made on both the relative and absolute fronts, albeit slowly.

Had this paper been prepared a decade ago, almost surely the prime focus would have been upon fluvial processes, then so vigorously pursued under stimulation of a group centered in the Water Supply Division of the US Geological Survey. The subject of fluvial processes is downplayed here simply because the results of that work have now become well known and are widely integrated into the professional literature. Anyone wishing to get a feel for the pertinence and impact of modern hydrological research should read the second chapter in Compton's (1977, pp. 19–46) physical geology book. Not only is this a fine example of pedogogy, it also nicely digests some fruits of that research.

FORMS AND PROCESSES ON OTHER PLANETARY SURFACES

Volcanic Features

Volcanology, always a fascinating subject in its own right, has received greatly increased attention owing to the space program, initially because of the similarity between impact and volcanic craters (Piero 1976, Roddy et al 1978), especially those created by gas-rich explosions, and subsequently because of the variety of volcanic features identified on other planetary surfaces. Planetological interest in volcanology is attested by conferences, field trips, and guidebooks, supported by the National Aeronautics and Space Agency (Greeley 1974, Greeley & King 1977).

Much attention has been given to large shield volcanos of the Hawaiian type (Macdonald & Abbott 1970) and to the associated caldera, because of the impressive development of similar features in the Tharsis region of Mars (Carr 1973, 1974b, Carr et al 1977b). A shield volcano is a cone of gentle slope (5°–15°) and usually of large size, formed primarily by extrusions of fluid lavas from a central vent or flank fissures. A caldera is a large flat-floored depression formed in part by collapse at the summit of such cones. The need to distinguish shield volcanos from other volcanic constructs, as viewed from an orbiting spacecraft at distances in excess of 1000 km, has caused volcanologists and planetary scientists to look more carefully at the pattern of forms created on the flanks of these cones by successive outpourings of lava. Volcanologists are familiar with lava channels, lobate tongues, flow units, lava levees, fissure eruptions, and similar features but have perhaps not fully appreciated the strongly lineated pattern of these forms when viewed from very high altitudes. The size of martian shields, especially Olympus Mons, roughly five times larger than any corresponding feature on Earth, has astounded terrestrial geologists. Smaller volcanic forms are also of interest (Greeley 1973, Malin 1977).

Distinguishing sheets of debris thrown out of an impact crater from sheets or lobes of material extruded from central volcanic vents, one means of differentiating impact from volcanic craters, has focused attention on the detailed morphology and characteristics of lava accumulations and pyroclastic deposits. Pyroclastics include all fragmental material ejected by volcanic explosions. Planetologists may not have appreciated fully the scale and mobility of such deposits or their possible role in creating landforms. On Earth, sheets of firmly welded pyroclastics form great, level, ponded areas within regions of otherwise rugged relief. Mackin's (1969, pp. 743–46) posthumous paper addressed to the possibility of pyroclastic materials on the lunar surface, based on wide experience with such rocks in southwestern Utah and southeastern Nevada, deserves more attention. Considerable thicknesses of pyroclastic debris can be emplaced over large areas by flow, and a thinner blanket is even more widely distributed by air fall. The mantling of topographic relief, possibly by pyroclastic materials, is recognized on Mars (West 1974, Malin 1979), but more attention could and should be given to the unusual forms created when such a blanket is partly removed by erosion.

Among the striking features of the lunar surface are narrow, deep, steep-walled, usually winding chasms, known as rilles, that start and stop abruptly. The abundance of volcanic rock on the lunar surface causes one to seek for an explanation of rilles in some type of volcanic process. This directed attention to tunnels and tubes within earthly lava flows, heretofore of interest as tourist attractions and for harboring perennial deposits of ice. Collapse of the roofs of such features has created small-scale forms resembling lunar rilles. Greeley (1970, 1971a, b) and Greeley & Hyde (1972) have been particularly active in this investigation, having at the same time increased understanding of the function of lava tubes within streams of flowing lava. This is a matter of considerable concern in areas of current volcanic activity, Hawaii for example.

The number, nature, size, and variety of channels on the martian surface go far beyond the category of lunar rilles and raise perplexing questions of origin (Milton 1973, Sharp & Malin 1975, Masursky et al 1977). Again, because of abundant martian volcanic features, consideration has been given to volcanic mechanisms of channel formation. Erosion by ash flows and glowing avalanches has long been posited, but such erosion appears limited to a narrow zone peripheral to the source of such ejections, usually a central vent. Erosion by flowing lava was hypothesized by Carr (1974a), but it seems capable of creating only small channels. Nonetheless, concern over the origin of martian channels has caused terrestrial volcanologists to look with a more critical eye at lava channels on Earth. Some lava channels on volcanic cones are the product of accretion rather than erosion. This occurs because the lateral margins of a narrow lobate lava tongue congeal while the central part flows away leaving a channel confined by natural levees of congealed rock.

Isolated, round-topped, steep-sided hills on the surface of Moon, Mars, and Mercury are something of an enigma. On Earth, similar features are commonly erosion residuals created by an episode of tectonic uplift followed by extensive subaerial erosion. This explanation cannot easily be invoked for many parts of other planetary surfaces where tectonic activity and erosion are minimal, so attention again shifts to volcanism and particularly to tholoids. A tholoid is a mass of highly viscous lava, often glassy, which has been extruded from a central vent, like mastic squeezed from a tube. Being too stiff to run away as a lava flow, it simply accumulates as a mound. Tholoids tend to be dome shaped, but many variations are recognized on Earth (Williams 1932, Macdonald 1972). Their possible abundance on other planetary surfaces, with characteristics largely unmodified by secondary processes, invites study. Volcanology can help immensely in understanding features and processes on other planetary surfaces, and volcanologists in turn have opportunity to inspect features created by volcanism where it is not so severely contested by tectonism and erosion as on Earth.

Eolian Erosion

Terrestrial geologists and engineers have not ignored eolian processes on Earth, but generally more attention has been given to depositional than to erosional products (Blackwelder 1931). However, interest in eolian erosion has been greatly increased by discovery of features on Mars possibly created by wind erosion (Figure 1). Expressions of this interest take the form of NASA support of a field conference on eolian processes (Greeley et al 1978), and laboratory investigations simulating martian features and conditions (Greeley et al 1974a, b, Iversen et al 1976, Wood et al 1974).

Wind is one of the few exogenic processes judged to be currently effective on the surface of Mars (Gifford 1964, Ryan 1964, Loomis 1965, Sagan & Pollack 1969, among others), and an assumption that it has long been a major agent shaping martian features seems justified. Evidence for both eolian deposition (Sagan et al 1971, Cutts & Smith 1973) and erosion (Arvidson 1972, McCauley 1973, Sagan 1973, Cutts 1973, Veverka 1975, Arvidson et al 1976) is advanced from study of orbiter photos. These processes were early hypothesized by McLaughlin (1954a, b, 1956), and the modern view is summarized by Mutch et al (1976, pp. 235–61).

On Earth, geological work by wind is hampered by vegetation, and the results of eolian erosion, even when it occurs, are usually rapidly modified or obscured by fluvial processes. The opportunity to study and evaluate the largely unmodified products of martian eolian erosion, continuing relatively uncontested for long intervals of time, has caused geologists to search more extensively and critically for features of eolian erosion on Earth, with profitable results.

Leaders in this effort have been J. F. McCauley and associates (1977) of the Astrogeology Branch of the US Geological Survey. Their attention has focused largely upon a feature long recognized as a product of eolian erosion, the yardang (Blackwelder 1934, p. 159). A yardang is an elongate streamlined ridge resembling an inverted ship's hull. Although usually of dimensions measurable in meters, some yardangs are up to 60 m high and a few are 2 km long and 1 km wide (McCauley et al 1977, p. 26, 50). Yardangs are recognized in the desert areas of Central China, Iran, Afghanistan, the Arabian Peninsula, Africa (Egypt, Libya, Sahara, Chad, and Namib), Peru, and North America. Although they occupy only a miniscule part of the terrestrial surface and require conditions of

Figure 1 Experienced arid-region geologists viewing this Viking 2 photomosaic would recognize that the martian surface shown has been deflated by wind. Rock fragments rest on the surface, not partly embedded within it. Large rock in center is about 30 cm high. NASA photos, Viking 2-14, P-17689.

high aridity, near-unidirectional winds, favorable material, and some assistance from weathering to form, the existence of yardangs in greater number and larger size in more areas than heretofore appreciated is significant. The McCauley group (1977, p. 169) feels that yardangs are more the product of deflation (removal of loosened material) than of abrasion (mechanical wear of coherent material). The erosional work of wind on Earth may have been greater than generally realized, although the magnitude of eolian erosion has long been a matter of debate (Ball 1927, Blackwelder 1931, Hobbs 1917, Keyes 1912, Tolman 1909).

Ground Ice and Permafrost

Prior to World War II, geologists in the United States, with some exceptions (Leffingwell 1919, Taber 1943, among others), had limited experience with perennially frozen ground compared to scientists in Canada and northern Eurasia, particularly Russia. Military operations during World War II required engineering treatment of frozen ground and ground ice and stimulated investigations and interest on the part of both geologists (Muller 1947) and engineers, which continued with strength into the post-war era. Large-scale operations such as the oil fields and pipelines of Alaska, and environmental considerations relating to them, have made extensive use of this experience and knowledge. The United States can now claim significant contributions and competent workers in the field (Anderson et al 1972, 1973, Black 1976, Corte 1969, Lachenbruch 1962, Péwé 1969, Stearns 1966, Washburn 1973, among many others). The exploration of Mars has created interest in this subdiscipline on the part of a wholly new clientele, the planetary scientists (Mutch et al 1976, Carr & Schaber 1977).

Permafrost is simply a condition of ground perennially at a sub-zero temperature. It may or may not be accompanied by ground ice, depending upon the presence or absence of moisture. Perennial ground ice exists in small intergranular interstices or as larger discrete segregated bodies of various shapes and dimensions. In terms of large-scale surface configurations, ground ice is more important than permafrost, although the former does not exist without the latter, except in a deteriorating phase. Unfortunately, in planetary-science literature the term permafrost is frequently used as synonymous with ground ice.

Mars has experienced some degree of differentiation and presumably, therefore, some degree of degassing, both on a scale more limited than Earth. Even limited degassing by Mars presumably produced water, and a troubling question is what has happened to the martian water? The amount in the atmosphere or locked up in polar ice caps (Smoluchowski 1968) is distressingly small. Water dissociates in the martian atmosphere, and the hydrogen escapes readily (McElroy 1972), so some water has wasted to outer space. Other water may be chemically combined in surface materials (Fanale 1976, Huguenin 1976). A further speculation is that much of the water never got to the surface but was captured as ground ice owing to the rigorous thermal environment of Mars, the current equatorial mean temperature being between 70° and 80°C below zero. The possibility of martian ground ice is supported by some terrain features.

Many of the unusual and striking forms of the present martian surface in equatorial regions appear to be the product of large-scale collapse, which has produced a jumbled topography featuring tilted blocks and a highly fractured crust, termed chaotic terrain (Figure 2). Issuing from

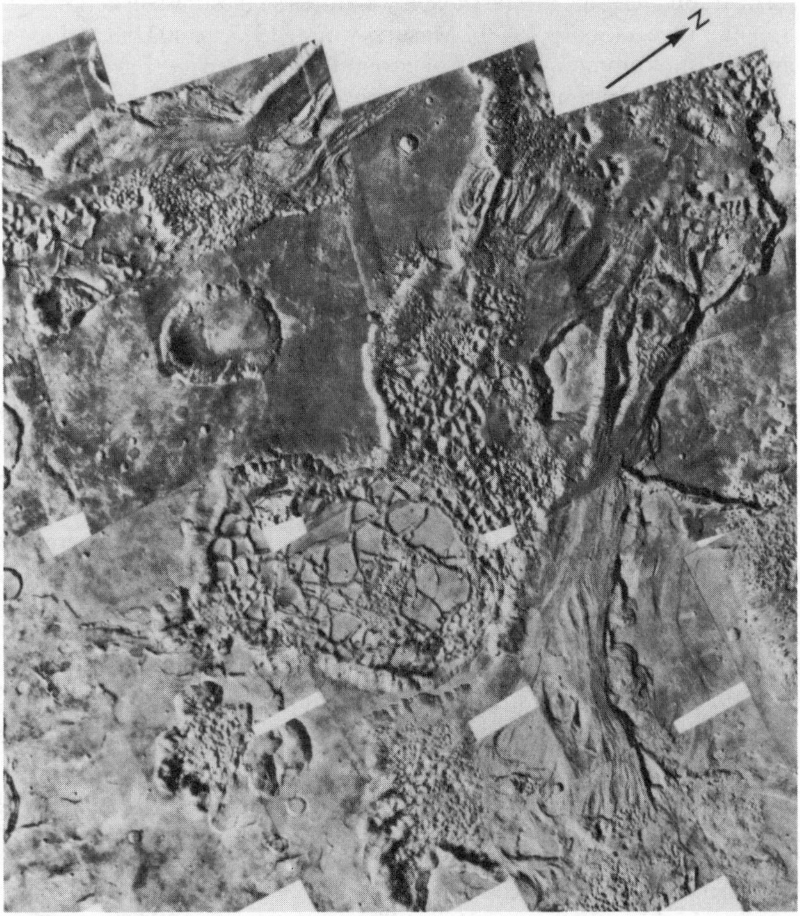

Figure 2 Photomosaic of collapsed chaotic terrain and large outflow channels displaying features resembling those created by huge terrestrial floods. Area viewed about 500 km wide. NASA photos, Viking 1-96, P-19131.

some areas of chaotic terrain are large outflow channels (Milton 1973, Sharp & Malin 1975, Masursky et al 1977) displaying features suggestive of huge floods (Baker & Milton 1974, Baker 1978) that resemble the products of the Spokane (Bretz 1923, Baker 1973, Baker & Nummendal 1978) and Bonneville (Malde 1968) floods of western United States. A reasonable although still highly speculative and not universally accepted interpretation (Wilson et al 1973) is that local melting of great masses of ground ice caused by geothermal anomalies or impacts (Maxwell et al 1973) generated large quantities of liquid water which broke forth to the surface making floods. Fluids other than water are more speculative (Milton 1974, Yung & Pinto 1978).

The fretted terrains of Mars display features suggesting the action of a sapping process (Sharp 1973b, Masursky et al 1977, p. 4022). On Earth, sapping is a common product of ground-water seepage; on Mars the melting or evaporation of ground ice exposed on the face of steep declivities may be the primary cause. Carr et al (1977a, p. 4062) speculate that the extended flow of debris lobes ejected from martian impact craters may have resulted from included water produced by impact melting of ground ice. Many features on the martian surface, possibly caused by ground-ice deterioration, occur on a scale far larger than on Earth.

Some of the smaller forms produced on Earth by the intense freeze and thaw activity that occurs in the surficial active layer of frozen ground may exist on Mars, but they are much too small to be seen on orbiter photos, and none is recognized in photos returned by the Viking landers. On Earth the exceptional mobility of the thin layer of seasonally thawed material above permafrost creates landforms of smoothed, subdued, and rounded configuration. Some parts of the martian surface display similar characteristics on a larger scale (Balsamo & Salsbury 1973). This is not absolute evidence for permafrost because a similar aspect can be created through mantling by pyroclastics or other air-borne debris. Convex rolls at the bottoms of some martian slopes could be the product of the flow and creep that occurs in the thin surficial debris mantle overlying permafrost on slopes. More impressive evidence of mass flow is provided by wide lineated debris streams (Figure 3) filling some martian valleys (Squyres 1978). These resemble the product of vigorous flow and creep produced by freeze and thaw and possibly testify to such activity on Mars, provided the martian environment at the time of their formation permitted liquid water at or near the surface.

The attempt to explain patterns of large polygonal cracks on parts of the martian surface by processes related to freeze and thaw and permafrost conditions (Masursky & Crabill 1976) has not been widely accepted because of the large scale of the cracks. They may be of tectonic origin (Mutch et al 1976, p. 232–34, Strom et al 1975, p. 2482).

Figure 3 Photomosaic of martian Milosyrtis region (34°N, 290°W), north to right, width of area covered about 100 km. Valleys, possibly created by sapping, now partly filled with lineated creeping debris. NASA photos, Viking 1-85, P-18086.

Mars is currently more than cold enough to have an extensive and deep permafrost. Whether or not it has the more important ingredient, ground ice, rests upon the controversial availability of liquid water. Without water, permafrost is more of a stabilizing factor than an active agent creating surface features. The role of CO_2 ice as a substitute for water ice remains speculative (Lambert & Chamberlain 1978). The fact that martian surface features resemble forms created by freeze and thaw and by conditions of permafrost and ground ice on Earth (Anderson et al 1967, 1972, 1973, Gatto & Anderson 1975, Wade & deWys 1968) could be employed as an argument favoring the existence of liquid water in significant quantities on or near the surface at some time during evolution of the planet.

Sapping as a Geomorphological Process

Planetary exploration, especially of Mars, has renewed interest in the efficacy and mechanisms of sapping, a surface process that causes steepening and recession of slopes by undermining at their base. This is a healthy influence, for sapping, although a long-recognized terrestrial process, has not fully received the attention it deserves.

The role of localized ground-water seepage (seepage sapping or spring sapping) in creating steep-head (box) canyons and amphitheaters on Earth is recognized (Jennings 1971, pp. 112–14). Cliff development and recession in areas of horizontally layered rock, sedimentary and volcanic, is also attributed in large part to sapping by seepage. Steep-walled, steep-headed gullies on gently rolling hillsides are more often the product of headward growth by seepage sapping than of surface runoff (Emmett 1968).

On Mars, the wide, flat-floored, steep-walled valleys, box-head canyons, amphitheaters, irregular cliffs, mesas, and buttes of fretted terrain all indicate the action of sapping (Sharp 1973b, Masursky et al 1977, p. 4022). The walls of the gigantic martian troughs (Sharp 1973a, Blasius et al 1977) display evidence suggestive of sapping and many of the abundant and huge martian landslides (Sharp 1973c, Luchitta 1978) have probably been set off by basal sapping. Dendritic canyons tributary to the large martian equatorial troughs and clearly controlled by sets of fractures in the crustal rocks (Figure 4) probably developed by some mechanism of headward growth, probably sapping. Spring sapping is judged a likely cause for other channels on Mars (Sharp & Malin 1975, Masursky et al 1977, p. 4022).

The variety, abundance, and wide distribution of such forms in low and middle martian latitudes suggest that sapping may have been a more effective process in landform genesis on Mars than on Earth, although the effectiveness of sapping on Earth may be underestimated because its products are so strongly overshadowed by the work of running water. Sapping on Earth is produced largely by ground water and only locally by deterioration of ground ice. On Mars the role of these two agents may be reversed.

Ancient Earth Terrains and Features

Renewed interest in states of the Earth's crust during its earliest evolution that may have been largely destroyed, masked, buried, or altered by subsequent events, conditions, and processes has been generated by

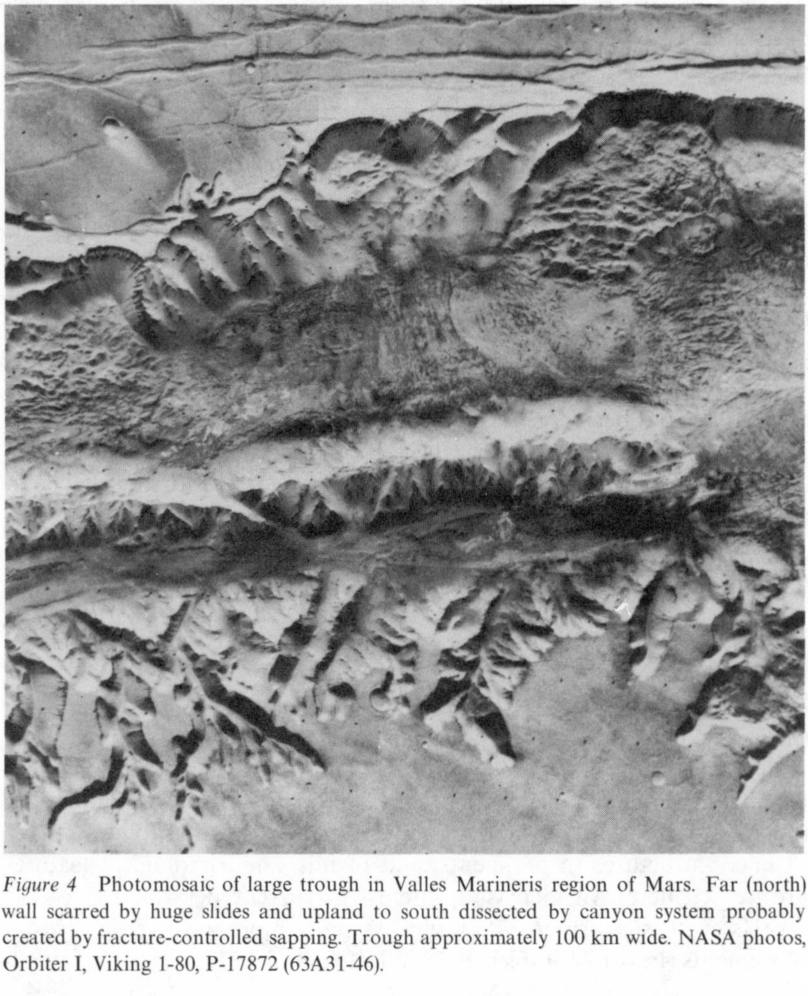

Figure 4 Photomosaic of large trough in Valles Marineris region of Mars. Far (north) wall scarred by huge slides and upland to south dissected by canyon system probably created by fracture-controlled sapping. Trough approximately 100 km wide. NASA photos, Orbiter I, Viking 1-80, P-17872 (63A31-46).

space exploration. The Earth's crust was almost surely subjected to the same initial heavy bombardment by meteoroidal bodies recorded by the surfaces of Moon, Mars, and Mercury, and it must have looked much like those planets in its earliest stage. Not only have the surface features of this bombarded crust been eliminated by weathering and erosion, which is not surprising, but also the deeper scars of large impacts are so far unrecognized. Scars of large, but younger, impacts may be represented by some of Earth's more enigmatic crustal structures (Dietz 1961, 1963, 1964). However, the evidence of very large early impacts, so spectacularly displayed on other planetary surfaces, seems to be gone, or so greatly modified as to be unrecognizable. The layer of impact brecciated rock, widely distributed within the lunar uplands and presumably also well developed on Mars and Mercury, has been so modified by metamorphic processes on Earth as to be unrecognizable.

The martian surface displays types of terrains or features with no known counterpart on Earth, present or past. Examples are the huge equatorial troughs of the Valles Marineris region or the complex rectilinear pattern of chasms in Labyrinthus Noctus. Earth does have limited areas of small-scale chaotic terrain, but nothing comparable in area or scale to the chaotic terrains of Mars, or like the etched terrain of the southern martian subpolar area. Martian fretted terrain has aspects similar to some terrestrial plateau country, but the rocks look different and the genetic processes may not have been the same.

This does not mean that Earth never had such terrains, but evidence of them has not been recognized. Especially intriguing is the evidence of major collapse of large areas on the martian surface. This may well have occurred on Earth, but so far we may not have looked for the effects on a large enough scale.

ON-GOING DEFORMATION OF EARTH'S CRUST

Current concern with earthquake hazards has created an awareness of the inadequacies of the short human record of seismic activity. Allen (1975) has shown how to extend this record backward into the Holocene (the last 11,000 years) by study of small-scale, subtle surface features along fault traces (micro-morphology) and by analysis of relationships within layered materials immediately underlying the surface in fault zones (micro-stratigraphy). Examples of the use of micro-stratigraphic sections, of micro-structures in sediments, and of micro-morphological features are treated here. An increasing interest in current deformation of the Earth's crust is clearly manifest, witness the symposia on recent crustal movements (Pavoni & Green 1975, Whitten et al 1979). Changes attending subsidence generated by withdrawal of underground fluids, often centered in heavily urbanized areas, also merit attention (Poland & Davis 1969, Geerstman 1973).

Seismic Activity and Micro-Morphology

Gilbert (1928) summarized observations, going back at least half a century, that demonstrated the value of scarplets in alluvium and other Holocene deposits for deciphering the history of recent activity along faults. The detailed forms of such scarplets have been accorded particular attention during the past decade.

Slemmons, Cluff, and associates (Slemmons 1967, 1969, 1975, Slemmons et al 1969, Oakeshott & Greensfelder 1972) in sparsely published studies have used a technique of low-altitude, low-sun-angle air photography to identify many previously unrecognized scarplets on the alluvial floor of Owens Valley just west of the Sierra Nevada in California. Particular attention has been given to scarplets with compound profiles (facets of different slope) as a means of identifying recurrent movements. Using all possible variations of time of day and season of the year, they have been able to photograph scarps under illumination-angle differences as great as 45°. Full use of this lighting spectrum has provided information on scarplet morphology not usually available from ground-based inspection. Scarps related to the 1872 Owens Valley earthquake proved to have a mean declinity of 34°, and two earlier families of scarps, or scarp facets, were distinguished, one sloping an average 15° and a still older group at 5–10°. Buckman & Anderson (1979) also address the matter of change in scarp height and slope with age.

Gary Carver (1970, 1975, Carver et al 1969) used Slemmons's methods in a study of recent tectonics within Owens Lake basin at the south end of Owens Valley. There scarps in well-consolidated glacial outwash were as steep as 70–80°. In addition to having gentler slopes, older scarps had rounded brinks, and many displayed a crestal bald strip, easily recognized from the air and presumably formed by rain beat and thread- and sheet-flow of water. Much of the surficial mantle bordering Owens Lake is a mixture of lacustrine and fluvial deposits, which are susceptible to sliding during seismic events. Thus, many of the surface scarplets and associated graben are regarded by Carver as secondary products of mass movement rather than primary tectonic forms.

Carver uses compound scarps with facets of different slope to identify separate episodes of movement, but he also gives attention to small alluvial fans at a scarp base. These fans and the layers of material composing them would appear to merit more careful study in other areas. Sieh (1978b, pp. 1428–34) made a careful study of three sets of small alluvial fans offset along the 1857 break on San Andreas fault just off Soda Lake Road, 3 km northwest of Highway 166 near the southern end of Carrizo Plain in Kern County, California. By measuring the offset of fan heads from source gullies, he demonstrates a 6.5-m right-lateral displacement in 1857 (Figure 5). Comparison of debris volumes of pre- and post-1857 fans suggests that the preceding major offset occurred here about 1470±40 A.D. This is consistent with data gathered elsewhere along the fault (Sieh 1978a).

Study of laterally offset fans can be made only on faults with a significant lateral-slip component of movement. However, investigation of deposits composing small fans, superimposed one on top the other, along the base of faults with only vertical displacements might prove equally rewarding in deciphering fault-movement history.

Wallace (1977), working in an area of historical faulting, Lake Lahontan shorelines, and related deposits in north-central Nevada, has refined and extended earlier studies of scarp morphology in relation to seismic history. Using principles of slope morphology and evolution, he identifies elements within the scarp as the free face, debris slope, and wash slope (Figure 6). Slopes of the free face usually range between 45° and overhanging, depending upon material, compaction, and binding by roots. Scarp brinks remain sharp as long as a free face exists, but they become rounded once the free face disappears. Debris slopes are at the angle of repose, mostly in a range of 34–37°, depending upon material. Wash slopes decline from 15° near their head to 3° near the base. If debris

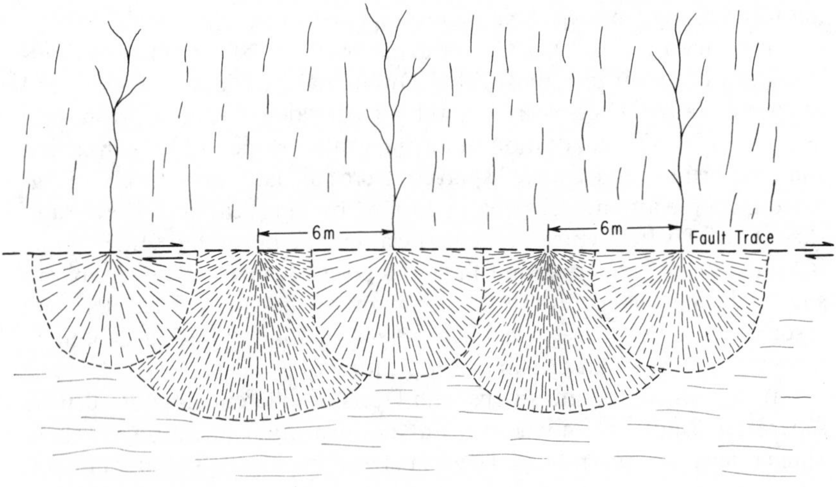

Figure 5 Planimetric sketch of small alluvial fans at hillslope base offset by 1857 lateral displacement on San Andreas fault, modified from Sieh (1978b).

composing the scarp face is heterogeneous, including coarse elements, armoring of slopes by larger fragments can become a significant factor. Gullies created by through-flowing drainage need to be treated separately from gullies originating on the scarp face as to their respective roles in scarp modification.

In general, changes on the scarp face occur primarily by gravity-controlled spalling (loosening and fall of individual fragments) and wash (rain beat, thread flow, sheet wash). Material composing the scarp is obviously a factor (Pease 1979), and climatic environment is important (Wallace 1979).

A scarp face progresses through a sequence of evolutionary stages ranging from an initial 100% gravity control to 90% wash control after a million years. Allowing for differences in material, a debris-controlled slope is usually the dominant feature after a century, and a combination of debris- and wash-controlled slopes appear in 1000–2000 years. Reduction in slope steepness occurs as wash control encroaches more strongly on the debris-controlled slope. Within 10,000 years, rounding of the scarp brink becomes significant. Some measure of the influence of material is shown by a 2-m recession in 20 years of the free face of an historical scarp in poorly indurated fanglomerate compared to a nearby scarp in well-indurated colluvium (ill-sorted slope mantle) which showed little significant modification in 60 years.

Histograms of historical scarp slopes in north central Nevada show bimodal peaks at about 30° and 85° within a widely dispersed range from 15–90°. The dispersal of values become narrower, 10–35°, and unimodal with age, as a shift to lower slope angles occurs. A near-symmetrical unimodal peak at about 25° develops in less than 1200 years.

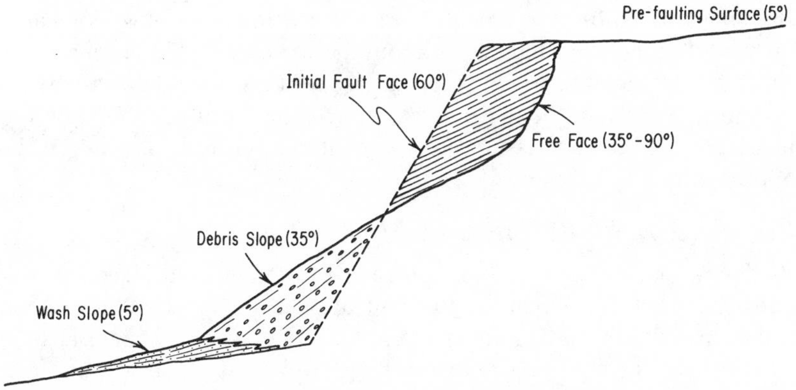

Figure 6 Essential morphological elements of a partly degraded recent fault scarplet, modified from Wallace (1977).

At greater than 12,000 years the spectrum, 5–25°, of slopes is asymmetrical but unimodal at roughly 15°. This peak shifts to about 12° for scarps much greater than 12,000 years. The bimodal histograms of historical scarps reflects the fact that a free face and a gravity-controlled debris slope are both present. The free face disappears within a few hundred, or at most a few thousand, years. The width of the crestal break becomes greater with age, and on scarps older than 12,000 years the break in slope at the toe can be greater than the break at the brink of the scarp.

Morphological features are also created by faulting in areas contiguous to but not directly along the line of faulting, and they can be used to establish the sequence and sometimes the dates of movements. Among such features are knick points (breaks in longitudinal stream profile) and terraces along drainage lines crossing the fault trace, as well as warped, tilted, or displaced shorelines or other normally planar features.

One objective of determining the detailed history of activity along a fault is to ascertain the frequency of seismic events, or the recurrence interval (Lamar et al 1974, Weber 1979). This is difficult through use of micro-morphological features alone because good time control is usually lacking. For example, Wallace (1968, 1969) and Sieh (1978b) working with laterally offset stream channels along San Andreas fault have successfully determined the amount of recurrent displacements, but accurate measurement of recurrence intervals largely eludes them. Some feel for relative values can be obtained from the delicate interplay between incision, alluviation, and bench- and terrace-formation related to individual seismic events (Wallace 1977, p. 1275). Such analyses usually involve subtle and sometimes complex interpretations.

Carver (1970, pp. 74–85) has attempted to use tilting of Owens Lake shorelines by faulting to calculate recurrence intervals of major earthquakes. He concludes that events about the size of the 1872 Owens Valley shock (8.3 M) have occurred in this region about every 1000 years, between extremes of 590 and 1540 years. Although the calculation involves an uncertain and sensitive assumption of constant tilt rate, it is an interesting approach.

Seismic Activity Recorded by Micro-Stratigraphy

Work by Clark et al (1972) on Borrego Mountain earthquake of 1968 and by Sieh (1978a) on San Andreas fault at Pallett Creek, both in southern California, shows that the micro-stratigraphy of surficial materials can extend an historical record of faulting backward by centuries, even millenia. Dating of layers within deposits is fundamental to successful micro-stratigraphic work. This can be accomplished by historical records,

^{14}C dating of associated organic materials, relationship to lake shorelines and associated deposits or glacial moraines and outwash, identifiable volcanic ash layers, tree rings, lichen, soil zones, and sedimentation rates among other methods.

Clark and associates excavated trenches across the 1968 break on Coyote Creek fault near Borrego Mountain in easternmost San Diego County to expose underlying surficial materials. Coyote Creek fault is a branch of San Jacinto fault, historically the most active seismic structure in southern California. The excavation showed progressively greater displacement of older strata with increasing depth along the fault plane as it cut through deposits of Holocene Lake Cahuilla. The oldest dated layer, 3000 years B. P. (before present), is offset vertically, including bending by drag, 3 times more (1.7 m vs 0.56 m) than the base of the youngest (860 years B. P.) dated layer. Abundant freshwater snail and clamlike shells provide ^{14}C dates at various levels. Reliability of these dates was checked with ^{14}C ages obtained from mussel shells, charcoal, and calcareous tufa (lakeshore $CaCO_3$) in Lake Cahuilla deposits at other sites. The data support a conclusion that the recurrence interval for seismic events of the approximate magnitude of the 1968 quake (6.8) is roughly 200 years $\pm$ 60 to 70 years.

Sieh studied, in painstaking detail, thicker accumulations of marsh deposits, silt, sand, and gravel with interlayers of peat, exposed in natural and artificial excavations along the bank of Pallett Creek on San Andreas fault north of San Gabriel Mountains in Los Angeles County. Age relationships were established by ^{14}C dating of peat and calculation of sedimentation rates for peat (1.4 mm/year) and silt (1.9 mm/year). Displaced strata and associated sand-boil structures provided evidence for 9 major seismic events since the Sixth Century AD. The latest was the 8.25 magnitude 1857 Fort Tejon earthquake, and 6 of the preceding 8 events were seemingly of similar size. The recurrence intervals of the 9 major events are bimodally distributed at 100 and 230 years, with an average of 160 years. Nine data points do not make a valid statistical sample, but they are better than just one (the historical record), and the limits of 100 to 230 years for the recurrence of large earthquakes along this reach of the San Andreas fault, inactive since 1857, is a useful parameter.

Intra-Stratigraphic Structures of Seismic Origin

Aside from actual displacement of strata along a fault plane, other structures generated by seismic shocks within soft sediments can be used to date earthquakes. Such structures have the advantage of wider geographic distribution, as they need not be confined to the fault zone.

Sims (1975) capitalized on the opportunity to study bottom sediments in Van Norman Reservoir when it was emptied following the San Fernando (California) earthquake of 1971 (magnitude 6.6). These sediments, accumulated during the 56-year life of the reservoir, displayed internal deformation at three levels. The structures included load casts (local depressions into an underlying layer by more competent materials), small intra-layer folds, heave ups, contorted laminae, and pseudo-nodules. This type of feature is most likely to develop by plastic deformation within a liquified host material, and Sims judged that they formed in a sediment layer close to the water interface.

Four moderate to large earthquakes (1930, 1933, 1952, and 1971) occurred within contiguous southern California during the life of the reservoir. From the thickness of material between deformed layers and estimated rates of accumulation, Sims concludes that the 1930, 1952, and 1971 events generated the 3 zones of deformation observed.

A search for deformed zones within Holocene sedimentary accumulations in other areas of historical seismic activity has revealed 5 deformed layers within Lake Cahuilla deposits near Brawley in Imperial Valley (California) and 14 deformed layers, of possible seismic origin, in finely layered clays of Puget Sound (Washington). Sieh (1978a) used sand-boil structures (mushroom-like extrusions of liquified sand formed during earthquakes) to establish the relative position of specific seismic events within the beds at Pallett Creek. Lamar et al (1979) have recognized possible seismogenic deformation within young sediments of now-drained Kern Lake in San Joaquin Valley, California. A continued and expanded search for zones of seismogenic deformation in other Holocene accumulations would seem a desirable endeavor.

Micro-Chronology

A necessary element in micro-stratigraphy and micro-morphology is a sense of time. The advent of ^{14}C dating was a great boon to this need, and other methods for measuring short time intervals and dates have been or are being developed and applied. These include $^{40}K/^{40}A$, ^{210}Pb, uranium daughter products, amino acids, fluorine impregnation, hydration rims on volcanic glass, tephrochronology, and paleomagnetism. Some of these procedures are secondary in the sense that they depend upon an absolute datum furnished by radioactive clocks. Tephrochronology and paleomagnetism are intimately tied to micro-stratigraphy and have the advantage of regional extent compared to spot measurements.

Tephra (pyroclastics) includes all fragmental material ejected by volcanic explosions. Layers of tephra incorporated within other Quaternary materials have long been used as a basis for correlation and relative dating (Swinford 1949, Wilcox 1965, D. G. W. Smith & Westgate 1969, Izett et al 1970, H. W. Smith et al 1977a, b, Porter 1978, as examples). Modern chemical and petrological analyses provide means of fingerprinting tephra layers from different sources more effectively (Borchardt et al 1971), and improvements in precision of $^{40}K/^{40}A$ analyses of younger materials have made absolute age assignments increasingly reliable.

Tephrochronology is most useful to micro-stratigraphers working near or downwind from volcanic centers active within the last one or two million years, but layers of fine volcanic ash have been singularly identified 1000 miles from their source. Tephrochronology appears destined for wider and more effective use in the future. Recent research shows that variations in the strength and reversals of Earth's magnetic field (paleomagnetism) can be used in correlation of young as well as older deposits (Opdyke et al 1977, Johnson 1975, Mörner 1977, Vitorello & van der Voo 1977, Manabe 1977, Davis et al 1977, Maenaka et al 1977).

The use of plants, mostly trees, in studies of surface processes and landform evolution, a technique long practiced by a few botanically inclined investigators (Cooper 1923, 1937, 1939, Heusser 1954, 1956, Lawrence 1950, 1958a, b, Sigafoos 1964, among others) is now enjoying renewed attention and wider application (Alestalo 1971, La Marche 1968, Page 1970, Shroder 1978, Zoltai 1975). La Marche & Wallace (1972) and Sieh (1978b) have recently used dendrochronology to date movements on San Andreas fault by asymmetry or compression of tree rings resulting from tilting or disturbance of trees.

Geodesy and Recent Deformation

Recently, a strong renaissance of interest in using geodesy as a means of detecting, measuring, and understanding current deformation of the Earth's crust has occurred (Kaula 1978, pp. 20–22). MacDoran (American Geophysical Union Meeting 1979) commented specifically upon exciting geophysical applications of the dusty old discipline of leveling.

The vast store of leveling and triangulation data gathered over many decades by public agencies and bureaus is proving to be a virtual gold mine of useful information (Castle et al 1974, 1975). An example is the dramatic discovery of southern California's "Palmdale Bulge" (Castle et al 1976), perhaps too hastily christened, as additional investigations (Bennett 1977, Castle et al 1977) show it to be an elongated ridge extending from Point Arguello on the coast at least 580 km eastward to the Arizona border with the point of greatest uplift, 45 cm, between Twenty-Nine Palms and Amboy, a considerable distance east of Palmdale. The bulge (or ridge) began to rise in the early 1960s, culminated in 1974, and had collapsed about 50 percent by 1977. Evidence of earlier upwarps in this general area are gleaned from still older surveys.

Analyses of old survey data from other parts of southern California are yielding useful tectonic information (Thatcher 1979) and may eventually produce additional surprises. Such endeavors will be interesting to monitor (Buchanan-Banks et al 1975). A study by Wood & Elliott (1979) reveals an uplift with westward tilting of the northern part of the Peninsular Ranges block involving elevation changes as great as 0.4 m within the earliest 20th Century. Thatcher (1976) shows how geodetic data may give premonitory warning of possible thrust-fault earthquakes (Castle et al 1975). Although the examples cited are drawn from southern California, similar investigations are under way in other parts of the US (Isachsen 1975, Brown & Oliver 1976, Reilinger et al 1977) and in many other parts of the World, including Alaska, Poland, Russia, India, Japan, New Zealand, Switzerland, and South America.

Aside from old methods and old data, geophysicists working on problems of crustal strain and seismicity (Meade 1975) have been quick to appreciate the value of various modern techniques of electronic surveying (Huggett & Slater 1975, Savage et al 1973, 1978, Prescott & Savage 1976) that appeared following World War II (Ewing & Mitchell 1970, p. 108, Tomlinson & Burger 1977). These newer systems include the use of modulated light waves (geodimeter), pulsed radio waves (tellurometer), laser ranging to satellites including Moon (Bender & Silverberg 1975), radio doppler, and various types of radio and radar altimeters. Recently, radio interferometry employing extra-galactic radio sources, so distant that they behave as fixed points, has been used over very long baselines (Coates et al 1975, Whitcomb 1976).

Some of these methods measure only distance, resulting in trilateration (Prescott & Savage 1976) in place of the older procedure of measuring angles (triangulation). However, ranging techniques yield three-dimensional data that can be used to determine elevation differences. The virtue of most new methods is the speed with which they can be executed and the long distances over which they can operate with high accuracy.

Earth scientists working in areas of current tectonic activity easily become preoccupied with faulting and its various manifestations. Warping is a much more subtle tectonic process, less dramatic and often less dangerous or damaging. Nonetheless, it is interesting, more extensive in area, and in urbanized regions economically significant.

Antecedent warping can be detected by geological relationships, if a suitable reference datum, such as the Funeral basalt of the Death Valley region, exists. Landforms, such as alluvial fans, can also record warping (Hooke 1972).

One good method of detecting and measuring current warping is by the time-proven procedure of tiltmetering. Long-period horizontal pendulum seismometers, borehole pendulums, wet tubes, telescopic spirit leveling, mercury pools, and tide gauges can all be used as tiltmeters (Rikitake 1976, pp. 126–49). The antecedent eastward tilting of the Death Valley block, indicated by geomorphological and lacustral evidence (Hunt 1966, pp. 46–47, Hunt & Maybe 1966, p. 100), was confirmed by installation of simple tiltmeters (Greene 1966). Tiltmeters are now being improved (Eaton 1959, Huggett et al 1976) and more widely deployed in areas of known or suspected crustal instability. They have long been used in Hawaii as part of the eruption-prediction program.

Like all human procedures, tiltmetering suffers from the brevity of the historical record, and geologists do well to look for natural tiltmeters of longer record. Ponded water bodies can be employed as tiltmeters (Sirén 1951), witness recent studies of Salton Sea in Imperial Valley (Wilson & Wood 1980). Using records of water-level staff gages maintained at 3 sites along the shore for over 25 years, a 1952-to-1972 downward tilting to the southeast of 11 cm over a 38-km distance was demonstrated. A reversal in direction of tilt set in after 1972. Hunt (1966, pp. 46–47) used the salt pan of a dessicated lake about 2000 years old to demonstrate tilting. Shoreline warping of more ancient vintage was long ago recognized by Gilbert (1890) for Lake Bonneville and for the glacial great lakes by Upham (1884) and Leverett & Taylor (1915), among others.

Geomorphologists and limnologists have probably not given sufficient attention to the micro-features of lake shorelines as possible indicators of ongoing tilting. A search for small-scale features indicating emergence along one reach of a lake shoreline and of submergence on the opposing shore may require patience and a sharp eye for detail but could be rewarding. Shores subject to ice shove may be unsatisfactory.

CONCLUSIONS

Interest in processes shaping the surface of Earth has been stimulated and broadened by exploration of other planetary bodies and by man's increasing desire to live as comfortably and economically as possible in harmony with the natural environment and in reasonable safety with respect to natural geological disasters, such as earthquakes, landslides, floods, volcanic eruptions, and related phenomena.

Since features seen on other planetary surfaces are best understood in terms of earthly analogues, space exploration has stimulated interest among planetologists in such terrestrial phenomena as volcanism, work of the wind, huge catastrophic floods, perennially frozen ground (permafrost), ground ice, sapping as an erosional process, and mass movements (landslides, creep, earthflows).

Increased interest in volcanology has been generated by the number and variety of volcanic features (cones, domes, rilles, lava fields, flow patterns, and above all craters and caldera) seen on the surfaces of other planetary bodies including Moon, and now the satellites of Jupiter. This interest has beneficially increased and broadened the study of volcanic forms on Earth. Not enough attention has yet been given to the possible role of tephra blankets (fragmented volcanic materials) in landscape evolution on extraterrestrial bodies. Such blankets may have been emplaced upon and partly or wholly stripped from underlying surfaces.

The likelihood of extensive eolian erosion on Mars has sparked a profitable search for more and larger features of eolian erosion on Earth. Abundant large erosional channels on Mars have created greater appreciation of phenomena related to huge catastrophic floods on Earth, such as the Spokane and Bonneville events of western United States.

Widespread evidences of collapse, extensive flooding, topographic sapping, and creeping surface debris on Mars have broadened interest in frozen ground, ground ice, and the freeze-thaw process, matters heretofore of concern principally to a small group of specialists. Recognition of the role of these processes and conditions in shaping terrestrial landscapes has been enhanced by interest in their possible effects on other planets.

The surface of Earth, in its earliest stage, must have looked much like the present surfaces of Moon, Mars, and Mercury, although vestiges of this early state, including a thick and widespread mantle of breccia (broken rock), have so far escaped detection owing to reconstitution by erosion, deposition, and metamorphism. The huge scale of features visible on other planetary surfaces should motivate geologists to look at Earth through wider eyes.

In the realm of man's desire to live in greater harmony with nature, micro-morphology, micro-stratigraphy, and micro-chronology are playing ever more creative roles. For example, such studies show the recurrence interval for 6–7 magnitude earthquakes on Coyote fault of the San Jacinto system (California) to be on the order of 200 ± 70 years, and on San Andreas fault, north of San Gabriel Mountains (California), to average 160 years within limits of 100 to 230 years for quakes of magnitude 8.

The deciphered history of displacements on active faults is being extended back into prehistoric times by study of fault-scarp forms (micro-morphology), of the succession of deposits laid down at the base of such scarps (micro-stratigraphy), and by the offset of features, such as small alluvial fans, along the traces of lateral-slip faults. Seismogenic deformation of soft sediments is proving useful in deciphering the history of fault displacements in areas remote from the actual fault zone.

Measurements of short intervals of time (micro-chronology) are essential to all micro-morphological and micro-stratigraphic investigations. Geochemical methods of short-time measurements, other than ^{14}C, are being developed and applied. Paleomagnetism and finger-printed layers of volcanic ash are providing means of correlating layers within young surficial materials.

Current deformation of segments of Earth's crust by warping and creep are probably more ubiquitous than usually perceived. Terra is not so firma. Modern geodectic techniques, as well as vast accumulations of older geodectic data, are leading to the discovery and determination of areas and rates of current tectonic deformation, witness the Palmdale Bulge.

The need, opportunity, and supporting techniques for study of features and processes on the terrestrial surface have never been greater than at present.

Literature Cited

Alestalo, J. 1971. Dendrochronological interpretation of geomorphic processes. *Fennia* 105:1–40

Allen, C. R. 1975. Geological criteria for evaluating seismicity. *Geol. Soc. Am. Bull.* 86:1041–57

American Geophysical Union. 1979. Southern California anomalies. *EOS, Trans. Am. Geophys. Union* 60:50

Anderson, D. M., Gaffney, E. S., Law, P. F. 1967. Frost phenomena on Mars. *Science* 155:319–22

Anderson, D. M., Gatto, L. W., Ugolini, F. 1972. An antarctic analog of martian permafrost terrain. *Antarctic J. US* 7:114–16

Anderson, D. M., Gatto, L. W., Ugolini, F. 1973. An examination of Mariner 6 and 7 imagery for evidence of permafrost terrain on Mars. In *Permafrost: The North American Contribution to the Second International Conference*, pp. 449–508. Natl. Acad. Sci. 783 pp.

Arvidson, R. E. 1972. Aeolian processes on Mars: Erosive velocities, settling velocities, and yellow clouds. *Geol. Soc. Am. Bull.* 83:1503–8

Arvidson, R. E., Coradini, M., Carusi, A., Coradini, A., Fulchignomi, M., Federico, C., Funiciello, R., Salomone, M. 1976. Latitudinal variation of wind erosion of crater ejecta deposits on Mars. *Icarus* 27:503–16

Baker, V. R. 1973. Paleohydrology and sedimentology of Lake Missoula flooding in eastern Washington. *Geol. Soc. Am. Spec. Pap. 144*. 79 pp.

Baker, V. R. 1978. The Spokane Flood controversy and the martian outflow channels. *Science* 202:1249–56

Baker, V. R., Milton, D. J. 1974. Erosion by catastrophic floods on Mars and Earth. *Icarus* 23:27–41

Baker, V. R., Nummendal, D., eds. 1978. *The Channeled Scabland*. Office of Planetary Geology Programs, NASA. 186 pp.

Ball, J. 1927. Problems of the Libyan Desert. *Geogr. J.* 70:32–34

Balsamo, S. R., Salsbury, J. W. 1973. Slope angle and frost formation on Mars. *Icarus* 18:156–63

Belcher, J. D., Veverka, J., Sagan, C. 1971. Mariner photography of Mars and aerial photography of Earth: Some analogies. *Icarus* 15:241–52

Bender, P. L., Silverberg, E. C. 1975. Present tectonic-plate motions from lunar ranging. *Tectonophysics* 29:1–7

Bennett, J. 1977. Palmdale "Bulge" update. *Calif. Geol.* 30:187–89

Black, R. F. 1976. Features indicative of permafrost. *Ann. Rev. Earth Planet. Sci.* 4:75–94

Blackwelder, E. 1931. The lowering of playas by deflation. *Am. J. Sci.* 21:140–44

Blackwelder, E. 1934. Yardangs. *Geol. Soc. Am. Bull.* 45:159–66

Blasius, K. R., Cutts, J. A., Guest, J. E.,

Masursky, H. 1977. Geology of the Valles Marineris: First analysis of imaging from the Viking I orbiter primary mission. *J. Geophys. Res.* 82:4067–91

Borchardt, G. A., Harvard, M. E., Schmitt, R. A. 1971. Correlation of volcanic ash deposits by activation analysis of glass separates. *Quat. Res.* 2:247–60

Bretz, J H. 1923. The channeled scabland of the Columbia Plateau. *J. Geol.* 31: 617–49

Brown, L. D., Oliver, J. E. 1976. Vertical crustal movements from leveling data and their relation to geologic structure in the Eastern United States. *Rev. Earth Space Phys.* 14:13–35

Buchanan-Banks, J. M., Castle, R. O., Ziony, J. I. 1975. Elevation changes in the central Transverse Ranges near Ventura, California. *Tectonophysics* 29: 113–25

Buckman, R. C., Anderson, R. E. 1979. Estimation of fault-scarp ages from a scarp-height–slope-angle relationship. *Geology* 7:11–14

Carr, M. H. 1973. Volcanism on Mars. *J. Geophys. Res.* 78:4049–62

Carr, M. H. 1974a. The role of lava erosion in the formation of lunar rilles and Martian channels. *Icarus* 22:1–23

Carr, M. H. 1974b. Tectonism and volcanism of the Tharsis region of Mars. *J. Geophys. Res.* 79:3943–49

Carr, M. H., Crumpler, L. S., Cutts, J. A., Greeley, R., Guest, J. E., Masursky, H. 1977a. Martian impact craters and emplacement of ejecta by surface flow. *J. Geophys. Res.* 82:4055–65

Carr, M. H., Greeley, R., Blasius, K. R., Guest, J. E., Murray, J. B. 1977b. Some martian volcanic features as viewed from the Viking orbiter. *J. Geophys. Res.* 82: 3985–4015

Carr, M. H., Schaber, G. G. 1977. Martian permafrost features. *J. Geophys. Res.* 82: 4039–54

Carver, G. A. 1970. Quaternary tectonism and surface faulting in the Owens Lake Basin, California. *Tech. Rep. AT-2.* Univ. Nevada, Mackay Sch. Mines. 103 pp.

Carver, G. A. 1975. Shoreline deformation at Owens Lake. *Calif. Geol.* 28:111

Carver, G. A., Slemmons, D. B., Glass, C. E. 1969. Surface faulting patterns in Owens Valley, California. *Geol. Soc. Am. Abstr. Programs.* 1:9–10 (Abstr.)

Castle, R. O., Alt, J. N., Savage, J. C., Balazs, E. I. 1974. Elevation changes preceding the San Fernando earthquake of February 9, 1971. *Geology* 2:61–66

Castle, R. O., Church, J. P., Elliott, M. R., Morrison, N. L. 1975. Vertical crustal movements preceding and accompanying the San Fernando earthquake of February 9, 1971: A summary. *Tectonophysics* 29: 127–40

Castle, R. O., Church, J. P., Elliott, M. R. 1976. Aseismic uplift in southern California. *Science* 192:251–53

Castle, R. O., Elliott, M. R., Wood, S. H. 1977. The southern California uplift. *EOS, Trans. Am. Geophys. Union* 58:495 (Abstr.)

Clark, M. M., Grantz, A., Rubin, M. 1972. Holocene activity of the Coyote Creek fault as recorded in sediments of Lake Cahuilla, *US Geol. Surv. Prof. Pap.* 787: 112–20

Coates, R. J., Clark, T. A., Counselman, C. C., Shapiro, I. I., Hinteregger, H. F., Rogers, A. E., Whitney, A. R. 1975. Very long baseline interferometry for centimeter accuracy geodectic measurements. *Tectonophysics* 29:9–18

Compton, R. R. 1977. *Interpreting the Earth.* New York: Harcourt. 554 pp.

Cooper, W. S. 1923. The recent ecological history of Glacier Bay, Alaska. *Ecology* 4:93–128, 223–46, 355–65

Cooper, W. S. 1937. The problem of Glacier Bay, Alaska: A study of glacier variations. *Geogr. Rev.* 27:37–62

Cooper, W. S. 1939. A fourth expedition to Glacier Bay, Alaska. *Ecology* 20:130–59

Corte, A. E. 1969. Geocryology and engineering. In *Reviews in Engineering Geology*, ed. D. J. Varner, G. Kiersch, pp. 119–85. Geol. Soc. Am. 350 pp.

Cotton, C. A. 1944. *Volcanoes as Landscape Forms.* Christchurch, New Zealand: Whitcombe & Tombs. 416 pp.

Cutts, J. A. 1973. Wind erosion in Martian polar regions. *J. Geophys. Res.* 78:4211–21

Cutts, J. A., Smith, R. S. U. 1973. Eolian deposits and dunes on Mars. *J. Geophys. Res.* 78:4139–54

Davis, P., Smith, J., Kukla, G. J., Opdyke, N. D. 1977. Paleomagnetic study at a nuclear power plant site near Bakersfield, California. *Quat. Res.* 7:380–97

Dietz, R. S. 1961. Astroblemes. *Sci. Am.* 205:50–58

Dietz, R. S. 1963. Astroblemes—ancient meteorite-impact structures on the earth. In *The Moon, Meteorites, and Comets*, ed. B. M. Middlehurst, G. P. Kuiper, pp. 285–300. Chicago, Ill: Univ. Chicago Press. 810 pp.

Dietz, R. S. 1964. Sudbury structure as an astrobleme. *J. Geol.* 72:412–34

Eaton, J. P. 1959. A portable water-tube tiltmeter. *Bull. Seismol. Soc. Am.* 49:301–16

Emmett, W. W. 1968. Gully erosion. In *The Encyclopedia of Geomorphology*, ed.

R. W. Fairbridge, pp. 517–19. New York: Reinhold. 1295 pp.

Ewing, C. E., Mitchell, M. M. 1970. *Introduction to Geodesy*. New York: Elsevier. 304 pp.

Fanale, F. P. 1976. Martian volatiles: The degassing history and chemical fate. *Icarus* 28:179–202

Frey, H. 1979. Martian canyons and African rifts: Structural comparisons and implications. *Icarus* 37:142–55

Gatto, L. W., Anderson, D. M. 1975. Alaskan thermokarst terrain and possible martian analog. *Science* 188:255–57

Geertsman, J. 1973. Land subsidence above compacting oil and gas reservoirs. *J. Petrol. Tech.* 25:734–44

Gifford, F. A. 1964. A study of martian yellow clouds that display movement. *Mon. Weather Rev.* 92:435–40

Gilbert, G. K. 1890. Lake Bonneville. *US Geol. Surv. Monogr. 1*. 438 pp.

Gilbert, G. K. 1928. *Studies of Basin and Range Structure. US Geol. Surv. Prof. Pap. 153*. 92 pp.

Greeley, R. 1970. Observations of actively forming lava tubes and associated structures, Hawaii. *Modern Geol.* 2:207–23

Greeley, R. 1971a. Lunar Hadley Rille: Considerations of its origin. *Science* 172: 722–25

Greeley, R. 1971b. Lava tubes and channels in the lunar Marius Hills. *Moon* 3:289–314

Greeley, R. 1973. Mariner 9 photographs of small volcanic structures on Mars. *Geology* 1:175–80

Greeley, R., ed. 1974. *Geologic Guide to the Island of Hawaii*. Office of Planetary Programs, NASA. 257 pp.

Greeley, R., Hyde, J. H. 1972. Lava tubes of the cave basalt, Mount St. Helens, Washington. *Geol. Soc. Amer. Bull.* 83: 2397–2418

Greeley, R., Iversen, J. D., Pollack, J. B., Udovich, N., White, B. R. 1974a. Wind tunnel studies of martian aeolian processes. *Proc. R. Soc. London Ser. A* 341: 331–60

Greeley, R., Iversen, J. D., Udovich, N., White, B. R. 1974b. Wind tunnel simulation of light and dark streaks on Mars. *Science* 183:847–49

Greeley, R., King, J. S., eds. 1977. *Volcanism of the Eastern Snake River Plain, Idaho*. Office of Planetary Programs, NASA. 308 pp.

Greeley, R., Womer, M. B., Papson, R. P., Sudis, P. D., eds. 1978. *Aeolian features of Southern California*. Office of Planetary Geology Programs, NASA. 264 pp.

Greene, G. W. 1966. Tiltmeter measurements. In Hunt, C. B., Maybe, D. R., 1966. *Stratigraphy and Structure, Death Valley, California*, pp. 112–14. *US Geol. Surv. Prof. Pap. 494-A*. 162 pp.

Hartmann, W. K. 1974. Geological observations of martian arroyos. *J. Geophys. Res.* 79:3951–57

Heusser, C. J. 1954. Glacier fluctuations, forest succession, and climatic variations in the Canadian Rockies. *Am. Philos. Soc. Yearb.* 355–65

Heusser, C. J. 1956. Postglacial environments in the Canadian Rocky Mountains. *Ecological Monogr.* 26:263–302

Hobbs, W. H. 1917. The erosional and degradational processes of deserts with especial reference to the origin of desert depressions. *Assoc. Am. Geogr. Ann.* 7: 43–60

Hooke, R. LeB. 1972. Geomorphic evidence for Late Wisconsin and Holocene tectonic deformation, Death Valley, California. *Geol. Soc. Am. Bull.* 83:2073–98

Huggett, G. R., Slater, R. E. 1975. Precision electromagnetic distance-measuring instrument for determining secular strain and fault movement. *Tectonophysics* 29: 19–27

Huggett, G. R., Slater, R. E., Pavlis, G. 1976. Precision leveling with a two-fluid tiltmeter. *Geophys. Res. Lett.* 3:754–56

Huguenin, R. L. 1976. Mars: Chemical weathering as a massive volatile sink. *Icarus* 28:203–12

Hunt, C. B. 1966. *Hydrologic Basin Death Valley, California. US Geol. Surv. Prof. Pap. 494B*. 138 pp.

Hunt, C. B., Maybe, D. R. 1966. *Stratigraphy and Structure Death Valley, California. US Geol. Surv. Prof. Pap. 494A*. 162 pp.

Isachsen, Y. W. 1975. Possible evidence for contemporary doming of the Adirondack Mountains, New York, and suggested implications for regional tectonics and seismicity. *Tectonophysics* 29:169–81

Iversen, J. D., Pollack, J. B., Greeley, R., White, B. R. 1976. Saltation threshold on Mars: The effect of interparticle force, surface roughness, and low atmospheric density. *Icarus* 29:381–93

Izett, G. A., Wilcox, R. E., Powers, H. A., Desborough, G. A. 1970. The Bishop ash bed, a Pleistocene marker bed in the western United States. *Quat. Res.* 1:121–32

Jennings, J. N. 1971. *Karst*. Cambridge, Mass.: MIT Press. 252 pp.

Johnson, N. M. 1975. Magnetic polarity stratigraphy of Pliocene-Pleistocene terrestrial deposits and vertebrate faunas, San Pedro Valley, Arizona. *Geol. Soc. Am. Bull.* 86:5–12

Kaula, W. M., Chairman. 1978. *Geodesy: Trends and Prospects*. Washington, DC: Natl. Acad. Sci. 86 pp.
Keyes, C. R. 1912. Deflative scheme of the geographic cycle in an arid climate. *Geol. Soc. Am. Bull.* 23:537–62
Komar, P. D. 1979. Comparisons of the hydraulics of water flows in martian outflow channels with flows of similar scale on Earth. *Icarus* 37:156–81
Lachenbruch, A. H. 1962. *Mechanics of Thermal Contraction Cracks and Ice Wedge Polygons. Geol. Soc. Am. Spec. Pap. 70*. 69 pp.
Lamar, D. L., Merifield, P. M., Proctor, R. J. 1974. Earthquake recurrence intervals on major faults in southern California. In *Geology, Seismicity and Environmental Impact*, ed. D. E. Moran, J. E. Slosson, R. O. Stone, C. A. Yelverton, pp. 265–76. Los Angeles, Calif: Spec. Publ. Assoc. Eng. Geol. 444 pp.
Lamar, D. L., Muir, S. G., Merifield, P. M. 1979. Possible earthquake deformed sediments in Kern Lake, Kern County, California. *Geol. Soc. Am. Abstr. Programs* 11(3):88 (Abstr.)
La Marche, V. C. 1968. Rates of slope degradation as determined from botanical evidence, White Mountains, California. *US Geol. Surv. Prof. Pap.* 352I:341–77
La Marche, V. C., Wallace, R. E. 1972. Evaluation of effects on trees of past movements on the San Andreas fault, northern California. *Geol. Soc. Am. Bull.* 83:2665–76
Lambert, R. St. J., Chamberlain, V. E. 1978. CO_2 permafrost and martian topography. *Icarus* 34:568–80
Lawrence, D. B. 1950. Estimating dates of recent glacier advance and recession rates by studying tree growth layers. *Trans. Am. Geophys. Union* 31:243–48
Lawrence, D. B. 1958a. Historic landslides of the Gros Ventre Valley, Wyoming. *Mazama* 40:10–20
Lawrence, D. B. 1958b. Glaciers and vegetation in southeastern Alaska. *Am. Sci.* 46:89–122
Leffingwell, E. deK. 1919. *The Canning River Region, Northern Alaska. US Geol. Surv. Prof. Pap. 109*. 251 pp.
Leverett, F., Taylor, F. B. 1915. *The Pleistocene of Indiana and Michigan and the History of the Great Lakes. US Geol. Surv. Monogr. 53*. 529 pp.
Loomis, A. A. 1965. Some geologic problems of Mars. *Geol. Soc. Am. Bull.* 76:1083–1104
Lucchitta, B. K. 1978. A large landslide on Mars. *Geol. Soc. Am. Bull.* 89:1601–9
Macdonald, G. A. 1972. *Volcanoes*, pp. 108–121. Englewood Cliffs, NJ: Prentice-Hall. 510 pp.
Macdonald, G. A., Abbott, A. T. 1970. *Volcanoes in the Sea*. Honolulu: Univ. Hawaii Press. 441 pp.
Mackin, J. H. 1969. Origin of lunar Maria. *Geol. Soc. Am. Bull.* 80:735–48
Maenaka, K., Yokoyama, T., Shiro, I. 1977. Paleomagnetic stratigraphy and biostratigraphy of the Plio-Pleistocene in the Kinki district, Japan. *Quat. Res.* 7:341–62
Malde, H. E. 1968. *The Catastrophic Late Pleistocene Bonneville Flood in the Snake River Plain, Idaho. US Geol. Surv. Prof. Pap. 596*. 52 pp.
Malin, M. C. 1974. Salt weathering on Mars. *J. Geophys. Res.* 74:3888–94
Malin, M. C. 1977. Comparison of volcanic features of Elysium (Mars) and Tibesti (Earth). *Geol. Soc. Am. Bull.* 88:908–19
Malin, M. C. 1979. Mars: Evidence of indurated deposits of fine materials. *2nd Int. Colloq. Mars, NASA Conf. Publ. 2072*, p. 54 (Abstr.)
Manabe, K.-I. 1977. Reversed magnetozone in late Pleistocene sediments from the Pacific Coast of Odaka, northeast, Japan. *Quat. Res.* 7:372–79
Masursky, H., Boyce, J. M., Dial, A. L., Schaber, G. G., Strobell, M. E. 1977. Classification and time of formation of martian channels based on Viking data. *J. Geophys. Res.* 82:4016–38
Masursky, H., Crabill, N. L. 1976. Search for the Viking 2 landing site. *Science* 194:62–68
Maxwell, T. A., Otto, E. P., Picard, M. D., Wilson, R. C. 1973. Meteorite impact: A suggestion for the origin of some stream channels on Mars. *Geology* 1:9–10
McCauley, J. F. 1973. Mariner 9 evidence for wind erosion in equatorial and mid-latitude regions of Mars. *J. Geophys. Res.* 78:4123–37
McCauley, J. F., Grolier, M. J., Breed, C. S. 1977. *Yardangs of Peru and Other Desert Regions. US Geol. Surv. Interagency Rep. Astrogeology 81*. 177 pp. See also *Geomorphology in Arid Regions*, ed. D. O. Doehring, pp. 233–69. Binghamton, N.Y.: Publ. in Geomorphology. 272 pp.
McElroy, M. B. 1972. Mars: An evolving atmosphere. *Science* 175:443–45
McLaughlin, D. B. 1954a. Volcanism and eolian deposition on Mars. *Geol. Soc. Am. Bull* 65:715–17
McLaughlin, D. B. 1954b. Changes on Mars as evidence of wind deposition and volcanism. *Astron. J.* 60:261–70
McLaughlin, D. B. 1956. The volcanic-eolian hypothesis of Martian features. *Publ. Astron. Soc. Pacific* 68:211–18

Meade, B. K. 1975. Geodectic surveys for monitoring crustal movements in the United States. *Tectonophysics* 29: 103–12
Milton, D. J. 1973. Water and processes of degradation in the Martian landscape. *J. Geophys. Res.* 78: 4037–47
Milton, D. J. 1974. Carbon dioxide hydrate and floods on Mars. *Science* 183: 654–56
Mörner, N.-A. 1977. The Gothenburg magnetic excursion. *Quat. Res.* 7: 413–27
Muller, S. W. 1947. *Permafrost or Permanently Frozen Ground.* Ann Arbor, Mich: Edwards Bros. 231 pp.
Mutch, T. A., Arvidson, R. E., Head, J. W., Jones, K. L., Saunders, R. S. 1976. *The Geology of Mars.* Princeton, NJ: Princeton Univ. Press. 400 pp.
Oakeshott, G. B., Greensfelder, R. W. 1972. 1872–1972, one hundred years later. *Calif. Geol.* 25: 55–61
Opdyke, N. D., Lindsay, E. H., Johnson, N. M., Downs, T. 1977. The paleomagnetism and magnetic polarity stratigraphy of the mammal bearing section of Anza Borrego State Park, California. *Quat. Res.* 7: 316–29
Page, R. 1970. Dating episodes of faulting from tree rings: Effects of the 1958 rupture of the Fairweather fault on tree growth. *Geol. Soc. Am. Bull.* 81: 3085–94
Pavoni, N., Green, R., eds. 1975. Recent crustal movements. *Tectonophysics* 29: 1–552
Pease, R. C. 1979. Fault scarp degradation in alluvium near Carson City, Nevada. *Geol. Soc. Am. Abstr. Programs* 11(3): 121
Péwé, T. L., ed. 1969. *The Periglacial Environment: Past and Present.* Montreal: McGill-Queens. 487 pp.
Piero, L. 1976. *Volcanoes and Impact Craters on the Moon and Mars.* New York: Elsevier. 432 pp.
Poland, J. F., Davis, G. H. 1969. Land subsidence due to withdrawal of fluids. In *Reviews in Engineering Geology*, ed. D. J. Varnes, G. Kiersch, pp. 187–269. 350 pp.
Porter, S. C. 1978. Glacier Peak tephra in the north Cascade Range, Washington: Stratigraphy, distribution, and relationship to late glacial events. *Quat. Res.* 10: 30–41
Prescott, W. H., Savage, J. C. 1976. Strain accumulation on the San Andreas fault near Palmdale, California. *J. Geophys. Res.* 81: 4901–8
Reilinger, R. E., Citron, G. P., Brown, L. D. 1977. Recent vertical crustal movements from precise leveling data in southwestern Montana, Western Yellowstone National Park, and the Snake River Plain. *J. Geophys. Res.* 82: 5349–59
Rikitake, T. 1976. *Earthquake Prediction.* Amsterdam: Elsevier. 357 pp.
Roddy, D. J., Peppin, R. O., Merrill, R. B., eds. 1978. *Impact and Explosion Cratering: Planetary and Terrestrial Implications.* New York: Pergamon. 1301 pp.
Ryan, J. A. 1964. Notes on martian yellow clouds. *J. Geophys. Res.* 69: 3759–70
Sagan, C. 1973. Sandstorms and eolian erosion on Mars. *J. Geophys. Res.* 78: 4155–61
Sagan, C., Pollack, J. B. 1969. Windblown dust on Mars. *Nature* 223: 791–94
Sagan, C., Veverka, J., Gierasch, P. 1971. Observational consequences of Martian wind regimes. *Icarus* 15: 253–78
Savage, J. C., Prescott, W. H., Kinoshita, W. T. 1973. Geodimeter measurements along the San Andreas fault. In *Proc. Conf. on Tectonic Problems of the San Andreas Fault System*, ed. R. L. Kovach, A. Nur, pp. 44–53. *Stanford Univ. Publ. Geol. Sci. 13.* 494 pp.
Savage, J. C., Prescott, W. H., King, M. L. N. 1978. Strain in southern California: Measured uniaxial north-south regional compression. *Science* 202: 883–85
Sharp, R. P. 1973a. Mars: Troughed terrain. *J. Geophys. Res.* 78: 4063–72
Sharp, R. P. 1973b. Mars: Fretted and chaotic terrain. *J. Geophys. Res.* 78: 4073–83
Sharp, R. P. 1973c. Mass movements on Mars. In *Geology, Seismicity, and Environmental Impact*, ed. D. E. Moran, J. E. Slosson, R. O. Stone, C. A. Yelverton, pp. 115–22. Dallas, Texas: Assoc. Eng. Geologists. 444 pp.
Sharp, R. P., Malin, M. C. 1975. Channels on Mars. *Geol. Soc. Am. Bull.* 86: 593–609
Shroder, J. F. 1978. Dendrochronological analysis of mass movements on Table Cliff Plateau, Utah. *Quat. Res.* 9: 168–85
Sieh, K. E. 1978a. Prehistoric large earthquakes produced by slip on the San Andreas fault at Pallett Creek, California. *J. Geophys. Res.* 83: 3907–39
Sieh, K. E. 1978b. Slip along the San Andreas fault associated with the great 1857 earthquake. *Bull. Seismol. Soc. Am.* 68: 1421–48
Sigafoos, R. S. 1964. *Botanical Evidence of Floods and Floodplain Deposition. US Geol. Surv. Prof. Pap. 485 A.* 35 pp.
Sims, J. D. 1975. Determining earthquake recurrence intervals from deformational structures in young lacustrine sediments. *Tectonophysics* 29: 141–52
Sirén, A. 1951. On computing the land uplift from the lake water level records in Finland. *Fennia* 73(5): 1–181
Slemmons, D. B. 1967. Pliocene and Quaternary crustal movements of the Basin-

and-Range province, U.S.A. *J. Geosci., Osaka City Univ.* 10:91–103

Slemmons, D. B. 1969. New methods of studying regional seismicity and surface faulting. *Trans. Am. Geophys. Union* 50: 397–98

Slemmons, D. B. 1975. Cenozoic deformation along the Sierra Nevada province and Basin and Range province boundary. *Calif. Geol.* 28:99–105

Slemmons, D. B., Carver, G. A., Cluff, L. S. 1969. Historic faulting in Owens Valley, California. *Geol. Soc. Am. Spec. Pap.* 121:559–60 (Abstr.)

Smith, D. G. W., Westgate, J. A. 1969. Electron probe technique for characterizing pyroclastic deposits. *Earth Planet. Sci. Lett.* 5:313–19

Smith, H. W., Okazaki, R., Knowles, C. R. 1977a. Electron microprobe data for the tephra attributed to Glacier Peak, Washington. *Quat. Res.* 7:197–206

Smith, H. W., Okazaki, R., Knowles, C. R. 1977b. Electron microprobe analysis of glass shards from tephra assigned to set W, Mount St. Helens, Washington. *Quat. Res.* 7:207–17

Smoluchowski, R. 1968. Mars: Retention of ice. *Science* 159:1348–50

Squyres, S. W. 1978. Martian fretted terrain: Flow of erosional debris. *Icarus* 34:600–13

Stearns, R. S. 1966. *Permafrost.* US Army Material Command, Cold Regions Res. & Eng. Lab. 77 pp.

Strom, R. G., Trask, N. J., Guest, J. E. 1975. Tectonism and volcanism on Mercury. *J. Geophys. Res.* 80:2478–508

Swinford, A. 1949. Source area of Great Plains Pleistocene volcanic ash. *J. Geol.* 57:307–11

Taber, S. 1943. Perennially frozen ground in Alaska, its origin and history. *Geol. Soc. Am. Bull.* 54:1433–548

Thatcher, W. 1976. Episodic strain accumulation in Southern California. *Science* 194: 691–95

Thatcher, W. 1979. Horizontal crustal deformation from historic geodectic measurements in southern California. *J. Geophys. Res.* 84:2351–70

Tolman, C. F. 1909. Erosion and deposition in the southern Arizona bolson region. *J. Geol.* 17:136–63

Tomlinson, R. W., Burger, T. C. 1977. *Electronic Distance Measuring Instruments.* Falls Church, Virginia: Am. Congr. Surv. Mapping. 79 pp. 3rd ed.

Trevena, A. S., Picard, M. D. 1978. Morphometric comparison of braided martian channels and some braided terrestrial features. *Icarus* 35:385–94

Upham, W. 1884. Lake Agassiz: a chapter in glacial geology. *Minnesota Geol. Surv. Ann. Rep.* 11:137–53

Veverka, J. 1975. Variable features on Mars V: Evidence for crater streaks produced by wind erosion. *Icarus* 25:595–601

Veverka, J., Liang, T. 1975. An unusual landslide feature on Mars. *Icarus* 24:47–50

Vitorello, I., van der Voo, R. 1977. Magnetic stratigraphy of Lake Michigan sediments obtained from cores of lacustrine clays. *Quat. Res.* 7:398–412

Wade, F. A., deWys, J. N. 1968. Permafrost features in the Martian surface. *Icarus* 9:175–85

Wallace, R. E. 1968. Notes on stream channels offset by the San Andreas fault, southern Coast Ranges, California. In *Geological Problems of the San Andreas Fault System*, ed. W. R. Dickinson, A. Grantz. pp. 6–21. *Stanford Univ. Publ. Geol. Sci. Ser. 11.* 374 pp.

Wallace, R. E. 1969. Earthquake recurrence intervals on the San Andreas fault. *Cordilleran Sec. Geol. Soc. Am. Abstr. Programs* 3:71–72 (Abstr.)

Wallace, R. E. 1977. Profiles and ages of young fault scarps, north-central Nevada. *Geol. Soc. Am. Bull.* 88:1267–81

Wallace, R. E. 1979. Degradation of the Hegben Lake and Madison Range fault scarps, Montana. *Geol. Soc. Am. Abstr. Programs* 11(3): 133 (Abstr.)

Washburn, L. R. 1973. *Periglacial Processes and Environments.* New York: St. Martins Press. 320 pp.

Weber, G. E. 1979. Recurrence interval for surface faulting along the Frijoles fault and the Ano Nuevo thrust fault of the San Gregorio fault zone, San Mateo County, California. *Geol. Soc. Am. Abstr. Programs* 11(3):134 (Abstr.)

West, M. 1974. Martian volcanism: Additional observations and evidence for pyroclastic activity. *Icarus* 21:1–11

Whitcomb, J. H. 1976. New vertical geodesy. *J. Geophys. Res.* 81:4937–44

Whitten, C. A., Green, R., Meade, B. K. 1979. Recent crustal movements, 1977. *Tectonophysics* 52:1–663

Wilcox, R. E. 1965. Volcanic-ash chronology. In *The Quaternary of the United States*, ed. H. E. Wright, D. G. Frey, pp. 807–16. Princeton, NJ: Princeton Univ. Press. 922 pp.

Williams, H. 1932. The history and character of volcanic domes. *Univ. Calif. Publ. Dept. Geol. Sci. Bull.* 21:51–146

Wilson, M. E., Wood, S. H. 1980. Tectonic tilt rates derived from lake-level measurements, Salton Sea, California. *Science* 207:183–86

Wilson, R. C., Harp, E. L., Picard, M. D.,

Ward, S. H. 1973. Chaotic terrain of Mars: A tectonic interpretation from Mariner 6 imagery. *Geol. Soc. Amer. Bull.* 84:741–48

Wood, G. P., Weaver, W. R., Henry, R. M. 1974. *The Minimum Free-Stream Wind Speed for Initiating Motion of Surface Material on Mars. Tech. Mem. X-71959.* NASA. 13 pp.

Wood, S. H., Elliott, M. R. 1979. Early 20th-Century uplift of the northern Peninsular Ranges province of southern California. *Tectonophysics* 52:249–65

Yung, Y. L., Pinto, J. P. 1978. Primitive atmosphere and implications for the formation of channels on Mars. *Nature* 273:730–32

Zoltai, S. C. 1975. Tree ring record of soil movements on permafrost. *Arct. Alp. Res.* 7:331–40

Ann. Rev. Earth Planet. Sci. 1980. 8: 263–301
Copyright © 1980 by Annual Reviews Inc. All rights reserved

GEOLOGIC PRESSURE DETERMINATIONS FROM FLUID INCLUSION STUDIES

×10133

E. Roedder and R. J. Bodnar[1]

959 US Geological Survey, Reston, Virginia 22092

Introduction

Fluid inclusions are small, usually microscopic, volumes of fluid trapped within crystals during their growth from the fluid. Although usually trapped as a homogeneous fluid at the temperature (and pressure) of growth, they are generally multiphase when examined at room temperature. Most commonly the two phases are vapor and liquid. Sorby (1858) explained that this phase separation was due to differential shrinkage on cooling of the "crystal bottle" and the contained fluid, and showed that it could be reversed on a heating microscope stage; thus, each microscopic inclusion is a recording geothermometer that provides us with data on the temperatures of past geologic events. In the following 120 years, inclusion thermometry has been the subject of numerous papers, in large part in the Soviet literature, and has provided a significant part of the data in many more.[2] The precision and accuracy of these thermometric data have been discussed and debated in many of these papers, and in the heat of this "debate" far too many partisan statements have been made, both pro and con, that are not supported by the available facts. The two aspects of inclusion study that are most commonly mistated or misunderstood are 1. the effect of the hydrostatic pressure at the time of trapping of a given inclusion on the thermometric results obtained, and 2. the use of inclusion data to obtain an estimate of this pressure. In this paper we show why some commonly used procedures for evaluating pressure from inclusion studies (i.e. geobarometers) are wrong, and can

[1] Current address: Ore Deposits Research Section, Pennsylvania State University, University Park, Pennsylvania 16802.

[2] For a summary of this extensive literature up to 1953, see Smith (1953); for abstracts of the more recent literature see Roedder (1968–onward, 1972).

0084-6597/80/0515-0263$01.00

yield very erroneous pressure values. We also review the variety of valid inclusion geobarometers, along with their precision, accuracy, limitations, and applications.

Data on the pressures existing during geologic processes are of far more than academic interest. Because they provide some evidence of the depth beneath the surface at which the process occurred, they can provide the exploration geologist with valuable information on the amount of cover that has been faulted or eroded away, the possible nature of the deposit, the former occurrence of boiling that can cause ore deposition, etc, and pressure differences may eventually also yield information on the direction of flow of the ore-forming fluids.

General Principles and Nature of Available Data

PRINCIPLES OF INCLUSION GEOTHERMOMETRY Most geobarometry based on inclusions considers first the data on the temperature of homogenization of fluid inclusions—normally the temperature at which the liquid and vapor phases become a homogeneous fluid on heating under the microscope.[3] There are five major assumptions on which all such geothermometry is based, and which must be evaluated for each sample studied. These are as follows (Roedder 1979a):

1. The fluid trapped upon sealing of the inclusion was a single, homogeneous phase.
2. The cavity in which the fluid is trapped does not change in volume after sealing.
3. Nothing is added or lost from the inclusion after sealing.
4. The effects of pressure are insignificant, or are known.
5. The origin of the inclusion is known.

The first assumption is essential to geothermometry, but as will be shown later, evidence of trapping of a *non*homogeneous mixture of two fluid phases, particularly liquid and vapor, can provide an excellent geobarometer.

[3] This is normally abbreviated T_H in the literature. We will also use the following:

T_T—temperature of trapping. (Some authors use T_F for temperature of formation).

T_H L-V—temperature of homogenization of the liquid and vapor phases on heating.

T_S—temperature of solution of a solid phase on heating. (Some authors use T_m for temperature of melting.)

T_S NaCl—temperature of solution of a solid crystal of NaCl.

T_{Frz}—temperature of freezing, i.e. the depression of the freezing point due to the solutes present, normally referring to aqueous inclusions. In actual practice it is the temperature of melting of the last ice crystal on warming a frozen inclusion (Roedder 1962).

T_D—temperature of decrepitation.

The second assumption is of relatively minor concern in most geobarometry because the most common volume change, that due to precipitation on the walls during cooling, is generally not only small but also easily reversed during heating in the laboratory except in silicate melts. Very significant errors may occur, however, if care is not used in the determination of T_H, because of permanent deformation ("stretching") of the host mineral walls from overheating, as is detailed in a later section. Serious stretching can also occur when an inclusion with a small vapor bubble is frozen, because of the expansion of water on freezing.

The cavity volume will be larger at T_H in the laboratory than it was in nature at the time of trapping. This occurs as a result of the expansion of the host mineral upon bringing it to the surface of the earth. The compressibility of most minerals (Birch 1966) is such that the inclusion volume will increase 0.1% per kilobar pressure decrease on bringing to surface pressures. This effect is still well below the errors from other sources and can generally be ignored at this time.

Assumption number three may be pertinent to geobarometry in several situations. The loss of hydrogen from inclusions by diffusion through the host mineral has been suggested as a possible explanation of some inclusion data (Roedder & Skinner 1968). As the hydrogen presumably comes from the dissociation of inclusion H_2O the volume change would be effectively controlled by the chemical behavior of the oxygen left behind. The O_2 may oxidize components of the fluid, such as sulfide to sulfate, or, in hydrous silicate melt inclusions, it may diffuse into the host (Anderson & Sans 1975, and personal communication). The volume changes within the inclusion in these examples might be difficult to predict.

Assumption four refers to the fact that the homogenization of the vapor and liquid phase of an inclusion establishes only a minimum value for the temperature of trapping (T_T). All inclusions of a given composition trapped along the isochore (line of constant volume) originating at that point on a *P-T* plot will have the same homogenization behavior, so either *P* or *T* must be known to determine the other. This interrelationship is examined in more detail below.

The fifth assumption, concerning the origin of the inclusion, causes problems in many inclusion studies. The inclusions in a given crystal generally have been trapped at more than one stage in the growth and subsequent fracturing and healing of the crystal, and there may be considerable ambiguity in the determination of the origin and sequence of formation (see, for example, Roedder 1976, Table II). This ambiguity, although frustrating to the student of inclusions, does not affect the validity of most of the geobarometers presented here but is of major concern to others.

Practically all use of fluid inclusions to provide estimates of either the temperature or the hydrostatic pressure at the time of formation of the host mineral requires two types of data: 1. the composition of the fluid phase (or phases) trapped, and 2. the phase behavior and *P-V-T-X* properties of that composition in the range involved. It is unfortunately all too true that these two main requirements place severe constraints on the accuracy of all pressure determinations based on inclusions.

COMPOSITION OF THE FLUID PHASE(S) TRAPPED The composition of the fluid phase that was trapped in a given inclusion is seldom known exactly, and there is no panacea for the problem of its determination. Most fluid inclusions represent the trapping of a small volume of an originally homogeneous fluid phase within a growing crystal, or within a fracture through a previously existing crystal that is in the process of healing. On cooling from T_T to that of observation, the originally homogeneous fluid phase generally undergoes one or more phase changes. Most commonly it splits into liquid and vapor phases, and frequently it forms other phases during this cooling, such as additional immiscible fluids, or various crystalline phases (i.e. "daughter minerals"). A variety of methods for estimating the composition of the original homogeneous fluid are available. These are mostly based on the use of microscopy to identify the phases present, and can give qualitative or even semiquantitative data on composition (Roedder 1972).

Some of these tests, such as the determination of the presence of noncondensable gases, are exceedingly sensitive ($\sim 10^9$ molecules; Roedder 1970) but low in accuracy. Others can be made very accurate, but yield data that are only indirect measures of composition, such as T_{Frz}. Some quantitative methods are exceedingly sensitive but very restricted in application. Thus, laser-activated Raman spectroscopy gives excellent analytical results, but is essentially limited in its application to certain polynuclear species such as CO_3, SO_4, CH_4 (Rosasco & Roedder 1979).

For many inclusion studies, the only readily available compositional data consist of a qualitative identification of the phases present (e.g. liquid water solution, NaCl or other daughter crystals, and vapor bubble) and a crudely quantitative estimate of the relative volumes and hence masses of these various phases. In the absence of a halite daughter mineral, the salinity of the water solution, normally expressed as wt. % NaCl equivalent, is usually obtained from T_{Frz}. In those inclusions containing an additional immiscible fluid phase such as liquid carbon dioxide, the relative volumes of these fluids can also be determined, but only rarely are such volume data highly accurate. Highly *precise* determinations of relative phase volumes can be made from area or length measurements on inclusions that have special configurations, such as very thin and flat or long and tubular, but the *accuracy* of such determinations is only as good as the assumption of a uniform third dimension. This assumption can result in gross errors. Thus, the vapor bubble in a simple two-phase (liquid + vapor) inclusion will always be found in the thickest part of a flattened but slightly wedge-shaped inclusion (thus minimizing surface energy), but measurements of this third dimension are generally inaccurate at best. Even in geometrically regular inclusions, optical distortion from the curvature of the several phase boundaries involved can yield erroneous volume estimates (e.g. Roedder 1972, Plate 11, Figures 8–9).

The fluids present around a growing or healing crystal are not always homogeneous. In many places, two immiscible fluids, such as H_2O and CO_2, or water and steam, were present. Fluid inclusions forming in such a heterogeneous fluid environment may trap only one of the fluids, or both of them in random ratios. Such compositionally divergent inclusions provide several of the geobarometers described below, but of course compositional data are required on both fluids. The use of data from more than one kind of inclusion to provide a geobarometer must be based on the assumption (often unstated) that the several inclusions used are truly contemporaneous (i.e. cogenetic). Validation of this assumption is a far from trivial matter, and has led to much erroneous data in the literature.

PHASE BEHAVIOR AND *P-V-T-X* PROPERTIES OF FLUID PHASE(S) TRAPPED

Any determination of pressure based on study of a single inclusion requires compositional data on the inclusion contents and *P-V-T-X* data on such a composition. The available data on the phase behavior and *P-V-T-X* properties of appropriate compositions in the range involved in fluid inclusion studies are meager, and even at best require considerable extrapolation. *P-V-T* data are most complete on the two pure compounds that are major components of many inclusion fluids, H_2O (Keenan et al 1969, Burnham et al 1969, Pollak 1974) and CO_2 (Kennedy 1954, Shmonov & Shmulovich 1974), and these data are frequently used in interpreting fluid inclusions. The natural fluids are not pure, however, because the water inclusions usually are at least 1-*m* in ionic solutes, and CO_2 inclusions frequently contain appreciable CH_4 and N_2, so the properties of the pure compounds are useful only as an approximation.

For water solutions, the most extensive experimental data available are for the system NaCl-H_2O, but even this system has not been studied in the entire range of interest (Potter & Brown 1977). In addition, most natural fluids are not simple solutions of NaCl and H_2O, but contain significant amounts of other solutes, such as K^+, Ca^{+2}, SO_4^{-2}, and HCO_3^-.

Although Potter & Clynne (1978) showed that many of the solutes present in natural inclusion fluids result in fluids that have thermodynamic properties rather close to those measured for simple NaCl-H_2O solutions with the same value for the depression of the freezing point, extrapolation to fluids containing significant quantities of CO_2 is hardly warranted. The available experimental data on the system NaCl-H_2O-CO_2 are even more limited (Ellis & Golding 1963, Takenouchi & Kennedy 1965).

DEPTH VS PRESSURE; HYDROSTATIC VS LITHOSTATIC Most pressures determined from fluid inclusions are stated to represent the "lithostatic" or the "hydrostatic" pressure, or some intermediate value. The local pressure at the site of the inclusion at the time of trapping is actually hydrostatic in any case because it is a fluid pressure, but the two terms are used to indicate the source of the pressure on the fluid. "Lithostatic" is generally used to refer to the pressure from a column of country rock of density and height appropriate for the depth of the sample below the surface at the time of trapping, and "hydrostatic" usually refers to the pressure of a column of fresh water of the appropriate temperature and hence density, corrected for depth to water table. Although the pressures are probably between these two limits in most natural situations, the actual values in nature may span a considerably wider range at both ends, and even within this range numerous factors may affect the specific pressure. Diagram *A* in Figure 1 represents the simplest possibility in which the hot fluid moves up an open fracture. The fluid will expand on pressure decrease, but the cooling will generally be much greater than adiabatic, as the fluid is losing heat to the cooler adjacent rock before flowing out on the surface as a hot spring. The pressure at the inclusion would be that from a column of fluid of appropriate composition and temperature. The actual integrated density of the fluid column would be higher than that of the fluid trapped in the inclusion because the temperature decreases toward the surface. A modification of this is shown in *B*, where the upper part of the column is ground water. Because density increases with both salinity and cooling, the pressure in *A* and *B* might be identical if the integrated density of the cool, dilute groundwater is the same as that of the hotter, more saline fluids over the same interval of the upper portion of the vein length. If the fluid in the vein above the inclusion is at the point of boiling throughout its length *C*, and the salinity is known, the data of Haas (1971) permit an unambiguous pressure determination. Unfortunately, however, there is a fourth and probably common possibility, *D*, which may result in pressures at the inclusion that are considerably less than simple "hydrostatic," i.e. diagram *A*, depending upon where in the column the switch from liquid to vapor occurs. A vapor-dominated system, with liquid above vapor, would be treated similarly. These four possibilities all involve a vein system that is assumed to be open to the surface. In *E* through *G*, a tight, almost closed vein or throttle is assumed. Where the host rocks are rigid, as shown in *E*, the pressure of a flowing fluid at the inclusion site will be the hydrostatic head, *h*, plus an unknown overpressure limited only by the vein constriction. This overpressure could result from the presence of an igneous body below, or from a hydrostatic load not directly above the vein, such as an artesian-type regional system. Diagram *F* illustrates an extreme condition, in which fault movements have increased the volume occupied by a relatively fixed volume of fluid; this increase causes boiling. The pressure at the inclusion trapped in this environment would then be simply the vapor pressure of the fluid at that temperature, even though this chamber might be deep in the earth. Although this condition, yielding boiling, is probably rare, the same mechanism has probably been operative in many ore deposits, but only enough to cause migration of fluids, and not enough to cause boiling. Thus, the well-known concen-

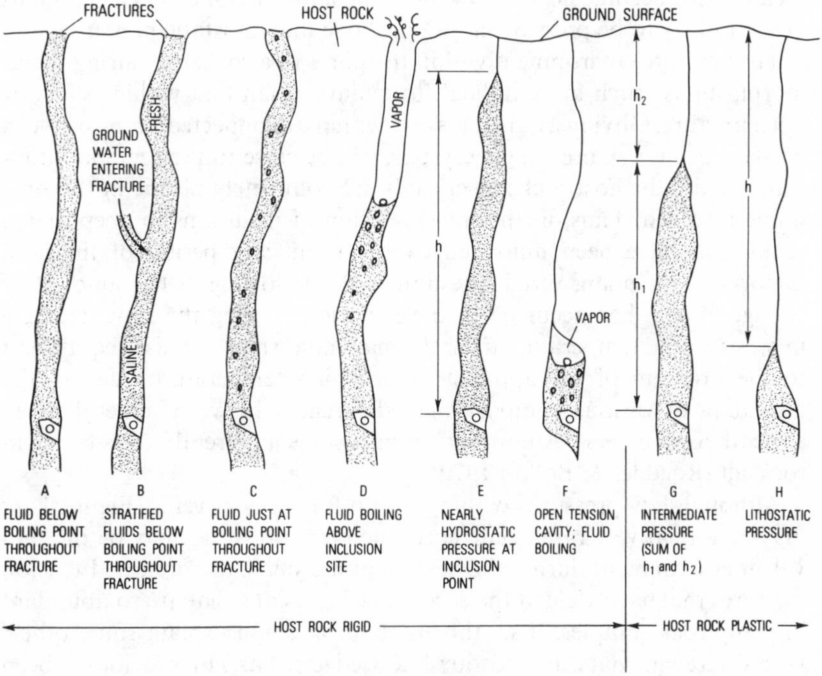

Figure 1 Diagram showing range of possible pressure conditions on a fluid inclusion trapped in a crystal growing freely into a fluid in a vein. See text for discussion.

trations of ore or vein matter into the apices of folds is frequently ascribed to the lower pressures there that are due to the mechanics of folding.

In diagrams *G* and *H* in Figure 1, depths are assumed to be such that the country rocks are plastic. In *G*, the pressure would be the sum of the hydrostatic pressure of fluid column h_1 plus the lithostatic pressure of rock column h_2; in *H*, it would be simply lithostatic for column *h*.

In any lithostatic pressure environment such as shown in diagram *H*, the true pressure at the inclusion can be considerably less than lithostatic if the country rocks are not completely plastic in the time frame involved. Also, the pressure in *H* can be more than lithostatic if the horizontal extent is small relative to the depth *h*, or if these country rocks are under compression from regional forces. The only situation in which the pressure will be truly "lithostatic" would be the rather rare occurrence of a horizontal vein with a lateral extent that is large relative to its depth. Such "overpressures" can also be caused by the increase in vapor pressure during crystallization of magmas; when the roof rocks yield suddenly, explosive volcanism may be the result. Similar overpressures can occur in sediments (Barker 1972), as formation waters trapped under impermeable beds become heated. Owing to the large lateral extent, such fluids are not likely to be pressurized much above the lithostatic pressures.

The pressure environment within the pores of a rock, e.g. during metamorphism, is much more difficult to quantify than that within a vein or open fracture. Obviously, the pressure within disconnected pores between crystals, or even within single crystals, will be close to the regional lithostatic load if the host rocks or crystals are completely plastic in the time frame involved. Thus, in the interpretation of inclusions in deep-seated rocks that have been unloaded over a significant period of time, an important but unanswered question is the following: what amount of change, if any, has occurred in the inclusions during the slow decrease in pressure and temperature from the maximum values? This is equivalent to the problem of the apparent "quenching temperature" during the annealing of various compositional differences between minerals used as geothermometers. "Annealing" of inclusions apparently does occur in rocksalt (Roedder & Bolkin 1979).

Although low pressures within tension fractures in veins (Figure 1, *F*) may be a relatively rare occurrence, such low pressures may be the prevalent environment during the healing of the small, disconnected tension fractures that have yielded the secondary inclusions that are so abundant in many rock samples. If so, this mechanism could explain some otherwise discrepant data, but to our knowledge it has not previously been suggested. Distinction between low-density inclusions formed at relatively great depths by this mechanism and similar low-density inclusions formed by boiling in effectively open fractures at much shallower depths would be impossible from the inclusions alone, and would require additional data.

GEOBAROMETRY BASED ON VAPOR PRESSURE OF SOLUTION If a group of two-phase, liquid-plus-vapor inclusions in a given sample all have the same composition and T_H, it is generally presumed that a homogeneous fluid was trapped. Thus, we can presume that the hydrostatic pressure at the time of trapping was greater than the vapor pressure of that particular solution at that temperature, or the fluid would have been boiling (i.e. two-phase). Such simple inclusions provide no evidence concerning how much the pressure might have exceeded the vapor pressure. Simple two-phase, liquid-plus-vapor inclusions are the most common by far, and hence this minimum value for the pressure is generally the only constraint on the pressure that can be obtained from the inclusions themselves. For ore deposits formed at low temperatures, this may not result in much of a constraint. Thus, only 37 m of hydrostatic head of cold water (at 1.0 $g \cdot cm^{-3}$) is required to prevent boiling of the typically 150°C fluids that formed the Mississippi Valley–type ore deposits (Roedder 1976).

If there are gases in solution, the vapor pressure of the mixed solution would be higher than that of the pure solvent liquid, but appropriate *P-V-T-X* data are generally lacking. If such inclusions are studied on the crushing stage (Roedder 1970), however, an estimation of the internal pressure, at room temperature, can be obtained. Such an estimate must be less, of course, than the pressure at T_T.

GEOBAROMETRY BASED ON COMPARISON OF T_H WITH AN INDEPENDENT GEOTHERMOMETER If the composition and T_H of a simple two-phase, liquid-plus-vapor inclusion are known, and another independent geothermometer is available to determine the (higher) temperature of formation of the host or associated minerals, the pressure may be determined from *P-V-T* data on the fluid of the inclusion. The uncertainty in the values derived from any such geobarometer is obviously at least equal to the sums of the uncertainties in the precision and accuracy of the determinations by the two geothermometers used, as well as other factors. The independent thermometer must be rather accurate in order to provide useful results. Thus, for low-temperature ore deposits, an error of 25°C in the temperature obtained from the independent geothermometer is equivalent to the hydrostatic pressure from >3-km depth of burial (Roedder 1971). Some published geobarometry data have been based on such poor "thermometric" data that the pressure values obtained (and the geologic speculation based in turn on them) are virtually meaningless. The limitations imposed by inadequate *P-V-T* data on the inclusion fluid can also be severe.

Essentially, the T_H for the fluid inclusion, along with knowledge of its composition, limits the possible conditions of trapping to a sloping line (isochore) on a *P-T* plot for the appropriate composition fluid. If another independent geothermometer-geobarometer can also be used on the same samples, and it yields a line with a different slope, the point of intersection of these two lines will be the *P* and *T* of formation; with experimental uncertainties in the determination of both lines, the intersection becomes an area. Bethke & Barton (1971) showed that the distribution of manganese between sphalerite and galena could thus be combined with fluid inclusion T_H to yield both *P* and *T* of the environment. Fortunately, the slopes of the data on manganese distribution and the inclusions on a *P-T* plot are strongly inclined to each other, thus reducing errors from this source. The accuracy of this determination would be limited not only by that of the experimental measurements involved, but also by the validity of the necessary assumption that the sphalerite and galena and the fluid inclusions were all truly cogenetic. Validation of this assumption is not a trivial matter (Barton et al 1963).

Coveney & Kelly (1970) described a similar technique for use on inclusions trapped in quartz that permits some limits to be placed on the *P-T* conditions at trapping. Quartz has two modifications, with a rapid inversion, at 573°C (at one atmosphere). Quartz that crystallized as the high-temperature (β) form can be recognized as such (by crystal morphology, etching, etc), even though it has inverted on cooling to the low-temperature (α) form. The inversion temperature is raised by pressure, but only $28°C \cdot kb^{-1}$. Most isochores for fluid inclusions in quartz are much flatter on a *P-T* plot, and hence they intersect the α–β line at a large angle. If a given inclusion were trapped by growth of an α-quartz crystal, the conditions of trapping would still have to lie along the isochore defined by its homogenization, but below the intersection of this isochore with the α–β-quartz inversion line. If the host crystal can be shown to have grown as β quartz, the growth conditions must lie along a different isochore at a higher *T* than that of the intersection. However, at the instant of inversion, the internal pressure in the inclusion will *decrease* owing to the $\sim 1\%$ volume increase on going from α to β. Because the thermal expansions of α and β quartz are quite different, the path through *P-T* space for an inclusion in quartz will not be a straight line and is not isochoric.

GEOBAROMETRY BASED ON SIMULTANEOUS TRAPPING OF TWO IMMISCIBLE FLUIDS Where two essentially immiscible fluids, each with known *P-V-T* properties, are present and inclusions of each fluid were trapped simultaneously, both *P* and *T* can be determined from the values for the T_H of the two inclusions. In addition to the requirement that the *P-V-T* properties of both fluids be known, it is important that the slopes of the appropriate isochores for the two fluids on a *P-T* plot be significantly different, or the accuracy of the determination will be poor. The bulk of the large number of geobarometry determinations that have been reported (mostly in the Russian literature, see Roedder 1968–onward) are based on the use of the two immiscible fluids CO_2 and H_2O. Unfortunately, these two fluids are not truly immiscible (see next section), and there are also severe problems in establishing the contemporaneity of trapping. Furthermore, many of the published reports do not make clear the all-important point of whether the CO_2 and H_2O phases studied were in separate inclusions or were two immiscible phases in a single inclusion.

The only pair of fluids that give promise of providing good geobarometric data by this method are the oil and brine inclusions found in some Mississippi Valley–type ore deposits (Roedder 1963, pp. 176–77), but the *P-V-T* data on the oil phase are unknown and can only be guessed at the present. Oil that shows post-entrapment degradation or maturation will not provide good data. Evidence that such changes have occurred is obvious in some oil inclusions from these Mississippi Valley–type ore deposits (Roedder 1972, Plate 9), but other oil inclusions in the same deposits show no visible evidence of change.

GEOBAROMETRY BASED ON SIMULTANEOUS TRAPPING OF TWO PARTLY IMMISCIBLE FLUIDS If two partly immiscible fluids, each saturated with respect to the other, are present at the time of trapping, inclusions of these two fluids can provide some constraints on the pressure, as long as the appropriate *P-V-T-X* data are available, and the mutual solubilities decrease from the conditions of trapping to those of observation. The method was originally proposed by Smith & Little (1959). The nature of the fluid-inclusion evidence on the phase condition at the time of trapping is crucial. If, for example, a series of inclusions all have the same ratio of the two fluids, presumably a homogeneous phase was trapped and the *P-T* conditions at trapping are limited to the appropriate one-phase area of the pertinent diagram. If different ratios are found in different coeval inclusions, the fluids were presumably immiscible at the time of trapping, but a variety of problems make the use of such data to obtain the pressure rather difficult. Even where it can be proved that the inclusions are coeval, there are problems. If necking down occurs after trapping, in inclusions that contain more than one phase (from either the trapping of immiscible fluids or subsequent phase separation), the resulting inclusions can be very misleading. Similarly, the changes in mutual solubilities between the temperature (and pressure) of trapping and those of observation can be rather complex. We believe that such problems may have invalidated many of the published geobarometric data sets where this method was used.

GEOTHERMOMETRY BASED ON TRAPPING OF BOILING FLUIDS The boiling of a fluid is merely a special example of immiscibility under the prevailing conditions. We use the term "immiscibility" in its most general sense (Roedder & Coombs 1967, p. 419) to refer to the existence, at equilibrium, of two or more noncrystalline polycomponent solutions (in this case fluid phases), differing in properties and generally in composition. Gas and liquid on the boiling curve in a one-component system thus become a special limiting case of immiscibility in which the composition variable is eliminated. If a "boiling" liquid and its coexisting vapor phase are trapped separately in a pair of inclusions, these two inclusions will generally each become two-phase on cooling: the liquid inclusion will develop a vapor bubble as the liquid contracts, and the vapor inclusion will develop a film of liquid by condensation. On reheating these two to the original temperature of trapping, these phase changes will be reversed: the expansion of the liquid will eliminate the vapor bubble in the inclusion that trapped liquid, and the liquid phase in the vapor inclusion will evaporate. These two homogenizations, "to the liquid phase" and "to the gas phase," respectively, must occur at the *same* temperature, and if the boiling curve is known for that fluid, the pressure can be determined from this T_H.

Experimental difficulties are involved (e.g. inaccurate T_H values may be recorded because the small amount of liquid phase in the vapor inclusion usually coats the walls as a film and hence is difficult to see), but more important, individual inclusions may have trapped a mixture of two phases, rather than a single, homogeneous phase. Thus, an inclusion trapping a gas bubble may enclose some liquid as well, and, more rarely, a liquid inclusion may trap a gas bubble along with the liquid. As the immiscibility "solvus" generally closes at higher temperatures, this trapping of a mixture will result in inclusions with higher T_H than would be obtained on inclusions that trapped only liquid or only gas. Thus, where a group of such inclusions are coeval, the *minimum* T_H for inclusion homogenization in vapor and in liquid is provided by inclusions trapping pure end members and should be equal and represent the trapping temperature. All other inclusions, from the trapping of mixtures, would yield higher, and spurious, T_H values.

The identity of these two homogenization temperatures is not only a necessary requirement, but also can be taken as relatively unambiguous proof of boiling. Because the pressure environment under which boiling can occur in nature may be rather variable and transient, and such pairs of inclusions are seldom *exactly* coeval, small differences may be expected. Of course, two separate fluids, one liquid and one vapor, could be trapped at different times and at a fortuitous combination of P and T values that would yield such similar homogenization temperatures.

It is not uncommon, however, to find individual inclusions grown from heterogeneous mixtures that have trapped only the dispersed phase (that which occurs as isolated droplets in the other, continuous phase). This is particularly expectable when the bulk of the growth of the host crystal occurs from the continuous phase, e.g. from a water solution that contains dispersed steam bubbles. In theory, it makes no difference whether the dispersed phase is the same composition as the continuous phase (as in true boiling of a one-component system) or whether it consists merely of bubbles of a minor, more volatile constituent in a solution. The method should work in either case, but solubility reversal and observational problems effectively preclude the latter.

Evidence of boiling provides us with some of the most accurate and unambiguous geobarometry data available, and has been reported in numerous ore deposits. However, the most important part of the evidence, the proof of contemporaneity of trapping of the two types of inclusions, is frequently poor or lacking. The mere existence of the two inclusion types is inadequate, as this can also come about by sequential trapping of different fluids at different times, by necking down, and by leakage (Roedder 1979a). Careful microscopy is necessary to minimize or eliminate these sources of ambiguity. The distinction is far more than merely academic, because several varieties of rich ore deposits ("bonanzas") are widely believed to have formed as a result of the gross change in the chemistry of the ore fluids upon boiling.

Not uncommonly one may find inclusion evidence of boiling at a given pressure, along with other inclusions, in the same deposit, indicating possibly higher pressures. In this situation, the low-pressure limit established by the boiling inclusions is most informative, because it places an upper (hydrostatic) depth limit on the crystal in a vein open to the surface (ignoring the interval of unsaturated ground above the water table). The higher pressures are not as meaningful because of mechanisms for generating lithostatic or greater pressures at shallow depths (Figure 1). The only way in which such an interpretation of the evidence of boiling could yield misleading depths is shown in the probably rather rare example F in Figure 1.

GEOBAROMETRY BASED ON INCLUSIONS CONTAINING DAUGHTER MINERALS

Many fluid inclusions contain one or more solid phases in addition to liquid and vapor; this occurrence indicates that on cooling the solution became saturated with respect to these solid phases, assuming that the inclusion trapped a homogeneous fluid. Upon heating, these daughter minerals will generally dissolve and the vapor bubble will decrease in size until, at some elevated temperature, the inclusion contains a homogeneous fluid phase. If the *P-V-T-X* properties of that particular salt-H_2O system are known, the phase-disappearance temperatures may be used to determine a pressure of formation for the fluid inclusion.

A fluid inclusion containing solid salt, saturated liquid, and saturated vapor at ambient temperature might follow any one of three possible paths to homogenization, indicating three different trapping pressures. If the vapor-bubble–disappearance temperature (T_H L-V) is higher than the temperature of solution of the salt (T_S salt), the inclusion had trapped

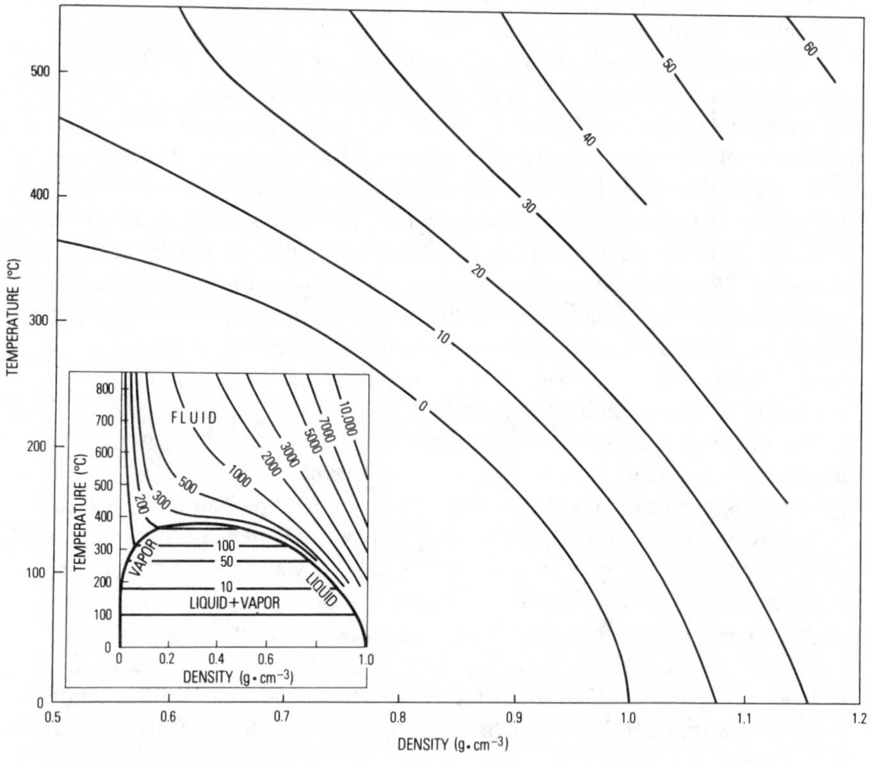

Figure 2 High-density portion of the temperature-density diagram for the system H_2O, from Keenan et al (1969). The boiling curves for NaCl solutions of stated weight percents are also shown, smoothed from the data of Haas (1976) and Urusova (1975). The inset shows the entire diagram for H_2O, along with isobars from Fisher (1976).

an unsaturated solution. The *minimum* temperature and pressure of trapping are T_H L-V and the pressure on the liquid-vapor curve at that temperature for a solution with a salinity determined from T_S. If T_S salt and T_H L-V are the *same*, that *is* the minimum temperature, and the minimum pressure is the pressure on the solid-liquid-vapor curve at that temperature. Furthermore, where one of these inclusion types coexists with inclusions homogenizing at the same temperature but in the vapor phase, then both were probably trapped from a boiling solution, and the temperature and pressure at homogenization equals that at trapping. Finally, if T_H L-V occurs below T_S salt, this latter provides a *minimum* value for T_T. The *minimum* pressure is the pressure along the solid-liquid-vapor curve at T_S.

Applications to Specific Inclusion Compositions

In this section we review the procedures that are most commonly used for geobarometry based on inclusions, but first we should emphasize that in many if not most inclusion investigations the pressure is *not* determined from the inclusions. Most inclusions have trapped fluids at pressures higher than their vapor pressures (i.e. in the one-phase "fluid" field above the boiling curve in Figure 2). All inclusions except those trapped from boiling fluids thus require that a *pressure correction* be added to T_H to obtain T_T.[4] Generally the pressure is estimated from independent evidence of the depth of cover at the time of trapping (e.g. from geological reconstructions of the thickness of material since removed by erosion or faulting), and then this pressure is used, along with *P-V-T* data on supposedly appropriate solutions, to calculate the pressure correction.

Obviously the difficulty in estimating the depth of cover, plus the uncertainty of hydrostatic vs lithostatic pressure described above, and the errors in estimations of the composition of the fluid and hence its *P-V-T* characteristics can cause relatively large uncertainties in the pressure correction. It is fortunate that absolute geothermometric accuracy is sometimes of minor concern, e.g. in current studies in ore deposition, where the relative values for various samples may provide the most valuable data. Furthermore, because dense fluids are relatively incompressible, the pressure corrections for inclusions homogenizing in the liquid phase at

[4] Two different sets of pressure-correction diagrams are in common use, those of Lemmlein & Klevtsov (1961), and those of Potter (1977). We recommend the use of Potter's diagrams because they are plotted from more recent data and are in a larger format; these features make interpolation more precise. Also, there is an apparent error in the high-pressure portion of the tabular data of Lemmlein & Klevtsov; graphical intercomparisons reveal that all data entries for pressures $\geqq 1500$ atm and degrees of fill $<85\%$ should have a degree of fill 2.5% larger than stated in the original Russian text.

low temperatures (shown on the lower right side of Figure 2) are generally small, in places only a few degrees, unless the pressures were very high. The lower-density fluids are much more compressible, and hence it is possible to have inclusions that homogenize in the liquid at 370°C, but that were trapped at perhaps 800°C, at moderately high pressures (2 kb), i.e. they have a 430°C pressure correction. The limitations are such that most commonly the geothermometric or geobarometric data from fluid inclusions are compared with independent evidence from other sources in an attempt to arrive at a consensus and to recognize which data or assumptions are invalid, and why.

Fluid-inclusion-pressure determinations are based on the volumetric properties of the inclusion fluid. Therefore, the composition of the fluid must be known and experimental *P-V-T-X* data *for that particular fluid* must be available before the results of heating/freezing runs can be used to calculate trapping pressures. Because most fluid inclusions contain an aqueous phase, we will begin our discussion of specific compositions with a review of the *P-V-T* properties of water, and proceed to more complex aqueous and CO_2-rich systems. Finally, we will consider pressure estimates from silicate-melt inclusions and their use in petrology.

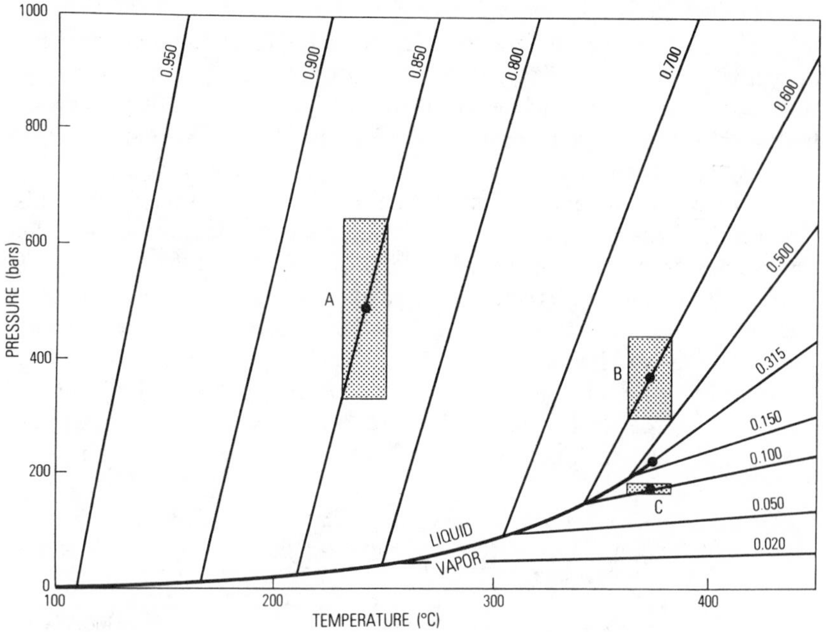

Figure 3 A portion of the H_2O phase diagram showing the densities ($g \cdot cm^{-3}$) of several isochores. Isochores *A*, *B*, and *C* correspond to the paths followed by three fluid inclusions discussed in text. Data are from Burnham et al (1969) and Keenan et al (1969).

PURE-WATER INCLUSIONS The volumetric properties of water are well known over a wide range of geologically applicable temperatures and pressures. Figure 3 shows the liquid-vapor curve for water from Keenan et al (1969) with selected isochores calculated from the data of Burnham et al (1969). At low temperatures, the liquid-field isochores are very steep, and become less steep as temperature increases, but the vapor-field isochores are relatively flat at all temperatures. Knowing the slopes of the isochores is important for pressure determinations because fluid inclusions represent essentially constant-volume systems and, therefore, follow isochoric paths. The effect that the slope of the isochore has on the precision of pressure determinations is shown by the *P-T* paths followed by three pure-water fluid inclusions upon heating. Inclusions *A* and *B* (Figure 3) homogenize to the liquid phase at 212° and 343°C, respectively, and inclusion *C* homogenizes to the vapor phase at 343°C. Let us assume that we have determined from an independent geothermometer that the actual trapping temperatures of all three inclusions were 30°C$\pm$10°C above their respective T_H and that we calculate trapping pressures using this information. The intersection of the appropriate isochore for the given temperature and mode of homogenization with the independently obtained trapping temperature defines the trapping pressure. As shown in Figure 3, inclusion *A*, having the steepest isochore, provides the least precise pressure determination (490$\pm$160 bars). As dP/dT decreases, the precision of the pressure determinations increases to 368$\pm$72 bars for inclusion *B*, and 172$\pm$6 bars for inclusion *C*. Therefore, as the slope of the isochore decreases, the precision of pressure determinations increases, but observational problems generally preclude accurate T_H determinations on inclusions such as *C*.

If we know the pressure at trapping from an independent geobarometer and we use this information to determine T_T, the inverse correlation applies. That is, the steeper the isochore, the more precisely T_T can be determined by the intersection of the isochore with the known pressure.

As described in a previous section, where fluid inclusions are trapped from boiling solutions, $T_H = T_T$. Also, the pressure on the liquid-vapor curve at T_H *is* the trapping pressure. Thus, if pure-water fluid inclusions having densities of 0.6 and 0.1 $g \cdot cm^{-3}$ were coeval, they were trapped from a boiling fluid at 343°C, and the trapping pressure would have been 152 bars.

Inclusions of essentially pure water are relatively rare, and are most commonly found in samples from low-pressure, near-surface hot-spring environments (Roedder 1977b), but are also found in samples from geothermal systems, formed at considerably higher pressures. In the latter, they may permit some reconstruction of the past *P-T* regime in the system (Browne et al 1976).

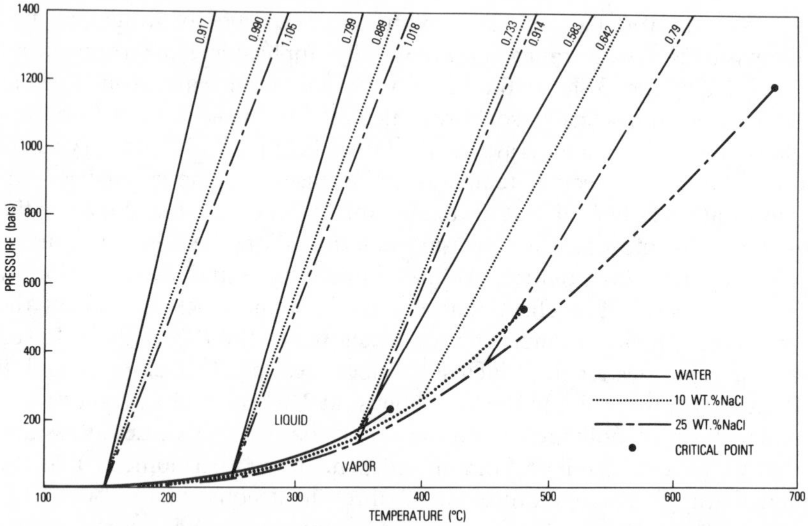

Figure 4 A portion of the phase diagrams for water and 10 and 25 wt. % NaCl solutions showing the liquid-vapor curves and several isochores ($g \cdot cm^{-3}$). Data for the 10 wt. % isochore originating at T_H 400° are from Urusova (1975), and for the 25 wt. % isochore originating at T_H 450° are extrapolated from Potter & Brown (1977). Other data from Sourirajan & Kennedy (1962), Burnham et al (1969), Keenan et al (1969), and Haas (1976).

LOW-TO-MODERATE SALINITY INCLUSIONS[5] To extend the methods of pressure determination described above to solutions of low or moderate salinity, we need to determine only the composition of the inclusion fluid and then use the volumetric properties of that fluid to calculate pressure. We obtain the composition of the inclusion by measuring T_{Frz} (Roedder 1962) and converting this temperature to salinity in weight percent NaCl equivalent, using the data of Potter et al (1978). Although fluid inclusions do not contain simple NaCl-H_2O solutions, Potter & Clynne (1978) showed that the *P-V-T-X* properties of brines in the system Na-K-Ca-Mg-Cl-Br-SO_4-H_2O can be estimated to within $\pm 1.0\%$ by the properties of an NaCl solution of the same T_{Frz}. Therefore, we can use the volumetric data for the system NaCl-H_2O, which are well known over a wide range of compositions, temperatures, and pressures (Potter & Brown 1977, Haas 1976, Urusova 1975), as a very good approximation of the *P-V-T* properties of the inclusion fluids.

The addition of NaCl to water lowers the vapor pressure along the liquid-vapor curve and extends this curve to higher temperatures relative

[5] For simplicity, we have arbitrarily defined low-to-moderate salinity inclusions as those which are unsaturated at room temperature, i.e. less than ~26 wt. % NaCl equivalent.

to water (Figures 2 and 4). This means that at a given temperature, an NaCl solution will boil at a lower pressure than pure water, and that at a given pressure, boiling, or the coexistence of two immiscible fluids, may occur at much higher temperatures for salt solutions than for water.

The addition of NaCl to water also changes the slopes of the isochores as shown in Figure 4. Note that, at low temperatures, isochores for pure water are steeper than those for salt solutions. However, as temperature increases, the water isochores gradually flatten until at ~325°C (at 10 wt. %) and ~300°C (at 25 wt. %) the water and salt-solution isochores have the same slopes. Above these temperatures, the salt-solution isochores are steeper than pure-water isochores. Of course, for any solution, as the critical point is approached, the isochores will become less steep and the slope of the critical isochore will be the same as, i.e. tangent to, the liquid-vapor curve at the critical point.

The above discussion indicates that low-temperature, high-salinity fluid inclusions provide more precise pressure estimates than their pure-water counterparts if the trapping temperature is known. In practice, however, errors in pressure estimates introduced by the variation in isochoric slope with salinity are minor compared to the precision of trapping temperatures obtained from other mineralogical geothermometers and the precision of T_H measurements obtained from inclusions homogenizing in the vapor phase.

HIGH-SALINITY, MULTIPHASE INCLUSIONS If a homogeneous fluid of sufficiently high sodium chloride concentration is trapped as an inclusion, the saturation limit may be exceeded as the fluid cools; the cooling causes precipitation of a daughter crystal of halite. When this multiphase inclusion is reheated to some elevated temperature, it once again becomes homogeneous. The pressure at homogenization depends on the temperature, salinity, and order of disappearance of the phases, as indicated by the three inclusions shown in Figure 5, all of which are assumed to be in the system NaCl-H_2O and to homogenize at 400°C.

Inclusion *A* follows the solid-liquid-vapor curve until the halite dissolves at 158°C (T_S NaCl), corresponding to a 30 wt. % NaCl solution (Keevil 1942). With continued heating the inclusion follows the liquid-vapor curve for a 30 wt. % NaCl solution (Haas 1976, Urusova 1975) until the vapor bubble disappears at 400°C (T_H L-V). At this final homogenization, the pressure in the inclusion, and thus the minimum trapping pressure, is 222 bars (Urusova 1975). This inclusion may have been trapped at any higher *P* and *T* along the isochore originating at *A*, shown on Figure 5.

Inclusion *B* follows the solid-liquid-vapor curve until, at 400°C, both the halite and the vapor bubble disappear simultaneously (T_S NaCl =

T_H L-V). At this temperature the inclusion contains a 46 wt. % NaCl solution (Keevil 1942) under a pressure of 182 bars (Sourirajan & Kennedy 1962), corresponding to the *minimum* trapping pressure. The inclusion could have been trapped at any higher *P* and *T* along the isochore originating at *B*, shown on Figure 5.

Inclusion *C* follows the solid-liquid-vapor curve until the vapor phase disappears at 310°C (T_H L-V) at only 66.4 bars (Haas 1976). From this temperature to homogenization at 400°C (T_S NaCl), the inclusion follows a solid-liquid curve shown schematically in Figure 5. The pressure at this point, which is the *minimum* trapping pressure, is calculated to be ~650 bars as shown below. Other ions in solution, such as Ca^{+2}, could reduce this minimum even further (Stewart & Potter 1979). The inclusion could

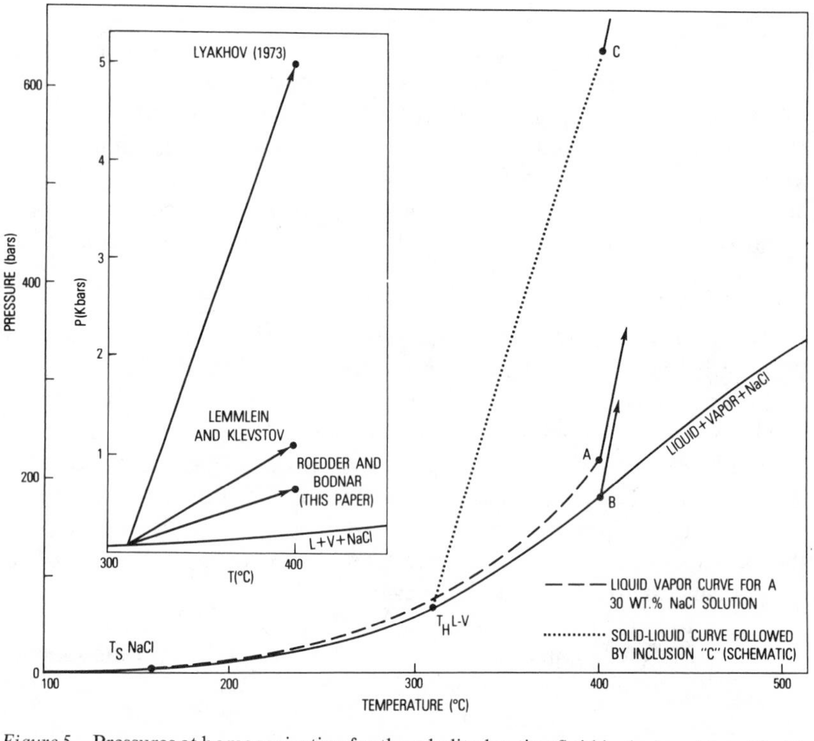

Figure 5 Pressures at homogenization for three halite-bearing fluid inclusions (*A*, *B*, *C*). All are assumed to be in the system NaCl-H_2O and to homogenize at the same temperature (400°C), but to exhibit differing modes of homogenization, resulting in different pressures at T_H. Data are from Sourirajan & Kennedy (1962), Urusova (1975), Haas (1976), and Potter & Brown (1977). The inset shows the pressures obtained for inclusion *C* using various methods described in the text.

have been trapped at any higher P and T along the isochore originating at C, shown on Figure 5.

Although fluid inclusions homogenizing by halite disappearance are common, there is much confusion concerning their origin and interpretation. Assuming that this behavior is not a result of kinetic effects (Chivas & Wilkins 1977, Eastoe 1978) or that the inclusions have not trapped the halite as a solid phase, as suggested by some data (e.g. Bodnar 1978, Nagano et al 1977, Wilson 1978), the most commonly used explanation is that the inclusions were trapped at "high pressure" (e.g. Piznyur 1968, Dolgov et al 1976, Kamilli 1978, Milovskiy et al 1978, and Andreev & Shvadus 1977), on the basis of what is generally called the "Lemmlein and Klevtsov method." We believe this method yields spurious results.

The "Lemmlein and Klevtsov method" was apparently first used by Klevtsov & Lemmlein (1959)[6] to determine minimum trapping pressures of fluid inclusions homogenizing by halite disappearance. The pressure was obtained by conducting experimental P-V-T studies on aqueous solutions thought to correspond to the composition of the inclusion fluids and then plotting a pressure-correction diagram from the data. Microchemical analyses indicated that the fluid composition could be approximated by an aqueous solution of 24 wt. % NaCl and 15 wt. % KCl; temperatures of phase changes in the inclusions (Lemmlein & Klevtsov 1955) verify this composition where compared to the data of Ravich & Borovaya (1949, 1950). However, we have found several errors in the experimental and theoretical data of Klevtsov & Lemmlein (1959) that invalidate their results.

Most important, the solutions used in their experiments were not representative of the inclusion fluids. Klevtsov & Lemmlein (1959) stated that the solutions used were 24 wt. % NaCl and 15 wt. % KCl. Recalculation from their original data (p. 662 in original ". . . 122.5 ml NaCl and 61.5 ml KCl per 1000 ml H_2O . . .") indicates that the solution *actually* used was 19 wt. % NaCl and 8.8 wt. % KCl. We are not certain how this error was introduced, but it may have resulted from their assumption that the total volume of the system can be calculated by adding the volumes of solid NaCl and KCl to the volume of water. This is clearly incorrect. For example, the volume of a 25 wt. % NaCl solution at 25°C may be approximated to within 2.6% of the actual volume if the two volumes of solid NaCl and liquid H_2O are assumed to be additive (Potter & Brown 1977). However, this small volume difference is equivalent to ~900 bars pressure. Thus, where pressures are calculated from volumes (or densities)

[6] A translation of this paper will appear in volume 10 of COFFI (Roedder 1968–onward).

as Klevtsov & Lemmlein (1959) have done, the assumption of additive volumes leads to highly inaccurate results.

As a result of use of the wrong solution, all data of Klevtsov & Lemmlein (1959) were collected in the one-phase fluid field and *not* along the solid-liquid curve as they imply. These workers then compared the isochores obtained in their study with those for a 30 wt. % NaCl solution (Lemmlein & Klevtsov 1956, 1961) and noted that they were very similar in slope. They then suggested that for any fluid inclusion homogenizing by halite disappearance, the *P-V-T* data for a 30 wt. % NaCl solution could be used to approximate the pressure, assuming that the difference between the vapor-bubble-disappearance and halite-dissolution temperatures corresponds to the "ΔT" values on the ordinate of Lemmlein & Klevtsov's (1956) pressure-correction diagram (p. 533, their Figure 4). Using this "pressure correction" (ΔT) and the "homogenization temperature" (actually T_H L-V) along the abscissa, they obtain a pressure. Thus for inclusion *C* (Figure 5) in which the vapor bubble disappears at 310°C and the halite dissolves at 400°C, an apparent homogenization pressure of ~1100 bars is obtained.

We believe that this method of calculating pressure is incorrect because the difference in the two phase-disappearance temperatures measured in these inclusions is *not* the same as the "ΔT" values on the pressure-correction diagrams. Lemmlein & Klevtsov's (1956, 1961) data were obtained by placing a known solution for 30% NaCl into a fixed-volume autoclave and heating until the autoclave contained a homogeneous fluid. Pressure was then measured as a function of temperature "in the one-phase liquid region of the system" (Lemmlein & Klevtsov 1961, p. 149). Obviously, the inclusions under consideration are not one phase and will not follow an isochoric path for a 30 wt. % NaCl solution as the method requires. The actual path along the solid-liquid surface cannot be determined because of the lack of volumetric data in this region. However, if the apparent molar volumes of NaCl are negative in this region, as Urusova's (1975) data indicate, the actual pressure change between the vapor-bubble- and halite-disappearance temperatures might be very small or even negative, though the temperature difference may be very large.

Lyakhov (1973) presented a different method of calculating pressures of inclusions homogenizing by halite disappearance, which is also based on the *P-V-T-X* properties of NaCl-H_2O solutions. Unfortunately, this method assumes a constant volume but allows the mass of water in the inclusion to vary as the density of the water changes. This results in the erroneous conclusion that above 320°C the solubility of NaCl decreases rapidly. In Lyakhov's method, the effect of this erroneous conclusion is that an inclusion following a path to homogenization described by inclusion *C* (Figure 5) would have been trapped at a fictitious minimum pressure of 5000 bars.

We hold that the "Lemmlein and Klevtsov" and "Lyakhov" methods of determining pressures are both invalid, not only for the theoretical reasons given above, but also because the pressures obtained from these methods are not consistent with other data. Thus, using the "Lemmlein and Klevtsov method," Kamilli (1978, p. 431) indicates that the inclusion data "require pressures much greater than any reasonable lithostatic load," and Bodnar & Beane (1977) reported pressures an order of magnitude higher than those obtained from other fluid-inclusion data and from geologic reconstruction. Also, pressures of 5000 bars (Lyakhov 1973) and 6000 bars (Dolgov et al 1976) in quartz and 2700 bars in fluorite (Erwood et al 1979) are in gross disagreement with measured pressures required to decrepitate inclusions in quartz (850 bars; Naumov et al 1966) and to stretch fluid inclusions in fluorite (100–700 bars; Bodnar & Bethke 1980).

We believe it is possible to calculate a good approximation of the pressure in such multiphase inclusions at homogenization. We will illustrate this method by calculating the pressure for the same inclusion *C* (Figure 5) in which the vapor bubble disappears at 310°C and the halite dissolves at 400°C. To simplify calculations, we assume that the inclusion contains 1000 g of H_2O and that the fluid properties are adequately represented by those of the NaCl-H_2O system. Assuming that NaCl solubility is independent of pressure (Adams 1931), we can obtain the bulk composition (46 wt. % NaCl) from T_S NaCl (400°C) and the solubility data of Keevil (1942). Hence, we need to calculate only the density of the homogeneous fluid and refer this value to the *P-V-T-X* measurements of Urusova (1975) to obtain the pressure at final homogenization (T_S NaCl).

At 310°C, the inclusion volume is simply the volume of saturated aqueous solution plus the volume of solid halite remaining. The solution volume is the total mass of solution divided by its density. A saturated solution on the liquid-vapor curve at 310°C is 10.8 molal NaCl and has a density of 1.07 $g \cdot cm^{-3}$ (Haas 1976) which, when combined with our previous assumption of 1000 g of H_2O and a molecular weight of NaCl of 58.44 $g \cdot mol^{-1}$, gives a solution volume of 1524 cm^3. The volume of halite remaining can be calculated from the difference in solubility between 310° and 400°C (Keevil 1942) and the molar volume at 310°C. Using a value for the solubility of NaCl at 400°C of 46 wt. % or 14.6 molal, calculated from Keevil (1942), we find that the difference in solubility is 3.8 moles/1000 g H_2O. The molar volume of halite at 25°C is 27.018 cm^3 (Robie et al 1966), which increases 3.92% upon heating from 25°C (Skinner 1966) to give a molar volume at 310°C of 28.08 cm^3. Therefore, the volume of halite at 310°C is 106.7 cm^3, and the total inclusion volume at 310°C is 1631 cm^3.

The inclusion volume at homogenization is the volume at 310°C plus the increase due to thermal expansion upon heating and dissolution of the inclusion walls. Where the host mineral is quartz, the inclusion volume will increase 0.54% or 8.8 cm^3 from thermal expansion when heated from 310° to 400°C (Skinner 1966). The increase in volume, due to dissolution of quartz from the walls when the inclusion is heated from 300° to 400°C, can be estimated from quartz solubility in water at 317° and 396°C and 500 bars (Kennedy 1950) and the apparent molal volume data for amorphous silica (Duedall et al 1976). The calculated volume increase, 2.5×10^{-2} cm^3 or $\sim 1.5 \times 10^{-3}$ volume %, is small, even when compared to the increase due to thermal expansion and will not be included in further calculations.

The inclusion volume will also increase as a result of the increase in internal pressure upon heating from 310° to 400°C. However, at the moment of entrapment, the external and internal pressures would have been equal and, assuming that the expansion from internal pressure is similar in magnitude to the compressibility under the same pressure, these volume changes would cancel one another. Previously these two volume changes were incorrectly thought to be additive (Roedder 1979a).

The density of the homogeneous fluid at 400°C calculated from the mass of solution (1853 g) and the inclusion volume (1639.8 cm^3) is 1.130 $g \cdot cm^{-3}$. Extrapolating Urusova's (1975) data at 400°C along isochores from 45 wt. % NaCl (her maximum concentration at 400°C) to 46 wt. % NaCl provides a pressure at final homogenization of 640 bars. This is compared with pressures obtained by the "Lemmlein and Klevtsov" method (~1100 bars), and the "Lyakhov" method (5000 bars), for this same inclusion (Figure 5).

The accuracy of our pressure determination is difficult to establish and, of course, depends on the accuracy of the *P-V-T-X* data used. In this respect, we note that, where the data of Urusova (1975) and Sourirajan & Kennedy (1962) overlap, Urusova's data are always higher by ~25 bars, and Sourirajan & Kennedy's data are known to be high because of the presence of hydrogen gas derived from corrosion (Liu & Lindsay 1972, J. Haas, Jr., personal communication, 1979). A complete analysis of the accuracy and precision of our method must await experimental volumetric data along the liquid-solid curve of the NaCl-H_2O system at high pressures and temperatures.

Although halite is by far the most common solid phase occurring in fluid inclusions, many other daughter minerals have been identified (Roedder 1972, pp. JJ19–JJ29). One of these, sylvite (KCl), is often found along with halite, especially in porphyry-type ore deposits. With Roedder's (1971) NaCl-KCl-H_2O ternary diagram and the data of Ravich & Borovaya (1949), these inclusions bearing sylvite and halite may be used to estimate fluid compositions and "minimum" pressures attending entrapment. Other ions in solution, such as Ca^{+2}, would significantly lower this "minimum." The vapor pressures of such "hydrosaline melts" are surprisingly low. Thus, Stewart & Potter (1979) showed that a fluid saturated with respect to NaCl, KCl, and $CaCl_2$ has a maximum vapor pressure of only ~25 bars, at 350°C.

Most other solid phases are so rarely found and their solubility and volumetric properties so poorly known that these inclusions provide little, if any, information concerning the pressure at trapping.

CARBON DIOXIDE–BEARING INCLUSIONS Carbon dioxide–bearing fluid inclusions occur in rocks from a wide range of geologic environments and many methods have been suggested for determining trapping pressures and temperatures from these inclusions. Considerable confusion exists, mainly from failure to consider the trapping conditions. The three most commonly used methods are described below.

Intersecting isochores in pure CO_2 and H_2O systems Kalyuzhnyi & Koltun (1953) applied a method of geobarometry, first described by Nacken (1921), which is applicable to *separate* CO_2 inclusions and H_2O inclusions trapped at the same *P-T* conditions, either at the same location at different times or at the same time but at different locations. By plotting the *P-V-T* diagrams for H_2O and CO_2 in the same plane, the pressure and temperature at trapping are defined by the intersection of the isochore corresponding to the CO_2 inclusion with that corresponding to the H_2O inclusion (Figure 6). Thus, a CO_2 inclusion homogenizing in the liquid phase at 11°C and an H_2O inclusion homogenizing in the liquid phase at 167°C would both have been trapped at 237°C and 1180 bars (*A*, Figure 6). This method of obtaining trapping conditions is valid if, and *only if*, the two inclusions were separately trapped as essentially pure components at the same temperature and pressure.

Pressure determinations from CO_2-bearing fluid inclusions are much more complex if the carbon dioxide and water were not physically separated at the time of trapping. The addition of CO_2 to water causes the critical locus to migrate to lower temperatures and higher pressures, reaching a minimum temperature of 266°C at 2450 bars and 41.5 mole % CO_2 (Todheide & Franck 1963). From there the critical temperature slowly rises, reaching 268°C at 43.5 mole % CO_2 and ~3600 bars; this rise results in a large two-phase field spanning a wide range of *P-T-X* conditions (Figure 7).

Naumov and Malinin graphical method Naumov & Malinin (1968) proposed a graphical technique for determining pressure that uses the T_H and T_D of inclusions containing both liquid water solution and CO_2. The method is based on a straightline extrapolation, in *P-T* space, from the point representing partial homogenization (liquid and gaseous CO_2), through T_D (assumed to be at 850 atm for inclusions in quartz), to the temperature, and thus pressure, at total homogenization. We question the validity of this technique because there is no basis for assuming that the pressure increases linearly with temperature between the temperature of partial homogenization (liquid and gaseous CO_2) and complete homogenization of this CO_2 fluid and the water solution. In fact, because the slopes of the "isochores" and mutual solubilities (see below) of the two phases are constantly changing as the system within the inclusion travels through *P-T* space, it would be quite surprising if the pressure did vary in such a simple fashion with temperature. Further, the assumption that decrepitation begins at 850 atm is based on experimental studies (Naumov et al 1966) using synthetic quartz crystals that normally contain very

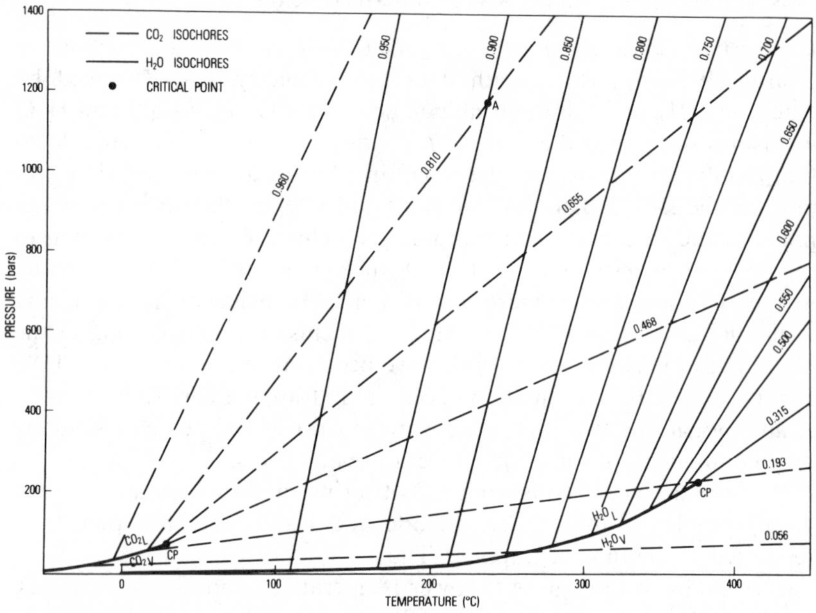

Figure 6 Combined *P-T* diagrams for CO_2 and H_2O illustrating the Kalyuzhnyi & Koltun (1953) method of geothermobarometry using separate CO_2 and H_2O inclusions trapped at the same temperature and pressure. Point *A* represents the common trapping temperature (237°C) and pressure (1180 bars) for a CO_2 inclusion and an H_2O inclusion homogenizing at 11° and 167°C, respectively, both in the liquid phase. Data are from Kennedy (1954), Burnham et al (1969), Keenan et al (1969), and Weast (1978).

large fluid inclusions relative to natural quartz and do not contain lower temperature secondary inclusions. As Naumov & Malinin (1968) pointed out, pressures necessary to decrepitate inclusions in synthetic quartz range from 850 to >3000 atm, depending on inclusion size. Therefore, if the

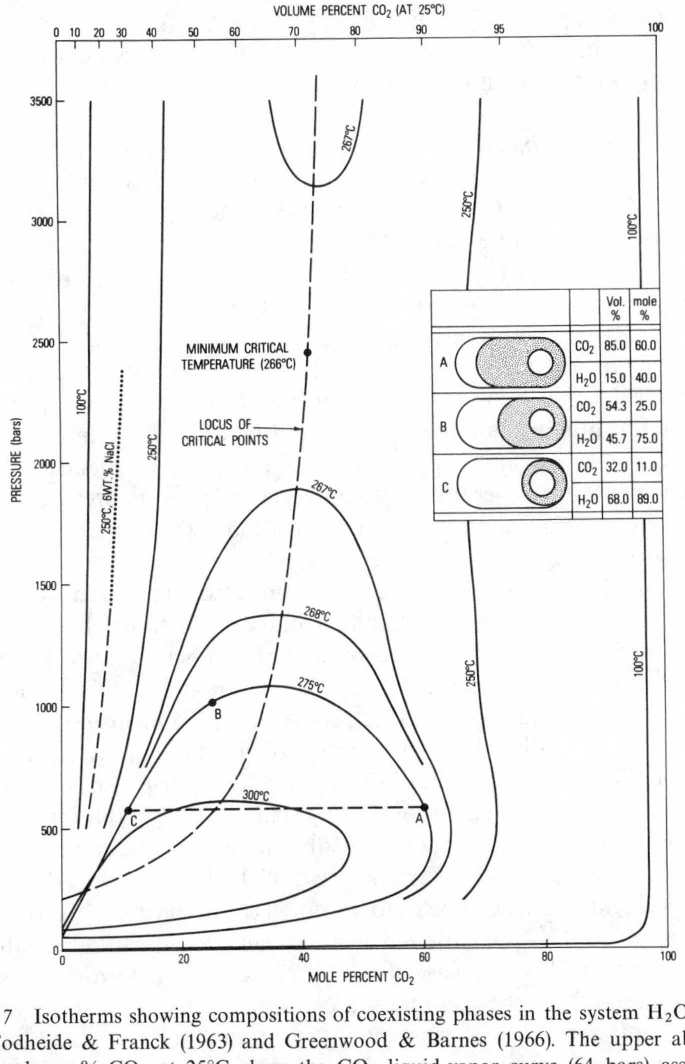

Figure 7 Isotherms showing compositions of coexisting phases in the system H_2O-CO_2, after Todheide & Franck (1963) and Greenwood & Barnes (1966). The upper abscissa shows volume % CO_2 at 25°C along the CO_2 liquid-vapor curve (64 bars) assuming densities of CO_2 liquid, CO_2 vapor, and H_2O liquid of 0.71, 0.24, and 1.0 g·cm^{-3}, respectively (Newitt et al 1956, Keenan et al 1969). The inset shows the two-dimensional appearance at the stated conditions for three cylindrical inclusions having compositions as given (liquid CO_2 shaded), which are also shown on the diagram. The 250°C isotherm for a 6 wt. % NaCl solution from Takenouchi & Kennedy (1965) is shown for comparison.

smallest inclusions in synthetic quartz (decrepitating at >3000 atm) correspond to the largest inclusions studied in natural quartz, as is possible, the onset of mass decrepitation of natural quartz samples will be at >3000 atm. Finally, and most important, at many of the *P-T-X* combinations obtained by means of this technique, water and CO_2 form two immiscible phases (Todheide & Franck 1963, Takenouchi & Kennedy 1964, Malinin 1959, Greenwood & Barnes 1966), which raises the possibility that the inclusions did not trap homogeneous fluids.

Mixed H_2O-CO_2 inclusions At ambient temperatures, inclusions consisting purely of carbon dioxide and water commonly contain three phases, liquid and gaseous CO_2, and liquid water. These phases are essentially pure, having solubilities of H_2O in the CO_2-liquid-rich phase and of CO_2 in the H_2O-rich phase of ~0.1 and ~2.1 mole %, respectively, at 25°C (Greenwood & Barnes 1966). Therefore, by measuring the volumes of the three phases at a known temperature and using the density data for CO_2 (Kennedy 1954) and H_2O (Burnham et al 1969), we may calculate the mole % CO_2 in the inclusion. Such a procedure was used to obtain the mole % CO_2 (and mole % H_2O) in three hypothetical inclusions shown in the inset of Figure 7. In practice, volume measurements are generally very imprecise for reasons previously mentioned and we urge caution when compositions are calculated by means of this technique.

The complexities of pressure determinations from inclusions containing both carbon dioxide and water might best be illustrated by examining the phase changes upon heating of three hypothetical pure H_2O-CO_2 fluid inclusions and referring these observational data to the CO_2 solubility diagram in Figure 7. First, consider a fluid inclusion containing 25 mole % CO_2 with volume percentages of phases as shown by inclusion *B* (Figure 7). With heating, the CO_2 liquid and vapor phases homogenize to a fluid that has a density of 0.672 $g \cdot cm^{-3}$ at 27°C (Quinn & Jones 1936, Kennedy 1954, Newitt et al 1956). The mode of homogenization is by vapor-phase disappearance, although, if the heating rate is too rapid such an inclusion will appear to exhibit critical phenomena (Roedder 1972, pp. JJ16–JJ17). Continued heating causes the mutual solubilities of CO_2 and H_2O to increase until, at 275°C, the inclusion contains a homogeneous fluid phase composed of 25 mole % CO_2 and 75 mole % H_2O (*B* in Figure 7). This means that the internal pressure,[7] and thus the *minimum* trapping pressure, is ~1000 bars. Note that homogenization at

[7] The high internal pressure developed during homogenization of many CO_2-H_2O inclusions is the reason that decrepitation commonly occurs before T_H is reached.

275°C and 1000 bars would also occur in an inclusion containing 45 mole % CO_2.

If the pressure on this fluid dropped below 1000 bars, carbon dioxide and water would no longer be completely miscible. Thus, at 275°C and 575 bars a CO_2-rich fluid containing 40 mole % water and a water-rich fluid containing 11 mole % CO_2 coexist (Figure 7), and two separate inclusions trapped at these conditions would have phase relations as shown by inclusions *A* and *C* in the inset on Figure 7. Upon heating from 25°C, the CO_2 phases in inclusions *A* and *C* would homogenize at 26° and 28°C, respectively, and complete homogenization would occur in both at 275°C and 575 bars. In most fluid inclusion work the composition is unknown or is imprecisely known from volume measurements. In this situation, the critical pressure along the 275°C isotherm determines a *maximum* trapping pressure of 1080 bars. Ypma (1963) pointed out, however, that it is possible to have noncoeval CO_2-H_2O inclusions, trapped under different *P-T* conditions, that can homogenize (fortuitously) at the same temperature.

Most CO_2-bearing fluid inclusions contain an aqueous salt solution rather than pure water, adding further complexity to pressure determination. With the addition of a salt to the CO_2-H_2O system, the critical solubility of CO_2 at a given temperature and pressure decreases (Takenouchi & Kennedy 1965, Ellis & Golding 1963, Malinin & Kurovskaya 1975), and thus the miscibility gap is widened. The 250°C isotherm for CO_2 solubility in a 6 wt. % NaCl solution, from Takenouchi & Kennedy (1965), is shown in Figure 7 for comparison with pure H_2O. Note that at 250°C, a 10 mole % CO_2 fluid at 750 bars would be just within the one-phase region if we assume a pure H_2O-CO_2 system, but well into the two-phase region if the H_2O phase actually is a 6 wt. % NaCl solution.

If a 10 mole % CO_2 inclusion that had a T_H of 250°C were trapped as a homogeneous fluid, Figure 7 shows that it would have had a *minimum* trapping pressure of 750 bars, if we assume that the inclusion consists of pure CO_2 and H_2O. However, if the H_2O phase contained 6 wt. % NaCl, the *minimum* trapping pressure would have been ~2500 bars, determined by extrapolating the 250°C, 6 wt. % NaCl isotherm of Takenouchi & Kennedy (1965) to higher pressure.

Methane is completely miscible with both liquid and gaseous CO_2 and is a common component of CO_2-bearing fluid inclusions in metamorphic rocks. Its presence is usually implied when the triple point for the CO_2-phases in these inclusions is found to be at a lower temperature than that for pure CO_2. Hollister & Burruss (1976) suggested that the addition of CH_4 to the H_2O-CO_2 system raises the top of the solvus to higher temperatures, and the miscibility gap is therefore widened. Swanenberg (1979) indicated that data from CO_2-CH_4 inclusions may be used in pressure determinations if the fluid density is expressed as an "equivalent CO_2 density." This method is based on the observation that, over the temperature range 200°–800°C, the isochoric pattern for the CO_2-CH_4 system is similar to that of the pure CO_2 system.

SILICATE-MELT INCLUSIONS Silicate-melt inclusions are trapped when crystals grow from a silicate melt (i.e. a magma or lava), in a manner quite similar to the trapping of aqueous fluids in other minerals. They may develop a shrinkage bubble on cooling, and they may form daughter minerals. The methods of study are similar in principle to ordinary (aqueous) fluid inclusions, but as the temperatures are much higher, the instrumentation is very different. Additional differences stem from the fact that relatively large amounts of host mineral may crystallize on the walls during cooling, and the high fluid viscosities commonly result in much metastability and sluggish equilibration (Roedder 1979b). Although there have been many determinations of T_H on silicate-melt inclusions (generally they fall in the range 800°–1200°C), pressure determinations based on such inclusions are normally not made. In large part this stems from the fact that silicate melts are relatively incompressible, so there is little need for a pressure correction to be applied to most such T_H determinations. Murase & McBirney (1973) determined the adiabatic compressibility of a series of silicate melts from their densities and longitudinal wave velocities, and found them to fall in the range of 2 to $7 \times 10^{-12}\,cm^2 \cdot dyne^{-1}$ at 1000–1200°C. Combining this with thermal expansion data on crystals and glasses (Skinner 1966), we find that the effect of 1 kilobar pressure at the time of trapping would correspond to $\sim$20°C, an amount far smaller than the probable experimental error alone on most determinations of silicate melt T_H (Roedder 1979b). Because most determinations of melt T_H are made on glassy inclusions, they are from rocks formed under intermediate or shallow depths, and hence the pressures are seldom >1 kb.

If the silicate melt inclusions give evidence of having been trapped from an immiscible mixture of silicate melt with another, more compressible fluid, pressure estimates become feasible. Here the silicate-melt inclusions, containing a relatively incompressible fluid, permit a fairly close estimate of the true temperature of trapping. This value can then be used, along with P-V-T data on the other, more compressible fluid, to obtain pressure. The major limitation in the practical application of the procedure lies in the difficulty of proving contemporaneity of the two inclusion types. Many of the published pressure determinations from inclusions in igneous minerals are based on such an assumption of immiscibility, but lack the necessary evidence to prove contemporaneity of trapping.

The abundant carbon dioxide inclusions found along with silicate-melt inclusions in the olivine of olivine nodules from basalt occurrences all over the world (Roedder 1965) provide an example of the application of the method. The silicate-melt inclusions in these samples, and actual observations on lavas, show that these inclusions were probably trapped at ~1200°C. If we assume this temperature of trapping, data on the density of the CO_2 inclusions (from their T_H values, normally in the range of 10–31°C), along with *P-V-T* data on CO_2, permit an estimate of 6 kb as the pressure at the time of trapping, corresponding to the hydrostatic pressure from ~20 km of liquid basalt. This is not a minimum or maximum value, but the actual pressure, under the given assumptions. Unfortunately this pressure estimate must be based on extrapolations from data in which maximum temperatures and pressures are 1000°C and 1400 bars (Kennedy 1954), or 707°C and 8000 bars (Shmonov & Shmulovich 1974). Other inclusions in these samples were presumably trapped at similar temperatures but greater depths; these had such high internal pressures that they have generally decrepitated upon eruption at the surface at ~1200°C. Therefore, the range of application of the method is limited (Roedder 1965).

Bilal & Touret (1977) reported homogenization temperatures as low as +10°C on presumably pure CO_2 inclusions in phenocrysts from a basalt, and estimated a pressure of 5 kb. However, using the molal volume data on CO_2 of Shmonov & Shmulovich (1974) and extrapolating to an assumed trapping temperature of 1200°C, we find that these inclusions indicate a trapping pressure well over 6 kb. Conversely, the trapping temperature required to yield inclusions with this density of CO_2 at 5 kb is only ~900°C.

Lower-pressure CO_2 inclusions in basaltic glass from submarine flows were studied by Moore et al (1977). They showed that there was a constant relationship between the pressure of CO_2 in gas vesicles (i.e. bubbles) and the known depth of water under which the eruption occurred. Using this method, we can estimate the pressure at the time of eruption of unknown basalt samples simply by piercing the bubbles under a liquid and measuring the volume expansion.

The above examples are based on the essentially immiscible pair of fluids, CO_2 and basalt. Inclusions from partly immiscible fluids, such as silicate melt and water, can also be used. If the second fluid phase is a hydrosaline melt, with >50% of NaCl and other salts, as in the Ascension Island granites (Roedder & Coombs 1967) or in the various hypabyssal granites reported in the extensive Soviet literature (see many entries in

Roedder 1968–onward, particularly those by A. I. Zakharchenko), inclusion data could, in theory, provide an estimate of pressure. This requires, however, that the composition of the aqueous fluid phase is known, and that *P-V-T* data on such fluids are available; neither of these requirements can be satisfactorily met at present.

Where the concentration of salts in the aqueous phase is low, the *P-V-T* data for H_2O can be used, and the accuracy of the resulting geobarometric values will increase. The limiting case involves silicate-melt inclusions containing water but without evidence of a second, volatile-rich phase. This requires that the pressure at the time of trapping was above that of the vapor pressure of water over that melt. Thus Anderson (1974), and Anderson & Sans (1975, and personal communications) devised several methods for estimating the water content of magmas from melt inclusions and found some high values, as much as $12 \pm 2\%$ H_2O (Roedder 1979b). Such inclusion data can be combined with experimental data on the vapor pressures of hydrous melts, when they become available, to provide at least a lower limit for the pressure at the time of trapping.

The Future

As in all science, we can expect that more accurate thermometric determinations on inclusions will certainly become available in the future. Will they help in geobarometry? *In theory,* it is possible to obtain both *P* and *T* from careful studies of the small differences that should exist between otherwise identical coeval inclusions in two host minerals with differing thermal expansions. Such an effort seems doomed, however, by the requirement of coeval inclusions; whenever detailed studies are made, it is apparent that the fluids present in most geological environments have varied in *P*, *T*, and *X* even during the formation of small parts of a single crystal (e.g. Roedder 1977a). Such data emphasize the difficulty of proving two inclusions to be truly coeval, and show that even small errors in such an assignment of origin can result in major errors in pressure by any method requiring coeval inclusions. Although immiscible fluid pairs now provide us with some of our best geobarometers, this same problem of finding truly coeval inclusions in material formed in a changing environment makes future refinements in inclusion thermometry on such inclusions of relatively little value for geobarometric determinations.

More accurate determinations of the composition of inclusions are also forthcoming, as a result of the application of a variety of new techniques. Although such improved data on the composition of the former fluid phase are invaluable in other investigations, such as the causes of ore deposition or the nature of the various reactions that have yielded metamorphic rock assemblages, they are of rather limited *immediate* value to geobarometry, because experimental *P-V-T-X* data are generally lacking for such compositions. Even the simple systems, such as NaCl-H_2O, are not known to the pressure, temperature, and concentration limits necessary, and as the complexity of the composition increases, the available data decrease very rapidly. After such *P-V-T-X* data become available, however, we expect that the accuracy of geobarometry based on inclusions that have trapped a single, homogeneous phase may exceed that from the two-fluid methods because the difficult requirement of coeval inclusions is not involved.

One interesting and important new facet of fluid-inclusion barometry that may well have more extensive application in the future lies in its use in evaluating a site for a nuclear reactor (Cunningham 1974). This was based on a study of fluid inclusions in euhedral crystals protruding into vuggy cavities along a fault that cut a proposed site. The important question was to date the last movement on the fault, and hence to evaluate the possibility of renewed movement. The occurrence of the crystals suggested that they must have grown since the last movement on the fault. An elevated pressure at trapping, shown by thermometric data on the inclusions in the crystals, established a minimum depth below the surface at the time of formation. Subsequent erosion had removed this much material, so with an estimate of the rates of denudation, a minimum time since the last movement was obtained.

Most single-fluid inclusion studies today involve the determination of T_H and a search for geologic evidence concerning depth of burial in order to estimate the pressure correction to be added to T_H to obtain T_T. Assuming that the composition of the inclusion can be determined, the three interrelated variables here are T_H, T_T, and P; if any two are known, the other can be obtained. T_H can be determined with relatively high accuracy at present. Estimates of pressure, however, have very large error bars that are the sum of the necessarily large uncertainties of the geologic reconstruction and the problem of "hydrostatic" vs "lithostatic" pressure mentioned in a previous section. Independent geothermometers for obtaining T_T are not always available, and are seldom very accurate. However, we believe that there is considerable possibility of using them in the future as these other geothermometers are developed, and hence of determining P from (T_T-T_H) and the *P-V-T-X* properties of the fluid with an accuracy at least comparable to that from geologic reconstruction.

The effects on an inclusion of its post-trapping history, although generally a source of possibly major error in both geothermometry and geobarometry of the original environment, may provide useful data on the pressures during this subsequent history. Although we do not believe that decrepitation of inclusions in the laboratory provides much quantitative thermometric data, natural decrepitation can prove useful. The rocks containing fluid inclusions, formed deep in the earth at elevated *P* and *T*, have dropped to present-day surface *P* and *T* via a generally unknown route on a *P*-*T* diagram for the appropriate fluid. If the path taken is such that the pressure within an inclusion exceeds the external pressure on the sample by some amount that varies with the sample, the inclusion may deform or rupture its host crystal. Evidence for such natural decrepitation has been reported by numerous investigators (see, for example, Voznyak & Kalyuzhnyi 1976, and various entries in Roedder 1968–onward), and may provide valuable information on depths of burial and rates of uplift and denudation; it should be looked for routinely. Perhaps the most serious problem lies in the development of petrographic criteria for assigning relative ages to the various generations of secondary inclusions present in many metamorphic samples that may reflect such decrepitation.

One aspect of natural decrepitation may hold considerable promise for clarifying the time sequence in certain geologically complex terrains, but has not been adequately applied in the past. This is the recognition of natural decrepitation from the pressure rise due to local heating, e.g. at the intersection of veins and dikes (Roedder 1977a). Similarly, some constraints on the *P*-*T* path taken by magmas during ascent and eruption may be obtained from data on those high-pressure CO_2 inclusions that have decrepitated, and those that have not (Roedder 1965).

During laboratory heating runs, some inclusions develop sufficient internal pressure to cause permanent deformation of the host mineral, but without decrepitation. Although this "stretching" can cause serious thermometric errors, it can be avoided by appropriate laboratory procedures (Larsen et al 1973). A recent study of the systematics of stretching of fluid inclusions when overheated (Bodnar & Bethke 1980) suggests that it may be possible to recognize inclusions that have stretched naturally. If so, we can use such naturally stretched inclusions as geobarometers. Thus, Bodnar & Bethke (1980) showed that the internal pressure necessary to initiate stretching of fluid inclusions in fluorite varies in a systematic manner as a function of inclusion size but that, once begun, the amount of stretching is a linear function of the internal pressure. Where a fluorite crystal has been overheated by some later thermal event, its larger inclusions might have been stretched but not other smaller inclusions, and heating tests would reveal an increase in filling temperature as inclusion size increases for inclusions exceeding a certain minimum size. Then, knowing the pressure necessary to initiate stretching of the smallest observed stretched inclusion, and assuming that this pressure represents the difference between the internal and external pressures existing when stretching occurred, we can calculate the external pressure. This pressure is the external pressure at the time of stretching and *not* necessarily the pressure during trapping of those particular inclusions. Furthermore, there is no verification at present that for a given pressure difference, the stretching is independent of the confining pressure.

Conclusions

We have discussed the numerous inclusion geobarometers that have been used in the past, and find that some are wrong in concept and can yield grossly inaccurate pressures, and others need an independent geothermometer to yield an estimate of the pressure. Single inclusions can yield only data that are functions of both P and T, but pairs of inclusions trapped from immiscible fluids (e.g. water and steam, or water and CO_2) can yield *both* P and T. Inclusions of an immiscible water (or CO_2) phase along with a silicate-melt phase are a special example, in that the silicate-melt inclusions provide, in effect, an independent thermometer, and hence the pair can provide a geobarometer.

The essential data needed for any determination of the pressure of formation from inclusions are the following:

1. Good evidence of the time of trapping of the inclusion relative to the geologic process being studied, from careful microscopy.
2. Good evidence of freedom from various secondary effects on the inclusion since trapping, both in nature and the laboratory (necking down, leakage, decrepitation, stretching, etc).
3. Good thermometric data (T_{Frz}, T_S, T_H) on the phases in the inclusion (and sometimes volumetric measures of the individual phases).
4. Good compositional data on the inclusion from thermometry and other procedures.
5. Good experimental P-V-T-X data on appropriate systems for the composition found, that covers the necessary range of conditions.

Of these five requirements, at the present time we believe that the most serious handicap to accurate geobarometry is the lack of good experimental P-V-T-X data.

Acknowledgments

The authors wish to thank the following for reviews of the manuscript: R. E. Beane, John Haas, Jr., W. C. Kelly, A. Kozlowski, R. W. T. Wilkins, and T. L. Woods. We have obviously profited from many discussions with our colleagues.

Literature Cited

Adams, L. H. 1931. Equilibrium in binary systems under pressure. I. An experimental and thermodynamic investigation of the system NaCl-H_2O. *J. Am. Chem. Soc.* 53: 3769–3813

Anderson, A. T. Jr. 1974. Evidence for a picritic, volatile-rich magma beneath Mt. Shasta, California. *J. Petrol.* 15(2): 243–67

Anderson, A. T., Sans, J. R. 1975. Volcanic temperature and pressure inferred from inclusions in phenocrysts (extended abstract). *Int. Conf. Geotherm. Geobarom. 5–10 Oct., 1975, Penn. State Univ., Extended Abstr.* University Park, Pa: Penn. State Univ. (Unpaginated)

Andreev, G. V., Shvadus, M. I. 1977. *P-T* conditions of formation of alkaline rocks of the Synnyr complex, on the basis of results of studies of the first found melt inclusion. *Dokl. Akad. Nauk SSSR* 235(4): 910–13

Barker, C. 1972. Aquathermal pressuring—Role of temperature in development of abnormal-pressure zones. *Am. Assoc. Petrol. Geol. Bull.* 56: 2068–71

Barton, P. B. Jr., Bethke, P. M., Toulmin, P. 3rd. 1963. Equilibrium in ore deposits. *Mineral. Soc. Am. Spec. Pap.* 1: 171–85

Bethke, P. M., Barton, P. B. Jr. 1971. Distribution of some minor elements between coexisting sulfide minerals. *Econ. Geol.* 66: 140–63

Bilal, A., Touret, J. 1977. Fluid inclusions in phenocrysts from basaltic lavas of Puy Beaunit (French Massif central). *Bull. Soc. Fr. Mineral. Cristallog.* 100: 324–28 (In French)

Birch, F. 1966. Compressibility; elastic constants. In *Handbook of Physical Constants*, ed. S. P. Clark Jr., *Geol. Soc. Am. Mem. 97*, pp. 97–173. Revised ed.

Bodnar, R. J. 1978. *Fluid inclusion study of the porphyry copper prospect at Red Mountain, Arizona.* MS thesis. Univ. Ariz., Tucson

Bodnar, R. J., Beane, R. E. 1977. Temperature variations in pre-intrusive cover over a buried pluton. *Geol. Soc. Am. Abstr. Programs* 9(7): 903–4

Bodnar, R. J., Bethke, P. M. 1980. Systematics of "stretching" of fluid inclusions as a result of overheating. *Econ. Geol.* In press

Browne, P. R. L., Roedder, E., Wodzicki, A. 1976. Comparison of past and present geothermal waters, from a study of fluid inclusions, Broadlands field, New Zealand. *Proc. Int. Symp. Water-Rock Interaction*, Czechoslovakia, 1974, ed. J. Cadek, T. Paces, pp. 140–49. Prague: Geological Survey

Burnham, C. W., Holloway, J. R., Davis, N. F. 1969. The specific volume of water in the range 1000 to 8900 bars, 20° to 900°C. *Am. J. Sci.* 267: 70–95

Chivas, A. R., Wilkins, R. W. T. 1977. Fluid inclusion studies in relation to hydrothermal alteration and mineralization at the Koloula porphyry copper prospect, Guadalcanal. *Econ. Geol.* 72: 153–69

Coveney, R. M., Kelly, W. C. 1970. Quartz as a geologic barometer. *Mich. Acad.* 3(3): 45–56

Cunningham, C. G. Jr. 1974. Geothermometry and geobarometry of fault plane mineralization—Ginna project, New York, Appendix C, pp. CIII1–CIV2, (17p.). In *Geologic and Geophysical Investigations, Ginna site, Rochester Gas & Electric Corp.* 593 pp. Prepared by Dames & Moore Corp. for Rochester Gas & Electric Corp. Submitted in support of proceedings for US Nuclear Regulatory Commission Docket No. 50-244, April 19, 1974

Dolgov, Y. A., Tomilenko, A. A., Chupin, V. P. 1976. Inclusions of salt melts—solutions in quartz of deep-seated granites and pegmatites. *Dokl. Akad. Nauk SSSR* 226(4): 938–41 (In Russian). Trans. in *Dokl. Akad. Nauk SSSR* 226(4): 206–9

Duedall, I. W., Dayal, R., Willey, J. D. 1976. The partial molal volume of silicic acid in 0.725 m NaCl. *Geochim Cosmochim. Acta* 40: 1185–89

Eastoe, C. J. 1978. A fluid inclusion study of the Panguna porphyry copper deposit, Bougainville, Papua New Guinea. *Econ. Geol.* 73: 721–48

Ellis, A. J., Golding, R. M. 1963. The solubility of carbon dioxide above 100°C in water and in sodium chloride solutions. *Am. J. Sci.* 261: 47–60

Erwood, R. J., Kesler, S. E., Cloke, P. L. 1979. Compositionally distinct, saline hydrothermal solutions, Naica Mine, Chihuahua, Mexico. *Econ. Geol.* 74: 95–108

Fisher, J. R. 1976. The volumetric properties of H_2O—A graphical portrayal. *J. Res. US Geol. Surv.* 4: 189–93

Greenwood, H. J., Barnes, H. L. 1966. Binary mixtures of volatile compounds. In *Handbook of Physical Constants*, ed. S. P. Clark Jr., *Geol. Soc. Am. Mem. 97*, pp. 385–400. Revised ed.

Haas, J. L. Jr. 1971. The effect of salinity on the maximum thermal gradient of a hydrothermal system at hydrostatic pressure. *Econ. Geol.* 66: 940–46

Haas, J. L. Jr. 1976. Physical properties of

the coexisting phases and the thermochemical properties of the H_2O component in boiling NaCl solutions. *US Geol. Surv. Bull. 1421-A*. 73 pp.

Hollister, L. S., Burruss, R. C. 1976. Phase equilibria in fluid inclusions from the Khtada Lake metamorphic complex. *Geochim. Cosmochim. Acta* 40:163–75

Kalyuzhnyi, V. A., Koltun, L. I. 1953. Some data on pressures and temperatures during formation of minerals in Nagol'nyy Kryazh, Donets Basin. *Mineral. Sb. Lvov Geol. Obshch.* 7:67–74

Kamilli, R. J. 1978. The genesis of stockwork molybdenite deposits: implications from fluid inclusion studies at the Henderson mine. *Geol. Soc. Am. Abstr. Programs* 10(7):431

Keenan, J. H., Keyes, F. G., Hill, P. G., Moore, J. G. 1969. *Steam Tables; Thermodynamic Properties of Water, Including Vapor, Liquid, and Solid Phases.* New York: Wiley. 162 pp.

Keevil, N. B. 1942. Vapor pressures of aqueous solutions at high temperatures. *J. Am. Chem. Soc.* 64:841–50

Kennedy, G. C. 1950. A portion of the system silica-water. *Econ. Geol.* 45:629–53

Kennedy, G. C. 1954. Pressure-volume-temperature relations in CO_2 at elevated temperatures and pressures. *Am. J. Sci.* 252:225–41

Klevtsov, P. V., Lemmlein, G. G. 1959. Determination of the minimum pressure of quartz formation as exemplified by crystals from the Pamir. *Zap. Vses. Mineral. Obshch.* 85(6):661–66 (In Russian)

Larsen, L. T., Miller, J. D., Nadeau, J. E., Roedder, E. 1973. Two sources of error in low temperature inclusion homogenization determination, and corrections on published temperatures for the East Tennessee and Laisvall deposits. *Econ. Geol.* 68:113–16

Lemmlein, G. G., Klevtsov, P. V. 1955. Physico-chemical analysis of liquid inclusions containing crystals of halite and sylvite. *Zap. Vses. Mineral. Obshch.* 84(1): 47–52 (In Russian)

Lemmlein, G. G., Klevtsov, P. V. 1956. Relationship of the thermodynamic parameters PTV of water and of 30 % solutions of NaCl. *Zap. Vses. Mineral. Obshch.* 85(4):529–34 (In Russian)

Lemmlein, G. G., Klevtsov, P. V. 1961. Relations among the principal thermodynamic parameters in a part of the system H_2O-NaCl. *Geokhimiya* 1961(2): 133–42 (In Russian). Transl. in *Geochemistry* 1961(2):148–58

Liu, C.-T., Lindsay, W. T. Jr. 1972. Thermodynamics of sodium chloride solutions at high temperatures. *J. Solution Chem.* 1: 45–69

Lyakhov, Y. V. 1973. Errors in determining pressure of mineralization from gas-liquid inclusions with halite, their causes and ways of eliminating them. *Zap. Vses. Mineral. Obshch.* 102(4):385–93 (In Russian)

Malinin, S. D. 1959. The system water-carbon dioxide at high temperatures and pressures. Transl. in *Geochemistry* 1959(3): 292–306 (From Russian)

Malinin, S. D., Kurovskaya, N. A. 1975. Solubility of CO_2 in chloride solutions at elevated temperatures and CO_2 pressures. *Geochem. Int.* 12(2):199–201

Milovskiy, G. A., Zlenko, B. F., Gubanov, A. M. 1978. Conditions of formation of scheelite ores in the Chorukh-Dayron mineralized area (as revealed by a study of gas-liquid inclusions). *Geokhimiya* 15(1):79–86 (In Russian). Transl. in *Geochem. Int.* 15(1):45–52

Moore, J. G., Batchelder, J. N., Cunningham, C. G. 1977. CO_2-filled vesicles in mid-ocean basalt. *J. Volcanol. Geotherm. Res.* 2:309–27

Murase, T., McBirney, A. R. 1973. Properties of some common igneous rocks and their melts at high temperatures. *Geol. Soc. Am. Bull.* 84:3563–92

Nacken, R. 1921. Welche Folgerungen ergeben sich aus dem Auftreten von Flüssigkeitseinschlüssen in Mineralien? *Centralbl. Mineral.* 1921:12–20, 35–43

Nagano, K., Takenouchi, S., Imai, H. 1977. Fluid inclusion study of the Mamut porphyry copper deposit, Sabah, Malaysia. *Min. Geol.* 27:201–12 (In English)

Naumov, V. B., Malinin, S. D. 1968. A new method for determining pressure by means of gas-liquid inclusions. *Geochem. Int.* 5:382–91 (English transl.)

Naumov, V. B., Balitskiy, V. S., Ketchikov, L. N. 1966. Correlation of the temperatures of formation, homogenization, and decrepitation of gas-fluid inclusions. *Dokl. Akad. Nauk SSSR* 171(1):146–48

Newitt, D. M., Pai, M. U., Kuloor, N. R. 1956. Carbon Dioxide. In *Thermodynamic Functions of Gases, Volume 1*, ed. F. Din, pp. 102–34. London: Butterworths

Piznyur, A. V. 1968. On pressure during the formation of the Zhireken copper-molybdenum deposit (East Transbaikalia). *Dokl. Akad. Nauk SSSR* 179(5):1186–88 (In Russian)

Pollak, R. 1974. *The thermodynamic properties of water represented by a canonical equation of state for the homogeneous and heterogeneous fluid states up to 1200 K and 3000 bar*. Dr.-Ing. thesis. Ruhr Univ., Bochum. (In German) Transl. by Central Electricity Generating Board, U.K., IUPAC, Thermodynamic Tables Project Centre Transl. No. PC/T(CEGB) 14. (1976) 113 pp.

Potter, R. W. II. 1977. Pressure corrections for fluid-inclusion homogenization temperatures based on the volumetric properties of the system NaCl-H_2O. *J. Res. US Geol. Surv.* 5(5): 603–7

Potter, R. W. II, Brown, D. L. 1977. The volumetric properties of aqueous sodium chloride solutions from 0° to 500°C and pressures up to 2000 bars based on a regression of available data in the literature. *US Geol. Surv. Bull. 1421-C.* 36 pp.

Potter, R. W. II, Clynne, M. A. 1978. Pressure correction for fluid inclusion homogenization temperatures. *Programs and Abstracts, Int. Assoc. Genesis Ore Deposits Symp., Snowbird, Alta, Utah, 1978*, p. 146 (Abstr.)

Potter, R. W. II, Clynne, M. A., Brown, D. L. 1978. Freezing point depression of aqueous sodium chloride solutions. *Econ. Geol.* 73: 284–85

Quinn, E. L., Jones, C. L. 1936. *Carbon Dioxide. Am. Chem. Soc. Monogr. 72.* New York: Rheinhold

Ravich, M. I., Borovaya, F. E. 1949. Phase equilibria in ternary water-salt systems at elevated temperatures. *Izv. Sekt. Fiz.-Khim. Anal., Inst. Obshch. Neorg. Khim., Akad. Nauk SSSR* 19: 68–81 (In Russian)

Ravich, M. I., Borovaya, F. E. 1950. Crystallization of melts of chlorides of K and Na in the presence of water vapor. *Izv. Sekt. Fiz.-Khim. Anal., Inst. Obshch. Neorg. Khim., Akad. Nauk SSSR* 20: 165–83 (In Russian)

Robie, R. A., Bethke, P. M., Toulmin, M. S., Edwards, J. L. 1966. X-ray crystallographic data, densities, and molar volumes of minerals. In *Handbook of Physical Constants*, ed. S. P. Clark Jr. *Geol. Soc. Am. Mem.* 97, pp. 27–74. Revised ed.

Roedder, E. 1962. Studies of fluid inclusions I: Low temperature application of a dual-purpose freezing and heating stage. *Econ. Geol.* 57: 1045–61

Roedder, E. 1963. Studies of fluid inclusions II: Freezing data and their interpretation. *Econ. Geol.* 58: 167–211

Roedder, E. 1965. Liquid CO_2 inclusions in olivine-bearing nodules and phenocrysts from basalts. *Am. Mineral.* 50: 1746–82

Roedder, E., ed. 1968–onward. *Fluid Inclusion Research—Proceedings of COFFI.* (An annual summary of world literature; vol. 1–5 (1968–1972) privately printed and available from the editor; vol. 6 (1973)–onward printed and available from the Univ. Mich. Press)

Roedder, E. 1970. Application of an improved crushing microscope stage to studies of the gases in fluid inclusions. *Schweiz. Mineral. Petrogr. Mitt.* 50: Pt. 1, 41–58

Roedder, E. 1971. Fluid-inclusion evidence on the environment of formation of mineral deposits of the southern Appalachian Valley. *Econ. Geol.* 66: 777–91

Roedder, E. 1972. The composition of fluid inclusions, Chapter JJ, In *Data of Geochemistry*, ed. M. Fleischer, 6th ed. *US Geol. Surv. Prof. Pap. 440 JJ.* 164 pp.

Roedder, E. 1976. Fluid-inclusion evidence on the genesis of ores in sedimentary and volcanic rocks. Chap. 4. In *Handbook of Strata-Bound and Stratiform Ore Deposits*, ed. K. H. Wolf, Vol. 2, Geochemical Studies, pp. 67–110. Amsterdam: Elsevier

Roedder, E. 1977a. Changes in ore fluid with time, from fluid inclusion studies at Creede, Colorado. *Problems of Ore Deposition, 4th IAGOD Symp., Varna, 1974, Vol. II*, pp. 179–85

Roedder, E. 1977b. Stable and metastable fluid inclusion data, Browns Canyon fluorspar district, Chafee County, Colorado, and similar epithermal and hot-spring (?) deposits. *Problems of Ore Deposition, 4th IAGOD Symp., Varna, 1974, Vol. II*, pp. 186–95

Roedder, E. 1979a. Fluid inclusions as samples of ore fluids. In *Geochemistry of Hydrothermal Ore Deposits*, ed. H. L. Barnes, pp. 684–737. New York: Wiley

Roedder, E. 1979b. Origin and significance of magmatic inclusions. *Bull. Soc. Mineral. Cristallog.* In press

Roedder, E., Bolkin, H. E. 1979. Fluid inclusions in salt from Rayburn and Vacherie domes, Louisiana. *US Geol. Surv. Open File Rep. 79–1675.* 25 pp.

Roedder, E., Coombs, D. S. 1967. Immiscibility in granitic melts, indicated by fluid inclusions in ejected granitic blocks from Ascension Island. *J. Petrol.* 8(3): 417–51

Roedder, E., Skinner, B. J. 1968. Experimental evidence that fluid inclusions do not leak. *Econ. Geol.* 63: 715–30

Rosasco, G. J., Roedder, E. 1979. Application of a new Raman microprobe spectrometer to nondestructive analysis of sulfate and other ions in individual phases in fluid inclusions in minerals. *Geochim. Cosmochim. Acta.* 43: 1907–15

Shmonov, V. M., Shmulovich, K. I. 1974. Molal volumes and equation of state of CO_2 at temperatures from 100 to 1000°C and pressures from 2000 to 10,000 bars. *Dokl. Akad. Nauk SSSR* 217: 935–38 (In

Russian) Transl. in *Dokl. Acad. Sci. USSR* 217:206–9 (1975).

Skinner, B. J. 1966. Thermal expansion. In *Handbook of Physical Constants*, ed. S. P. Clark Jr., *Geol. Soc. Am. Mem. 97*, pp. 75–96. Revised ed.

Smith, F. G. 1953. *Historical Development of Inclusion Thermometry*. Toronto, Canada: Univ. Toronto Press. 149 pp.

Smith, F. G., Little, W. M. 1959. Filling temperatures of H_2O-CO_2 fluid inclusions and their significance in geothermometry. *Can. Mineral.* 6(3): 380–88

Sorby, H. C. 1858. On the microscopic structure of crystals, indicating the origin of minerals and rocks. *Geol. Soc. London Q. J.* 14: Pt. 1, 453–500

Sourirajan, S., Kennedy, G. C. 1962. The system H_2O-NaCl at elevated temperatures and pressures. *Am. J. Sci.* 260(2): 115–41

Stewart, D. B., Potter, R. W. II. 1979. Application of physical chemistry of fluids in rock salt at elevated temperature and pressure to repositories for radioactive waste. *Scientific Basis for Nuclear Waste Management, Vol. 1*, ed. G. J. McCarthy, pp. 297–311. New York: Plenum

Swanenberg, H. E. C. 1979. Phase equilibria in carbonic systems, and their application to freezing studies of fluid inclusions. *Contrib. Mineral. Petrol.* 68: 303–6

Takenouchi, S., Kennedy, G. C. 1964. The binary system H_2O-CO_2 at high temperatures and pressures. *Am. J. Sci.* 262: 1055–74

Takenouchi, S., Kennedy, G. C. 1965. The solubility of carbon dioxide in NaCl solutions at high temperatures and pressures. *Am. J. Sci.* 263: 445–54

Todheide, K., Franck, E. U. 1963. Das Zweiphasengebiet und die kritische Kurve in System Kohlendioxid-Wasser bis zu Drucken von 3500 bar. *Z. Phys. Chem.* 37: 388–401

Urusova, M. A. 1975. Volume properties of aqueous solutions of sodium chloride at elevated temperatures and pressures. *Zh. Neorg. Khim.* 20: 3103–10 (In Russian) Transl. in *Russ. J. Inorg. Chem.* 20(11): 1717–21

Voznyak, D. K., Kalyuzhnyi, V. A. 1976. Use of decrepitation of inclusions to reconstruct *P-T* conditions of mineral formation (using quartz from Volyn pegmatites as an example). *Mineral. Sb.* 2(30): 31–40 (In Russian)

Weast, R. C., ed. 1978. *Handbook of Chemistry and Physics, 1977–1978*. Cleveland, Ohio: Chemical Rubber Publ. Co.

Wilson, J. C. 1978. Ore-fluid-magma relationships in a vesicular quartz latite porphyry dike at Bingham, Utah. *Econ. Geol.* 73: 1287–1307

Ypma, P. J. M. 1963. *Rejuvenation of ore deposits as exemplified by the Belledonne metalliferous province*. PhD thesis. Univ. Leiden. 212 pp.

Ann. Rev. Earth Planet. Sci. 1980. 8:303–42
Copyright © 1980 by Annual Reviews Inc. All rights reserved

SATELLITE SENSING OF OCEAN SURFACE DYNAMICS

×10134

John R. Apel

Pacific Marine Environmental Laboratory, Environmental Research Laboratories, National Oceanic and Atmospheric Administration, Seattle, Washington 98105

INTRODUCTION

The prospect of making planetary-scale observations of the surface of the sea from satellites has been an intriguing one for oceanographers ever since the early spacecraft missions returned color photographs of the ocean that showed surprising amounts of structure. However, it has only been in the past several years that the quality, type, and amount of satellite data have allowed the process of acquiring scientific information from space to resemble "remote measurement." As a result, it is now becoming possible in certain cases to study the dynamics of the ocean surface, insofar as they are represented by changes in surface temperature, color, set-up, or roughness variations.

Remote measurement is certainly not the stand-alone panacea for oceanographers that its most enthusiastic supporters have made it out to be. However, if skilfully used as a new tool in the enlarging tool kit now available, it can be made to yield information of a type and extent that should be of much value to the advance of science. When utilized in conjunction with in-water measurements and an adequate theory, it is often valuable for interpolation between and extrapolation beyond the regions of surface observations, for determining the spatial variability of the surface phenomenon under study, and for deducing scales, speeds, amplitudes, and the like.

This paper is a representative (rather than exhaustive) review of selected results on dynamics of the sea surface as obtained from satellite remote measurement and published through mid-1979, coupled with a few theoretical interpretations, where possible. Examples are presented for methods of arriving at quantitative values for certain of the parameters of the theory, in the hope of stimulating investigators into thinking in conceptual, interpretative terms rather than in descriptive, gestalt ones. In

0084-6597/80/0515-0303$01.00

being broad rather than deep, the paper reports on a number of isolated, single results rather than carrying out an in-depth, continuing study of a particular process. However, as we accumulate larger bodies of data, better methods of interpretation, and more confidence in the results, such studies are more likely to be worth the investigator's time and trouble, and more papers are expected to appear.

This review is divided into three parts, determined by the scales of the phenomena—dynamics on planetary scale, mesoscale, and small scale are treated. These correspond roughly to physical events in the ocean: large-scale circulation and its variability; mesoscale eddy and upwelling motions; small-scale fronts, internal waves, and surface gravity and capillary waves. The paper includes some preliminary results from the dedicated oceanic satellite, Seasat, where appropriate. It concludes with observations on the future and exhortations on the practice of remote sensing. No discussion of the type, performance, or accuracies of the various remote sensing instruments has been given; instead, results obtained from visible, infrared, and microwave devices are cited. None of the various space platforms is discussed. These two aspects have been reviewed recently and the papers probably retain their usefulness (Apel 1976, 1979a). More detailed, quantitative papers on the radio-science side of ocean remote sensing are to be found in Apel (1972), Stewart (1978), and Gower (1980).

PLANETARY-SCALE PHENOMENA

On the planetary scale there are dynamic events such as variations in western boundary currents, equatorial currents, and large-scale fronts. Historically, the Gulf Stream was the earliest such phenomenon observed from space, so an inordinate portion of this section will be devoted to it. We begin with the Gulf of Mexico Loop Current, followed by discussion of the motions of the Stream along the southeast US coast, and then on to the meander and eddy region east of Cape Hatteras.

The Gulf Stream System

GENERAL The temperature, salinity, and color of the Gulf Stream distinguish it from the adjacent waters through which it flows, over the entire geographical region discussed. Particularly in the winter and spring, the surface temperature gradients delineate the boundary of the Stream through a reasonably close association between the subsurface baroclinic velocity field, $\mathbf{u}_h$, and the location of the maximum horizontal cross-stream temperature gradient, $\nabla_h T$. It has been demonstrated on several occasions that the (x, y) locus of the maximum surface temperature gradient overlies the location of the (x, y) locus of a selected subsurface isotherm at a fixed depth, to within an error of order 10 or 20 km, as long as anomalous wind conditions do not exist (Hansen & Maul 1970, also see below).

The temperature field is related to the velocity field by the equation for geostrophic balance,

$$(\mathbf{f} \times \mathbf{u})_{\rm h} = (-1/\rho)\nabla_{\rm h}p,$$

where $\mathbf{f}$ is the Coriolis vector, ρ the density field, and p the pressure; both p and ρ are functions of temperature, depth, and salinity.

In the Gulf of Mexico, the locus of the 22°C isotherm at 100 m is a good indicator of the boundary of the Loop Current (Leipper 1970, Maul 1977), while in the Gulf Stream meander region, the northwest wall of the discontinuity is classically taken as the location of the 18°C isotherm at 200 m.

Figure 1 shows a vertical temperature section across the Gulf Stream

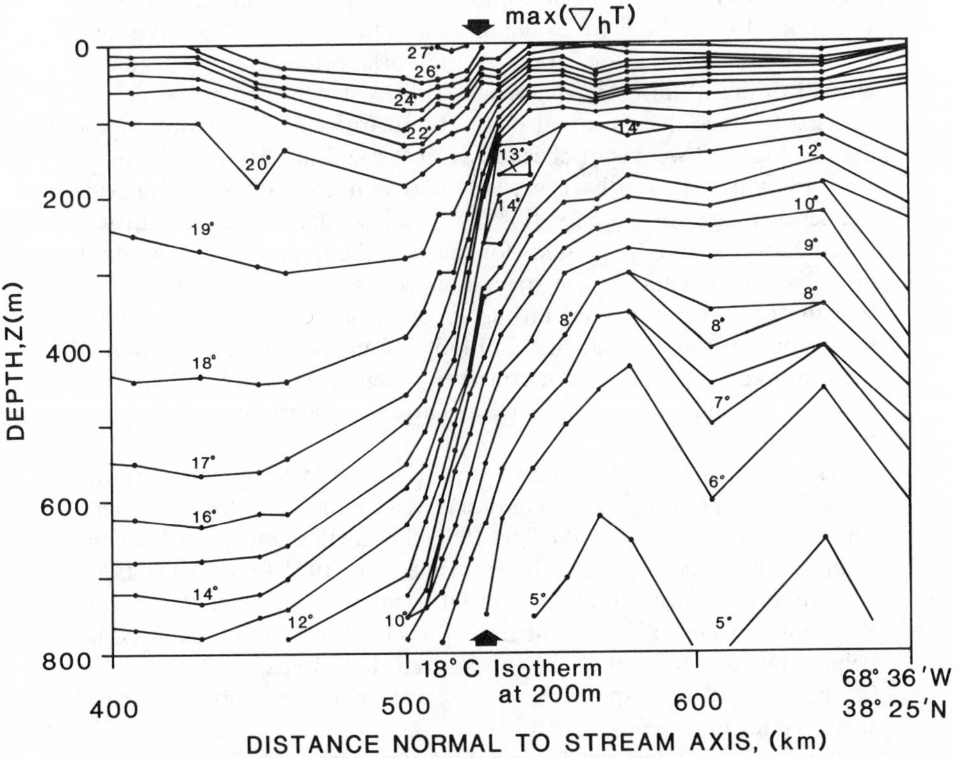

Figure 1 Temperature section across Gulf Stream near 38°N, 69°W, on June 23, 1976, taken under conditions of very light winds for the preceding several days. Locations are shown for the maximum horizontal temperature gradient and the 18°C isotherm at 200-m depth. These usually coincide to 10–20 km. From Apel (unpublished).

nearly parallel to $68\frac{1}{2}$°W, taken with XBT's between 23 and 24 June 1976, under conditions of light to zero winds. The vertical arrows identify $\max(\nabla_h T)$ at the surface, and the location of the 18°C isotherm at 200-m depth. These agree to within approximately 10 km; they also agree with the position of the edge, as deduced from the NOAA-4 satellite thermal infrared imagery, to within 5 to 10 km—which is about the same as our ability to interpolate the motion of the boundary that occurs during a 24-hour interval separating two satellite observations. Thus, according to the limited information available, if high wind-shear conditions are avoided, the maximum surface temperature gradient is a reasonable indicator of the location of the maximum underlying velocity gradient of western boundary currents.

THE LOOP CURRENT Maul (1977) has traced out the motion of the Gulf of Mexico Loop Current between August 1972 and September 1973 using hydrographic measurements in conjunction with data from the Landsat 1 spacecraft. Figure 2 illustrates the motion of the Loop Current over that year (Maul 1976). In the summer, the surface temperature gradients virtually vanish, but color differences still exist which in principle can be used to trace the evolution of the Loop Current. The hydrographic data of Figure 2 were spot-checked against Landsat data (which covers only 184-km-wide swaths every 18 days and thus does not provide frequent enough coverage for this task) and satisfactory agreement was noted between the two methods for locating the boundary (G. A. Maul, private communication). The space-time series obtained by Maul show that the Loop Current goes through a cycle of growth and decay at a north-south speed approaching 250 km in four months. The motion of the Loop apparently does not conserve potential vorticity. It had been observed to split off from the main current at least once during the previous five years.

Winter intrusions of the Loop Current were studied by Molinari and coworkers (1977) using a combination of XBT data and satellite imagery. They conclude that from 1965 to 1973 the northernmost reach of the Loop Current occurred in summer, while from 1974 to 1977 the Loop was found farthest north of 26°N in late winter or spring. These results cast doubt on the validity of the annual cycle of intrusions advanced by Leipper (1970). Molinari et al hypothesized that events upstream in the Caribbean may be responsible for the variability, but no evidence is advanced for that view.

Huh et al (1978) have used thermal infrared data to study the seasonal cycle of the horizontal temperature distribution in the Gulf of Mexico. Starting in the fall, typically October, polar air masses successively extract heat from the upper ocean. By winter, the only strong temperature signature remaining is that of the Loop Current, which stands out in contrast to the cooler slope water and coldest shelf water until well into the spring. Their paper contains useful illustrations of both the annual march of the cooling cycle and the development of small-scale eddies along the cyclonic front. The origin of the eddies is unknown but may be due to baroclinic or shear-flow instability of the Loop Current.

Maul et al (1974) have reported summer observations of small-scale eddies (12- to 32-km diameter) in the Gulf Loop Current, as viewed by Skylab with multispectral photography at visible wavelengths. The eddies described by them were cyclonic and appeared to be imbedded in the main current system, whereas the eddies observed by Huh et al were along the edge of the current. The precise location of the north wall of the current was not known at the time of the spacecraft overpass, and in retrospect it appears more likely that the eddies represented either a current boundary effect or advection of surface oils by turbulent winds.

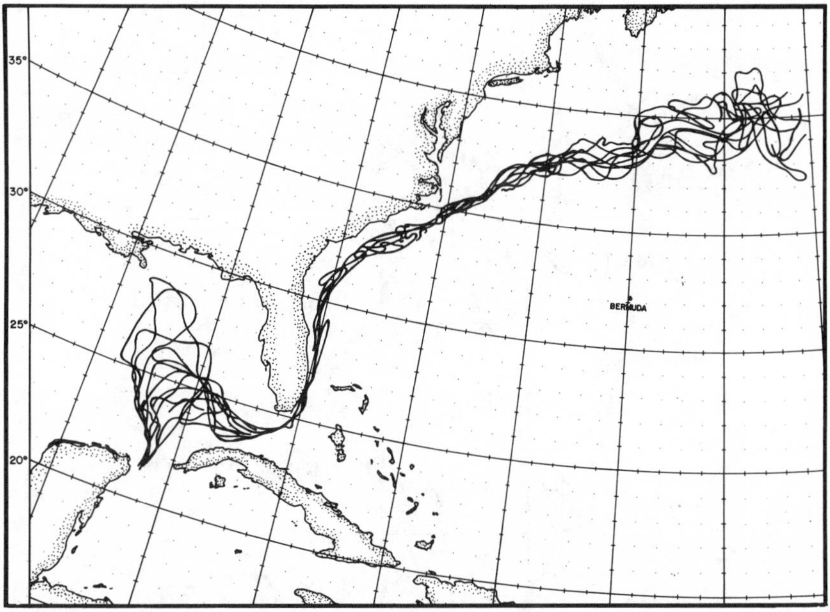

Figure 2 Locus of the boundary of the Gulf Stream system as deduced from ship and satellite measurements. Lines on this chart are of the left-hand side of the flow, facing downstream. The current has several names, depending on the locale. Data used were obtained at near-monthly intervals by research ships in the Gulf of Mexico and northeast of Cape Hatteras; satellite data were used between the Straits of Florida and Cape Hatteras. The lines depict the annual variability of the current system. In addition, many eddies are known to exist throughout the Sargasso Sea, east of New England, and in the Gulf of Mexico. From Maul (1976).

THE FLORIDA CURRENT The view that eddies lie along the region of velocity shear is reinforced in studies by Legeckis (1975, 1979) of the Florida Current on the eastern side of that peninsula. Data were obtained from the Synchronous Meteorological Satellite, SMS, or its operational equivalent, the Geostationary Operational Environmental Satellite, GOES. These spacecraft return full-earth-disc imagery at visible (0.55–0.75-μm) and thermal infrared (10.5–12.6-μm) wavelengths from their geosynchronous orbits in the earth's equatorial plane every 30 minutes. Up to 40 images per day are available from the VISSR instrument, a multi-

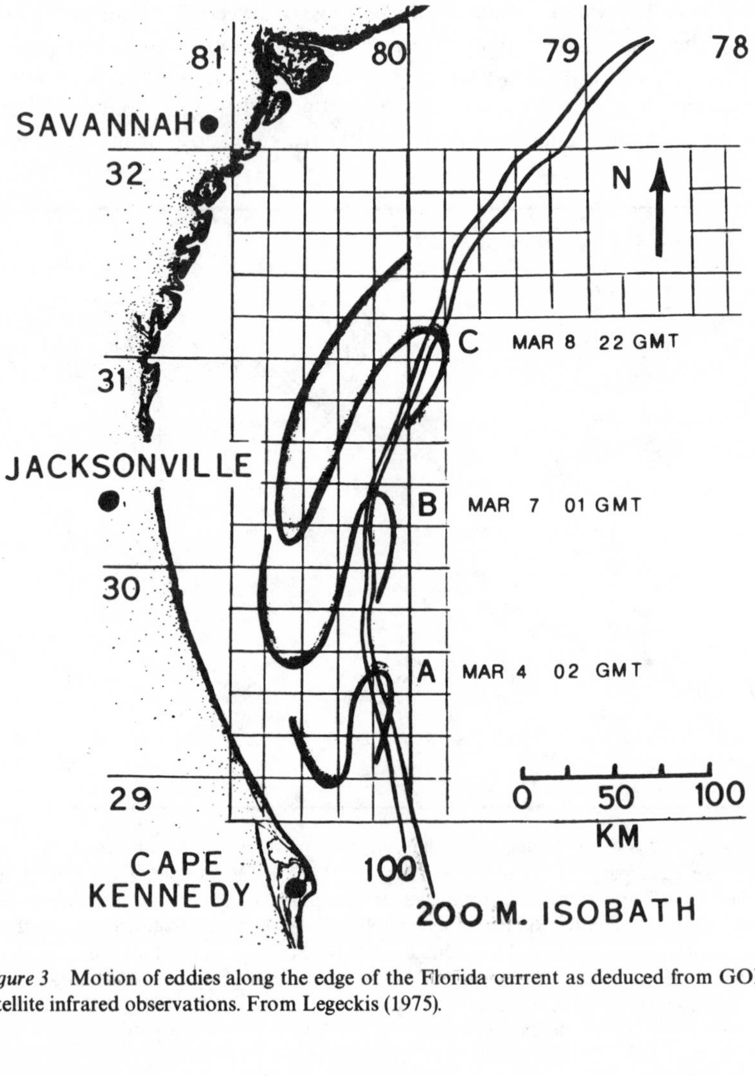

Figure 3 Motion of eddies along the edge of the Florida current as deduced from GOES satellite infrared observations. From Legeckis (1975).

spectral imaging radiometer, which has a 7-km resolution and a noise-equivalent-temperature of 0.5°C in the IR. Because of the very large differences in the characteristic times of motion for clouds as compared with those for the oceans, it is possible to clearly differentiate between thermal signatures due to atmosphere and ocean. This is accomplished by processing a series of images into a cinema loop and, from it, studying the slow motion of oceanic thermal boundaries beneath the rapidly moving clouds. Figure 3 shows a sequence of frontal positions for the dates 4–8 March 1975 constructed from such a loop. These data yield a northward-propagating, eddylike tongue having an average speed of 39 cm s^{-1}, which is approximately one-fourth of the surface speed of the Gulf Stream in that region: 150 cm s^{-1}. Using this speed and the observed spatial wavelength of about 276 km, an average period of 7.8 d is deduced. These results are discussed in the light of a number of observational and theoretical investigations of shelf waves, but insufficient data were available for a detailed intercomparison. It seems likely to the present author, however, that because of the length and time scales observed, the eddy development is due to either 1. a barotropic continental shelf wave which is rendered unstable by the horizontal velocity shear along the current edge (Niiler & Mysak 1971, LeBlond & Mysak 1978), or 2. a baroclinic instability in which both potential and kinetic energies contribute to the wave motion (Orlanski 1969, Orlanski & Cox 1973). The former seems more likely as an explanation of the phenomenon observed by Legeckis, because of the values of wavelength, period, and group speed calculated for the Blake Plateau region using the Niiler & Mysak theory (LeBlond & Mysak 1978, p. 435): 140 km, 10 d, and 36 cm s^{-1}, respectively. However, the analysis is quite sensitive to the numerical values of the flow and the topographic features. The validity of the period is reinforced through findings by Düing et al (1977), who report on in-water measurements of the Florida Current having energy in this same frequency region. The problem is deserving of much more satellite observational work and a subsequent detailed comparison between theory and experiment.

Progressing on downstream in the Gulf Stream, but still remaining south of Cape Hatteras, a short but interesting sequence of eddies during 12–18 February 1974 has been reported by Stumpf & Rao (1975), using infrared imagery from the polar orbiting satellite NOAA-2. These eddies seem phase-locked to the prominent cuspate features of Cape Romain and Cape Fear, but within a period of 8 d they evolved in amplitude to a characteristic multivalued configuration similar to those shown in Figure 3. No dynamical information was given other than to cite a wavelength of 180 km and peak-to-trough amplitude of 65 km. The present author speculates that the eddy positions and lengths may be set by interactions with the Carolina cape structures of Romain, Fear, Lookout, and Hatteras; however, this is derived from casual observation of a sequence of infrared images and not from an orderly investigation.

A somewhat more detailed study of Gulf Stream meanders in the region between Florida and North Carolina has been made by Bane & Brooks (1979) using the *Experimental Ocean Frontal Analysis* charts published by the US Naval Oceanographic Office (NAVOCEANO 1975) for the period 5 January 1977 to 29 March 1978. From a compilation of 14 surface temperature sections across the Stream boundary, they define the mean and standard deviation for the amplitude of the surface thermal front. The rms meandering grows from approximately 4 km off central Florida to a maximum of 34 km southeast of Cape Fear, North Carolina. The large-scale meandering commences downstream of a rise in the bottom topography off Charleston, South Carolina. At a latitude of 31°59′N, the Current has been deflected approximately east due to conservation of potential vorticity. The subsequent meandering is felt to be a baroclinic topographic Rossby wave propagating downstream, possibly induced by the Charleston bump (Brooks & Mooers 1977).

GULF STREAM MEANDER REGION Once the Gulf Stream departs the continental shelf, it is well known that it breaks into spatially growing oscillations that evolve into detached rings and eddies. There is much literature on this subject, but two representative papers on the in-water measurements are those by Fuglister & Worthington (1951) and Hansen (1970). The motions of the northwest wall of the Stream as published by Hansen are also shown on Figure 2; these were obtained from towed thermistor measurements during nine cruises made at 30-day intervals during 1965–1966.

In order to illustrate how repeated satellite data can give information equivalent to ship data, an analysis is presented of the positions of the northwest boundary using thermal infrared imagery from NOAA-4 Very High Resolution Radiometer (VHRR) for the period 9 April to 21 May 1975 (Apel & Apel 1976). Such imagery, which has a spatial resolution of 0.8 km, is acquired twice a day and is available in film format archived by the NOAA Environmental Data and Information Service. Figure 4 shows one such image made on 12 May 1975 over the region between Cape Hatteras, North Carolina, and Newfoundland, Canada. Lighter shades correspond to warmer temperatures; the black features are clouds. The Gulf Stream, two warm-core anticyclonic rings, and the colder slope and shelf water are readily seen. This figure has been contrast-enhanced to bring out oceanic features at the expense of land and atmospheric information. Approximately two months of such image data were analyzed

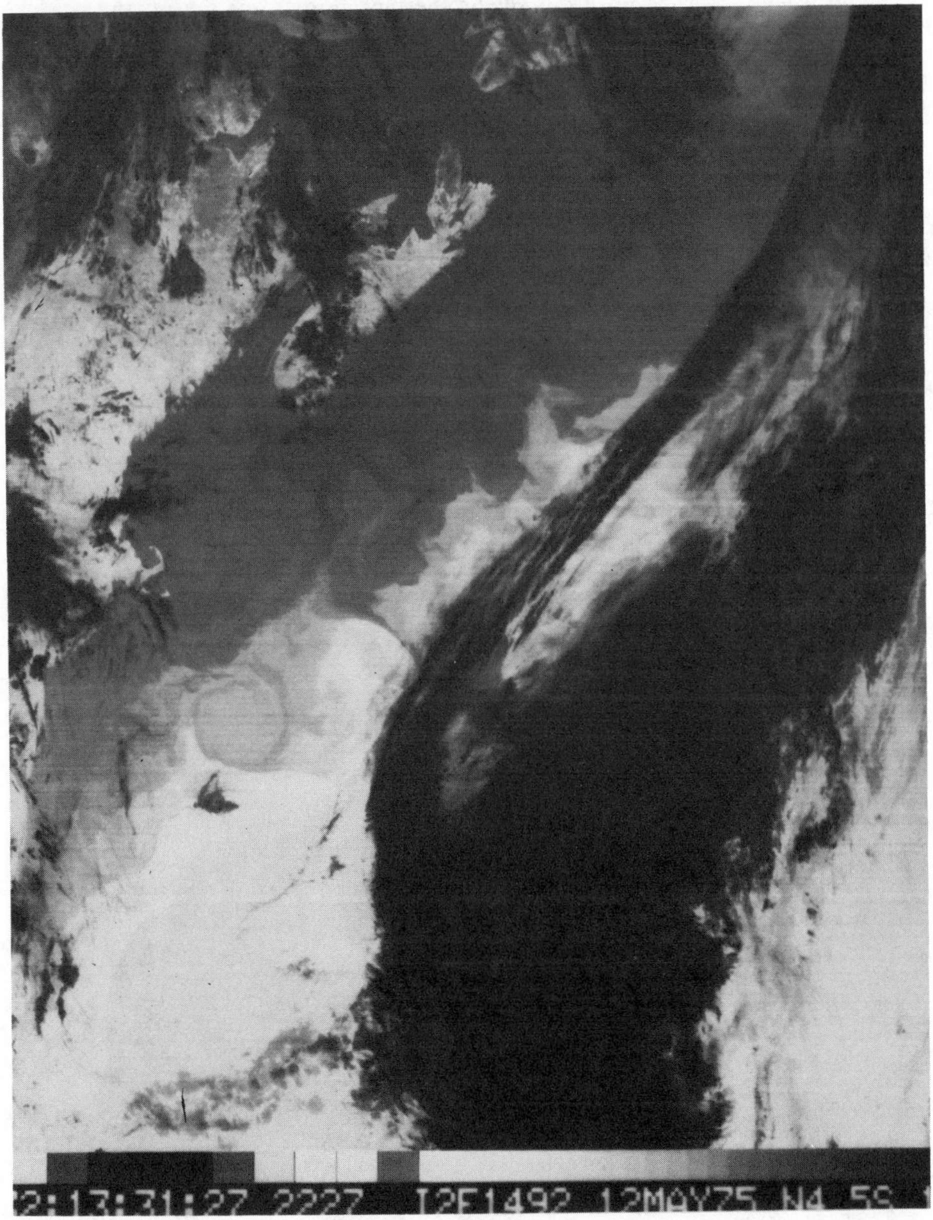

Figure 4 NOAA-4 thermal infrared image of the Gulf Stream region for May 12, 1975, shown with warm temperatures in light tones and cold temperatures in dark. The black features are clouds. National Oceanic and Atmospheric Administration.

and weekly average positions of the thermal gradient of the northwest boundary were plotted. This averaging was necessary because of obfuscation of surface information by clouds. In addition, the motions of the two large detached rings to the north of the wall were tracked. The resultant data are shown on Figure 5. Wavelengths and phase speeds for the most coherent oscillations in the small amplitude region directly east of Cape Hatteras were obtained from sequences of daily images, to avoid smearing out due to the weekly averaging.

An analysis of the frontal motion was made in terms of a spatially unstable oscillation about the mean edge of the Stream of the form

$$A(x, y, t) = A_{\omega} e^{-k_i s} \sin(k_r s - \omega t),$$

where the location of the mean for the Stream edge is given by $s(x, y)$; A is the lateral displacement of the boundary about that mean; $k_r = 2\pi/\lambda$ is the real part of the complex along-axis wave number; k_i is the imaginary part; and $\omega = 2\pi/\tau$ is the radian frequency. Phase speed, $c = \omega/k_r$, is directly measured from the images by the propagation of points of constant phase. This form approximately describes the coherent motion for about three wavelengths northeast of Cape Hatteras. Table 1 contains values for average wavelength, period, phase speed, and spatial growth

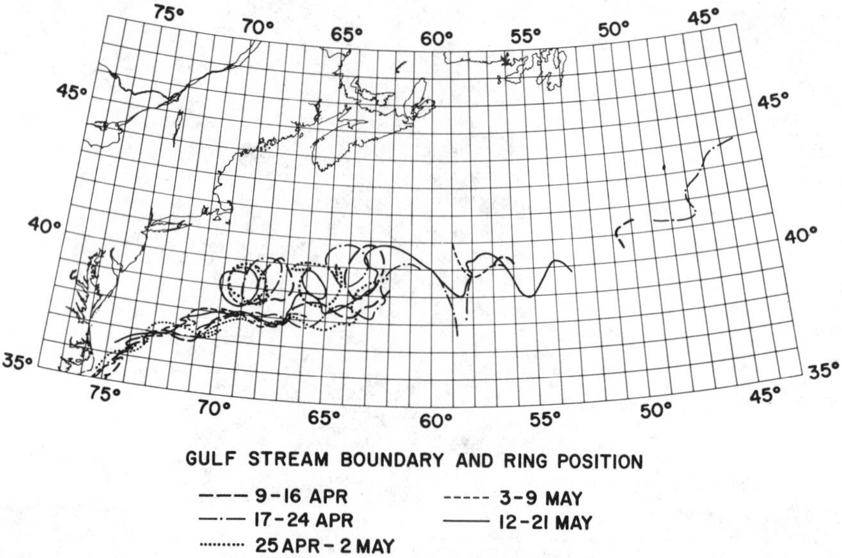

Figure 5 Positions of northwest front of the Gulf Stream and detached rings as deduced from NOAA-4 satellite thermal infrared data during two months in 1975. Weekly averages are presented. Compare with Figure 4 for an instantaneous image and with Figure 2 for equivalent shipboard data. From Apel & Apel (1976).

Table 1 Gulf Stream meander characteristics: 60° to 75°W

Parameter	Apel & Apel 1976	Hansen 1970
Wavelength, λ (km)	310	300
Phase speed, ω/k_r (m s^{-1})	0.28	0.05–0.10
Period, τ (days)	13	42
Spatial growth rate, $-k_i$ (km^{-1})	1.8×10^{-3}	2×10^{-3}
Amplitude near detached anticyclonic rings, $2A$ (km)	160	—
Ring diameter, $2R$ (km)	160	—
Ring speed, $\bar{U}$ (m s^{-1})	0.055–0.073	—

rate from this analysis; the Table also includes Hansen's 1970 data (compare Figures 5 and 2). It can be seen from Table 1 that the wavelength and spatial growth rates agree very well, but that the phase speeds and hence the periods are in poor agreement. Since the satellite-derived phase speeds are obtained from 14 separate observations of highly coherent phase advances having an rms spread of $\pm 51\%$, it is unlikely that the differences are statistical. It is concluded that during the two-month interval under study, the dominant motion is correctly estimated by the satellite data on Table 1. The dominant period obtained from this record is approximately 13 days. This same period has also been reported in studies by Robinson et al (1974), Düing et al (1977), and Legeckis (1975). In an extensive analysis of GOES infrared satellite data, Maul and co-workers (1978) report periods centered at 4, 8, 14, 18–21, 29, 36, and 45 days near the region under consideration. Düing attributes the 13-day period to atmospheric forcing.

Thus, the NOAA-4 imagery suggests that the roughly biweekly oscillations seen at a variety of locations upstream and ascribed to atmospheric stress may become baroclinically unstable upon the departure of the Gulf Stream from the US east coast. While propagating at a speed of $\omega/k_r u_h = 28/150 \simeq 20\%$ of the mean current, the waves grow exponentially at a rate of $2\pi k_i/k_r \simeq 56\%$ per wavelength for a few cycles and then become highly nonlinear. Much more work is needed to firmly establish this, however.

In a recent in-depth study, Halliwell & Mooers (1979) have made a statistical analysis of two years' worth of NOAA satellite data, working from the weekly-averaged *Experimental Ocean Frontal Analysis* produced by the US Navy (NAVOCEANO 1975). (Their analyses include in-water data and are therefore of greater value than the satellite data by themselves.) They examine three surface thermal features: the boundary between the shelf and slope waters; the Gulf Stream front; and the warm-core, anticyclonic eddies (see below) in the region between Hatteras and approximately 64°W. Figure 6 summarizes the results of a spectral analysis of the Gulf Stream front, assuming a cross-stream sinusoidal motion of the form

$$A(s,t) = \sum_n C_n(s) \cos \omega_n t + S_n(s) \sin \omega_n t$$

where, as above, $s(x, y)$ denotes the mean position of the boundary. The quantity displayed in Figure 6 is the propagating wavenumber-frequency autospectrum of the amplitudes C_n and S_n, but multiplied by $10^4\ \omega k_r$. Level surfaces of this spectral function are contoured and regions of northeastward propagation are shown unshaded.

The dominant propagating motion observed during this two-year interval has a period near eight weeks and a wavelength slightly greater than 300 km, but there are significant concentrations of energy in the bands given on Table 2. The "phase speed" ω/k_r, as evaluated at the maximum of the autospectrum that corresponds to this motion, is approximately 7 cm s^{-1}; this agrees well with Hanson's 1970 estimate. The high frequency peak at a two-week period and 200-km wavelength is approximately the same as for the motion cited above in Table 1,

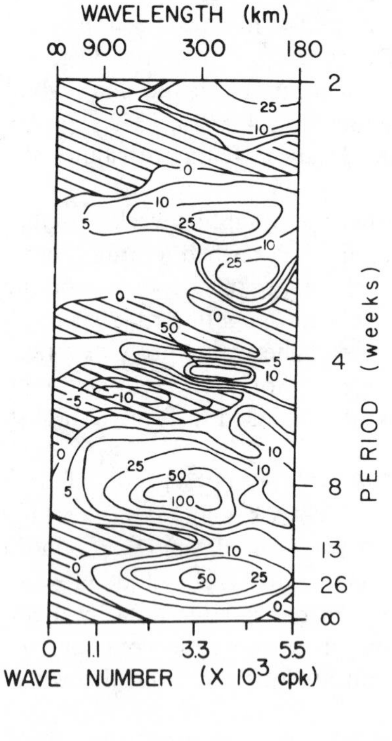

Figure 6 Wavenumber-frequency autospectrum for propagating Gulf Stream meanders as observed from two years of NOAA IR satellite data for region northeast of Cape Hatteras. Dominant motion is near eight weeks and 320 km. From Halliwell & Mooers (1979).

Table 2 Wavenumbers and frequencies of dominant Gulf Stream meander motions, 64° to 75°W[a]

Wavenumber, k_r (cyc km^{-1})	Frequency, ω (cyc yr^{-1})	Speed, ω/k_r (cm s^{-1})
3.4×10^{-3}	2.0	—
2.8	6.3	7.2
3.7	12.5	—
4.4	17.0	—
3.8	19.2	—
4.9	26.0	17.0

[a] Halliwell & Mooers 1979.

except that the calculated speed ω/k_r from Table 2 is one-half of the phase speed of 28 cm s^{-1} as deduced by Apel & Apel (1976). A pervasive characteristic is that, for all propagating modes, the wavelengths decrease and the amplitudes increase with distance downstream. This effect is also visible in Figure 2. There is also evidence of standing waves along the boundary.

EDDIES AND RINGS Regarding the motion of the Gulf Stream detached eddies, the two rings to the north of the Stream in Figure 5 moved to the west-southwest during the observation period. The easternmost one, which was seen in the process of detaching itself, had a mean speed of about 7.3 cm s^{-1} and the westernmost ring had a mean speed of 5.5 cm s^{-1}. These speeds are entirely consistent with earlier reports of ring drift rates (Fuglister & Worthington 1951).

Halliwell & Mooers (1979) also report on ring motion to the north of the Stream and derive much more complete data and statistics than those presented by Apel & Apel. Fourteen rings were observed during the 104-week observation period. These had a mean diameter of 100 km and a mean southwestern propagation speed of 6 cm s^{-1}. They broke off between 500 and 1000 km downstream from Hatteras and on occasion were observed to be recaptured by the Stream close to the continental shelf. The severe winter of 1977 was a time of enhanced transport and the rate of eddy generation doubled, from two to four per year. The authors hypothesize that large-scale atmospheric circulation may thereby drive shelf circulation through the indirect process of enhanced eddy production.

Cyclonic, cold-core eddies to the south of the Gulf Stream have been frequently reported in satellite data and studies of their drifts made from both ship and thermal infrared data (Stumpf et al 1973). Their mean drift rates were of order 2–4 cm s^{-1} in the region between Cape Hatteras and Grand Bahama Island, and their surface temperature contrasts ranged from 0–6°C. Even after their surface temperature signatures have eroded, there is often evidence of subsurface circulation via the baroclinicity of the density field.

The generation of warm-core rings to the north of the Gulf Stream and cold-core eddies to the south is a process that reduces the average equator-to-pole temperature gradient in the ocean and is apparently one means by which excess solar heat energy deposited in the tropics is redistributed.

SATELLITE ALTIMETRY Another means of making measurements of intense current systems from space, satellite altimetry, represents a much less well-developed capability under the present circumstances. Skylab, GEOS-3, and Seasat satellites each carried a precision radar altimeter; the latter two could measure the vertical distance from spacecraft to sea surface to an rms precision of ±70 cm and ±10 cm, respectively, even in the

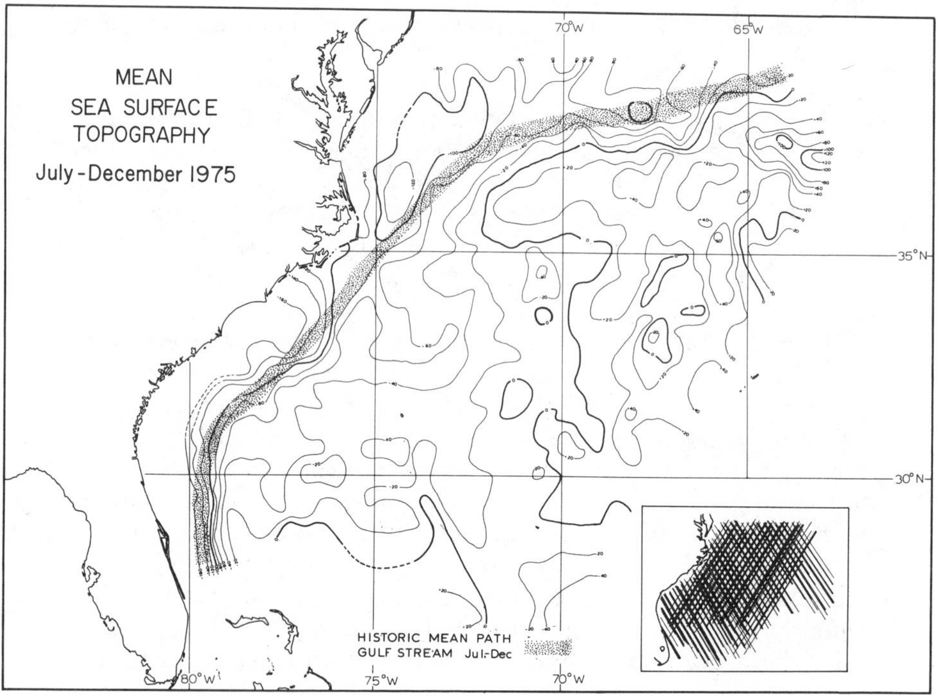

Figure 7a Mean sea surface topography for western Atlantic as derived from GEOS-3 satellite altimetry data, averaged over 6 months. A best-estimate geoid has been removed. Contours are in 20-cm intervals. Insert shows satellite orbits used. From Huang & Leitao (1978).

presence of ocean waves of much larger amplitude. If these altitude measurements could be coupled with equally precise determinations of both the satellite orbit and the equipotential surface (or geoid) of the ocean, it is estimated that the surface current speeds could be measured to an accuracy approaching 20–30 cm s^{-1}; this assumes errors in altitude, orbit, and geoid are each of order ± 10 cm (Kaula 1970, Apel 1972, Apel & Byrne 1974, Apel & Siry 1974). This would be done by relating the slope, δ, of the sea surface relative to the geoid, to the speed of the surface geostrophic current, u_h, via an alternate form of the equation for geostrophic balance,

$$u_h = (g/f) \tan \delta.$$

Here g is the acceleration of gravity and f is the Coriolis parameter. Slopes of order 10^{-5} and set-ups of about one meter are estimated for the Gulf Stream. Workers at NASA Wallops Flight Center have derived the ocean surface topography in the western Atlantic from GEOS-3 data (Leitao et al 1978, 1979, Huang & Leitao 1978). By compositing

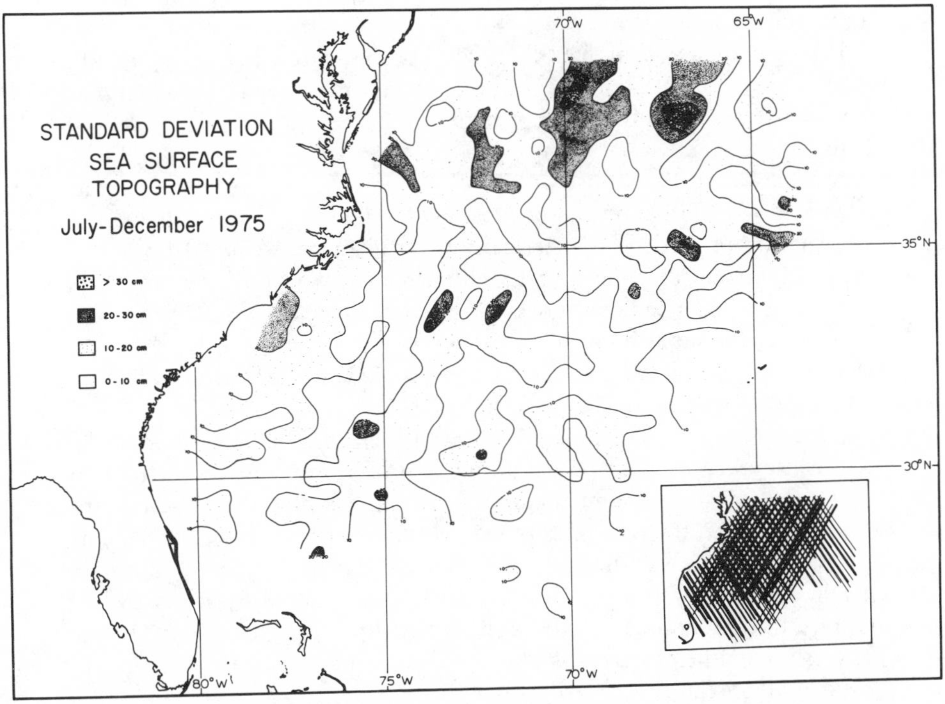

Figure 7b Standard deviation of sea surface topography for the same case as 7*a*. Contours are on 10-cm intervals. The shaded regions are ascribed to 300-km-long meanders. From Huang & Leitao (1978).

and statistically analyzing altimetric data from 163 passes between 1 July and 31 December 1975, topographic maps were obtained that are estimated to be significant above the ± 20 cm level. Figure 7*a*, from Huang & Leitao (1978), shows the mean topography contoured in 20-cm intervals; independent best estimates of the geoid were used to subtract out that equipotential surface. The satellite orbits are shown in the insert. The set-up due to the Gulf Stream is apparent as a rise of approximately 110 cm in the area historically traversed by that current system. Figure 7*b* illustrates the standard deviation of the same data, with three large, statistically significant contours spaced at about 300 km and located near the north wall of the historical boundary. These are attributed to the same type of 300-km-long meanders visible in Figures 2 and 5. No values of surface current have been cited from the above measurements because the averaging process spreads out the horizontal scale of the frontal gradient. However, these investigators also show maps of topographic contours derived from only one month of data, and the map for November 1975 exhibits a cross-stream horizontal gradient at 38°N, 67°W of 120 cm in 100 km. This slope indicates a geostrophic surface velocity $u_h \simeq 1.2 \text{ m s}^{-1}$, a reasonable value.

Huang et al also demonstrate tracking a cold-core eddy at one-month intervals by both its thermal infrared and altimetric surface signatures. The drift rate so obtained is approximately 2.5 cm s^{-1}, which agrees well with the speeds cited by Stumpf.

A somewhat different use of GEOS-3 altimetric data over the sea has been made by Mather and associates (1979). They demonstrate that useful information on ocean-wide steric anomalies can be extracted from altimetric data, and that highs and lows correlated with Gulf Stream rings and eddies appear in the altimetric signals from the spacecraft. However, because of the large noise figure of ± 50–± 70 cm, typical oceanic surface set-up and elevations are barely resolved in the Mather analyses under the best of circumstances.

Pullen & Byrne (1978) and Byrne (1979) analyzed data from the Kuroshio taken with the Seasat altimeter and have compared them with near-simultaneous dynamic height measurements made from shipboard while that satellite was in a precisely repeating orbit. Measurements across the current set-up east of Japan were made from the altimeter height variations and dynamic heights were calculated from the hydrography. Both techniques yielded a surface set-up of approximately 110 cm. The estimated height error is of order ± 20 cm for both techniques and the overall agreement is very good.

The Seasat altimeter appears to be precise enough to arrive at semiquantitative estimates of the total horizontal pressure gradient at the sea surface, provided the equipotential surface can be found by independent means. This pressure gradient drives both the baroclinic and the barotropic circulation, of course, and its measurement would represent a determination of the total surface current speed, independent of assumptions of a level of no motion.

An assessment of the ability of the Seasat altimeter to observe ocean surface features is presented by Tapley et al (1979). With an rms noise figure of ± 5–8 cm in sea states under 4-m significant wave height, the instrument allows the 100-cm set-up of the Gulf Stream to be identified easily. As of this writing, there have not been enough data analyzed to study ocean dynamics via time changes in sea surface topography.

SUMMARY OF GULF STREAM OBSERVATIONS In order to summarize the satellite-derived measurements made over the Gulf Stream system, the following observations are made:

1. High resolution thermal infrared imagery taken twice per day from polar orbiters gives quality spatial and thermal data, but is plagued by cloud problems. Nevertheless, dynamical information that agrees with surface values appears to be derivable for time scales in excess of approximately one week and space scales in excess of about 10 km. A large archive of data exists for the past several years for near-US waters and other selected areas. Halliwell & Mooers have exploited a portion of these data in their useful study cited above.
2. Lower resolution thermal-IR images obtained from geosynchronous satellites up to 40 times per day can be made into cinema loops. From these, temperature patterns may be studied between openings in cloud cover. In the hands of Legeckis, this technique is yielding useful dynamics. In principle, motions having time scales in excess of a few days and space scales exceeding a few tens of kilometers may be deduced. However, the archived photographic images are not suitable for this purpose and only a few such loops have been produced from computer-enhanced images.
3. Very high precision altimetric measurements show promise for measuring surface geostrophic speeds over the high-speed portions of the current at intervals of several days and on space scales of order of the satellite orbit separations. The archived GEOS-3 data span approximately three years, while the more precise measurements from Seasat exist only for the interval between June and October, 1978, due to the loss of power on the latter satellite. However, enough information probably exists for proof of the concept and for an analysis of short-term dynamical effects at a wide variety of locations around the

world. No oceanographically useful information has yet emerged from such data, but the potential clearly exists.

4. The Coastal Zone Color Scanner (CZCS) on the Nimbus-7 satellite is probably more nearly optimum for observing the large- and intermediate-scale dynamics of the Current system. This spectral radiometer operates with four channels of visible light and one channel of thermal infrared at 10.5–12.3-μm wavelength; it yields images having 0.8-km resolution, which cover 1200-km-wide swaths every three to four days. In the summer, when the thermal contrast of the Gulf Stream is reduced to very small values at the more southerly latitudes, color contrast across the boundary should still be visible. However, as of this writing, no analyzed Nimbus-7 CZCS data are available to demonstrate the technique. Thus, the use of color to map current systems has barely been touched.

Equatorial Current Systems

Other major oceanic current systems have received much less attention from remote sensing scientists than the Gulf Stream, so little quantitative work has appeared in the literature. One example of a major circulation feature deserving much more study is the equatorial current, countercurrent, and undercurrent system. These alternating bands of zonal flows in the vicinity of the Equator are of importance in global circulation and heat balance and are clear candidates for investigation by satellite techniques, especially because of their large scale.

Legeckis (1977a) has presented evidence of long-wave meandering of the thermal front between the South Equatorial Current (SEC) and the North Equatorial Countercurrent (NECC), as derived from infrared imagery from the GOES geostationary satellite during 1975. The SEC may be considered as the zonal extension of the Peru Current; in the eastern tropical Pacific, it flows approximately along the Equator (see Figure 8*a*). The thermal contrast across the SEC-NECC front is of order 2–4°C under normal conditions. The GOES observations showed long-length waves moving westward at an average phase speed of 47 cm s^{-1}. As shown in Figure 8*b*, the waves ranged in length between 800 and 1200 km, with the shorter lengths generally found to the east; on the average, they were about 1000 km long and had double amplitudes of order 300 km. During 1976 (an El Niño year), no such waves were visible in the GOES thermal-IR imagery. At that time, sea surface temperatures in the eastern tropical Pacific were higher than the previous year by 4–6°C and the waves may have been obscured by the high surface temperature anomalies.

Somewhat analogous motions of the Atlantic Equatorial Undercurrent (EUC) have been reported by Düing and others (1975), based on multiple ship operations during the 1974 GATE Equatorial Experiment. The eastward-flowing undercurrent is a subsurface jet centered near 90-m depth and having speeds approaching 100 cm s^{-1}. During GATE, its meanders took the form of a westward-propagating wave with lengths approximating 2660 km and periods of order 16 days. It surfaced on occasion, which implies that the cooler subsurface water would be visible as a thermal anomaly to a satellite infrared sensor.

Philander (1976) has modeled the stability of the various equatorial current components in the beta-plane approximation and concludes that the Atlantic SEC-NECC system should be unstable to barotropic shear flow. He predicts a westward-propagating unstable wave of 2000-km length having periods of order two to three weeks. However, according to this analysis, the Atlantic EUC appears to be stable under the conditions existing during GATE. If this is true, then the SEC-NECC surface current instability may drive the undercurrent and thus the meandering motion of the EUC may be part of a phenomenon having a much larger latitudinal scale. Obviously, well-developed satellite observations can play an important role in delineating these motions.

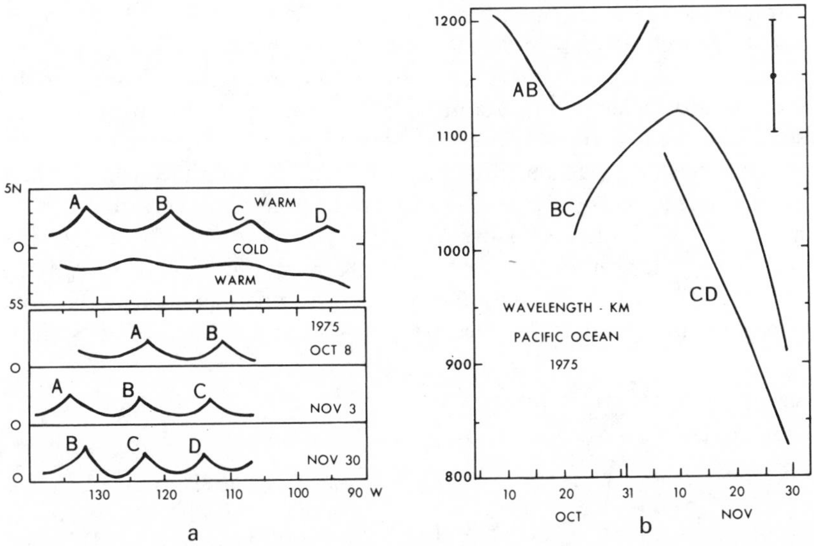

Figure 8 (*a*) Waves in the eastern equatorial Pacific current systems as derived from GOES infrared imagery. Four wave peaks (*A*, *B*, *C*, *D*) are identified as moving westward at the frontal boundary between the South Equatorial Current and the North Equatorial Countercurrent. (*b*) The daily averaged distances between adjacent wave peaks (*AB*, *BC*, and *CD*) as a function of time in October–November 1975. The probable error is indicated by the error bar. Wavelengths average 1000 km and phase speeds 47 cm s^{-1}. From Legeckis (1977a).

INTERMEDIATE-SCALE PHENOMENA

Grouped under the title of intermediate-scale phenomena are such isolated processes as coastal eddies, upwellings, and continental shelf waves. They are generally characterized as being due to a localized forcing or manifestation and not necessarily to some ocean-wide system.

Coastal Eddies

A good example of a combined satellite and in-water study of a localized system is one by Bernstein et al (1977) on eddy formation in the California Current. This system is characterized by mean flows of order 10 cm s^{-1} directed toward the south and by intermittent events resembling mesoscale eddies or tongues with speeds up to 40 cm s^{-1}. During February and March, 1975, a data set was obtained consisting of 1. NOAA-3 VHRR imagery showing an anticyclonic eddy off Point Conception, California, and 2. dynamic height observations made from a hydrographic survey of the same area during the same interval. Figure 9 shows a superposition of dynamic heights (in dynamic meters relative to the 500-dbar level) and infrared sea surface temperature fronts; data were taken on 21 February and 17 March, 1975. The density of hydrographic stations, as shown by the dots, is not great enough to resolve the fine structure seen in the satellite surface temperature contours, but the general agreement between shapes of the two types of patterns is very good. The feature is an anticyclonic meander, i.e. an intrusion of warm, low-salinity offshore water having speeds of order 10 to 15 cm s^{-1} and surface temperature contrasts near 7°C. Additional data presented in their paper reinforce the view that the surface temperature patterns are manifestations of underlying dynamical events. Bernstein interprets the meandering as being due to a baroclinic instability in a current system having a vertical velocity shear

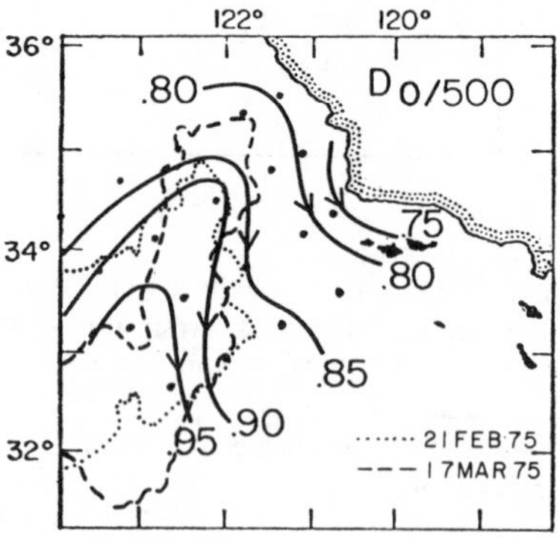

Figure 9 Manifestations of a mesoscale meander in the California Current off Point Conception, California. Solid lines give heights in dynamic meters relative to a 500-dbar surface, while the broken lines give the position of a sea surface temperature anomaly from the NOAA-3 VHRR infrared radiometer. The agreement between the two methods is considered to be very good. From Bernstein et al (1977).

of $du/dz \simeq 0.08/200\ s^{-1}$ and a Brunt Väisälä frequency of $N \simeq 1.5 \times 10^{-2}$ rad s^{-1}. The application of simple instability theory then yields a growing disturbance with the most rapidly growing mode having a wavelength of 230 km and a phase speed of 10 cm s^{-1}; these agree with the observations to within a factor of 2 to 3. The authors point out that the correctness of their interpretations depends heavily on the availability of subsurface data collected by conventional means.

Other investigators have reported on the usefulness of satellite infrared imagery in determining the evolution of near-shore dynamical features. Düing has studied a monsoon feature, the "great whirl" in the Somali Current off the Horn of Africa, which is a pronounced anticyclonic eddy of dimensions near 400 × 600 km having transports of order 70 Sverdrups (Düing 1978). The surface temperatures range from below 18°C to above 23°C over distances of order 90 km. The drift of such a large anomaly can be readily studied using a sequence of images, as was the case above involving detached Gulf Stream eddies.

Upwelling

Stumpf & Legeckis (1977) conducted a month-long study of mesoscale anticyclonic gyres in the Gulf of Tehuantepec off the southwest coast of Mexico. These gyres are generated by polar high-pressure systems with winds that blow across the narrow Mexican peninsula at speeds of 10–20 m s^{-1}, thereby causing strong upwellings. Figure 10 is a NOAA-4 VHRR image of the area from 5 February 1976 and illustrates cold water in light shades and warm water in dark ones. A 280-km-diameter circular gyre is plainly visible. A similar feature located to the east-southeast was observed to propagate due west at an average speed of 15 cm s^{-1}. This is compared with the theoretical group speed of a nondispersive baroclinic Rossby wave,

$$c = -\beta r_d^2,$$

where β is the planetary vorticity ($f = f_0 + \beta y$) and $r_d = (g\Delta\rho h/\rho f^2)^{1/2}$ is the baroclinic Rossby radius of deformation in a two-layer system of density contrast $\Delta\rho/\rho$ and an upper-layer depth h. For conditions extant at the time and place of Figure 10, $r_d = 80$ km and $c \simeq 14$ cm s^{-1}, which is very close to the drift speed of the eddy. Further theoretical study is required in order to make these order-of-magnitude calculations more firm, especially because a gyre is the extreme nonlinear limit of a Rossby wave. The calculation for group speed holds for small amplitudes only.

Legeckis has also published a survey of satellite images that show temperature fronts at a wide variety of locations around the world (Legeckis 1977b, 1978). He discusses data availability, problems with atmospheric opacity, image processing, and the general utility of the infrared imagery. His 1978 paper serves to illustrate that the types of features presented in the imagery above are not unusual events but rather are widespread features of the ocean, occurring at many times and places.

In essentially all of the above observations of planetary or mesoscale

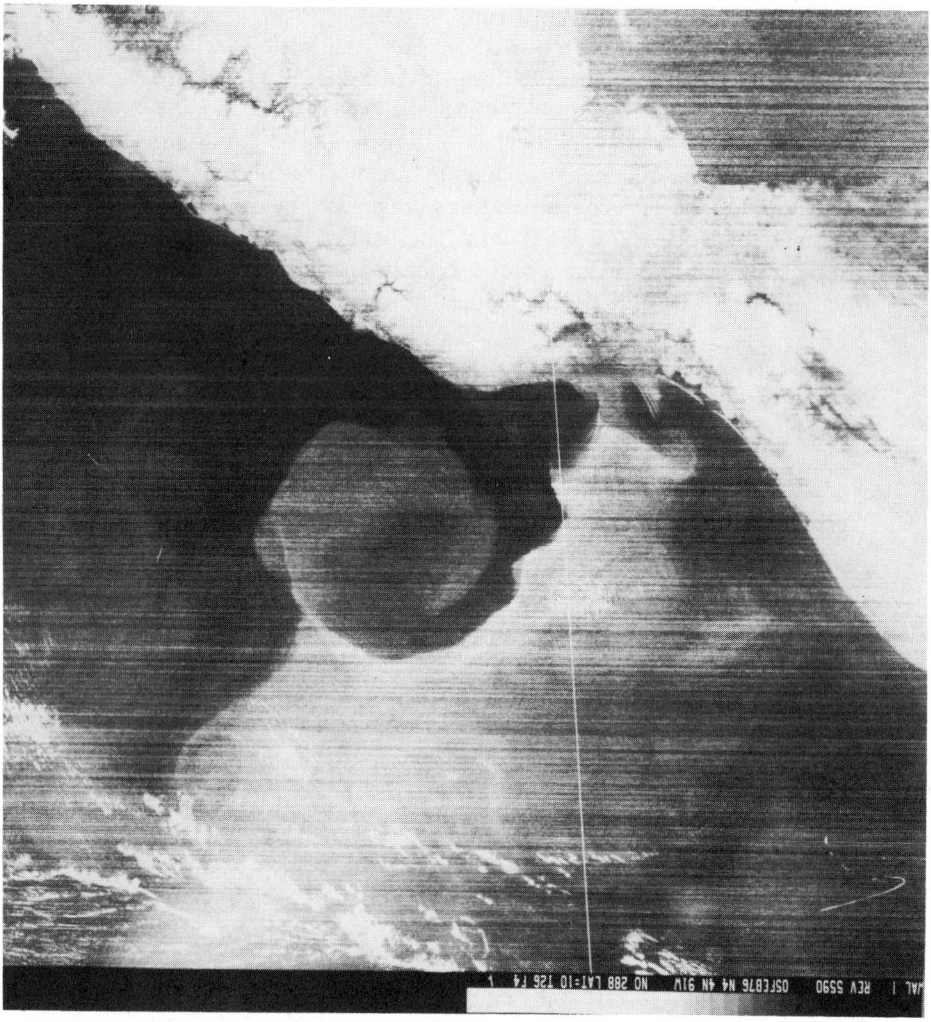

Figure 10 NOAA-4 VHRR image for February 5, 1978, off the Gulf of Tehunatepec near Acapulco, Mexico. The cooler water is due to wind-induced upwelling across the Central American isthmus. The large gyre has a diameter of 280 km. From Stumpf & Legeckis (1977).

current systems, the appearance of meandering and eddy development is a frequent feature. Where theoretical interpretations are made, they generally involve a baroclinic Rossby wave or instability. Such large-scale variations are necessarily influenced by the change in Coriolis parameter or bottom depth over their scales of motion, and the conservation of potential vorticity must play an important role thereby. The satellite data should be of considerable value in establishing spatial, velocity, and amplitude scales for the oscillations.

A useful compendium of imagery obtained from environmental satellites is found in a recent publication by the National Oceanic and Atmospheric Administration (NOAA 1979). This report contains many illustrations of oceanic phenomena along with explanations and references.

Coastal and Estuarine Fronts

The use of satellite sensors having fine resolution and operating in the visible portion of the spectrum allows one to study small-scale dynamics associated with turbidity, sediment, pollutants, or chlorophyll. These often serve as water mass tracers and delineate coastal, estuarine, or tidal flow and diffusion.

The literature on this subject is broad but scattered. Much of the reported work has made use of the Landsat multispectral scanner, MSS, for studies of advective and diffusive effects. Landsat data are obtained over the entire US and selected portions of the remainder of the world every 18 days, with the images covering a swath of 184 km at a resolution of 70 m. The MSS provides radiometrically calibrated images in four channels, with wavelengths in the green, red, and near-infrared.

The satellite is in a sun-synchronous orbit that crosses the equator at 9:30 AM local time and completes 13 & 13/18ths revolutions per 24 hours. The orbit also precesses westward at about 1.4° per day, thereby providing an overlap between adjacent swaths separated by precisely 24 hours amounting to approximately 40% of the image width at midlatitudes. This allows the area of overlap to be observed on two successive days at essentially the same local time. A given area is also revisited at intervals of 18 × 24 hours (cf NASA 1976). Thus, tidally driven circulation, which is forced predominantly at $12\frac{1}{2}$- and 25-hr periods, can be observed at all phases of the tidal cycle during the course of time. This feature makes Landsat especially useful for near-shore circulation studies.

A representative paper that shows the utility of visible imagery in coastal investigations is a recent one by Feely and associates (1979), which reports on studies of sediment distributions in the Gulf of Alaska. Figure 11 illustrates a Landsat MSS-4 image for the area under study; its dimensions are approximately 150 × 150 km^2. Sediment is entrained in the near-shore waters and delineates an anticyclonic, baroclinic gyre of approximately 50-km diameter to the southwest of Kayak Island, Alaska. It is clear that surface water motions are traced out by the suspended sediments, which serve as natural dye markers for the satellite. The authors gathered nephelometer data to deduce the suspended sediment concentration at fixed locations. It appears possible to use this to calibrate the upwelling radiances observed by Landsat in terms of

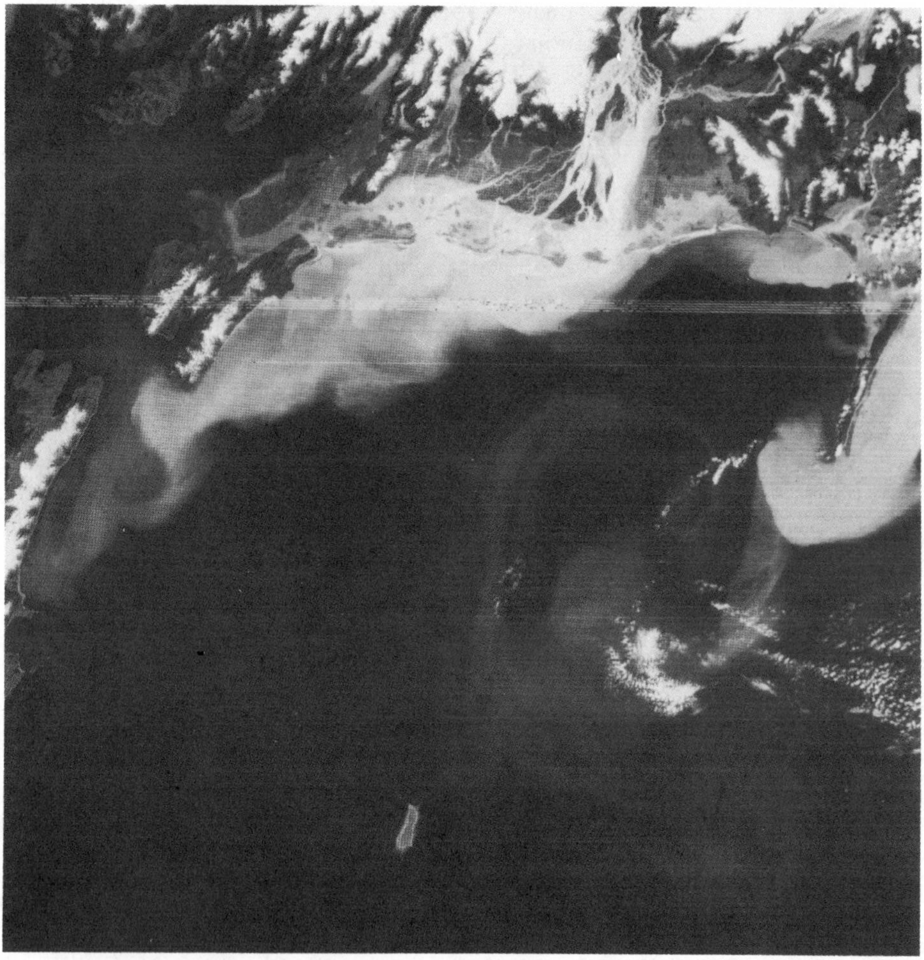

Figure 11 Landsat MSS-4 visible image from August 14, 1974, taken over the Gulf of Alaska. The image shows the advection of sediments by coastal currents, which trace out a large baroclinic gyre to the southwest of Kayak Island, shown at center-right. Image ID #-1387-20281-4. From Feely et al (1979).

near-surface concentration of sediment (E. T. Baker, private communication). This procedure is probably acceptable for a specific sediment type in a specific area, but cannot be generalized to arbitrary materials or areas without such calibration. The general utility of the method obviously depends on the existence of sources of sediment in the region of interest.

The Landsat MSS was not intended for marine measurements, even though it has proven to be of some value to oceanographers. An instrument that has been specifically designed for quantitative determination of oceanic chlorophyll and sediment concentrations is the Coastal Zone Color Scanner, CZCS, on the Nimbus-7 satellite. The narrow-band channels on CZCS view 1000-km-wide swaths with a resolution of 840 m and are expected to yield useful data for chlorophyll in the range of approximately 0.1–20 mg m^{-3} and for sediment in concentrations of about 0.2–100 mg m^{-3}. As of this writing, no published results from CZCS are available, but the preliminary assessments of its capabilities are favorable (W. Hovis, private communication).

SMALL-SCALE MOTIONS

For the most part, the small-scale motions discussed here are internal, surface gravity, and capillary waves.

Internal Waves

Surface signatures of coherent internal waves occur under limited but frequently realized conditions of the ocean surface and have been observed with Landsat (Apel et al 1975a, b, 1976). The waves that are visible are apparently generated by tidally driven flow over sills, continental shelf edges, or other major variations in underwater topography. If the significant surface wave height is less than approximately a meter, the internal wave currents modulate the small-scale surface roughness by focusing surface wave energy and surface oils into narrow regions parallel to the internal wave crests. This leads to paired, narrow bands of water that are alternately rougher than average and smoother than average; the pairs are separated by wider intervals equal to the distance between internal wave troughs (Elachi & Apel 1976, Proni et al 1978). Sunlight or radar energy is diffusely scattered from the rough regions and specularly reflected from the smooth ones, thereby rendering the underlying internal wave pattern visible to remote sensors under light-to-moderate wind conditions.

The surface signatures are also visible with an imaging radar (Elachi & Apel 1976). Figure 12 shows an image taken off the west coast of Baja California, Mexico, made with the Seasat Synthetic Aperture Radar, SAR, on 7 July 1978 (Apel 1980). Several packets of internal waves are visible in the radar image, which has a resolution of 25 m; these are shown schematically in the line drawing of Figure 12*a*. The average length of the waves is 390 m, and the average interval between packets is 19 km. The wave period is estimated to be approximately 10 min. They are thought to be generated every $12\frac{1}{2}$ hours by tidal flow over the offshore banks visible on Figure 12*a*. The waves then propagate towards shore, increase in amplitude, and finally dissipate in shallow water through breaking and turbulence. They have clear nonlinear features described below.

Sawyer & Apel (1976) have compiled an atlas of internal wave surface signatures as seen by Landsat off the North American east coast during the summers of 1972 to 1974, inclusive. The data clearly show the continental shelf edge to be a major source of coherent waves. Many of the packets propagate more rapidly than predicted by calculations of the linear phase and group velocities, which is thought to be a nonlinear effect. In general, within a given packet, the waves at the front have the longest lengths, longest crests, and largest amplitudes, with all of these quantities diminishing toward the trailing edge. During the time of passage of the packet, the average mixed layer depth is significantly depressed (Proni et al 1978). The surface pattern is usually convex, as if radiating away from a small source, with the leading edge of the packet extremely well defined. Up to six packets have been observed in one image, implying the waves have lifetimes on the shelf of at least three days (Apel et al 1975a).

Osborne and co-workers (1978) have made moored current meter and thermistor measurements of very large amplitude internal waves in the Andaman Sea west of the Malay Peninsula and have described the vertical displacements and currents in terms of a train of solitary waves. Their measured characteristics are the most dramatic reported to date, with peak-to-trough amplitudes of order 100 m, wavelengths of 10–20 km, current speeds near 150 cm s^{-1}, and group speeds of 250 cm s^{-1}. The sea surface accompanying each wave trough is cited as having approximately a 2-m significant wave height, with the roughened region confined to a 1000-m-wide band overlying the descending portion of the internal wave phase (A. R. Osborne, private communication). Similar surface signatures were photographed on the Andaman Sea by the US Apollo-Soyuz astronauts on another occasion (Apel 1979b). The photographs suggest that there are two sources, one of which is to the west of the area of observation near the Nicobar Islands, and the other in the Strait of Malacca.

An analytical expression approximately describing the major proper-

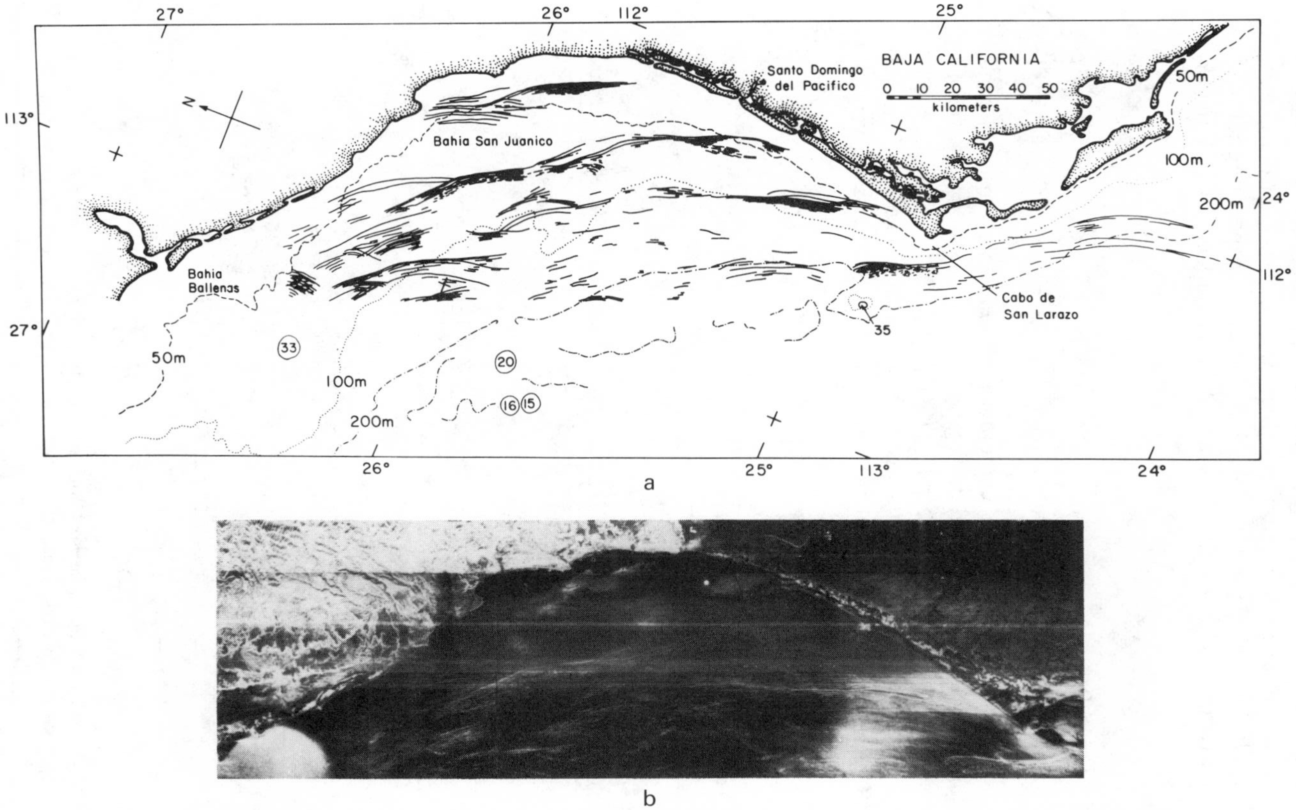

Figure 12 (*a*) Map of the area off Baja California showing bottom topography and a line drawing interpretation of surface signatures of internal waves. (*b*) Seasat Synthetic Aperture Radar image of continental shelf internal waves off Baja California, taken at about 0400 local time on July 7, 1978. The wavelengths are approximately 390 m and they occur in packets separated by about 19 km. Tidal flow over the offshore shallows is thought to be the source of the waves. From Apel (1980).

ties of coherent internal waves is given by the equation below. Assume a stratified system undergoing nonlinear, vertical oscillatory displacements from equilibrium of $\eta(x, y, z, t)$; then for a packet propagating along a horizontal arc $s = s(x, y)$ in the (x, y) plane, the subsurface displacement is (see Figure 13)

$$\eta(x, y, z, t) \simeq \begin{cases} 0, \; s < 0, \\ -2\eta_0 R_c(x, y) W(\alpha z) \exp(-s/s_1) cn_m^2\{\tfrac{1}{2}[k_0(1+s/s_2)s - \omega t]\}, \; s > 0. \end{cases}$$

This model describes an internal cnoidal wave train $cn_m^2(\phi)$ abruptly commencing at $s = 0$, with an initial downgoing pulse of total amplitude $2\eta_0$, a linearly increasing wavenumber $k = k_0(1+s/s_2)$, and an exponentially decreasing amplitude toward the rear of the packet. The curvature of the packets in the horizontal plane is described by $R_c(x, y)$; the vertical

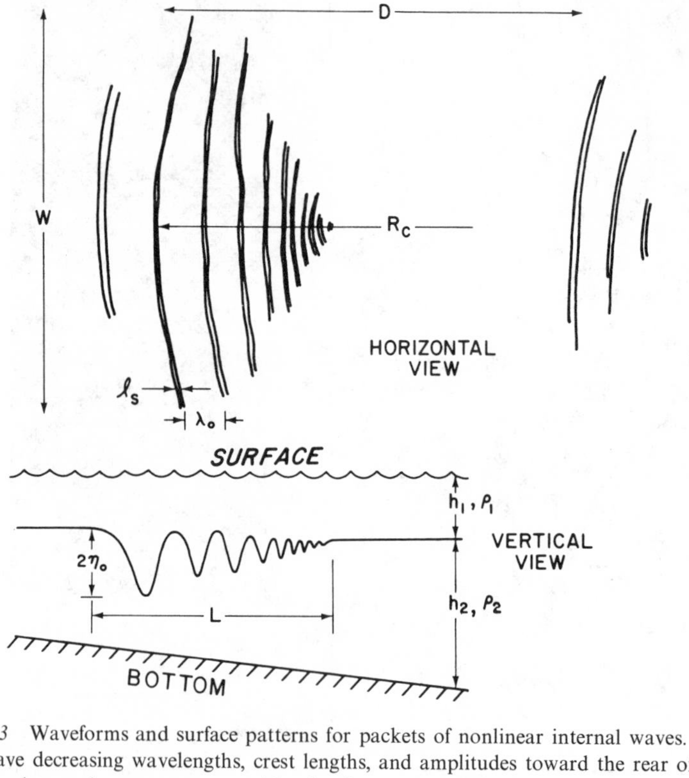

Figure 13 Waveforms and surface patterns for packets of nonlinear internal waves. The waves have decreasing wavelengths, crest lengths, and amplitudes toward the rear of the packet, and occur in groups separated by the distance traveled in one tidal period. From Apel et al (1976).

eigenmode structure function is $W(\alpha z)$. The parameter m gives the degree of nonlinearity, with $m = 0$ corresponding to sinusoidal waves and $m = 1$ to a solitary wave; s_1, s_2, and α^{-1} give scales to the waveform. This approximates a solution to the Korteweg–de Vries equation but is not itself a solution. (See Whitham 1974 for a discussion of nonlinear wave characteristics.) It is possible to evaluate all of these parameters from the combined remote sensing and in-situ data. A sequence of such packets is assumed to be generated by the repeating semidiurnal and diurnal tides.

Two mechanisms have been advanced to explain the generation process. First, Tsai & Apel (1979) have invoked a linear instability in a stratified, tidally driven shear flow that has been rendered highly baroclinic by bottom topography. The Richardson number

$$\mathrm{Ri} = N^2/(\partial u/\partial z)^2$$

is assumed to drop below 1/4 for a few hours during the peak tidal current, thereby generating a finite-length train or packet, typically numbering about ten waves. As the tidal currents rotate through the tidal ellipse, the instability starts up and shuts down every $12\frac{1}{2}$ hours. This theory provides good agreement between observed and calculated wave lengths, periods, and vertical eigenmodes, but remains to be directly verified in the field.

Second, Maxworthy (1979) has experimentally demonstrated the generation of nonlinear internal wave trains downstream of a submerged barrier in a stratified flow in a two-layer system of depths h_1 and h_2. If the flow speed over the barrier, u, exceeds the shallow water internal wave phase speed, $c = [(g\Delta\rho h_1 h_2/\rho(h_1+h_2)]^{1/2}$, an internal undulatory bore is formed as a train of solitary lee waves. When the mean flow ceases, the internal bore propagates opposite to the former direction of flow as a freely moving packet. This mechanism adequately explains the formation of internal wave packets in narrow fjords (Farmer & Smith 1978) but fails on the continental shelf in its present formulation. Because of the generally monotonic decrease in the depth of the shelf towards shore, the theory implies that the flow speed, u, would first exceed c in the shallowest water near shore and form a bore there and not at the shelf break, where the wave packets are first observed. However, the solitary-wave nature of many of the packets on the continental shelf argues for some such mechanism as a source. It appears possible to conduct a definitive experiment in the field that would resolve the question.

In summary, because of their coherence and clearly defined characteristics, continental shelf internal waves, when viewed from spacecraft, have been relatively easy to interpret and analyze. The results have been more quantitative than those obtained from larger-scale circulation studies. However, much remains to be learned from increasingly sophisticated intercomparisons between theory and observation; the latter will without doubt include satellite imagery.

Surface Waves

The quantity often desired in measurements of surface wave properties is the two-dimensional height spectrum, $S(k_x, k_y)$, as a function of horizontal wave vector (k_x, k_y). Current techniques allow only a partial satisfaction of this requirement.

GRAVITY WAVE SPECTRA Images of surface waves obtained with sufficiently high resolution may be used to deduce $S(k_x, k_y)$ for the longer wavelength portion of the spectrum, assuming the wave field is spatially homogeneous over the field of view. While the application of a Nyquist sampling theorem for a two-dimensional field would imply that two samples per wavelength are required in order to resolve a given component, in practice it appears that three to five samples are needed.

In this regard, the use of Landsat imagery, with its 70-m resolution, appears possible for ocean surface waves in excess of approximately 200-m length. Because of the near-constant viewing and illumination angles over a given area, such imagery can be Fourier-transformed to obtain estimates of the dominant wave vector components, if not the spectrum itself (Apel et al 1975a). Obviously the technique is limited to conditions of clear skies and long wavelengths, a combination that does not occur frequently. In addition, the size of the image needs to be at least large enough to contain of order 100 waves of the longest length present to avoid broadening of the spectrum due to finite record lengths. If the field is not statistically homogeneous (as in a region of wave refraction), then a spectral description is unsatisfactory and the image itself must be directly used.

Imaging radars have also been shown to give useful information on wave spectra, although the mechanisms by which microwave radar energy is backscattered and modulated to produce ocean wave images are not well understood (Brown et al 1976, Gonzalez et al 1979). The Seasat Synthetic Aperature Radar (SAR) has produced approximately 1.5×10^6 km of radar imagery having a swath width of 100 km and a resolution of 25 m. Much of this is over the ocean, and some of it was gathered during hurricanes and other storms, to whose effects the radar is virtually immune (the radar wavelength of 25 cm is little affected by rainfall or clouds). Figure 14*a* shows a SAR image containing waves of 275-m length and propagating at an angle of about 45° with respect to the image edge. North is directed upward 24° clockwise from the top of the image. The

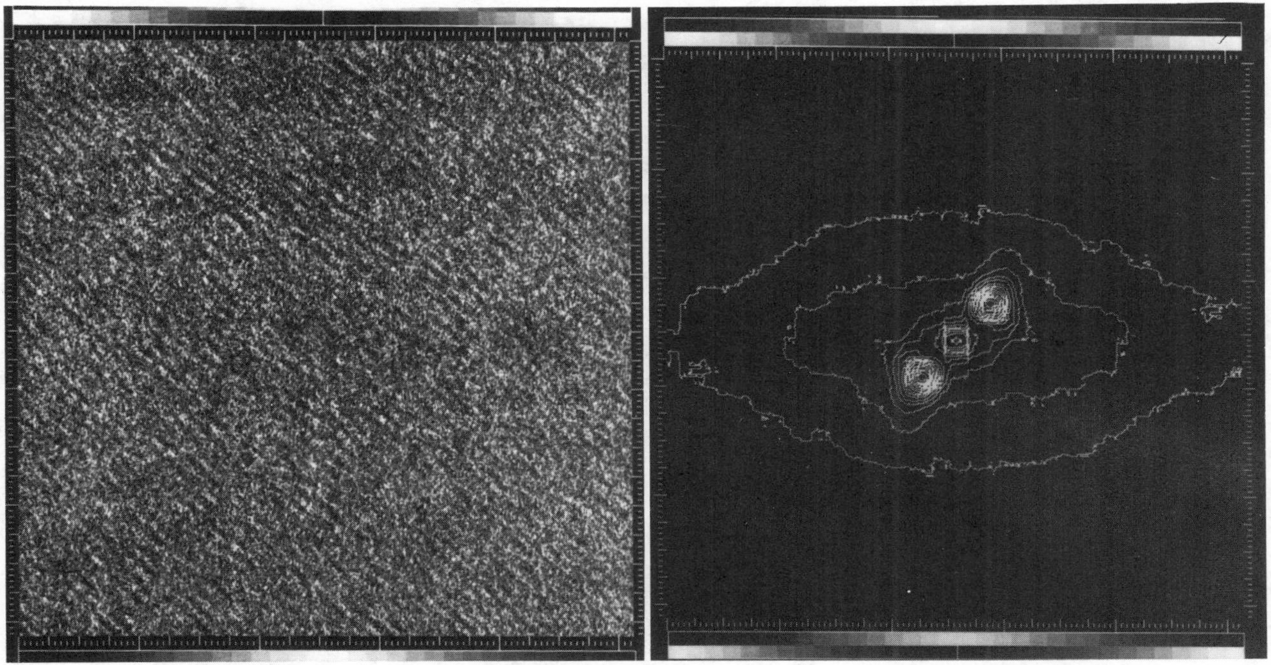

Figure 14 (*a*) Seasat Synthetic Aperture Radar image taken over Gulf of Alaska September 16, 1978, at 1753 GMT. Location 42.8°N, 129.4°W. From a NOAA data buoy, $H_{1/3} = 2.7$ m, $\lambda = 318$ m. Smallest increment, 125 m. Image size is 12.8 km. North is upward 24° clockwise from the top edge. Wavelengths deduced from the image have $\lambda = 275$ m. (*b*) Digital Fourier transform of the wave image in (*a*). Smallest increment = 0.078 cyc/km. Contour intervals arbitrary but equally spaced; origin at center. Concentrations of energy at 45° and 225° represent the waves visible in (*a*). Jet Propulsion Laboratory, Pasadena, California.

significant wave height, $H_{1/3}$, defined via

$$H_{1/3}^2 = \int\int_{-\infty}^{\infty} S(k_x, k_y)\, dk_x\, dk_y,$$

is estimated at approximately 2.7 m from near-coincident surface data. (Amplitude information is not currently derivable from SAR imagery.) Figure 14*b* is a digital Fourier transform made from this image, showing the dominant spectral components as increases in intensity disposed anti-symmetrically with respect to the origin, $k_x = k_y = 0$, which is at the center of the figure (F. I. Gonzalez and W. Stromberg, personal communication).

Preliminary comparisons between Seasat SAR and surface data indicate that the agreement between satellite and surface measurements is good to approximately ±15% in wavelength and ±25° in angle. Significant wave heights in excess of 1 to 2 m appear to be required in order to form images (Gonzalez et al 1979).

WAVE HEIGHTS Another radar technique has been shown to yield good quality wave amplitude measurements along the subsatellite track. The radar altimeters on GEOS-3 and Seasat emit a very short pulse of length τ, in order to obtain their fine altitude resolution. The pulse length is 12.5 ns for GEOS-3 and 3 ns for Seasat; the equivalent widths in space are given by $c\tau/2$, where c is the velocity of light; the widths are 1.9 and 0.45 m, respectively. If such a pulse is reflected from a sea surface having significant wave height (SWH), $H_{1/3}$, it will be broadened out in space by the distribution of reflecting heights represented by the waves. The broadening will be measurable as a stretching of the shape of the received pulse in time and can be discerned if $H_{1/3} \gtrsim c\tau/2$. For the two satellites above, wave heights in excess of about 2.0 and 0.5 m are measurable (Barrick 1972, Walsh et al 1978). If the sea is gaussian, the leading edge of the pulse shape is in the form of an error function, erf(x), so that the received power as a function of time and sea state is

$$P(t, H_{1/3}) = P_0 \operatorname{erf}(t/t_p),$$

where the time origin is at the half-power point, $P = P_0/2$, and the time scale t_p is related to the pulse width and wave height, viz (Fedor & Barrick 1978)

$$t_p = [(c\tau)^2/\ln 2 + 2H_{1/3}^2]^{1/2}/2c.$$

GEOS-3 has been shown to yield wave heights to within an rms value of ±43% of surface measurements (Fedor & Barrick 1978). Recently, the Seasat altimeter has been subjected to a much more extensive analysis that indicates that it will provide SWH measurements accurate to ± 0.5 m over a range of at least 0.8–3.5 m (Tapley et al 1979).

The designed upper limit to the Seasat wave height capability is $H_{1/3} \simeq$ 20 m; to date, measurements have been reported to 12-m SWH in the 40°–50°S latitude bands (W. F. Townsend, private communication).

The technique depends only on determination of pulse shape, is conceptually simple, and appears to be reasonably well founded; the range and accuracy have not been fully established as yet. It is likely, however, that the system will reach its design capability and that the measurements can be regarded with confidence over the range $0.5 \lesssim H_{1/3} \lesssim$ 20 m. It is also likely that it will be implemented on operational satellites in the next few years.

SHORT WAVES The remote measurement of very short features, i.e. ultragravity and capillary waves, and small-scale roughness, is currently possible using microwave radar whose wavelength λ_μ is in the centimeter range. The physical basis for this technique is resonant Bragg scattering of microwaves from the capillary waves. The first-order Bragg condition is $\lambda_\mu = 2\lambda \sin\theta$, where λ is the capillary wave length and θ is the angle of incidence measured from the vertical. At an angle of 30°, $\lambda_\mu = \lambda$; thus, a radar operating at 15 GHz, for example, will be Bragg-scattered from 2-cm capillaries. These waves are in near-instantaneous equilibrium with the short-term surface wind field, so the strength of the radar return will depend on the capillary wave amplitude and direction of illumination relative to the wind direction (Jones & Schroeder 1978).

This physical phenomenon has been exploited in the design of the Seasat wind scatterometer system, SASS, an instrument intended to measure vector surface winds. The range and precision specified for wind speed are 2 to perhaps 25 m s^{-1} $\pm 25\%$, while the range for wind direction is 0–360° with a precision of $\pm 20°$; these quantities are to be derived on a grid spacing of 25 km. The wind so obtained is referenced to a standard level of 19.5 m above sea level.

Preliminary results from Seasat are reported by Jones et al (1979) and indicate that the specifications have been met or exceeded for winds up to approximately 17 m s^{-1}. Some measurements taken over hurricanes indicate the useful range of the instrument may be as high as 35–40 m s^{-1} (P. Black, private communication).

Figure 15 displays results obtained from one pass of the Seasat scatterometer over the Gulf of Alaska, where much of the preliminary verification work has been carried out (Jet Propulsion Laboratory 1979). This figure illustrates a GOES image of the west coast of North America from the Alaska North Slope to San Francisco, taken on 14 September 1979 at

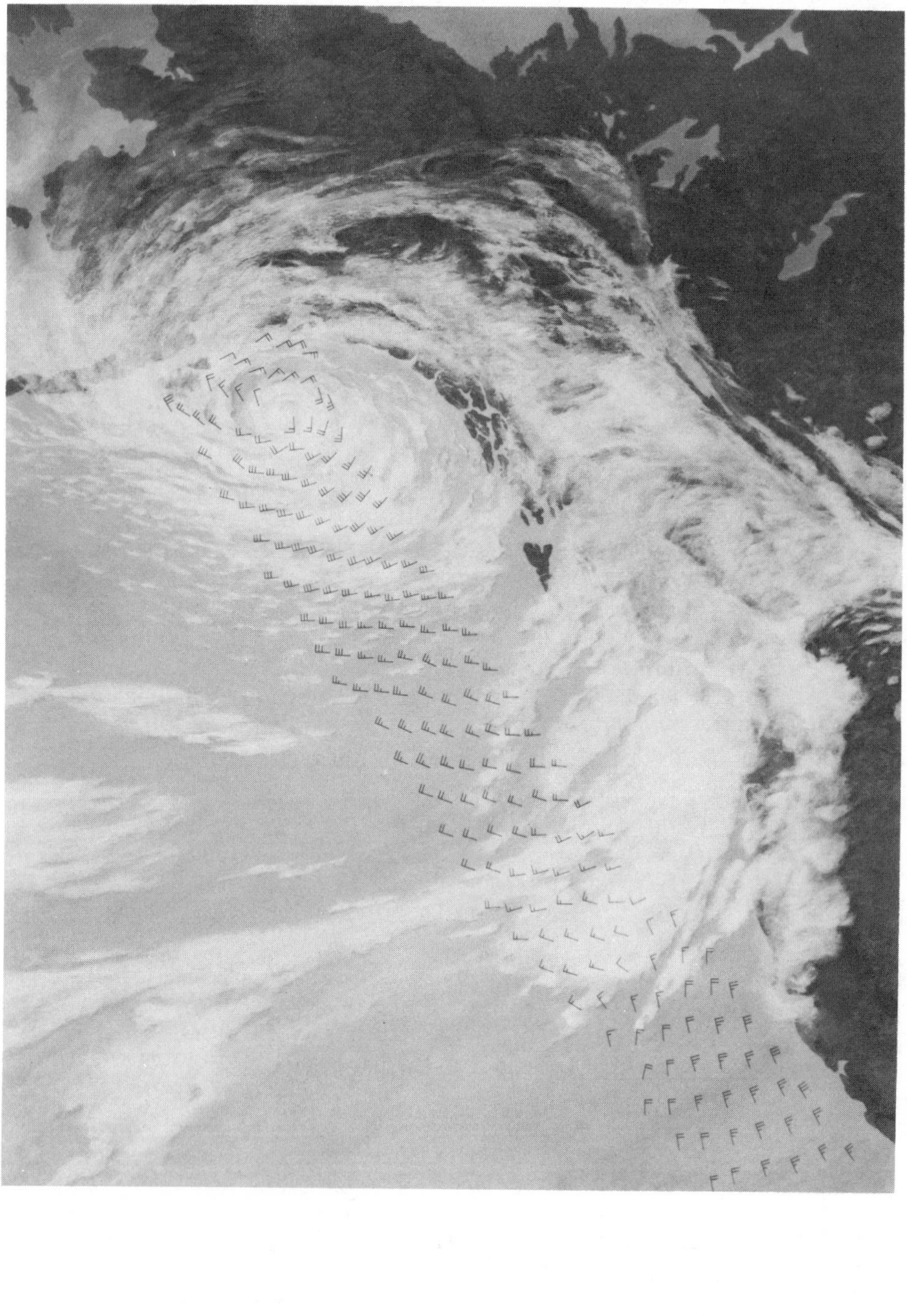

1715 GMT; a cyclonic low-pressure system is seen over the Gulf of Alaska. Superimposed are wind barbs derived from the SASS for this pass, with the highest surface winds indicated being 30 knots (15 m s^{-1}). The lower pressure cell is well resolved by the SASS, as is the westerly jet to the south and the unusually strong northerly flow off the California coast.

If the SASS functions as well as the early analyses indicate, in the future it should be possible for satellites to yield high resolution (25 to 50 km) surface wind fields over the world's oceans on a daily basis. Such data would be of great value in studies of the generation of waves, air-sea interaction, and wind-driven circulation. It would appear possible, for instance, to test hypotheses on Eckman and Sverdrup transports, generation of El Niño events, and other wind-associated oceanic processes. Such studies must await the establishment of future operational satellites equipped with radar systems, however.

Another microwave instrument that is designed to provide wind speeds on a 150-km grid on the ocean surface is the scanning multichannel microwave radiometer, SMMR; this device has flown on Seasat and Nimbus-7. The instrument functions by measuring the enhanced microwave emission association with increased small-scale surface roughness and foam. The increased emission is approximately linearly related to wind speed from roughly 5 to as much as 30–35 m s^{-1}. The early analyses from Seasat indicate a precision of ± 2 m s^{-1} may be attainable under non-precipitating conditions, but considerable work remains before surface fields of the desired accuracy may be routinely achieved. The SMMR is also expected to yield surface temperatures to a precision of approximately ± 1.5°C (Lipes et al 1979). Further evaluations will be obtained from the Nimbus-7 instrument in the near future.

CONCLUDING REMARKS

Discussion and Summary

CURRENTS A number of individual studies of the Gulf Stream have been reviewed that collectively offer a broad view of its behavior near the US

← *Figure 15* Ocean surface winds derived from Seasat microwave scatterometer data for September 14, 1979, 1715 GMT. Winds are indicated at each point to the nearest 5 knots (full bar, about 10 knots, half barb, about 5 knots). The data identify the storm center in the Gulf of Alaska, the westerly jet associated with it, the ridge of high pressure and weak winds to the south, and the unusually strong northerly flow off the California coast. Cloud structure is taken from Western GOES imagery. Jet Propulsion Laboratory, Pasadena, California. *Science* 204: cover.

east coast. The use of satellite infrared sensors is shown to delineate successfully the locations of the north and west boundary and to yield data that are useful additions to the understanding of western boundary current dynamics. While these patterns of surface flow do not yield current speeds, it appears possible to arrive at such speeds with altimetric measurements. The limited altimeter data set available from Seasat will probably suffice to prove the concept, but extensive studies must await a future generation of altimetric satellites.

In regions of less intense flows, especially those having reduced thermal contrast at the surface, the use of infrared data becomes more difficult. There are reasonable indications that equatorial waves and upwellings, eastern boundary currents, and wind-induced coastal upwelling may be studied with imaging infrared sensors but much less work has been done on these phenomena. To the author's knowledge, no studies of such circulatory features that utilize altimetric data have been carried out to the point of being published. The area is ripe for exploitation since data exist that would allow original research to be done.

For surface currents below 20–30 cm s^{-1}, space measurements offer little direct benefit at present. A possible exception might be the use of high-resolution visible imagery in coastal regions where material suspended in the water might serve as dye markers for near-shore transport.

SEA SURFACE TEMPERATURE It has only been since the launch of TIROS-N in 1978 that sufficiently high quality infrared imagery has been available to allow the measurement of sea surface temperature to an accuracy approaching 1°C. Earlier IR data, when composited into sea surface temperature maps, has had accuracies of order 2–2½°C when compared with surface measurements. The TIROS-N data may allow a new class of observations of dynamic processes to be made, especially those for which surface heat transport is important.

OCEAN COLOR The quantitative measurement of ocean color from space has only been possible since late 1978, since the launch of Nimbus-7, and no results are available for publication. However, color contrast may be as useful as temperature contrast in observing current systems, especially in tropical regions where surface temperatures tend to be uniform.

WAVES It is clear that dominant surface gravity wave heights, lengths, and directions may be determined from radar instrumentation on satellites when such waves exceed 150–200 m in length. These measurements are especially interesting in storms, where essentially all conventional means of arriving at the wave properties fail. Images of refracting wave patterns in shoal areas are possible to obtain under storm conditions or cloud cover.

Coherent internal waves on the continental shelves may often be observed and their propagation characteristics determined when surface wave conditions allow.

WINDS A preliminary assessment of the wind-measuring capabilities of microwave-equipped satellites suggests that wind speeds from perhaps 5–35 m s^{-1} can be determined; some angular information is available for the lower speeds.

Exhortations on the Practice of Remote Sensing

In the study of ocean dynamics, as elsewhere, a sensible approach is to use whatever tools are appropriate for the task at hand. It is becoming clear that remote sensing has certain limited but unique and valuable contributions to make to oceanography. In another discipline, astrophysics, the power of remote sensing has been most impressive—nearly all that we know about planetary and stellar dynamics has been obtained by a combination of remote measurement and the subsequent interpretation of the data using physical laws. Nevertheless, it is certain that if it were possible for astrophysicists to make direct, in-situ measurements of stellar atmospheres, for example, they would do so, for the use of remote measurement alone is frought with difficulties.

If the process being studied admits of indirect as well as direct measurement, it would seem most fruitful to merge in-situ and remote measurement with the interpretative apparatus of physical theory. Even for well-defined dynamical processes visible on the surface, the satellite data is best supported by in-water measurements that portray the underlying fields in a convincing manner.

The practice of remote sensing is becoming more and more quantitative with time, and this is an evolution that is absolutely essential if further progress is to be made in its application to oceanography. For dynamical studies, the practitioners require training in electromagnetic theory and photogrammetry, and access to digital data and computers, as well as a thorough grounding in the theory of geophysical fluid dynamics. By these means the remote sensing oceanographer can arrive at the point where formulation and testing of scientific hypotheses constitutes the major activity at hand. Oceanographers have been hard put to gain the overview of their domain required to understand large-scale processes in the sea. The combination suggested here promises to provide a vantage point for viewing important phenomena occurring near the surface of the sea.

ACKNOWLEDGMENTS

I am indebted to H. M. Byrne for assistance in preparation and review of some of the material, to the authors who provided the figures included here, and to Ms. Mary Jensen and Ms. Sally Dong for aid in preparation and typing of the manuscript.

Literature Cited

Apel, J. R., ed. 1972. Sea surface topography from space, I & II. *NOAA Tech. Rep. ERL 228-AOML 7*

Apel, J. R. 1976. Ocean science from space. *EOS, Trans. Am. Geophys. Union* 57:612–24

Apel, J. R. 1979a. Past, present and future capabilities of satellites relative to the needs of ocean science. *Bruun Mem. Lect., 1977 Intergov. Oceanogr. Tech. Ser.* 19: 7–39. Paris: UNESCO

Apel, J. R. 1979b. Observations of internal wave surface signatures in ASTP photographs. *Apollo-Soyuz Test Project, Summary Sci. Rep. NASA SP-412*

Apel, J. R. 1980. Oceanic internal wave signatures as observed from the Seasat imaging radar. To be published

Apel, J. R., Apel, J. J. 1976. Gulf Stream meander and eddy dynamics as observed from ship and spacecraft. Conf. Atmosphere and Oceanic Waves and Stability. *Bull. Am. Meteorol. Soc.* 57:99

Apel, J. R., Byrne, H. M. 1974. Oceanography and the marine geoid, *Proc. Int. Symp. Applications of Marine Geodesy*, pp. 59–66. Washington DC: Marine Technology Society

Apel, J. R., Siry, J. W. 1974. Synopsis of Seasat-A scientific contributions. *Seasat-A Sci. Contrib. 3-30*, Washington DC: Natl. Aeronaut. Space Adm.

Apel, J. R., Byrne, H. M., Proni, J. R., Charnell, R. L. 1975a. Observations of oceanic internal and surface waves from the Earth Resources Technology Satellite. *J. Geophys. Res.* 80:865–81

Apel, J. R., Proni, J. R., Byrne, H. M., Sellers, R. L. 1975b. Near-simultaneous observations of intermittent internal waves on the continental shelf from ship and spacecraft. *Geophys. Res. Lett.* 2:128–31

Apel, J. R., Byrne, H. M., Proni, J. R., Sellers, R. L. 1976. A study of oceanic internal waves using satellite imagery and ship data. *Remote Sensing Environ.* 5:125–35

Bane, J. M. Jr., Brooks, D. A. 1979. Gulf Stream meanders along the continental margin from the Florida Straits to Cape Hatteras. *Geophys. Res. Lett.* 6:280–82

Barrick, D. E. 1972. Determination of mean surface position and sea state from the radar return of a short-pulse satellite altimeter. In *Sea Surface Topography from Space 1*, ed. J. R. Apel, 16-1 to 16-19. *NOAA Tech. Rep. ERL 228-AOML 7*

Bernstein, R. L., Breaker, L., Whritner, R. 1977. California current eddy formation: Ship, air and satellite results. *Science* 195:353–59

Brooks, D. A., Mooers, C. N. K. 1977. Free, stable continental shelf waves in a sheared, barotropic boundary current. *J. Phys. Oceanogr.* 7:380–88

Brown, W. E. Jr., Elachi, C., Thompson, T. W. 1976. Radar imaging of ocean surface patterns. *J. Geophys. Res.* 81: 2657–67

Byrne, H. M. 1979. Radar altimetry for use in measuring sea surface slope. *Natl. Radio Sci. Meet. Program*, June 1979, p. 237. Int. Union Radio Sci.

Düing, W. O. 1978. Spatial and temporal variability of major ocean currents and mesoscale eddies. *Boundary-Layer Meteorol.* 13:7–13

Düing, W. O., Hisard, P., Katz, E., Meincke, J., Miller, L., Moroshkin, K. V., Philander, G., Ribnikov, A. A., Voigt, K., Weisberg, R. 1975. Meanders and long waves in the equatorial Atlantic. *Nature* 257:280–84

Düing, W. O., Mooers, C. N. K., Lee, T. N. 1977. Low-frequency variability in the Florida current and relations to atmospheric forcing from 1972 to 1974. *J. Mar. Res.* 35:129–61

Elachi, C., Apel, J. R. 1976. Internal wave observations made with an airborne synthetic aperture imaging radar. *Geophys. Res. Lett.* 3:647–50

Farmer, D., Smith, J. D. 1978. Nonlinear internal waves in a fjord. In *Hydrodynamics of Estuaries and Fjords*, ed. J. Nihoul. New York: Elsevier

Fedor, L. S., Barrick, D. E. 1978. Measurement of ocean wave heights with a satellite radar altimeter. *EOS, Trans. Am. Geophys. Union* 59:843–47

Feely, R. A., Baker, E. T., Schumacher, J. D., Massoth, G. J., Landing, W. M. 1979. Processes affecting the distribution and transport of suspended matter in the Northeast Gulf of Alaska. *Deep-Sea Res.* 26:445–64

Fuglister, F. C., Worthington, L. V. 1951. Some results from a multiple ship survey of the Gulf Stream. *Tellus* 3:1–14

Gonzalez, F. I., Beal, R. C., Brown, W. E., DeLeonibus, P. S., Sherman, J. W. III, Gower, J. F. R., Lichy, D., Ross, D. B., Rufenach, C. L., Schuchman, R. A. 1979. Seasat synthetic aperture radar: Ocean wave detection capabilities. *Science* 204: 1418–21

Gower, J. F. R., guest ed. 1980. *Boundary-Layer Meteorol.* 18, Nos. 1, 2, & 3

Halliwell, G. R. Jr., Mooers, C. N. K. 1979. The space-time structure and variability of the shelf water/slope water and Gulf Stream surface temperature fronts and associated warm-core eddies. *J. Geophys. Res.* 84: In press

Hansen, D. V. 1970. Gulf Stream meanders between Cape Hatteras and the Grand Banks. *Deep-Sea Res.* 17:495–511

Hansen, D. V., Maul, G. A. 1970. A note on the use of sea surface temperature for observing ocean currents. *Remote Sensing Environ.* 2:161–64

Huang, N. E., Leitao, C. D. 1978. Large scale Gulf Stream frontal study using GEOS-3 radar altimeter data. *J. Geophys. Res.* 83:4673–82

Huh, O. K., Wiseman, W. J. Jr., Rouse, L. J. Jr. 1978. Winter cycle of sea surface thermal patterns, northeast Gulf of Mexico. *J. Geophys. Res.* 83:4523–29

Jet Propulsion Laboratory. 1979. Ocean surface winds derived from Seasat microwave scatterometer data for 14 September 1979, 17:15 G.M.T. *Science* 204: cover

Jones, W. L., Black, P. G., Boggs, D. M., Bracalente, E. M., Brown, R. A., Dome, G., Ernst, J. A., Halberstram, I. M., Overland, J. E., Peteherych, S., Pierson, W. J., Wetnz, F. J., Woiceshyn, P. M., Wurtele, M. G. 1979. Seasat Scatterometer: Results of the Gulf of Alaska Workshop. *Science* 204:1413–15

Jones, W. L., Schroeder, L. C. 1978. Radar backscatter from the ocean: Dependence on surface friction velocity. *Boundary-Layer Meteorol.* 13:133–49

Kaula, W. M., ed. 1970. *The Terrestrial Environment: Solid Earth and Ocean Physics.* Cambridge, Mass: MIT Press

LeBlond, P. H., Mysak, L. A. 1978. *Waves in the Ocean*, pp. 431–37. Amsterdam: Elsevier

Legeckis, R. V. 1975. Application of synchronous meteorological satellite data to the study of time dependent sea surface temperature changes along the boundary of the Gulf Stream. *Geophys. Res. Lett.* 2:435–38

Legeckis, R. V. 1977a. Long waves in the eastern equatorial Pacific Ocean: A view from a geostationary satellite. *Science* 197:1179–81

Legeckis, R. V. 1977b. Oceanic polar front in the Drake Passage—Satellite observations during 1976. *Deep-Sea Res.* 24:701–4

Legeckis, R. V. 1978. A survey of worldwide sea surface temperature fronts detected by environmental satellites. *J. Geophys. Res.* 83:4501–22

Legeckis, R. V. 1979. Satellite observations of the influence of bottom topography on the seaward deflection of the Gulf Stream off Charleston, South Carolina. *J. Phys. Oceanogr.* 9:483–97

Leipper, D. F. 1970. A sequence of current patterns in the Gulf of Mexico. *J. Geophys. Res.* 75:637–57

Leitao, C. D., Huang, N. E., Parra, C. G. 1978. Ocean current surface measurements using dynamic elevations obtained by the GEOS-3 radar altimeter. *J. Spacecraft Rockets* 15:43–49

Leitao, C. D., Huang, N. E., Parra, C. G. 1979. A note on the comparison of radar altimetry with IR and in-situ data for the detection of the Gulf Stream surface boundaries. *J. Geophys. Res.* 84:3969–73

Lipes, R. G., Bernstein, R. L., Cardone, V. J., Katsaros, K. B., Njoku, E. G., Riley, A. L., Ross, D. B., Swift, C. T., Wentz, F. J. 1979. Seasat scanning multichannel microwave radiometer: Results of the Gulf of Alaska workshop. *Science* 204:1415–17

Mather, R. S., Rizos, C., Coleman, R. 1979. Remote sensing of surface ocean circulation with satellite altimetry. *Science* 205: 11–17

Maul, G. A. 1976. Variability in the Gulf Stream system. *Gulfstream* 2 (Oct. 1976): 6–7. Washington DC: NOAA, Natl. Weather Serv.

Maul, G. A. 1977. The annual cycle of the Gulf Loop Current Part I: Observations during a one-year time series. *J. Mar. Res.* 35:29–47

Maul, G. A., Norris, D. R., Johnson, W. R. 1974. Satellite photography of eddies in the Gulf Loop Current. *Geophys. Res. Lett.* 1:256–58

Maul, G. A., deWitt, P. W., Yanaway, A., Baig, S. R. 1978. Geostationary satellite observations of Gulf Stream meanders: Infrared measurements and time-series analysis. *J. Geophys. Res.* 83:6123–35

Maxworthy, T. 1979. A note on the internal solitary waves produced by tidal flow over a three-dimensional ridge. *J. Geophys. Res.* 84:338–46

Molinari, R. L., Baig, S., Behringer, D. W., Maul, G. A., Legeckis, R. V. 1977. Winter intrusions of the Loop Current. *Science* 198:505–7

National Aeronautics and Space Administration. 1976. *Landsat Data Users Handbook, Doc. No. 76SDS4258.* Greenbelt, Maryland

National Oceanic and Atmospheric Administration. 1979. *Oceanic and Related Atmospheric Phenomena as Viewed from Environmental Satellites.* Washington DC: NOAA, Off. Oceanic Atmos. Serv. April 1979

NAVOCEANO. 1975. *Experimental Ocean Frontal Analysis.* Washington DC: US Naval Oceanographic Office

Niiler, P. P., Mysak, L. A. 1971. Barotropic waves along an eastern continental shelf. *Geophys. Fluid Dyn.* 2:273–88

Orlanski, I. 1969. The influence of bottom topography on the stability of jets in a baroclinic fluid. *J. Atmos. Sci.* 26:1216–32

Orlanski, I., Cox, M. D. 1973. Baroclinic instability in ocean currents. *Geophys. Fluid Dyn.* 4:297–332

Osborne, A. R., Burch, T. L. Scarlet, R. I. 1978. The influence of internal waves on deep-water drilling. *J. Petrol. Tech.* Oct. 1978:1497–504

Philander, S. G. H. 1976. Instabilities of zonal equatorial currents. *J. Geophys. Res.* 81:3725–35

Proni, J. R., Apel, J. R., Byrne, H. M., Sellers, R. L., Newman, F. C. 1978. Oceanic internal waves from ship, aircraft and spacecraft: A report on the New York-to-Bermuda remote sensing experiment. *USGPO 1978-796-417-116-10.* Natl. Oceanic Atmos. Admin.

Pullen, P., Byrne, H. M. 1978. Preliminary comparison of Seasat altimetry data and oceanographic data across the Kuroshio. *EOS, Trans. Am. Geophys. Union* 59:1095

Robinson, A. R., Luyten, J. R., Fuglister, F. C. 1974. Transient Gulf Stream meandering, I: An observational experiment. *J. Phys. Oceanogr.* 4:237–55

Sawyer, C., Apel, J. R. 1976. Satellite images of oceanic internal wave signatures. *NOAA S/T 2401,* Sept. 1976. Natl. Oceanic Atmos. Adm.

Stewart, R. H., guest ed. 1978. *Boundary-Layer Meteorol.* 13

Stumpf, H. G., Legeckis, R. 1977. Satellite observations of mesoscale eddy dynamics in the eastern tropical Pacific Ocean. *J. Phys. Oceanogr.* 7:648–58

Stumpf, H. G., Rao, P. K. 1975. Evolution of Gulf Stream eddies as seen in satellite infrared imagery. *J. Phys. Oceanogr.* 5: 388–93

Stumpf, H. G., Strong, A. E., Pritchard, J. 1973. Large cyclonic eddies of the Sargasso Sea. *Mariners Weather Log* 17:208–10

Tapley, B. D., Born, G. H., Hagar, H. H., Lorell, J., Parke, M. E., Diamante, J. M., Douglas, B. C., Goad, C. C., Kolenkiewicz, R., Marsh, J. G., Martin, C. F., Smith, S. L. III, Townsend, W. F., Whitehead, J. A., Byrne, H. M., Fedor, L. S., Hammond, D. C., Mognard, N. M. 1979. Seasat altimeter calibration: Initial results. *Science* 204:1410–12

Tsai, J. J., Apel, J. R. 1979. Tidally induced shear flow instability as a source of internal waves on the continental shelf. *NOAA Pacific Marine Environ. Lab. Contrib. No. 415*

Walsh, E. J., Uliana, E. A., Yaplee, B. S. 1978. SWH measurement with a high-resolution pulse-limited radar altimeter. *Boundary-Layer Meteorol.* 13:263–78

Whitham, G. B. 1974. *Linear and Nonlinear Waves.* New York: Wiley

Ann. Rev. Earth Planet. Sci. 1980. 8: 343–70
Copyright © 1980 by Annual Reviews Inc. All rights reserved

QUATERNARY DEEP-SEA BENTHIC FORAMINIFERS AND BOTTOM WATER MASSES

✕10135

Detmar Schnitker
Department of Oceanography, University of Maine–Orono, Walpole, Maine 04573

INTRODUCTION

The floor of the ocean constitutes the world's largest single biotope, but also one of the least known. Its vastness, inaccessibility, and the high cost of its exploration have contributed to the slow growth of our understanding of its biota and its abiotic environment. In this apparently unchanging, cold, and dark environment an abundant and species-rich fauna of benthic foraminifers has evolved. The existence of this rich fauna was revealed by the first systematic investigation of the oceans, the CHALLENGER expedition of over 100 years ago. H. B. Brady's report (1884) on the foraminifers of that expedition still represents the high-water mark for thoroughness and comprehensiveness of deep-water foraminiferal investigations and is unmatched in the sheer beauty of its illustrations. Many national oceanographic expeditions followed that of the CHALLENGER and their reports on the foraminifers slowly increased global coverage and added detail to the taxonomy and systematic description of the deep-water species. This "taking of the inventory" failed to produce recognition of biogeographic differentiation, and as late as 1953, Phleger, Parker & Peirson stated in their report on the foraminifers of the Swedish Deep Sea Expedition that "the deep-sea benthonic species are widely distributed in the cores indicating that the conditions influencing them on the ocean floor are essentially uniform."

Several workers, notably Bandy and his collaborators, and F. B. Phleger in this country, Saidova in the USSR, and A. Caralp and M. Pujos-Lamy in France established depth zonation schemes for benthic foraminifers. Most of these zonations proved to be of local significance only, and the existing depth-differentiation, in particular for the deep-water species,

0084-6597/80/0515-0343$01.00

could not be explained by any "significant" change in water or substrate attributes.

Apparently, familiarity with the extreme and manifold differentiation of shallow-water foraminifers that can so readily be associated with differences of the environment had led past investigators to expect similar clear-cut relationships in the deep oceans. Failing to find these, researchers declared the deep-water foraminiferal faunas "cosmopolitan" and then largely left them alone. This attitude is apparently changing. A fresh look has been taken at the deep-water foraminifers and their physical and chemical environment; differentiation and systematic correlations have been detected, however subtle. At the present time only a handful of investigators are pursuing this subject and the number of published reports is no larger.

This review summarizes the accomplished work, drawing upon the small published record as well as upon oral presentations made at scientific meetings, personal communications, and, to a considerable degree, as yet unpublished research results. The work that has been accomplished to date does not constitute much more than an extensive feasibility study; the bulk of the work still remains to be done. This review, therefore, is less a retrospect than a look ahead.

THE DEEP-WATER SPHERE

The 4°C isotherm, as suggested by Bruun (1957), probably serves better than specific depths to separate the abyssal environment from the overlying bathyal zone. In the Atlantic Ocean the 4°C isotherm is most commonly encountered at a depth of about 2000 meters; it is somewhat shallower in the Indian and Pacific Oceans. The deep waters have their origin in high polar latitudes and are primarily driven by differences of density: the thermohaline circulation. Only the North Atlantic, specifically the Norwegian and Greenland Seas, is presently a northern hemisphere locus of deep-water formation. Probably almost one half of all deep water originates there (Worthington 1976). Most of the southern hemisphere deep water is produced in the Weddell Sea, with additional deep water coming from the Ross Sea and minor contributions from other regions of Antarctica (Carmack 1977). These deep waters can be identified readily by their respective temperatures and salinities. As shown in Figure 2, the Antarctic water mass, as it occurs in the Atlantic Ocean, is characterized by very low temperatures (<2°C) and relatively low salinity. The Norwegian Sea Overflow Water, on the other hand, stands out for its high salinity, reflecting the fact that it was formed by cooling of salty North Atlantic Drift Water.

These water masses lose their characteristic "signature" of temperature and salinity only very slowly by some mixing as they flow, either as fast western boundary contour-following currents or as slow bodies of water in the open ocean basins. Salinity and, to a lesser extent, temperature are "conservative" properties of ocean water. However, many chemical species that occur in ocean water are not conservative, particularly those that participate in biological processes. These chemical species are also very useful in identifying water masses and in tracing their movement through the ocean basins. Oxygen, which can be replenished only at the sea surface, is slowly consumed by the deep-sea biota, thus the "older" the deep-water mass, the lower its oxygen content. Conversely, nutrients such as silica, phosphate, nitrite, and also $\sum CO_2$ are accumulated along the course of flow; the "older" the water mass, the higher its dissolved nutrient content. These deep-water masses appear to be rather unvarying features of the deep oceans; no changes have been noted throughout the history of oceanographic investigation. The abyssal environment is the most constant environment in existence: perpetually dark, without seasons, with unvaryingly cold temperatures and unchanging salinities and other chemical or physical attributes. In this unique environment a highly specialized biota evolved of which the foraminifers probably constitute the most abundant and ecologically important component (Bernstein et al 1978).

RECENT DEEP-WATER FORAMINIFERS

Two kinds of benthic foraminifers exist: those with calcium carbonate tests and those that agglutinate their shells from various kinds of foreign mineral matter. The agglutinated specimens are often large and very fragile and have a tendency to fragment during sample preparation so that specimen counts become either impossible or meaningless. The tests of most agglutinating species also disintegrate on the seafloor so that, as already noted by Phleger et al (1953), these species cannot be found much below the top 10 or 20 centimeters of sediment cores. Calcareous foraminifers, on the other hand, are limited to the ocean floor above the carbonate compensation depth. It is therefore desirable that the agglutinated species of foraminifers be reappraised to gain distributional and ecological information on those that appear to be "preservable." The present investigations, being limited to the calcareous foraminifers, essentially stop at the carbonate compensation depth, leaving the vast areas of red clay deposits as *terra incognita.*

As the early work on deep-water foraminifers has shown, nearly all species of this large group are truly cosmopolitan, occurring in all major ocean basins. Only very few exceptions have been noted to date. It is common that 80 to 90 different species of calcareous foraminifers can be identified from relatively small (~30 ml) seafloor samples. Exhaustive search of large samples routinely produces about 140 to 160 species. Replicate sampling of the same site (i.e. within about 1/4 mile, due to ship's drift) may produce an additional 20 species, indicating some small-scale heterogeneity of species distribution on the seafloor. But, as illustrated in Figure 1, a few species usually dominate the fauna, the remainder being quantitatively unimportant. Such frequency distribution is very unusual: the species richness is typical of very stable environments, whereas the great dominance of only one or two species is characteristic of unstable or stressed environments. In order to detect the differentiation that exists in such faunas, quantitative analyses are mandatory.

North Atlantic Faunal Distribution

Two regional studies of deep-water North Atlantic foraminifers have been made to date.

A Q-mode factor analysis (Klovan & Imbrie 1971) of the foraminiferal counts of Phleger et al (1953) by Streeter (1973) resolved the existence of three faunal associations. Figure 2 shows the relationship of these faunal groups to the water masses on the temperature/salinity plot of Worthington & Wright (1970). Faunal association I, dominated by *Osangularia umbonifera*, occurs in samples that were under the influence of Antarctic Bottom Water; faunal association II, characterized by *Planulina wuellerstorfi* and *Epistominella exigua*, occurs in samples from North Atlantic Deep Water influence; and faunal association III, dominated by *Uvigerina peregrina*, occurs in conjunction with relatively warm and saline water that is in large part of Mediterranean origin. The *Osangularia umbonifera* fauna occurs in the deep basins of both the western and eastern North Atlantic,

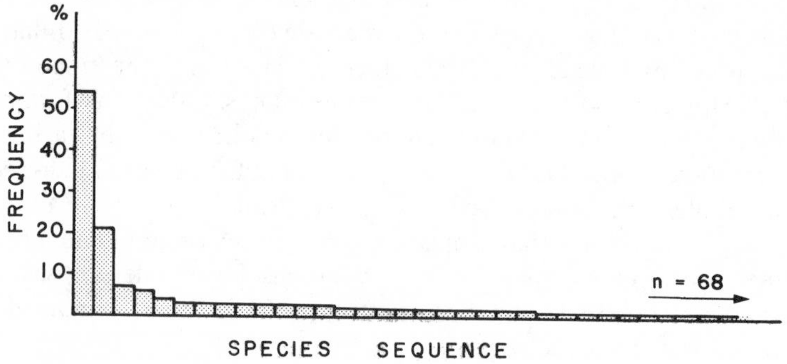

Figure 1 Frequency distribution of foraminiferal species in a typical deep-sea sample (TR111-15, 4430 m).

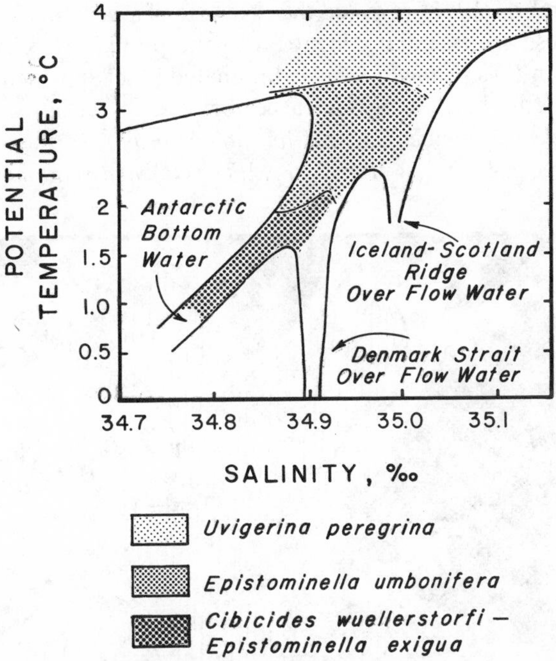

Figure 2 North Atlantic deep-water masses (after Worthington & Wright 1970). Superimposed are temperature-salinity ranges in which three distinct benthic foraminiferal faunas occur (after Streeter 1973).

the *Planulina/Epistominella* fauna on the mid-Atlantic Ridge and the European-African continental margin, while the *Uvigerina peregrina* fauna occurs on relatively shallow bottom, the crest of the mid-Atlantic Ridge, and Rockall Bank.

Schnitker (1974) found a geographically coherent distribution of recurring faunal associations within the western North Atlantic. As shown in Figure 3, a fauna which is dominated by the species *Epistominella exigua* occurs in the northern half of the deep basin, whereas the southern half is occupied by a fauna of which *Osangularia umbonifera* is the dominant species. The dominance of each of the nominal and its accessory species is greatest in the northernmost and southernmost samples, respectively. Towards the center of the deep western North Atlantic basin (~35°N latitude) the distinction between these two faunas is essentially lost. Along the continental margin of North America and along the crest of the mid-Atlantic Ridge a fauna occurs in which *Uvigerina peregrina*, *Hoeglundina elegans*, and *Gyroidina* spp. are variously dominant.

A Q-mode factor analysis of the faunal data, using the program of

Klovan & Imbrie (1971), revealed the existence of four faunal associations. Three of these are essentially identical to the previously selected faunal groupings: an *Epistominella exigua*–dominated fauna in which *Oridorsalis umbonatus* is the most important accessory species; a fauna dominated by *Osangularia umbonifera*, with *Cibicidoides bradyi* as most important accessory species; and a fauna in which *Globocassidulina subglobosa*

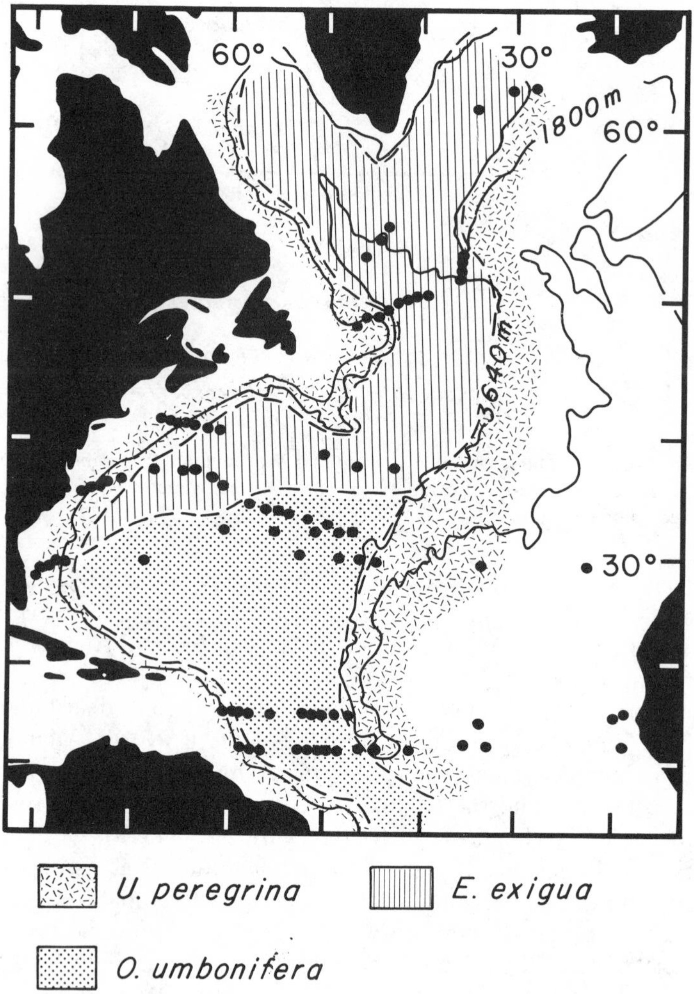

Figure 3 Areal distribution of three associations of Recent foraminifers (after Schnitker 1974).

usually dominates, with *Uvigerina peregrina* as subdominant species. A new grouping of species emerged as a fourth factor in which *Elphidium incertum* dominates, with *Uvigerina peregrina* again as subdominant species. When mapped (Figure 4), these faunal associations essentially duplicate the distribution pattern of the dominant faunas (Figure 3). The important differences are that Q-mode factoring places the division be-

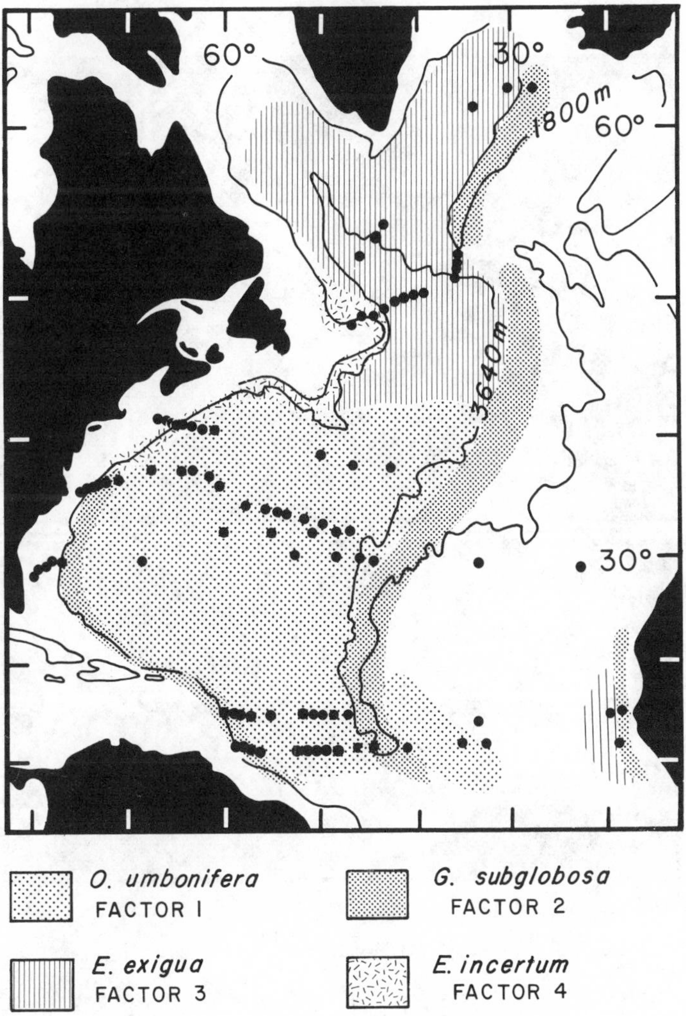

Figure 4 Areal distribution of factor dominance. The four factors resulted from Q-mode factor analysis of foraminiferal counts that were also the basis for Figure 3.

tween the northern *Epistominella exigua* fauna and the southern *Osangularia umbonifera* fauna 10° further north, at about 45°N latitude. The emergence of a separate faunal group, the *Elphidium incertum* fauna, along the continental margin of Canada and New England was expected: *E. incertum* is usually considered to be a shallow-water, even estuarine species and had been excluded as a "contaminant" from the analyses of Schnitker (1974). Finding of many live specimens (= stained with Rose

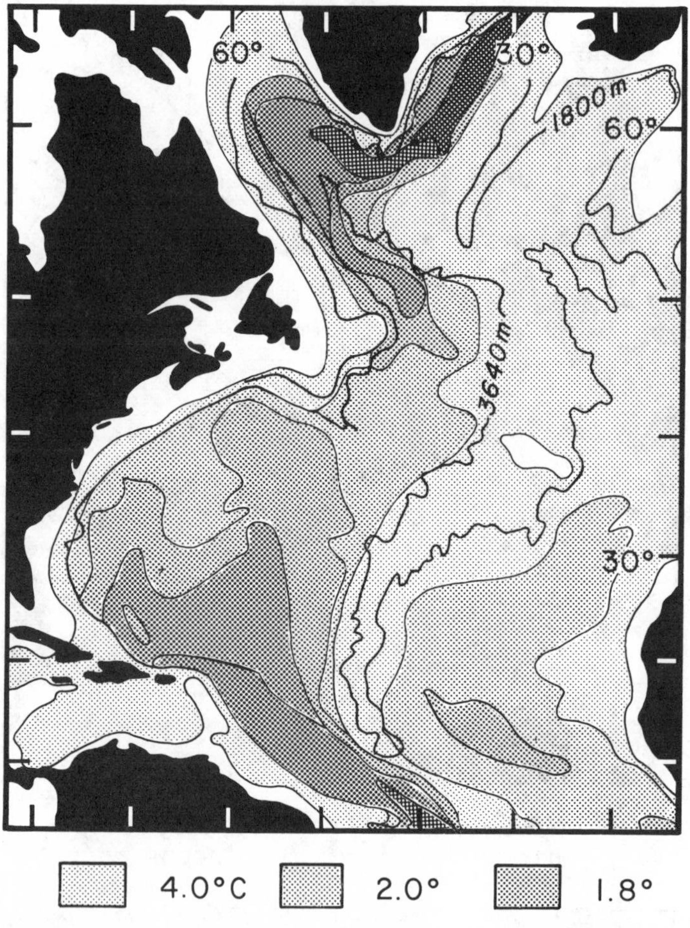

Figure 5 Temperatures of bottom water (after Worthington & Wright 1970) showing the influx of Norwegian Sea Overflow Water and of Antarctic Bottom Water.

Bengal) of this species off the Grand Banks at 840- and 1935-m depth suggests that *Elphidium incertum* may well be part of the deep-sea foraminiferal biota.

These faunal distribution patterns have a striking resemblance to the distribution of the deep- and bottom-water masses (Figure 5) that was described and mapped by Worthington & Wright (1970). The *Epistominella exigua* fauna occurs in areas overlain by Norwegian Sea Overflow Water and approximately within the 1.9°C potential temperature contour. The *Osangularia umbonifera* fauna is best developed within the 1.4°C potential temperature contour of the Antarctic Bottom Water, which enters the area from the south. From the 1.4°C to the 1.9°C potential temperature contours of the Antarctic Bottom Water, the *Osangularia umbonifera* fauna becomes less distinct (*O. umbonifera*–dominance decreases, the abundance of accessory species increases) so that between 40°N and 50°N latitude, exactly where the two bottom-water masses mingle, the two foraminiferal faunas are no longer distinguishable. Overlying these bottom-water masses is the North Atlantic Deep Water with potential temperatures between 2°C and 4°C. Where this water mass is in contact with the seafloor, along the continental margin and along the mid-ocean ridge, are the *Globocassidulina subglobosa/Elphidium incertum/Uvigerina peregrina* faunas. This close association between the three water masses and foraminiferal faunas argues strongly that the relationship is one of cause and effect, that the distribution of the foraminiferal faunas is controlled by the occurrence of distinct water masses.

South Atlantic Faunal Distribution

The studies by Lohmann (1978a) and Gofas (1978), instead of searching for regional distribution patterns of deep-water benthic foraminifers, investigated their distribution over a great depth range in small areas where large differentiation and sharp contrasts between deep-water masses exist. Both authors found that on the slopes of the Rio Grande Rise the most outstanding feature of the benthic foraminiferal faunas is their distinct twofold division: 1. an extraordinary dominance of *Osangularia umbonifera* below about 3500-m to 4000-m depth, and 2. a much more diversified and equitable fauna above. This sharp and fundamental division finds its immediate hydrographic explanation: the Rio Grande Rise intercepts Antarctic Bottom Water on its way north and North Atlantic Deep Water on its way south through the Vema and Hunter Channels. As was the case in the North Atlantic, the *Osangularia umbonifera* fauna again is associated here with the occurrence of Antarctic Bottom Water, while the overlying heterogenous fauna occurs in conjunction with North Atlantic Deep Water. High abundances of *Uvigerina*

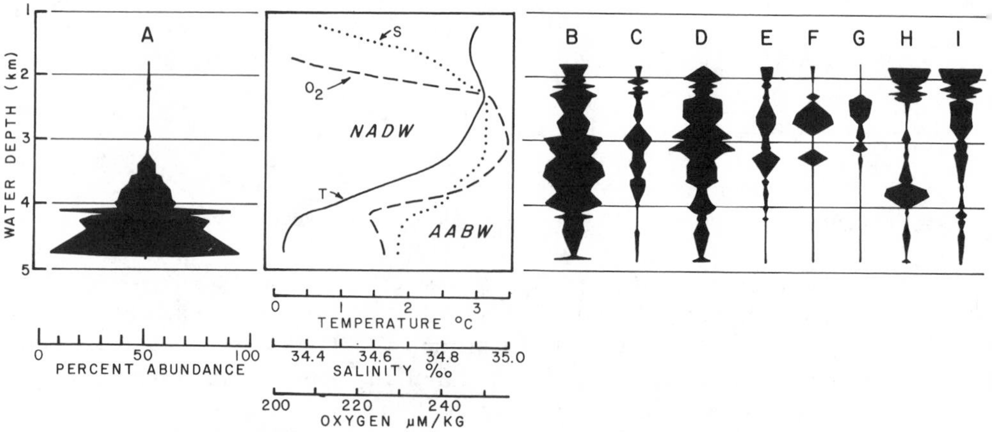

Figure 6 Depth distribution of several common benthic foraminifers and of principal water qualities on the Rio Grande Rise. (*A*) *Osangularia umbonifera*; (*B*) *Planulina wuellerstorfi*; (*C*) *Cibicidoides kullenbergi*; (*D*) Pyrgo ssp.; (*E*) *Quinqueloculina* sp.; (*F*) Nummuloculina irregularis; (*G*) *Hoeglundina* elegans; (*H*) *Uvigerina peregrina*; (*I*) *Globocassidulina subglobosa* (after Lohmann 1978a).

peregrina and also of *Globocassidulina subglobosa* occur above 2000-m depth, clearly associated with Circumpolar Deep Water (Figure 6).

On the Walvis Ridge in the eastern South Atlantic, Gofas (1978) did not detect any significant occurrence of *Osangularia umbonifera* (only 13% at one station at 5060-m depth). However, north of the Walvis Ridge the foraminifers *Pyrgo*, *Planulina wuellerstorfi*, and several others are abundant to below 5000-m depth, whereas the same species become insignificant below 4000-m depth on the Rio Grande Rise. *Uvigerina peregrina* displays a distinct twofold distribution: it is common above 4000-m depth, but on the north flank of the ridge it disappears below that depth whereas on the south flank of the ridge it becomes the dominant species of the foraminiferal fauna below 4000-m depth.

The Walvis Ridge does not lie in the pathway of Antarctic Bottom Water. The Angola Basin to the north of the Walvis Ridge is filled with North Atlantic Deep Water, with only a minor influx of Antarctic Bottom Water through the Romanche Fracture zone of the mid-Atlantic Ridge to the north. The slight increase of *Osangularia umbonifera* abundance at 5060 m at the northern base of Walvis Ridge coincides with the southernmost vestiges of Antarctic Bottom Water in the Angola Basin. The close correlation between water mass and fauna is exemplified by the very deep extension of both the North Atlantic Deep Water and the *Pyrgo* spp./*P. wuellerstorfi* fauna on the northern flank of the Walvis Ridge. The abundance of *Uvigerina peregrina* on the crest of the Walvis Ridge and its dominance on the southern flank coincides with the occurrence there of Circumpolar Deep Water.

Central Pacific Faunal Distribution

The one recent study of seven samples from the central Pacific (Samoa Passage) by Gofas (1978) can be regarded more as a confirmation that the observations made elsewhere hold true for the Pacific as well, rather than as a demonstration of Pacific deep-water foraminiferal distributions.

For the central and northern Pacific, which does not have a northern source of deep and bottom water, the Samoa Passage is a principal conduit for the influx of cold and dense water from the south. The northward-flowing Pacific Deep Water is composed near the bottom mostly of Antarctic Bottom Water, but toward its upper levels contains significant amounts of North Atlantic Deep Water (Reid & Lynn 1971, Hollister et al 1974).

The faint dichotomy in water masses is strongly reflected in the foraminiferal distribution. *Osangularia umbonifera* largely dominates the deeper samples, with *Globocassidulina subglobosa* as subdominant species. While never relinguishing its dominance, *O. umbonifera* and also *G. subglobosa* become less abundant in shallower samples, where *Planulina wuellerstorfi* and *Favocassidulina favus* increase significantly in abundance. No samples from the overlying Pacific Deep Water were studied.

Southeast Indian Ocean Faunal Distribution

Corliss (1979a) studied the foraminifers contained in 64 samples from the southeast Indian Ocean. Q-mode factor analysis of these data revealed the existence of two faunal assemblages in the area (Figure 7). The first faunal assemblage is dominated by *Osangularia umbonifera*, particularly south of the Southeast Indian Ridge where the lower and colder Antarctic Bottom Water is present. *Globocassidulina subglobosa*, the subdominant species of the first faunal assemblage, assumes a larger role in certain samples from north of the Southeast Indian Ridge in apparent conjunction with the upper and slightly warmer Antarctic Bottom Water. The second faunal assemblage, dominated by *Uvigerina* spp. and *Epistominella exigua*, is found on the western Southeast Indian Ridge where it is associated with Indian Bottom Water.

The close tie between bottom-water mass and foraminiferal fauna led Corliss to assume that these water masses were also present in those areas where the respective faunas occur but where data on water characteristics are inadequate. He thus concluded that the band of the *Osangularia umbonifera* fauna, which extends along the northern flank of the Southeast Indian Ridge and thence around the southern flank of

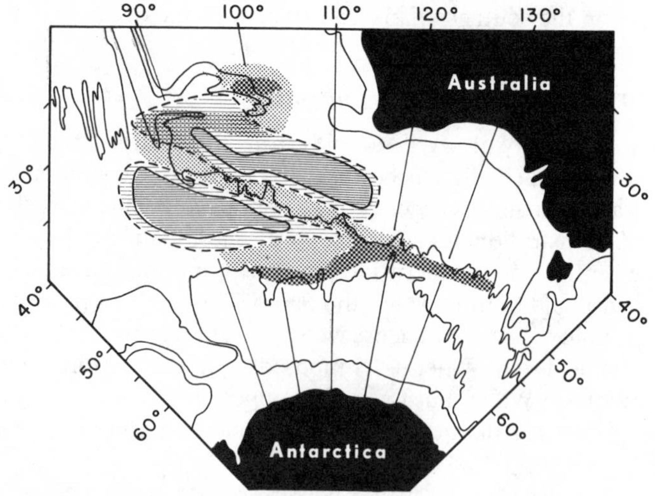

Figure 7 Areal distribution of factor dominance in the southeast Indian Ocean. (*stippled area*) dominance of AABW associated fauna; (*ruled area*) dominance of IBW associated fauna (after Corliss 1979a).

Broken Ridge into the Wharton Basin, is an expression of a contour-following western boundary current of Antarctic Bottom Water. Similarly, the presence of Indian Bottom Water is inferred in the western portion of the south Australian Basin where a well-developed *Uvigerina* spp./*Epistominella exigua* fauna was found.

TOWARD AN ECOLOGICAL EQUATION

Two recurring observations stand out from the previous presentation: 1. wherever Antarctic Bottom Water is present, *Osangularia umbonifera* is likely to be the dominant species of the benthic foraminiferal fauna, and 2. systematic patterns of faunal distribution exist on the deep-sea floor, in apparent correlation with the distribution patterns of deep-water masses.

The first observation leads to the expectation that "index species" can be singled out, whose presence above a certain background level indicates the presence of a specific "water mass."

The second observation leads to the expectation that perhaps for any combination of physical and chemical parameters of the bottom environment there develops a specific combination of benthic foraminiferal species. The alternate side of this expectation is obviously of basic significance to paleoecology, i.e. that a specific faunal composition represents a unique combination of the physical and chemical environment in which it lived. As there are many more species of deep-water benthic foraminifers than there are known and commonly measured environmental attributes, this expectation, from a statistical point of view at least, does not appear unjustified.

A basic prerequisite for such attempts is the standardization of foraminiferal taxonomy and, above all, techniques of investigation. Neither has been achieved to date.

Index Species

A tentative listing of "index species" is given in Table 1. There are many uncertainties associated with these assignments. For example, a technical problem is the use of various sieve sizes by different authors. Schnitker (1974) used the >125-μm fraction, Streeter (1973) and Corliss (1979a,b) the >150-μm fraction, and Lohmann (1978a) and Gofas (1978) the >250-μm fraction for faunal analyses. One of the consequences of sieve size differences may be the different levels of significance that *Epistominella exigua* assumes in the various studies. The average largest diameter of this species lies around 100–120 μm. It is the dominant species in many samples from the western North Atlantic (Schnitker 1974), much less common in the eastern North Atlantic (Streeter 1973) and also the southeast Indian Ocean (Corliss 1979a), and is not at all recorded from the South Atlantic by Lohmann (1978a) and Gofas (1978). A cursory check of one sample from the South Atlantic (D. Schnitker, unpublished) shows the species indeed to be present there. The recordings of other species certainly were affected in similar fashion.

Statistical Analysis

The next question is what aspects of the physical and chemical environment determine, alone or in combination, the success or failure of a foraminiferal species. Powerful statistical techniques are necessary to analyze the large and varied set of data provided by the foraminiferal counts and the environmental variables that accompany them.

The techniques described by Imbrie & Kipp (1971), or some modifications thereof, are the most commonly employed methods of analysis. Briefly, they consist of a Q-mode principal component analysis (factor analysis, if subsequent rotation of principal components is made) of the faunal census data. This usually produces a smaller number of independent composite variables, which correspond to faunal associations. Multiple regression analysis then establishes the correlation between these faunal associations and individual environmental variables and expresses them as "ecological equations." In addition, it is necessary to check the

Table 1 Water mass and dominant species associations

Water mass[a]	Western N. Atlantic	Eastern N. Atlantic	Western S. Atlantic	Eastern S. Atlantic	Central Pacific	Southeast Indian Ocean
AABW	O. umbonifera	O. umbonifera	O. umbonifera	O. umbonifera	O. umbonifera	O. umbonifera
ABW	E. exigua	—	—	—	—	—
lower NADW	E. exigua	P. wuellerstorfi	P. wuellerstorfi	P. wuellerstorfi	—	—
upper NADW	U. peregrina	U. peregrina	—	—	—	—
CPW	—	—	U. peregrina	U. peregrina	—	—
PDW	—	—	—	—	P. wuellerstorfi	—
IBW	—	—	—	—	—	U. peregrina

[a] Water masses:
AABW: Antarctic Bottom Water
ABW: Arctic Bottom Water
NADW: North Atlantic Deep Water
CPW: Circum Polar Water
PDW: Pacific Deep Water
IBW: Indian Bottom Water

independence or degree of intercorrelation of the environmental variables, because if they are highly covariant, variations of one may not be discriminated from variations of another on the basis of the faunal data.

For example, both Streeter (1973) and Schnitker (1974) noted the abundance of *Uvigerina peregrina* in samples from the upper and relatively warm North Atlantic Deep Water. As a consequence, the "ecological equations" developed for the western North Atlantic (D. Schnitker, unpublished) indicated high temperatures in conjunction with high scores of factors 3 and 4, the *Uvigerina*-rich faunal associations. At the same time, the oxygen content of the western North Atlantic water varies roughly inversely with temperature, but because the range of dissolved oxygen values is small, and the quality of their observations not very high, the "oxygen equation" allowed for only minor variation in the level of dissolved oxygen and had a rather low correlation coefficient.

Applications of these equations led to very erroneous interpretations of fossil communities (Schnitker 1975) which could be rectified only after the covariance of oxygen and *Uvigerina* and the independence of this species from temperature had been established outside of the North Atlantic (Lohmann 1978a, Schnitker 1979).

Lohmann (1978a) found that in the Rio Grande Rise area the nine variables of depth, temperature, salinity, oxygen, silica, PO_4, NO_3, alkalinity, and CO_2 were strongly intercorrelated. The correlation between these environmental variables and the principal components of the foraminiferal fauna (Figure 8) permitted resolution of temperature, O_2, and salinity as one group and alkalinity, silica, CO_2, PO_4, depth, and NO_3 as another group of variables on the basis of the second principal component, while oxygen was independently correlated with the third principal component. Only two ecological equations could therefore be written from the available data. The first equation estimated temperature in some combination with the correlated variables such as silica or alkalinity. The second equation estimated oxygen content. The precision of the oxygen content estimates increased sharply when separate equations were formulated for the Antarctic Bottom Water and the North Atlantic Deep Water (Figure 9).

Corliss (1979a) found significant correlations between the faunal associations (factors) and several environmental variables. Factor 1 is negatively correlated with salinity and the log benthonic foraminiferal number (a carbonate dissolution indicator). Factor 2 is negatively correlated with oxygen and positively correlated with temperature and salinity. He refrained from determining "ecological equations" because none of the correlations explained more than 50% of the variance of the faunal data. It is probable, however, that exclusion from the analysis of the south Australian Basin and the north flank of the southeast Indian Ridge, where hydrographic data are inadequate, would have tightened the correlations and made the formulation of an ecological equation possible.

Rather unique deep-water conditions exist in the Gulf of Mexico

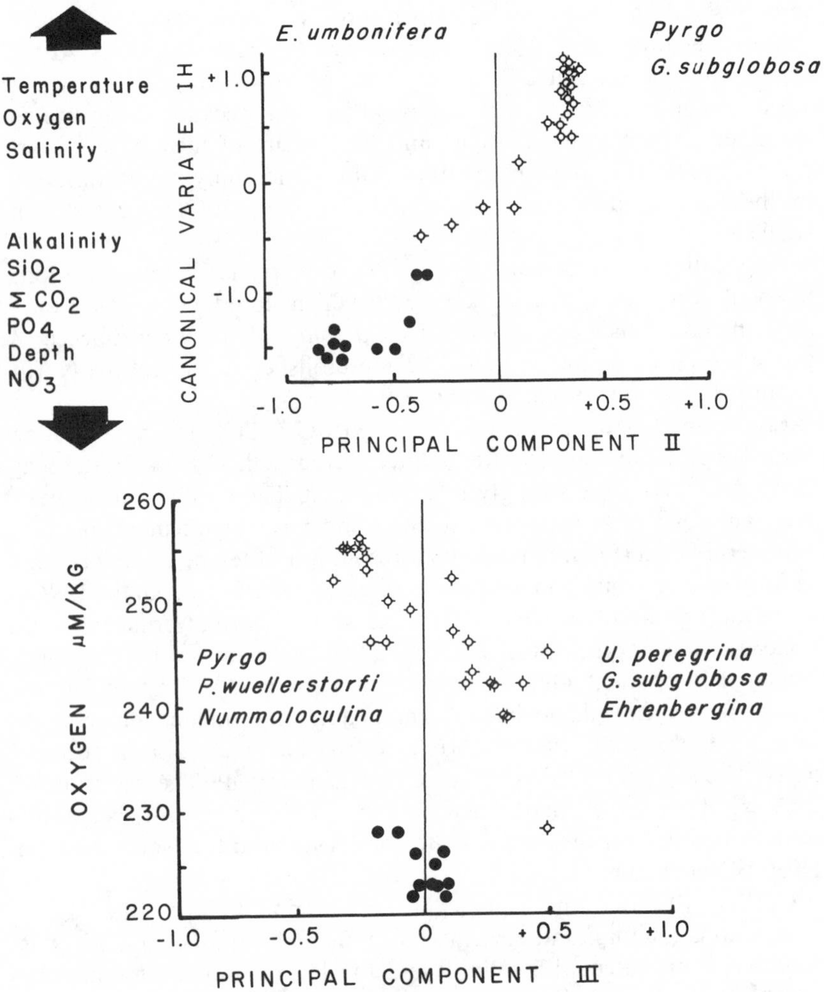

Figure 8 Variations of principal component II in relation to a set of unresolved water properties, and variations of principal component III in relation to dissolved oxygen: (*dots*) samples from AABW covered areas; (*stars*) samples from NADW covered areas (after Lohmann 1978a).

where below the sill depth of about 1500 m, only hardly perceptible gradients of nearly all environmental attributes exist. Dignes (1979), however, has demonstrated the presence of a rich and distinctively differentiated foraminiferal bottom fauna below 1500 m. He could not detect any significant correlation between any of the measured environmental attributes and the existing faunas, which therefore must be sensitive to aspects of the deep-water environment that have either not been measured or whose existence is as yet unsuspected.

Despite the complexity of the deep-sea environment, "ecological equations" can apparently be written. The greatest obstacle that needs to be overcome is the high degree of intercorrelation that exists between so many environmental properties in any one area, and also the existence of environmental properties that have yet to be measured and made available for analysis. Discrimination of individual environmental attributes may be possible because intercorrelations are not the same in the different ocean basins, nor in the different compartments of the same ocean.

Exit Depth Zonation

These findings explain why previous attempts to calibrate the bathymetric occurrence of foraminifers are bound to meet with limited success and that it is not permissible to apply them to paleodepth interpretations. The occurrence of foraminiferal species is tied to water-mass charac-

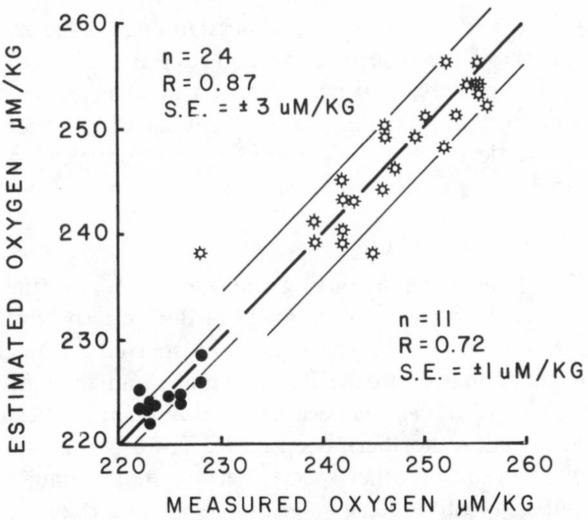

Figure 9 Measured versus estimated oxygen contents: (*dots*) samples from AABW; (*stars*) samples from NADW (after Lohmann 1978a).

teristics, and only insofar as ocean water masses tend to be stratified and widespread will foraminiferal distributions give the impression of stable depth sequencing. However, on a regional or oceanic scale, water masses and, with them, foraminiferal faunas are definitely depth-transgressive.

FROM PALEONTOLOGY TO PALEOCEANOGRAPHY

Benthic foraminifers, together with the much scarcer ostracodes, are the only members of the deep-sea-bottom biota that, upon death or reproduction, leave their shells to be incorporated into the sediment as testimony of their presence and of former living conditions. Assuming that faunal response to ecological conditions has not changed through time, these fossil faunas can be interpreted in terms of past deep-water oceanography. To the extent that the ecological response of these organisms to their environment is properly understood, deep-water paleoceanography may be described in terms of various "water masses" or it may progress to a more detailed description of specific physical and chemical properties of these deep-water masses.

Two modes of analysis and representation are commonly employed in paleontological-paleoceanographic studies, "time-slice" and "time-series." Because of its similarity to the presentation of modern data, time-slice presentation of past faunal distributions is perhaps the more illustrative and intuitively comprehensible of the two, whereas time-series analysis is best suited for an understanding of the evolution of specific conditions. Time-slice analysis requires the availability of a large number of sediment cores and necessitates a great investment of effort in preparatory stratigraphic analysis. It is not surprising, therefore, that most studies of deep-water paleontology and paleoceanography are time-series analyses.

The North Atlantic Trigger

Schnitker (1974) showed that during the late glacial maximum, about 18,000 years ago, the benthic foraminifers of the western North Atlantic formed somewhat different associations and displayed drastically altered distributions as compared to the Recent faunas and their distributions (Figure 10). *Uvigerina peregrina* became the dominant species of a fauna that invaded the entire northern deep basin. Towards the south *Epistominella exigua* and several other species became more significant so that south of about 35°N latitude *E. exigua* and *U. peregrina* were codominant and the mapping of a proper faunal association is warranted. From about 20°N latitude south the importance of *Uvigerina peregrina* decreased while the abundance of *Osangularia umbonifera* increased, creating a new *Epistominella exigua/Osangularia umbonifera* fauna. Approximately 120,000 years ago, during the previous interglacial, the faunal assemblages and their distributions were nearly identical to those of today (Figure 11). Such rearrangement of the nature and distribution of the fauna strongly

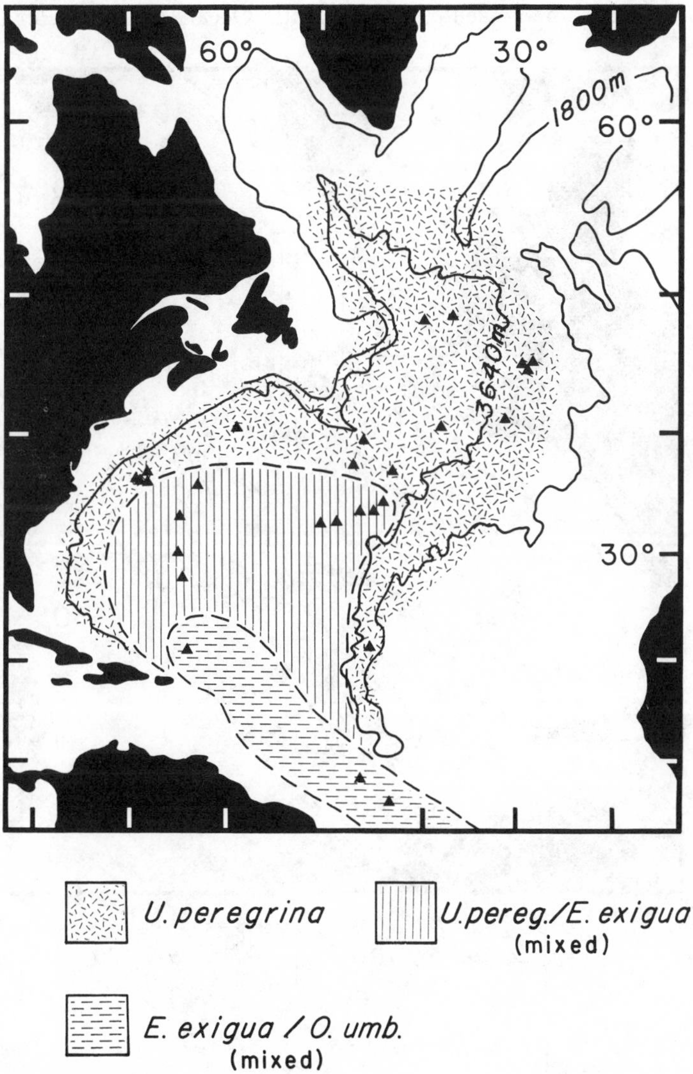

Figure 10 Areal distribution of three associations of late Glacial (18,000 yr B.P.) foraminifers (after Schnitker 1974).

suggests that a change in the nature of the deep and bottom water and a rearrangement of its circulation had taken place. Streeter (1973) had also noted that in cores from the Swedish Deep Sea Expedition the species *Uvigerina peregrina* became very common in glacial-age samples, occasionally even dominant.

Because of the apparent close correlation of *Uvigerina peregrina* and temperature in the present North Atlantic Ocean, both Streeter (1973)

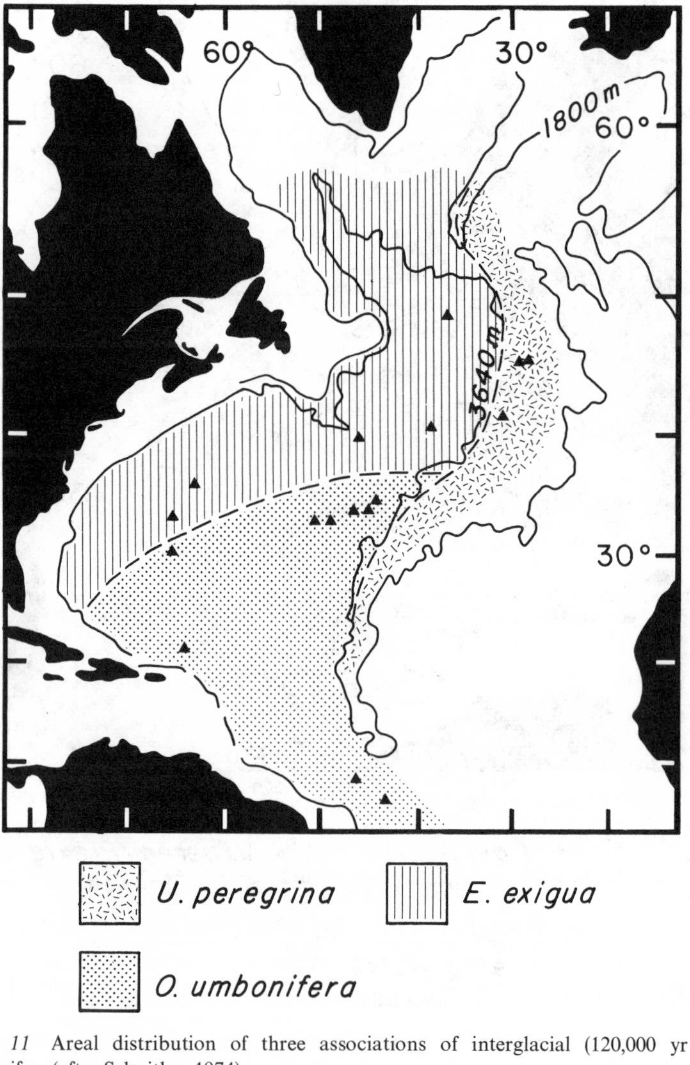

Figure 11 Areal distribution of three associations of interglacial (120,000 yr B.P.) foraminifers (after Schnitker 1974).

and Schnitker (1974) concluded that glacial-age North Atlantic bottom water must have been warm, with Streeter (1973) presenting a temperature estimate of 4°C. These glacial-age faunal distribution patterns and their initial interpretation seemed to be explained in part by Weyl's (1968) hypothesis of glacial-age oceanic circulation changes. As shown in Figure 12, Weyl (1968) proposes that during glacial periods the sites of present-day deep- and bottom-water formation were ice-covered or occupied by relatively low density water, so that glacial bottom water could be formed only from relatively warm but very saline central North Atlantic surface water. Weyl (1968) proposed that such water had a temperature of about 5°C and a salinity of approximately 35.9‰. The presence of the *Osangularia umbonifera*/*Epistominella exigua* mixed fauna in the southwest North Atlantic during glacial times, however, suggested that a modified Antarctic Bottom Water continued to enter the North Atlantic (Schnitker 1974).

When it became apparent that outside of the North Atlantic Ocean *Uvigerina peregrina* was existing abundantly in waters that were much

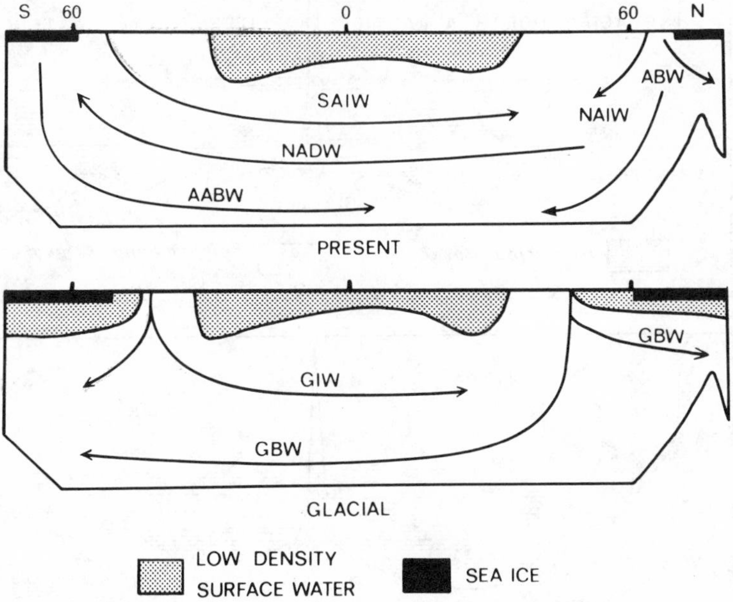

Figure 12 Schematic circulation of the Atlantic Ocean at present and during glacial conditions. AABW = Antarctic Bottom Water, ABW = Arctic Bottom Water, NAIW = North Atlantic Intermediate Water, NADW = North Atlantic Deep Water, SAIW = South Atlantic Intermediate Water, GBW = Glacial Bottom Water, GIW = Glacial Intermediate Water (after Weyl 1968).

colder than the present-day upper North Atlantic Deep Water, Streeter (1976) believed the *Uvigerina peregrina* peaks in his cores to indicate peaks in the flux of organic matter to the seafloor. Schnitker (1976), while still considering that glacial-age North Atlantic bottom water had been relatively warm and saline, suggested that during glacial ages the North Atlantic had not been a site of bottom-water formation and that instead the area was occupied by oxygen-deficient glacial Antarctic Bottom Water. Streeter (1977) showed that this "stagnation" was observable in the entire North Atlantic and in 1979 Streeter & Shackleton demonstrated that in core V29-179 from the mid-Atlantic Ridge the *Uvigerina peregrina* fauna, which indicates "old," oxygen-deficient water, had been present for most of the past 150,000 years. Only during short intervals did *Uvigerina peregrina* disappear, which suggests that formation of deep water again took place in the North Atlantic.

A series of three cores, taken from different depths of the western

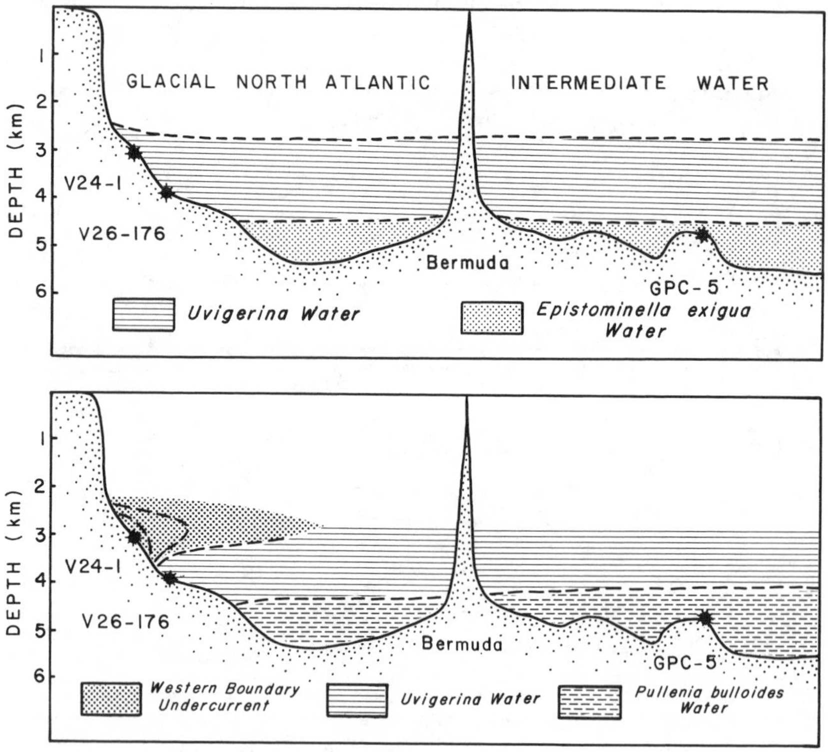

Figure 13 Possible configurations of late glacial North Atlantic deep-water structures: (*a*) about 17,000 yr B.P.; (*b*) 10,000 yr B.P. (after Schnitker 1979).

North Atlantic basin, permitted a detailed time-series study of the foraminiferal faunal changes that occurred there during the past 24,000 years (Schnitker 1979). These faunas indicate that the deep waters of the western North Atlantic were stratified during late glacial as well as during Holocene times. The presence of an *Epistominella exigua*–dominated fauna below 4500 m and of *Uvigerina peregrina*–dominated faunas above suggests that during late glacial times the deepest part of the western North Atlantic basin was filled with cold, well-oxygenated bottom water, probably of local western North Atlantic origin. However, most of the late glacial deep water was "old" and oxygen-deficient, probably of Antarctic origin (Figure 13*a*). Very remarkable is the two-step transition from this glacial mode to the interglacial (present) mode of deep bottom-water distribution. The first basin-wide change occurred about 12,500 years ago: the deepest fauna (*Epistominella exigua*) changed to a *Pullenia bulloides*–dominated fauna, suggesting that this bottom water was no longer renewed and had lost oxygen. The disappearance of *Uvigerina peregrina* from the continental slope and its replacement by a *Globobulimina auriculata*/*Eponides bradyi* fauna at about 3000 m and by an *Epistominella exigua* fauna at 4000 m indicates the return of well-oxygenated water, probably marking the reestablishment of the Western Boundary Undercurrent. However, soon afterwards, "old" oxygen-deficient water reoccupied the 4000-m level (Figure 13*b*). A second abrupt change occurred shortly after 9000 years B.P., marking the return of nearly modern faunas and, presumably, modern deep-water conditions. The deep basin below 4500 m became filled with Arctic Bottom Water as well, as indicated by an *Epistominella exigua*–dominated fauna. Modern Antarctic Bottom Water, as indicated by a dominance of *Osangularia umbonifera*, arrived at the center of the basin less than 5000 years ago.

The first appearance of "new" water on the continental slope 12,500 years ago correlates with the initial retreat of the polar front into the Norwegian Sea. The return of "old" water 2000 years later occurs precisely at the time of a temporary and brief southward readvance of the polar front. The final "flushing" of bottom water, about 9000 years ago, coincides with the final retreat of the polar front into the Norwegian Sea (Ruddiman & McIntyre 1977, Kellogg 1977).

This close correlation between the position of the polar front (and hence the possible locus of bottom-water formation) and the actual arrival of Arctic Bottom Water in the western North Atlantic suggests that the North Atlantic is a very critical and also unstable control point for what is presently one half of the world ocean's deep and bottom water.

The rapidity with which these faunal changes were accomplished approaches the residence time of Arctic Bottom Water in the present North Atlantic Ocean (Stuiver 1976): most changes were 80% complete in less than 700 years, the change at 9000 yr B.P. in core V26-176 took less than 500 years. Because of the dispersing effects of bioturbation, the actual changes will have been faster.

The Southern Domain

In the southwest Atlantic, on the Rio Grande Rise, sedimentation rates are low, so that the cores studies by Lohmann (1978b) did not permit high resolution of a glacial-interglacial sequence of bottom-water change. But on the other hand, they provided a record of change spanning the past 700,000 years.

As expected, the three faunas that are presently associated with the Antarctic Bottom Water, the North Atlantic Deep Water, or the Circumpolar Deep Water were present on the Rio Grande Rise during the past 700,000 years as well. Also as expected, the distribution of these faunas

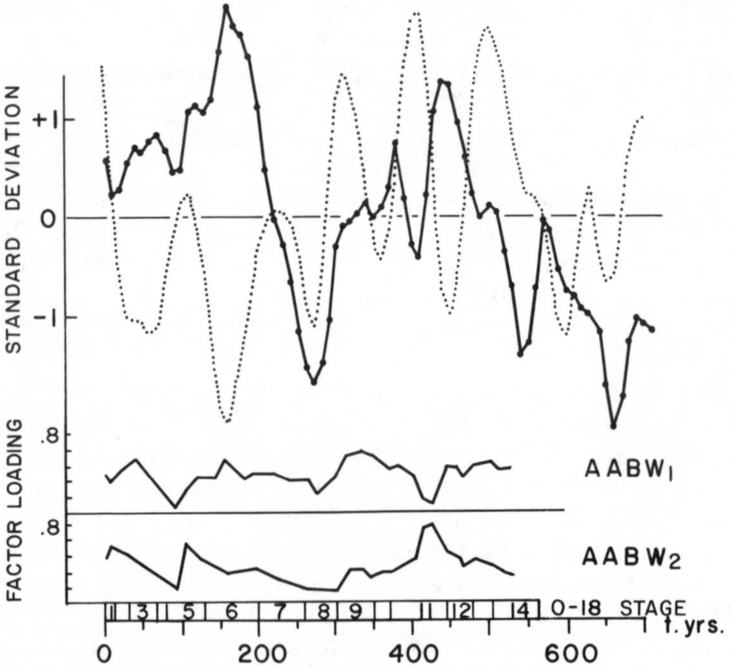

Figure 14 Variations of the significance of Antarctic Bottom Water—associated faunas. Top: Rio Grande Rise (after Lohmann 1978b), bottom: Southeast Indian Ocean (after Corliss 1979b). Carbonate content (*dotted line*) fluctuates synchronously with global surface climate.

showed large variations during the course of time. The really unexpected and perhaps most significant result of this study is the realization that deep-water faunas, and thus deep-water masses, varied quite independently of the glacial-interglacial cycles, that apparently there has been no consistent response of the deep sea to ice ages.

As shown in Figure 14, the significance of the *Osangularia umbonifera* fauna on the Rio Grande Rise and, by inference, the amount of Antarctic Bottom Water within the area has changed rather drastically. While the fluctuations of the calcium carbonate content (*dotted line*) follow closely the glacial-interglacial climatic cycles, the main variations of Antarctic Bottom Water do not. During some glacials (~90,000, 150,000, 450,000 years B.P.) Antarctic Bottom Water covered more of the Rio Grande Rise than it normally does; during other glacials, particularly about 270,000 and 660,000 years ago, much less. Other yet unpublished data (G. P. Lohmann, personal communication, July 1979) indicate that North Atlantic Deep Water was relatively important during the past interglacials, particularly above 4000-m depth; but this water was "turned off" during most glacial periods. Much less regular was the occurrence of Circumpolar Deep Water, which attained great significance during only four of the past seven glacial periods, about 80,000, 270,000, 550,000, and 660,000 years ago.

Four cores have been studied from the southeast Indian Ocean (Corliss 1979b). The benthic foraminifers from these cores indicate that Indian Bottom Water has been present on the Southeast Indian Ridge during the past 300,000 years, and also that Antarctic Bottom Water entered the South Australian Basin during most of the past 500,000 years. As shown in Figure 14, the abundance of those faunas that indicate Antarctic Bottom Water varied strongly, which suggests that the flow of Antarctic Bottom Water across the Southeast Indian Ridge underwent large fluctuations. As in the southwest Atlantic, the changes in Antarctic Bottom Water flow do not coincide with the glacial-interglacial climatic cycles that so dominate surface conditions.

The time-series observations by Gofas (1978) from the central Pacific are twice handicapped: 1. all cores are from below 4500 m where faunal contrasts are apt to be minimal, and 2. sedimentation rates were low so that the three relatively short cores penetrated well into sediments of Pliocene age. Nevertheless, significant changes in the composition of the bottom faunas did occur, particularly in the contribution of the *Planulina wuellerstorfi* fauna at the shallowest site. This fauna may possibly be an indicator of the North Atlantic Deep Water contribution to the Pacific Bottom Water.

Comparison of the indices for Antarctic Bottom Water from the Rio

Grande Rise and from the Southeast Indian Ridge suggests that during most of the past 500,000 years the Antarctic Ocean acted as a single unit. The curve of $AABW_1$ (first Antarctic Bottom Water assemblage) factor scores (Figure 14) indicates the significance on the Southeast Indian Ridge of the *Globocassidulina subglobosa* fauna, which at present occurs there in association with the upper, and warmer, Antarctic Bottom Water, while $AABW_2$ (second Antarctic Bottom Water assemblage) represents the significance of *Osangularia umbonifera* fauna which is closely associated with the lower Antarctic Bottom Water. Of these two, the $AABW_1$ curve matches best the variations in Antarctic Bottom Water on the Rio Grande Rise. Nearly all highs and lows of these two curves match, although not in magnitude. Mismatches occur during oxygen isotope stages nine and ten, about 370,000 to 310,000 years ago. The great peaks of $AABW_2$ during isotope stages 5 and 11 would indicate an extraordinarily high level, and thus rate of production, of Antarctic Bottom Water. No expression of this is noted on the Rio Grande Rise. The South Atlantic maxima of Circumpolar Water at 80,000 and 270,000 years (G. P. Lohmann, personal communication) are also matched by pronounced maxima of Indian Bottom Water.

Search for an Explanation

How did these deep water changes come about? The model of glacial-interglacial modes of deep water circulation that was proposed by Weyl (1968) fails in two ways: the North Atlantic did not provide saline, warm bottom water to the world ocean, nor did the Antarctic Bottom Water formation come to a stop during glacial times as required by the model.

Worthington (1968) suggested that the buildup of continental ice sheets increased the ocean's salinity by 1‰. In principle, this would not have altered the thermohaline circulation except during deglacial periods, when meltwater diluted the ocean's surface water and formed a stable stratification. Deep-water circulation could only resume after geothermal heating had warmed the saline bottom water sufficiently to equalize densities, which may have taken 15,000 years. A "stagnation" of 15,000 years would probably not have rendered the deep oceans anoxic, but certainly would have reduced the oxygen content appreciably. The occasional great preponderance of the *Uvigerina*-rich Circumpolar Water fauna on the Rio Grande Rise and of the equally *Uvigerina*-rich Indian Bottom Water fauna on the Southeast Indian Ridge may indicate times when the ocean's deep-water circulation resembled Worthington's deglacial circulation stage.

For the remaining glacial times, the faunal evidence suggests that Antarctic Bottom Water continued to form, which could have happened only if the loci of formation were not fixed in place and could migrate with the expanding Antarctic ice shelf during glacial periods.

SUGGESTIONS

Understanding how the deep-water circulation behaves and how it evolved is without doubt of greatest importance because of the significant role of the ocean in shaping the world's climate, and of the deep-water circulation in the oceanic mass, nutrient, and energy budget. Any additional study, whether it be a grand overview or a detail of space or time, is bound to make a significant contribution to our sparse knowledge of deep-ocean paleontology and paleoceanography.

ACKNOWLEDGMENTS

Many of the ideas presented here were brought into focus during discussions with Drs. G. P. Lohmann and B. H. Corliss. Dr. G. P. Lohmann kindly provided me with some of his unpublished research results. Dr. Les Watling deserves my gratitude for his critical reading of the manuscript. I thank Patrice Rossi for her help with the drawings and Lois Lane for typing the manuscript. The National Science Foundation gave financial support through grant DES-75-13855.

Literature Cited

Bernstein, B. B., Hessler, R. R., Smith, R., Jumars, P. A. 1978. Spatial dispersion of benthic foraminifera in the abyssal central North Pacific. *Limnol. Oceanogr.* 23:401–16

Brady, H. B. 1884. Report on the foraminifera dredged by H.M.S. *Challenger*, during the years 1873–1876. *Rep. Sci. Res. Voy. Challenger, Vol. 9* (*Zool.*) 814 pp.

Bruun, A. F. 1957. Deep sea and abyssal depth. *Mem. Geol. Soc. Am.* 67(1):641–72

Carmack, E. C. 1977. Water characteristics of the southern ocean south of the Polar Front. *A Voyage of Discovery*, ed. M. Angel, pp. 15–41. Oxford & New York: Pergamon

Corliss, B. H. 1979a. Recent deep-sea benthonic foraminiferal distributions in the southeast Indian Ocean: inferred bottom water routes and ecological implications. *Mar. Geol.* 31:115–38

Corliss, B. H. 1979b. Quaternary Antarctic Bottom Water history: deep-sea benthonic foraminiferal evidence from the southeast Indian Ocean. *Quat. Res.* 12: 271–89

Dignes, T. B. 1979. *Late Quaternary abyssal foraminifera of the Gulf of Mexico.* PhD thesis. Univ. Maine, Orono. 175 pp.

Gofas, S. 1978. Une approche du paléo-environnement océanique: Les foraminifères benthiques calcaires, traceurs de la circulation abyssale. Thèse doct. Univ. de Bretagne Occidentale, Brest, France. 149 pp.

Hollister, C. D., Johnson, D. A., Lonsdale, P. F. 1974. Current-controlled abyssal sedimentation; Samoa Passage, Equatorial West Pacific. *J. Geol.* 82:275–300

Imbrie, J., Kipp, N. G. 1971. A new micropaleontological method for quantitative paleoclimatology: application to a Late Pleistocene Caribbean core. In *The Late Cenozoic Glacial Ages*, ed. K. K. Turekian, pp. 71–181. New Haven: Yale Univ. Press

Kellogg, T. B. 1977. Paleoclimatology and paleoceanography of the Norwegian and Greenland Seas: the last 450,000 years. *Mar. Micropal.* 2:235–49

Klovan, J. E., Imbrie, J. 1971. An algorithm and Fortran IV program for large scale Q-mode factor analysis. *Int. Assoc. Math. Geol. J.* 3:61–78

Lohmann, G. P. 1978a. Abyssal benthonic foraminifera as hydrographic indicators in the western South Atlantic ocean. *J. Foram. Res.* 8:6–34

Lohmann, G. P. 1978b. Response of the deep sea to ice ages. *Oceanus* 21(4): 58–64

Phleger, F. B., Parker, F. L., Peirson, J. F. 1953. North Atlantic Foraminifera. *Rep. Swed. Deep Sea Exped.* 7:1–122

Reid, J. L., Lynn, R. J. 1971. On the influence of the Norwegian-Greenland and Weddell seas upon the bottom waters of the Indian and Pacific oceans. *Deep-Sea Res.* 18:1063–88

Ruddiman, W. F., McIntyre, A. 1977. Late Quaternary surface ocean kinematics and climatic change in the high latitude North Atlantic. *J. Geophys. Res.* 82:3877–87

Schnitker, D. 1974. West Atlantic abyssal circulation during the past 120,000 years. *Nature* 248:385–87

Schnitker, D. 1975. Dynamics of the late Pleistocene thermohaline circulation in the western North Atlantic. *Geol. Soc. Am., Abstr. Programs*, p. 1261

Schnitker, D. 1976. Structure and cycles of the western North Atlantic bottom water, 24,000 yrs. B.P. to present. *EOS, Trans. Am. Geophys. Union* 57:257–58

Schnitker, D. 1979. The deep waters of the western North Atlantic during the past 24,000 years, and the re-initiation of the Western Boundary Undercurrent. *Mar. Micropal.* 4:265–80

Streeter, S. S. 1973. Bottom water and benthonic foraminifera in the North Atlantic—glacial-interglacial contrasts. *Quat. Res.* 3:131–41

Streeter, S. S. 1976. Deep water benthic foraminiferal faunas in the Atlantic during the late Pleistocene—the significance of uvigerinid peaks. *EOS, Trans. Am. Geophys. Union* 57:258

Streeter, S. S. 1977. Deep circulation of the Atlantic Ocean over the last 150,000 years—two contrasting modalities. *Geol. Soc. Am. Abstr. Programs*, p. 1192

Streeter, S. S., Shackleton, N. J. 1979. Paleocirculation of the deep North Atlantic: 150,000-year record of benthic foraminifera and oxygen-18. *Science* 203: 168–71

Stuiver, M. 1976. The ^{14}C distribution in West Atlantic abyssal waters. *Earth Planet. Sci. Lett.* 32:322–30

Weyl, P. K. 1968. The role of the oceans in climatic change: a theory of the ice ages. *Meteorol. Monogr.* 8:63–67

Worthington, L. V. 1968. Genesis and evolution of water masses. *Meteorol. Monogr.* 8:63–67

Worthington, L. V. 1976. *On the North Atlantic Circulation.* Baltimore: Johns Hopkins Univ. Press. 110 pp.

Worthington, L. V., Wright, W. R. 1970. North Atlantic Ocean atlas of potential temperature and salinity in the deep water, including temperature, salinity and oxygen profiles from the Erika Dan cruise of 1962. *Woods Hole Oceanogr. Inst. Atlas Ser. II.* 24 pp., 58 plates

Ann. Rev. Earth Planet. Sci. 1980. 8: 371–406
Copyright © 1980 by Annual Reviews Inc. All rights reserved

RARE EARTH ELEMENTS IN PETROGENETIC STUDIES OF IGNEOUS SYSTEMS

×10136

Gilbert N. Hanson
Department of Earth and Space Sciences, State University of New York, Stony Brook, New York 11794

Introduction

Petrogenetic studies of igneous rocks involve determining the history of the sources of melts, the conditions of melting, the mineralogical and chemical composition of the source during melting, the extent of the melting processes involved, and how the melt is modified by assimilation, metasomatism, differentiation, and fluids. In order to evaluate the importance of each, a detailed knowledge of the geochemistry of systems involving fluids, minerals, and melts would be required. This knowledge, however, is not available. Thus, it is necessary to make inferences based on the information we have. The development of precise analytical techniques for the analysis of the individual rare earth elements (REE) as well as the development of quantitative trace element modeling using single element distribution coefficients (K_d)[1] in mineral-melt systems have made the REE particularly valuable for placing limits on the applicability of proposed petrogenetic models. Realizing the great potential of the REE many laboratories have analyzed wide ranges of rock types and determined K_d's based on natural samples and laboratory experiments. Although many more K_d determinations are needed, K_d's are available for a large number of minerals over a wide range of temperatures and

[1] A mineral-melt K_d is the measured weight or mole fraction of a given element in the mineral divided by the weight or mole fraction of the element in the melt. For precise petrogenetic modeling K_d data should be given in mole fractions because the K_d's in weight fractions will vary with the atomic weight of the melt or phases considered. This can be considerable, for example in a suite consisting of komatiites and basalts. In order to model K_d's in mole fractions, however, it is necessary to have a complete major and minor element analysis.

0084-6597/80/0515-0371$01.00

compositions making it possible to quantitatively model many igneous systems.

The REE are a group of 15 elements with atomic numbers from 57 lanthanum (La) to 71 lutetium (Lu); 14 of these elements occur naturally. The REE with lower atomic numbers are generally referred to as the light REE, those with higher atomic numbers as the heavy REE, and those with intermediate atomic numbers as the middle REE. The REE include those elements that have two electrons in the 6s energy level and one electron in the 5d level; with increasing atomic number the additional electrons are added to the 4f energy level. Lanthanum (La) has none in the 4f level, Cerium (Ce) has one, and lutetium (Lu) has fourteen and a full 4f level. As the 4f energy level fills the ionic radii are systematically reduced, a result called lanthanide contraction (Table 1).

The REE are particularly useful in petrogenetic studies because they are geochemically very similar. Except for Eu and Ce, the REE are trivalent under most geologic conditions, and there is a regular change in the ionic radii for the trivalent REE in octahedral coordination from 1.03 Å for La to 0.861 Å for Lu (Shannon 1976). Thus, a given REE has geochemical characteristics very similar to those of its nearest atomic neighbor, but differing systematically from those of the REE with greater or smaller atomic numbers. Thus the REE have the relatively greatest abundance in minerals with sites preferring an element with a cation radius of about 0.9 to 1.0 angstrom, for example hornblende. If a mineral does not have such a site it will generally have relatively low abundances in the REE, for example, olivine where the cation sites have a radius of much less than 0.9 to 1.0 angstrom and plagioclase where the cation sites have a radius much greater than 0.9 to 1.0 angstrom. Eu is both divalent and trivalent in igneous systems, the ratio depending on the fugacity of oxygen (f_{O_2}). Divalent Eu is geochemically very similar to strontium (Sr). Ce may be tetravalent under highly oxidizing conditions such as during weathering or hydrothermal alteration. Haskin et al (1966, 1968) review the REE geochemistry and the composition of a wide spectrum of rocks.

In order to graphically compare REE for different rocks, it is necessary to eliminate the Oddo-Harkins effect, which is the occurrence of higher concentrations for the elements with even atomic numbers as compared to the concentrations for the elements with odd atomic number. Thus, concentrations for the individual REE are generally normalized to their abundance in chondrites by dividing the concentration of a given element in the rock by the concentration of the same element in chondritic meteorites. This smooths out the concentration variations from element to element. Chondrites have been used because they are primitive solar material which may have been the parental material of the earth. Laboratories use different values for normalizing. Two sets of values are given in Table 1. In general the values used are within a range of about 25%, which reflects the variations within chondrites as well as analytical bias and uncertainties (see, for example, Evensen et al 1978 and Masuda et al 1973). In a cogenetic sequence of rocks it may be advantageous to normalize the REE to one of the rocks in the sequence rather than to chondrites. A sample closest to the parent melt composition would be best to use for normalizing.

The development of the Sm-Nd dating technique and the application of initial Nd isotope ratios in a manner similar to that for initial Sr and Pb isotopes allows an estimate to be made of the long-term Sm-Nd ratio

Table 1 Two sets of chondrite normalizing values and octahedral ionic radii for the trivalent rare earth elements (REE)

Atomic number	Element		Chondrite normalizing values[a] (ppm)		Ionic radii[a] (Å)
			1	2	3
57	Lanthanum	La	0.315	0.330	1.032
58	Cerium	Ce	0.813	0.88	1.01[b]
59	Praseodymium	Pr	—	0.112	0.99
60	Neodymium	Nd	0.597	0.60	0.983
61	Promethium[c]	Pm	—	—	—
62	Samarium	Sm	0.192	0.181	0.958
63	Europium	Eu	0.0722	0.069	0.947[d]
64	Gadolinum	Gd	0.259	0.249	0.938
65	Terbium	Tb	—	0.047	0.923
66	Dysprosium	Dy	0.325	—	0.912
67	Holmium	Ho	—	0.070	0.901
68	Erbium	Er	0.213	0.200	0.890
69	Thulium	Tm	—	0.030	0.880
70	Ytterbium	Yb	0.208	0.200	0.868
71	Lutetium	Lu	0.0323	0.034	0.861

[a] Column identification:
1. REE values for Leedey chondrite (Masuda et al 1973) divided by 1.20. These analyses were by isotope dilution.
2. REE values for a composite of nine chondrites, Haskin et al (1968). These analyses were by neutron activation analysis.
3. Ionic radii in angstroms for octahedral coordination of trivalent REE from Shannon (1976).

[b] Ce^{4+} has an octahedral ionic radius of 0.87 Å.

[c] Promethium has no stable isotopes and does not occur in nature.

[d] Eu^{2+} has an octahedral ionic radius of 1.17 Å comparable to that of Sr^{2+} with an ionic radius of 1.18 Å.

in the source of igneous rocks. O'Nions et al (1979) have reviewed the application of Nd isotopes to interpretation of geochemical systems.

Rather than presenting a review of the REE patterns found in a wide range of rock types, the main emphasis here is the calculation of REE patterns for petrogenetic interpretations. The following sections present the basic concepts for trace element modeling using single-element K_d's; a discussion of the mineral-melt K_d's for REE; and calculated REE patterns for melts and solids that result by melting of the mantle, melting of basalts, melting of continental rocks, differentiation, assimilation and interaction with carbonate melts and H_2O and CO_2 fluids, as well as calculated patterns for coexisting immiscible liquids.

Distribution Coefficients

Since the middle 1960s there has been a shift from using trace elements as "finger prints" in comparative studies to the use of trace elements in petrogenetic studies where the main emphasis is the origin and evolution of a suite of igneous rocks. This advance has been made possible through the use of single element distribution coefficients (K_d) in quantitative models. The K_d's for a given element are dependent on the temperature, pressure, and composition of the mineral and melt. The ideal temperature dependence of a K_d for a given element, holding all other variables constant, is (McIntire 1963)

$$\ln K_d = (C_1/T) + C_2 \qquad (1)$$

where C_1 and C_2 are constants. C_1 is a function of the difference in free energies of the standard states of the trace element in the two phases and C_2 is a function of the logarithm of the ratio of the activity coefficients for the element in the two phases (Banno & Matsui 1973). Generally the mineral/melt K_d's increase with decreasing temperature. Although the mineral/melt K_d's are also dependent on pressure, the dependence is apparently not as great as it is for temperature. The compositional dependence of mineral/melt K_d's is related to the structures of both the minerals and melts. Until adequate models for the structures of melts are available, however, it is not possible to quantitatively evaluate this dependence. Theoretical derivations of quantitative modeling and approaches to application of trace elements in igneous systems have been presented, for example, by Neumann et al (1954), Schilling & Winchester (1967), Gast (1968), Shaw (1970, 1978), Greenland (1970), Albarede & Bottinga (1972), Treuil & Varet (1973), Paster et al (1974), Hertogen & Gijbels (1976), Arth (1976), Allegre et al (1977), Minster et al (1977), Langmuir et al (1977, 1978), Allegre & Minster (1978), and Hanson (1977, 1978).

Equations for Melting

Unless melting takes place in an inert container such as in the laboratory, all melting of rocks is partial melting. This is because in a homogeneous, multiphase system containing a zone of complete melting and a zone with no melting there must also be a zone of partial melting and the derived melt will reflect an intermediate extent of melting for the system. If mixing or diffusion homogenizes the melt prior to separation, the melt will reflect the average extent of melting. For consideration of the effects of diffusion during melting see Hanson (1978) and Magaritz & Hofmann (1978).

There are two end member models for melting: batch melting and fractional fusion. In *batch melting* or *equilibrium fusion* (Presnall 1969) the trace elements in the melt and solid residue are in equilibrium as melting proceeds until enough melt accumulates so that it can migrate away from the zone of melting, melting then ceases. In *fractional melting* or *fractional fusion*, as melting takes place small fractions are continuously and completely removed as melting proceeds (Bowen 1928, Presnall 1969). *Continuous melting* is an intermediate process in which fractions of melt are continuously, but not completely, removed from the solid phases as melting proceeds (Langmuir et al 1977). *Zone refining* is a melting process in which a melt passes pervasively through a solid and in which each of the individual crystals through which the melt has passed has interacted and equilibrated with the melt (Harris 1974).

BATCH MELTING During batch melting simple mass balance requires that for a given element

$$C_0 = F\,C_L + (1-F)\,C_s \tag{2}$$

where F is the fraction of melt, C_0 is the initial concentration of the element in the system, C_L is the concentration of the element in the melt, and C_s is the concentration of the element in the solid residual phases.

Given that

$$D = \sum X\,K_d = C_s/C_L \tag{3}$$

where D is the bulk solid-melt distribution coefficient at the time of removal of the melt, and X is the fraction of each of the minerals in the solid residue, we can then substitute (3) into Equation (2). Rearranging gives the batch melting equation (Schilling 1966)

$$C_L/C_0 = 1/(D(1-F)+F). \tag{4}$$

This formulation shows that at the time of removal of a melt from a solid residue the concentration of a given element in the melt is dependent only on C_0, D, and F. Because it is an equilibrium process, the resulting concentration in the melt is not dependent on the path of melting, and so Equation (4) is also applicable to equilibrium crystallization. Thus, for batch melting, if a mineral was originally present during melting but has subsequently melted or reacted out, it has no influence on the concentration of that element in the melt or solid phases. Shaw (1970) and Hertogen & Gijbels (1976) present derivations that take into account variations in K_d's and mineral proportions during melting. However, use of these equations requires knowledge of the phase equilibria and K_d's as a function of T, P, and composition during melting. Since this information is poorly known, it is more appropriate to consider only the possible parent composition, the mineral content of the residue at the time of melt removal, and the extent of melting, which can be done by using Equation (4). The trace element concentration in the solid residue at the time of removal of the melt is given from Equation (3) where $C_s = DC_L$.

The ratio C_L/C_0 more closely approaches $1/D$ as F decreases. As D approaches zero, C_L/C_0 approaches $1/F$ and approaches it more closely as F increases.

FRACTIONAL FUSION For fractional fusion Equation (4) may be used incrementally by giving new values for D and C_0 for each increment of melting. C_0 would be the value of C_s for the previous melting increment. Shaw (1970) derived a continuous relation for fractional fusion for modal and non-modal melting. During modal melting (Equation 5) the minerals melt in the same proportion as they occur in the solid. During non-modal melting the minerals melt in a proportion different from that in which they occur in the original solid (Equation 6). Melting is generally non-modal.

$$C_L/C_0 = (1/D)(1-F)^{[(1/D)-1]} \tag{5}$$

$$C_L/C_0 = (1/D_0)(1-PF/D_0)^{[(1/P)-1]} \tag{6}$$

In Equation 6, D_0 is the value for D for the first melt and P accounts for the proportion of minerals which go into the melt for non-modal melting.

$$P = \sum p\,K_d \tag{7}$$

where p is the proportion of a given mineral in the melt.

CONTINUOUS MELTING For continuous melting Equation (4) may be used by calculating a new C_0 and giving new values of D after each increment.

$$C_0 = [(1-F)\,C_s + (F - Y\,F)\,C_L]/(1 - Y\,F) \tag{8}$$

where F is the extent of melting for each increment of melting, Y is the proportion of F that escapes, and C_s and C_L are the concentrations from the previous increment of melting. For example, if each increment of melting represents 4% melting, half of which remains in the residue, a new C_0 for each increment is given by

$$C_0 = (96/98)\, C_s + (2/98)\, C_L.$$

Since melting will generally be non-modal, it is necessary to continually change the mineralogic composition of the residue.

ZONE REFINING For zone refining Harris (1974) gives

$$C_L/C_0 = (1/D) - (e^{-DN})[(1/D)+1] \qquad (9)$$

where N is the number of equivalent volumes of solid through which the given volume melt has passed. For example, if the melt passes through a volume of solid five times that of the melt, $N = 5$. Zone refining produces melts whose trace element patterns are similar to those produced by batch melting. The steady state trace element concentration for any trace element during zone refining occurs when a relatively large number of equivalent volumes of solid have melted, i.e. N is large and e^{-DN} approaches zero. As can be seen from Equation (9) the steady state concentration is $1/D$. As for batch melting the limit of enrichment for any trace element with D less than one is $1/D$.

Equations for Crystallization

There are two end members for crystallization: equilibrium crystallization and fractional crystallization. During equilibrium crystallization the melt and crystals continually re-equilibrate, whereas in fractional crystallization as soon as an infinitesimal fraction of crystals form they are removed from the melt and will not re-equilibrate with any later melt. Because most natural systems cool relatively rapidly, which does not allow equilibration between the minerals and melt, fractional crystallization is probably a more realistic model for most crystallizing bodies. However, the actual process may be intermediate to these two end members. Equilibrium crystallization is described by Equation (4).

During fractional crystallization the Rayleigh fractionation law (Rayleigh 1896) as applied by Neumann et al (1954) can be used to describe the concentration of a given trace element in the melt, C_L, relative to the parent melt C_0.

$$C_L/C_0 = F^{(D-1)} \qquad (10)$$

where F is the fraction of melt left and D is the bulk distribution coefficient for the minerals settling out of the melt. Greenland (1970) presents equations that take into account variations in the proportions of minerals or variations in K_d as a function of temperature and composition. Equation (10) can also be used incrementally to take into account the variations in mineral proportions and K_d's. Unlike batch melting, where the former presence of a mineral has no influence on the composition of the melt, during fractional crystallization the composition of a derived melt depends on the presence, proportions, and sequence of crystallization of any and all phases present during the crystallization process. For fractional crystallization, as D approaches zero, C_L/C_0 approaches $1/F$. If D is larger than 1, fractional crystallization rapidly depletes that element from the melt.

For crystallization or batch melting it is possible to determine the extent of the process by selecting a trace element that is suspected of having a low D (i.e. an incompatible element), because as D approaches zero, $C_L/C_0 = 1/F$ for both processes, which gives the maximum allowable extent of melting or the maximum fraction of melt remaining due to fractional crystallization.

It is possible to distinguish whether a suite of cogenetic rocks is related by different extents of melting or fractional crystallization by considering those trace elements with D's significantly greater than 1. If the concentration remains relatively constant throughout the suite, the process is different extents of melting. If the element with a large K_d shows a strong depletion consistent with greater extents of fractional crystallization, the process is fractional crystallization. Within a suite of rocks variations in both processes may be expected to be important and discernable.

Types of Elements in Mineral-Melt Systems

Hanson & Langmuir (1978) proposed for quantitative modeling in mineral-melt systems a classification of elements into trace elements, essential structural constituents (ESC), and intermediate elements. The classification is based on how the elements are distributed among the phases, as determined by the stoichiometric constraints that the individual phases place on the distribution and the types of solution in the various phases.

A trace element is an element whose concentration is so low that there are no stoichiometric constraints on its abundance in any of the phases in the system. It is in ideal solution or follows Henry's law or the activity coefficients vary systematically in the phases involved. Thus, for the given conditions, the K_d for a trace element does not vary over the range of abundances considered. Unless a mineral occurs that has the REE as a major component in a given mineral-melt system, the REE can generally be treated as trace elements.

An essential structural constituent (ESC) is an element that totally fills one site in any one of the phases in the system. For any other phase in the system, the concentration of the ESC is fixed and will vary only if the K_d varies. Because the REE have very similar geochemical properties it would be highly unlikely that any one REE would be an ESC in a phase in a mineral system.

An intermediate element has characteristics between those of a trace element and an ESC. An intermediate element is abundant enough in one of the phases in the system so that there are stoichiometric constraints. The abundance of an intermediate element in a solid phase can be calculated using K_d's, but the abundance can be varied only with the limits of the stoichiometric constraints. That is, you cannot allow the concentration of elements in a site to be greater than the site can hold. For non-ideal solutions, an element may be of low enough abundance in the phases so that it is not constrained stoichiometrically, yet it may not fulfill the solution criteria for trace elements and K_d will vary as the abundance of the element varies.

When attempting petrogenetic modeling it is extremely important to understand the constraints placed on an element. In order to treat an element as a trace element (i.e. K_d is a function of T, P, and the composition of the system, but not the abundance of the element) the element cannot be an ESC or intermediate element in any phase in the system. If in any phase in the system the element is an ESC, the element considered must be modeled as an ESC. If the element is not an ESC, but is an intermediate element in any phase, it must be modeled as an intermediate element. The concentration of all intermediate elements occupying the same site in a phase must be calculated together so that the site is fully occupied and the stoichiometric constraints are satisfied.

Mineral-Melt K_d*'s for REE*

Although the values of the mineral/melt K_d's for the REE for a given mineral vary widely as a function of temperature or composition, in general the ratios of the elements, which determine the shape of the pattern for the K_d's for a given mineral, are consistent (Philpotts 1978). A given mineral will have a characteristic effect on the REE pattern for a melt which allows identification of that mineral during melting or fractional crystallization, although it may not be possible to calculate the fraction of that mineral in the residue. For most igneous minerals the mineral/melt K_d's increase with decreasing T and are greater for rhyolitic compositions, whose K_d's are greater than for dacitic compositions, whose K_d's are greater than those for basaltic composition. The K_d's determined on natural systems (i.e. phenocryst/matrix or phenocryst/glass K_d's) reflect a range in temperature and the effects of diffusion and disequilibrium

(Albarede & Bottinga 1972, and Hart & Brooks 1974). In modeling a sequence of rocks it is necessary to use K_d's, whether based on natural or experimental systems, that are appropriate for the compositions of the rocks considered and the conditions under which the processes took place. In the study of a differentiated sequence of volcanics, it is advantageous to determine and use the phenocryst/matrix K_d's within those volcanics for modeling. However, great care is necessary in using the abundances of trace elements in phases in plutonic rocks for K_d's. If the phase of interest is definitely a liquidus phase of low abundance and forms phenocryst in a fine-grained matrix it might be useful to use the mineral/matrix ratio for a given trace element as a K_d. This approximation may not be appropriate for relatively equigranular holocrystalline rocks because the composition of a given phase may depend on where in the paragenetic sequence it crystallized. This is because in a small closed system the melt will continually change composition as the rock crystallizes and the minerals which crystallize later in the paragenetic sequence will reflect the changing composition of the melt.

Arth & Hanson (1975) reviewed the K_d's based on natural systems and Irving (1978) reviewed the K_d's based on experimental systems. Due to extensive experimental studies under way more data are continually becoming available which will allow more constraints during modeling.

In this section individual minerals are considered with respect to how they qualitatively affect the concentration of the REE and the shape of the REE pattern in the melts. The magnitude of the effect of a mineral is directly related to both the relative abundance of the mineral and the magnitude of the K_d's. Table 2 and Figure 1 give some typical mineral-melt K_d's for the REE.

Feldspar The feldspars have low K_d's for the REE and large positive europium (Eu) anomalies. The plagioclase/melt K_d's for the REE other than Eu are not strongly dependent on temperature or composition (Drake & Weill 1975). The plagioclase Eu anomaly decreases with increasing f_{O_2} and increasing T, but will be significant in almost all known terrestrial igneous systems (Drake 1975). Feldspar has but a minor effect on the REE pattern of the melt, except the large positive Eu anomaly in the K_d pattern will contribute to a negative Eu anomaly in the melt.

Garnet Garnet has very low K_d's for the light REE and increasingly larger K_d's for the heavy REE; the K_d's decrease by an order of magnitude from rhyolitic to basaltic systems. The presence of garnet leads to depletion of the heavy REE and generally contributes to a positive Eu anomaly in the melt.

Table 2 Mineral–silicate melt K_d's used in calculations[a]

	Melting of mantle				Melting of basalts						Melting of sediments								
	1 CPX	2 OPX	3 OL	4 GAR	5 AP	6 PLAG	7 CPX	8 GAR	9 HBLDE	10 SPH	11 OR	12 OP	13 CPX	14 HBLDE	15 AP	16 ZR	17 PLAG	18 BIOT	19 GAR
Ce	0.098	0.0030	0.0005	0.021	16.6	0.20	0.30	0.35	0.899	53.3	0.044	0.15	0.50	1.52	34.7	2.64	0.27	0.32	0.35
Nd	0.21	0.0068	0.0010	0.087	21.0	0.14	0.65	0.53	2.80	88.3	0.025	0.22	1.11	4.26	57.1	2.20	0.21	0.29	0.53
Sm	0.26	0.010	0.0013	0.217	20.7	0.11	0.95	2.66	3.99	102	0.018	0.27	1.67	7.77	62.8	3.14	0.13	0.26	2.66
Eu	0.31	0.013	0.0016	0.320	14.5	0.73	0.68	1.50	3.44	101	1.13	0.17	1.56	5.14	30.4	3.14	2.15	0.24	1.50
Gd	0.30	0.016	0.0015	0.498	21.7	0.066	1.35	10.5	5.48	102	0.011	0.34	1.85	10.0	56.3	12.0	0.097	0.28	10.5
Dy	0.33	0.022	0.0017	1.06	16.9	0.055	1.46	28.6	6.20	80.6	0.006	0.46	1.93	13.0	50.7	45.7	0.064	0.29	28.6
Eu	0.30	0.030	0.0015	2.00	14.1	0.041	1.33	42.8	5.94	58.7	0.006	0.65	1.66	12.0	37.2	135	0.055	0.35	42.8
Yb	0.28	0.049	0.0015	4.03	9.4	0.031	1.30	39.9	4.89	37.4	—	0.86	1.58	8.4	23.9	270	0.049	0.44	39.9

[a] Column identification:

1. Clinopyroxene-melt K_d's average of 2 from Grutzeck et al (1974).
2. Orthopyroxene-melt K_d's determined by using clinopyroxene-orthopyroxene K_d's from peridotite nodule GSFC 20 Philpotts et al (1972) and clinopyroxene-melt K_d's from 1.
3. Olivine-melt K_d's are estimated using clinopyroxene-olivine K_d's and clinopyroxene-melt K_d's from 1.
4. Garnet-melt K_d's from Shimizu & Kushiro (1975).
5. Apatite-melt K_d's from Nagasawa & Schnetzler (1971).
6. Plagioclase-melt K_d's from Schnetzler & Philpotts (1970).
7. Clinopyroxene-melt K_d's from Schnetzler & Philpotts (1970).
8. Garnet-melt K_d's from Schnetzler & Philpotts (1970).
9. Hornblende-melt K_d's from Arth & Barker (1976).
10. Sphene-melt K_d's from Simmons & Hedge (1978).
11. Orthoclose-melt K_d's from Schnetzler & Philpotts (1970).
12. Hypersthene-melt K_d's average of four from Nagasawa & Schnetzler (1971) as reported in Arth & Hanson (1975).
13. Clinopyroxene-melt K_d's average of two from Nagasawa & Schnetzler (1971) and Schnetzler & Philpotts (1970) as reported in Arth & Hanson (1975).
14. Hornblende-melt K_d's average of three from Nagasawa & Schnetzler (1971) as reported in Arth & Hanson (1975).
15. Apatite-melt K_d's average of four from Nagasawa (1970) as reported in Arth & Hanson (1975).
16. Zircon-melt K_d's average of two from Nagasawa (1970) as reported in Arth & Hanson (1975).
17. Plagioclase-melt K_d's average of four from Nagasawa & Schnetzler (1971) as reported in Arth & Hanson (1975).
18. Biotite-melt K_d's from Higuchi & Nagasawa (1969).
19. Garnet-melt K_d's from Schnetzler & Philpotts (1970).

Low-Ca *pyroxene* Hypersthene and pigeonite generally have K_d's which are less than one and strongly dependent on composition. The presence of low-Ca pyroxene may lead to slightly more enrichment of the light REE than the heavy REE in the melt and may contribute to a positive Eu anomaly.

High-Ca *clinopyroxene* The K_d's for high-Ca clinopyroxene show a strong dependence on composition, but are generally not much greater than one. The presence of clinopyroxene leads to more enrichment of the light than middle or heavy REE and contributes to a positive Eu anomaly in the melt.

Hornblende The K_d's for hornblende show a strong dependence on composition and may be greater than 10 for the middle REE in rhyolitic systems. The presence of hornblende will lead to depletion of the middle REE, less so the heavy REE, and contribute to a positive Eu anomaly in the melt.

Olivine The K_d's for olivine are so low that the presence of olivine leads to essentially equivalent enrichment of all of the REE.

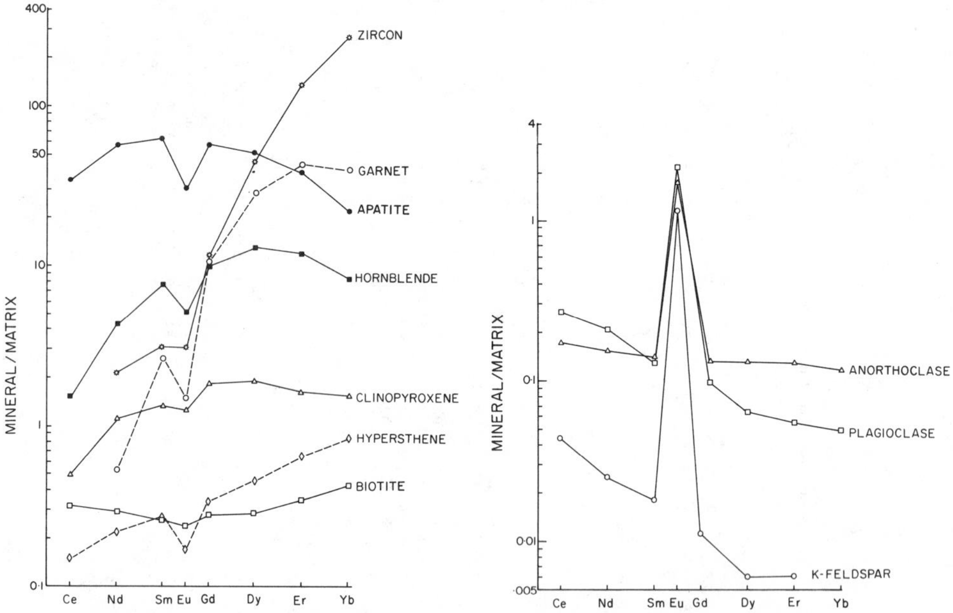

Figure 1 Mineral-melt K_d's for REE from dacites and rhyolites from Hanson (1978). These are the average values from Arth & Hanson (1975), which were selected from Nagasawa & Schnetzler (1971), Higuchi & Nagasawa (1969), Nagasawa (1970), and Schnetzler & Philpotts (1970). The data for the anorthoclase are from Sun & Hanson (1976).

Biotite Biotite generally has low K_d's for the REE and its presence should have little effect on the REE pattern of the melt.

Perovskite Perovskite has larger K_d's for the light and middle REE than the heavy REE. The presence of significant perovskite can lead to depletion or less enrichment of the light and middle REE relative to the heavy REE (Irving 1978).

Zircon Zircon has very large K_d's (on the order of 100's) for the heavy REE. However, its low abundance (generally less than 0.1%) leads to only a minor depletion of the heavy REE in the melt.

Apatite and sphene Apatite and sphene have similar K_d patterns for the REE with K_d's greater than one for all of the REE. Their generally low abundance reduces the effect of their large K_d's. Both apatite and sphene lead to enrichment of the light and heavy REE relative to the middle REE. The presence of apatite could contribute to a positive Eu anomaly.

In general the K_d's for the light REE are less or even much less than one for the major rock-forming minerals. Apatite, sphene, perovskite, and most likely monazite and allanite have K_d's much greater than one for the light REE and may deplete the light REE in the melt if they are present in abundance during melting or fractional crystallization. The middle REE in a melt will be less enriched or even depleted relative to the light and heavy REE principally by the occurrence of apatite, sphene, and hornblende. The heavy REE in a melt will be less enriched or depleted relative to the light or middle REE by the presence of garnet, zircon, pyroxene, and hornblende. A positive Eu anomaly may result from the presence in the melt of hornblende, apatite, clinopyroxene, and garnet. The presence of feldspar will contribute to a negative Eu anomaly in the melt. Equal proportions of plagioclase and clinopyroxene or twice as much plagioclase as hornblende, however, will produce melts with negligible Eu anomalies.

A common misconception is that the trace element composition of igneous rocks can be determined by the relative proportions of minerals present using the mineral/melt K_d's. For example, this type of thinking would suggest that a plagioclase-rich rock will have low total REE and a positive Eu anomaly. Except for cumulates, this is not necessarily the case. The mineral composition of a melt is determined by the major element composition and the conditions of crystallization. The trace element concentration of a melt is dependent on the trace element concentration of the parent, the extent of melting, the processes of melting involved, the solid phases remaining at the time of removal of the melt, any differentiation prior to complete crystallization, and any possible interaction with rocks, melts, or fluids. When the melt finally crystallizes, the minerals distribute the available trace elements according to the K_d's and their paragenesis. Thus, the minerals do not control the trace element content of the crystallized melt. The history of the melt prior to crystallization does. For example, if a basaltic melt with an essentially flat REE pattern crystallizes without differentiating at shallow conditions, it will be a plagioclase-clinopyroxene rock. If the same basaltic melt crystallizes without differentiating in the field of stability of eclogite, it will be a garnet-clinopyroxene rock. Both rocks will have the same REE pattern reflecting their prior history, not the physical conditions of crystallization.

Trace minerals with large abundances of REE and thus potentially large K_d's may modify the REE patterns calculated based only on the major rock-forming minerals. Due to the generally low abundances of the REE the trace mineral will probably make up less than 0.1 % of the residue. Thus, to play a significant role in most systems where F is greater than 5 %, the mineral/melt K_d must be much greater than 100. For example, zircon has a K_d of about 300 for Ytterbium (Yb) in granitic systems. Zr has an abundance of about 100–200 ppm in most rocks. Thus, if the parent had 200 ppm Zr, zircon would make up 0.04 % of the parent. If no Zr went into the melt during melting, which is highly unlikely, the effect on the D for Yb is $XK_d = (0.0004)(300) = 0.12$, which would be equivalent to the addition of 8 % clinopyroxene, 1.5 % hornblende, or 0.3 % garnet to the residue. Thus, the presence of zircon will affect the abundance of Yb, but the extent is also dependent on the abundance and K_d's of the other minerals present and on the extent of the process considered.

Analytical Techniques

In modeling petrogenetic processes it is generally found that higher precision analyses are always more useful. Even though the K_d's are not known precisely, the effects of a process will usually cause a definite change in the REE abundances and patterns. The higher the precision, the easier it is to detect the effect of a process. Precise REE analysis adequate for petrogenetic studies is possible by a number of techniques, including isotope dilution, spark source mass spectrography, x-ray fluorescence, and neutron activation.

Isotope dilution requires chemical separation of the REE for analysis. Usually they are divided into two or more fractions to separate the lighter and heavier REE in order to reduce the interference of REE metals and oxides on the surface emission mass spectrometer (see, for example, Hanson 1976). All of the REE can be analyzed, except for Praseodymium (Pr), Terbium (Tb), Holmium (Ho), and Thulium (Tm) as they have but one stable isotope each. Potentially the precision possible by this technique is 0.1%, but most of the better analyses are at the 1 to 2% level. Lanthanium (La), Gadolinium (Gd), and Lutitium (Lu) may have a somewhat worse precision (Hanson 1976, Hooker et al 1975, Masuda et al 1973).

In *spark sources mass spectrography* all of the REE can be analyzed to a precision and accuracy of about 3%. No chemical separation is required; however, the sample size is small, less than 50 mg (Taylor 1965, Taylor & Gorton 1977). An advantage is that a number of other elements may be analyzed at the same time.

X-ray fluorescence analysis of the REE on a thin film is possible if the REE are first separated as a group. The technique works reasonably well for the lighter REE or those REE with abundances greater than 2 ppm for which precision of about 10% is possible (Fryer & Edgar 1977, Eby 1972).

Neutron activation analysis for the REE may be done on a whole rock without chemical preparation (instrumental neutron activation analysis) or on the chemically separated REE (radiochemical neutron activation analysis). Most of the REE can be analyzed and the precision may be as good as 2–3% for some of the REE, e.g. La and Sm. Others may be difficult to do with a precision better than 10% (see, for example, Jacobs et al 1977). A number of other elements may also be analyzed at the same time.

For any technique the precision is dependent on the competence of the analyst, the amount of REE available for analysis, and the type of rock. An indication of the quality can be based on the analyses of rock standards (for example, USGS standard BCR-1), which gives an indication of the accuracy, and on replicate analyses for the same sample, which gives the precision. Generally high quality analyses when normalized to high quality chondrite data[2] will produce a smooth pattern. Chondrite-normalized data that plot with random peaks and valleys in going from one REE to another (excluding Eu) have an analytical uncertainty probably not much better than the difference between the peaks and valleys divided by the total abundance.

Rock types in a sequence may superficially appear to have a common origin and evolution, whereas they may actually have quite unique histories. Thus, it is very important for petrogenetic interpretations to have analyses for single well-selected samples rather than for composite samples, and to use analyses for well-described individual rock samples rather than for average compositions.

[2] For example, the analysis of Leedy chondrite by Masuda et al (1973) has a precision and accuracy of better than 2% for the analyzed REE.

An important consideration in the analysis of any granitoid rock is that the majority of the REE in the rock may reside in trace mineral phases such as apatite, sphene, monazite, zircon, xenotime, etc. (Gromet & Silver 1978). This can lead to two sources of uncertainty. The first is that homogeneity of the rock sample is difficult to achieve when a trace element is associated with a trace phase. This requires that the sample be well mixed, a very fine-grained powder, and that samples be large enough to be representative of the powder (on the order of 100's of mg). The second is that for those analyses which require dissolution of the sample for chemical separation of the REE as a group or individually, it is important to realize that minerals such as zircon, monazite, and xenotime are generally not soluble in HF with HNO_3 or $HClO_4$ in open beakers. Although zircon is soluble in HF in a pressurized vessel, monazite is not. At Stony Brook we have found that fusion with lithium metaborate, $LiBO_2$, is the most reliable method for dissolution of the major as well as trace minerals generally found in granitoid and basaltic rocks.

Effects of Alteration and Metamorphism on REE Abundances

An important consideration in the application of REE to petrogenetic studies is the mobility of the REE during metamorphism, hydrothermal alteration, and weathering, as most rocks found at the surface of the earth may have been affected by one or even all of these processes. Unless these processes are obviously severe, they do not cause a major change in the patterns or abundances for the REE (see, for example, Sun & Nesbitt 1978, Condie et al 1977, Martin et al 1978, Herrmann et al 1974, Vocke et al 1979, Ludden & Thompson 1979). This would be consistent with the experimental studies of Cullers et al (1973), Zielinski & Frey (1974), Mysen (1978), and Flynn & Burnham (1978), who have shown that aqueous fluids have very large silicate melt– and mineral–aqueous fluid K_d's. Thus, only if water were flushed many times through a system would the REE abundances be significantly affected. It would appear that most igneous rocks that have undergone essentially static metamorphism or only very limited hydrothermal alteration or weathering should give REE patterns and abundances indicative of the original rock.

Wendlandt & Harrison (1979) have shown that anhydrous carbonate melts or vapor at 5 Kb and 1200°C may be able to mobilize the REE enough to significantly affect the REE abundance of silicate systems. Flynn & Burnham (1978), however, report that their experiments suggest that mineral/fluid K_d's for a mixture of equal amounts of CO_2 and H_2O at 4 Kb and 800°C are not significantly different from those for pure water. During granulite grade metamorphism, mantle-derived carbonate melt or vapors may be involved (Goldsmith 1976), which Collerson & Fryer (1978) suggest would result in the relative depletion of the heavy REE abundances of the rocks. CO_2 vapor would deplete the light REE relative to the heavy REE (Table 3). A carbonate melt that can deplete the heavy relative to the light REE, which has no REE to start with, and represents 1% of a tonalitic rock would reduce the abundance of Yb in the tonalite by only 10%. CO_2, however, makes up much less than one percent of the crust including sediments in which most of the CO_2 resides (Rubey 1955). Also the CO_2 would have a significant mantle-derived REE content before invading the granulite grade lower crust (see the discussion in the section on the effect of CO_2 and H_2O on mantle-derived melts). Mantle-derived carbonate melts may have such a high REE content that, rather than causing a depletion of the REE, they may actually add REE to the lower crust. Therefore, CO_2 should have a significant effect on the REE abundances of rocks only locally where large quantities of CO_2 with low initial concentration of REE may be available.

Modeling Melting Processes

In this section we will consider the effects upon the REE composition of melts of (*a*) different melting processes, (*b*) varying conditions of melting, (*c*) residual trace minerals with large K_d's for the REE, (*d*) the inclusion of residual phases in melts, and (*e*) the loss of aqueous and carbon dioxide fluids prior to, during, and after melting. Rather than review the types of REE patterns found for each different rock type and interpreting the individual sets of rocks, we present calculated REE patterns for the conditions and processes considered using mineral-melt K_d's in Table 2

Table 3 Carbonate melt–, H_2O fluid–, and CO_2 vapor–silicate melt K_d's used in calculations

Element	20 Kb[a] 1075°C F-H_2O	5 Kb[a] 1075°C F-H_2O	5 Kb[b] 1200°C V-CO_2	20 Kb[c] 1200°C L-CO_3	5 Kb[c] 1200°C L-CO_3	4 Kb[d] 800°C F-H_2O
Ce	7.8	0.0035	45.2	3.20	2.60	0.0666
Sm	4.8	0.0035	9.7	4.43	3.71	—
Eu	—	—	—	—	—	0.159
Gd	—	—	—	—	—	0.0455
Tm	4.0	0.0035	27.2	8.32	7.69	—
Yb	—	—	—	—	—	0.0303

[a] Selected fluid water–silicate melt K_d's from Wendlandt & Harrison (1979).
[b] CO_2 vapor–silicate melt K_d's from Wendlandt & Harrison (1979).
[c] Carbonate melt–silicate melt K_d's from Wendlandt & Harrison (1979).
[d] Selected fluid water–silicate melt K_d's from Flynn & Burnham (1978).

and carbonate melt–, CO_2 vapor–, and H_2O fluid–silicate melt K_d's in Table 3. Some of the plots of REE patterns are normalized to a parent composition, C_L/C_0, and others are normalized to chondrites.

MELTING OF MANTLE If the earth originally had a chondritic REE abundance and then differentiated to form the early mantle and core, the mantle would have retained the REE because they are lithophile. This would result in a mantle with a greater abundance of REE and a parallel pattern relative to chondrites. Melting of the mantle to form basalts leads to a relative depletion of the light REE compared to the middle and heavy REE suggesting that the evolving mantle should become relatively depleted in the light REE. A chondritic to slight light REE depleted mantle is generally substantiated by Nd isotope studies. Some of the papers that relate the REE abundances in basalts to melting in the mantle include Schilling & Winchester (1967), Frey et al (1968, 1974, 1978), Gast (1968), Kay et al (1970), Kay & Gast (1973), Schilling (1973), Sun & Hanson

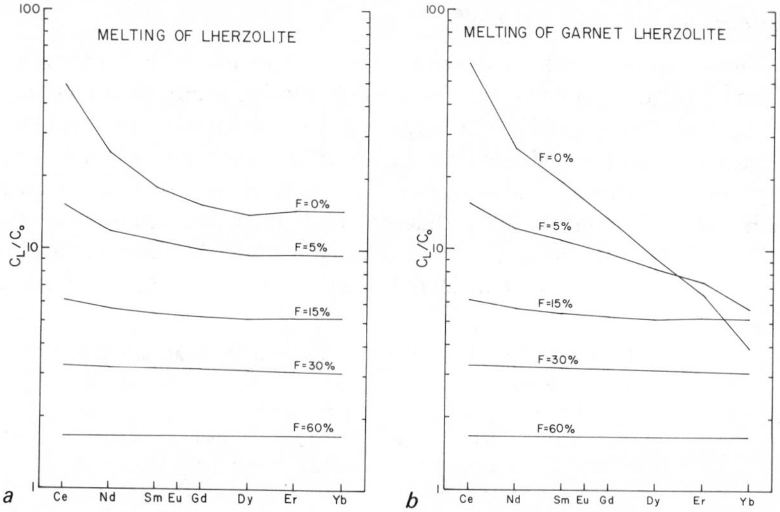

Figure 2 (*a*) Calculated REE patterns for melts derived by batch melting of lherzolite consisting originally of 55% olivine, 25% orthopyroxene, and 20% clinopyroxene. The data are normalized to the original REE composition of the lherzolite. The calculations were made using the K_d's in Table 2. As melting proceeds it uses 20% olivine, 25% orthopyroxene, and 55% clinopyroxene. (*b*) Calculated REE patterns for melts derived by batch melting of garnet lherzolite consisting originally of 55% olivine, 25% orthopyroxene, 15% clinopyroxene, and 5% garnet. The data are normalized to the REE composition of the original garnet lherzolite. The calculations were made using the K_d's in Table 2. At 5% melting there is 58% olivine, 26% orthopyroxene, 14% clinopyroxene, and 2% garnet in the residue. There is no garnet in the residue at 15% melting or greater and the residue is the same as that in Figure 2*a*.

(1975), and Langmuir et al (1977). Assuming the mantle has a lherzolitic composition the mantle may be melted at shallow levels leaving a residue of olivine ± orthopyroxene ± clinopyroxene or at depths great enough so that garnet may also be a residual solid phase. Figure 2*a* shows calculated REE patterns normalized to the parent mantle composition (C_L/C_0) for batch melting of a mantle under shallow conditions with a range of 0–60% melting. Figure 2*b* shows calculated REE patterns for batch melting of a mantle under conditions where garnet occurs as a

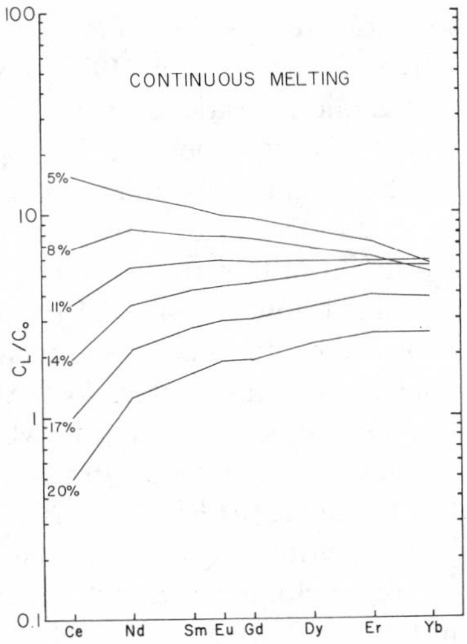

Figure 3 Calculated REE patterns for melts derived by continuous melting of the garnet lherzolite in Figure 2*b*. The data are normalized to the REE composition of the original garnet lherzolite. The calculations were made using the K_d's in Table 2. As melting proceeds it uses 40% garnet and 60% clinopyroxene until the garnet is used up. It then uses 50% clinopyroxene, 25% orthopyroxene, and 25% olivine.

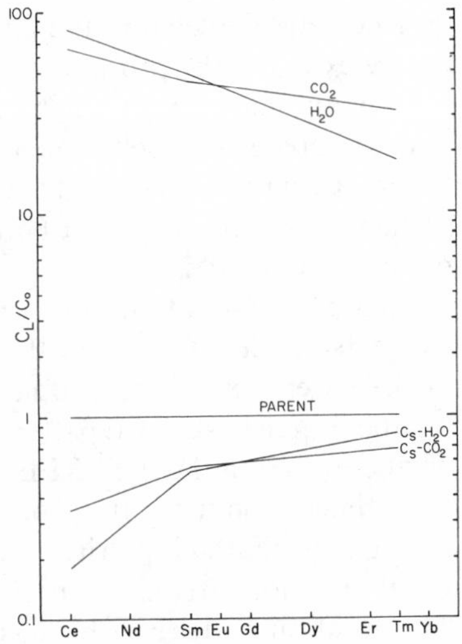

Figure 4 Calculated REE patterns for carbonate melt and H_2O fluid for $F = 1\%$ and a garnet lherzolite residue (C_s) after removing 1% carbonate melt and 1% H_2O fluid. The data are normalized to the original REE composition of the garnet lherzolite. The calculations were made using the carbonate melt– and H_2O fluid–silicate melt K_d's in Table 3 for 20 Kb and the mineral-melt K_d's in Table 2 for mantle melting.

residual solid phase during low fractions ($F < 15\%$) of melting. For low fractions of melting both models will show essentially the same enrichment in the light REE. The main difference is in the heavy REE. The presence of garnet leads to less enrichment of the heavy REE and a significant increase in the Ce/Yb ratio relative to melts derived from a mantle without garnet.

In both Figures 2*a* and 2*b* it can be seen that more than about 15% melting results in patterns for the melt that are parallel to subparallel to that of the parent. Thus it would be expected that the light REE ratios for tholeiites, picrites, and komatiites would be essentially those of the parent. Langmuir et al (1977), however, found that in the FAMOUS area on the mid-Atlantic Ridge that the tholeiitic basalts from within a couple of km of each other had REE patterns that crossed, suggesting that if batch melting were the main process that more than one source was involved. The incompatible element and radiogenic isotope ratios for basalts from this area, however, indicated a homogeneous source. They explained how crossing REE patterns may develop by a process called dynamic melting. Dynamic melting occurs in a continuously upwelling adiabatic mantle source. As the mantle rises it intersects its solidus and creates a zone of melting in which both batch melting and continuous melting are occurring. With continuous melting it is possible to significantly change the REE pattern as compared to patterns developed during batch melting. Figure 3 shows REE patterns for melts derived by continuous melting of a garnet lherzolite in which whenever there is 5% melt in the source region 60% of that melt is removed. During continuous melting, it is possible to derive melts with patterns that are both slightly enriched and depleted in the light REE relative to the parent depending on the extent of the process.

Dynamic melting may help explain how it is possible to find in an Archean greenstone belt tholeiites with essentially flat REE patterns and komatiites with depleted light REE patterns (Hanson & Langmuir 1978). Dynamic melting may also be important for other light REE depleted ocean ridge basalts.

Figure 2*b* shows that at 0% melting, when maximum enrichment for elements with low *D*'s occurs, the melt has a Ce abundance of 60 times that of the parent, even using some of the lower values for the K_d's. If the mantle source has a REE pattern parallel to that of chondrites or is somewhat depleted in the light REE and if the source has a Ce abundance of three times chondrites, this would produce melts with a maximum abundance of 180 times chondrites. Yet nephelinites and potassic basalts have Ce abundances greater than 200 times chondrites (Kay & Gast 1973). The addition of REE-rich phases such as apatite or perovskite to the residue would only exacerbate this problem because of their high K_d's for the light REE. Kay (1975), Green & Liebermann (1976), and Harris (1974) have suggested that basalts with very high light REE concentrations may be derived through zone refining of the mantle. However, the REE patterns for melts derived by zone refining cannot be enriched for elements with low D's by more than $1/D$, the $F = 0\%$ pattern in Figure 2*b*. Therefore Sun & Hanson (1975) and Frey et al (1978) have argued that the sources for alkali basalts and nephelinites do not have REE patterns parallel to those of chondrites, but they are light REE enriched. Nd isotope ratios, however, suggest that alkali basalts and nephelinites are derived from sources which, on a chondrite-normalized basis, have been on average light REE depleted or flat through the history of the earth. This would suggest that the source was enriched in light REE shortly prior to melting or that the light REE abundance in the melts is not explained by the standard melting models.

THE EFFECT OF CO_2 AND H_2O ON MANTLE-DERIVED MELTS Under the conditions of melting in the mantle H_2O may be present as a supercritical fluid whose ability to dissolve is mainly a function of its density which is controlled by T and P. As the density of a supercritical fluid approaches that of the liquid its properties as a solvent approach that of the liquid phase. In the following discussion, supercritical water will be referred to as a fluid, i.e. neither liquid nor vapor. A fluid phase may be involved prior to melting of the mantle and if the fluid phase is immiscible in the silicate melt a fluid may exist during melting. Mysen (1978) and Wendlandt & Harrison (1979) report K_d's for carbonate melts and H_2O and CO_2 as fluid phases for Ce, Sm, and Tm, i.e. light, middle and heavy REE, respectively. Table 2 compares representative values for H_2O fluid–, carbonate melt–, and CO_2 vapor–silicate melt K_d's. These are for very restricted silicate compositions and may not be appropriate for all compositions. Mineral-fluid K_d's are calculated by dividing mineral–silicate melt K_d's by fluid–silicate melt K_d's selected for the appropriate conditions.

The distribution of trace elements in fluid-mineral, fluid-melt, or fluid-mineral-melt systems can be calculated in the same fashion as the distribution of trace elements for mineral-melt systems. In fluid-mineral-melt systems where the fluid is removed in one step, the silicate melt can be considered as one more phase in the residue for calculating D for batch melting (Equation 4). If the fluid is streaming off the mineral-melt system as it forms, the equations for fractional fusion during modal or non-modal melting should be used (Equations 5 and 6).

From Table 3 it can be seen that, at 5 Kb and 1075°C, fluid water has a much lower concentration of the REE than a silicate melt in equilibrium with it. Thus, removal of a small fraction of water at 5 Kb and 1075°C will have little effect on the REE abundances of either derived melts or solid residual phases. At 20 Kb and 1075°C, water has a REE content greater than that of a silicate melt. Removal of one per cent water from the system will have a rather strong effect on the residual solids (Figure 4). Carbonate melt has a high REE concentration relative to silicate melts at either 20 Kb or 5 Kb, and the heavy REE are enriched over the light. Carbonatites, however, are strongly enriched in the light REE suggesting that they represent much less than 1% melting of garnet-bearing sources or come from sources enriched in light REE. Separation of a CO_2 vapor phase from a carbonate melt while approaching the surface will reduce the Ce more than the Sm and Tm concentrations of the carbonate melt, i.e. it will tend to flatten the REE pattern for the melt. Introduction of H_2O, carbonate, or CO_2 derived from a garnet-bearing source can significantly enrich the light REE over the heavy REE in metasomatized mantle (see, for example, Lloyd & Bailey 1975 for a discussion of the geochemistry and petrography of metasomatized lherzolite nodules).

MELTING OF A BASALTIC PARENT Melts derived from melting of rocks of basaltic composition may be important whenever large volumes of basalt may be brought to temperatures and pressures where significant melting can take place. Under high water pressures this may occur at temperatures as low as 700°C (Helz 1976). Melting of basaltic rocks is potentially important in an arc environment where ocean floor basalts are subducted. In this environment melting of basalts may take place along the slab, at the base of the crust, or between the crust and subducting slab. A scenario for the basalt in the subducting slab is that it is amphibolite until a depth of some 100 km when it converts to eclogite (Ringwood 1974). The conversion may be associated with melting.

The different conditions for melting of a basalt considered here are (*a*) relatively shallow and wet conditions, which would produce dacitic melts and leave an amphibolite residue (Helz 1976); (*b*) relatively shallow and dry conditions, producing andesitic melts and granulite residue (Green & Ringwood 1968); (*c*) moderate depths producing anorthositic gabbro melts, and granulite residue (Simmons & Hanson 1978); and (*d*) relatively greater depths producing dacitic melts and leaving an eclogitic residue (Green & Ringwood 1968). Potential parents of melts would probably be light REE depleted ocean floor basalts. However, light REE enriched basalts may also be present. In all of the calculations for this section the REE are normalized to the parent composition (C_L/C_0) and not to possible chondrite normalized compositions. Discussion of the effects on REE patterns for melts derived by melting of basalts have been presented by Hanson & Goldich (1972), Arth & Hanson (1972, 1975), Gill (1974), Kay (1977), Dostal et al (1977), Lopez-Escobar et al (1977), and Simmons & Hanson (1978).

A well-documented study of the melting of basalts at low fractions is that of Helz (1976) who melted a continental tholeiite, an olivine tholeiite, and an alkali basalt at 5 Kb water pressure with controlled oxygen fugacity. The main difference between these experiments and melting of wet ocean floor basalts is that these basalts have between 0.5 and 1% K_2O whereas most ocean floor basalts would have about 0.1% K_2O. Figure 5 shows a plot of REE patterns calculated using the relative fractions of melt and minerals for melting of a Picture Gorge tholeiite, at 5 Kb p_{H_2O} from 700°C where $F = 10\%$ to 1000° where $F = 63\%$. All of the melts have similar REE patterns. As melting proceeds amphibole becomes more abundant in the residue leading to the most depleted REE pattern occurring at $F = 40\%$. For increased extents of melting clinopyroxene becomes a more important residual phase. If the K_d's were dependent upon temperature, the effect on the K_d's for hornblende would be the most important because hornblende has the largest K_d's and is the most abundant residual mineral. Presumably the K_d's would decrease as T increased. This would lead to either greater spread between the samples or even less spread depending on which temperature was appropriate for the K_d's used in the calculation. In any case the patterns would be expected to be very similar to those shown.

Figures 5 and 6 both show for comparison the plot of a melt derived by 20% melting of a basalt leaving 60% clinopyroxene and 40% plagioclase in the residue. This melt would be appropriate for melting either at shallow or moderate depths leaving a granulite residue forming andesite or anorthositic gabbro respectively. Figure 6 also shows a melt derived by 20% melting of basalt leaving 55% clinopyroxene, 25% garnet, and 10% quartz in the residue. To show the effect of sphene in the residue, a separate REE pattern, which results from the addition of 2% sphene to the residue, is shown. Two per cent is a very large percentage especially when it is considered that although sphene may be stable under lower pressure conditions it may not be stable in an eclogitic residue (Hellman & Green 1979). Because sphene has fairly large K_d's for all of the REE, but the K_d's for garnet and clinopyroxene are low for the light REE, the main effect is to reduce the abundance of the light REE for the melt with less effect on the middle REE and the least effect on the heavy REE.

Loss of water from any of these rocks prior to melting will lead to moderate depletion of Ce relative to Sm. The effect will be less than that for the mantle, because of the higher K_d's for the residual minerals in a basaltic residue as compared to a mantle residue. Likewise, several per cent water from either the mantle or other basalts added to the basalts prior to melting could lead to moderate enrichment of the light REE.

Melting a basalt produces melts of differing major element composition and of differing REE concentrations and patterns. Melts leaving a dominantly hornblende residue should have relative to the parent basalt composition a concave upward pattern, strong positive Eu anomaly, with moderate depletion of the middle and heavy REE. These melts will probably be nearly water-saturated and would probably occur mainly as tonalitic to granodioritic intrusions rather than as extrusions.

Melts leaving a granulite residue will have REE patterns relative to the parent basalt composition essentially flat in the middle to light REE and moderately enriched in the light REE. These melts may be moderately enriched in H_2O and may occur as either intrusions (dioritic, gabbroic, or anorthositic complexes) or extrusions (andesites). Andesites derived by partial melting of basalt leaving a granulite residue would be difficult to distinguish from melts derived from the mantle or by fractional crystallization of a mantle-derived tholeiitic basalt melt.

Melts leaving an eclogite residue will have relative to the parent basalt

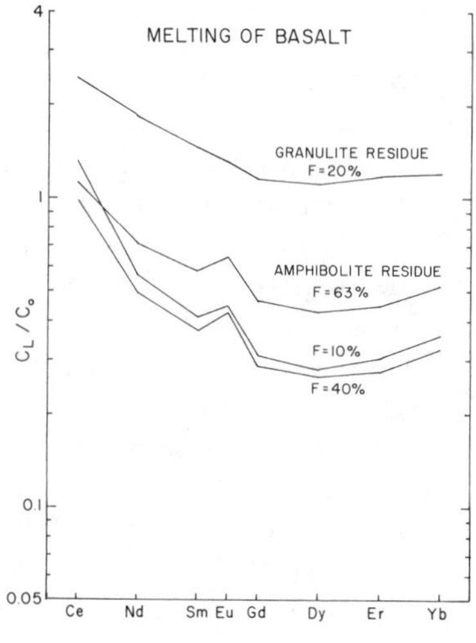

Figure 5 Calculated REE patterns for melts derived by batch melting of basalt leaving a granulite residue and amphibolite residue. The data are normalized to the REE composition of the original basalt. The K_d's used in the calculations are given in Table 2. The melt leaving a granulite residue is a result of 20% melting leaving a residue consisting of 60% clinopyroxene and 40% plagioclase. The melts leaving amphibolite residue were calculated using the proportions of melt and residual minerals given by Helz (1976) for melting of a Picture Gorge basalt at $P_{H_2O} = 5$ Kb. The melt derived by 10% melting at 700°C left a residue consisting of 0.9% apatite, 32% plagioclase, and 59% hornblende. The melt derived by 40% melting at 825°C left a residue consisting of 1.3% apatite, 7% plagioclase, and 88% hornblende. The melt derived by 63% melting at 1000°C left a residue consisting of 0.3% apatite, 67% hornblende, and 26% clinopyroxene. In each case the residue has significant quartz or oxides which have very low K_d's for the REE.

composition a strong depletion of the heavy REE. Although there may be some upturn in the heaviest REE, it will not be as dramatic as it is for the melts leaving an amphibolitic residue. Melts leaving an eclogitic residue may be relatively dry and may occur as shallow intrusions or extrusions.

It is not the mineralogy of a parent prior to melting that controls the REE patterns. It is, rather, the mineralogy of the residue at the time of removal of melt, which may be drastically different from the parent mineralogy at the onset of melting. For example, the parent may originally be an amphibolite, but the residue may be eclogitic or granulitic. Also, the residual mineralogy will depend on the major element composition of the residue, not on the composition of the original parent. Melting of a basalt will lead to a residue depleted in SiO_2 and alkalies and enriched in Ca, Fe, and Mg relative to that of the parent.

MELTING OF SEDIMENTS AND CONTINENTAL CRUST The main products of melting sediments and continental crust are leucocratic granites (quartz monzonites) and granodiorites. Because we are used to seeing and mapping the details of sedimentary, igneous, and metamorphic terrains, it is easy to think of melting as occurring in small well-defined lithologic units. Yet the parent may be large and consist of a heterogeneous collection of rocks. For example, a relatively small homogeneous granitic intrusion may have dimensions of 5 km on a side or a volume of 125 km^3. If the pluton is derived by 20% melting of sedimentary rocks or the basement, the parent would have had a volume of 625 km^3 which would be a block 8.5 km on a side. Due to the large dimensions involved, the parent

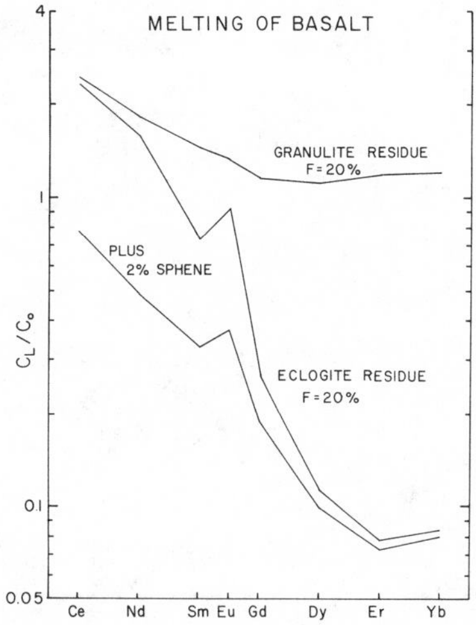

Figure 6 Calculated REE patterns for melts derived by batch melting of basalt leaving eclogite residues and for comparison with the melt in Figure 5 leaving a granulite residue. The K_d's used in the calculation are given in Table 2. The data are normalized to the REE composition of the original basalt. The melts leaving eclogite residues result from 20% melting leaving 55% clinopyroxene, 35% garnet, and 10% quartz; and 55% clinopyroxene, 35% garnet, 8% quartz, and 2% sphene.

undergoing melting may consist of a number of rock types. The melt may, however, be derived from similar least refractory portions of a heterogeneous parent. Thus, although a given source may be variable in composition, large volumes susceptible to melting may have similar mean compositions.

Nance & Taylor (1976) show that, although there are some variations as a function of age, all analyzed graywackes and shales have similar REE patterns and concentrations. They conclude that these patterns and concentrations are representative of the continental crust. Shale plus graywacke are often abundant in thick sedimentary sequences and are most susceptible to melting, so that in melting various sedimentary-rich sequences the parent REE compositions should be similar and the main source of variability should be a result of the conditions of melting and the amount of the solid residue that accompanies the melt. If melting occurs where the water pressure equals total pressure, the melt will not rise much above the site of melting before crystallizing (Harris et al 1970) and will probably lose an aqueous fluid or vapor phase prior to or during crystallization. If melting occurs with high water pressure there will be an amphibolite grade solid residue, whereas if the water pressure is relatively low there will be granulite grade residue.

Figure 7 presents the REE patterns (normalized to chondrites) for average graywacke (Nance & Taylor 1976), the residue after melting the graywacke, the melt, and a 70% melt + 30% residue mixture. This example of melting is similar to that given by Arth & Hanson (1975). The 40% melting occurs during upper amphibolite grade metamorphism

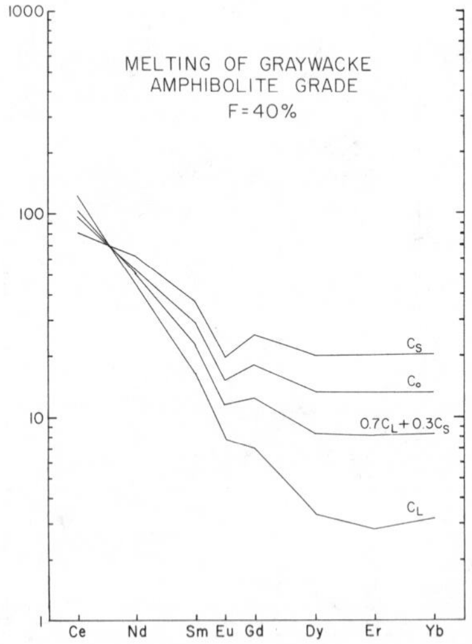

Figure 7 Calculated REE patterns for melt (C_L), residue (C_s), mixture of 70% melt and 30% residue ($0.7C_L+0.3C_s$) and the original graywacke composition which underwent 40% melting leaving an amphibolite residue. The data are normalized to chondrites. The residue consists of 10% quartz, 55% plagioclase, 20% hornblende, 11% garnet, 0.5% apatite, and 0.02% zircon. The K_d's used in the calculation are given in Table 2.

leaving a residue of 10% quartz, 55% plagioclase, 20% hornblende, 11% garnet, 0.5% apatite, and 0.02% zircon. Except for zircon and apatite the mineralogy of the residue has been determined by an inversion procedure in which the mineral compositions are selected from the literature for the appropriate conditions and the melt is a minimum melt composition (Vocke & Hanson 1980). Since the D for Ce is 0.67, small changes in the fraction of melting will have but little effect on the light REE abundances. Reduction in the extent of melting, however, may lead to higher REE abundances in the melt as the residue at lower extents of melting may have more biotite and less garnet.

To a first approximation the REE patterns for melting of a graywacke-shale composition leaving a granulite residue are the same as that for an amphibolite residue. Pyroxene plus garnet has approximately the same effect on a melt as amphibole plus garnet. Since the REE pattern for graywackes approximates that of the continental crust, if large volumes of continental crust are involved in melting, the parent will probably have a REE composition similar to that of graywacke. Thus, the melts derived from graywacke at amphibolite grade may be hard to distinguish on the basis of REE from melts derived from melting of large volumes of heterogeneous continental crust at granulite grade.

Chappell & White (1974) have suggested that during melting a significant

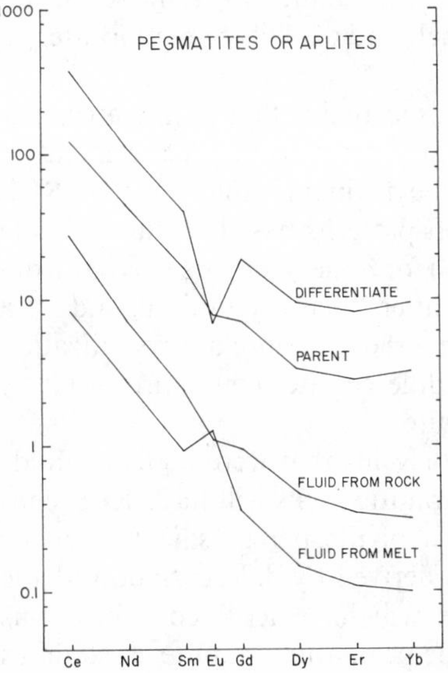

Figure 8 Calculated REE patterns for pegmatites or aplites derived from the parent granite, which has the same REE composition as the melt in Figure 7. The data are normalized to chondrites. The K_d's used in the calculations are the fluid H_2O–silicate melt K_d's at 800°C and 4 Kb in Table 3 and the mineral–silicate melt K_d's for melting of sediments in Table 2. The "fluid from melt" represents the REE composition for H_2O fluids representing 3% of the mass of the granite body. The fluid separates before the melt crystallizes. The "fluid from rock" represents the REE composition for H_2O fluid representing 3% of the mass of the crystalline granite (parent) which consists of 30% quartz, 30% plagioclase, 30% orthoclase, 10% biotite, 0.5% apatite, and 0.05% zircon. The K_d data have been interpolated for some of the REE. The "differentiate" represents 80% fractional crystallization or 97% equilibrium crystallization or 3% batch melting of the crystalline granite.

proportion of the residue will accompany the rising melt. For drier systems a melt rising adiabatically may reach temperatures above its liquidus and melt the included residue, so that the residue may not be apparent as xenoliths. For the example of melting graywacke, the shape of the parent and melt are similar, except for the heavy REE, so that even relatively large fractions of residue added to the melt may be difficult to distinguish from the effects of varying the proportions of hornblende and garnet in residual phases.

Those parts of the residue concentrated in high density garnet or hornblende may be preferentially left behind. The melt-residue mixture would then contain more of the low density phases, for example, quartz and plagioclase. This would lead to a reduction of the trivalent REE in the mixture due to the low K_d's for plagioclase and quartz. The addition of residual plagioclase would, however, lead to a relative increase in the Eu content.

The effect on the REE in the melt plus accompanying residue due to settling out of a high density phase such as garnet will depend on how the garnet settles out. If the garnet settles rapidly or in large xenoliths so that the garnet is not in direct communication with the melt, the effect on the REE is an unmixing process. If the garnet settles out very slowly while remaining in equilibrium with the melt, it has the potential of reducing the heavy REE in a manner similar to, but not identical to, fractional crystallization. Fractional crystallization involves the crystallization of an infinitesimal fraction of a mineral and its immediate removal. In this case the garnet is already crystallized and the mineral is removed in infinitesimally small fractions while the melt and crystals re-equilibrate.

APLITES AND PEGMATITES Frequently associated with granitic intrusions are aplite and pegmatite dikes. These may form from silicate melts or aqueous solutions. These origins should be distinguishable based on REE abundances. Shown in Figure 8 are REE patterns based on fluid/silicate melt K_d's from Flynn & Burnham (1978) for aqueous fluids derived from the melt, (C_L) in Figure 7, from the melt after it has crystallized, and for a differentiate derived by 97% equilibrium or 80% fractional crystallization of a granitic melt. The pattern marked "differentiate" could also result by 3% batch melting of the crystalline granite.

Late stage aplite or pegmatitic dikes or veins that were aqueous fluids derived from melts or from crystalline granitic rocks will have REE concentrations significantly lower than that of the parent silicate melt or rock, whereas aplite or pegmatite veins derived by differentiation of the granitic melt or melting of the granite will have REE concentrations greater than that of the parent. Even if the pegmatite or aplite crystallized from an aqueous fluid and the fluid escaped, the pegmatite or aplite should not have higher REE concentrations than the parent.

Assimilation

The effects of assimilation upon the abundance and shapes of the REE patterns for a melt or melt plus assimilated material will depend on how the assimilation takes place. If the rocks are totally assimilated, then a simple mass balance relation similar to that of Equation (2) describes the result. C_0 is the melt plus assimilated material. C_L is the concentration of a given element in the melt. C_s is the concentration of a given element in the rock assimilated, F is the proportion of melt, and $(1-F)$ the proportion of the rock assimilated. If there is melting of the wall rock, assimilation may involve mixing of the original melt with the melt from the wall rock.

Assimilation may also involve incomplete equilibrium of phases that do not remain with the melt because of incomplete interaction with the wall rock or because the rock settles out. In either case, the results can be estimated by calculating the bulk distribution coefficients for the rock (D) and the composition of a melt (C_L) in equilibrium with the rock. From Equation (3)

$$C_L = C_s/D \tag{11}$$

where C_s is the concentration of a given element in the wall rock. If the C_L calculated is less than the concentration of the element in the melt, the rock will take on the element in order to decrease the concentration. Likewise, if C_L is greater than the concentration of the melt, the rock will give up the element to increase the concentration of the melt. The mineralogy that develops is particularly important. For example, if a pelitic rock interacting with the melt were crystallizing mica, feldspar, and quartz, it would probably donate some of each REE to a granitic melt. If, however, the same rock were crystallizing a relatively large proportion of garnet, it may donate some of its light REE to the melt and remove some heavy REE from the melt.

Differentiation

The effects of differentiation may be noted in both intrusive and extrusive rocks. Gabbroic plutonic bodies, particularly those that have undergone convection, may include the cumulates of fractional crystallization as well as an intercumulus liquid with the composition of the differentiated melt (Paster et al 1974). Volcanics derived from a magma chamber in which differentiation is taking place will generally include a predominance of the differentiating melt plus the phenocrysts that represent those phases that are crystallizing (Zielinski & Frey 1970). In a granitic pluton or a pluton that is cooling rapidly so that convection is not an important process, there may be incremental equilibrium crystallization in which the body cools from the outside-inward with the outer portions involving the higher temperature or near liquidus phases as well as significant melt (McCarthy & Hasty 1976).

If D for a given element approaches zero and F is fairly large, it does not matter whether the batch melting equation (Equation 4) or the Rayleigh fractionation law (Equation 10) is used for calculating the effects of differentiation. As D approaches one or greater or F approaches zero, it is important to decide which equation may be more appropriate. If the solid phases are actually being removed instantaneously, the Rayleigh Fractionation Law is appropriate. If, however, crystallization is taking place very slowly and all of the phases remain in equilibrium with each other until the melt is completely crystallized, then the batch melting equation is appropriate. If crystallization is somewhere between these two extremes, it may be more appropriate to use the batch melting equation in an appropriate number of steps changing C_0 discontinuously. To more easily visualize the effect of the crystallizing process it is best to normalize the cumulate-rich rocks or differentiates to the most primary liquid.

One of the large number of possible models of differentiation is shown in Figure 9 in which the REE abundances are normalized to the initial melt abundances. Figure 9 shows the initial melt, cumulate plagioclase

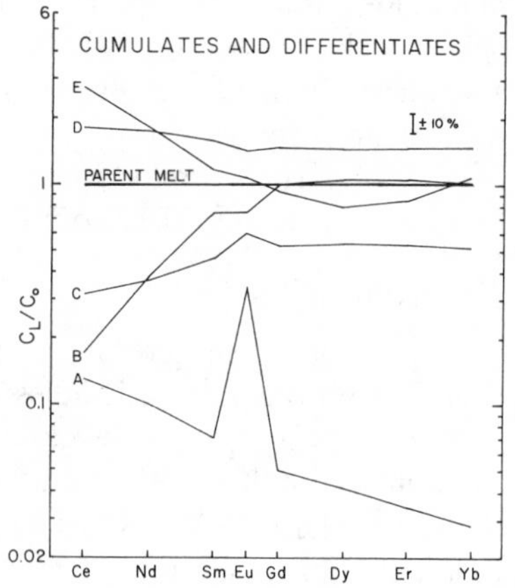

Figure 9 Calculated REE patterns for cumulates and differentiates from a basaltic parent melt. The data are normalized to the parent melt. REE composition *A* represents a pure plagioclase cumulate from the parent melt. *B* represents a pure clinopyroxene cumulate from the parent melt. *C* represents 50% plagioclase with the composition of *A* plus 30% clinopyroxene with the composition of *B* plus 20% intercumulus parent melt. *D* represents the REE composition for a differentiate derived by fractional crystallization of 50% crystals consisting of 60% plagioclase (*A*) and 40% clinopyroxene (*B*) from the parent melt. *E* represents a melt derived by a further 25% fractional crystallization of 70% plagioclase and 30% hornblende with the K_d's given in Table 2 under melting of basalt.

(*A*), cumulate clinopyroxene (*B*), a rock consisting of 50% cumulate plagioclase, 30% cumulate clinopyroxene and 20% of parent melt (*C*), a melt derived by removal of 50% fractional crystallization of clinopyroxene and plagioclase in the ratio of 2/3 (*D*), and a melt derived by a 25% further crystallization of hornblende and plagioclase in the ratio of 3/7 (*E*). It can be seen that the first 50% crystallization involving clinopyroxene and plagioclase produces a melt whose REE pattern is somewhat enriched and subparallel to the original pattern. Error bars for an uncertainty of $\pm 10\%$ are indicated near the melt composition. It can be seen that the light REE enrichment indicative of clinopyroxene and the negative Eu anomaly indicative of plagioclase are indiscernable after 50% crystallization if the uncertainty is 10%.

For the melt (*E*) involving fractional crystallization of hornblende and plagioclase from melt (*D*), the light REE are enriched, but the middle REE are depleted and the heavy REE less so, leading to a concave upward pattern in the middle to heavy REE. If apatite were crystallizing, it would have little effect on the shape of the derived melt but would reduce all of the REE by approximately the same extent. For example, 1% apatite reduces the REE content by 5–15%. The main, but subtle, indicator that apatite may be present is that the lightest REE are not as enriched as much as might be expected by the presence of only hornblende and plagioclase.

In treating cumulate sequences it would be best to (*a*) analyze the cores of cumulate minerals or mono-mineralic cumulate rocks (e.g. pure plagioclase or pure clinopyroxene rocks) to give an indication of how the liquid composition and K_d's are changing, then (*b*) compare these analyses with those of nearby rocks rich in intercumulus liquid.

In volcanic sequences the phenocryst minerals may be used to calculate mineral-melt K_d's for the minerals involved in fractional crystallization. It is very important to be cautious about using possible xenocrysts for the determination of K_d's.

Immiscible Silicate Liquids

Watson (1976) and Ryerson & Hess (1978) report K_d's for immiscible ferro-basaltic and granitic melts in which they find that the ferrobasalts are enriched in the REE relative to the granitic melts. The ferrobasalt/granitic melt K_d's are essentially the same for the trivalent REE so that although the concentrations of the REE are quite different the patterns for the two melts are parallel. The ferrobasalt/granitic melt K_d for Eu^{2+} is, however, about 1, so that if the melts are highly reduced one or both may have significant Eu anomalies. The extent to which the REE are enriched in the ferrobasalt is proportional to the P_2O_5 content of the melts, i.e. the higher the P_2O_5 content the more the REE prefer the ferrobasalts.

Figure 10 shows two examples of how the REE are distributed in melts for equal amounts of both ferrobasalt and granite (patterns labeled 50% G) and for 85% granite and 15% basalt (patterns labeled 85% G) using a K_d of 10 for the trivalent REE and a K_d of 1 for Eu. If the ferrobasalt makes up a small fraction of the system, it will have a strong negative Eu anomaly, because the REE are concentrated in the ferrobasalt. At equal amounts of ferrobasalt and granite the ferrobasalt has a small negative Eu anomaly and the granite has a large positive Eu anomaly. If there is significant trivalent Eu this will reduce the extent of the Eu anomalies proportional to the fraction of Eu^{3+}.

Conclusion

The main emphasis of this paper has been the use of REE in petrogenetic studies of igneous rocks, i.e. determining the chemical and mineralogical composition of the parent at the time of melting, the extent of melting, the *T* and *P* conditions during melting, and the modification of the melt due to mixing of melts, assimilation, metasomatism, differentiation, and interaction with fluids. Although the REE are a very powerful tool for petrogenetic studies, they can only make up part of such a study, which must also include field observations, careful selection and preparation of samples, petrography, and major, minor, and trace element analysis. When possible, studies should also include geochronology and radiogenic

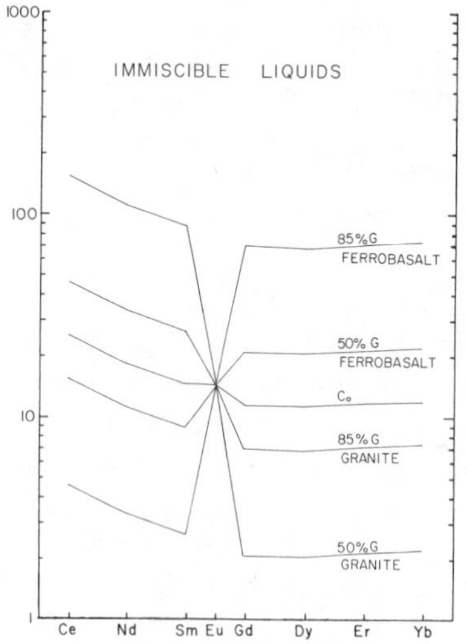

Figure 10 Calculated REE patterns for immiscible granites and ferrobasalts and the composition of the total system (C_0). The data are normalized to chondrites. The K_d's are based on the data of Ryerson & Hess (1978), and Watson (1976). There are two examples shown. The first (patterns labeled 85% G) is for immiscible liquids consisting of 85% granite and 15% ferrobasalt. The second (patterns labeled 50% G) is for immiscible liquids consisting of 50% granite and 50% ferrobasalt.

and stable isotope analysis. Thus, for maximum petrogenetic information it is important that REE analyses be coordinated with ongoing projects or undertaken in areas that have been subject to intensive study. It is also very important in order to obtain the maximum information that data of the highest possible precision and accuracy are available for interpretation.

ACKNOWLEDGMENTS

Financial support was provided by National Science Foundation Grant No. EAR 76-13354 (Geochemistry) and OCE 78-20058 (Submarine Geology). J. F. Bender, K. L. Cameron, C. H. Langmuir, and P. I. Nabelek gave useful suggestions for improving the manuscript.

Literature Cited

Albarede, F., Bottinga, Y. 1972. Kinetic disequilibrium in trace element partitioning between phenocrysts and host lava. *Geochim. Cosmochim. Acta* 36:141–56

Allegre, C. J., Minster, J. F. 1978. Quantitative models of trace element behavior in magmatic processes. *Earth Planet. Sci. Lett.* 38:1–25

Allegre, C. J., Treuil, M., Minster, J. F., Minster, B., Albarede, F. 1977. Systematic use of trace elements in igneous processes. Part I: Fractional crystallization processes in volcanic suites. *Contrib. Mineral. Petrol.* 62:57–75

Arth, J. G. 1976. Behavior of trace elements during magmatic processes—a summary of theoretical models and their applications. *J. Res. US Geol. Surv.* 4:41–47

Arth, J. G., Barker, F. 1976. Rare earth partitioning between hornblende and dacitic liquid and implications for the genesis of trondhjemetic-tonalitic magmas. *Geology* 4:534–36

Arth, J. G., Hanson, G. N. 1972. Quartz diorites derived by partial melting of eclogite or amphibolite at mantle depths. *Contrib. Mineral. Petrol.* 37:161–74

Arth, J. G., Hanson, G. N. 1975. Geochemistry and origin of the early Precambrian crust of Northeastern Minnesota. *Geochim. Cosmochim. Acta* 39:325–62

Banno, S., Matsui, Y. 1973. On the formulation of partition coefficients for trace elements distribution between minerals and magma. *Chem. Geol.* 11:1–15

Bowen, N. L. 1928. *The Evolution of Igneous Rocks.* Princeton, N.J.: Univ. Press. 334 pp.

Chappell, B. W., White, A. J. R. 1974. Two contrasting granite types. *Pac. Geol.* 8: 173–74

Collerson, K. D., Fryer, B. J. 1978. The role of fluids in the formation and subsequent development of early continental crust. *Contrib. Mineral. Petrol.* 67:151–67

Condie, K. C., Viljoen, M. J., Kable, E. J. D. 1977. Effects of alteration on element distributions in Archaean tholeiites from the Barberton greenstone belt, South Africa. *Contrib. Mineral. Petrol.* 64:75–89

Cullers, R. L., Medaris, L. G., Haskin, L. A. 1973. Experimental studies of the distribution of rare earth as trace elements among silicate minerals and liquids and water. *Geochim. Cosmochim. Acta* 37: 1499–1512

Dostal, J., Zentilli, M., Caelles, J. C., Clark, A. H. 1977. Geochemistry and origin of volcanic rocks of the Andes (26°–28°S) *Contrib. Mineral. Petrol.* 63:113–28

Drake, M. J. 1975. The oxidation state of europium as an indicator of oxygen fugacity. *Geochim. Cosmochim. Acta* 39: 55–64

Drake, M. J., Weill, D. F. 1975. Partition of Sr, Ba, Ca, Y, Eu^{2+}, Eu^{3+} and other REE between plagioclase, feldspar and magmatic liquid: an experimental study. *Geochim. Cosmochim. Acta* 39:689–712

Eby, G. N. 1972. Determination of rare earth, yttrium and scandium abundances in rocks and minerals by an ion exchange X-ray fluorescence procedure. *Anal. Chem.* 44:2137–43

Evensen, N. M., Hamilton, P. J., O'Nions, R. K. 1978. Rare-earth abundances in chondritic meteorites. *Geochim. Cosmochim. Acta* 42:1199–1212

Flynn, R. T., Burnham, C. W. 1978. An experimental determination of rare earth partition coefficients between a chloride

containing vapor phase and silicate melts. *Geochem. Cosmochim. Acta* 42:682–701
Frey, F. A., Haskin, M. A., Poetz, J. A., Haskin, L. A. 1968. Rare earth abundances in some basic rocks. *J. Geophys. Res.* 73: 6085–98
Frey, F. A., Bryan, W. B., Thompson, G. 1974. Atlantic Ocean floor: Geochemistry and petrology of basalts from Legs 2 and 3 of the Deep-Sea Drilling Project. *J. Geophys. Res.* 79:5507–27
Frey, F. A., Green, D. H., Roy, S. D. 1978. Integrated models of basalt petrogenesis: A study of quartz tholeiites to olivine melilites from southeastern Australia utilizing geochemical and experimental petrological data. *J. Petrol.* 19:463–513
Fryer, B. J., Edgar, A. P. 1977. Significance of rare earth distribution in coexisting minerals of peralkaline undersaturated rocks. *Contrib. Mineral. Petrol.* 61:35–48
Gast, P. W. 1968. Trace element fractionation and the origin of tholeiitic and alkaline magma types. *Geochim. Cosmochim. Acta* 32:1057–86
Gill, J. B. 1974. Role of underthurst oceanic crust in the genesis of a Fijian calc-alkaline suite. *Contrib. Mineral. Petrol.* 43:29–45
Goldsmith, J. R. 1976. Scapolites, granulites, and volatiles in the lower crust. *Geol. Soc. Am. Bull.* 87:161–68
Green, D. H., Liebermann, R. C. 1976. Phase equilibria and elastic properties of a pyrolite model for the oceanic upper mantle. *Tectonophysics* 32:61–92
Green, T. H., Ringwood, A. E. 1968. Genesis of the calc-alkaline igneous rock suite. *Contrib. Mineral. Petrol.* 18:163–74
Greenland, L. P. 1970. An equation for trace element distribution during magmatic crystallization. *Am. Mineral.* 55:455–65
Gromet, L. P., Silver, L. T. 1978. Implications of rare earths distribution among minerals in a granodiorite, Peninsular Range batholith, Southern California. *EOS, Trans. Am. Geophys. Union* 59:399–400
Grutzeck, M. W., Kridelbaugh, S. J., Weill, D. F. 1974. The distribution of Sr and REE between diopside and silicate liquid. *Geophys. Res. Lett.* 1:273–75
Hanson, G. N. 1976. Rare earth element analysis by isotope dilution. *Natl. Bur. Stand. Spec. Publ.* 422:937–49
Hanson, G. N. 1977. Geochemical evolution of the suboceanic mantle. *J. Geol. Soc.* 134:235–53
Hanson, G. N. 1978. The application of trace elements to the petrogenesis of igneous rocks of granitic composition. *Earth Planet. Sci. Lett.* 38:26–43
Hanson, G. N., Goldich, S. S. 1972. Early Precambrian rocks in the Saganaga Lake–Northern Light Lake area, Minnesota–Ontario, Part II: Petrogenesis. *Geol. Soc. Am. Mem.* 135:179–92
Hanson, G. N., Langmuir, C. H. 1978. Modeling of major elements in mantle-melt systems using trace element approaches. *Geochim. Cosmochim. Acta* 42:725–41
Harris, P. G. 1974. Origin of alkaline magmas as a result of anatexis. *The Alkaline Rocks*, ed. H. Sorenson, pp. 427–36. London: Wiley
Harris, P. G., Kennedy, W. Q., Scarfe, C. M. 1970. Volcanism versus plutonism—the effect of chemical composition. *Mechanism of Igneous Intrusion*, ed. G. Newall, N. Rast, pp. 187–200. Liverpool: Gallery Press
Hart, S. R., Brooks, C. 1974. Clinopyroxene-matrix partitioning of K, Rb, Cs, Sr, and Ba. *Geochim. Cosmochim. Acta* 38:1799–1806
Haskin, L. A., Frey, F. A., Schmitt, R. A., Smith, R. H. 1966. Meteoritic, solar and terrestrial rare earth distributions. *Phys. Chem. Earth* 7:167–321
Haskin, L. A., Haskin, M. A., Frey, F. A. 1968. Relative and absolute terrestrial abundances of the rare earths. In *Origin and Distribution of the Elements*, ed. L. H. Ahrens, pp. 889–912. Oxford: Pergamon
Hellman, P. L., Green, T. H. 1979. The role of sphene as an accessory phase in the high pressure partial melting of hydrous mafic compositions. *Earth Planet. Sci. Lett.* 42:191–201
Helz, R. T. 1976. Phase relations of basalts in their melting ranges at $P_{H_2O} = 5$ Kb. Part II. Melt compositions. *J. Petrol.* 17:139–93
Herrmann, A. G., Potts, M. J., Knake, D. 1974. Geochemistry of the rare earth elements in spilites from the oceanic and continental crust. *Contrib. Mineral. Petrol.* 44:1–16
Hertogen, J., Gijbels, R. 1976. Calculations of trace element fractionation during partial melting. *Geochim. Cosmochim. Acta* 40:313–22
Higuchi, H., Nagasawa, H. 1969. Partition of trace elements between rock-forming minerals and the host volcanic rocks. *Earth Planet. Sci. Lett.* 7:281–87
Hooker, P. J., O'Nions, R. K., Pankhurst, R. J. 1975. Determination of rare earth elements in U.S.G.S. standard rocks by mixed solvent ion exchange and mass-spectrometric isotope dilution. *Chem. Geol.* 16:189–96
Irving, A. J. 1978. A review of experimental

studies of crystal-liquid trace element partitioning. *Geochim. Cosmochim. Acta* 42:743–70

Jacobs, J. W., Korotev, R. L., Blanchard, D. P., Haskin, L. A. 1977. A well tested procedure for instrumental neutron activation analysis of silicate rocks and minerals. *J. Radioanal. Chem.* 40:93–114

Kay, R. W. 1975. Chemical zonation of the oceanic mantle. *EOS, Trans. Am. Geophys. Union* 56:1077

Kay, R. W. 1977. Geochemical constraints on the origin of Aleutian magmas. In *Island Arcs, Deep Sea Trenches and Back-Arc Basins*, ed. M. Talwani, W. C. Pitman. *Am. Geophys. Union, Maurice Ewing Series* 1:229–42

Kay, R. W., Gast, P. W. 1973. The rare earth content and origin of alkali-rich basalts. *J. Geol.* 81:653–82

Kay, R. W., Hubbard, N. J., Gast, P. W. 1970. Chemical characteristics and origin of oceanic ridge volcanic rocks *J. Geophys. Res.* 75:1585–1614

Langmuir, C. H., Bender, J. B., Bence, A. E., Hanson, G. N., Taylor, S. R. 1977. Petrogenesis of basalts from the FAMOUS area: Mid-Atlantic Ridge. *Earth Planet. Sci. Lett.* 36:133–56

Langmuir, C. H., Vocke, R. D., Hanson, G. N., Hart, S. R. 1978. A general mixing equation with applications to Icelandic basalts. *Earth Planet. Sci. Lett.* 37:380–92

Lloyd, F. E., Bailey, D. K. 1975. Light element metasomatism of the continental mantle: evidence and consequences. *Phys. Chem. Earth* 9:389–416

Lopez-Escobar, L., Frey, F. A., Vergara, M. 1977. Andesites and high alumina basalts from the central-south Chile high Andes: geochemical evidence bearing on their petrogenesis. *Contrib. Mineral. Petrol.* 62:199–228

Ludden, J. N., Thompson, G. 1979. An evaluation of the behaviour of the rare earth elements during the weathering of sea-floor basalt. *Earth Planet. Sci. Lett.* 43:85–92

Magaritz, M., Hofmann, A. W. 1978. Diffusion of Eu and Gd in basalt and obsidian. *Geochim. Cosmochim. Acta* 42:847–58

Martin, R. F., Whitley, J. E., Wooley, A. R. 1978. An investigation of rare-earth mobility: fenitized quartzite Borralan complex, N.W. Scotland. *Contrib. Mineral. Petrol.* 66:69–73

Masuda, A., Nakamura, N., Tanaka, T. 1973. Fine structures of mutually normalized rare-earth patterns of chondrites. *Geochim. Cosmochim. Acta* 37:239–48

McCarthy, T. S., Hasty, R. A. 1976. Trace element distribution patterns and their relationships to the crystallization of granitic melts. *Geochim. Cosmochim. Acta* 40:1351–58

McIntire, W. L. 1963. Trace element partition coefficients—a review of theory and applications to geology. *Geochim. Cosmochim. Acta* 27:1209–64

Minster, J. F., Minster, J. B., Treuil, M., Allegre, C. J. 1977. Systematic use of trace elements in igneous processes, Part II. Inverse problem of the fractional crystallization process in volcanic suites. *Contrib. Mineral. Petrol.* 61:49–77

Mysen, B. O. 1978. Experimental determination of crystal-vapor partition coefficients for rare earth elements to 30 Kbar pressure. *Carnegie Inst. Washington Yearb.* 77:689–95

Nagasawa, H. 1970. Rare earth concentrations in zircons and apatites and their host dacites and granites. *Earth Planet. Sci. Lett.* 9:359–64

Nagasawa, H., Schnetzler, C. C. 1971. Partitioning of rare-earth, alkali and alkaline earth elements between phenocryst and acidic igneous magma. *Geochim. Cosmochim. Acta* 35:953–68

Nance, W. B., Taylor, S. R. 1976. Rare earth element patterns and crustal evolution—I. Australian post-Archean sedimentary rocks. *Geochim. Cosmochim. Acta* 40:1539–51

Neumann, H., Mead, J., Vitaliano, C. J. 1954. Trace element variations during fractional crystallization as calculated from the distribution law. *Geochim. Cosmochim. Acta* 6:90–99

O'Nions, R. K., Carter, S. R., Evensen, N. M., Hamilton, P. J. 1979. Geochemical and cosmochemical application of Nd isotopes. *Ann. Rev. Earth Planet. Sci.* 7:11–38

Paster, T. P., Schauwecker, D. S., Haskin, L. A. 1974. The behaviour of some trace elements during the solidification of the Skaergaard layered series. *Geochim. Cosmochim. Acta* 38:1549–77

Philpotts, J. A. 1978. The law of constant rejection. *Geochim. Cosmochim. Acta* 42:909–20

Philpotts, J. A., Schnetzler, C. C., Thomas, H. H. 1972. Petrogenetic implications of some new geochemical data on eclogite and ultrabasic inclusions. *Geochim. Cosmochim. Acta* 36:1131–66

Presnall, D. C. 1969. The geometric analysis of partial fusion. *Am. J. Sci.* 267:1178–94

Rayleigh, J. W. S. 1896. Theoretical considerations respecting the separation of gases by diffusion and similar processes. *Philos. Mag.* 42:77–107

Ringwood, A. E. 1974. The petrological

evolution of island arc systems. *J. Geol. Soc. London* 133:193–204

Rubey, W. W. 1955. Development of the hydrosphere and atmosphere. In *The Crust of the Earth. Geol. Soc. Am. Spec. Pap.* 62:631–50

Ryerson, F. J., Hess, P. C. 1978. The implications of liquid-liquid distribution coefficients to mineral-liquid partitioning. *Geochim. Cosmochim. Acta* 42:921–32

Schilling, J.-G. 1966. *Rare earth fractionation in Hawaiian volcanic rocks.* PhD thesis. Mass. Inst. Tech., Cambridge, Mass.

Schilling, J.-G. 1973. Iceland mantle plume: geochemical study of Reykjanes Ridge. *Nature* 242:565–71

Schilling, J.-G., Winchester, J. W. 1967. Rare earth fractionation and magmatic processes. In *Mantles of the Earth and Terrestrial Planets,* ed. S. K. Runcorn, pp. 267–83. London: Interscience

Schnetzler, C. C., Philpotts, J. A. 1970. Partition coefficients of rare-earth elements between igneous matrix material and rock-forming mineral phenocrysts, II. *Geochim. Cosmochim. Acta* 34:331–40

Shannon, R. D. 1976. Revised effective ionic radii and systematic studies of interatomic distances in halides and chalcogenides. *Acta Crystallogr.* A32:751–67

Shaw, D. M. 1970. Trace element fractionation during anatexis. *Geochim. Cosmochim. Acta* 34:237–43

Shaw, D. M. 1978. Trace element behaviour during anatexis in the presence of a fluid phase. *Geochim. Cosmochim. Acta* 42: 933–43

Shimizu, N., Kushiro, I. 1975. Partitioning of rare earth elements between garnet and liquid at high pressures: preliminary experiments. *Geophys. Res. Lett.* 2:413–16

Simmons, E. C., Hanson, G. N. 1978. Geochemistry and origin of massif type anorthosites. *Contrib. Mineral. Petrol.* 66:119–35

Simmons, E. C., Hedge, C. E. 1978. Minor-element geochemistry and Sr-isotopes of Tertiary stocks, Colorado mineral belt. *Contrib. Mineral. Petrol.* 67:379–96

Sun, S. S., Hanson, G. N. 1975. Origin of Ross Island basanitoids and limitations upon the heterogeneity of mantle sources for alkal basalts and nephelinites. *Contrib. Mineral. Petrol.* 52:77–106

Sun, S. S., Hanson, G. N. 1976. Rare earth element evidence for differentiation of McMurdo Volcanics, Ross Island Antartica. *Contrib. Mineral. Petrol.* 54:139–55

Sun, S. S., Nesbitt, R. W. 1978. Petrogenesis of ultrabasic and basic volcanics: evidence from rare earth elements. *Contrib. Mineral. Petrol.* 65:301–25

Taylor, S. R. 1965. Geochemical analysis by spark source mass spectrography. *Geochim. Cosmochim. Acta* 29:1243–61

Taylor, S. R., Gorton, M. P. 1977. Geochemical application of spark source mass spectrography-III. Element sensitivity, precision and accuracy. *Geochim. Cosmochim. Acta* 41:1375–80

Treuil, M., Varet, J. 1973. Critères volcanologiques, petrologiques et geochimiques de la genèse et de la differenciation des magmas basaltiques, exemple de l'Afar. *Soc. Geol. Fr. Bull. 7ième sér.* 15:506–40

Vocke, R. D., Hanson, G. N. 1980. U-Pb zircon ages and petrogenetic implications for two basement units from Victoria Valley, Antarctica. *Joint AGU/ARS volume on DVDP.* In preparation

Vocke, R. D., Hanson, G. N., Gruenenfelder, M. 1979. The effects of low temperature metasomatism on REE distributions in the Roffna Gneiss, Switzerland. *EOS, Trans. Am. Geophys. Union* 60:425

Watson, E. B. 1976. Two-liquid partition coefficients: experimental data and geochemical implications. *Contrib. Mineral. Petrol.* 56:119–34

Wendlandt, R. F., Harrison, W. J. 1979. Rare earth element partitioning between immiscible carbonate and silicate liquids and CO_2 vapor: Results and implications for the formation of light rare earth–enriched rock. *Contrib. Mineral. Petrol.* 69:409–19

Zielinski, R. A., Frey, F. A. 1970. Gough Island: Evaluation of a fractional crystallization model. *Contrib. Mineral. Petrol.* 29:242–54

Zielinski, R. A., Frey, F. A. 1974. An experimental study of a rare earth element (Gd) in the system diopside aqueous vapor. *Geochim. Cosmochim. Acta* 38:545–65

Ann. Rev. Earth Planet. Sci. 1980. 8:407–24
Copyright © 1980 by Annual Reviews Inc. All rights reserved

EVOLUTIONARY PATTERNS IN EARLY CENOZOIC MAMMALS

×10137

Philip D. Gingerich
Museum of Paleontology, The University of Michigan, Ann Arbor, Michigan 48109

INTRODUCTION

Life in the past was different. This is one of the most important empirical generalizations we can draw from three centuries of inquiry into the nature of fossils. By 1859, when Charles Darwin published *The Origin of Species*, evolution itself was not in question. Darwin's contribution was in understanding the mechanism: natural selection. Earlier efforts of Lamarck, Smith, Cuvier, Lyell, and Darwin himself had already established beyond doubt that life changed through time. They had no *a priori* reason to expect this but all were at some time in their career paleontologists, and all had observed firsthand the petrified remains of forms now extinct. William Smith's principle of faunal succession and correlation lies at the heart of biostratigraphy—it is also at the core of empirical evidence for evolution.

The study of organic evolution is both a geological and a biological subject. Evolution means change, change implies time, and the great sweep of life history is recorded in sedimentary rocks and measured in geological time. There are two components of organic evolution: earth history records a general increase in both the *complexity* and the *diversity* of life. Each corresponds to a mode of speciation. The first can be explained by progressive change within a single lineage. The second requires a multiplication of lineages. We know that the study of genetics, ecology, behavior, biogeography, and morphology of living animals is all necessary for understanding the evolutionary process, but this is not sufficient. Fundamental questions persist about the origin and nature of species that can only be answered by more detailed study of the historical record.

In this paper I want to review the nature of the early Cenozoic record of mammalian evolution because it illustrates both the potential and limitation of the fossil record for contributing to a better under-

0084-6597/80/0515-0407$01.00

standing of evolution at the species and faunal level. Since the most detailed record of Paleocene and Eocene mammalian evolution discovered to date is in the Clark's Fork Basin and the Big Horn Basin of Wyoming, most of the following discussion will be illustrated by examples documented there.

OUTLINE OF FAUNAL HISTORY

Dinosaurs became extinct at the end of the Cretaceous and their role in terrestrial ecosystems was rapidly filled by mammals as the Cenozoic "age of mammals" began. Paleocene mammalian faunas are dominated by archaic orders: Multituberculata, Proteutheria, Condylarthra, and plesiadapiform Primates. In contrast, Eocene mammalian faunas are dominated by modern orders: Insectivora, Chiroptera, Rodentia, Artiodactyla, Perissodactyla, Cetacea, and modern Primates. Oligocene faunas differ from those of the Eocene in being composed not only of modern orders but principally of modern families as well.

Paleogeography and Paleoclimates

Geographically, most of the major continents were approximately in their present positions by the beginning of the Cenozoic (Smith & Briden 1977). Exceptions with an important bearing on mammalian distributions were 1. a high latitude land connection between North America and Europe severed in the early Eocene by final opening of the North Atlantic, 2. partial or complete separation of eastern from western Asia by the epicontinental Obik Sea, with a high latitude land connection between eastern Asia and North America, 3. isolation of Indo-Pakistan in the present Indian Ocean, and 4. contiguity of Antarctica and Australia at high latitudes in the southern hemisphere.

Within the early Cenozoic, there were several major climatic fluctuations. The first half of the Cenozoic (Paleocene-Eocene) was characterized by an equable climate and mean annual temperatures some 10°C warmer than those of the less equable second half of the Cenozoic (Oligocene to Recent). The first half of the Paleocene was relatively warm, with climates in western North America being "subtropical." The second half of the Paleocene was cooler, with "warm temperate" climates in western North America. The lowest mean annual temperature appears to have been in the early or middle part of the late Paleocene. From this point through the early Eocene there was a general if variable increase in mean annual temperature, accompanied by a pronounced drying trend, restoring "subtropical" climates to western North America (Wolfe 1978). About 33 Ma before present (at the end of the Eocene or in the early Oligocene, depending on which boundary definition is followed) there was a major climatic change with a geologically sudden worldwide decrease in mean annual temperature and a major decrease in equability. (Ma refers to a specific date; m.y. to duration.) This event had a profound effect on marine (Savin et al 1975) and continental temperatures, and on floras (Wolfe & Hopkins 1967) and faunas (Lopez & Thaler 1975). In Europe the profound terminal Eocene or early Oligocene break in mammalian faunas is termed the *Grande Coupure* (Stehlin 1909). In North America this break corresponds to an apparently less dramatic change from forest to savanna faunas between the middle and upper parts of the Chadron Formation (Clark & Beerbower 1967).

Faunal History

The combined effect of changing paleogeography and changing climate on early Cenozoic mammalian faunas was complex. Late Mesozoic and early Cenozoic high latitude land connections such as those between North America and Asia, North America and Europe (McKenna 1975), or South America-Antarctica-Australia (Cox 1973, Tedford 1974) were easily crossed only during periods of warm climate and low sea level, and the latter two routes were subsequently broken by major tectonic rifting opening the northern part of the North Atlantic and the southern Indian Ocean, respectively. Retreat of the epicontinental Obik Sea connected eastern and western Asia. Finally, joining a land mass as large as the Indo-Pakistan subcontinent to the remainder of Asia must have had a profound effect on mammalian faunas either in releasing a mammalian fauna long evolving in isolation (if India was previously colonized by mammals) or by opening an enormous "empty" land mass for colonization by Asian mammals, with the rapid evolutionary radiations and specializations that probably accompany such an invasion.

This discussion highlights the potential complexity of early Cenozoic mammalian evolution. At the same time, however complex, a better knowledge of the early Cenozoic is essential for understanding the radiation and diversification of mammals, our own order Primates included. Early Cenozoic mammalian faunas are still poorly known or entirely lacking for much of the world, exploration continues, and several important new faunas are discovered literally every year. It is also possible to study previously known areas more intensively to document patterns of faunal evolution in greater detail. The most complete record of mammalian evolution from the middle Paleocene through early Eocene is in the Clark's Fork Basin of Wyoming. This is a particularly interesting period in mammalian history because it spans the major late

Paleocene climatic cooling and subsequent warming, and it records the first appearance of most of the modern orders that dominate mammalian faunas today. Geologically this is an interesting period as well, because it coincides with the Laramide orogeny in western North America.

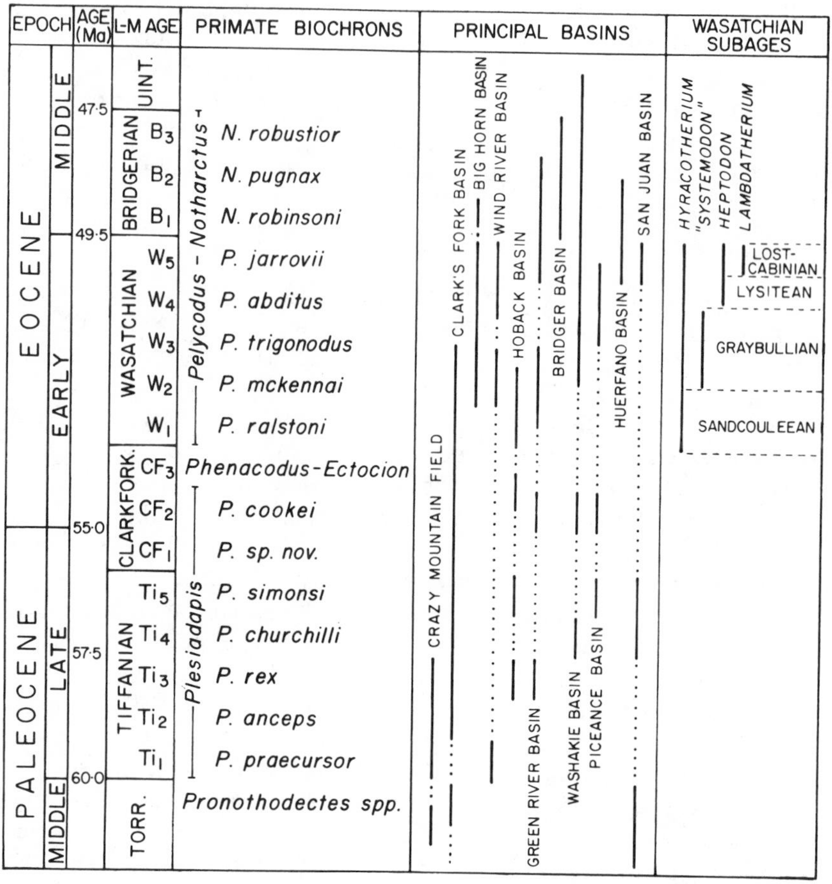

Figure 1 Biochronology of the Paleocene-Eocene transition in North America. Radiometric ages in million years before present (Ma) are from Berggren et al (1978). Tiffanian land mammal ages discussed in Gingerich (1976b), Clarkforkian extensively documented in Rose (1979), Wasatchian discussed by Savage (1977), Bridgerian reviewed by West (1976). Primate zones and biochrons documented in Figures 2 and 3. Subdivisions of land mammal ages corresponding to biochrons are listed in land mammal age column. Also shown are chronological ranges of sediments and faunas documented to date in the principal basins of the Western Interior. Right hand column shows approximate equivalence of Wasatchian subages based on first appearance of perissodactyl genera in the Clark's Fork, Big Horn, and Wind River basins.

Study of mammalian faunas in the Clark's Fork Basin is still in progress and the results presented here are incomplete, but they nevertheless serve to illustrate the potential contribution of detailed biostratigraphical studies for understanding evolution at the species level and at the faunal level.

Biostratigraphy

The Cenozoic time scale is subdivided into Epochs, and these are further divided into Ages. In North American continental sediments, land mammal "ages" have been defined based on faunas found primarily in the Western Interior (Wood et al 1941, Savage 1977). Each land mammal age has a duration of from 2 to 7 million years (Berggren et al 1978). Thus the land mammal ages themselves do not provide adequate resolution for detailed evolutionary study. In an attempt to provide a more detailed chronology for study of faunal evolution across the Paleocene-Eocene boundary, the Tiffanian, Clarkforkian, Wasatchian, and Bridgerian land mammal ages have been subdivided into biochrons based on successive species of rapidly evolving lineages of the primates *Plesiadapis*, *Pelycodus*, and *Notharctus* (Gingerich 1976b, 1979a, Gingerich & Simons 1977). These primate biochrons are listed in Figure 1. There is one gap between the ranges of *Plesiadapis* and *Pelycodus*, and the *Phenocodus-Ectocion* biochron is based on a zone of the same name in the upper Clarkforkian (Rose 1979).

The relationship of *Plesiadapis* biochrons to stratigraphy is illustrated in Figure 2. All of the *Plesiadapis* species shown, except *P. praecursor*, are now known from correlated measured sections in the Fort Union (Polecat Bench) and Willwood Formations on Polecat Bench and in the Clark's Fork Basin. *P. praecursor* is interpolated based on its occurrence between *Pronothodectes gidleyi* and *Plesiadapis anceps* in the Fort Union Formation of the Crazy Mountain Field of Montana, a northwesterly extension of the Clark's Fork Basin. Carpolestidae have also been used to refine the chronology of the middle and late Paleocene (Rose 1977), and an interpretation of carpolestid phylogeny is shown in Figure 2. In this case several of the species have been interpolated based on association with *Plesiadapis* in other basins (*Elphidotarsius shotgunensis* and *Carpodaptes hobackensis*, for example, have not been found in the Clark's Fork Basin or the Big Horn Basin).

The stratigraphic record of *Pelycodus* in the central Big Horn Basin and the sequence of *Notharctus* species in the Bridger Basin (Figure 3) provide the basis for the *Pelycodus-Notharctus* biochrons.

The total number of biochrons listed in Figure 1 is 16, and the time represented is approximately 12.5 m.y. (60 Ma to 47.5 Ma). Thus the

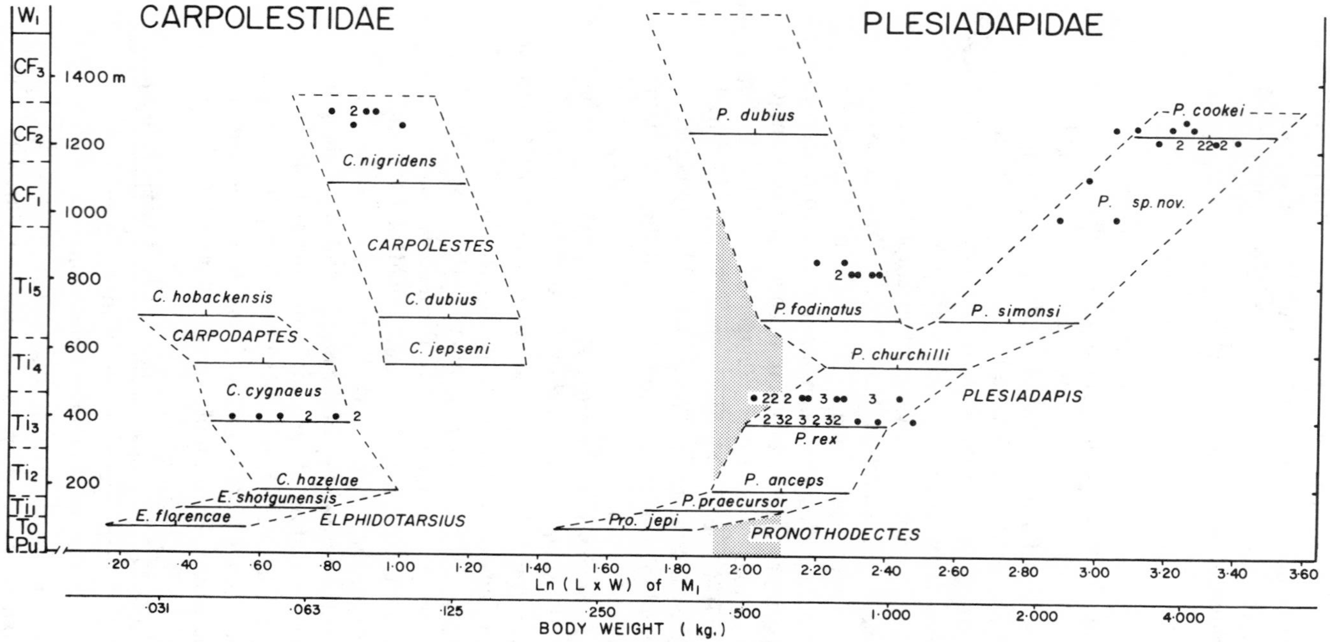

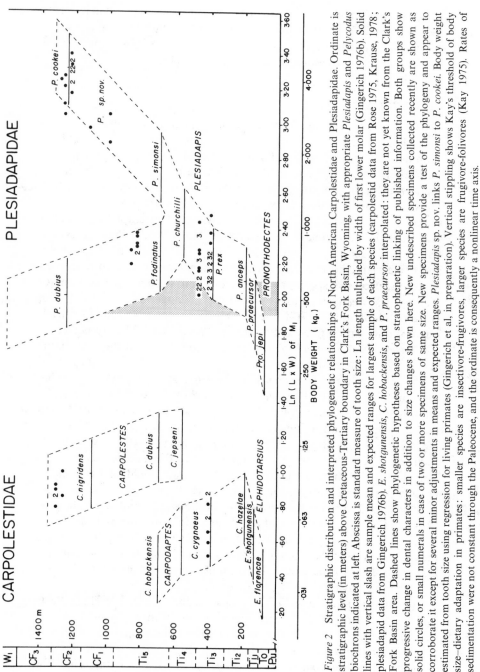

Figure 2 Stratigraphic distribution and interpreted phylogenetic relationships of North American Carpolestidae and Plesiadapidae. Ordinate is stratigraphic level (in meters) above Cretaceous-Tertiary boundary in Clark's Fork Basin, Wyoming, with appropriate *Plesiadapis* and *Pelycodus* biochrons indicated at left. Abscissa is standard measure of tooth size: Ln length multiplied by width of first lower molar (Gingerich 1976b). Solid lines with vertical slash are sample mean and expected ranges for largest sample of each species (carpolestid data from Rose 1975, Krause, 1978; plesiadapid data from Gingerich 1976b). *E. shotgunensis*, *C. hobackensis*, and *P. praecursor* interpolated: they are not yet known from the Clark's Fork Basin area. Dashed lines show phylogenetic hypotheses based on stratophenetic linking of published information. Both groups show progressive change in dental characters in addition to size changes shown here. New undescribed specimens collected recently are shown as solid circles, or small numerals in case of two or more specimens of same size. New specimens provide a test of the phylogeny and appear to corroborate it except for several minor adjustments in means and expected ranges. *Plesiadapis* sp. nov. links *P. simonsi* to *P. cookei*. Body weight estimated from tooth size using regression for living primates (Gingerich et al, in preparation). Vertical stippling shows Kay's threshold of body size–dietary adaptation in primates: smaller species are insectivore-frugivores, larger species are frugivore-folivores (Kay 1975). Rates of sedimentation were not constant through the Paleocene, and the ordinate is consequently a nonlinear time axis.

average duration of each biochron is about 0.8 m.y. or 800,000 years. Although there was undoubtedly some variation in the length of time represented by different biochrons, there is close agreement of ages interpolated using biochrons with radiometric ages at the bases of Ti_4, CF_2, and B_1. As a result, it is possible to achieve much greater chronological resolution using *Plesiadapis-Pelycodus-Notharctus* biochrons than is possible using land mammal ages. Obviously mammalian biochrons are not a substitute for good radiometric ages, but they can be recognized in a suite of continental sediments not usually amenable to radiometric dating and the two approaches are complementary.

There are a number of important applications of improved chronological resolution using *Plesiadapis-Pelycodus-Notharctus* biochrons in

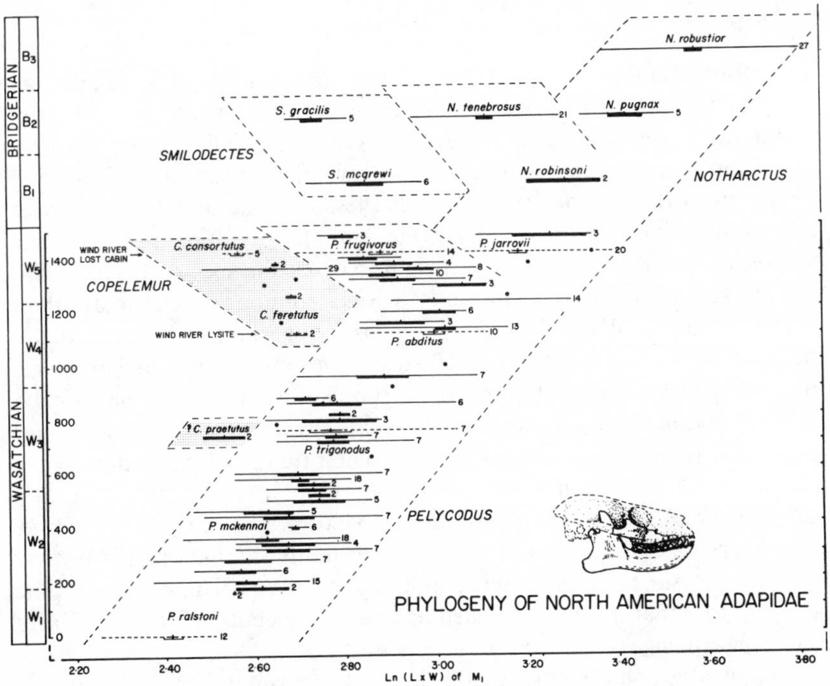

Figure 3 Stratigraphic distribution and interpreted phylogenetic relationships of North American Adapidae. Ordinate is stratigraphic level (in feet) above base of Willwood Formation in central Big Horn Basin (Gingerich 1976a), with superimposed Bridgerian faunal zones from the Bridger Basin. Abscissa is measure of tooth size. Horizontal line is observed range, vertical slash is mean, horizontal bar is standard error, and numeral at right of range is number of specimens in sample. Solid circles are individual specimens. Data from Gingerich & Simons (1977) and Gingerich (1979a). Dashed lines show pattern of stratophenetic linking and hypothesis of phylogenetic relationships.

the temporal interval spanning the Paleocene-Eocene boundary: 1. improved correlations between basins in the Western Interior (e.g. Gingerich 1976b, Figure 15; Krause 1978, Figure 11), 2. improved understanding of phylogenetic relationships in other less abundantly represented taxa, such as Carpolestidae (Figure 2) and Esthonychidae (Gingerich & Gunnell 1979), 3. dating of tectonic events, such as the movement of major thrust slices in the Wyoming thrust belt (Dorr & Gingerich 1979). Finally, as fossiliferous sections are studied in more detail it should be possible to compare types and rates of sedimentation within and between sedimentary basins of the Western Interior and thus gain a better understanding of the detailed chronology of many aspects of earth history as well as life history.

EVOLUTIONARY PATTERNS

The more detailed the stratigraphic sampling of a lineage or clade, the greater the power of resolution of events occurring in its evolution. Stratigraphic detail depends on several factors, especially on the thickness of section spanning a given temporal interval and the spacing of samples drawn from the section. In Figure 2, Plesiadapidae and Carpolestidae are known from approximately one sample per biochron, i.e. one sample every 0.8 m.y., which is sufficient to order the species in succession but not sufficient to document the transition from one species to another. The evolution of Wasatchian *Pelycodus* is known in much greater detail because there are an average of 10 successive samples per biochron, i.e. one every 80,000 years (Figure 3). In the central Big Horn Basin each foot of section represents about 2500 years on average, and each meter of section represents about 8000 years. The rate of sedimentation in the Clark's Fork Basin during the early Wasatchian was more rapid, and each meter of section represents about 3300 to 3400 years on average. The Clark's Fork Basin is not yet as thoroughly sampled as the central Big Horn Basin, but it potentially should yield a Wasatchian stratigraphic record more than twice as detailed as that in the central Big Horn Basin as a consequence of more rapid sedimentation.

The stratigraphic record of the lemur-like primate *Pelycodus* in the Clark's Fork Basin is shown in Figure 4. The pattern is exactly the same as that shown in Figure 3, with gradual increase in size in a single lineage. Measured in darwins [defined by Haldane (1949) as change by a factor of *e* per m.y.], the rate of increase in size in *Pelycodus* was approximately 0.20. At the same time, in the same stratigraphic section and localities, tarsier-like *Tetonoides* and *Tetonius* show a very different evolutionary pattern. *Tetonoides* is present through biochron W_1 and the

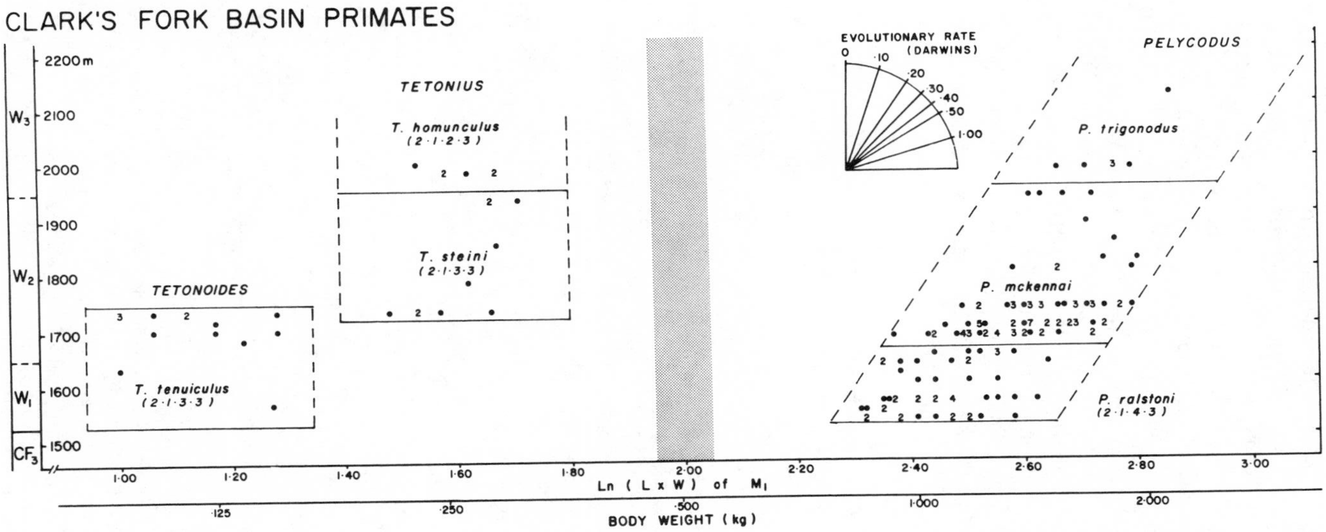

Figure 4 Evolutionary patterns in early Eocene Primates. Abscissa, ordinate, and stippling as in Figure 2. Individual and multiple specimens shown with solid circles and numerals, respectively. Note replacement of *Tetonoides* by a larger *Tetonius* with the same dental formula (2.1.3.3). Evolution of *Pelycodus* in the Clark's Fork Basin is identical to that in the Big Horn Basin (Figure 3). *Tetonoides* and *Tetonius* show little change in size within species lineages, whereas tooth size in *Pelycodus* increased at a rate of about 0.20 darwins.

lower part of W_2. It exhibits no change through time, and apparently disappeared at about level 1750 m. Just before *Tetonoides* disappeared, the closely related larger genus *Tetonius* appeared abruptly as an immigrant. *Tetonius* exhibits little or no change in tooth size through time (although the loss of one tooth, P_2, and enlargement of the central incisor distinguish *T. homunculus* from *T. steini*).

The stratigraphic record of the Clarkforkian and Wasatchian tillodont *Esthonyx* is shown in Figure 5. *Esthonyx* is a tillodont the size of a small pig and it was also probably pig-like in dental adaptation. One lineage of *Esthonyx* appears at or near the beginning of the Clarkforkian and it became larger through time, evolving at a rate of 0.50 darwins. At the beginning of the Wasatchian a second lineage of *Esthonyx* appeared and it also became larger through biochron W_1 and part of W_2. Initially the

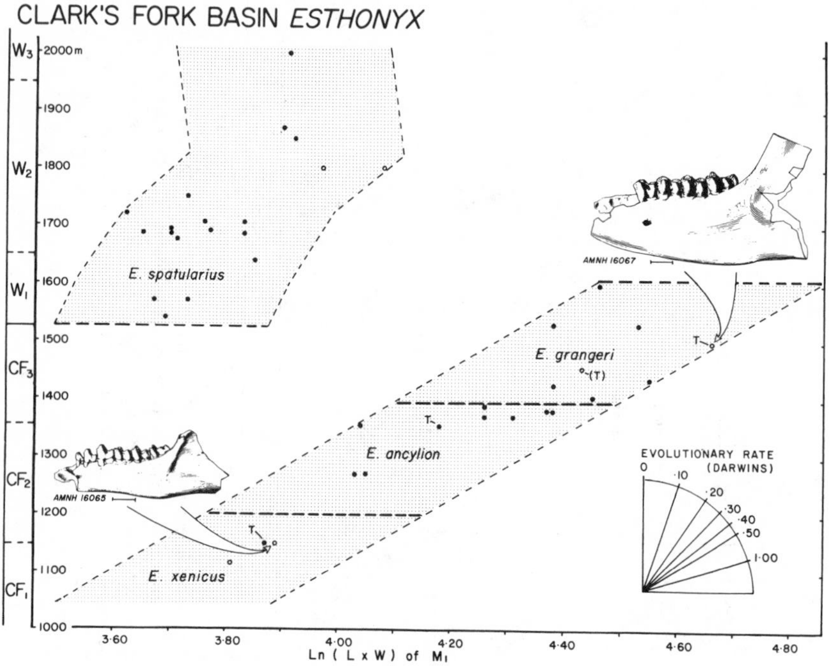

Figure 5 Evolutionary patterns in early Eocene Tillodontia. Abscissa and ordinate same as Figure 2. One lineage of *Esthonyx* appeared in the early Clarkforkian (CF_1) and evolved rapidly toward larger size (tooth size increased at about 0.50 darwins). This lineage was replaced in the early Wasatchian by another lineage that initially became larger and then stabilized. It is possible that the transition from *E. spatularius* in lower W_2 to the species present in upper W_2 and W_3 was discontinuous but critical evidence is lacking. Figure from Gingerich & Gunnell (1979).

rate of size evolution in *E. spatularius* was about 0.20 darwins, and then it stabilized and changed little through the last half of biochron W_2 and W_3. Both lineages are found in the same fossil localities in biochron W_1. The transition from one lineage to the other, like that from *Tetonoides* to *Tetonius*, is not preserved in the Clark's Fork Basin.

The fossil record of the horse has been an evolutionary classic for over a century (Drake 1978). The earliest well-dated record of horses in North America is in the Clark's Fork Basin, where *Hyracotherium* first appeared at the beginning of the Wasatchian [intensive collecting in the vicinity of Princeton Quarry has failed to corroborate Jepsen & Woodburne's (1969) record of a late Tiffanian *Hyracotherium* in the Clark's Fork Basin]. *Hyracotherium* first became larger, then smaller, and then larger again, with *H. grangeri* giving rise to *H. aemulor* (Figure 6). The maximum rate of evolution in this lineage was about 0.50 darwins. Near the beginning of biochron W_3 a second lineage represented by *H. pernix* appeared abruptly and both lineages subsequently evolved with little change. The tapiroid *Homogalax* is represented in the Clark's Fork Basin by two species, the earlier of which appeared near the base of the Wasatchian and became smaller at a rate of about −0.30 darwins. It was replaced by E. D. Cope's Big Horn Basin "*Systemodon*," or

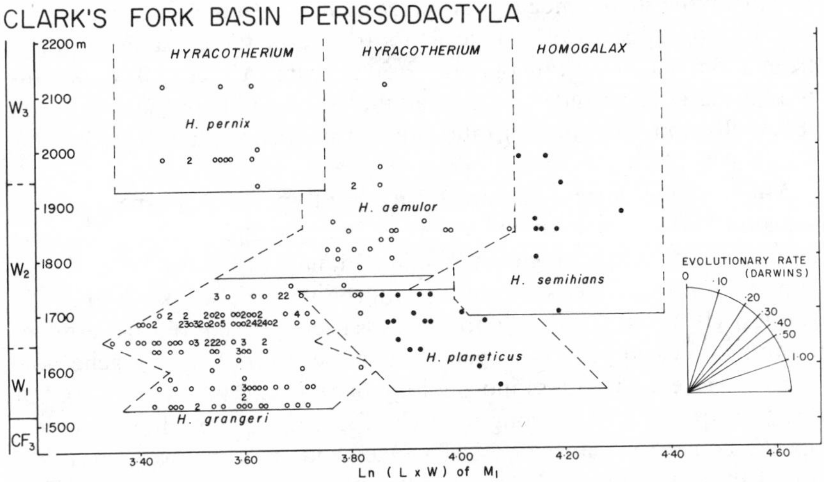

Figure 6 Evolutionary patterns in early Eocene Perissodactyla. Abscissa and ordinate same as Figure 2. Note zigzag pattern of gradual evolution in *Hyracotherium grangeri* leading to the large *H. aemulor*. Maximum rate of change in tooth size in this transition was about 0.50 darwins. *Homogalax semihians* immigrated into the Clark's Fork Basin, replacing *H. planeticus*. *Hyracotherium pernix* also appears to be an immigrant. Figure from Gingerich (1979b).

Homogalax semihians. The two species are both found together in the interval from 1700 to 1750 m. Following its abrupt appearance, *H. semihians* apparently changed very little during subsequent evolution.

Species Level Evolution

The detailed species level evolutionary patterns discussed here represent only six genera in an early Wasatchian fauna containing approximately 50 or more mammalian genera, most of which remain to be analyzed. These examples clearly do not adequately represent all possible patterns, but they do document a spectrum of evolutionary rates and two distinct modes of appearance of new species.

Cope's Rule is an evolutionary generalization sometimes cited as law: animals tend to evolve toward larger body size (Stanley 1973). Considering the direction of evolution in the lineage patterns shown in Figures 2–6, a total of ten lineage segments (45%) evolved toward larger tooth and body size, eight (36%) evolved toward smaller size, and four (18%) remained unchanged. Thus there is a slight tendency in these examples for evolution in the direction of larger body size to predominate over evolution toward smaller size, but since it occurs in less than 50% of the examples this is hardly a general rule. The simple historical fact that mammals originated at relatively small body size is probably sufficient to explain the slight tendency for increases in size to predominate.

Evolutionary rates of change in tooth size in these examples range from 0 to a maximum rate of about 0.50 darwins. A rate of 0.50 darwins is an order of magnitude greater than the average rate for mammals, but it is well within the range of rates previously documented (Kurten 1960, Van Valen 1974).

Much of the current disagreement about species and speciation is semantic. It is possible to define species in a given place at a given time as clusters of morphologically similar animals separated by real discontinuities from other such clusters (morphological, or "skin and skull" species). It is also possible to define species based on patterns of reproduction and gene flow (genetical species). If patterns of gene flow were congruent with gross morphology there would be no problem, but the two are not always congruent (Endler 1973). Some morphological species consist of many genetical species and the converse is possible. The fact that both concepts are included under the term "species" is confusing. *Deme* is a term meaning an "interbreeding community" and it is an appropriate name for the "genetical species." Normally many demes make up a morphological species, and restriction of the term *species* to the larger "skin and skull" concept would be consistent with historical precedent going back to Linnaeus. Natural selection probably operates at many levels in evolution, selectively favoring individuals, demes, and even species (Stanley 1975). Knowledge of evolution at each level is important for understanding the general process. Patterns of evolution at the individual and deme level are beyond the power of resolution of virtually all paleontological sampling, and the patterns described here illustrate evolution at the level of the Linnaean species.

There are three principal components of evolutionary space: morphology, time, and geography. These components are all portrayed in Figures 2–6. Morphology is on the horizontal x-axis, time is on the vertical y-axis, and geography is (theoretically) on the z-axis going into the page. In a sense, the different figures each represent different positions of the morphological component as well, since each includes only genera that are closely similar. All of the figures illustrate evolution in one geographical area, and consequently values on the z-axis are constant and all changes can be shown in the morphology-time plane. This is not to say that geography is unimportant in evolution or that it should purposely be ignored. Geography is another important component necessary for a full understanding of species and speciation, but the level of our knowledge of the geographical distribution of Paleocene and Eocene taxa is not yet sufficient to adequately characterize geographical distributions.

Some Wasatchian species are found from Wyoming to New Mexico, while others appear to be confined to one region or the other. Before a real geographical component can be added to Figures 2–6, vertical successions in other sedimentary basins will have to be studied in the detail shown here. Detailed study of any one basin is a large undertaking, and we are not likely to have sufficient knowledge to discuss geographical distributions of species in the near future. On a related problem, the "subspecies" is a geographical concept clearly violated by use in a temporal sense, with the unnecessary burden of writing a trinomial when a binomial conveys exactly the same information. Taking the evolution of *Esthonyx* shown in Figure 5 as an example, reference of *Esthonyx ancylion* and *E. xenicus* to the previously named *E. grangeri* as subspecies (*E. grangeri ancylion*, *E. grangeri xenicus*, with *E. grangeri grangeri*) conveys no additional information over writing *E. grangeri*, *E. ancylion*, and *E. xenicus*. In either case, relationships must be shown in a diagram.

Species, as portrayed in Figures 2–6, are groups of similar individuals. They are segments of evolving lineages bounded by morphological and temporal (and potentially geographical) discontinuities from other such lineages. Apparent branching points like the one shown in *Plesiadapis* (Figure 2) or in *Pelycodus* (Figure 3) also bound species lineages temporally at the point where one lineage becomes two. Within some lineages there is sufficient change in size or other characteristics to require subdivision of the lineage into two or more successive species. In practice, the amount of size change separating successive easily recognizable species is a shift in mean size of one-half of the expected range of size variation, or a shift of 0.2 units on the scale of tooth size used in Figures 2–6. Species are naturally bounded in most dimensions, but this is the one case where an arbitrary boundary must be drawn between species, similar to the arbitrary boundary that separates successive hours in our daily lives. Successive hours are different and worthy of delimitation, and successive species in a single lineage are also worthy of delimitation if the lineage is changing significantly.

Two modes of species appearance are evident in Figures 2–6. Some new species arise by gradual evolution within lineages (anagenesis). Appearance of most of the successive species of *Carpodaptes*, *Plesiadapsis*, or *Pelycodus* are examples of speciation by gradual evolution. Other species appear abruptly in the record as immigrants from elsewhere. *Pelycodus ralstoni*, *Esthonyx spatularius*, and *Hyracotherium grangeri* are three examples of species that first appear by immigration at the beginning of the Wasatchian. In several cases an immigrant replaces a closely related species previously present, for example *Tetonius steini* (Figure 4), *Esthonyx spatularius* (Figure 5), or *Homogalax semihians* (Figure 6). This pattern conforms closely to a prediction of the "punctuated equilibria" model of speciation (Gould & Eldredge 1977), although the geographical perspective necessary to confirm rapid speciation in a small peripheral isolate is still lacking. Leaving out ambiguous cases, there are 24 examples of species appearing by gradual evolution and 14 examples of species appearing by immigration shown in Figures 2–6. The relative frequencies of the two modes may change as additional components of Paleocene and Eocene faunas are studied, but this evidence is sufficient to establish that both modes are present. It is, of course, possible that species first appearing as immigrants in the Clark's Fork Basin evolved gradually elsewhere.

Faunal Evolution

Definitive study of evolution at a faunal level will not be possible until the systematics of all of the component species are well understood. This is not yet the case. Nevertheless, it is possible to make some preliminary observations on patterns of faunal evolution across the Paleocene-Eocene boundary. Some species evolved rapidly while others evolved more slowly. Some appeared abruptly by immigration and disappeared by emigration or extinction. Our work to date has shown that the net result is relatively slow gradual evolution of the mammalian fauna as a whole, interrupted by several episodes of major faunal turnover.

The best documented episode of turnover is at the beginning of the Wasatchian (W_1), when Artiodactyla, Perissodactyla, modern Primates, and hyaenodontid Creodonta first appeared. Another faunal turnover occurred at the beginning of the Clarkforkian (CF_1), when Rodentia, Tillodontia, miacine Carnivora, and coryphodontid Pantodonta first appeared (Rose 1979). A possible episode of faunal turnover ocurred in the Tiffanian near the end of the *Plesiadapis rex* biochron or beginning of the *P. churchilli* biochron (Ti_3 or Ti_4), when a number of archaic primate and other genera first appeared. This turnover may have been more gradual than the others, and it is not yet well documented. Several elements with possible South American affinities (an edentate, a notoungulate, and *Probathyopsis*) first appeared in the late Tiffanian (Ti_4 or Ti_5), possibly as part of this turnover. Finally, as Figures 4–6 suggest, there may have been another distinct episode of faunal turnover in the early Wasatchian (W_2) between levels 1650 and 1750 m in the Clark's Fork Basin: *Tetonius* replaced *Tetonoides, Esthonyx grangeri* disappeared, and *Homogalax semihians* replaced *H. planeticus*. Evolution at a faunal level across the Paleocene-Eocene boundary included periods of episodic turnover as well as gradual evolution.

Rose (1979) has made an extensive study of faunal composition and diversity in mammalian faunas spanning the Torrejonian, Tiffanian, Clarkforkian, and early Wasatchian, and he shows a clear relationship of mammalian faunal diversity with changing climate. Diversity was high in the Torrejonian. It was relatively low in the middle Tiffanian (Ti_2 and Ti_3), corresponding to the low in mean annual temperature discussed above. Diversity then increased gradually though the early Wasatchian as mean annual temperature increased.

Climate and high latitude land connections between the northern continents can be combined in a tentative hypothesis to explain the episodes of faunal turnover across the Paleocene-Eocene boundary. A warming climate would expand the potential geographic range of subtropically and tropically adapted animals. In the late Tiffanian this may explain the first appearance of mammals with Central American or South American affinities in northwestern Wyoming (Sloan 1969, Gingerich 1976b). The initial Clarkforkian episode of faunal turnover includes immigrants with Asian affinities, suggesting that climates had warmed sufficiently for these mammals to cross into North America from Asia. The immigration at the beginning of the Wasatchian includes no taxa clearly indicating area of origin, but there is suggestion of European (*Hyracotherium*) or African (adapid primates, hyaenodontid creodonts) affinities. It is probable that this immigration was the only major one to cross the high latitude route between Europe and North America after late Paleocene climatic cooling. The route was closed permanently by further rifting of the North Atlantic in the early Wasatchian. Considering present knowledge, every detail of this hypothesis is tentative and subject to change with future discoveries. One critical gap, the nature of eastern Asian mammalian faunas, is rapidly being filled by colleagues in China, Mongolia, and the U.S.S.R. Other major gaps remain, but the hypothesis gives some structure for further inquiry and it is one possible explanation for documented episodes of faunal turnover.

SUMMARY AND CONCLUSIONS

Historical records are essential for deciphering past events. Geology, stratigraphy, and paleontology contribute uniquely to understanding evolution in its temporal dimension. The stratigraphic record of early Cenozoic mammals is remarkably complete in some areas, although we certainly still lack an adequate geographical perspective. Faunal changes across the Paleocene-Eocene boundary are documented in a 2000 m stratigraphic section spanning the Tiffanian, Clarkforkian, and early Wasatchian land mammal ages in the Clark's Fork Basin of Wyoming. Here it is possible to trace individual mammalian lineages through time. In the illustrated examples ten species lineages become larger, eight become smaller, and four exhibited little or no change. The maximum rate of change in tooth size is about 0.50 darwins. In the examples where mode of appearance of new species is clearly shown, 24 originated by gradual evolution and 14 show a "punctuated" pattern of first appearance by immigration into the Clark's Fork Basin from elsewhere.

At a faunal level, the net result of dynamic change in individual lineages is slow gradual evolution of the mammalian fauna as a whole, interrupted by several episodes of faunal turnover. Interpreted in light of early Cenozoic paleoclimates, mammalian faunal diversity appears to be closely related to climate, and episodes of faunal turnover may coincide with the successive introduction of South American, Asian, and European-African elements, respectively, into North America. The latter two introductions possibly appear episodic rather than gradual because they resulted from a threshold effect of immigration across high latitude land bridges opened by progressively warming climates.

Mammalian paleontology has reached a stage where progress in understanding early Cenozoic faunal evolution, including the origin and diversification of Primates and other modern orders of mammals, depends on detailed biostratigraphic documentation. Fortunately there are many sedimentary basins in the Western Interior where mammalian remains are abundantly preserved in stratigraphic context, and in future these promise to yield a better perspective on evolution at the species level as well.

ACKNOWLEDGMENTS

Results presented here reflect the combined efforts of a research group studying mammalian paleontology, stratigraphy, and taphonomy in the Clark's Fork Basin. I am indebted to K. D. Rose, D. W. Krause, D. A. Winkler, G. F. Gunnell, and many others for discussion and assistance in compiling data. E. H. Lindsay and R. F. Butler supplied preliminary results of work on paleomagnetic stratigraphy. I also thank E. Peer, K. Steelquist, G. Paulson, and A. Benson for their help in completing this paper. Supported by NSF grant DEB 77-13465.

Literature Cited

Berggren, W. A., McKenna, M. C., Hardenbol, J., Obradovich, J. D. 1978. Revised Paleogene polarity time scale. *J. Geol.* 86:67–81

Clark, J., Beerbower, J. R. 1967. Geology, paleoecology, and paleoclimatology of the Chadron Formation. *Fieldiana Geol. Mem.* 5:21–74

Cox, C. B. 1973. Systematics and plate tectonics in the spread of marsupials. *Spec. Pap. Paleontol.* 12:113–19

Dorr, J. A., Gingerich, P. D. 1979. Early Cenozoic mammalian paleontology, geologic structure, and tectonic history in the overthrust belt near LaBarge, western Wyoming. *Contrib. Geol. Univ. Wyoming.* In press

Drake, E. T. 1978. Horse genealogy: the Oregon connection. *Geology* 6:587–91

Endler, J. A. 1973. Gene flow and population differentiation. *Science* 179:243–50

Gingerich, P. D. 1976a. Paleontology and phylogeny: patterns of evolution at the species level in early Tertiary mammals. *Am. J. Sci.* 276:1–28

Gingerich, P. D. 1976b. Cranial anatomy and evolution of early Tertiary Plesiadapidae (Mammalia, Primates). *Univ. Mich. Pap. Paleontol.* 15:1–140

Gingerich, P. D. 1979a. Phylogeny of middle Eocene Adapidae (Mammalia, Primates) in North America: *Smilodectes* and *Notharctus*. *J. Paleontol.* 53:153–63

Gingerich, P. D. 1979b. Evolution of early Eocene Perissodactyla (Mammalia) in the Clark's Fork Basin of Wyoming, with a systematic review of North American *Hyracotherium* and *Homogalax*. *Contrib. Mus. Paleontol. Univ. Mich.* In press

Gingerich, P. D., Gunnell, G. F. 1979. Systematics and evolution of the genus Esthonyx (Mammalia, Tillodontia) in the early Eocene of North America. *Contrib. Mus. Paleontol. Univ. Mich.* 25:125–53

Gingerich, P. D., Simons, E. L. 1977. Systematics, phylogeny, and evolution of early Eocene Adapidae (Mammalia, Primates) in North America. *Contrib. Mus. Paleontol. Univ. Mich.* 24:245–79

Gould, S. J., Eldredge, N. 1977. Punctuated equilibria: the tempo and mode of evolution reconsidered. *Paleobiology* 3: 115–51

Haldane, J. B. S. 1949. Suggestions as to quantitative measurement of rates of evolution. *Evolution* 3:51–6

Jepsen, G. L., Woodburne, M. O. 1969. Paleocene hyracothere from Polecat Bench Formation, Wyoming. *Science* 164:543–47

Kay, R. F. 1975. The functional adaptations of primate molar teeth. *Am. J. Phys. Anthropol.* 43:195–216

Krause, D. W. 1978. Paleocene primates from western Canada. *Can. J. Earth Sci.* 15:1250–71

Kurten, B. 1960. Rates of evolution in fossil mammals. *Cold Spring Harbor Symp. Quant. Biol.* 24:205–15

Lopez, N., Thaler, L. 1975. Sur le plus

ancien lagomorphe européen et la "grande coupure" oligocène de Stehlin. *Palaeovertebrata* 6:243–51

McKenna, M. C. 1975. Fossil mammals and early Eocene North Atlantic land continuity. *Ann. Mo. Bot. Gard.* 62:335–53

Rose, K. D. 1975. The Carpolestidae, early Tertiary primates from North America. *Bull. Mus. Comp. Zool.* 147:1–74

Rose, K. D. 1977. Evolution of carpolestid primates and chronology of the North American middle and late Paleocene. *J. Paleontol.* 51:536–42

Rose, K. D. 1979. *The Clarkforkian land-mammal "age" and mammalian faunal composition across the Paleocene-Eocene boundary*. PhD thesis. Univ. Mich., Ann Arbor. 601 pp.

Savage, D. E. 1977. Aspects of vertebrate paleontological stratigraphy and geochronology. In *Concepts and Methods of Biostratigraphy*, ed. E. G. Kauffman, J. E. Hazel, pp. 427–42. Stroudsburg, PA: Dowden, Hutchinson & Ross. 658 pp.

Savin, S. M., Douglas, R. G., Stehli, F. G. 1975. Tertiary marine paleotemperatures. *Bull. Geol. Soc. Am.* 86:1499–1510

Sloan, R. E. 1969. Cretaceous and Paleocene terrestrial communities of western North America. *Proc. N. Am. Paleontol. Conv.* E:427–53

Smith, A. G., Briden, J. C. 1977. *Mesozoic and Cenozoic Paleocontinental Maps*. Cambridge, England: Cambridge Univ. Press. 63 pp.

Stanley, S. M. 1973. An explanation for Cope's Rule. *Evolution* 27:1–26

Stanley, S. M. 1975. A theory of evolution above the species level. *Proc. Natl. Acad. Sci. USA* 72:646–50

Stehlin, H. G. 1909. Remarques sur les faunules de mammifères des couches éocènes et oligoènes du Bassin de Paris. *Bull. Soc. Géol. France Sér. 4* 9:488–520

Tedford, R. H. 1974. Marsupials and the new paleogeography. *Soc. Econ. Paleont. Mineral. Spec. Publ.* 21:109–26

Van Valen, L. 1974. Two modes of evolution. *Nature* 252:298–300

West, R. M. 1976. Paleontology and geology of the Bridger Formation, southern Green River Basin, southwestern Wyoming, part 1, history of field work and geological setting. *Contrib. Biol. Geol. Milwaukee Publ. Mus.* 7:1–12

Wolfe, J. A. 1978. A paleobotanical interpretation of Tertiary climates in the northern hemisphere. *Am. Sci.* 66:694–703

Wolfe, J. A., Hopkins, D. M. 1967. Climatic changes recorded by Tertiary land floras in northwestern North America. *Symp. 11th Pacific Sci. Congr.* 25:67–76

Wood, H. E., Chaney, R. W., Clark, J., Colbert, E. H., Jepsen, G. L., Reeside, J. B., Stock, C. 1941. Nomenclature and correlation of the North American continental Tertiary. *Bull. Geol. Soc. Am.* 52:1–48

Ann. Rev. Earth Planet. Sci. 1980. 8: 425–87

ORIGIN AND EVOLUTION OF PLANETARY ATMOSPHERES[1]

✕10138

James B. Pollack
Space Science Division, NASA-Ames Research Center, Moffett Field, California 94035

Yuk L. Yung
Division of Geological and Planetary Sciences, California Institute of Technology, Pasadena, California 91125

INTRODUCTION

Spacecraft and groundbased observations of the atmospheres of solar system objects have provided a definition of their present characteristics and have yielded clues about their past history. Table 1 presents a summary of our current knowledge of the atmospheric properties of all the planets, except Pluto, and several satellites. The masses of these atmospheres range from the very miniscule values for the Moon, Mercury, and Io, to the more substantial values for the Earth, Venus, Mars, and Titan, to the very large values for the giant planets, where the atmosphere constitutes a significant fraction of the total planetary mass. The compositions of these atmospheres encompass ones dominated by rare gases (the Moon and Mercury), ones containing highly oxidized compounds of carbon, nitrogen, and sulfur (the outer three terrestrial planets and Io), and ones with highly reduced gases (Titan and the giant planets). What factors account for this enormous diversity in properties?

There are reasons for thinking that in the past these atmospheres were quite different from their present state. For example, it is considered highly unlikely, if not impossible, for the chemical steps that led to the origin of life on the Earth to have occurred with the current, highly oxidized atmosphere. Rather, an early, somewhat reducing atmosphere needs to be postulated (Miller & Orgel 1974). In the case of Mars, certain

[1] The US Government has the right to retain a nonexclusive, royalty-free license in and to any copyright covering this paper.

Table 1 Properties of the atmospheres of solar system objects

Object	$\bar{\rho}$[a]	g[a]	P_s[a]	T_s[a]	Major gases[a]	Minor gases[a]	Aerosols[a]
Mercury	5.43	3.95×10^2	$\sim 2 \times 10^{-15}$	440	He(~0.98), H(~0.02)[b]	—	—
Venus	5.25	8.88×10^2	90	730 (~230)	CO_2(0.96), N_2(~0.035)	H_2O(20–5000), SO_2(~150), Ar(20–200), Ne(4–20), CO(50)[c], HCl(0.4)[c], HF(0.01)[c]	Sulfuric acid (~35)
Earth	5.52	9.78×10^2	1	288 (~255)	N_2(0.77), O_2(0.21), H_2O(~0.01), Ar(0.0093)	CO_2(315), Ne(18), He(5.2), Kr(1.1), Xe(0.087), CH_4(1.5), H_2(0.5), N_2O(0.3), CO(0.12), NH_3(0.01), NO_2(0.001), SO_2(0.0002), H_2S(0.0002), O_3(~0.4)	Water (~5) Sulfuric acid (~0.01–0.1)[d] Sulfate, sea salt Dust, organic (~0.1)[d]
Mars	3.96	3.73×10^2	0.007	218 (~212)	CO_2(0.95), N_2(0.027), Ar(0.016)	O_2(1300), CO(700), H_2O(~300), Ne(2.5), Kr(0.3), Xe(0.08), O_3(~0.1)	Water ice (~1)[e] Dust (~0.1–10)[e] CO_2 ice (?)[e]
Moon	3.34	1.62×10^2	$\sim 2 \times 10^{-14}$	274	Ne(~0.4), Ar(~0.4), He(~0.2)	—	—
Jupiter	1.34	2.32×10^3	$\gg 100$[f]	(129)	H_2(~0.89), He(~0.11)	HD(20), CH_4(~2000), NH_3(~200), H_2O(1?), C_2H_6(~5)[f], CO(0.002), GeH_4(0.0007), HCN(0.1), C_2H_2(~0.02)[f], PH_3(0.4)	Stratospheric "smog" (~0.1) Ammonia ice (~1) Ammonium hydrosulfide (~1) Water (~10)

Saturn	0.68	8.77×10^2	$\gg 100$[f]	(97)	$H_2(\sim 0.89)$, $He(\sim 0.11)$	$CH_4(\sim 3000)$, $NH_3(\sim 200)$, $C_2H_6(\sim 2)$[f]	Same aerosol layers as for Jupiter
Uranus	1.55	9.46×10^2	$\gg 100$[f]	(58)	$H_2(\sim 0.89)$, $He(\sim 0.11)$	CH_4	Same aerosol layers as for Jupiter, but thinner smog layer plus possibly methane ice
Neptune	2.23	1.37×10^3	$\gg 100$[f]	(56)	$H_2(\sim 0.89)$, $He(\sim 0.11)$	CH_4	Same aerosol layers as for Jupiter, plus possibly methane ice
Titan	~ 1.4	$\sim 1.25 \times 10^2$	$2 \times 10^{-2} \rightarrow \sim 1$	~ 85	$CH_4(0.1–1)$	$C_2H_6(\sim 2)$	Stratospheric "smog" (~ 10)
Io	3.52	1.79×10^2	$\sim 1 \times 10^{-10}$	~ 110	$SO_2(\sim 1)$	—	—

[a] Reading from left to right, these variables are the following: the object's mean density, in units of gm/cm^3; acceleration of gravity, in cm/s^2; surface pressure, in bars; surface temperature, in K—the numbers in parentheses are values of effective temperature; major gas species—the numbers in parentheses are volume mixing ratios; minor gas species—the numbers in parentheses are fractional abundance by number in units of ppm; aerosol species—the numbers in parentheses are typical values of the aerosols' optical depth in the visible. The numbers in this table were derived from the following sources: Blanco & McCuskey (1961), Hunten (1977), Jaffe et al (1979), Kumar (1976), Pearl et al (1979), Pollack & Black (1979), Pollack et al (1979b), Rages & Pollack (1979), Ridgway et al (1976), Rossow (1978), and Weidenschilling & Lewis (1973).

[b] These mixing ratios refer to typical values at the surface.

[c] These mixing ratios pertain to the region above the cloudtops.

[d] The sulfuric acid aerosol resides in the lower stratosphere, while the sulfate, etc. aerosols are found in the troposphere, especially in the bottom boundary layer.

[e] The ice clouds are found preferentially above the winter polar regions. Dust particles are present over the entire globe.

[f] These mixing ratios pertain to the stratosphere.

classes of ancient channels appear to have been formed by running water and may require a much more massive atmosphere for their genesis, one capable of creating a strong greenhouse warming (Pollack 1979). What factors control these and other changes in the properties of planetary and satellite atmospheres?

In this paper, we review our current understanding of the origin and evolution of the atmospheres of solar system objects and assess our ability to answer the above questions. In the first part of the paper, we describe physical processes that control this evolution in an attempt to develop a set of general principles that can help to guide studies of specific objects. In the second part, we discuss the planetary and satellite atmospheres of the inner solar system objects and critically consider in each case current hypotheses on their origin and evolution, including a few new ones of our own. We regret that space permits only a cursory discussion of the atmospheres of objects in the outer solar system. The giant planets have been adequately treated elsewhere (Cameron & Pollack 1976, Pollack et al 1977, 1979a).

Several review articles touch on some of the topics discussed here. Among these are papers by Shemansky & Broadfoot (1977) on the exospheric atmospheres of Mercury and the Moon; Hunten & Donahue (1976) on the escape of light gases from the atmospheres of the terrestrial planets and Titan; Pollack (1979) on climatic changes on the Earth, Mars, and Venus; Cameron & Pollack (1976) on the formation and evolution of the giant planets; Hunten (1977) on Titan's atmosphere; and McElroy (1976) on photochemical processes in planetary atmospheres; and books by Walker (1977) and Holland (1978a) that treat the geochemical cycles that control the Earth's atmospheric composition (Walker's book also contains some discussion of the atmospheres of the other terrestrial planets).

PHYSICAL PROCESSES

In this section, we discuss the physical processes that determine the chemical composition and mass of a planetary or satellite atmosphere at a given epoch and the ones that cause long term changes in these properties.

Sources of Atmospheric Gases

There are four major sources of the gases in planetary and satellite atmospheres: the primordial solar nebula, the solar wind, colliding bodies, and the object's interior. Table 2 summarizes the time period during which each of these sources is expected to be most effective and the dominant volatile-forming elements they contain, in approximate ranked order.

Table 2 Nature of the sources of planetary and satellite atmospheres

Source	Elemental composition[a]	Time period of greatest potency[b]
Primordial solar nebula	H > He > O > C > N > Ne > Mg > Si > Fe > S > Ar	4.6
Solar wind	H > He > O > C > N > Ne > Mg > Si > Fe > S > Ar	~4.6[c]
Colliding bodies	H_2O > S > C > N	~4.6–4.0
Interior	H_2O > C > Cl > N > S	~4.6–4.0[d]

[a] The compositions given here are based on solar abundances, carbonaceous chrondritic meteorites, and terrestrial "excess volatiles" for the primordial solar nebula and solar wind, colliding bodies, and the interior, respectively (Cameron 1970, and Walker 1977). Only the elements that help to constitute key volatiles are considered here.

[b] Time in aeons from the present, with 4.6 taken as the time of solar system formation.

[c] However, it is the present day solar wind that is relevant for the atmospheres of Mercury and the Moon.

[d] Later, episodic outgassing can also be important. This is particularly true in the case of Io, where the composition of the outgassed material may have changed with time.

Justification for the entries in Table 2 will be given below. All four sources are most potent during the early history of the solar system.

1. PRIMORDIAL SOLAR NEBULA The term "primordial solar nebula" refers to the diffuse cloud of gas and dust that formed the source material from which all solar system objects, including the Sun, were formed. Presumably, the elemental composition of the solar nebula was very similar to that characterizing the outer portions of the current Sun. Radioactive dating of meteorites, lunar samples, and terrestrial rocks indicates that planet formation took place about 4.6 billion years ago, with the epoch probably spanning less than several hundred million years (e.g. Wasserburg et al 1977). Although the solar nebula is the ultimate source of all planetary and satellite constituents, including volatiles, we consider here only the direct way in which it may have contributed to the formation of planetary atmospheres.

First consider satellites and terrestrial planets. After they had completely formed, their gravitational fields may have caused a nearby concentration of solar nebula gases. A portion of the material may then have been retained after the solar nebula dissipated. Whether this scenario was in fact followed depends on the relative timing of the completion of satellite and planetary formation and the disappearance of the solar nebula and on the effectiveness of the nebular dissipation process in stripping away such an early atmosphere. This type of atmosphere, as well as one due to an early intense solar wind, is frequently referred to as a "primary" atmosphere.

The solar nebula is the only plausible source for the massive atmospheres of the giant planets. The origin of these planets and their atmospheres will be discussed in the next section of this paper.

2. SOLAR WIND The solar wind is a magnetized and fully ionized gas that originates in the uppermost portions of the Sun's atmosphere and flows outward into interplanetary space. At the orbit of the Earth, typical values for the solar wind's temperature, velocity, number density, and flux are 2×10^5 K, 500 km/s, 5 particles/cm^3, and 2×10^8 particles cm^{-2} s^{-1}, respectively, for quiescent conditions (Walker 1977). At other distances from the Sun, the flux scales crudely as the inverse of the distance squared. The composition of the solar wind is approximately, but not exactly, the same as the Sun's. In the case of an object that lacks both an appreciable atmosphere and intrinsic magnetic field, such as the Moon, the solar wind flows down to the object's daytime surface, before being deflected around the object. In the case of objects with an appreciable magnetic field, such as the Earth and Mercury, or objects with an appreciable atmosphere and hence ionosphere, such as Venus, the solar wind is deflected well above the object's surface, although some material may leak in through the interface between the solar wind and the object's magnetic field or ionosphere.

Consider first objects such as Mercury and the Moon that lack appreciable atmospheres. In such cases, temporary capture of solar wind material occurs as the result of charged particles striking surface material, being temporarily held by the surface material, becoming neutralized, and ultimately being released back into the atmosphere. Once neutralized, the solar wind derived gases can no longer be removed immediately by being caught on magnetic field lines within the solar wind and swept away. According to the classical model of this capture process, the surface grains quickly become saturated with solar wind derived neutrals so that, as new material is added to the surface, an equal amount of older gases leave and enter the atmosphere with a Maxwell-Boltzmann distribution of kinetic energies and a temperature equal to that of the surface (Hodges 1975). Important modifications to the above first order description are given in Hodges (1975) and Shemansky & Broadfoot (1977).

In principle, the solar wind could also be an important source of gases present in more massive atmospheres, such as those of Venus, the Earth, and Mars. For example, if all the solar wind incident upon Venus was added to its atmosphere, a very unlikely situation, the number of H atoms added cumulatively over the age of the solar system would equal 6×10^{25} atoms/cm^2, equivalent to an H/C ratio of 4×10^{-2}. In this case, the solar wind could constitute an important source of H_2 in Venus' atmosphere. Similarly, the solar wind could be an important source of He and Ne. However, proceeding in an analogous manner as for H, we find that the current solar wind fails by orders of magnitude to supply the amount of C, N, and Ar known to be present in Venus' atmosphere. Analogous comparisons hold for the Earth and Mars.

The very early solar wind may have been much more intense. A popular theme in the literature has been the dissipation of the solar nebula with a "T-Tauri" type solar wind, by analogy with the high outflow rates that characterize some of the earliest stages of stellar evolution (see, for example, Cameron & Pollack 1976). Such a wind could also add much material to the atmospheres of solar system objects during the earliest history of the solar system.

3. COLLISIONS WITH ASTEROIDS AND COMETS The highly cratered surfaces of the Moon, Mars, and Mercury attest to the large cumulative mass brought into them and into other solar system bodies through impacts with much smaller stray objects, such as asteroids and comets, whose initial orbits were perturbed into planet- and satellite-crossing orbits. Some asteroids in the asteroid belt and some with Earth-crossing orbits resemble, in their spectral characteristics, carbonaceous chondrites, a volatile-rich type of meteorite (Morrison 1977). When they enter the inner solar system, comets are extremely volatile-rich. However, because the orbital period of a short period comet ($\sim$several years) is much smaller than its residence time in the inner solar system prior to impact ($\sim 10^8$ years; Wetherill 1975), a "dead" comet nucleus looses much of its volatiles prior to impact. Nevertheles, it still may be as volatile-rich at this point as a carbonaceous chondrite. Thus, a significant amount of volatiles may have been added to inner solar system bodies through impacts with small stray objects. Small stray objects may be even more volatile-rich when they strike bodies in the outer solar system, since little evaporation takes place at these large distances prior to impact. At the typical velocities of impact (~ 10 km/s), all the volatiles are released into the impacted body's atmosphere.

Comparison of crater densities on various areas of the Moon with the age of the terranes, as established from returned lunar samples, indicates that there was a much higher flux of impacting objects during the Moon's first 700 million years than at subsequent times (Wetherill 1975). Since gravitational perturbations by Venus and the Earth change the orbits of stray bodies into ones that, in most cases, cross the orbits of all the terrestrial planets (Wetherill 1975), a similar bombardment history must have been experienced by all bodies in the inner solar system.

We can make a crude estimate of the importance of small body collisions as a source of planetary atmospheres subsequent to planet formation by comparing the mass of water brought in by these bodies over the lifetime of the Earth with the amount of water contained in the Earth's oceans. The observable crater population on the moon's surface was produced by impacting bodies having a cumulative mass of about 10^{21} gm, with most of the mass contained in the largest impacting objects, such as the one that produced the Imbrium basin (Wetherill 1975). For the moment, let us neglect the highly uncertain, additional mass of objects which formed craters that were subsequently obliterated. Note that obliterated basins are of greatest interest in this regard. Therefore, allowing for differences in surface area and impact probability per unit area (Wetherill 1975), we estimate that a cumulative mass of about 2×10^{22} gm impacted the Earth over its lifetime. If we assume that the impacting objects, on the average, had a water content of 10%, a value typical of the more volatile-rich carbonaceous chondrites (Mason 1962), then these impacts added about 2×10^{21} gm of water. The above water content could well be a gross overestimate. Nevertheless, even with this generous assumption, the above estimate of water brought in by impacts is about three orders of magnitude smaller than the mass of water contained in the Earth's oceans (Turekian & Clark 1975). Similar differences occur when the carbon brought in by impacts is compared with the amount of carbon-containing material that resided at some time in the Earth's atmosphere (Turekian & Clark 1975). Therefore, impacting bodies may not represent an important source of the major volatiles for at least the Earth's atmosphere subsequent to the Earth's formation. But, in view of the crudeness of the above calculation, such as our neglect of obliterated lunar basins, it is premature to make a definitive judgment at this point. Also, impacting objects might be an important source of certain volatiles in cases where the planetary interior is greatly depleted in these volatiles, such as perhaps S and H_2O in the case of Venus (Lewis 1974).

4. INTERIOR SOURCES The grains that accreted to form planets and satellites may have contained volatile-forming compounds. After the planet or satellite formed, heating of its interior in conjunction with associated volcanism and tectonic activity resulted in the release of some of the stored volatiles into the atmosphere. Some release also occurred during planetary accretion. Below, we discuss the composition of the volatile-forming compounds, the sources and timing of interior heating, and the composition of the outgassed volatiles.

Meteorites provide useful clues on the chemical composition of the material that was initially present inside terrestrial planets and satellites, since certain classes of meteorites have undergone relatively little thermal or chemical evolution since their formation. However, it is probably a mistake to model quantitatively inner solar system bodies after certain classes of meteorites since meteorites originated in only limited regions of the solar system. Table 3 summarizes the principal chemical compounds found in meteorites that contain volatile-forming elements. Also given in the table are the atmospheric gases that contain the volatile element and the temperature within the solar nebula at which it is thermodynamically possible to form the chemical compound. Except for the rare gases, the volatile elements form an integral part of the crystal structure of their associated chemical compounds. Thus, heating to high temperatures is required for the release of these volatile elements, an act which destroys the host mineral. Also note that, except for the rare gases and water of crystallization, none of the volatile elements appear with the same chemical makeup as their corresponding atmospheric gases. Such conversions can take place in the air spaces of magma chambers. For example, C originating in organic compounds can obtain O needed to form CO_2

Table 3 Chemical compounds in meteorites that contain volatile-forming elements[a]

Volatile gas	Volatile elements	Chemical compound	Formation temperature
Water	H	Organics	$\lesssim$750 K
		Hydrated minerals (water of crystallization and hydroxyl groups)	~150–450 K
Carbon dioxide	C	Organics	$\lesssim$750 K
Methane		Graphite	
Carbon monoxide		Carbonate[b]	
Nitrogen	N	Organics	$\lesssim$750 K
Ammonia		Ammonium salts[b]	
		Nitrides	
Sulfur dioxide	S	Troilite (FeS)[c]	500–700 K
Hydrogen sulfide		Sulfates[b]	
		Free sulfur	
		Organics	$\lesssim$750 K
Hydrogen chloride	Cl	Apatite ($Ca_5(PO_4)_3Cl$)	
		Lawrencite ($FeCl_2$)	
		Organics	$\lesssim$750 K
Rare gases	rare gases	Occluded gases	
		Radioactive elements[d]	

[a] Based on data given in Mason (1962) and Lewis (1974).
[b] Probably an alteration product.
[c] Dominant sulfur compound.
[d] ^{40}K is a source of ^{40}Ar and U and Th are sources of helium.

from either the oxygen released by dissociated water vapor or the oxygen contained in ferro magnesian silicates and magnetite. It is interesting to observe that organic compounds contain all the volatile elements given in Table 3. Finally, in addition to the rare gases found occluded within minerals, the decay of radioactive elements also serves as a source of certain rare gases (see footnote d on Table 3).

The above discussion has relevance for objects in the inner solar system and possibly for Io. In the case of satellites of the outer solar system, such as Titan, temperatures below about 200 K within the circum-planetary nebula of their birth may have permitted ices to condense and therefore to be incorporated within the satellite-forming materials (Lewis 1974, Cameron & Pollack 1976). As the temperature is lowered, first water ice becomes thermodynamically stable, followed by ammonia hydrate and then methane clathrate ($NH_3 \cdot H_2O$, $CH_4 \cdot 6H_2O$). These interior ices can serve in a rather direct fashion as sources of atmospheric water vapor, ammonia, and methane.

There are three major epochs of strong heating during which volatiles were released into planetary and satellite atmospheres: accretion, global internal differentiation, and local volcanism. Especially during the later stages of planetary and satellite formation, accreted material may be heated to high temperatures due to ablation in the object's very early atmosphere or high velocity impact with its surface (Jakosky & Ahrens 1979). However, if the best current values for the time scale of planetary accretion are used ($\sim 10^8$ years), then the surface temperature of the outer three terrestrial planets remained low enough to permit the re-incorporation of the released H_2O and CO_2 into these bodies due to chemical reactions with surface minerals, followed by burial during later accretion (Jakosky & Ahrens 1979). Rare gases and nitrogen may have remained in the atmosphere. Below, we will give some evidence that the subsurface temperatures may have been quite high during accretion. If so, complete reincorporation may not have occurred.

During the later stages of accretion and/or thereafter, terrestrial planets and large satellites underwent global internal differentiation, resulting in the appearance of a low density crust and a high density core. Opinions on the degree of initial chemical zonation have ranged the full gamut from a totally homogeneous mixture of low and high temperature condensates to a fully inhomogeneous model with iron and other core-forming materials already located near the object's center (see, for example, Jakosky & Ahrens 1979, Walker 1977). In our opinion, while materials of different volatility classes may have formed at different times from the primordial solar nebula, accretion into a completely segregated body seems unlikely. Hence, some core formation probably took place during the internal differentiation event that generated the crust and, most importantly, some metallic iron was initially present in the upper mantle.

Age dating of returned lunar samples indicates that the Moon's sialic crust was formed during the first several hundred million years (Wasserburg et al 1977). In order to create a 60-km thick crust, as implied by seismic measurements, a "magma" ocean with a minimum depth of several hundred kilometers had to be present close to the surface during this epoch of internal differentiation. Portions of the surface of Mars and Mercury exhibit a high density of craters, comparable to that of the lunar highlands (Toksöz et al 1978, Wetherill 1975). Since the lunar highlands record the early epoch of intense meteorite bombardment and since the bombardment flux is not expected to vary greatly throughout the inner solar system (Wetherill 1975), crustal formation and associated outgassing took place during the first billion years of the history of Mars, Mercury, and, by implication, Venus and the Earth.

Age dated lunar samples also provide a good definition of the timing of subsequent igneous activity on the Moon (Wasserburg et al 1977). Mantle-derived magmas reached the lunar surface, locally and episodically, from about 4.4 to 3.1 billion years ago, but no sizeable expression of igneous activity on the surface occurred at later times, due probably to lithospheric thickening. Currently, this rigid outer region of the Moon extends to a depth of 700–1000 km. Extensive lithospheric thickening is expected to be delayed to progressively later times for progressively larger objects than the Moon (Toksöz et al 1978). Hence, basaltic volcanism and associated outgassing span larger time intervals for larger objects.

Figure 1 illustrates the results of thermal history models of an initially homogeneous Moon, Mars, Mercury, and Venus that are consistent with the above and other relevant constraints (Toksöz et al 1978). In order to have global internal differentiation occur at early times, the temperature of the outer layers of these bodies was assumed to be at or close to the melting point. Consequently, it is quite conceivable that much of the volatiles released during the later stages of accretion were never effectively reincorporated back into the body's interior. Furthermore, given the extensive melting associated with accretion and internal differentiation, much of the juvenile volatiles may have been released during the first billion years of the solar system for inner solar system objects. Finally, the degree of early outgassing may have been size-dependent, with the smaller bodies outgassing less effectively due to a smaller amount of heating per unit mass by accretional processes and core formation (Toksöz et al 1978, Jakosky & Ahrens 1979).

The next major period of outgassing is associated with later partial melting of portions of the mantle. For small objects such as the Moon, this epoch ended several billion years ago as the lithosphere became too thick. For objects of the size of the Earth, this epoch is still occurring and may have spanned much of the object's history. Although most of the juvenile volatiles may have been released during the first two epochs, this last epoch is important not only for supplying new material, but also in recycling volatiles that were outgassed, but subsequently incorporated into surface rocks.

An understanding of the thermal history of the satellites of the outer solar system is at a much more primitive stage than the study of inner solar system objects. However, it appears that Io and Titan, among others, are probably large enough to have undergone extensive differen-

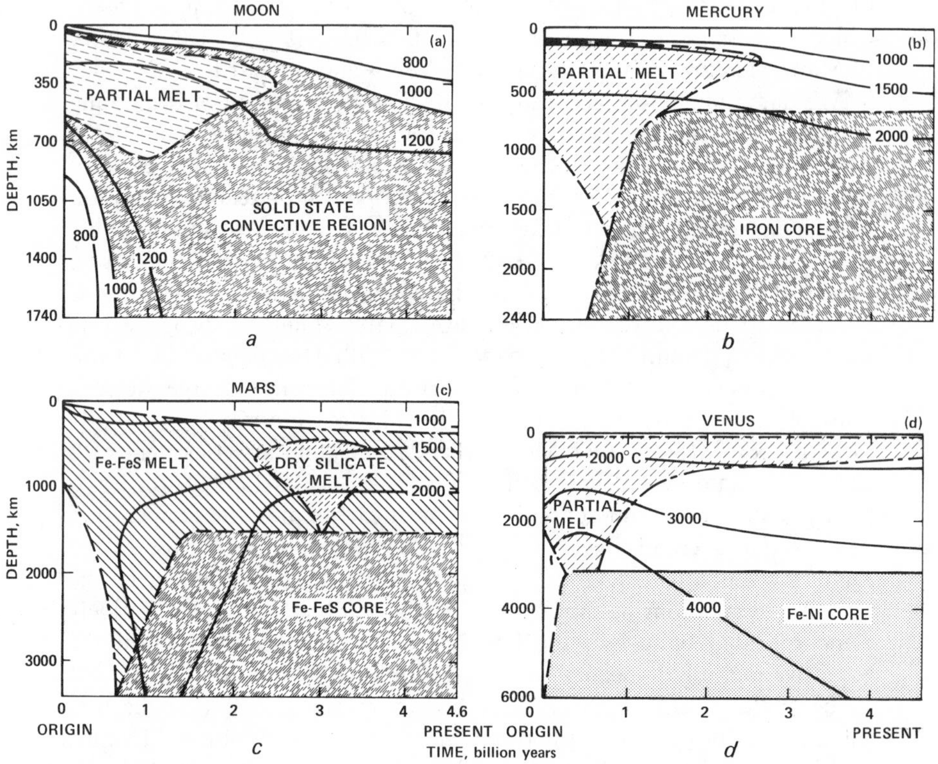

Figure 1 Thermal evolution of the interiors of the Moon (*a*), Mercury (*b*), Mars (*c*), and Venus (*d*) as a function of time from the completion of planetary formation (0 years). The curves in these figures are isotherms, with their associated numbers being temperatures in °C. In all cases, the model planets were assumed to have a homogeneous composition at the initial epoch. From Toksöz et al (1978).

tiation (Consolmagno & Lewis 1977). In the case of Io, intense tidal heating throughout most of its history appears to have resulted in a very thin lithosphere and extensive active volcanism (Peale et al 1979, Smith et al 1979).

Finally, we discuss the chemical composition of the outgassed volatiles. Considerations of thermodynamic equilibrium provide a useful first approximation to the state of oxidation of gases vented from the interior (Holland 1962). It is useful to distinguish between two stages at which thermodynamic equilibrium may apply. During stage 1, the gases come to thermodynamic equilibrium with their source magma at partial melting temperatures (1500 K for mantle silicates). Volatiles outgassed into the atmosphere have a composition similar to this equilibrium situation. Once outgassed, the volatiles cool to temperatures similar to that of the surface and some gases are severely depleted due to condensation. During stage 2 gases may achieve a new state of thermodynamic equilibrium appropriate to the revised set of conditions.

In order to illustrate these concepts more clearly, we consider a terrestrial planet having a rock and volatile content similar to that of the Earth (Holland 1962). Thus, ferro magnesian silicates are abundant and water is the chief volatile. According to Holland (1962), the state of oxidation of iron in the volatile source region controls the oxidation state of the gases during stage 1. At temperatures characteristic of silicate magma, water vapor partially dissociates into hydrogen and oxygen. But the oxygen partial pressure is severely limited by the iron-containing rocks and hence a net reduced state with some free hydrogen but little oxygen results for the volatiles. Suppose first that some metallic iron is present, as might be the case prior to the completion of core formation. In this case, the hydrogen partial pressure approximately equals that of water vapor. Under these conditions CO is the dominant C-containing gas, H_2S the dominant S gas, and N_2 the dominant N gas. During stage 2, there is more molecular hydrogen, at least initially, than carbon-containing gases. As a result of this fact and the much lower temperatures, methane may become the dominant carbon gas. Also, some N_2 may be converted to ammonia. H_2S remains the dominant S gas. Figure 2 illustrates the dependence of the oxidation state of C and N gases on the hydrogen partial pressure for room temperature conditions, when only thermodynamic equilibrium is considered.

Next we consider a situation more appropriate to the current oxidation state of the Earth's mantle. No metallic iron is present, the rocks contain mostly FeO, but they also contain some Fe_2O_3. Under these conditions, stage 1 is characterized by a hydrogen partial pressure of about 10^{-2} that of water vapor, and CO_2, SO_2, and N_2 are the dominant C, S, and N

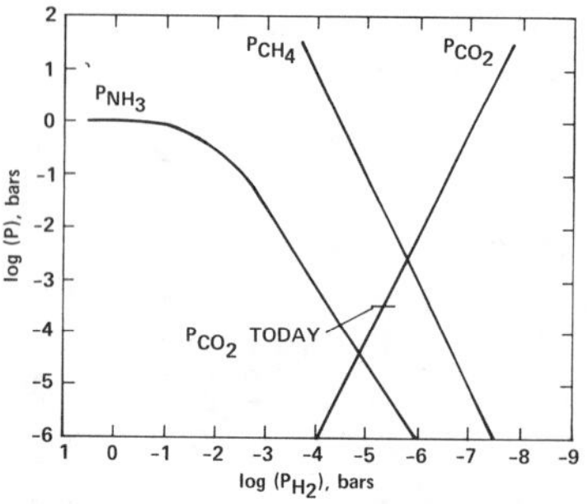

Figure 2 Relationship between the partial pressure of H_2, P_{H_2}, and the partial pressures of CH_4, CO_2, and NH_3, P, under conditions of thermodynamic equilibrium at room temperature. From Holland (1962).

gas species, respectively. During stage 2, some but not all CO_2 and CO may be converted to CH_4, the conversion being limited by the abundance of H_2. Also CO is rapidly converted to graphite and CO_2. If, as is likely, there is more H_2 than SO_2, all the SO_2 may be converted to H_2S. Alternatively, SO_2 is largely eliminated by surface weathering reactions. Essentially all the N volatiles are in the form of N_2.

It is quite conceivable that kinetic limitations may prevent thermodynamic equilibrium from being achieved during stage 2. In this case, even if much of the outgassing from the interior took place while metallic iron was present in the upper mantle, only trace amounts of CH_4 and NH_3 would be present initially in the atmosphere. Thus, both the uncertainty in possible kinetic inhibitions during stage 2 and in the fraction of time metallic iron was present in the upper mantle lead to very large possible ranges for the oxidation state of C and N in early atmospheres.

A further complication is created by the distinct possibility that much of the initial release of volatiles may have occurred during the accretion phase. In this case, both possibly elevated temperatures, reflecting those in the surrounding solar nebula, and sizeable partial pressures of hydrogen, originating from the nebula, may have created a more favorable circumstance for the creation of an early, highly reducing atmosphere.

Sinks of Atmospheric Gases

Light gases can escape to space, while many chemically reactive gases as well as the more easily condensible ones can be severely depleted from the atmosphere by being incorporated into near-surface reservoirs under conditions of moderate or low surface temperature. In the case of the satellites and the terrestrial planets, only the rare gases of intermediate molecular weight, such as neon, argon, and krypton, appear to be largely immune to such loss processes.

1. ESCAPE TO SPACE Gases can escape from the top of planetary atmospheres through a variety of processes that include hydrodynamical flow, thermal evaporative loss (Jeans escape), photochemical escape, and loss by charge or energy exchange. Each of these processes is discussed below. The reader is referred to the review article by Hunten & Donahue (1976) for a more detailed presentation of this material.

Hydrodynamical loss involves mass outflow of atmospheric gases into interplanetary space at supersonic speeds. Thus, there is no mass fractionation of the gases present in the upper atmosphere, but there may still be a fractionation between the gas species present in the lower and upper atmospheres due to diffusive separation below the altitudes of supersonic flow. Hydrodynamical loss occurs when $H_c > 1/2\ r_c$, where H_c and r_c are the scale height of the lightest gas and the distance from the body's center to the critical level (Hunten 1973). Critical level or base of the exosphere refers to the place at which the mean free path for collisions equals the scale height. For almost all atmospheres, $H_c < 1/2\ r_c$ and hence hydrodynamical escape does not occur. Important exceptions to this conclusion include the solar wind; ion flow in the polar regions of the Earth; and the blowoff of a hypothetical hydrogen-rich atmosphere of Titan (Hunten 1973).

In the absence of hydrodynamical loss, light gases can escape to space through thermal evaporation at and above the critical level. To first order, all atoms having an upward-directed velocity greater than the gravitational escape velocity and lying within one mean free path of the top of the atmosphere are lost to space. The fraction of atoms having velocities in excess of the escape velocity can be calculated approximately from the Boltzmann velocity distribution at the exospheric temperature, although some depletion in the high velocity tail of the distribution results from escape (Chamberlain 1963). The classical Jeans equation for the escape flux, F_J^i, for constituent i may be written as (Hunten 1973):

$$F_J^i = n_c^i W_c^i = n_c^i \frac{U}{2\sqrt{\pi}} (1 + \lambda_c^i) \exp(-\lambda_c^i) \tag{1}$$

where

$$U = (2kT/m^i)^{1/2}, \quad \lambda_c^i = r_c/H_c^i, \quad H_c^i = kT_c/m^i g_c$$

and n, W, k, T, m, H, and g denote number density, "effusion" velocity, Boltzmann's constant, temperature, molecular weight, scale height, and acceleration of gravity, respectively. Because $\lambda_c^i \sim m^i$ and $F_J^i \sim \exp(-\lambda_c^i)$, only the lightest gases escape at appreciable rates through thermal evaporation.

There are three possible controls on the value of the thermal escape rate, F_J^i: the exospheric temperature, T_c; the rate at which the escaping atom diffuses to the exosphere from the atmosphere beneath; and the rate at which the escaping gases are supplied chemically. T_c is the controlling factor when it is sufficiently low.

Now, suppose that T_c is sufficiently large that diffusion from the lower atmosphere limits the escape rate (Hunten 1973). In this case, the diffusion rate and hence the escape rate are approximately equal to the limiting flux F_l^i, evaluated at the homopause:

$$F_\mathrm{l}^\mathrm{i} \cong b_\mathrm{h}^\mathrm{i} f_\mathrm{h}^\mathrm{i} / H_\mathrm{h}^\mathrm{a} \quad (\text{if} \quad m^\mathrm{i} \ll m^\mathrm{a} \quad \text{and} \quad f_\mathrm{h}^\mathrm{i} \ll 1) \tag{2}$$

where b_h^i and f_h^i are the binary collision coefficient and the number mixing ratio of species i to the rest of the atmosphere, respectively. Scripts h and a refer to the homopause level and the rest of the atmosphere, respectively. Homopause refers to the level at which the eddy mixing coefficient, K, equals the diffusion coefficient of species i, D^i. More general expressions are given in Hunten (1973). The corresponding diffusion velocity, W_l, is given by $D^\mathrm{i}/H_\mathrm{h}^\mathrm{a} = K/H_\mathrm{h}^\mathrm{a} = W_\mathrm{l}$. Diffusion rather than exospheric temperature is the rate-limiting process when $W_\mathrm{l} < W_\mathrm{c}$. In such a situation, the number density at the exospheric level, n_c^i, is adjusted so as to insure that $F_\mathrm{J}^\mathrm{i} = F_\mathrm{l}^\mathrm{i}$. In the more general situation of several escape processes contributing to the loss at the top of the atmosphere, but diffusion being the rate-limiting step, Equation (2) remains

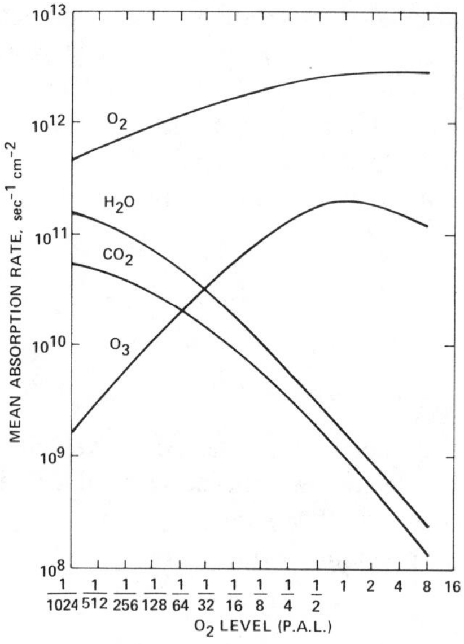

Figure 3 Mean absorption rates of UV photons capable of dissociating water vapor for various atmospheric species as a function of the oxygen level relative to today's level (=1 P.A.L.). From Brinkman (1969).

valid and F_l^i is equal to the sum of the loss fluxes at the top of the atmosphere.

In evaluating Equation (2) at the homopause, f_h^i may be set equal to the mixing ratio of all i-containing species. For example, if atomic hydrogen, but not molecular hydrogen, escapes efficiently from the exosphere and molecular hydrogen is the chief H-containing species at the homopause, f_h^i is simply twice the molecular hydrogen mixing ratio at the homopause. In the absence of condensation, the mixing ratio of all i-containing species remains constant below the homopause, despite photochemical transformations. Hence f_h^i can be evaluated by using data on abundances in the stratosphere.

As indicated by Equation (2), $F_l^i \sim f_h^i$. As f_h^i increases, eventually a point is reached at which the loss rate becomes limited by the rate at which gas species containing species i can be converted by photochemistry into species i. In this case, the escape rate of species i may be limited by the flux of solar photons needed for the rate-determining photochemical step. For example, Figure 3 illustrates the photon-limited rate at which water vapor is photodissociated into hydrogen and oxygen as a function of oxygen content for an atmosphere containing 5 ppm H_2O and 330 ppm CO_2 above the tropopause and for the current solar UV flux at the Earth (Brinkman 1969). The ozone content has been assumed to be proportional to the oxygen abundance. For other choices of water vapor or carbon dioxide mixing ratio, the appropriate curve will scale approximately linearly with mixing ratio, under the constraint that the sum of all curves for a fixed oxygen level is an invariant.

When the velocity distribution is dictated by thermodynamic equilibrium as is appropriate for Jeans escape, only the lightest atoms escape at appreciable rates. However, it is possible to generate medium weight atoms with speeds in excess of the escape velocity by highly exothermic photochemical reactions. When these reactions occur above the base of the exosphere, about half of the fast atoms (the ones with upward-directed velocities) are lost to space. There are three types of photochemical reactions that can lead to photochemical escape: dissociative recombination, photodissociation, and dissociation by photoelectrons. Examples of each of these types, which are relevant for the escape of C, O, and N atoms from Mars, are (McElroy 1972, McElroy et al 1976)

$$O_2^+ + e \rightarrow O + O, \tag{3}$$

$$CO + h\nu \rightarrow C + O, \tag{4}$$

$$e + N_2 \rightarrow N + N + e. \tag{5}$$

Photochemical escape provides an important loss mechanism for C, N, and O in the case of Mars (McElroy 1972) and for H in the case of Venus (Kumar et al 1978).

Another important loss process involves charge exchange between a very energetic or "hot" proton and a neutral atom (Hunten & Donahue 1976). In the case of unmagnetized planets, such as Venus and Mars, the solar wind is the source of the hot proton. The slow atom, which acquires a charge through the charge-exchange reaction, is trapped around magnetic field lines in the solar wind and is swept away from the planet. The very "hot" proton, which is neutralized by the reaction and therefore is no longer tied to the solar wind's magnetic field lines, escapes to space either directly, if its motion is away from the planet, or indirectly, following collisions with atmospheric atoms. Thus, there is a net loss of the neutral atom as a result of the charge-exchange reaction. Photoionization can also convert a neutral atom to an ion that can be picked up by the solar wind (Kumar 1976).

The capacity of the solar wind to pick up atmospheric atoms is limited (Cloutier et al 1969). If it adds a mass more than $\sim 9/16$ of its initial mass, a shock layer forms and the solar wind flow is diverted further from the planet so that this limiting mass factor is not exceeded. Assuming solar wind sweeping operates at maximum efficiency, one may write this escape flux, $F_{\text{s.w.}}$, as (Hunten & Donahue 1976)

$$F_{\text{s.w.}} \cong 0.17\, \phi\, H/r, \tag{6}$$

where ϕ, H, and r are the flux of solar wind protons, atmospheric scale height at the level of the solar wind, and distance from the center of the planet to the interface with the solar wind near the limb, respectively. Solar wind sweeping represents an important loss process for C and N from Mars (McElroy 1972), for H from Venus (Kumar et al 1978), and for all the constituents of the lunar atmosphere (Kumar 1976).

In the case of planets with magnetic fields, such as the Earth, the solar wind is deflected far above any significant atmosphere, but charge-exchange reactions within the magnetic cavity can still be important. In particular, charge-exchange reactions between atoms in the uppermost region of the exosphere and very hot protons trapped by the planetary magnetic field, a component of the plasmasphere, result in a fast atom, which escapes, and a trapped thermal ion. The escape flux in this case is given by Equation (7) of Hunten & Donahue (1976). This process provides an important loss mechanism for H from the Earth at present (Hunten & Donahue 1976). Since the solar wind provides the energy to heat the plasmasphere, there is again a limit associated with the loss rate, although a higher one than in the unmagnetized planet case because of a higher cross section for interaction. A similar mechanism for escape of oxygen from the Earth's atmosphere has recently been proposed by Torr et al (1974). Energetic O^+ ions produced in magnetic storms could charge-transfer with a thermal oxygen atom and result in producing a "hot" O atom capable of escaping from the Earth's gravitational field.

The environment of the satellites of the outer planets may be dominated by the magnetosphere of their parent planets rather than the solar wind. In this case, potentially important loss processes include charge-exchange reactions with the magnetospheric plasma and energy exchange with high energy magnetospheric particles. In the latter case, a given magnetospheric particle may be capable of ejecting a number of atmospheric particles by elastically scattering them to energies far in excess of the escape energy, with the scattered particles eventually escaping after suffering a number of collisions and energy losses with other atmospheric particles. We note that it may suffice for the escaping particles to simply reach the magnetosphere proper, where they can become ionized and swept away by the magnetic field.

2. LOSS TO SURFACE AND INTERIOR RESERVOIRS In the case of satellites and terrestrial planets, a large proportion of many gas species that were present at some time in the atmosphere currently reside in near-surface reservoirs as a result of condensation, adsorption, and chemical weathering. In the case of the outer planets, certain gas species may be partially removed from the observable atmosphere to the deep interior because of the onset of fluid immiscibility.

When the partial pressure of a gas reaches its saturation vapor pressure close to the surface, the gas begins to condense, with the condensate ultimately being removed to the surface. The key variable in such cases is the surface temperature since the saturation vapor pressure depends exponentially on it. Such a dependence implies that only modest changes in surface temperature can result in quite substantial changes in the amount of condensible gases in the atmosphere. Water is a good example of a condensible gas. In the case of the Earth, there is about five orders of magnitude more water in the oceans than in the atmosphere. Other examples include H_2O and CO_2 in the case of Mars, CH_4 in the case of Titan, and SO_2 in the case of Io.

When fine-grained sedimentary deposits are present on the surface, some gases, particularly ones with large polarizabilities, become physically adsorbed onto the surface of the grains. For a given sediment, the amount of gas of a given type that is adsorbed onto a unit volume of the sediment is a function of both the sediment's temperature and the gas's partial pressure. Figure 4 illustrates this dependence between CO_2 gas and both a fine-grained clay (nontronite) and a coarse basalt powder, for conditions appropriate for Mars. For a fixed temperature and partial pressure, the total amount of gas adsorbed onto surface deposits is proportional to the "effective" area of the grains and the total volume of the sedimentary layer in contact with the atmosphere. Thus, finer-grained deposits are more effective reservoirs than coarser-grained ones (cf Figure 4). The fine-grained regolith of Mars may contain much more adsorbed CO_2 than is present in the atmosphere (Fanale & Cannon 1979). In the case of the Earth, more xenon may reside in shales than in the atmosphere as a result of adsorption (Fanale & Cannon 1971).

Sedimentary, metamorphic, and igneous rocks that are brought to a body's surface by tectonic processes are generally not in thermodynamic equilibrium with their new environment, one characterized by much lower temperatures than those of the body's interior, a variety of atmospheric gases, and, in the case of the Earth, liquid water. Consequently, some of the mineral constituents of surface rocks undergo chemical reactions with atmospheric gases, which convert the former to more stable phases. This "chemical weathering" represents an important sink for many atmospheric gases. For example, calcium-bearing silicates react with CO_2 to produce calcium carbonate, which may be represented schematically as (Urey 1952)

$$CaSiO_3 + CO_2 \rightarrow CaCO_3 + SiO_2. \quad (7)$$

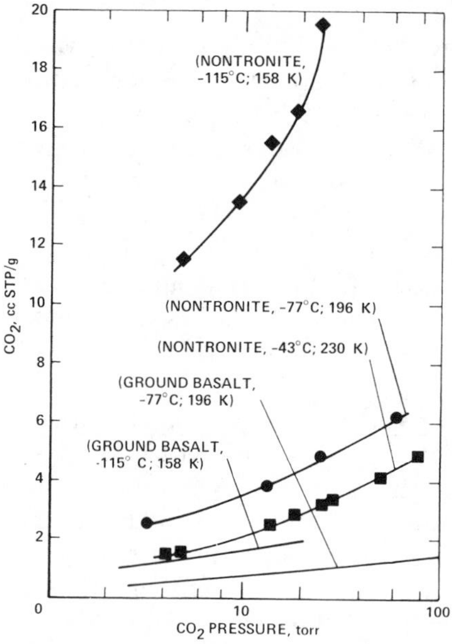

Figure 4 Amount of adsorbed CO_2 (vertical ordinate) as a function of the partial pressure of CO_2 gas for two types of powdered materials and several temperatures. From Fanale & Canon (1979).

Ideally, the CO_2 pressure in the atmosphere is adjusted until thermodynamic equilibrium is achieved. However, as explained in a later subsection on the Earth, the true situation can be much more complicated on a planet that has liquid water and active tectonism.

The rate at which atmospheric gases are lost as a result of chemical weathering is a function of the erosion rate, the fraction of rock material that undergoes chemical weathering, and the degree of completion of the weathering reactions. Quite possibly, the Earth has a uniquely high erosion rate among solar system bodies due to the occurrence of rainfall and various biological processes. Chemical weathering is an important loss process for CO_2 and O_2 in the case of the Earth. Its importance for other objects with atmospheres is less clear.

Conversion Processes

PHOTOCHEMISTRY Photochemical reactions within the atmosphere result in the creation of many trace constituents; the conversion of some gases to aerosols, which are subsequently removed rapidly from the atmosphere; the conversion of some gases irreversibly to other gases; and the production of light constituents, especially atomic hydrogen, which can escape from the top of the atmosphere.

Trace constituents are of particular interest if they affect the solar or thermal radiation. An obvious and very important example is ozone in the Earth's atmosphere, which prevents biologically harmful solar UV radiation from reaching the Earth's surface. Ozone is generated through the photodissociation of molecular oxygen. Recent calculations of its abundance have taken into account not only the chemistry of O, O_2, and O_3, but also that of hydrogen-, nitrogen-, and chlorine-containing compounds, which exert important catalytic control over the ozone destruction rate (Logan et al 1978).

The conversion of gases to particles is an effective mechanism for reducing the gas content of an atmosphere. The very low abundance of SO_2 in the Earth's atmosphere is partly the result of its rapid conversion into sulfate particles, the conversion being initiated by reactions with such radicals as OH. Also, as discussed later, the production of aerosols in Titan's atmosphere may constitute a long term sink for their precursor gases.

As an atmosphere evolves, gas species may become irreversibly converted into other gas species. For example, as an atmosphere becomes more highly oxidized, photochemical reactions can irreversibly convert ammonia into nitrogen, with this process being initiated by ammonia photodissociation (Kuhn & Atreya 1979). Such a change in oxidation state is driven by the loss of atomic hydrogen from the top of the atmosphere, with photochemistry being responsible for converting other hydrogen-containing species into atomic hydrogen in the upper atmosphere (Hunten & Donahue 1976).

LIFE Within the solar system, the Earth has an apparently unique set of transformation processes, ones involving its biosphere. As discussed in the next section, photosynthesis by green plants is chiefly responsible for the large amount of free oxygen in the Earth's atmosphere. Living organisms are also important sources of such trace gases as nitrous oxide and methane.

Sources and Sinks of Aerosols

For our purposes, aerosols are an important part of an atmosphere both because they can strongly influence the radiation budget and because their formation may represent a sink of atmospheric gases, as discussed above. There are three important sources of atmospheric aerosols: winds, condensation, and photochemistry. When winds near the surface exceed a threshold speed, sand-sized particles are set into motion and dust-sized particles are put into suspension (Iversen et al 1976). Rising motions in the atmosphere cause a cooling of air parcels which may induce certain gas species to condense and form clouds. Photochemical conversion has been discussed above.

Aerosols generally remain in an atmosphere for only a short period of time ($\sim$day–year). Removal occurs by sedimentation, scavenging, evaporation, and precipitation (Rossow 1978). Gravity causes all particles to have a net downward velocity, with the fall velocity increasing monotonically with particle size and eddy mixing generally enhancing the net removal rate. Small particles may serve as nucleation centers for the formation of larger condensate particles, with the net result being an enhanced removal rate of the small particles because of a higher fall velocity. This process is responsible for the efficient removal of non-water aerosols from the Earth's troposphere and possibly the preferential deposition of dust particles in the Martian polar regions (Pollack et al 1979b). If a condensate particle falls to a region having unsaturated air, it evaporates, as occurs for small-sized water particles in the Earth's atmosphere. But rapid growth of particles within cloud regions through coalesence can produce large "precipitation"-sized particles, which fall quickly enough to survive to the surface.

Table 1 summarizes the principal aerosol constituents of planetary and satellite atmospheres. A much more detailed description of the above physical processes and their application to these atmospheres is given in Rossow (1978).

Drives for Climatic Change

The radiative properties and the general circulation of an atmosphere help to determine the climatic conditions at the surface and within the atmosphere. These aspects of an atmosphere are in turn related to the concentration of optically active gases, condensible gases, aerosols, and the total atmospheric pressure. Several external factors exert a strong control on these properties of an atmosphere. These factors include the solar input of energy at the top of the atmosphere, the interior heat flux, and tectonic and associated interior processes. Changes in climate and atmospheric properties result from changes in these factors (Pollack 1979).

In discussing the amount of solar energy incident at the top of an atmosphere, it is useful to distinguish between the total energy emitted by the Sun and the effect of astronomical factors, including an object's orbital and axial characteristics. Over the lifetime of the solar system, almost all models of the Sun's evolution predict that its luminosity has increased by several tens of percent, with the percentage increase ranging from about 25–40% (Sagan & Mullen 1972). [See, however, Canuto & Hsieh (1978) on the possible influence of cosmological factors on these results.] This leads to interesting questions concerning the generation of favorable climates on the Earth and Mars in their early history, despite the decreased solar luminosity then. These questions are discussed in the next section. On much shorter time scales of years to centuries, climatically significant, but much smaller changes in solar luminosity and/or spectral characteristics, especially in the UV, are known or suspected to have occurred (Eddy 1977).

On time scales of 10^4–10^6 years, the orbital eccentricity, axial tilt, and axial precession of the Earth and Mars undergo quasi-periodic variations (Ward et al 1974). Such changes result in seasonal and latitudinal alterations in the amount of incident sunlight. These variations may have played a role in causing ice ages on the Earth during the last several million years (Hays et al 1976) and in producing the layered terrain in the polar regions of Mars (Cutts 1973, Pollack 1979).

Internal heat fluxes are trivial compared to the solar input in the case of satellites and terrestrial planets. But even today the internal heat fluxes from Jupiter, Saturn, and Neptune are comparable to the solar inputs (Hubbard 1979). At earlier times, the internal heat fluxes may have been substantially higher (Pollack et al 1977).

A variety of internal processes affects the atmospheres and climates of satellites and planets. With the exception of the giant planets, the most important of these are ones related to the outgassing of juvenile and recycled volatiles. In the case of the giant planets, internal phase changes can influence the atmospheric composition, as discussed earlier.

Finally, the atmosphere itself is not only affected by climate change, but, through loss processes to space, geochemical interactions with the surface, and regulation of surface temperature, it has a dynamic role. The complicated interrelationships between the atmosphere, surface, and interior make rigorous calculations of climate and atmospheric evolution impossible at the present time, although useful approximate methods have proved to be interesting and provocative. They form part of the subject matter of the next section of this paper.

ATMOSPHERES OF SOLAR SYSTEM OBJECTS

In this section, we review in sequential fashion current concepts on the origin and evolution of the atmospheres of the solar system objects listed in Table 1.

Origin of the Atmospheres of Venus, Earth, and Mars

Here, we consider relevant data that permit preliminary assessments of the major sources of the atmospheres of the outer three terrestrial planets and of the spatial distribution of the initial volatile endowment. The starting point for such an assessment is an estimate of the composition and amount of gases that, at some time, have been present in a planetary atmosphere—the volatile inventory. In the case of the Earth, reasonably good determinations of its volatile inventory can be obtained from the known composition of its atmosphere, in conjunction with estimates of the amount of volatiles stored in surface, crustal, and mantle reservoirs (e.g. the oceans and carbonate rocks; Turekian & Clark 1975). The high surface temperature of Venus implies that, to first approximation, all of its volatile inventory is presently in its atmosphere. Thus, atmospheric compositional measurements made in situ from the Pioneer Venus and Venera spacecraft provide good estimates of Venus' volatile inventory, except perhaps for water, as discussed in the subsection on Venus. Because of the low surface temperatures on Mars and a lack of information on the amount of volatiles present in surface and subsurface reservoirs, our knowledge of its volatile inventory is incomplete, despite the compositional measurements made in situ from the Viking spacecraft. Nevertheless, the abundance of rare gases, except possibly Xe, in the Martian atmosphere probably closely matches that of its volatile inventory (Anders & Owen 1977). Also, the observed $^{15}N/^{14}N$ isotopic ratio, which exceeds the terrestrial value by about 62%, provides useful constraints on the total abundance of N_2 (McElroy et al 1977). Table 4 summarizes current estimates of the volatile abundances of Venus, Earth, and Mars (Pollack & Black 1979). Also given are values appropriate for two classes of meteorites and the Sun, which, however, refer to the bulk content of these objects.

Several attempts were made to estimate the "unseen" components of the Martian volatile inventory on the basis of the Viking results at a time when the Venus results had not yet been obtained. In essence, each of these studies selected one of the components of the Martian volatile inventory that was known, and determined the remaining components by appropriately scaling the volatile inventory of the Earth or a class of meteorites, which was assumed to have the same *relative proportion* of volatiles as Mars. Rasool & LeSergeant (1977) used the Martian ^{36}Ar abundance and LL chondrites for this purpose, with the meteoritic material forming a 100-km thick veneer. Although they concluded that this model was consistent with the Earth's volatile inventory, more recent comparisons by Bogard & Gibson (1978) indicate that the $^{36}Ar/C$ ratio of LL chondrites, as well as other classes of chondrites, differs substantially from the terrestrial value, and hence scaling on the basis of ^{36}Ar can be dangerous (see Table 4). Anders & Owen (1977) also favored a veneer source, due either to inhomogeneous accretion or to asteroidal/cometary sources, but they considered a carbonaceous chrondrite of type 3V to be a more appropriate analogue. They also employed a very sophisticated model in which elements are divided into five volatility classes, with ^{36}Ar belonging to the class of highest volatility and ^{40}Ar, since it is derived from the decay of ^{40}K, belonging to the class of second highest volatility. According to them, the higher value of $^{40}Ar/^{36}Ar$ for the Martian atmosphere, as compared to the Earth's, indicates that Mars was more poorly endowed with volatiles than the Earth. The analysis of Owen & Anders is subject to the same criticism as that of Rasool & LeSergeant. Clark & Baird (1979) have used the Martian value of ^{40}Ar in conjunction with the relative abundances of the Earth's volatile inventory to estimate other components of the Martian volatile inventory. The chief weakness with this approach is that ^{40}Ar is built up in the interior only gradually with time and thus its atmospheric abundance depends chiefly on outgassing events that occurred after planetary differentiation, whereas all the other volatiles can be released into the atmosphere chiefly at earlier epochs (see Table 2). Finally, McElroy et al (1977) used their estimate of the N_2 abundance, inferred from the $^{15}N/^{14}N$ ratio, together with terrestrial volatile abundances, to infer the amount of CO_2 and H_2O in the Martian volatile inventory. These amounts were consistent with the lower bound of 1 bar of oxygen in the form of either CO_2 or H_2O, as

Table 4 Volatile abundances (gm/gm) and isotopic ratios[a]

Object	Primordial Ar	N_2	CO_2	H_2O	$^{40}Ar/^{36}Ar$	$^{20}Ne/^{36}Ar$	$^{84}Kr/^{36}Ar$	$^{36}Ar/^{14}N$	$^{15}N/^{14}N$	N/C
Venus	6×10^{-10} to 9×10^{-9}	(2.0–2.2) $\times 10^{-6}$	(9.4–9.6) $\times 10^{-5}$	1×10^{-9} to 6×10^{-8}	1.1 to 1.3	0.2 to 0.8	0.005 to 0.015	8×10^{-5} to 9×10^{-4}	2.6×10^{-3} to 4.6×10^{-3}	0.05 to 0.09
Earth	4.6×10^{-11}	2.4×10^{-6}	1.6×10^{-4}	2.8×10^{-4}	292	0.5	0.036	5×10^{-6}	3.6×10^{-3}	0.05
Mars	(1.9–2.5) $\times 10^{-13}$	4×10^{-8} to 4×10^{-7}	$> 3.5 \times 10^{-8}$	$> 5 \times 10^{-6}$	2500 to 3500	0.2 to 0.9	0.01 to 0.05	2×10^{-7} to 2×10^{-6}	5.6×10^{-3} to 6.6×10^{-3}	—
Ordinary chondrites	3×10^{-12} to 1×10^{-9}	—	—	—	—	~0.2	0.015 to 0.029	1×10^{-7} to 7×10^{-7}	—	0.01 to 0.05
Carbonaceous chondrites	4×10^{-10} to 1×10^{-9}	—	—	—	—	~0.2	0.015 to 0.029	1×10^{-7} to 7×10^{-7}	—	0.01 to 0.05
Sun	1×10^{-4}	1.3×10^{-3}	1.3×10^{-2}	9.5×10^{-3}	—	31	0.00027	2.7×10^{-2}	—	0.31

[a] Adapted from Pollack & Black (1979).

implied by an essentially identical $^{18}O/^{16}O$ ratio for the Earth and Mars (McElroy et al 1977). McElroy et al also noted that their approach implied that the relative proportions of rare gases and other volatiles, such as N_2, were different for the two planets. As discussed next, the Venus data support these conclusions.

Pollack & Black (1979) made use of the new Venus data, in conjunction with the volatile abundances for the other objects listed in Table 4, to draw tentative conclusions about the origin of the atmosphere of the outer three terrestrial planets. In accord with Table 4, they noted four key trends: 1. the absolute abundances of N_2 and CO_2 are essentially the same for Venus and the Earth, but Mars has a lower abundance of N_2 (absolute abundance means the ratio of the mass of a given volatile to that of the planet); 2. the absolute abundance of ^{36}Ar and the ratio $^{36}Ar/^{14}N$ decrease systematically by several orders of magnitude from Venus to the Earth to Mars; 3. a similar sharp increase occurs for $^{40}Ar/^{36}Ar$; and 4. the ratios of *primordial* rare gas species (e.g. $^{20}Ne/^{36}Ar$) are very similar for the terrestrial planets and chrondritic meteorites, but differ considerably from solar values. Sources involving the solar nebula and solar wind are inconsistent with trend 4, while the one involving meteorites and comets is incompatible with trends 1 and 2. The release of intrinsic volatiles can account for all these trends, if one assumes that volatiles were incorporated into the planet-forming material at times when there were large pressure differences among the regions where Venus, the Earth, and Mars were forming, but little temperature differences. These constraints on conditions in the solar nebula follow from the differences in the mode of incorporation of rare gases and other volatiles into planet-forming material, as discussed in the previous section. Rare gases are occluded, but C, N, and H_2O are incorporated into chemical compounds. Hence, the amount of rare gases incorporated per gram of other volatiles increases linearly with pressure and decreases exponentially with temperature. Also, the ratio of one incorporated rare gas species to a second varies with temperature. Hence, trend 4 implies a nearly constant temperature, while trend 2 implies a strong pressure gradient. Note that these conditions need not occur simultaneously, but rather apply to whenever volatiles were formed in different regions. Whether such conditions are physically realizeable needs to be carefully assessed. This conclusion that the atmospheres of the three terrestrial planets are derived from the release of intrinsic volatiles is consistent with the rough estimates of source strengths given in the previous section.

Pollack & Black (1979) were able to quantitatively fit all the data of Table 4 with a model that allowed for differences among the three planets in nebular pressure and differences in the efficiency of the release of volatiles at the times of accretion and interior differentiation and at later times. Note again that little ^{40}Ar comes out during the early epochs of outgassing. If P, e_1, and e_2 represent nebular pressure and efficiency of outgassing for the early and later epochs, all normalized to a value of 1 for the Earth, they find $P = 20 \rightarrow 200$, $e_1 = 1$, and $e_2 = 1/15 \rightarrow 1/1.5$ for Venus and $P = 1/40 \rightarrow 1/20$, $e_1 = 1/5 \rightarrow 1/20$, and $e_2 = 1/20$ for Mars. These estimates of e_1 and e_2 implicitly assume that Venus, the Earth, and Mars were all endowed with the same proportion of N_2- and CO_2-forming volatiles. If not, e_1 and e_2 must include differences in initial abundances as well as differences in outgassing efficiency.

Information on bulk potassium content, crustal potassium content, and rare gas abundances in rocks allows us to place some further constraints on the efficiency of volatile release. The bulk K content of the Moon is about a factor of 2 smaller than that of the Earth, as inferred from surface heat flow measurements and K/U and K/Th ratios in rocks for these two bodies (Anders & Owen 1977). If the global abundance of K for Mars lies intermediate between that of the Earth and Moon, then differences in volatile endowment can account for no more than a factor of 1/2 in the value of e_2 for Mars and, thus, this low value is chiefly due to a much more incomplete outgassing of ^{40}Ar from the Martian interior.

The above small spread in values of bulk K content implies that sizeable variances in crustal K abundances may provide a qualitative guide to the efficiency of volatile release, with more highly differentiated objects presumably having higher efficiencies. The crustal K abundances of the Earth and Venus are quite similar, that of Mars is about a factor of 4 smaller, and that of the Moon about a factor of 20 smaller than the terrestrial value (Anders & Owen 1977). Hence, comparable release efficiencies are expected for the Earth and Venus, with progressively smaller values for Mars and the Moon. This expectation is consistent with the above estimates of the relative efficiency factors e_1 and e_2.

The absolute efficiency of later volatile release for the Earth can be obtained from comparisons of bulk K content and atmospheric ^{40}Ar abundance. Values ranging from 25% (Hart et al 1979) to 50% (Stacey 1979) have been obtained in this manner, with the spread in values reflecting differing treatments of mantle heat transport and differing assumptions on the similarity of the K/U and K/Th ratios for the crust and mantle. Rocks derived from the upper mantle of the Earth ($\lesssim$400 km) have very high $^{40}Ar/^{36}Ar$ ratios ($\sim 10^4$) compared to the atmospheric value ($\sim$300) and low absolute abundances of ^{36}Ar, while rocks derived from deeper depths have $^{40}Ar/^{36}Ar$ ratios ($\sim$450) similar to the atmospheric value and much higher absolute abundances of ^{36}Ar (Stacey

1979). These results imply that the absolute efficiency for outgassing primordial rare gases is greater than that for ^{40}Ar; that the Earth has incompletely outgassed both types of rare gases; and that its initial endowment of volatiles extended to sizeable depths in the mantle, perhaps encompassing the entire mantle.

Earth

Here, we consider controls on the present composition of the Earth's atmosphere, its long term evolution, and the ability of the early atmosphere to generate an enhanced greenhouse effect and hence counteract the lower solar luminosity then. In discussing the present and past atmospheres of the Earth, we will concentrate on water vapor, hydrogen, oxygen, ozone, and carbon-, nitrogen-, and sulfur-containing gases. Table 1 summarizes the concentration of these species in the Earth's current, highly oxidizing atmosphere. The dominant C, N, and S species are CO_2, N_2, SO_2, and H_2S.

As discussed in the previous section, the amount of atmospheric water vapor is buffered by the huge ocean reservoir, with the H_2O concentration in the troposphere being limited by its saturation vapor pressure at ambient temperature and its concentration at higher altitudes being regulated by the tropopause cold trap. However, the very efficient removal of water vapor from the troposphere by condensation and precipitation keeps the time-averaged relative humidity somewhat below unity (Rossow 1978). The biosphere converts nitrogen into such "fixed" compounds as nitrates, but it also efficiently recycles fixed nitrogen back into the atmosphere as N_2 and N_2O (Holland 1978a). As a result, much of the N of the Earth's volatile inventory is present as atmospheric N_2 rather than as subsurface nitrate. The biosphere is also responsible for the trace amounts of CH_4, NH_3, and H_2S in the atmosphere, despite the large amount of free O_2 (Walker 1977). Photodissociation of these gases and water vapor in the upper atmosphere helps to generate H_2. The very low abundance of S-containing gases is due to the photochemical conversion of H_2S to SO_2 and the efficient removal of SO_2 by dissolution in water cloud particles and their subsequent precipitation and by sulfate aerosol formation. Finally, the trace amount of ozone in the atmosphere is ultimately governed by the oxygen abundance through a complex set of photochemical reactions (Logan et al 1978).

The amounts of CO_2 and O_2 in the present atmosphere are thought to be governed by complex geochemical cycles that involve the transfer of C and O compounds between the atmosphere and subsurface rock reservoirs (Walker 1977, Holland 1978a). The carbon cycle is illustrated in Figure 5*a* (Walker 1977). The primary reservoir is the crustal rocks,

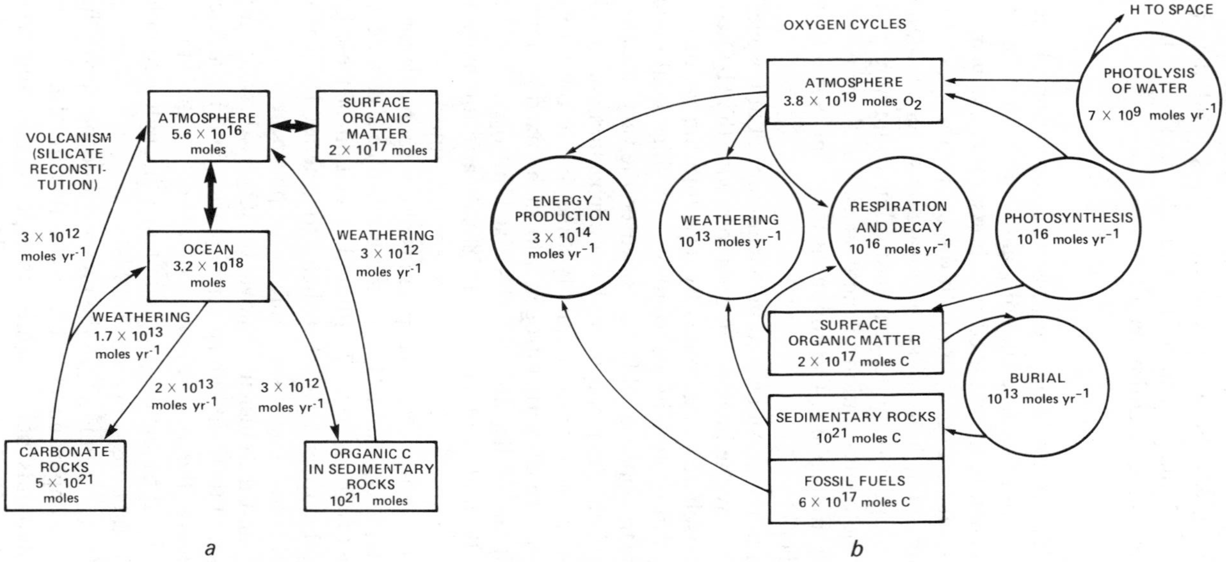

Figure 5 Geochemical cycles of carbon (*a*) and oxygen (*b*). The rectangular boxes show the amount of material stored in major reservoirs, while the numbers within circles in (*b*) and along arrows in (*a*) indicate the fluxes between reservoirs. From Walker (1977).

with the oxidized C-containing rock reservoir, i.e. carbonate rocks, being substantially larger than the reduced C (organic) rock reservoir. Conceivably, the atmospheric CO_2 content is determined by a dynamic equilibrium between the loss of CO_2 through the weathering of silicate rocks to carbonate rocks and the addition of CO_2 due to the metamorphism of buried carbonate rocks and the outgassing of the resulting CO_2 (Walker 1977) as well as possibly juvenile CO_2 (Holland 1978a). The amount of CO_2 lost by weathering probably varies monotonically with the atmospheric partial pressure, but in a rather complicated way (Walker 1977): First, cations must be extracted from silicate rocks by rainwater containing dissolved CO_2 and then the CO_3^{2-} anion content of seawater must be high enough for the cations to precipitate in the sea as carbonates. At present, almost all the Ca leached from silicate rocks is precipitated as calcite, with this being the dominant carbonate rock. At higher partial pressures of CO_2, other carbonates may be produced in abundant amounts [e.g. $CaMg(CO_3)_2$] and, at lower pressures, not all of the leached Ca^{2+} may be precipitated as calcite (Holland 1978a).

Figure 5*b* illustrates the oxygen cycle (Walker 1977). As for the C cycle, subsurface rocks constitute the largest reservoir and the amount of atmospheric O_2 may be determined by a dynamic equilibrium between the rate at which O_2 is being added to and lost from the atmosphere. The principle source is the tiny fraction ($\sim 0.1\%$) of organic matter produced by photosynthesis that is buried in sea sediments without being oxidized. The dominant contributor to this net source of O_2 is organic matter produced in the sea by phytoplankton. The O_2 generated in this manner constitutes a source that is a thousand times more potent than the source due to water photolysis followed by the escape to space of H. Oxygen is lost from the atmosphere as the result of the chemical weathering of rocks containing reduced compounds, with old organic sediments constituting the dominant sink material. Almost complete oxidation occurs during chemical weathering. Hence, control on the O_2 level in the atmosphere is exerted through the dependence of the photosynthetic source on the O_2 partial pressure. For example, increasing this pressure causes an increase in the amount of O_2 dissolved in seawater and a reduction in the fraction of buried organic matter that is not oxidized.

We next discuss the early atmosphere of the Earth and its evolution to its present state. Our previous discussion of thermal history models indicated that much of the release of volatiles occurred during the earliest epochs ($\sim 1 \times 10^9$ years), in conjunction with accretion and global internal differentiation. Later degassing of the remaining juvenile volatiles may have occurred episodically throughout the Earth's history in conjunction with more localized volcanism. A rather key feature in the history of the

Earth's volatiles is the extensive recycling of them between the atmosphere and subsurface, as illustrated in Figure 5. On the basis of the size of the dominant C reservoirs and the fluxes to them given in Figure 5*a*, we estimate a characteristic recycling time of about 3×10^8 years.

The above history of volatile release implies that sizeable oceans were present over almost all of the Earth's lifetime. As at present, atmospheric water vapor in the past was buffered by the oceans, with its abundance being a sensitive function of surface temperature.

Upper bounds can be placed on the amount of solar nebular derived H_2 present in the earliest atmosphere. Using the present amount of atmospheric neon and solar elemental abundances, Walker (1977) shows that the H_2 pressure could have been no more than about 10^{-1} bars. Quite likely it was much less since most of the terrestrial neon was derived from gases trapped in planet-forming material.

After the solar nebula was dissipated, the fractional abundance of hydrogen by number, f_{H_2}, was controlled by a balance between its rate of release from accretional and interior sources and the escape rate of H from the top of the atmosphere. If we assume that the exospheric temperature T_{ex} exceeded 500 K, as it does today ($T_{ex} \sim 1500$ K), the escape rate is equal to the diffusion limiting flux, as given by Equation (2). To obtain a very crude estimate of the initial rate of release of H_2, we suppose that half of the juvenile volatiles were released during the first 3×10^8 years and that $H_2/H_2O = 0.1$ in the released gases, a value intermediate between ones appropriate for silicate melts containing metallic iron and ones containing only FeO and Fe_2O_3 in their present proportions (Holland 1962). Equating the release and escape rates of H_2, we find $f_{H_2} \cong 10^{-3}$. Thus, the partial pressure of H_2 was $\sim 10^{-3}$ bars during this initial phase, if the total pressure was comparable to its present value.

Recycling of water through the Earth's interior at the current characteristic time scale for C compounds would result in a similar mixing ratio and partial pressure of H_2 throughout subsequent epochs, until the time that significant amounts of free oxygen began to accumulate in the atmosphere. Even higher H_2 amounts could have been realized at the earliest times as a result of metallic iron being present in the melt and/or a possibly much shorter time scale associated with the accretional formation of the Earth.

It is very difficult to estimate the oxidation state of the C-, N-, and S-containing gases in the Earth's early atmosphere. For the full range of possible oxidation states of iron, the released volatiles contained only trace amounts of CH_4, and NH_3, with CO, CO_2, N_2, and H_2S representing the dominant C, N, and S species (Holland 1962). For the higher oxidation states of iron, sizeable, even dominant amounts of SO_2 were released. Thus, significant amounts of CH_4 and NH_3 could have resulted only from the hydrogenation of the more oxidized forms of these compounds, once the latter had been placed in the atmosphere. From the standpoint of thermodynamic equilibrium, a sizeable conversion of N_2 to NH_3 and a complete reduction of CO and CO_2 to CH_4 were favored by the comparatively low atmospheric temperature (~ 300 K) and moderate H_2 partial pressure ($\sim 10^{-3}$ bars; Figure 2).

However, the degree to which thermodynamic equilibrium was approached may have been limited by both the very slow reaction rates, especially for NH_3, and by the total amount of H_2 available. With regard to the latter, H_2/CO and H_2/CO_2 ratios of at least 3/1 and 4/1, respectively, are required to fully reduce these gases to CH_4. But $H_2O/CO_2 \cong 5$ in the Earth's volatile inventory (Table 4). Hence, either H_2/H_2O had to approach the value appropriate for a metallic iron-containing melt or water had to be preferentially recycled to meet the above stoichiometric constraints on complete reduction.

It has been traditional to attempt to constrain the oxidation state of the Earth's early atmosphere by the requirement that it favor the generation of complex organic compounds, ones capable of leading to the origin of life. As there is evidence for living organisms in the geological record over at least the last 3 billion years (Windley 1977), the origin of life presumably occurred during the first billion years of the Earth's history. Early experiments by Miller and Urey showed that complex organic molecules, including a variety of amino acids, are produced when a highly reduced atmosphere, dominated by NH_3, CH_4, H_2, and H_2O, is subjected to electrical discharges (Miller & Orgel 1974). Subsequent experimentation has yielded comparable results for a variety of energy sources, including UV light, and for less reducing mixtures of gases (Miller & Orgel 1974). However, in all cases, reducing gases have constituted a significant fraction of the starting materials and it has been necessary to quickly remove the organic molecules from the energy source in order to preserve their integrity.

Theoretical calculations by Pinto et al (1979) suggest that at least formaldehyde can be produced in copious amounts in an atmosphere that is only mildly reducing. In these calculations, the atmosphere was assumed to have its present abundance of N_2, H_2O, and CO_2, with only trace amounts of H_2 ($\sim 10^{-3}$ bars) and CO ($\sim 10^{-6}$ bars) being present. The H_2 abundance is consistent with the estimates given above, while the CO amount is determined by CO_2 and H_2O photochemistry. Formaldehyde is generated through a series of photochemical reactions in which either H_2 or CO are the ultimate reducing agents. According to the model calculations, about 1% of the formaldehyde reaches the ocean before being destroyed by UV sunlight. Pinto et al (1979) find that the net production rate is adequate to generate a 10^{-2} molar concentration of CH_2O in the ocean in $\sim 10^8$ years. Whether more complicated organic molecules would result is unclear. Baur (1978) has even suggested that organic molecules could be produced in an atmosphere containing no reducing gases, with Fe^{2+}-containing clays acting as the reducing agent. In any event, few constraints on the CH_4 and NH_3 content of the early atmosphere appear to be placed by our current understanding of chemical steps leading to the origin of life.

Due to changes in outgassing, photochemical conversion processes, and photosynthesis by green plants, the atmosphere became more oxidizing with increasing time. The early disappearance of metallic iron and the subsequent progressive oxidation of interior Fe by biologically produced O_2 caused the H_2/H_2O ratio in volcanic effluent to change from about 1 to its present value of about 10^{-2}. Also, a decline in the interior heat flux by about a factor of 3 over the age of the Earth may have led to a decline in the outgassing rate (Holland 1978b). The above changes resulted in a decrease in the mixing ratios of H_2, NH_3, and CH_4.

Photodissociation by UV radiation at wavelengths less than about 0.23 μm converts NH_3 very efficiently into N_2 (Kuhn & Atreya 1979). For example, with f_{NH_3} maintained at 10^{-5} or higher, the current atmospheric amount of N_2 can be photolytically generated from NH_3 in only 10^7 years in an atmosphere containing only N-gas species (Kuhn & Atreya 1979). This rapid destruction of NH_3 may be slowed down by the presence of H_2 in the atmosphere and by the presence of H_2O in the upper atmosphere. Water absorbs in the same range of UV wavelengths as NH_3 (Brinkmann 1969). Conceivably the NH_3 abundance in the upper troposphere and higher atmosphere may be greatly reduced due to its reaction with H_2S to form NH_4SH clouds in the upper troposphere, by analogy to the cloud layers of the giant planets (Weidenschilling & Lewis 1973).

Photolysis may also tend to destroy the CH_4 component of the atmosphere. Photochemical reactions create minor amounts of hydrocarbons in the upper atmosphere (Lasaga et al 1971). The higher order hydrocarbons may polymerize and hence form hydrocarbon aerosols that are quickly removed from the atmosphere to form an "oil slick" at the ocean interface. Lasaga et al (1971) estimate that a 1-bar atmosphere of CH_4 would be totally polymerized in only 10^6 to 10^7 years. Conceivably, much longer time scales may apply when allowance is made for the presence of some H_2 and for the thermal decomposition of the polymers.

Not even N_2 is "safe" in a primitive atmosphere. Lightning and subsequent photochemistry can convert the current atmospheric amount of

N_2 in about 5×10^8 years into nitrates, which dissolve in the oceans (Yung & McElroy 1979).

Due to buffering by Fe and FeO, only negligible amounts of O_2 are contained in volcanic effluents. Furthermore, the O_2 generated from the photodissociation of H_2O and the subsequent escape of H is completely removed by the reducing gases emanating from the interior. The escape rate of H resulting from H_2O photodissociation is controlled by diffusion (Hunten & Donahue 1976). For H_2O mixing ratios in the stratosphere comparable to today's value, the photolytic production of O_2 is limited to a rate of about 5×10^7 molecules/cm^2/s. This production rate is about two orders of magnitude smaller than the volcanic input of H_2 given above for the early Earth. Even the current volcanic input of reducing gases is probably sufficient to overwhelm the production of O_2 from water (Kasting & Donahue 1979).

Thus, the introduction of nontrivial quantities of O_2 into the Earth's atmosphere awaited the development of the first green plants capable of photosynthesizing CO_2 and H_2O into organic matter and releasing O_2 as a waste product. Studies of the geological record suggest that the first oxygen-producing organisms appeared about 2.3×10^9 years ago and were similar to blue-green algae of today (Windley 1977).

However, the buildup of atmospheric oxygen may have been delayed somewhat after the first appearance of O_2-producing organisms by the oxidation of reduced compounds in seawater, such as FeO. It has been suggested that this oxidation process generated the prominent banded iron formations of the epoch centered around 2.1×10^9 years ago (Cloud 1972). The delay in the buildup of atmospheric O_2 may have given biota time to evolve oxygen-mediating enzymes that permitted the safe handling and eventual utilization of the otherwise poisonous O_2. Prior to 2×10^9 years ago, the occurrence of reduced minerals of uranium and iron suggests that the atmospheric O_2 partial pressure was quite low (Holland 1978b). However, subsequently, the widespread occurrence of red beds—ferric oxide containing sediments that were formed *on land*—suggests a buildup in the O_2 pressure.

During much of the Earth's history, all lifeforms were single-celled organisms called prokaryotes. Their cells lacked internal organelles, such as a nucleus. The first microscopic forms of life with intracellular organelles—eukaryotes—appeared about 1.4×10^9 years ago (Windley 1977). Their more sophisticated cell organization permitted the occurrence of multicellular organisms. This key evolutionary development may be related to the buildup of atmospheric O_2: almost all current-day eukaryotes are capable of respiration, i.e. the utilization of O_2 for the metabolic release of energy. Furthermore, current-day microorganisms switch from fermentation to respiration when the oxygen partial pressure exceeds 1% of the present atmospheric level (PAL). Thus, the first occurrence of eukaryotes may coincide with the time when the atmospheric O_2 level reached 1% PAL. The replacement of fermentation by respiration permitted organisms to obtain a much larger energy release during metabolism.

Further buildup in the oxygen level may have encouraged the development of more sophisticated means of handling respiration, including the advent of highly specialized systems, such as early versions of the respiratory, circulatory, and nervous systems. The first megascopic forms of life appeared about 7×10^8 years ago (Windley 1977).

The increase in atmospheric oxygen content also led to an increase in atmospheric ozone, which more and more effectively prevented potentially lethal UV radiation from reaching the surface (Berkner & Marshall 1965, Kasting & Donahue 1979). This buildup of atmospheric ozone may have permitted certain organisms to live near the top of the oceans and set the stage for the invasion of the continents by plants about 4×10^8 years ago (Berkner & Marshall 1965). Unfortunately, there is inadequate information on the doses of UV radiation that can be tolerated by various organisms, and hence it is difficult to define at present critical ozone levels and hence critical oxygen levels (L. Margulis, private communication).

Carbonate rocks are present throughout the geological record, which spans the last 3.8×10^9 years (Windley 1977). Hence, some atmospheric CO_2 existed over this entire period. Calcite—$CaCO_3$—is the dominant constituent of carbonate sediments formed during recent times. However, there is some evidence that increasing amounts of dolomite—$CaMg(CO_3)_2$—are present in progressively older carbonate sediments (Holland 1978b). This trend may be related to a higher heat flux and hence a higher outgassing rate in the past, which could imply a higher atmospheric CO_2 amount in the past (Holland 1978b). However, the absence of significant amounts of Na- and K-containing carbonates in the geological record probably implies that the CO_2 partial pressure was always less than $\sim 10^{-1}$ bars over the last 3.8×10^9 years (H. D. Holland, private communication).

Finally, we consider attempts to reconcile the lower solar luminosity in the past with the occurrence of moderate surface temperatures then, as implied by the occurrence of life and oceans for at least the last 3.8×10^9 years (Sagan & Mullen 1972). Greenhouse calculations indicate that the average surface temperature would have fallen below the freezing point of seawater about 2×10^9 years ago if the atmospheric composition and mass were the same then as today (Sagan & Mullen 1972). Hence, the atmosphere must have been different then. Sagan & Mullen considered two types of modified atmospheres, which can produce large enough greenhouse effects to successfully counteract the trend in the solar luminosity. One involved an atmosphere nearly the same as today's, but with NH_3 also present at the level of several parts per million. The other involved the first modified atmosphere plus 1 bar of H_2. Hart (1978) constructed evolutionary models of the Earth's atmosphere, with a large number of adjustable parameters. He successfully overcame the solar luminosity problem with a model in which the CO_2 pressure steadily declined with time, CO_2 was always the dominant C-gas species, and small amounts of NH_3 were present for the first 2×10^9 years. Subsequently, Owen et al (1979) performed much better greenhouse calculations with Hart's prescribed CO_2 time history and found that adequate greenhouse effects occurred at all times with only CO_2 and H_2O opacity, due to the occurrence of larger amounts of CO_2 at earlier times.

It is of interest to compare the above greenhouse models with the discussion given earlier in this subsection of the evolutionary history of the Earth's atmosphere. Even during earliest times, hydrogen escapes too rapidly for it to ever become the dominant constituent of the atmosphere, and hence Sagan & Mullen's H_2-rich atmosphere seems implausible. The occurrence of small amounts of NH_3 during the first several billion years, as invoked in their other model and Hart's model, is not unreasonable, but the rapid photolytic destruction of NH_3 presents a potentially serious problem. The steadily increasing amounts of CO_2 with time into the past, called for in the Owen et al and Hart models, is not inconsistent with our understanding of the CO_2 geochemical cycle, as explained above, but more work is needed on this subject.

Mars

In this section, we discuss the evolution of the mass and composition of the Martian atmosphere and the atmospheric and climatic changes involved in generating its fluvial channels and polar layered terrain. According to Pollack & Black's (1979) model for the origin of the atmospheres of the terrestrial planets, approximately 1–3 bars of CO_2 (or equivalently 0.3–1 bars of CH_4), 50–100 mb of N_2 (or NH_3), and an equivalent depth of 80–160 meters of water have been outgassed from the interior of Mars over the planet's lifetime. If Mars has the same proportion of volatiles as the Earth, the above figures represent no more than 20% of the planet's total volatile content. The inefficient degassing of Mars, as compared to that of the Earth ($e_1 \lesssim 1/5$), may be due to a combination of a smaller amount of accretional heating and a smaller energy release per unit mass during core formation, with both of these differences arising from Mars' smaller mass relative to that of the Earth. According to Figure 1 and Pollack & Black's model, a large fraction of the total degassing took place over the first billion or so years of history, during the period of accretion and global internal differentiation. However, significant additional degassing ($\sim$20–50% of the total) took place at later times in association with major volcanic events. These volcanic events are spread over much of Martian history, with perhaps a decline occurring during the last billion years due to a thickening of the lithosphere (Toksöz et al 1978). Thus, the present degassing rate is expected to be substantially smaller than the time-averaged value, in accord with McElroy et al's (1977) estimate that it is at least a factor of 20 times smaller than the average value on the basis of evolutionary models of the $^{15}N/^{14}N$ ratio.

We next consider the evolution of the oxidation state of the atmosphere. Following the same type of analysis as for the Earth, we estimate that during the early phase of volatile release $f_{H_2} \cong 10^{-3}$, with this mixing ratio declining at subsequent times towards its current value of 4×10^{-6}. The current mixing ratio of hydrogen is due entirely to H_2O dissociation. Once again, it is very difficult to estimate the state of oxidation of C and N compounds in Mars' early atmosphere, with there being some possibility of significant proportions of CH_4 and NH_3 in addition to CO_2 and N_2.

Photochemical processes may have greatly reduced the CH_4 and NH_3 contents of the early partially reducing atmospheres and eventually led to the current atmosphere that contains only CO_2 and N_2. As discussed in the subsection on the Earth, photochemical conversion of NH_3 to N_2 can occur so rapidly as to thwart the establishment of thermodynamic equilibrium concentrations of NH_3. Also, creation of higher hydrocarbons from methane acts as a sink for CH_4 since the higher hydrocarbons can drop out of the atmosphere and collect at the surface. According to calculations by Yung & Pinto (1978), a 1-bar atmosphere of CH_4 could be lost in this way in only about 10^8 years. Once this occurs, oxygen generated from the photodissociation of water vapor and the escape of H oxidizes the hydrocarbons, eventually resulting in the production of a CO_2 atmosphere on a time scale of 10^8–10^9 years (Yung & Pinto 1978). While the eventual production of a CO_2, N_2 atmosphere seems to be well established, the reader is again cautioned that additional work is needed before the stability of CH_4- and NH_3-containing atmospheres is adequately understood.

Through a series of loss processes, almost all of the outgassed volatiles have been removed from the atmosphere. At the low temperatures on Mars, saturation and condensation hold the water vapor content of the atmosphere to a very low level ($\sim 10^{-5}$ m). Principal near-surface reservoirs of water include the fine-grained regolith, subsurface permafrost layers, and the polar layered terrain, with probably only small amounts being present in the permanent polar caps (Pollack 1979). According to Clark & Baird (1979), the fine-grained regolith may have removed atmospheric water through the irreversible formation of clays and salts, the reversible hydration of salts, and reversible adsorption, with a net removal of perhaps as much as several tens of meters of water. Under present conditions, subsurface permafrost *in contact with the atmosphere* is stable down to a latitude of about 40° (Toon et al 1979). A modest amount of water ($\sim$3 m) has been lost over Mars' lifetime due to the joint escape of H and O from the top of the atmosphere (McElroy 1972), with the H escaping by thermal evaporation and the oxygen by photochemical escape. In fact, the coupling of these processes through the interaction of O_2 and H_2 in the lower atmosphere insures that at present precisely two H escape for every O lost (McElroy 1972).

While CO_2 is the dominant constituent of Mars' present atmosphere, a much larger amount of CO_2 (1–3 bars) resides in near-surface reservoirs. At one time, it was thought that the *permanent* polar caps of Mars might be made of CO_2 and contain large amounts of CO_2 ice. However, Viking data show that the northern cap is made of H_2O ice, while the southern cap might contain CO_2, at least during some years, but probably not a large amount (Kieffer et al 1977). The two major sinks for CO_2 are thought to be reversible adsorption onto the fine-grained regolith and the irreversible formation of carbonate rocks. According to Fanale & Cannon (1979), the regolith may at present contain 20 times as much CO_2 as the atmosphere, i.e. 100 mb. The loss of CO_2 from the atmosphere to these reservoirs may have occurred gradually over the course of time, as weathering produced carbonate rocks and a progressively deeper regolith and as the regolith was moved preferentially towards the colder subpolar regions. Thus, at earlier times, the pressure may have been much larger. Finally, through a combination of photochemical escape and solar wind sweeping, Mars has lost an amount of CO_2 comparable to that currently residing in the atmosphere (McElroy 1972).

Nitrogen has also been gradually lost from the atmosphere due mostly to the photochemical formation of HNO_2 and HNO_3, which reacted with surface minerals to form nitrates. Also, some N_2 (1–2 mb) has been lost by photochemical escape (McElroy et al 1977). Settle (1979) has studied the fate of sulfur gases (H_2S and SO_2) vented into the atmosphere. He finds that on a time scale of a few years, under current conditions, these gases are converted to sulfuric acid aerosols, which are subsequently removed from the atmosphere by scavenging and sedimentation. The acid reacts with surface minerals to produce sulfate salts.

The surface of Mars has been excised by a number of channels, whose morphology suggests that they were formed by running water. Some classes of these fluvial channels, particularly the numerous gullies, appear to require surface temperatures above the freezing point of water for their formation, although others, particularly the large outflow channels, might be formed under current Martian climatic conditions (Pollack 1979). The gullies were probably formed after the end of the first two stages of volatile release and towards the beginning of the third stage, i.e. several billion years ago (Pollack 1979). At first glance, it is a rather remarkable inference that clement conditions existed on Mars several billion years ago, in view of its current cold climate and a lower solar luminosity then.

Attempts have been made to explain the clement conditions needed for channel formation by invoking atmospheres capable of generating a large greenhouse effect. Within this context, Pollack (1979) has studied fully reducing atmospheres, with NH_3, H_2O, CH_4, and H_2 as the sources of infrared opacity; Sagan (1978) has examined mildly and very reducing atmospheres, with CO_2, NH_3, H_2, and H_2O opacity sources; and Pollack (1979) and Cess et al (1979) have investigated fully oxidized atmospheres, with CO_2 and H_2O opacity sources. Figures 6*a* and *b* illustrate Pollack's results for fully oxidized and reduced atmospheres, respectively. In Figure 6*a*, curves of surface temperature are displayed as a function of insolation factor f for fixed values of surface pressure P. The horizontal line indicates the melting temperature of water ice. The parameter f is defined as the ratio of the actual solar energy absorbed to the globally averaged amount of insolation that is currently being absorbed by Mars at its mean distance from the Sun. The value of f depends on the solar luminosity, the eccentricity, and orbital position of Mars at the time of interest, and the geographical position of interest (Pollack 1979). For simplicity, consider an orbital position close to perihelion when Mars had a large eccentricity value (see below). Then the orbital factors offset the lower solar luminosity several billion years ago and $f \cong 1$. In this case, a surface pressure of about 2 bars is required to reach the melting point of water for CO_2-H_2O atmospheres according to Figure 6*a*. More precise calculations by Cess et al (1979) suggest a "critical" pressure of 1 bar. Values of the "critical" pressure for other situations, that imply other values of f, are given in Pollack (1979). Thus, the fully oxidized models require that most of the CO_2 outgassed by Mars be present in the atmosphere to generate the required greenhouse effect. This condition is not unreasonable in light of the earlier discussion of the evolution of the surface pressure, but whether the surface reservoirs were sufficiently underdeveloped at this earlier time is still an open question. It is interesting to note that the very conditions that permitted a strong greenhouse effect may have led to its demise through a much enhanced rate of regolith and carbonate rock formation at times when liquid water was available (Pollack 1979).

Figure 6*b* shows corresponding greenhouse calculations for fully reduced atmospheres having varying proportions of NH_3 and H_2, with CH_4 being the principal gas and NH_3 and H_2 having mixing ratios of 0.03 and 0.1 in the nominal case. For all the curves, $f = 1$. These calculations indicate that H_2 opacity is unimportant, but NH_3 opacity is very important. These fully reduced atmospheres provide the needed greenhouse effect with total pressures not in excess of the amount of outgassed C gases ($CH_4 \lesssim 1$ bar) only when nonnegligible amounts of NH_3 are present. Thus, it is necessary to find a way of preventing the rapid photolytic destruction of NH_3, if they are to be viable options. Sagan (1978) required about 1 bar of CO_2 and an ammonia mixing ratio of about $10^{-4.5}$ to generate the required surface warming with his mildly reducing atmospheres, with $f \cong 0.8$. Again, the stability of NH_3 could constitute a serious problem.

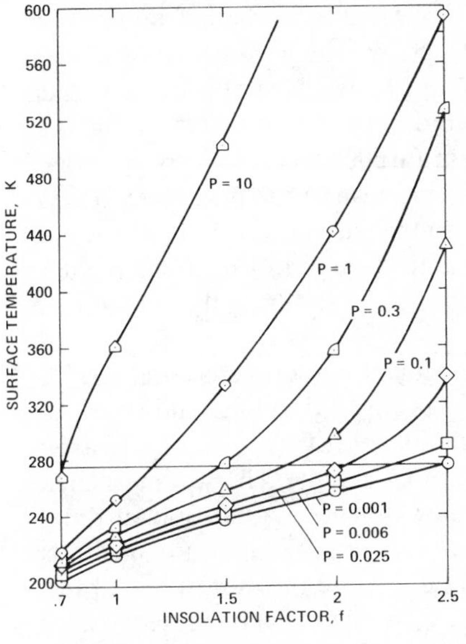

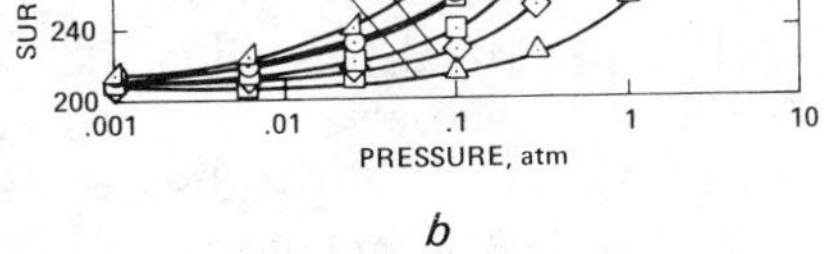

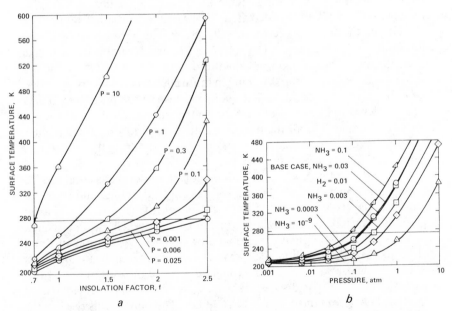

Figure 6 (*a*) Surface temperature of Mars as a function of insolation factor f for CO_2-H_2O-N_2 atmospheres. The curves show results for various choices of surface pressure, P, in bars. In all cases CO_2 is the major atmospheric constituent and $N_2/CO_2 = 0.03$ (*b*) Surface temperature of Mars as a function of surface pressure for CH_4-NH_3-H_2O-H_2 atmospheres of varying composition. In all cases, CH_4 is the dominant gas species. From Pollack (1979).

A portion of the polar regions of Mars consists of a layered sedimentary deposit (Cutts 1973). This laminated terrain is many layers thick, with individual layers having an approximately uniform thickness of 30 m. The layers are thought to be composed of a mixture of fine-grained dust and water ice, which were transported by winds and preferentially deposited in the polar regions. Quasi-periodically varying climatic conditions are believed to be responsible for the layered nature of this terrain (Cutts 1973, Pollack 1979). The most obvious drive for these climatic variations are the quasi-periodic changes of orbital eccentricity, axial tilt, and axial precession, which have periods of 10^5–10^6 years (Ward et al 1974). The amplitudes of these variations for Mars are much larger than for the Earth, with the eccentricity changing from 0 to 0.14 and the obliquity varying from 15 to 35°. Some direct evidence in favor of the astronomical climate model was provided by Pollack et al (1979b), who inferred a time scale of $\sim 10^5$ years to form a lamina from observed aerosol mass loadings and their temporal variations above the Viking lander sites.

One question connected with the astronomical theory of the laminated terrain is the mechanism for preferential aerosol deposition in the polar regions. Cutts (1973) suggested that this was due to the net downward velocity of the atmosphere during the polar winter resulting from the formation of the seasonal CO_2 polar cap. However, Pollack (1979) has pointed out that scavenging of dust and water ice by CO_2 ice in the winter polar region is probably a much more effective agent of preferential deposition. In addition, dust that migrates towards the polar region acts as a condensation center for water ice, so that in general both water and dust are contained in all the particles that settle out of the winter polar atmosphere and form the layered terrain (note that during the subsequent summer, the CO_2 ice sublimates).

A second question connected with the formation of the layered terrain is the manner in which astronomical variations lead to the building of the individual layers. Cutts emphasized the joint role of eccentricity and axial precession variations in modulating dust storm activity. This was motivated by the tendency of present-day global dust storms to occur only at times close to orbital perihelion. Perhaps an even more important factor is obliquity, whose variations can strongly modulate the atmospheric pressure (Pollack 1979). As discussed above, there may be a large reservoir of adsorbed CO_2 contained within the regolith. The bulk of this reservoir is at high latitudes. Hence, at times of high obliquity, the annually averaged polar temperatures are higher and so the atmospheric pressure increases. Calculations by Fanale & Cannon (1979) and Toon et al (1979), based on Figure 4, suggest that the atmospheric pressure varies from less than 1 mb at obliquity minimum to about 20 mb at obliquity maximum. At the lowest pressures, global dust storms are probably impossible as supersonic wind speeds would be required, while at the highest pressures global dust storms could occur all year round (Pollack 1979). In this way, obliquity variations modulate the dust loading of the atmosphere and, hence, the dust deposition rate in the polar regions. Fanale & Cannon (1979) and Toon et al (1979) also find that permanent polar caps of CO_2 ice form at obliquity minimum, largely at the expense of the CO_2 contained in the regolith. Obliquity variations can also modulate the amount of water ice deposited in the polar regions by increasing temperatures at equatorial and middle latitudes and decreasing them in the polar regions at times of low obliquity. According to Toon et al (1979), these temperature variations cause a poleward retreat of the midlatitude boundary of subsurface permafrost and an increase in water ice deposition in the polar regions at times of low obliquity. The above discussion implies that variations in both the amount of dust and water ice deposited in the polar regions and their relative proportions occur over an obliquity cycle.

The above discussion has focused on the manner in which the polar sedimentary deposits have been built up over time. At present, there may also be a significant recycling of material within the polar laminated terrain, which is caused by the joint effects of eccentricity and axial variations acting on the water ice component (Toon et al 1979). These modulations may cause a continual migration of large scale features in the laminated terrain (Toon et al 1979).

Venus

In this subsection, we consider the reasons that the atmospheres of the Earth and Venus are so different, the question of whether Venus' surface temperature was always as high as it is today, and the compositional evolution of the atmosphere. As shown in Table 1, Venus' surface temperature is elevated by about 500 K over the effective temperature at which it radiates to space, in contrast to a modest elevation of 33 K in the case of the Earth. Related to this difference is the presence in Venus' atmosphere of the vast bulk of its outgassed volatile inventory, whereas much of the Earth's inventory is stored in near-surface reservoirs. According to the discussion of the subsection on the origin of the terrestrial atmospheres, rare gases and perhaps other volatiles were incorporated into the outer three terrestrial planets at times when the nebular temperatures were very similar in the various regions of planet formation. The similar abundances of outgassed CO_2 and N_2 for Venus and the Earth may therefore imply similar abundances of outgassed water (Pollack & Black 1979). If this is true, then the current differences in atmospheric mass and surface temperature between Venus and the Earth cannot be due to differences in the composition of their outgassed volatiles.

The above differences between the two planets can be understood in terms of the extreme sensitivity of surface temperature to the amount of sunlight absorbed by a planet (Ingersoll 1969, Rasool & de Bergh 1970, Pollack 1971). This sensitivity arises from the control exerted by near-surface reservoirs on the amount of H_2O and CO_2 in the atmosphere, with these gas abundances depending exponentially on temperature. Since the amount of greenhouse warming depends on the amount of CO_2 and H_2O in the atmosphere, there is a strong positive feedback between surface temperature and gas amount. As illustrated in Figure 7, increasing the insolation at the top of a H_2O-N_2 atmosphere from the current value at Earth to the current value at Venus leads to a very sharp rise in surface temperature and the emplacement of all the volatiles into the atmosphere, a process termed a "runaway greenhouse" (Pollack 1971). Similar results are found when CO_2 and carbonate rocks are considered together with H_2O vapor and oceans (Rasool & de Bergh 1970).

Figure 7 also illustrates another interesting circumstance. Venus' surface may have had a moderate temperature during early epochs when the solar luminosity was considerably lower. When allowance is made for the presence of CO_2 at levels consistent with thermodynamic equilibrium

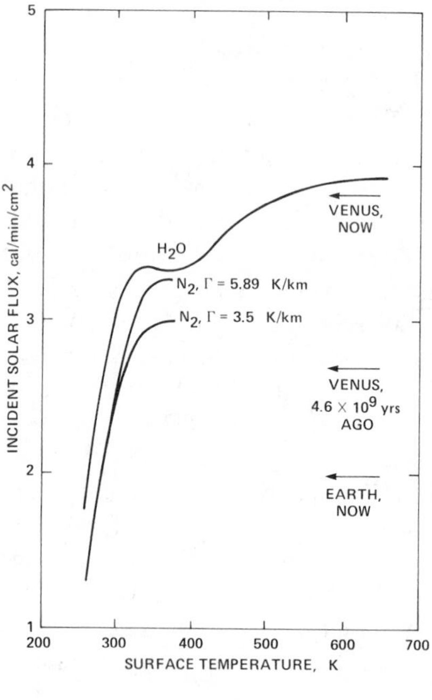

Figure 7 Surface temperature as a function of solar flux for N_2-H_2O atmospheres having 1 bar of N_2 partial pressure. Γ is the tropospheric lapse rate. From Pollack (1971).

with carbonate rocks, this conclusion remains unaltered (Pollack 1979). Thus, the "runaway greenhouse" may not have occurred on Venus until an intermediate epoch, at a time when the solar luminosity reached a critical level. This deduction is still quite tentative since thermodynamic-equilibrium considerations are at best a crude indicator of the atmospheric CO_2 content (Walker 1977, Holland 1978a). However, if this conclusion is correct, one could further speculate that life may have arisen on Venus as it did on the Earth, but that the Venusian biota were ultimately wiped out by the rising surface temperature. Future high resolution photographs of Venus' surface, as obtained from ground-based and spacecraft radar, might provide constraints on the planet's early environment by, for example, finding ancient river beds.

The evolution of Venus' atmosphere from a mildly reducing state to a fully oxidized state should have been similar to that for the Earth and so will not be repeated here. Instead, we focus on the matter of how much water vapor may have been eliminated from Venus' atmosphere over the course of time. If Pollack & Black (1979) are correct, Venus initially may have had an amount of water with an equivalent depth of about 1 km, while today there is at most 1 m of H_2O in the atmosphere (Table 1).

When H_2O is a major atmospheric constituent, the tropopause cold trap ceases being effective and the mixing ratio of H_2O in the stratosphere is comparable to its value in the lower troposphere (Ingersoll 1969). As a result, water vapor has access to almost all the available photons that can dissociate it (Figure 3) and it is fragmented at a rate approaching 10^{13} molecules/cm^2/s. This rate is four orders of magnitude larger than the dissociation rate of water in the Earth's atmosphere today (Figure 3). Furthermore, the abundant H and H_2 in the uppermost atmosphere, resulting from water dissociation, elevate the exospheric temperature far above its current cold value ($\lesssim 400$ K). In this case, the rate of escape of H is not controlled by the exospheric temperature, but by either diffusion or photon limitations (Walker 1975). Since the limiting flux is proportional to the abundance of H-containing species in the stratosphere, it is much larger than for the present Earth ($\sim 10^8$ H/cm^2/s) and is comparable to the rate at which water is photodissociated. If the escape rate of H were equal to 10^{13} atoms/cm^2/s, an ocean of water could be lost in only 30 million years (Walker 1975).

After an initial rapid loss of water vapor from the atmosphere, the rate of loss slowed down considerably for several reasons. First, with a diminishing water vapor mixing ratio in the troposphere and an increasingly effective tropopause cold trap, the diffusion flux decreases and becomes the limiting factor (Walker 1975). Also, O_2 becomes a major atmospheric constituent and its elimination becomes the central problem, both to prevent it from shielding H_2O too effectively from dissociating photons and from rapidly converting H and H_2 back to H_2O. Walker (1975) suggests that O_2 can be eliminated from the atmosphere by the chemical weathering of the surface. But he requires a weathering rate comparable to that for the Earth, despite a lack of liquid water to abet the erosion and to remove the already oxidized material. The basic difficulty with Walker's type of model is the need to get rid of large amounts of hydrogen and oxygen *at the same time*. Let us assume that, to first order, the escape of hydrogen is proportional to the mixing ratio of total hydrogen ($H+H_2$), f_H, and that the consumption of oxygen is proportional to the mixing ratio of oxygen f_{O_2}. In order to escape large amounts of hydrogen and bury equivalent amounts of oxygen we need high values for both f_H and f_{O_2}. Based on preliminary calculations by Yung & McElroy (1979) for the Earth's primitive atmosphere, it is found that high values for f_H and f_{O_2} cannot coexist in a H_2O-CO_2 atmosphere. For instance, if $f_H \gtrsim 10^{-3}$, then $f_{O_2} \lesssim 10^{-10}$. The exact calculations have not been performed for the evolution of the atmosphere of Venus, but it is easy to see that this is a fundamental difficulty. An interesting aspect of this model for the evolutionary loss of water is the possibility of a transient, but long-lived epoch when O_2 was a major constituent of the Venus atmosphere (Walker 1975).

Another possible sink for water arises from its vigorous recycling between the atmosphere and interior. During one cycle, water is removed from the atmosphere and water and hydrogen are added back, with the excess oxygen being taken up by the interior. The emitted hydrogen readily escapes to space. Hence, the net loss of water is equal to the oxygen consumed in the interior or equivalently the hydrogen emitted. The amount of water consumed in a given cycle could be as much as 10% of the amount entering the interior, if the mantle lacks both metallic and ferric iron (Holland 1962). Thus, recycling of water could serve as an important loss mechanism if there is rapid and efficient recycling of atmospheric water.

Much water can also be lost to the interior in an irreversible manner. Fricker & Reynolds (1968) have pointed out that water is highly soluble in silicate melts, with the weight percentage of dissolved water increasing with increasing H_2O partial pressure. Thus, a large initial H_2O pressure might be reduced substantially as a result of H_2O dissolving in surface volcanic flows, which are not recycled back through the interior. Later flows can also remove some atmospheric water, although they would be less effective due to the lowered H_2O pressure. At moderate surface temperatures and hence low H_2O pressures, as is the case of the Earth, this process is not important.

Finally, redox reactions between H_2O and the reduced components of the early atmosphere can result in the elimination of some H_2O. Of particular interest is the oxidation of CO to CO_2 and the loss of the resulting H_2 through the exospheric escape of H. Clearly, only an amount of H_2O equal to the amount of CO can be eliminated in this manner. Depending on the oxidation state of iron in the volatile-bearing material, the released gases are characterized by a CO/CO_2 ratio ranging from $\sim 1/40$ to 4 (Holland 1962). Hence, an amount of water ranging from a few percent to almost 100 percent of the amount of CO_2 in Venus' present atmosphere could have been removed in this manner. While this mechanism is not effective in eliminating a full terrestrial amount of water, it could remove all the water present in a more modest, but still water-rich volatile inventory. However, this scenario is not the full story since a rather precise near equality of initial H_2O and CO is required to generate the rather low H_2O and CO mixing ratios of Venus' present atmosphere. We can also use the very low mixing ratio of CO in Venus' present atmosphere in comparison to its value in the released volatiles to argue that $H_2O/CO_2 \gtrsim 1/40$ in the released volatiles. Therefore, the material that formed Venus had at least 1% of the Earth's water endowment.

The possible exospheric escape of large amounts of hydrogen in the past that is hypothesized in the water photodissociation, interior recycling, and CO oxidation scenarios described above would have left a characteristic signature in the D/H ratio. Future measurements of the ratios of species such as D/H, HD/H_2, HDO/H_2O, DCl/HCl, and DF/HF can provide definitive evidence bearing on these models of water loss.

Moon and Mercury

We discuss the sources of the current exospheric atmospheres of the Moon and Mercury and the possibility of early denser ones. It seems likely that the very thin atmospheres of the Moon and Mercury are derived from a combination of three sources: interior outgassing, accretion of solar wind ions, and accretion of atoms from the interstellar medium (Hodges 1975, Shemansky & Broadfoot 1977). The isotopic ratio $^{40}Ar/^{36}Ar$ in the lunar atmosphere is on the order of 10, whereas the solar value is much less than 1 (Kumar 1976). Thus, ^{40}Ar is almost entirely derived from the decay of ^{40}K in the lunar interior and its subsequent release to the surface. Using estimates of the bulk K content of the Moon, the lifetime of Ar in the lunar atmosphere and surface, and the observed atmospheric number density of ^{40}Ar, Hodges (1975) deduced a source strength for ^{40}Ar that is equivalent to the outgassing of 8% of the current interior production rate. This is a surprisingly large outgassing efficiency in view of the 800-km depth of the lunar lithosphere (Toksöz et al 1978). Another major component of the lunar atmosphere is He. Since the helium concentration shows a distinct correlation with solar activity indices, the solar wind is thought to be a major source, although radioactive decay of ^{238}U and ^{232}Th may supply about 10% of the observed He (Hodges 1975). The solar wind may also be responsible for Ne, another major component of the lunar atmosphere, and for trace amounts of CH_4, NH_3, and CO_2, with the latter being produced by chemical reactions between solar wind derived atoms on soil grains (Hoffman & Hodges 1975).

In contrast to the situation for the Moon, the solar wind usually does not reach the surface of Mercury, but is deflected at an altitude of about 1500 km by the planet's magnetic field. Nevertheless, some solar wind material is expected to be mixed into the magnetosphere (Kumar 1976). Furthermore, at times of intense solar activity, the enhanced solar wind can reach the surface directly. The H in Mercury's atmosphere is derived from a combination of solar wind protons and interstellar atoms (Shemansky & Broadfoot 1977), whereas interior outgassing and/or the solar wind represent the principal sources for the atmospheric He (Kumar 1976).

The components of the lunar atmosphere are lost chiefly by ionization, followed by solar wind sweeping (Kumar 1976). However, Jeans escape represents the chief loss mechanism for the H and He components of Mercury's atmosphere. Heavier components of Mercury's atmosphere are lost by photoionization, followed by magnetospheric convection into the solar wind (Kumar 1976).

Using upper limits on the H_2O, CO_2, and CO abundances in Mercury's atmosphere and estimates of their lifetimes, Kumar (1976) inferred that the current outgassing rate of H_2O and CO_2 is at least four orders of magnitude lower for Mercury than the average rate for the Earth. In part, this extreme difference may be due to the great depth of Mercury's lithosphere at present (Figure 1), as well as a deficiency in Mercury's endowment of volatiles.

Our earlier discussion of thermal history models and lunar chronology indicated that extensive outgassing could be expected early in the history of the Moon and Mercury, in association with accretion, global internal differentiation, and somewhat later basaltic volcanism. Most of the outgassing probably took place during the first two of these epochs, which spanned about the first two hundred million years of lunar history and a somewhat longer period for Mercury (Figure 1). We now inquire as to whether an appreciable atmosphere might have developed then. The formal criterion for having a substantial early atmosphere, but a very thin one at present, may be written as

$$L\Delta t/t < S < L, \tag{8}$$

where L is the assumed time-independent loss rate, S is the average outgassing rate during an intense outgassing epoch of duration Δt, and t is the planet's lifetime.

We can obtain a very crude estimate of S for the Moon. According to Pollack & Black (1979), Mars has outgassed about 2 bars of CO_2, with an efficiency for outgassing that is about 10% of that for the Earth. Comparison of the composition of terrestrial and lunar rocks indicates that the Moon's interior is depleted by a factor of about 30 in the most volatile elements in comparison with the Earth's interior (Anders & Owen 1977). Allowing for both this depletion factor and a somewhat lower outgassing efficiency than for Mars, we estimate that the Moon outgassed about 1 mb of CO_2. According to Equation (6), solar wind sweeping can eliminate about 3 mb of atmosphere over the age of the solar system. By analogy to the situation for Mars (McElroy 1972), a comparable amount can be lost by photochemical escape. However, an early lunar atmosphere dominated by CO_2 can be expected to have an exospheric temperature of about 400 K, by analogy to the exospheric temperature of Venus and Mars, and hence little loss of medium weight atoms occurs by thermal evaporation. The above estimates imply that as much as 6 mb of a CO_2-dominated atmosphere can be lost over the age of the solar system. Hence, if about 1 mb of CO_2-N_2-Ar has been outgassed from the Moon's interior in toto, only an exospheric atmosphere is expected at present. But if $\Delta t/t \sim 0.04$, then, according to Equation (8), it is possible that there was an early dense lunar atmosphere. Naturally, S is too poorly known to make a definitive statement on this interesting possibility.

Outgassed water on the Moon might also have had a long transient existence, but it too was eventually eliminated. Using the same reasoning as for CO_2, we estimate that the Moon may have outgassed about 0.1 m of H_2O. Initially, most of this condensed. The removal of water vapor through photodissociation, escape of H, and crustal loss of O, was impeded by diffusion while the CO_2 atmosphere was present. Nevertheless, even in this case, by analogy to Mars, 0.1 m of water could have been eliminated in just a few times 10^8 years. Conceivably, this early atmosphere may have left clues about its existence in the form of alteration products, such as carbonates, ferric oxides, and hydroxylated minerals.

Similar considerations apply to Mercury. If the planet's magnetic field was established prior to or during the chief epochs of volatile release, conditions may have been more favorable, compared to the Moon, for establishing a transient atmosphere, since sweeping by the solar wind would have been far less efficient. Without the field, Mercury's closer distance to the Sun may have made things less favorable. Once established, the transient atmosphere may have been dissipated through thermal evaporation of CO_2 (Belton et al 1967), photodissociation of CO_2 into CO and O and their thermal evaporation (Belton et al 1967), photochemical loss processes, and ionization and magnetospheric convection into the solar wind.

Finally, the above discussion of the rate at which an early lunar atmosphere was dissipated can be used to set limits on the contribution made by comets/asteroids to the volatile inventory of inner solar system bodies. In particular, this source could not have added more than about 6 mb of CO_2 to the Moon over the age of the solar system. This constraint is consistent with Pollack & Black's (1979) inference that the cometary/asteroid source was not the dominant source of the atmospheres of Venus, Earth, and Mars.

Titan

Titan, Saturn's largest moon, is the only satellite in the solar system with a substantial atmosphere. Here, we discuss the factors that gave rise to the atmosphere and the effect of loss processes on its gaseous composition and total mass. There is a large quantity of aerosols in Titan's atmosphere, which are optically thick at visible wavelengths, but optically thin at infrared wavelengths. Deposition of most of the sunlight in the aerosol layer results in a hot stratosphere (~ 160 K) and a cool surface (~ 80 K) (Danielson et al 1973, Jaffe et al 1979). Solar UV light and/or precipitated high energy particles from Saturn's magnetosphere act as energy sources for transforming some of the major atmospheric gases, such as methane, into more complicated compounds, ones capable of polymerizing. The polymerized material forms the source material of the aerosols (Podolak et al 1979, Hunten 1977).

As discussed in the subsection on interior sources, Titan's interior may have initially contained large quantities of water ice, ammonia hydrate, and methane clathrate. Interior and accretional heating resulted in the outgassing of H_2O, NH_3, and CH_4. Almost all of the outgassed material condensed on the surface, whose average temperature prior to condensation was ~ 85 K; these volatiles eventually migrated preferentially toward the colder polar regions.

Consider first the nature of a hypothetical, very early atmosphere that contained no aerosols. At summer solstice, an ice-covered region near the summer pole achieved a peak surface temperature of ~ 75 K, with a resulting partial pressure of ~ 6 mb of methane and much lower values for NH_3 and H_2O, where we have assumed an ice albedo of 0.8 and an axial obliquity equal to Saturn's (27°). The evaporated gas flowed rapidly to the winter polar region, where it recondensed. Thus, ~6 mb represents an upper limit to the mean surface pressure at the optimum season for this very early atmosphere and substantially lower values may have applied at other seasons.

A key factor in the evolution of Titan's atmosphere from the incipient state described above to its present state was the development of an optically thick aerosol layer, resulting from solar UV/high energy particle induced chemical transformations. Since the aerosols are highly absorbing and their optical depth in the visible is at least 5 at present (Rages & Pollack 1979), it is likely that thermal radiation from the aerosols totally dominates the energy input to the surface. In this case, the average surface temperature is simply equal to $T_e/\sqrt[4]{2} \cong 73$ K, where T_e is the effective temperature. Furthermore, latitudinal and seasonal temperature variations in the atmosphere and over the surface are almost entirely eliminated, since the dynamical time scale for an air parcel to travel from the equator to the pole is short compared to the aerosol layer's radiative cooling time scale (Leovy & Pollack 1973). Thus, the partial pressure of methane in the atmosphere assumes a seasonably independent value of about 4 mb, as dictated by vapor pressure equilibrium at 73 K.

While the above end state represents an approximately time-invariant state, some modulation of the atmosphere's characteristics can be expected, due to variations in the input of energy that converts gases to aerosols. Lockwood & Thompson (1979) find that Titan's albedo shows a marked (~10%) secular variation on the time scale of a decade, with the phase of the variation correlating closely with the phase of solar activity. Pollack et al (1980) propose that these secular brightness variations are caused by solar related changes in the input of energy needed for aerosol production (e.g. UV light below 0.16 μm). Such changes in input energy result in perturbations to the aerosols' characteristics (e.g. mean size) that determine the satellite's albedo. For our purposes, these solar-related changes in albedo are important because they modify the amount of visible sunlight absorbed by Titan by ~2% and hence change the surface temperature and methane partial pressure by ~0.5 K and ~10%, respectively (Pollack et al 1980). The long term trend in the solar luminosity has effects analogous to the much shorter term variations discussed above. If the solar luminosity was 30% smaller 4.6×10^9 years ago than it is today, the solar input to Titan's atmosphere was about 30% less, the surface temperature was ~6 K cooler and the methane pressure was reduced by a factor of ~4. In the future Titan's atmosphere will become more massive as the solar luminosity increases.

Due to gravitational sedimentation, the aerosols remain in the atmosphere for only ~1 year (Podolak & Bar-Nun 1979). Since it is unlikely that they can be converted back to their precursor gases at the cold surface temperature, their continued production represents a steady sink for the atmospheric gases, such as methane. On the basis of inferred properties of Titan's aerosols, their lifetime in the atmosphere, and constraints placed by the photolytic production rate of polymeric ethylene and acetylene, Podolak & Bar-Nun (1979) obtained a "conservative" value of 3.5×10^{-14} gm/cm^2/s for the mass loss rate of both aerosols and methane from the atmosphere. Over the age of the solar system, there is a net loss of about 5×10^3 gm/cm^2 of CH_4, which corresponds to a pressure of ~1 bar. Thus, much more methane than currently resides in the atmosphere may have been lost over Titan's lifetime due to aerosol formation. Naturally, the methane ice buffer keeps the atmospheric pressure at a constant value. But the above estimate implies that Titan's surface has acquired an amount of CH_4, which is at least comparable to Mars' volatile inventory of C. Thus, outgassing from the interior seems to be the most likely source of Titan's atmosphere. We also see that polymerization of methane represents a significant potential effect for hypothesized early, highly reducing atmospheres of the terrestrial planets, as was discussed earlier in this section on the basis of theoretical considerations. At present, there may be a sedimentary blanket of organic polymers covering the surface of Titan to an average depth of ~50 m.

Photolysis of gases in Titan's atmosphere may have produced additional gaseous constituents. On the basis of the rate at which methane is being converted by solar UV radiation to unsaturated hydrocarbons (which are eventually polymerized), Hunten (1977) estimates that $\sim 9 \times 10^9$ molecules of H_2/cm^2/s are being produced. Balancing this production rate against the diffusion-limited rate of H escape, he finds a mixing ratio of $\sim 5 \times 10^{-3}$ for H_2. Even somewhat larger steady-state abundances of H_2 might result from the possible presence of H_2S and PH_3 in Titan's atmosphere. The escape of large amounts of H_2, as implied by the above discussion, seems to be consistent with the detection of an emission cloud around Titan by the long wavelength channel of the UV photometer experiment aboard the Pioneer 11 spacecraft (D. Judge and R. Carlson, private communication). Potentially, photolysis of NH_3 vapor could generate huge amounts of N_2 (~20 bars) (Hunten 1977, Atreya et al 1978). However, the actual production of N_2 is probably much lower than this figure since the low surface temperature implies a very small partial pressure of NH_3; the aerosol layer intercepts many of the NH_3 dissociating photons; and N_2H_4, an intermediate product in the production of N_2 from NH_3, condenses at temperatures below ~150 K.

Io

Io, the innermost Galilean satellite of Jupiter, displays an unprecedented level of active volcanism, even by terrestrial standards (Smith et al 1979). Here we investigate the implications of Io's volcanism for the evolution of its volatile inventory. The high level of volcanic activity is thought to be caused by tidal heating due to Jupiter (Peale et al 1979). Strong tidal heating occurs, despite the satellite's synchronous rotation, as a result of Io's large orbital eccentricity, which is forced by resonant interactions with Europa and Ganymede. If this explanation is correct, Io may have experienced a comparable level of volcanic activity throughout most of its history. As a result of the tidal heating, the interior is molten, except for a thin crust, whose thickness is $\sim$20 km or 1% of the satellite's radius (Peale et al 1979).

Sulfur compounds dominate Io's current volatile inventory. SO_2 gas has been identified as a major component of Io's thin atmosphere (Peale et al 1979); SO_2 ice and solid elemental S are major constituents of its surface (Fanale et al 1977, 1979). At the same time, there are very stringent upper bounds on the amount of water, ammonia, carbon dioxide, and nitrogen ice and bound water on Io's surface (Pollack et al 1978, Pearl et al 1979). The very low values of the average atmospheric pressure, 10^{-8}–10^{-11} bars (McElroy & Yung 1975), imply that very little N_2 or ^{40}Ar has accumulated in its atmosphere.

The mean density of Io is comparable to that of the Moon (cf Table 1). Hence, little if any ice was incorporated into its interior during satellite formation and its intrinsic volatiles may bear a crude resemblance to those of the terrestrial planets. Since S gases constitute some of the more abundant volatiles vented from terrestrial volcanoes, their association with volcanism on Io is not surprising. However, the Earth has outgassed considerably more H_2O vapor and C- and Cl-containing gases than S-containing gases and comparable amounts of N-containing gases (Walker 1977). Hence, the total dominance of S volatiles on Io is, at first glance, a puzzle. This puzzle remains even if we assume little S was incorporated into Io's core, in contrast to the situation for the Earth. Below, we explore the ability of efficient loss processes to explain the current absence of volatiles, aside from S compounds, on Io's surface and in its atmosphere.

Let us first consider the loss processes for water. Suppose that Io formed with the same fractional abundance of water as the Earth and hence outgassed $\sim$100 m of water. Because of Io's greater distance from the Sun and its low surface temperature ($\sim$100 K), the dominant loss process for the terrestrial planets—photodissociation followed by loss of H—may operate too slowly on Io for it to lose 100 m of water over the age of the solar system. For example, using the photodissociation time constant for water (Kumar 1976) and vapor pressure constraints on the amount of water vapor in Io's atmosphere at the average surface temperature, we find that no more than about 0.01 m of water can be lost in this way over 4.6×10^9 years. An attractive alternative mechanism arises from the partial dissociation of water into H_2 at typical silicate lava temperatures. Estimates of the current resurfacing rates on Io imply that an amount of mass equal to ~0.1–10 times the total mass of Io has been extruded volcanically over its lifetime (Johnson et al 1979). If crustal material is preferentially used, then condensible, outgassed volatiles may be recycled ~10–1000 times. According to Holland (1962), $H_2/H_2O \cong$ 0.01–1 in outgassed material, depending on the oxidation state of the iron in the melt. Hence, through repeated outgassing, a very large fraction of the water may have been converted to H_2. Provided that the exospheric temperature, T_{ex}, was greater than about 200 K, as seems reasonable (McElroy & Yung 1975), Jeans escape of H_2 proceeds rapidly enough so that the escape rate would have been limited by diffusion to the exosphere. For almost any Io atmosphere, escape at the limiting flux rate proceeds very rapidly, with all but a minute amount of the volcanic H_2 being eliminated. We also note that the O liberated from H_2O in the magma chamber might have partially been used to oxidize other gases and so may have been the source of some of the O now present as SO_2.

Much water may also have been lost through explosive volcanism. At present, volcanic plumes have been observed to attain altitudes as high as 300 km above the surface (Johnson et al 1979). Thus, while none of the volcanic ejecta has been seen to have velocities above the escape velocity and gases would adiabatically expand and cool to too low a temperature to permit Jeans escape, occasionally this material reaches exospheric altitudes and perhaps even passes through the nearby Jovian magnetosphere. At present S-containing gases probably power the explosive volcanism (Smith et al 1979). Conceivably, at earlier times, when water was abundant, it may have powered the explosive ejecta to velocities above the escape velocity, in which case much water was lost. However, such a mechanism would result in an indiscriminate loss of all materials, including S gases, and thus would not naturally result in a totally water-depleted object that still had large amounts of S volatiles. Similarly, if S was always the dominant volatile and loss occurred only through high velocity volcanic explosions, it would be very difficult to so completely eliminate such atmospheric constituents as ^{40}Ar without greatly depleting the S volatiles. In conclusion, explosive volcanism is not the only loss mechanism for volatiles, although it could be a contributing factor.

We next consider the elimination of N_2. Again, if a terrestrial analogue is used for Io's volatile inventory, then ~ 30 mb of N_2 would be outgassed, which would remain in the atmosphere for *average* surface temperatures on Io. In order to lose this amount of N_2 in less than 4.6×10^9 years, the escape rate must be at least 3×10^7 molecules/cm^2/s. This requirement probably far exceeds the rates that can be expected from the photochemical escape processes that were effective on Mars. In particular, even with CO_2 as the dominant gas species to maximize the loss rate, we estimate that the maximum photochemical loss rate of N_2 is $\sim 10^6$ molecules/cm^2/s and it would most likely be much less. In order for Jeans escape to eliminate N_2 at the desired rate, T_{ex} must be greater than 1000 K, which far exceeds estimates of T_{ex} for a predominant N_2 atmosphere (McElroy & Yung 1975). A more promising possibility arises from the interaction of Io's atmosphere with Jupiter's magnetosphere. First, consider volcanic plumes. Occasionally, the material in plumes may penetrate into the magnetosphere. In this case, some of the gases are ionized through interaction with the magnetospheric plasma and are then trapped along magnetic field lines and swept away. For reasons given above in the discussion of explosive volcanism, this scenario is probably not the whole story. Second, consider gases that are continually present in the atmosphere. Elastic scattering by the high energy protons of the Jovian magnetosphere results in the ejection of gases close to the exosphere into the magnetosphere, where they are ionized and swept away. Furthermore, a buildup in the heavy ion component of the magnetosphere due to gases derived from Io sets up a positive feedback situation, with these heavy ions also ejecting gases from Io's atmosphere. Preliminary calculations of these processes indicate that they can generate substantial loss rates, ones adequate to eliminate terrestrial amounts of N_2 over Io's lifetime (J. B. Pollack & F. C. Witteborn, in preparation).

The above magnetospheric loss mechanism can equally well remove significant amounts of atmospheric ^{40}Ar and CO_2. The current dominance of SO_2 and S as Io's major volatiles may be due in part to their volatility being lower than ^{40}Ar, N_2, and CO_2, which would lead to the latter's preferential occurrence in the atmosphere and hence preferential loss at earlier times. We note that both the volcanic loss mechanism for water and the magnetospheric mechanism for the other gases would eliminate them in less than 4.6×10^9 years, even had we chosen a higher initial volatile content for Io than indicated by terrestrial standards.

At the present time, S and O atoms are being transferred from Io's atmosphere into the surrounding torus at a rate of about 10^{10} atoms/cm^2/s (Johnson et al 1979), perhaps by means of the magnetospheric processes described above. If this loss rate of S has been sustained over a significant fraction of Io's lifetime, then a total of about 1×10^{27} S atoms/cm^2 have been lost. But this amount of S exceeds by more than two orders of magnitude the amount in Io's entire volatile inventory that would be expected by a strict analogy to the Earth's volatile inventory (Walker 1977). Hence, a much smaller fraction of S has been segregated into Io's core than into the Earth's core.

The Giant Planets

These objects are thought to possess cores made of rocky and icy materials and gaseous envelopes that have an approximately solar elemental composition, but with perhaps some enrichment of core-type material. The envelope constitutes approximately 95, 85, 20, and 15% of the mass of Jupiter, Saturn, Uranus and Neptune, respectively (Podolak & Cameron 1974, Slattery 1977). In this section, we consider the possible relationships between the evolution of these massive gaseous envelopes and the capture of these planets' irregular satellites, the composition of their regular satellites, and changes in the composition of their observable atmospheres. Irregular satellites have orbits with large eccentricities and/or inclinations and they sometimes travel in a retrograde direction around the planet.

The giant planets may have formed in one of two ways: either the core formed first and grew to a sufficient mass to permanently capture their envelopes, perhaps by inducing a hydrodynamical instability in the nearby solar nebula (Perri & Cameron 1974); or, alternatively, a local gas instability occurred within the solar nebula, resulting in the gravitational contraction of a large gaseous envelope, which subsequently accumulated a core (Bodenheimer 1974). In either case, a gaseous envelope with dimensions of a hundred to a thousand times the current planetary size was present initially for times on the order of 10^7 years; then the envelope underwent a hydrodynamical collapse to several times its present size on a time scale of years; and subsequently has slowly contracted to its present size. During this latter phase, the large intrinsic luminosity, built up by early rapid contraction, decreased by several orders of magnitude up until the present (Pollack et al 1977, Bodenheimer et al 1979).

During the earliest epochs, when the envelopes were very distended, the giant planets had very large cross sections and may have captured small solid planetesimals ($\lesssim$10–100 km) through gas drag interactions (Pollack et al 1979a). In general, the orbits of the captured bodies decayed very rapidly, through continued gas drag, to the center of the planet, thus contributing to the core. However, objects captured just prior to the hydrodynamical collapse phase may have been able to remain in the outer portions of the planet's system, thereby becoming irregular satellites.

The regular satellites may have begun to form towards the end of the hydrodynamical collapse phase, when the outermost portions of the envelope spread out into a disk (Cameron & Pollack 1976). At this time, the planet's intrinsic luminosity may have been sufficiently high, particularly in the case of Jupiter and Saturn, to inhibit the condensation of ices close to them, thereby creating a compositional gradient within their regular satellite system (Pollack & Reynolds 1974, Pollack et al 1977). Conceivably, the systematic decrease of the mean density of the Galilean satellites with increasing distance from Jupiter was created in this way. Note that ices constitute a significant fraction of the interior of many satellites of the outer solar system. By way of contrast, ices were probably not incorporated into bodies in the inner solar system and atmosphere-forming volatiles represent only a very small fraction of their total mass.

During most of their history, the envelopes have undergone a progressive cooling. When the interior temperatures became sufficiently low, chemical constituents of the envelope may have begun to become immiscible and started to separate. A particularly interesting possibility in this regard is the immiscibility of helium in metallic hydrogen when the temperature falls below $\sim 10^4$ K (Stevenson & Salpeter 1976). Metallic hydrogen zones occur at interior pressures above several megabars. The temperatures inside Saturn may have reached this threshold, in which case some He within the metallic hydrogen zone has sunk towards the planet's center (Pollack et al 1977). Through mixing, the He depletion of the metallic zone may also lead to a helium depletion in the observable atmosphere. The gravitational energy released by the separation of He may be providing a significant fraction of Saturn's intrinsic luminosity at the present time (Pollack et al 1977). The envelopes of Uranus and Neptune are probably too small to have metallic H zones.

CONCLUSIONS

The dominant source of the atmospheres of Earth, Venus, Mars, Io, and Titan appears to be the release of intrinsic volatiles. Due to substantial heating during accretion and global internal differentiation in the case of the terrestrial planets and perhaps Titan and by tidal forces in the case of Io, most of the juvenile gases were vented into the atmosphere during the first 10^8–10^9 years. Additional outgassing of juvenile volatiles may have occurred episodically at later times in association with volcanic and tectonic events. The Earth and Venus have outgassed a significant fraction of their initial endowment of volatiles, while Mars may have outgassed no more than about 20% of its endowment, due to its smaller mass. Since most volatiles were probably incorporated into planet-forming material at very similar temperatures for the outer three terrestrial planets the relative proportion of volatiles should be quite similar, with the dominant outgassed volatiles being H_2O (and associated H_2) and C-, Cl-, N-, and S-containing gases, in approximate order of abundance. However, the Moon and Mercury were apparently endowed with a much smaller fractional amount of volatiles.

Io's juvenile volatiles may have had a similar composition to those of the terrestrial planets, although the S-gas species abundances may have been substantially higher than for the Earth. Titan's outgassed volatiles were probably dominated by H_2O, NH_3, and CH_4, which were derived from ices. Although Titan represents an end-member of satellites of the outer solar system in having a substantial atmosphere, it is qualitatively similar to many of these satellites: volatiles constitute a significant fraction of Titan's mass, in contrast to the much lower fractional abundance of volatiles for inner solar system objects. Yet its low surface temperature acts to constrain severely both the mass and composition of its atmosphere.

The solar wind, interior outgassing of radiogenic gases, and perhaps the interstellar medium constitute the dominant sources of the present exospheric atmospheres of the Moon and Mercury. However, in their early history, outgassing of modest amounts of Earth-type volatiles probably occurred and may even have led to the formation of transient atmospheres that were much more massive than their current ones. The solar nebula is the only reasonable source for the extremely massive atmospheres of the giant planets.

The early atmospheres of the Earth, Venus, Mars, and perhaps Io were at least mildly reducing. These atmospheres evolved to their current oxidized states as a result of juvenile gases being less reducing at later times, photodissociation of water followed by the escape to space of H, other photochemical processes, and, in the case of the Earth, green plant photosynthesis followed by the burial of organic matter. Indeed, photosynthesis gave rise to large amounts of free oxygen in the Earth's atmosphere. Recycling of water through its interior, dissolution of water in magmas that were not recycled, and redox reactions with CO may have enabled Venus to lose enormous quantities of water over its lifetime. The relative ease with which H can escape from the atmospheres of satellites and the terrestrial planets has prevented H_2 from ever being the dominant constituent of these atmospheres, in contrast to the situation for the giant planets. Photochemical escape has led to the elimination of modest amounts of N and O from Mars' atmosphere over its lifetime and modest amounts of H from Venus' atmosphere in recent times.

The Moon and Mercury lack substantial atmospheres at the present time since they have outgassed volatiles at a slower rate, averaged over their lifetimes, than the rate at which these gases were lost to space. In the case of the Moon, solar wind sweeping is the dominant loss process. In the case of Mercury, solar wind sweeping is also an important loss process, but in a more indirect fashion—ionization followed by magnetospheric convection to the solar wind. Thermal evaporation is also an important loss process for the lighter constituents of Mercury's atmosphere. Elastic scattering of constituents in Io's upper atmosphere by high energy magnetospheric particles may have led to the elimination of species more volatile than SO_2 over Io's lifetime. In particular, large amounts of outgassed ^{40}Ar, N_2, and CO_2 may have been lost in this manner. At present, significant amounts of SO_2 are being lost by Io to its circumjovian torus, presumably chiefly as a result of this process.

Large amounts of outgassed volatiles have been lost to surface and subsurface reservoirs as a result of condensation, adsorption, and chemical weathering in the case of the Earth, Mars, Titan, and Io. Most of the Earth's water and CO_2 reside in its oceans and carbonate rocks, respectively; most of Mars' water, CO_2, and N_2 have been lost to ice, regolith, and rock reservoirs; and most of Titan's CH_4 and Io's SO_2 reside in surface and subsurface ice deposits. However, in many of the above cases, such losses are not irreversible. For all the above objects, some reservoirs, e.g. the Earth's oceans, Mars' regolith, and Titan and Io's ice deposits, act to buffer the atmospheric content of their associated volatiles. Also volatiles contained in buried sediments have been cycled through the interiors of the Earth, Io, and Venus, and eventually outgassed back into their atmospheres. In the case of Io, such recycling has occurred many times over its lifetime as a result of tidal heating and may have played a key role in eliminating H_2O through the partial dissociation of H_2O into H_2 and O_2. This process also may have helped eliminate H_2O from Venus' past atmosphere, as mentioned earlier. In the case of the Earth, recycling of volatiles may lead to a partial kinetic control on the amounts of O_2 and CO_2 in the atmosphere.

An increase in the amount of sunlight incident on an Earth-like planet from that at the Earth's distance from the Sun to that at Venus' results in a "runaway" greenhouse, due to the very strong positive feedback between surface temperature and atmospheric water and CO_2. However, conversely, the Earth did not ice over, but had oceans and life in its early history and water flowed across the martian surface several billion years ago, despite the solar luminosity being several tens of percent lower then. The answer to these apparent paradoxes may lie in altered early atmospheres that were capable of stronger greenhouse effects. Smaller CO_2 reservoirs and hence much larger amounts of CO_2 in Mars' past atmosphere seem to offer the most promising possibility for achieving the needed greenhouse effect for that planet, while a similar solution, perhaps in conjunction with trace amounts of NH_3 at early times, may be a viable solution for the Earth. The lower solar luminosity in the past may have permitted Venus to have a moderate surface temperature and maybe even life in its early history and may have led to a reduced partial pressure of CH_4 in Titan's past atmosphere.

Quasi-periodic variations of orbital eccentricity and axial obliquity and orientation may have modulated the atmospheres and climates of the Earth and Mars on time scales of 10^4–10^6 years. These variations are much larger for Mars and may have led, particularly through the obliquity oscillations, to large variations in atmospheric pressure, dust storm activity, occasional presence of permanent CO_2 ice caps, and transfer of water between polar and midlatitude sinks, whose occurrence is recorded in the polar laminated terrain. These same astronomical variations appear to be important modulators of the Pleistocene succession of ice ages on the Earth.

Finally, the large dimensions of the early giant planets may have enabled them to capture core material and irregular satellites by gas drag effects. Their early high luminosity may have set up a compositional gradient among their regular satellites. Cooling of the metallic hydrogen region of Saturn may have reached a point at which He has begun to segregate from H_2, perhaps resulting in a He depletion in its observable atmosphere.

ACKNOWLEDGMENTS

We are very grateful to Ray Reynolds and Brian Toon for their careful reading of this paper and their helpful suggestions.

Literature Cited

Anders, E., Owen, T. 1977. Mars and earth: origin and abundance of volatiles. *Science* 198:453–65

Atreya, S. K., Donahue, T. M., Kuhn, W. R. 1978. Evolution of a nitrogen atmosphere on Titan. *Science* 201:611–13

Baur, M. E. 1978. Thermodynamics of heterogeneous iron-carbon systems: implications for the terrestrial reducing atmosphere. *Chem. Geol.* 22:189–206

Belton, M. J. S., Hunten, D. M., McElroy, M. B. 1967. A search for an atmosphere on Mercury. *Astrophys. J.* 150:1111–24

Berkner, L. V., Marshall, L. C. 1965. On the

origin and rise of oxygen concentration in the earth's atmosphere. *J. Atmos. Sci.* 22:225–61

Blanco, V. M., McCuskey, S. W. 1961. *Basic Physics of the Solar System.* Reading, Mass: Addison-Wesley

Bodenheimer, P. 1974. Calculations of the early evolution of Jupiter. *Icarus* 23:319–25

Bodenheimer, P., Grossman, A. S., DeCampli, W. M., Marcy, G., Pollack, J. B. 1979. Calculations of the evolution of the giant planets. *Icarus.* In press

Bogard, D. D., Gibson, E. K. Jr. 1978. The origin and relative abundance of C, N, and the noble gases on the terrestrial planets and in meteorites. *Nature* 271: 150–53

Brinkmann, R. T. 1969. Dissociation of water vapor and evolution of oxygen in the terrestrial atmosphere. *J. Geophys. Res.* 74:5355–68

Cameron, A. G. W. 1970. Abundances of the elements in the solar system. *Space Sci. Rev.* 15:121–46

Cameron, A. G. W., Pollack, J. B. 1976. On the origin of the solar system and of Jupiter and its satellites. In *Jupiter*, ed. T. Gehrels, pp. 61–84. Tucson: Univ. Ariz. Press

Canuto, V., Hsieh, S. H. 1978. Scale covariant cosmology and the temperature of the earth. *Astron. Astrophys.* 65:389–91

Cess, R. D., Ramanathan, V., Owen, T. 1979. The martian paleoclimate and enhanced atmospheric carbon dioxide. *Icarus.* In press

Chamberlain, L. W. 1963. Planetary coronae and evaporation. *Planet. Space Sci.* 11: 901–60

Clark, B. C., Baird, A. K. 1979. Volatiles in the Martian regolith. *Geophys. Res. Lett.* 6:811–14

Cloud, P. 1972. A working model of the primitive earth. *Am. J. Sci.* 272:537–48

Cloutier, P. A., McElroy, M. B., Michel, F. C. 1969. Modification of the Martian ionosphere by the solar wind. *J. Geophys. Res.* 74:6215–28

Consolmagno, G. J., Lewis, J. S. 1977. Preliminary thermal history models of icy satellites. In *Planetary Satellites*, ed. J. A. Burns, pp. 492–500. Tucson: Univ. Ariz. Press

Cutts, J. A. 1973. Nature and origin of layered deposits of the Martian polar regions. *J. Geophys. Res.* 78:4231–49

Danielson, R. E., Caldwell, J. J., Larach, D. R. 1973. An inversion in the atmosphere of Titan. *Icarus* 20:437–43

Eddy, J. A. 1977. Climate and the changing Sun. *Climatic Change* 1:173–90

Fanale, F. P., Cannon, W. A. 1971. Physical adsorption of rare gas on terrigenous sediments. *Earth Planet. Sci. Lett.* 11:362–68

Fanale, F. P., Cannon, W. A. 1979. Mars: CO_2 adsorption and capillary condensation of clays—significance for volatile storage and atmospheric history. *J. Geophys. Res.* In press

Fanale, F. P., Johnson, T. V., Matson, D. L. 1977. Io's surface composition: observational constraints and theoretical considerations. *Geophys. Res. Lett.* 4:303–6

Fanale, F. P., Brown, R. H., Cruikshank, D. P., Clark, R. N. 1979. Significance of adsorption features in Io's infrared reflectance spectrum. *Nature* 280:761–63

Fricker, P. E., Reynolds, R. T. 1968. Development of the atmosphere of Venus. *Icarus* 9:221–30

Hart, M. H. 1978. The evolution of the atmosphere of the earth. *Icarus* 33:23–39

Hart, R., Dymond, J., Hogan, L. 1979. Preferential formation of the atmosphere–sialic crust system from the upper mantle. *Nature* 278:156–59

Hays, L. D., Imbrie, J., Shackelton, N. J. 1976. Variation in the earth's orbit: pacemaker of the ice ages. *Science* 194: 1121–32

Hodges, R. R. Jr. 1975. Formation of the lunar atmosphere. *The Moon* 14:139–57

Hoffman, J. H., Hodges, R. R. 1975. Molecular gas species in the lunar atmosphere. *The Moon* 14:159–67

Holland, H. D. 1962. Model for the evolution of the earth's atmosphere. In *Petrologic Studies: A Volume in Honor of A. F. Buddington*, ed. A. E. Engle, H. L. James, B. F. Leonard, pp. 447–77. New York: Geol. Soc. Am.

Holland, H. D. 1978a. *The Chemistry of the Atmosphere and Oceans.* New York: Wiley

Holland, H. D. 1978b. The evolution of seawater. In *The Early History of the Earth*, ed. B. F. Windley, pp. 559–68. New York: Wiley

Hubbard, W. B. 1979. Intrinsic luminosities of the jovian planets. *Rev. Geophys. Space Phys.* In press

Hunten, D. M. 1973. The escape of light gases from planetary atmospheres. *J. Atmos. Sci.* 30:1481–94

Hunten, D. M. 1977. Titan's atmosphere and surface. In *Planetary Satellites*, ed. J. A. Burns, pp. 420–37. Tucson: Univ. Ariz. Press

Hunten, D. M., Donahue, T. M. 1976. Hydrogen loss from the terrestrial planets. *Ann. Rev. Earth Planet. Sci.* 4:265–92

Ingersoll, A. P. 1969. The runaway greenhouse: a history of water on Venus. *J. Atmos. Sci.* 26:1191–98

Iversen, J. D., Greeley, R., Pollack, J. B. 1976. Windblown dust on Earth, Mars, and Venus. *J. Atmos. Sci.* 33:2425–29

Jaffe, W., Caldwell, J., Owen, T. 1979. The brightness temperature of Titan at 6 cm from the very large array. *Astrophys. J. Lett.* In press

Jakosky, B. M., Ahrens, T. J. 1979. The history of an atmosphere of impact origin. *Proc. 10th Lunar Sci. Conf.*, p. 3043. New York: Pergamon

Johnson, T. V., Cook, A. F., Sagan, C., Soderblom, L. A. 1979. Volcanic resurfacing rates and implications for volatiles on Io. *Nature* 280:746–50

Kasting, J. F., Donahue, T. M. 1979. Evolution of oxygen and ozone in the earth's atmosphere. Conference on Life in the Universe, Moffett Field, NASA

Kieffer, H. H., Martin, T. Z., Peterfreund, A. R., Jakosky, B. M., Miner, E. D., Palluconi, F. D. 1977. Thermal and albedo mapping of Mars during the Viking primary mission. *J. Geophys. Res.* 82:4249–92

Kuhn, W. R., Atreya, S. K. 1979. Ammonia photolysis and the greenhouse effect in the primordial atmosphere of the Earth. *Icarus* 37:207–13

Kumar, S. 1976. Mercury's atmosphere: a perspective after Mariner 10. *Icarus* 28: 579–91

Kumar, S., Hunten, D. M., Broadfoot, A. L. 1978. Non-thermal hydrogen in the Venus exosphere: the ionospheric source and the hydrogen budget. *Planet. Space Sci.* 26:1063–75

Lasaga, A. C., Holland, H. D., Dwyer, M. J. 1971. Primordial oil slick. *Science* 174: 53–55

Leovy, C. B., Pollack, J. B. 1973. A first look at atmospheric dynamics and temperature variations on Titan. *Icarus* 19:195–201

Lewis, L. 1974. Volatile element influx on Venus from cometary impact. *Earth Planet. Sci. Lett.* 22:239

Lockwood, G. W., Thompson, D. T. 1979. A relationship between solar activity and planetary albedos. *Nature* 280:43–45

Logan, L. A., Prather, M. L., Wofsy, S. C., McElroy, M. B. 1978. Atmospheric chemistry: response to human influence. *Philos. Trans. R. Soc. London Ser. A* 290:187–234

Mason, B. 1962. *Meteorites.* New York: Wiley

McElroy, M. B. 1972. Mars: an evolving atmosphere. *Science* 175:443–45

McElroy, M. B. 1976. Chemical processes in the solar system: a kinetic perspective. In *Chemical Kinetics*, pp. 127–211. London: Butterworth

McElroy, M. B., Yung, Y. L. 1975. The atmosphere and ionosphere of Io. *Astrophys. J.* 196:227–50

McElroy, M. B., Yung, Y. L., Nier, A. O. 1976. Isotopic composition of nitrogen: implications for the past history of Mars' atmosphere. *Science* 194:70

McElroy, M. B., Kong, T. Y., Yung, Y. L. 1977. Photochemistry and evolution of Mars' atmosphere: a Viking perspective. *J. Geophys. Res.* 82:4379–88

Miller, S. L., Orgel, L. E. 1974. *The Origins of Life on Earth.* Englewood Cliffs, NJ: Prentice-Hall

Morrison, D. 1977. Asteroid sizes and albedos. *Icarus* 31:185–220

Owen, T., Cess, R. D., Ramanathan, V. 1979. Early Earth: an enhanced carbon dioxide greenhouse to compensate for reduced solar luminosity. *Nature* 277:640–42

Peale, S. J., Cassen, P., Reynolds, R. T. 1979. Melting of Io by tidal dissipation. *Science* 203:892–94

Pearl, J., Hanel, R., Kunde, V., Maguire, W., Fox, K., Gupta, S., Ponnamperuma, C., Raulin, F. 1979. Identification of gaseous SO_2 and new upper limits for other gases on Io. *Nature* 280:755–58

Perri, F., Cameron, A. G. W. 1974. Hydrodynamic instability of the solar nebula in the presence of a planetary core. *Icarus* 22:416–25

Pinto, J. P., Gladstone, G. R., Yung, Y. L. 1979. Photochemical production of formaldehyde in the Earth's primitive atmosphere. *Science.* In press

Podolak, M., Bar-Nun, A. 1979. A constraint on the distribution of Titan's atmospheric aerosol. *Icarus* 39:272–76

Podolak, M., Cameron, A. G. W. 1974. Models of the giant planets. *Icarus* 22:123–48

Podolak, M., Noy, N., Bar-Nun, A. 1979. Photochemical aerosols in Titan's atmosphere. *Icarus* 40:193–98

Pollack, J. B. 1971. A nongrey calculation of the runaway greenhouse: implications for Venus' past and present. *Icarus* 14: 295–306

Pollack, J. B. 1979. Climatic change on the terrestrial planets. *Icarus* 37:479–553

Pollack, J. B., Black, D. C. 1979. Implications of the gas compositional measurements of Pioneer Venus for the origin of planetary atmospheres. *Science* 205:56–59

Pollack, J. B., Reynolds, R. T. 1974. Implications of Jupiter's early contraction history for the composition of the Galilean satellites. *Icarus* 21:248–53

Pollack, J. B., Grossman, A. S., Moore, R., Graboske, H. C. Jr. 1977. A calculation of Saturn's gravitational contraction history. *Icarus* 30:111–28

Pollack, J. B., Witteborn, F. C., Erickson, E. F., Strecker, D. W., Baldwin, B. J., Bunch, T. E. 1978. Near-infrared spectra of the Galilean satellites: observations and compositional implications. *Icarus* 36: 271–303

Pollack, J. B., Burns, J. A., Tauber, M. E. 1979a. Gas drag in primordial circumplanetary envelopes: a mechanism for satellite capture. *Icarus* 37:587–611

Pollack, J. B., Colburn, D. S., Flasar, F. M., Kahn, R., Carlston, C. E., Pidek, D. 1979b. Properties and effects of dust particles suspended in the Martian atmosphere. *J. Geophys. Res.* 84:2929–45

Pollack, J. B., Rages, K., Toon, O. B., Yung, Y. L. 1980. On the possible relationship between secular brightness changes of Titan and solar variability. *Geophys. Res. Lett.* In press

Rages, K., Pollack, J. B. 1979. Titan aerosols: optical properties and vertical distribution. *Icarus*. In press

Rasool, S. I., de Bergh, C. 1970. The runaway greenhouse and accumulation of CO_2 in the Venus atmosphere. *Nature* 226:1037–39

Rasool, S. I., LeSergeant, L. 1977. Implications of the Viking results for volatile outgassing from Earth and Mars. *Nature* 266:822–23

Ridgway, S. T., Larson, H. P., Fink, U. 1976. The infrared spectrum of Jupiter. In *Jupiter*, ed. T. Gehrels, pp. 384–417. Tucson: Univ. Ariz. Press

Rossow, W. B. 1978. Cloud microphysics: analysis of the clouds of Earth, Venus, Mars, and Jupiter. *Icarus* 36:1–50

Sagan, C. 1978. Reducing greenhouses and the temperature history of Earth and Mars. *Nature* 269:224–26

Sagan, C., Mullen, G. 1972. Earth and Mars: evolution of atmospheres and surface temperature. *Science* 177:52–56

Settle, M. 1979. Formation and deposition of volcanic sulfate aerosols on Mars. *J. Geophys. Res.* In press

Shemansky, D. E., Broadfoot, A. L. 1977. Interaction of the surfaces of the Moon and Mercury with their exospheric atmospheres. *Rev. Geophys. Space Phys.* 15: 491–99

Slattery, W. L. 1977. The structure of the planets Jupiter and Saturn. *Icarus* 32: 58–72

Smith, B. A., Shoemaker, E. M., Kieffer, S. W., Cook, A. F. II. 1979. Volcanism on Io—the role of SO_2. *Nature* 280:738–43

Stacey, F. D. 1979. The cooling Earth: a reappraisal. *Phys. Earth Planet. Inter.* In press

Stevenson, D. J., Salpeter, E. E. 1976. Interior models of Jupiter. In *Jupiter*, ed. T. Gehrels, pp. 88–112. Tucson: Univ. Ariz. Press

Toksöz, M. N., Hsui, A. T., Johnston, D. H. 1978. Thermal evolutions of the terrestrial planets. *The Moon and the Planets* 18:281–320

Toon, O. B., Pollack, J. B., Ward, W., Burns, J., Bilski, K. 1979. The astronomical theory of climatic change on Mars. *Icarus*. In press

Torr, M. R., Walker, J. C. G., Torr, D. G. 1974. The escape of fast oxygen from the atmosphere during geomagnetic storms. *J. Geophys. Res.* 79:5267–71

Turekian, K. K., Clark, S. P. Jr. 1975. The non-homogeneous accumulation model for terrestrial planet formation and the consequences for the atmosphere of Venus. *J. Atmos. Sci.* 32:1257–61

Urey, H. C. 1952. *The Planets, Their Origin, and Development.* New Haven: Yale Univ. Press

Walker, J. C. G. 1975. Evolution of the atmosphere of Venus. *J. Atmos. Sci.* 32:1248–56

Walker, J. C. G. 1977. *Evolution of the Atmosphere.* New York: Macmillan

Ward, W., Murray, B. C., Malin, M. C. 1974. Climatic variations on Mars. 2. Evolution of carbon dioxide atmosphere and polar caps. *J. Geophys. Res.* 79: 3387–95

Wasserburg, G. J., Papanastassiou, D. A., Tera, F., Huneke, J. C. 1977. Outline of a lunar chronology. In *The Moon, A New Appraisal*, pp. 7–22. London: Royal Soc.

Weidenschilling, S. J., Lewis, J. S. 1973. Atmospheric and cloud structure of the Jovian planets. *Icarus* 20:465–76

Wetherill, G. W. 1975. Late heavy bombardment of the moon and terrestrial planets. *Proc. Lunar Sci. Conf.* 6:1539–61

Windley, B. F. 1977. *The Evolving Continents.* Chapter 19. New York: Wiley

Yung, Y. L., McElroy, M. B. 1979. Fixation of nitrogen in the prebiotic atmosphere. *Science* 203:1002–4

Yung, Y. L., Pinto, J. P. 1978. Primitive atmosphere and implications for the formation of channels on Mars. *Nature* 273:730–32

Ann. Rev. Earth Planet. Sci. 1980. 8: 489–525
Copyright © 1980 by Annual Reviews Inc. All rights reserved

FRACTURE MECHANICS APPLIED TO THE EARTH'S CRUST

✖10139

John W. Rudnicki
Department of Theoretical and Applied Mechanics, University of Illinois, Urbana, Illinois 61801

INTRODUCTION

Fracture mechanics concerns the study of stress concentrations caused by sharp-tipped flaws and the conditions for the propagation of these flaws. Although this approach has been very successful at explaining many features of the brittle and ductile failure of technological materials (metals, ceramics, glasses, polymers) under tensile loads, its application to geophysics has not been widespread. This is due, at least in part, to a historical preference for an approach that employs results from the theory of dislocations in elastic solids. With regard to the analysis of the stress or displacement induced by relative displacement on a surface of discontinuity, the two approaches are essentially coincident. They differ, however, with respect to the criteria for growth or spreading of the surface of discontinuity. In the dislocation approach, the relative motion on the surface of discontinuity is assumed a priori whereas, in fracture mechanics, a criterion based on plausible physical grounds or experimental evidence is used to determine whether the region of relative displacement will spread. The usefulness of the fracture mechanics approach is largely due to the success of simple fracture criteria in describing the failure of many materials. This article will focus on fracture criteria for the earth's crust.

The reluctance to adopt the fracture mechanics approach may also be due, in addition to historical reasons, to the attitude that so little is known about the details of the failure process in the earth's crust that the pursuit of studies based on assuming a fracture criterion is not worthwhile. There is, of course, good reason for this objection: seismological data is typically collected at large distances from the faults and is complicated by propagation effects; geological data is mostly limited to the earth's surface and is often obtained a long time after faulting occurs; and experimental data

0084-6597/80/0515-0489$01.00

is limited to size and time scales vastly different from those in situ. Moreover, seismologists have been very successful in using simple dislocation models to describe the transient motion due to earthquakes. In spite of these difficulties, the increasing sophistication of seismic studies of the transient motion near earthquake sources (strong motion studies; see, for example, Heaton & Helmberger 1978) and the direct observation of medium scale faulting in mines (McGarr et al 1979) suggest that these studies can be used with experimental and theoretical work to obtain a better understanding of failure processes in the earth's crust. Indeed, even an elementary understanding of the vast variety of earthquake-related phenomena, foreshocks, swarms, fault creep, aftershocks, etc. seems impossible without better knowledge of a fracture criterion.

The next section will review the fundamentals of fracture mechanics and discuss some aspects of fracture criteria. Although fracture mechanics has developed to the point where several texts (Knott 1973, Broek 1974, Lawn & Wilshaw 1975)[1] as well as extensive treatises (Liebowitz 1968) are available, some of the essential ideas are not well understood. The relationship between models that idealize the crack-tip as a singular stress field and those that include a cohesive zone will be discussed. In particular, the application of the Palmer & Rice (1973) cohesive zone model to shear faulting will be discussed in detail. Succeeding sections will review data from laboratory experiments, including some recent experiments on stress corrosion cracking in rock, and data inferred from observations of crustal faulting. Some results for fracture propagation in viscoelastic and fluid-infiltrated solids will be reviewed and the implications of these studies for fault creep and migration of seismic activity will be discussed. Finally, some aspects of rupture in the earth's crust, which are not possible to treat within the present framework of fracture mechanics, will be mentioned. Freund (1979) has recently reviewed several aspects of dynamic fracture mechanics relevant to fault propagation. Consequently, in this article inertial effects will, for the most part, be neglected and any growth of the fracture will be assumed to take place quasi-statically.

FUNDAMENTALS OF FRACTURE MECHANICS

Linear Elastic Crack-Tip Stress Fields

Consider a body which is infinite in extent in one direction (plane strain conditions) and which, except for the presence of the crack, is linearly elastic, isotropic, and homogeneous. The stresses near the crack-tip then

[1] Because these texts and treatises are available, no effort will be made to reference the original source when citing a result from fracture mechanics.

have the following form [e.g. Rice 1968a, Equations (78–83)]:

$$\sigma_{ij} = K_{\mathrm{L}}(2\pi r)^{-1/2} f_{ij}^{\mathrm{L}}(\theta) + O(1) \quad (\mathrm{L} = \mathrm{I, II, III}), \tag{1}$$

where r is the distance from crack-tip, θ is the angle measured from the plane ahead of the crack, and $O(1)$ denotes terms which are bounded at the crack-tip. The coefficients K_{L} are the stress intensity factors and the index L refers to the type of loading: loading which tends to cause separation of the crack faces is designated as mode I; loading which tends to cause relative slip is called mode II (plane strain shear) or mode III (anti-plane shear) depending on whether the slip is perpendicular or parallel to the crack edge. The functions $f_{ij}(\theta)$ are the same for all loadings of a given type (mode I, II, or III) and, thus, all details of the geometry and applied loading are contained in K. The stress intensity factor is proportional to the magnitude of the applied loading and to the square root of a characteristic length.

As examples of specific forms for the stress intensity factor, consider the mode II loadings, which are shown in Figure 1, and which have been discussed by Rice & Simons (1976). In Figure 1a a crack of length l is loaded at infinity by uniform shear stresses τ_{∞}. If the crack encloses no net dislocation so that the displacement field is single-valued outside the fault, the stress intensity factor is (Paris & Sih 1965)

$$K_{\mathrm{II}} = \tau_{\infty}(\pi l/2)^{1/2}. \tag{2}$$

If there is an entrapped dislocation of magnitude sufficient to eliminate the singularity at one end of the crack, the stress intensity factor at the

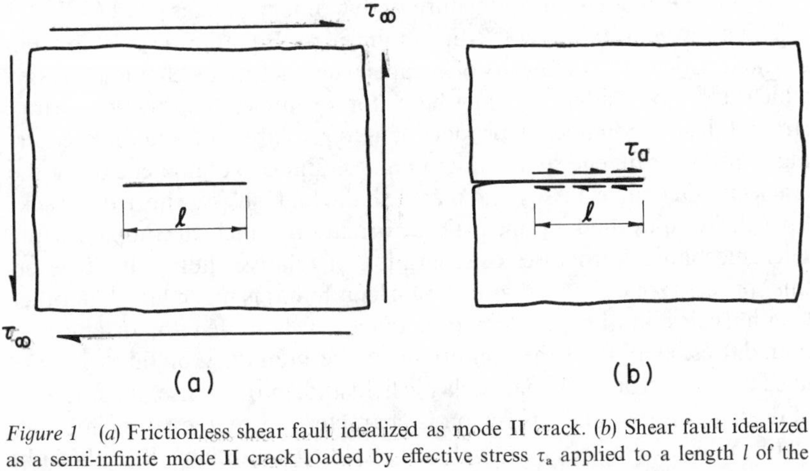

Figure 1 (a) Frictionless shear fault idealized as mode II crack. (b) Shear fault idealized as a semi-infinite mode II crack loaded by effective stress τ_a applied to a length l of the crack surfaces.

other end is [Rice 1968a, Equation (318)]

$$K_{\mathrm{II}} = \tau_{\infty}(2\pi l)^{1/2}. \tag{3}$$

If the crack models a fault that has a resistive frictional stress τ_f, then appropriate values for K_{II} are given by Equations (2) and (3) with the difference $\tau_{\infty} - \tau_f$ replacing τ_{∞}. Often the analysis is simplified by treating a crack or fault as semi-infinite with loading applied over a finite portion of the surface, as illustrated in Figure 1*b*. The stress intensity factor for this geometry is [Rice 1968a, Equation (98)]

$$K_{\mathrm{II}} = \tau_a(8l/\pi)^{1/2} \tag{4}$$

where τ_a may again be chosen as the excess of a remote stress over the fault friction. The length over which the loading is applied can be adjusted so that stresses near the fault-tip are the same as those in Figure 1*a*.

There is a large number of solutions for crack problems of various loadings and geometries (Paris & Sih 1965, Tada, Paris & Irwin 1973) and these provide a powerful means of stress analysis near faults and other acute geometries. The solutions in the form of (1) pertain strictly to plane strain conditions, but it is evident that such conditions are approached asymptotically near the edge of a crack in a three-dimensional body. Kassir & Sih (1966) verified this result directly for elliptical cracks and it is likely to apply for all cracks for which the edge has a smoothly turning tangent. The number of three-dimensional solutions is somewhat limited, but Weaver (1977) has recently presented an expedient numerical procedure for solving three-dimensional crack problems and has used it to study stresses near the edge of a rectangular fault [Budiansky & Rice (1979) have extended this procedure to dynamic problems].

The stresses near the crack-tip, as given by Equation (1), exhibit the characteristic $r^{-1/2}$ singularity of linear elastic fracture mechanics. Stresses which are more singular than these (for example, the stresses near a straight dislocation line are proportional to r^{-1}) are prohibited by requiring that the strain energy be finite in any bounded volume enclosing the crack-tip. Of course, in any real material, the high stresses near the crack-tip will be alleviated by inelastic deformation which accompanies the micromechanical processes of separation or relative sliding. If, however, this "process zone" has a characteristic length that is much less than other length scales in the problem (e.g. crack or fault length, distance to boundaries, etc.), then the conditions at the boundary of this zone are fixed by the surrounding linear elastic field and, in this sense, the intensity of deformation within this region is characterized by the stress intensity factor. When these so-called small scale yielding conditions (Rice 1968a, b) prevail, it is possible to ignore the details of inelastic processes near the crack-tip, which are often poorly known, and to treat the problem as fully elastic with the understanding that K characterizes the actual near tip field. Conversely, a boundary layer formulation (Rice 1968b, Edmunds & Willis 1976, Simons 1977) may be employed to study the details of the near-tip region for more realistic material models by imposing the condition that the deformation field approach the elastic solution as distance from the crack-tip becomes large.

Relation to Elastic Dislocations

In geophysics, a more popular approach to stress analysis near faults has made use of the theory of dislocations in elastic solids (e.g. Steketee 1958). Although this approach may be intuitively appealing to some because it reflects more directly the notion of idealizing faulting as slip on a planar surface, it is essentially equivalent to the approach taken in fracture mechanics. To illustrate this equivalence, consider a straight edge dislocation at a point $x = \xi$ on the axis $y = 0$ with a Burger's vector b in the x direction. The shear stress on the x axis is given by

$$\tau_d = \mu b/[2\pi(1-\nu)(x-\xi)] \tag{5}$$

where ν is Poisson's ratio and μ is the shear modulus. If there is a continuous distribution of dislocations with infinitesimal Burger's vectors over a segment S of the x axis, then the resulting stress is

$$\tau_d = \frac{\mu}{2\pi(1-\nu)} \int_S \frac{B(\xi)}{x-\xi}\,d\xi \tag{6}$$

where $B(\xi) = -\partial b/\partial \xi$ is the dislocation density. If S is assumed to represent a fault with resistive stresses $\tau(x)$ applied to the fault surfaces and a constant stress τ_∞ applied remotely, then the condition

$$\tau(x) = \tau_\infty + \tau_d \tag{7}$$

yields an integral equation for determining $B(\xi)$. As an example, consider the special case in which $\tau(x) = \tau_f$, a constant, as for the configuration shown in Figure 1*a*. If the crack encloses no net dislocation, $B(\xi)$ satisfies the condition [Equation (115) in Rice (1968a)]

$$\int_{-l/2}^{l/2} B(\xi)\,d\xi = 0, \tag{8}$$

and, in this case, the solution of (7) is

$$B(\xi) = 2\mu^{-1}(1-\nu)(\tau_\infty - \tau_f)\,\xi[(l/2)^2 - \xi^2]^{-1/2}. \tag{9}$$

The relative displacement of the fault surface is

$$\delta(x) = 2\mu^{-1}(\tau_\infty - \tau_f)[(l/2)^2 - x^2]^{1/2}, \tag{10}$$

but this can, of course, be determined by other methods [for example, see Equations (22) and (96) of Rice (1968a)].

A thorough discussion of this method has been given by Bilby & Eshelby (1968). This approach is often advantageous when S is regarded as a frictional surface and the stress sustained by the surface is related nonlinearly to the relative slip on the surface (e.g. Cleary 1976, Rudnicki 1979, Stuart & Mavko 1979). However, when the stress on S is regarded as given, these problems can often be solved more directly by other techniques (Rice 1968a). Alternatively, the stress on the fault may be regarded as unknown and chosen to satisfy some other condition in the problem. For example, Dmowska (1973) and Freund & Barnett (1976) have used this procedure to estimate the distribution of stress on dip-slip faults [Barnett & Freund (1975) treated the strike-slip case] by comparing calculated with observed surface displacements.

Energy Release Rate

During an infinitesimal increase in the length of a crack in a linearly elastic body, the work done by the tractions on any contour surrounding the crack-tip exceeds the increase of strain energy of the material within the contour. This difference, denoted G, is the energy released per unit area of crack advance. If the boundary displacements are fixed during the crack advance then G equals the *decrease* of potential energy per unit crack advance. Rice (1966) has shown that G can also be calculated as the work which is done in unloading the newly created crack surfaces. Because the near-tip stress field has the universal form (1) in a linearly elastic solid, G can be expressed in terms of the stress intensity factors as

$$G = [K_{\rm I}^2 + K_{\rm II}^2 + K_{\rm III}^2(1-\nu)^{-1}](1-\nu)(2\mu)^{-1} \tag{11}$$

for any combination of plane strain and anti-plane shear loadings. Consequently, for small scale yielding conditions, G is an alternative to K as a parameter that characterizes the intensity of the near-tip deformation. However, as Rice (1966, 1979a) has emphasized, it is implicit in (11) that the energy release can be calculated from the elastic singularity independently of the actual microstructural breakdown processes. In other words, in this purely continuum viewpoint the release of energy is due to the motion of the singularity in the elastic stress field which accompanies crack advance.

A viewpoint that has proved to be extremely useful is that extension of the fracture occurs when G is equal to some critical value, say $G_{\rm crit}$. In the case of purely brittle fracture considered by Griffith (1920), $G_{\mathrm{crit}} = 2\Gamma_0$ where Γ_0 is the specific surface energy. More generally, however, G_{crit} may be taken to represent the energy required for the microstructural processes of breakdown near the crack-tip. Thus, the fracture criterion is

$$G = G_{\mathrm{crit}}. \tag{12}$$

Because of (11), it is clear that the criterion (12) is equivalent to the condition that fracture occurs, for a given type of loading, when the stress intensity factor attains some critical value. Although this criterion is very simple, it often is an improvement over other simple criteria that are commonly used. For example, Abou-Sayed, Bretchel & Clifton (1978) employed the fracture criterion (12) in a study of stress determination by hydraulic fracturing. They concluded that the maximum horizontal stress was overestimated by an analysis based on a maximum tensile stress criterion because the effect of loading the faces of pre-existing cracks was neglected.

It is worth repeating that the left-hand side of (12) is determined by a calculation which assumes that the crack-tip can be idealized as a singularity in the elastic stress field whereas the right-hand side is determined from a model which makes more detailed assumptions about the nature of processes at the crack-tip. Such an uncoupling is not always realistic. Indeed, Rice (1979a) has stressed that all attempts to use such an approach as a fracture criterion for inelastic solids have led to physically unacceptable results. (Examples relevant to geophysics will be given in a subsequent section.)

J Integral

Another useful tool in fracture mechanics is the J integral which was first applied extensively to fracture mechanics by Rice (1968b) although Cherepanov (1968) and Eshelby (1956) had earlier introduced similar integrals. For a coordinate system chosen so the crack line coincides with the x direction and the y direction is perpendicular to the crack surface, the J integral is defined as (Rice 1968b)

$$J = \int_C (W n_x - n_i \sigma_{ij} \partial u_j / \partial x)\, \mathrm{d}s \qquad (i,j = x, y) \tag{13}$$

where σ_{ij} and u_j are the stress and displacement components, respectively;

$$W(\varepsilon_{ij}) = \int_0^{\varepsilon_{ij}} \sigma_{ij}\, \mathrm{d}\varepsilon'_{ij} \tag{14}$$

is the strain energy density; s is the arc length and the n_i are the components of the unit normal to the contour C which begins and ends on the crack surfaces and encloses the crack-tip. Generalizations of this integral have been discussed by Eshelby (1970), Knowles & Sternberg (1972), and Budiansky & Rice (1973).

The utility of J arises from the fact that it is path-independent in linear or nonlinear elastic materials which are homogeneous in the x direction (Rice 1968b). [If the crack surfaces are not traction-free, the value of J does depend on the endpoints but is the same for all paths with the same endpoints (Palmer & Rice 1973).] Thus, evaluation of J on a remote contour and one near the crack-tip makes it possible to relate conditions at the crack-tip to parameters characterizing the applied loads. (An example of the application of J will be given later.) Moreover, in the case of small scale yielding, Rice (1968b; see also Palmer & Rice 1973) has shown that J is equal to the energy release rate G. Thus, for small scale yielding, J, in addition to G, is an alternative to K as a parameter that characterizes the intensity of the near-tip deformation. Indeed, it is a feature of linear elastic fracture mechanics that a single parameter, of which J, G, and K are three possibilities, characterizes the near-tip field. J, however, has the additional advantage of possessing a precise meaning for nonlinearly elastic behavior (or for time-independent inelastic behavior if unloading is not allowed) and, thus, there is the possibility of using J to correlate experiments in which substantial amount of inelasticity occurs.

Cohesive Force Models for Shear Faulting

The microstructural processes of breakdown near the crack-tip can be included directly by assuming that they give rise to cohesive forces in a zone ahead of the crack-tip. These forces oppose the action of the applied loads so as to eliminate the crack-tip singularity. Several cohesive force models have been suggested for tensile cracks (Barenblatt 1962, Dugdale 1960, Bilby, Cottrell & Swindon 1963) and Ida (1972) has proposed a cohesive force model for steady propagation of an anti-plane shear fault. However, treatment here will follow the formulation proposed by Palmer & Rice (1973), hereafter abbreviated as PR, for shear band propagation in clay slopes and applied by Rice (1979b) and by Rice & Simons (1976) to shear faulting.

The PR model is shown schematically in Figure 2. In Figure 2*a*, τ_f is the residual friction stress on the fault surface, but the stress rises to τ_p near the end of the fault because of microstructural processes which cause resistance to slip initiation. In Figure 2*b*, the stress on the fault surface is shown as a function of the relative sliding $\delta(x)$. An amount of sliding δ^* is necessary to reduce τ_p to τ_f. In general, the end zone size is not known a priori. If, however, the end zone is small (small scale yielding), then its length can be determined by requiring the cohesive forces to cancel the singularity that would be caused by the applied loads in the absence of the end zone. As an example, consider the semi-infinite fault in Figure 1*b* with a small end zone and a constant distribution (Figure 3*a*) of stress in the end zone. The end zone size is given by [see Equation (10) of Rice & Simons (1976) and Equation (44) of PR]

$$R = \pi K^2[8(\tau_p - \tau_f)^2]^{-1} \tag{15}$$

where l in Figure 1*b* is the total length over which the loads are applied (including the end zone), and K is the stress intensity factor caused by the applied loads in the absence of the end zone. Because K for this geo-

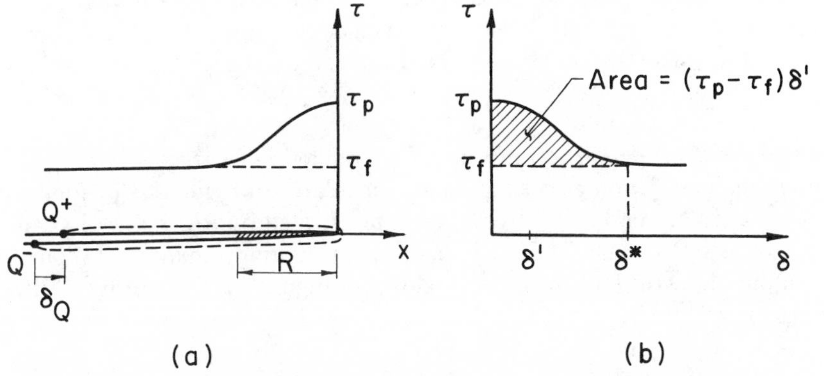

Figure 2 Palmer & Rice (1973) model of a cohesive zone. (*a*) Fault surface stress as a function of distance back from the fault-tip. Dashed line denotes contour for J integral used in deriving Equation (20). (*b*) Fault surface stress as a function of relative slip on the fault.

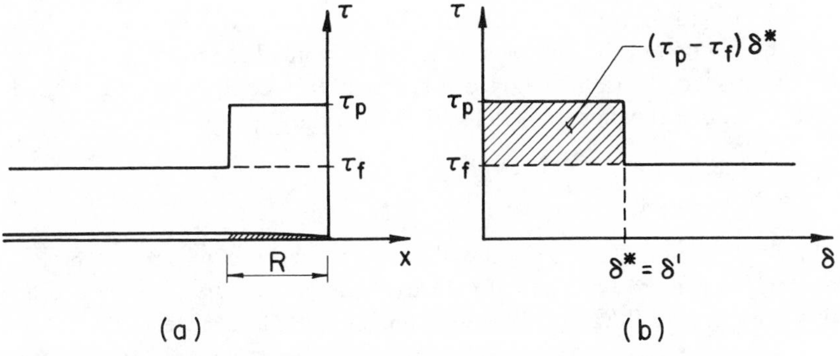

Figure 3 Palmer & Rice (1973) cohesive zone model with constant end zone stress.

metry is given by (4), Equation (15) can be rewritten as

$$R = l[(\tau_\infty - \tau_f)/(\tau_p - \tau_f)]^2. \tag{16}$$

If the stress in the end zone decreases linearly from τ_p to τ_f, instead of being constant, the expressions in (15) and (16) should be multiplied by 9/4. The relative slip in the end zone can be obtained by a standard calculation of fracture mechanics (Rice 1968a). For a small end zone, the result can be expressed as

$$\delta(x) = (1-\nu) K^2 f(|x|/R)[2\mu(\tau_p - \tau_f)]^{-1} \tag{17}$$

where $f(0) = 0$ and $f(1) = 1$ and, for the case of constant end zone stress

$$f(\lambda) = \lambda^{1/2} - \tfrac{1}{2}(1-\lambda) \log [(1+\lambda^{1/2})/|1-\lambda^{1/2}|]. \tag{18}$$

A fracture criterion which is appropriate for this model is that fault growth occurs when the relative displacement at the end of the breakdown zone is sufficient to reduce τ_p to τ_f or, more concisely, when

$$\delta(x = -R) = \delta^*. \tag{19}$$

The work expended in accomplishing this reduction, in excess of the work done against friction τ_f, is simply the shaded area in Figure 2*b* (or Figure 3*b* for constant end zone stress). As demonstrated by PR, this work can be expressed in terms of the *J* integral by evaluating Equation (13) on a contour along the fault surfaces as shown in Figure 2*a*. The result is [PR, Equation (7)]

$$J_Q - \tau_f \delta_Q = \int_0^{\delta(x=-R)} (\tau - \tau_f)\, d\delta \tag{20}$$

where Q denotes the position of the contour endpoints (assumed to be outside the end zone) and δ_Q is the relative slip of the fault surfaces at Q. [Note that $n_x = 0$ for this contour so that only the second term under the integral in (13) contributes to (20). The form of (20) then results from recognizing that in the end zone τ is a single-valued function of δ.] When (19) is satisfied, the right-hand side of (20) can be recognized as the shaded area in Figure 2*b* where δ' is defined by the relation

$$(\tau_p - \tau_f)\,\delta' = \int_0^{\delta^*} [\tau(\delta) - \tau_f]\, d\delta. \tag{21}$$

The right side of (21) is the work that is necessary to overcome the cohesive forces and advance the fracture. For a vanishingly small end zone this work *defines* the critical value of the energy release rate,

$$(\tau_p - \tau_f)\,\delta' = G_{crit}, \tag{22}$$

and, in this small end zone case, the Griffith criteria of Equation (12) for a singular crack-tip model is equivalent to the criteria expressed by (19) for a cohesive zone model. Willis (1967) first demonstrated this equivalence by a direct calculation based on complex variable methods and Rice (1968b) confirmed the result using the J integral. Formally, the equivalence can be rationalized by reducing the end zone size to zero while allowing $\tau_p - \tau_f$ to become unbounded in a manner such that the right-hand side of (21) remains finite and equal to G_{crit}. The size of the breakdown zone R can now be expressed in terms of δ' by using Equation (11) (with $K = K_{II}$ and $K_I = K_{III} = 0$) and Equation (22) in Equation (15). The result is

$$R = \pi\mu\delta'[4(\tau_p - \tau_f)(1-\nu)]^{-1}. \tag{23}$$

Cohesive zone models have been widely used in studies of dynamic fault propagation (e.g. Ida 1973a, Andrews 1976a,b, Das & Aki 1977a,b) because they avoid the numerical difficulties associated with treating a singular crack-tip stress field. In the case of dynamic crack propagation, the Griffith fracture criteria (based on an energy balance at the tip of a crack at which stress is singular) is not, however, exactly equivalent to a criteria based on the achievement of a certain relative displacement in the end zone [as expressed, for example, by Equation (19)] even if the end zone is small. As Freund (1976a, 1979) has emphasized, the two criteria are equivalent for small end zones only if, in addition, conditions are completely steady as viewed by an observer moving with the crack tip. Because results are more dependent on the choice of fracture criteria than in the quasi-static case, it is often difficult to distinguish general features of the solution from those which are peculiar to the particular model and method of discretization. Ida (1972) introduced a cohesive zone model for dynamic propagation of a mode III crack and he considered several forms for the decrease of stress with relative slip in the cohesive zone. His treatment was, however, limited to steady state conditions and small end zones. Andrews (1976a,b) studied the propagation of anti-plane strain and plane strain shear faults for a cohesive zone model in which the stress decreased linearly with slip. For plane strain fault propagation (Andrews 1976b), he found that the propagation speed could exceed the Rayleigh wave velocity, in contrast to predictions based on singular crack-tip models. The study of Das & Aki (1977a) is purportedly based on the criterion that propagation occurs when the dynamic stress intensity factor attains a certain level. The result of the discretization is, however, a criterion that requires that a certain stress level be achieved over a critical length. Freund (1979) has given an example demonstrating that predictions based on this criterion differ from those based on a critical value of the stress intensity factor. In the discretized fracture criterion of Das & Aki (1977a) the critical distance has no apparent physical basis and is simply identified as the grid spacing in the finite difference calculation. Das & Aki (1977b) used this criterion in the study of fault propagation on a plane with barriers, that is, areas of the fault plane where the critical stress level was locally higher. One of their results is the prediction that the propagating fault can leave areas behind the tip on which no relative slip occurs. In the absence of a physical basis for the critical length in the fracture criterion, it is, however, impossible to evaluate whether the predictions of this model are realistic.

EXPERIMENTAL VALUES OF MATERIAL PARAMETERS

Experimental Determination of K_{IC}

If any inelastic behavior near the tip of a crack is confined to a sufficiently small region and if the onset of crack growth is unstable (i.e. results in abrupt loss of load bearing capacity), then the value of the crack-tip stress intensity factor at fracture may be identified as a material parameter characterizing failure. For tensile (mode I) loading this critical value of K is usually denoted K_{IC}. Because of the success of linear elastic fracture mechanics in describing the brittle failure of technological materials, much attention has been given to establishing standard experimental procedures for measuring K_{IC}. There have recently been attempts to apply these procedures to the determination of K_{IC} for rocks and an ASTM subcommittee (E24.07) on the Fracture Testing of Brittle Non-metallic Materials was formed in 1977. Although K_{IC} has obvious relevance for tensile fracture as, for example, in hydraulic fracturing (Abou-Sayed, Brechtel & Clifton 1978), there may be some question about the applicability of these results to shear faulting. The macroscopic inelastic behavior of brittle rock does, however, result from microcracking at the tips of pre-existing fissures where the stress fields can be locally tensile even when all the applied principal stresses are compressive (Tapponnier & Brace 1976, Kranz 1979a,b). Macroscopic shear failure appears, at least in laboratory experiments, to result from the rapid growth and link-up of these fractures and the direct observation of fracture surfaces associated with seismic events in mines (McGarr et al 1979) suggests that this process may be prevalent on a larger scale. At the very least, studies of K_{IC} are likely to improve the understanding of inelastic processes and fracture in rocks.

Investigators have measured K_{IC} for a variety of rocks using a number of different tests (Clifton et al 1976, Abou-Sayed 1977, Abou-Sayed, Brechtel & Clifton 1978, Kobayashi & Fourney 1978, Schmidt 1976, 1977, Schmidt & Huddle 1977, Schmidt & Lutz 1979, Atkinson 1979a) and

Atkinson (1979a) has tabulated many of the existing results. Most of the rocks tested were carbonates or sandstones and a representative value of K_{IC} seems to be 1 $MNm^{-3/2}$. Schmidt & Huddle (1977) have, however, found that K_{IC} for Indiana limestone increased with confining pressure from 930 $kNm^{-3/2}$ at atmospheric pressure to 4.20 $MNm^{-3/2}$ at 62 MPa. Measurements of K_{IC} for granite have been less frequent: Schmidt & Lutz (1979) obtained a preliminary value of 2.5 $MNm^{-3/2}$ for Westerly granite whereas Atkinson & Rawlings (1979a) and Swanson & Spetzler (1979) reported a lower value of 1.74 $MNm^{-3/2}$. Atkinson (personal communication, 1979) has also obtained $K_{IC} = 1.66$ $MNm^{-3/2}$ for a "coarse orthoclase feldspar granite" and Atkinson & Rawlings (1979b) obtained $K_{IC} = 2.9$ $MNm^{-3/2}$ for a gabbro. These values of K_{IC} can be converted to critical values of the energy release rate by using Equation (11) with $K_{II} = K_{III} = 0$. For $\mu = 20$ GPa (1 GPa $= 10^9$ $Nm^{-2} = 10^4$ bars) and $v = 0.2$, K_{IC} values of 1, 2.5, and 4 $MNm^{-3/2}$ yield $G_{crit} = 20$, 125, and 320 J m^{-2}, respectively.

In the testing of metals, an empirical criterion that ensures that the inelastic zone at the crack-tip is sufficiently small has been found to be (Brown & Srawley 1966)

$$\text{crack length} > 2.5\,(K_I/\sigma_y)^2 \tag{24}$$

where σ_y is the yield stress (the stress at which the first departure from linearly elastic behavior occurs). If this condition is met, the value of K_I may be identified as the material parameter K_{IC}. This criterion seems to apply approximately in brittle rocks when the ultimate (peak) stress σ_u is substituted for σ_y. However, the right-hand side of (24) is based on a formulation for metal plasticity and because the inelasticity of brittle rocks results primarily from microcracking rather than dislocation motion, this criterion may not be appropriate. For a crack in a 12.9-mm-thick plate of Westerly granite, Kobayashi & Fourney (1978) observed thin cracks extending 13 mm and 20 mm ahead of the main crack when the applied value of K_I was 660 $kNm^{-3/2}$ and 990 $kNm^{-3/2}$, respectively. For $\sigma_u =$ 21 MPa (Jaeger & Cook 1976, Brace 1964), using $(K/\sigma_u)^2$ as an estimate for the size of the inelastic zone yields values too small by a factor of 10. A clear interpretation is prevented by the fact that the thinness of the plate is likely to have caused departure from plane strain conditions.

Wilkening (1978) has suggested that measurements of K_{IC} may be invalid because a significant amount of stable crack growth occurs prior to fracture. Consequently, he has employed the J integral, which may be applied in situations of inelastic loading, if unloading is not permitted, as a parameter characterizing conditions at fracture. On the basis of his direct measurements of J for Barre granite, Wilkening (1978) concluded that calculations of J based on linear elasticity and the identity $J = G$ for small end zones underestimate the actual value by a factor of 2 to 3. On the other hand, Schmidt & Lutz (1979) found that measurements of K and J for Westerly granite were generally in agreement. Unfortunately, there are not enough data available to resolve these differences but, given the nonlinear behavior of rock prior to fracture, more direct measurements of J would seem to be a worthwhile underatking.

Stress Corrosion Cracking

It is well known that the presence of water or other pore fluid in either liquid or vapor form can enhance microcrack growth in brittle rocks (e.g. Martin 1972, Scholz 1972, Swolfs 1972, Martin & Durham 1975, Anderson & Grew 1977). In particular, surface chemical effects can cause slow stable crack growth at values of the applied stress intensity factor (K_I) which are well below the critical value for unstable growth (K_{IC}). Such time-dependent crack growth has been observed in glass and ceramics (see Wachtman 1974 for a review), but only a few detailed results exist for rock. The results for glass and ceramics are often summarized in a logarithmic plot of crack velocity (V) versus the applied stress intensity factor (K_I) such as that shown schematically in Figure 4. Slow crack growth in region I is thought to be controlled by surface chemical effects. In region

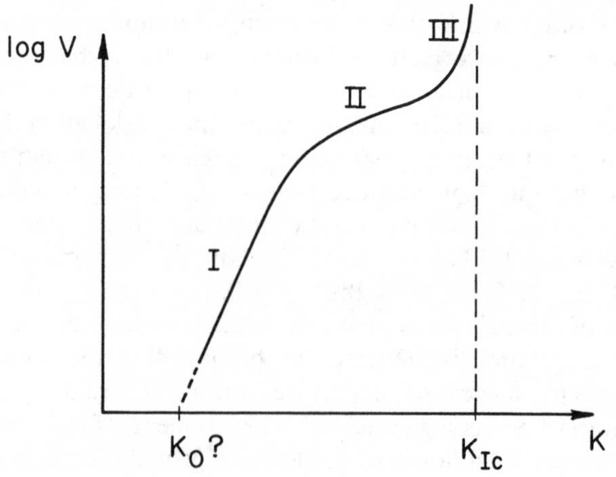

Figure 4 Schematic illustration of typical results on stress corrosion cracking. Crack velocity V is shown as a function of Mode I stress-intensity factor K_I. K_{IC} is the critical value of K_I for rapid fracture and K_0 is the possible threshold value below which no crack growth occurs.

II the speed of crack growth is limited by the rate at which the corrosive species can move to the crack-tip. In region III, crack growth is rapid and independent of environmental effects. The limiting value of the stress intensity factor for rapid growth is K_{IC} but whether there exists a threshold value K_0 below which no propagation occurs is unknown.

Atkinson (1979b, 1980) has studied slow crack growth (region I) in several rocks for the range of speeds from roughly 10^{-10} m s^{-1} to 10^{-3} m s^{-1} and he has found the crack speeds were proportional to K_I^n. For quartz crystals, $n = 12$ or 20 depending on orientation; for Arkansas novaculite, a fine-grained quartz rock, $n = 25$; and for a gabbro, $n = 32$. For Westerly granite, Atkinson & Rawlings (1979a,b) found $n = 39$ for deformation in air at 30% relative humidity and $n = 34$ in liquid water (in both cases at 20°C). [Swanson & Spetzler (1979) have also reported some preliminary results for environmentally assisted crack growth in Westerly granite for crack velocities ranging from approximately 10^{-7} m s^{-1} to 10 m s^{-1}.] In all cases the crack velocity at a given stress intensity factor increased with relative humidity and with temperature. The existence of a stress corrosion limit of $K_0 \simeq 0.52\,K_{IC}$ was inferred for the quartz, but no limit was observed for either the novaculite at crack velocities down to 10^{-10} m s^{-1} (corresponding to $K = 0.6\,K_{IC}$) or for the gabbro at crack velocities down to approximately 10^{-9} m s^{-1} (corresponding to $K \simeq 0.5\,K_{IC}$). For the novaculite, region I extended to $K_I = 0.94\,K_{IC}$ at 20°C so that regions II and III of Figure 4 may not be present.

Rice (1979b) has discussed the importance of these studies as an ingredient in more complete constitutive descriptions of the macroscopic time-dependent behavior of brittle rock and, as an example, he interprets the following observation of Wawersik & Brace (1971): when the stress on an axisymmetric compression sample of Westerly granite was held constant at a value slightly below peak stress (as determined by a test at a strain rate of about 10^{-5} s^{-1}), the sample deformed at constant stress, "tunnelled" through the peak of the stress-strain curve, and rejoined the post-peak curve. Rice (1979b) rationalized this behavior as caused by an increase in the microcrack density owing to slow crack growth at local values of the stress intensity factors which are below the levels for rapid growth. A rapid increase of stress at essentially constant microcrack density then brought the stress intensities to levels appropriate for rapid growth as the post-peak curve was rejoined. As a further example, Atkinson (1980) has used his V vs K_I data to predict the reduction of tensile fracture stress for Arkansas novaculite, as determined by bend experiments, from 72 MPa at a strain rate of 6×10^{-5} s^{-1} to 39 MPa at 10^{-10} s^{-1}. In addition, he suggests that his results support the contention of Rutter & Mainprice (1978) that stable sliding of wet pre-faulted Tennessee sandstone may, under certain conditions, be controlled by environmentally assisted cracking of near-fault grains. Dieterich & Conrad (1978) have also reported some observations which suggest that the time dependence of friction (Dieterich 1972, 1978) may, at least in part, be due to environmental effects. More specifically, the presence of a humid environment both increases the rate of growth of asperity contacts and decreases the shear strength of the contacts.

Because of the slow rate of tectonic loading and the typically long recurrence intervals for earthquakes, at least by comparison to laboratory time scales, time-dependent crack growth may be important in determining long term strength levels for crustal faulting. Crack speeds of 10^{-9} m s^{-1} may be very slow on laboratory time scales but this corresponds to 3 cm yr^{-1} which is comparable to observed rates of relative slip on the San Andreas fault in central California (Savage & Burford 1973).

A recent analysis by Rice (1978a), which takes proper account of thermodynamic restrictions on the brittle growth of cracks, seems to offer a promising theoretical framework for the study of environmental effects in crack growth. He finds that thermodynamic considerations require crack growth to be compatible with the inequality

$$(G-2\Gamma_0)V > 0 \tag{25}$$

where V is the crack speed (positive for extension) and the quantity in parenthesis is the difference between the energy release rate and the energy required to create new crack surface. This difference is the "force" which is thermodynamically conjugate to V. The condition (25) includes the Griffith criterion $G = 2\Gamma_0$ when $V = 0^+$, but Rice (1978a) has discussed the role of environmental effects in altering Γ_0. An important implication of (25) for earth faulting is that crack healing is not prohibited; that is V may be negative when $G < 2\Gamma_0$. Because all models of earthquake instability, at least those consistent with the principles of mechanics, require strength loss upon faulting [see Stuart (1979) for a review], the observation of repeated earthquakes in the same location indicates that some mechanism of strength recovery is necessary. As noted by Rice (1978a) the small stress drops (10–100 bars = 1–10 MPa) in earthquakes and the extremely brittle nature of cracking on the microscale in rock make crack healing an attractive possibility. Although apparent healing of microcracks has been observed by use of a scanning electron microscope in studies of microcrack growth in rock (e.g. Kranz 1979a), to the author's knowledge no systematic study of microcrack healing has been carried out.

Shear Fracture and Friction Experiments

Most seismic faulting in the earth's crust is thought to involve processes of shear fracture and frictional sliding rather than tensile fracture, although, as mentioned earlier, tensile fractures may play an important role in the formation of a macroscopic throughgoing fault. In the laboratory, however, shear fractures are unstable in the sense that they tend to propagate out of their original plane [e.g. Brace & Bombolakis (1963) and Ingraffea et al (1977); Cotterell & Rice (1979), in a reexamination of the solution for a slightly kinked crack, have shown that out-of-plane propagation occurs when the traction parallel to the crack-line is compressive]. Consequently, there are no measurements of critical values of K, G, or J for shear fracture analogous to those for mode I loading. On the other hand, there have been numerous laboratory investigations of frictional sliding [see Evernden (1977) for recent work relevant to earth faulting]. Although these experiments are frequently interpreted as analogues of crustal shear faulting, Rice (1979b) has used the PR model, which was discussed earlier, to point out that the possibility for laboratory friction experiments to simulate field conditions is critically dependent on the size of the breakdown zone, that is, the distance behind the fault-tip over which the fault surface stress exceeds the residual friction level. In order for laboratory friction experiments to be geometrically similar to slip zones in the field, the breakdown zone must be small by comparison to the length of the slipping zone.

For the PR model with a small end zone and a linear decrease of stress in the end zone, the size of the end zone is obtained by multiplying Equation (23), which is for constant end zone stress, by 9/4:

$$R = (9/16)\pi\mu\delta'[(\tau_p - \tau_f)(1-\nu)]^{-1}. \qquad (26)$$

The quantity δ', which is defined by Equation (21), is a measure of the slip necessary to reduce the stress in the end zone from the peak value τ_p to the residual level τ_f. If $\mu = 2 \times 10^4$ MPa, $\nu = 0.3$, and $\tau_p - \tau_f$ is taken to be the same order as a representative stress drop (10 MPa), then $R = (4.42 \times 10^3)\delta'$. Rice (1979b) has inferred values of $\delta' = 2.5$ mm from data of Coulson (1972) (as quoted by Barton 1973) on a natural joint in coarse-grained granite and $\delta' = 2.5$ μm from Dieterich's (1979) sliding experiments on flat ground specimens of granite. These yield $R = 1$ m and 11 mm, respectively. Rudnicki (1979), however, has suggested, based on observations of Barton (1971), that larger values of δ', for instance, on the order of centimeters, may be more representative of in situ conditions. If $\delta' = 2.5$ cm, $R = 110$ m. Barton (1971) has suggested that δ' may increase with the length of the sliding surface. In model studies designed to simulate test specimens of different sizes, he found that peak and residual strengths for rough joints were reached after a displacement of about 1% and 10% of the test length, respectively. If this result can be extrapolated to faults in situ, the corresponding size of the end zone will be enormous. However, this increase presumably reflects the presence of larger wavelength surface irregularities on larger sliding surfaces which may not exist on a fault surface that has undergone a large amount of relative slip. In contrast, the resistance to sliding in Dieterich's (1979) experiments on flat ground samples is apparently due to adhesion at asperity contacts rather than to sliding over geometric irregularities.

Obviously δ' depends strongly on conditions at the surface: the nature of asperity contacts; grain size; pore pressure; whether deformation at asperity contacts occurs by brittle cracking, plastic flow, or, at elevated temperatures, by local melting and pressure solution; whether gouge or moisture is present. Despite the evident importance of this size effect in extrapolating friction experiments to in situ conditions, there have been relatively few investigations of the precise variation of frictional stress for small values of the relative offset. If the values cited for δ' are indeed representative, geometric similarity will necessitate large laboratory specimens and/or preparation of the sliding surface to produce small values of δ', as in the experiments of Dieterich (1979).

CRITICAL ENERGY RELEASE RATES FOR CRUSTAL FAULTING

Although, as mentioned in the last section, shear fractures in the laboratory tend to propagate out of their original plane, many crustal earthquakes obviously do propagate in a way that can be idealized as a shear crack. Strong kinematic constraints or the effects of pre-existing zones of weakness may be responsible for preventing out-of-plane propagation of shear fractures on this size scale. In any case, this section reviews and discusses some estimates of the critical energy release rate for crustal faulting. Also, the critical energy release rate at the onset of the 1857 California earthquake is estimated by using the J integral, current geodetic data, and Sieh's (1978a) estimate of the recurrence time for such earthquakes.

Husseini et al (1975) have estimated G_{crit} corresponding to the arrest of seismic faulting. They inferred values of 1–10^4 J m^{-2} associated with frictional sliding and values of 10^4–10^6 J m^{-2} associated with fresh fracture. Ida (1973b) estimated a value of 10^7 J m^{-2} based on maximum values of ground surface displacements and acceleration. Aki (1978) has estimated

$G_{crit} = 8 \times 10^5$ J m^{-2} corresponding to arrest of the 1966 Parkfield earthquake and $G_{crit} = 10^8$ J m^{-2} associated with fracture barriers during the 1857 California earthquake. Rice & Simons (1976) obtained $G_{crit} = 2.6 \times 10^2$ J m^{-2} from observations by King, Nason & Tocher (1973) of a creep event on the San Andreas fault. Of these estimates, the last is likely to be the most accurate because, as Rice & Simons note, the observed displacement profile for the creep event was very nearly elliptical in agreement with the prediction of Equation (10) for the relative surface displacements of a fault with constant stress applied to the fault surfaces and no net enclosed dislocation. However, because this estimate was for a creep event the value of G_{crit} is probably lower than that for a typical seismic rupture.

The estimates of Aki (1978) tend to be higher than the others because they are based on an interpretation of faulting in terms of the "barrier" model introduced by Das & Aki (1977b). As mentioned earlier, the discretization of the fracture criteria in this work introduces a characteristic length whose physical significance has not yet been established. Aki (1978), however, actually estimates G_{crit} based on the dynamic version of Equation (11) (see Freund 1976a,b, 1979) for the propagation of a fault with a sharp tip. For comparison Husseini et al (1975) estimated $G_{crit} = 1.1 \times 10^2$ J m^{-2} for arrest of the Parkfield earthquake, a value 7000 times smaller than that obtained by Aki (1978) for the same event. The difference occurs because Aki (1978) used a stress drop (5 MPa) inferred from accelerogram data, which presumably reflects a local value, and an effective fault length of 3 km corresponding to a "barrier interval" inferred from the aftershock data, whereas the fault length inferred from the surface rupture was 35 km.

An application of the J integral to the idealized geometry in Figure 5 can be used to estimate G_{crit} for initiation of the 1857 California earthquake. Figure 5 is meant to idealize a long strike-slip fault which is locked to the right of point T, but which is undergoing steady relative displace-

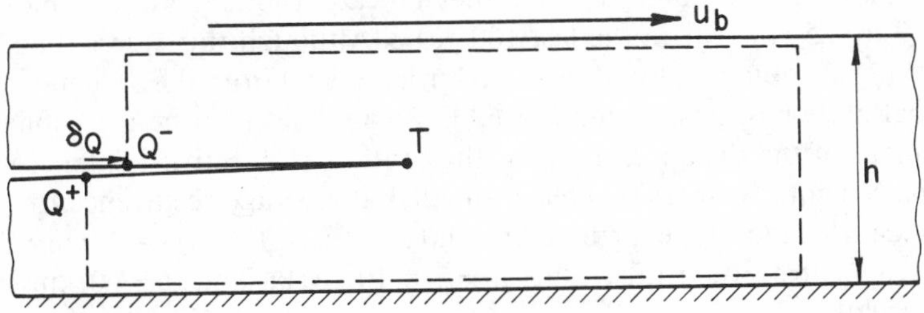

Figure 5 Idealization of a long strike-slip fault loaded by displacements u_b. Relative slip at point Q is δ_Q.

ment to the left of T. At point Q the relative displacement is δ_Q. The fault is loaded by displacements u_b at a distance $h/2$ from the fault. These displacements are assumed to be imposed by the large scale relative plate motion. Calculation of the J integral [Equation (13)] for the contour shown in Figure 5 follows exactly the calculation of PR for a long shear box apparatus and, consequently, details are omitted. It is, however, clear that only the vertical end segments contribute to the integral because of the imposed displacements on the top and bottom. Path independence of J requires that the value for this contour be equal to that for a contour along the fault surfaces (with the same endpoints). The result is [from Equation (17) of PR, after using their (16)]

$$J_Q - \tau_f \delta_Q = (h/2)\mu(\gamma^+ - \gamma^-)^2 \tag{27}$$

where $\gamma^+ = u_b/h$ is the uniform strain well ahead of point T and $\gamma^- = \gamma^+ - \delta_Q/h$ is the uniform strain in the material adjacent to the fault near point Q. Use of the expression for γ^- in (27) yields

$$J_Q - \tau_f \delta_Q = \mu \delta_Q^2/(2h) \tag{28}$$

where, from Equation (20) and Equation (21), the left-hand side is equal to the work necessary to overcome the cohesive forces and advance the fracture. For a small end zone, this work is equal to the energy release rate.

The San Andreas fault is approximately straight in central California and Figure 5 is a reasonable idealization of the deformation pattern. The fault tip (point T) is in the vicinity of Cholame. To the northeast (left in Figure 5), the fault is creeping and to the southwest the fault is locked, although in the area of Parkfield-Cholame, both creep and seismic faulting are intermittently observed (Sieh 1978b). The observed relative displacement rate on the creeping portion is about 3 cm/year (Savage & Burford 1973). Thatcher (1979) has discussed geodetic measurements from a net which has a width of 60 km and includes the fault. The relative displacement rate measured across the net indicates right lateral plate motion of 3.3 to 4.5 cm/year in the last 100 years. Although this is less than the 5.6 cm/year obtained by Minster & Jordan (1978) from the kinematics of global plate motions, taking $h = 60$ km would seem to be a reasonable estimate of the distance at which the motion is effectively imposed by displacements. Sieh (1978a) has estimated an average recurrence time of 160 years for 1857-type earthquakes, and thus, $\delta_Q = (160 \text{ yr}) \times (3 \text{ cm/yr}) = 480$ cm. Substituting this value, $\mu = 2 \times 10^4$ MPa and $h = 60$ km into (28) yields

$$J_Q - \tau_f \delta_Q = G_{\text{crit}} = 3.8 \times 10^6 \text{ J m}^{-2}. \tag{29}$$

This calculation must, of course, be regarded as an order of magnitude estimate but it would seem to be at least as accurate as the other estimates of G_{crit} except for that of Rice & Simons (1976). It should also be emphasized that the calculation pertains to the onset of rupture whereas a much smaller value of G may be necessary to continue rupture. A model with more detailed assumptions about the distribution of fault surface stress near the end of the fault might provide a mechanical basis for the foreshock sequence which Sieh (1978b) suggests is characteristic of these earthquakes.

The range of values for G_{crit} which have been estimated from field observations is so large and the estimates are, in general, so rough that it is difficult to know what significance should be attached to them. Some perspective may be obtained by comparison of these values with those corresponding to laboratory tensile fracture (although there is no evident reason why these values should be related) and with estimates of seismic energy radiated in earthquakes. The smallest values of G_{crit} obtained by Husseini et al (1975) for frictional sliding (1–10^2 J m^{-2}) and the value obtained by Rice & Simons (1976) for fault creep are comparable in magnitude to those obtained in laboratory tests on tensile fracture. More generally, however, the values estimated from field observations are much greater than those obtained from laboratory measurements. Nevertheless, even the largest values of G_{crit}, when multiplied by the fault area, yield an energy much less than empirical estimates of the seismic energy released by earthquakes. For example, the 1857 California earthquake had a rupture length of about 275 km (Sieh 1978a) and if it is assumed that the rupture extended to 10-km depth, the fracture energy is about 10^{15} J for $G_{crit} \simeq 10^6$ J m^{-2}. Use of Sieh's (1978a) estimate of the surface wave magnitude for this earthquake $M_S = 8.25$ in the Gutenberg-Richter relationship [p. 366 of Richter (1958) with correction noted by Kanamori & Anderson (1975)]

$$\log E = 11.8 + 1.5\, M_S \tag{30}$$

yields a value for the seismic energy of 10^{21} J. For smaller earthquakes, the fracture energy may be more comparable to the total seismic energy. For $M_S = 6.0$ Equation (30) yields $E \simeq 10^{17}$ J. The data plotted in Figure 5 of Kanamori & Anderson (1975) suggest a fault area of about 100 km^2 for $M_S = 6.0$ and, if $G_{crit} = 10^6$ J m^{-2}, the corresponding fracture energy is 10^{15} J.

It is, of course, important to remember that such estimates reflect values that are averaged over the fault plane and the critical value of the energy release rate needed to extend the fault is likely to be extremely nonuniform. More specifically, G_{crit} is likely to be very high where "asperities" lock the fault and low where the resistive stress is at the residual friction level. The higher values of G_{crit}, which were estimated by Aki (1978), may correspond to the values at asperities. Although the present estimates of G_{crit} are of limited usefulness, the study of the critical energy release rate would seem to be a worthwhile approach to obtaining more detailed information about the earthquake rupture process in general, and the nature of nonuniform rupture in particular. There has been a large amount of empirical data accumulated on earthquakes that seems to be consistent with simple physical models (e.g. Kanamori & Anderson 1975). It would be useful to investigate the possibility of similar relationships involving G_{crit}. For example, if, as suggested above, the fracture energy is a larger proportion of the seismic energy for smaller earthquakes, it is unlikely that small earthquakes could be considered "similar" to larger events.

TIME-DEPENDENT EFFECTS IN FAULTING

Although earth faulting is typically a very brittle event, observations of fault creep (Goulty et al 1978), slow precursors (Kanamori & Cipar 1974), tsunami earthquakes (Fukao 1979), slow earthquakes (Sacks et al 1977), and migration of aftershocks (Melosh 1976) all suggest that deviations from purely brittle behavior are frequent. Moreover, because observations of very slow faulting events that do not generate extensive seismic waves are limited to areas well-instrumented with creep or strain meters, such events may be more prevalent than present observations suggest. Time-dependent effects may arise from a number of sources: stress corrosion cracking, plastic creep, or coupling of the deformation with diffusion of an infiltrating pore fluid are examples. Moreover, for very large earthquakes, the coupling between the asthenosphere and the lithosphere will introduce time-dependent effects into the response. In this section, some results for the spreading of shear faults in viscoelastic and fluid-infiltrated porous elastic solids will be reviewed. In addition to their relevance to the geophysical phenomena noted above, these examples will illustrate the fundamental inadequacy of the Griffith fracture criterion when applied to nonelastic solids.

In the earlier discussion of fracture criteria, it was emphasized that the cohesive force model was equivalent to the Griffith approach if the end zone size was small and the material was elastic. The Griffith approach is convenient in that the actual microstructural processes of fracture that occur in the breakdown zone can be ignored in favor of regarding the crack-tip as a singularity in the elastic stress field. Energy changes of the body can then be calculated from the singular elastic field and the microstructural processes of breakdown enter only in defining a critical value of the energy release rate (or, equivalently, a critical value of K, J, or any other parameter characterizing the deformation near the crack-tip, since these are all related for a small end zone) at which fracture occurs. Rice (1966), however, drew attention to some difficulties with application of the Griffith approach to fracture in nonelastic solids and Rice (1979a) has emphasized that all attempts to employ an energy balance of the Griffith type in nonelastic materials have led to physically unacceptable results. As Rice (1979a) explains, the difficulty is not, of course, with the notion that energy should balance, but rather with the assumption, which is implicit in the Griffith criteria, that energy changes due to crack advance can be calculated independently of the microstructural processes of breakdown. This issue has been discussed in detail by Rice (1979a) for a number of material models, but for the time-dependent materials discussed here, the effect is straightforward. The material response introduces a characteristic time, say t^*, which is associated with the breakdown processes. If the fault is spreading at speed V, then the cohesive force model will be equivalent to the Griffith model only if the size of the breakdown zone is small by comparison to Vt^*. This is generally not the case.

In both of the examples reviewed here attention will be restricted to steady state propagation of a mode II shear fault. The fault will be taken as semi-infinite with the loading applied over a portion of the fault surface in order to reflect, as discussed earlier, the finite length of the actual fault. Because of the anticipated inadequacy of the Griffith energy balance model, the PR cohesive zone model is employed at the outset. The Griffith model of a singular crack-tip will be recovered as a special case.

Viscoelastic Effects in Fault Propagation

A number of authors (e.g. Kostrov & Nikitin 1970, Barenblatt, Entov & Salganik 1970, Mueller & Knauss 1971, Knauss 1974, Schapery 1975) have examined fracture in viscoelastic solids, but the treatment here follows Rice's (1979a) synopsis of his earlier analysis (unpublished lecture notes for graduate course in Fracture Mechanics at Brown University, Providence, R. I., 1973). For steady state fault propagation, the results can be obtained directly from the linear elastic results by means of the correspondence principle of linear viscoelasticity (Fung 1965, Knauss 1974). For the PR model with constant end zone stress, the linear elastic result for the slip in the end zone is given by (17) with $f(\lambda)$ as in (18). If the viscoelastic solid is characterized by a compliance $M(t)$ appropriate to tensile stress application to a body constrained to respond in plane strain, then the relative slip at the end of the breakdown zone is

$$\delta(x=-R) = [K^2/(\tau_p - \tau_f)] \int_0^1 M[(1-\xi)R/V] f'(\xi)\, d\xi \tag{31}$$

where V is the propagation velocity and R is still given by (15). Evidently, if the short-time response of the viscoelastic solid is elastic, comparison with Equation (17) reveals that $M(0) = (1-\nu_0)/2\mu_0$ where ν_0 and μ_0 are the appropriate Poisson's ratio and shear modulus. Similarly, if the limiting long-time response is also elastic, $M(\infty) = (1-\nu_\infty)/2\mu_\infty$. The fracture criterion (19) can now be applied to Equation (31) where it is assumed that δ^* is independent of time. The result is

$$K^2 \int_0^1 M[(1-\xi)R/V] f'(\xi)\, d\xi = G_{crit} \tag{32}$$

where Equation (22) has been used with $\delta^* = \delta'$ for constant end zone stress. The limiting result for very slow fault propagation ($V = 0^+$) is easily extracted from (32):

$$K^2 M(\infty) = G_{crit}. \tag{33}$$

This relation defines a threshold value of the stress intensity factor $K_s = (G_{crit}/M(\infty)]^{1/2}$ below which no fault propagation is possible. For very fast propagation ($V = \infty$), (32) reduces to the expression

$$K^2 M(0) = G_{crit} \tag{34}$$

and $K_f = [G_{crit}/M(0)]^{1/2}$ defines the limiting stress intensity factor for dynamic propagation. For intermediate values of K, fault propagation occurs at the velocity for which (32) is satisfied, as illustrated schematically in Figure 6.

The predictions of the Griffith energy balance model can be recovered in the limit of vanishing end zone size. Letting $R \to 0$ in (32) yields Equation (34) independently of the velocity of propagation. Because G_{crit}

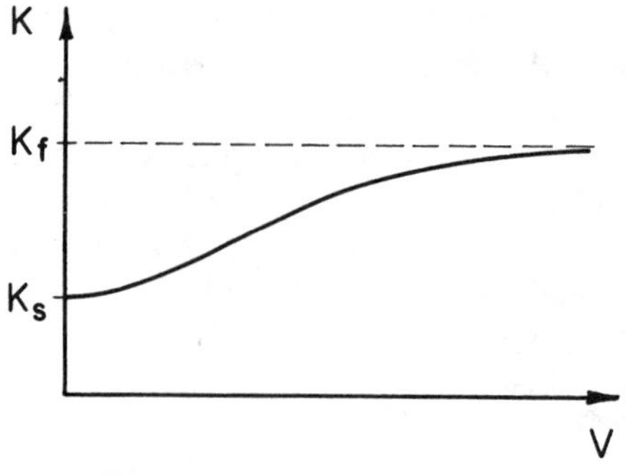

Figure 6 Applied stress intensity factor K vs fault propagation velocity V for a linear viscoelastic solid. K_f and K_s are the critical values for fast ($V \to \infty$) and slow ($V = 0^+$) propagation, respectively.

is assumed to be a fixed material parameter, the limit $R \to 0$ corresponds to simultaneously letting $\tau_p - \tau_f \to \infty$ so that the shaded area in Figure 3b is constant. As mentioned earlier, the limit $R \to 0$ is appropriate only when $R \ll Vt^*$ where t^* is a characteristic time for the response of the end zone material.

Although this model is very simple, its relevance to the geophysical phenomena mentioned earlier is clear. For applied loads that induce a stress intensity factor $K < K_s$, no fault propagation occurs, whereas for $K \simeq K_f$, dynamic (seismic) faulting is predicted. However, for $K_s < K < K_f$ fault growth is quasi-static. Furthermore, in this range of K values, fault propagation is stable in the sense that increases in K (which may result from either increases in the magnitude of the applied loads or an increase in the length of fault surface that is loaded) are necessary to increase the velocity.

The example can be carried further by assuming a specific form for $M(t)$. If the viscoelastic response is, for simplicity, assumed to be described by a linear Maxwell model, then

$$M(t) = M_0(1 + t/t_m) \tag{35}$$

where $M_0 = M(0)$ and t_m is the Maxwell relaxation time. The Maxwell time can be expressed as η/M_0 where η is a viscosity. For a Maxwell solid $M(\infty)$ is unbounded so that the threshold K value for slow fault growth is zero. Although it is often stated that for $t \gg t_m$ a Maxwell solid "behaves like a viscous fluid," the imprecision of this statement when acute geometries are present is evident. In contrast to the behavior of a viscous fluid in which no crack growth would be expected, growth of any finite size crack will occur in a Maxwell solid (although it may be very slow) at any level of the applied load. Substituting (35) into (32) and performing the integration yields

$$K^2 M_0[1 + R/(3Vt_m)] = G_{crit}. \tag{36}$$

This equation can be rearranged to yield the following expression for the velocity after using (16) for R and the definition of K_f:

$$V = (l/3t_m)[(\tau_\infty - \tau_f)/(\tau_p - \tau_f)]^2[(K_f/K)^2 - 1]^{-1}. \tag{37}$$

Unfortunately, good estimates for the parameters in this equation are not available. If, however, $K = 0.9K_f$, $t_m = 5$ yr, $l = 100$ km, and the first term in square brackets is unity, then Equation (37) yields $V = 60$ km/yr. This velocity is typical of observed rates for migration of seismic activity along large transform fault systems (Savage 1971). For predominantly "brittle" behavior and low stress drops K and $(\tau_\infty - \tau_f)$ may differ only slightly from K_f and $(\tau_p - \tau_f)$, respectively. A length of 100 km seems to be an appropriate size scale for phenomena associated with large earthquakes. A Maxwell time of five years corresponds to a shear modulus of 6×10^4 MPa and a viscosity of 10^{19} Poise, values that may be representative of the earth's mantle. There is, of course, disagreement about appropriate values for the viscosity of the mantle and, moreover, whether a linearly viscous model is adequate (Melosh 1976, 1978, Savage & Prescott 1978). If viscosity does vary nonlinearly with stress, the high stress near the fault tip would presumably result in a lower effective viscosity and shorter Maxwell times. In addition, laboratory tests on rock yield values ranging down to 10^{13} Poise (e.g. Handin 1966) so that shorter Maxwell times may be appropriate for faulting in the crust. In any case, Equation (37) illustrates the type of behavior that might be expected.

Coupled Deformation-Diffusion Effects in Fault Propagation

Fluid infiltration of a porous elastic solid introduces a time dependence into the response because the deformation of the solid is coupled to the diffusion of the infiltrating pore fluid. Rice & Simons (1976; also, see Simons 1977) investigated the steady-state propagation of a plane strain (mode II) shear fault in such a fluid-infiltrated solid. The presence of diffusion introduces a length scale c/V where c is the diffusivity and V is the propagation speed. Different regimes of behavior are obtained depending on the relative magnitudes of c/V, the end zone size, and the fault length. Rice & Simons demonstrated that fault propagation was stabilized by coupled deformation diffusion effects for a range of velocities, which is consistent with the range of observed creep events on the San Andreas fault. Ruina (1978) has analyzed the corresponding problem of a tensile (mode I) crack and has discussed the implications of his results for the retardation of hydraulic fractures. A complementary stabilizing effect due to coupling of pore fluid diffusion with dilatancy in the end zone has been treated by Rice (1979b; also see Rice & Cleary 1976) but will not be discussed here.

The equations governing the response of a linear fluid-infiltrated porous elastic solid were established by Biot (1941) but Rice & Cleary (1976) recently achieved a propitious rearrangement of them. For plane strain conditions (deformation is independent of z; $i,j = x, y$ in the following equations), the strain of the solid matrix ε_{ij} and the fluid mass content per unit volume of porous medium m are related to the total stress σ_{ij} and the alteration of pore fluid pressure p (from some ambient value) by

$$2\mu\varepsilon_{ij} = \sigma_{ij} - \nu(\sigma_{xx}+\sigma_{yy})\delta_{ij} + 3\{(\nu_u - \nu)/[B(1+\nu_u)]\}\delta_{ij}p \tag{38}$$

$$m - m_0 = \{3\rho_0(\nu_u - \nu)/[2\mu B(1+\nu)(1+\nu_u)]\}\cdot[(1+\nu_u)(\sigma_{xx}+\sigma_{yy})+(3/B)p] \tag{39}$$

where m_0 is the value of m in the unstressed state and ρ_0 is the density of the pore fluid. The material parameters in these equations are best explained by considering the limiting cases of deformation that is very rapid and very slow by comparison to the time scale of pore fluid diffusion. In the latter case, which is said to be "drained," the deformation is slow enough so that alterations in pore fluid pressure can be alleviated by diffusive mass flux and, consequently, the pore fluid pressure is constant in time. Examination of (38) for $p = 0$ then reveals that the response is that of an ordinary linear elastic solid with μ and ν as the shear modulus and Poisson's ratio. In the alternative short-time limit, which is said to be "undrained," the deformation is too rapid to allow time for fluid mass to diffuse out of material elements. Thus, $m = m_0$ and B can be identified from (39) as the ratio of the pore fluid pressure induced by undrained response to the mean normal stress. Substitution of this value for p into (38) discloses that the undrained response is also that of a linear elastic solid, but with Poisson's ratio ν_u (where the subscript u denotes "undrained") rather than ν. Because $\nu \leqq \nu_u \leqq 1/2$ (Rice & Cleary 1976), the response of the fluid-infiltrated solid is elastically stiffer for undrained conditions than for drained conditions and this increased stiffness is the source of the stabilizing effect in fault propagation. One additional constitutive equation is needed to describe the fluid mass flux. This is Darcy's law

$$q_i = -\rho_0 \kappa\, \partial p/\partial x_i \tag{40}$$

where q_i is the mass flow rate in the x_i direction per unit area and κ is a permeability.

The formulation is completed by the set of field equations that express stress equilibrium, strain compatibility, and fluid mass conservation. The equilibrium equation is

$$\partial \sigma_{ij}/\partial x_i = 0. \tag{41}$$

The conditions of strain compatibility and mass conservation yield, after some rearrangement, the following two equations:

$$\nabla^2(\sigma_{xx}+\sigma_{yy}+2\eta p) = 0 \tag{42}$$

where $\eta = 3(\nu_u - \nu)/2B(1+\nu_u)(1-\nu)$ and

$$(c\nabla^2 - \partial/\partial t)[\sigma_{xx}+\sigma_{yy}+(2\eta/\xi)p] = 0 \tag{43}$$

where $\xi = (\nu_u - \nu)/(1-\nu)$ and the diffusivity is

$$c = \frac{2\mu\kappa B^2(1-\nu)(1+\nu_u)^2}{9(1-\nu_u)(\nu_u-\nu)}. \tag{44}$$

Rice & Simons (1976) first solved these equations for the case of a steadily propagating fault (at a speed V) with no end zone. The fault was assumed to be semi-infinite and loaded by stresses $\tau_\infty - \tau_f$ over a length l of its surfaces, as in Figure 1*b*. For this geometry, the stress intensity factor for an elastic solid is given by (4). Because no material constants enter this expression, one might anticipate that the stress intensity factor for the fault in the fluid-infiltrated solid would also be given by (4) (as was the case for the linear viscoelastic solid discussed in the last section). The solution of Rice & Simons demonstrates, however, that the stress intensity factor is given by

$$K = K_{\text{nom}} h(Vl/c) \tag{45}$$

where K_{nom} is given by (4) and the function h decreases monotonically from unity at $Vl/c = 0$ to $\beta^{-1} = (1-\nu_u)/(1-\nu)$ as $Vl/c \to \infty$. (Because $\nu < \nu_u < 1/2$, $\beta^{-1} < 1$.) Thus K_{nom} is the value of the stress intensity factor that would be caused by the applied loads in the absence of pore fluid effects. For fault propagation in a fluid-infiltrated solid, this value is attained only for zero propagation speed and the stress intensity factor for any nonzero speed is less than K_{nom}. Rice & Simons (1976) rationalized this result by pointing out that for a fault without an end zone, a region near the fault-tip responds in drained fashion for any finite propagation speed whereas the surrounding material further from the fault responds in undrained fashion and, consequently, is elastically stiffer. Because the region near the fault-tip is drained, the energy release rate corresponding to (45) can be calculated from Equation (11) using the drained value of Poisson's ratio. The result is

$$G = G_{\text{nom}} h^2(Vl/c) \tag{46}$$

where

$$G_{\text{nom}} = (1-\nu)K_{\text{nom}}^2/2\mu \tag{47}$$

is the energy release rate for the same applied loads but in the absence of coupled deformation diffusion effects.

If fault growth is assumed to occur when G attains some critical value, say G_{crit} (where G_{crit} may be defined by (22) from a more detailed model of the fault-tip), then the fracture criterion is

$$G_{\text{nom}} h^2(Vl/c) = G_{\text{crit}} \tag{48}$$

This relation is shown schematically in Figure 7 by the curve labeled

$R/l = 0$ corresponding to the idealization of a vanishingly small end zone. G_{nom} can be interpreted as the "driving force" supplied by the applied loads. In the absence of porous media effects the driving force would be sufficient to cause dynamic ($V \to \infty$), that is, seismic fault propagation when

$$G_{\text{nom}} = G_{\text{crit}}. \tag{49}$$

The coupling of deformation with the diffusion is, however, a stabilizing effect and dynamic fault propagation is predicted for the fluid-infiltrated solid only when the driving force rises to

$$G_{\text{nom}} = \beta^2 G_{\text{crit}}. \tag{50}$$

Moreover, fault propagation is stable in the sense that an increase in driving force (G_{nom}) is necessary to increase the speed of propagation. This result is, however, mitigated by the inclusion of a small, but finite size end zone.

Rice & Simons reanalyzed the problem for an end zone of length R and a constant value of the end zone stress (see Figure 3). In this case, the results differ dramatically from those for a vanishingly small end zone. As shown in Figure 7, fault propagation is stable only for a range of values for Vl/c. The results of Equations (48–50) are recovered from the analysis for a finite size end zone in the limit $VR/c = 0$, but for nonzero VR/c there are two additional limiting cases of interest. If $Vl/c \to \infty$, but

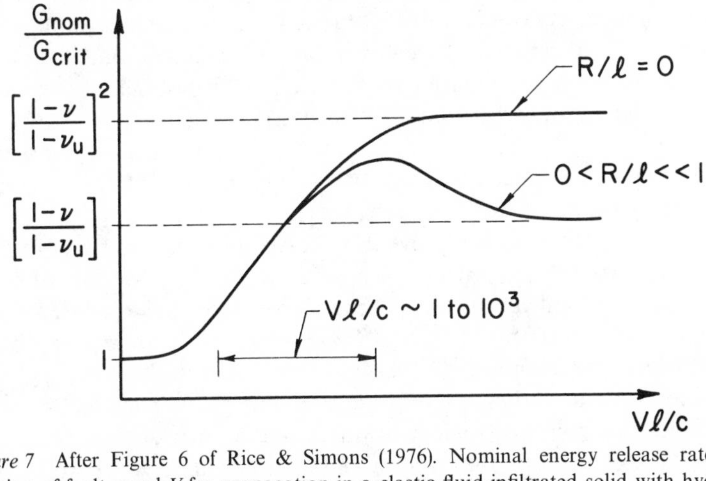

Figure 7 After Figure 6 of Rice & Simons (1976). Nominal energy release rate as a function of fault speed V for propagation in a elastic fluid-infiltrated solid with hydraulic diffusivity c and Poisson's ratios ν (drained) and ν_u (undrained). Effective fault length is l and end zone size is R. Arrow denotes range of creep events observed on the San Andreas fault by King, Nason & Tocher (1973).

VR/c is finite, the response on the overall size scale is undrained but the response on the size scale of the end zone is intermediate between drained and undrained. This limit corresponds to the falling portion of the curve in Figure 7, and thus G_{nom} decreases with increasing V. If both VR/c and Vl/c approach infinity with R/l fixed, the response is undrained on both the overall size scale and the size scale of the end zone and

$$G_{\text{nom}} = \beta G_{\text{crit}}. \tag{51}$$

The magnitude of the stabilizing effect can be judged by the increase in driving force (G_{nom}) required to propagate the fault over that value (G_{crit}) necessary to propagate the fault in the absence of coupled deformation diffusion effects. The maximum value of $G_{\text{nom}}/G_{\text{crit}}$ lies between β and β^2. For brittle rock, $\nu = 0.2$ and $\nu_u = 0.3$–0.4 are typical (Rice & Cleary 1976) and the corresponding values of β range from 1.14–1.33. For clay soils this stabilizing effect can be even more significant as $\nu = 0.15$ and $\nu_u = 0.5$ are typical values and, thus, $\beta = 1.667$.

The existence of the falling portion of the curve in Figure 7 predicts that there is a definite range of velocities for which stable creep events are possible. If the driving force corresponding to the peak in the curve is exceeded, the fault will accelerate and propagate dynamically. From the observations of creep events by King, Nason & Tocher (1973), Rice & Simons (1976) estimate slip lengths l in range 0.1–10 km and velocities V in the range 1–10 km/day. Anderson & Whitcomb (1975), on the basis of a variety of field observations, have suggested a value of the diffusivity $c = 1.0\,\text{m}^2/\text{s}$ and Rice & Simons (1976) have argued that this value is reasonable for fissured rock. Rice (1979b) estimated a value of $c = 0.1\,\text{m}^2/\text{s}$ from measurements by Kovach et al (1975) of water level changes in wells. For these ranges of numerical values, the corresponding values of Vl/c do indeed fall on the rising portion of the curve, as predicted by the analysis. This agreement, although suggestive, does not, of course, mean that all creep events involve stabilization by pore fluid effects. The viscoelastic effects discussed in the last section offer an alternative mechanism and a mechanism based on time-dependent friction, as observed by Dieterich (1972, 1978), is another possibility. Nevertheless, the result of Rice & Simons and its consistency with observation is an indication of the role the mechanical coupling of pore fluid diffusion with deformation can play in processes of crustal faulting. Unfortunately, the overemphasis and ensuing controversy over the possible role of pore fluid in altering seismic wave travel times has been detrimental to an appreciation of this effect. It does appear to be worthy of further study.

PROBLEMS OUTSIDE THE REALM OF FRACTURE MECHANICS

This article has focused on the usefulness of fracture mechanics as an approach to problems of faulting and fracture of the earth's crust. In particular, the fracture criteria was emphasized as an important means for understanding the mechanical conditions that cause the onset of rupture and that distinguish different types of rupture, for example, stable creep versus seismic faulting. It would, however, be misleading to conclude the article without mentioning some difficulties with this approach and briefly reviewing some results for problems of rupture that do not fit within the present framework of fracture mechanics. There is much that is not understood about failure of the earth's crust, and, although fracture mechanics can be helpful in approaching many problems, it will not, of course, be advantageous or even possible to apply it to all problems.

The results reviewed here were, for the most part, obtained for fractures that sustained a uniform stress, except possibly in a small end zone. If the fault surface stress is very nonuniform, the situation is much more complicated and there may be difficulty in even defining a discrete end zone. Also, if the stress on the entire fault depends on the relative slip, then, in effect, the entire fault is in the breakdown zone. In this case the dislocation approach discussed earlier is advantageous and has been employed by, for example, Stuart & Mavko (1979) and Cleary (1976).

More generally, departures from elastic or linear behavior will not be confined to a narrow zone that can be idealized as a crack. Although faults are often observed to be relatively narrow features, these may be merely the final result of processes of inelastic deformation in a larger volume. Rudnicki (1977a) has suggested an inclusion model of faulting in which an ellipsoidal zone of inelastically deforming material is considered to be embedded in an infinite nominally elastic rock mass. The rock mass is loaded by far-field stresses or strains and these drive the inclusion material past peak stress and into the strain softening regime. Eventually, the inclusion material softens sufficiently that the response to an increment of far-field loading is no longer quasi-static and a dynamic (seismic) "runaway" of inclusion strain occurs. Rudnicki (1977a) derived conditions for the onset of the runaway instability in terms of the geometry of the inclusion, the moduli of the surrounding elastic material, and parameters describing the inelastic response of the inclusion material. Rice & Rudnicki (1979) used a solution of Rice, Rudnicki & Simons

(1978) for the deformation of a spherical inclusion in an elastic fluid-infiltrated solid to examine the transient stabilization of runaway instability by coupled deformation-diffusion effects.

Rudnicki (1977a) also used the results of a study by Rudnicki & Rice (1975) to determine conditions for which the deformation in the inclusion would become concentrated into a narrow zone. He found that these conditions are generally met prior to the onset of runaway instability. Although such localization of deformation may result from the growth of local inhomogeneities or the propagation of crack-like flaws, Rudnicki & Rice (1975) adopted the point of view, which was suggested by Rice (1973), that this phenomenon could be explained in terms of the constitutive description of homogeneous deformation. More specifically, conditions were sought for which the constitutive behavior allowed a bifurcation from homogeneous deformation corresponding to nonuniform deformation in a narrow band. A general review of this approach has been given by Rice (1976) and by Cleary & Rudnicki (1976) in the context of rupture in geological materials. Rudnicki (1977a,b,c) has argued that the predictions of Rudnicki & Rice (1975) are consistent with faulting of brittle rock in laboratory specimens, at least if instability is not caused by the testing machine and inhomogeneities introduced by the end constraints are minimized.

Faulting in the earth's crust seems to occur most frequently in areas where the deformation has already been localized due to some past history of faulting. Nevertheless, for earthquakes that fracture fresh or largely rehealed rock, localization of deformation may be important in processes leading up to rupture. Indeed, recent observations of the seismicity patterns prior to the San Fernando earthquake (Ishida & Kanamori 1977) are suggestive of localization of deformation. Moreover, Rice (1978b) has suggested that the difference reported by Lindh, Fuis & Mantis (1978) in the orientations for fault plane solutions of foreshocks as compared with the mainshocks is consistent in magnitude and sense with predictions of Rudnicki & Rice (1975).

ACKNOWLEDGMENTS

Support from the Department of Theoretical and Applied Mechanics of the University of Illinois at Urbana-Champaign is gratefully acknowledged. A portion of this review was prepared while I was visiting Geophysicist at the US Geological Survey at Menlo Park, California, during June, 1979.

Literature Cited

Abou-Sayed, A. S. 1977. Fracture toughness K_{IC} of triaxially loaded Indiana Limestone. In *Energy Resources and Excavation Technology, Proc. US Symp. Rock Mech., 18th, Keystone, Colo. 1977*, ed. F. Wang, G. B. Clark, pp. 2A3-1–2A3-8

Abou-Sayed, A. S., Brechtel, C. E., Clifton, R. J. 1978. In-situ stress determination by hydro-fracturing: a fracture mechanics approach. *J. Geophys. Res.* 83:2851–62

Aki, K. 1978. Origin of the seismic gap: what initiates and stops a rupture propagation along a plate boundary? In *Proceedings of Conference VI, Methodology for Identifying Seismic Gaps and Soon-to-Break Gaps*, convened under auspices of National Earthquake Hazards Reduction Program, 25–27 May 1978 (USGS open file report 78-943), pp. 3–46

Anderson, D. L., Whitcomb, J. H. 1975. Time-dependent seismology. *J. Geophys. Res.* 80:718–32

Anderson, O. L., Grew, P. C. 1977. Stress corrosion theory of crack propagation with applications to geophysics. *Rev. Geophys. Space Phys.* 15:77–104

Andrews, D. J. 1976a. Rupture propagation with finite stress in anti-plane strain. *J. Geophys. Res.* 81:3575–82

Andrews, D. J. 1976b. Rupture velocity of plane strain shear cracks. *J. Geophys. Res.* 81:5679–87

Atkinson, B. K. 1979a. Fracture toughness of Tennessee sandstone and Carrara marble using the double torsion testing method. *Int. J. Rock Mech. Min. Sci. Geomech. Abstr.* 16:49–53

Atkinson, B. K. 1979b. A fracture mechanics study of subcritical tensile cracking of quartz in wet environments. *Pure Appl. Geophys.* 117: In press

Atkinson, B. K. 1980. Stress corrosion and the rate dependent tensile failure of a fine-grained quartz rock. *Tectonophysics* 64: In press

Atkinson, B. K., Rawlings, R. D. 1979a. *Acoustic emission during subcritical tensile cracking of gabbro and granite.* Presented at Spring meet. Am. Geophys. Union

Atkinson, B. K., Rawlings, R. D. 1979b. Acoustic emission characteristics of subcritical and fast tensile cracking in rock. *Proc. Inst. Acoustics.* In press

Barenblatt, G. I. 1962. Mathematical theory of equilibrium cracks in brittle fracture. *Adv. Appl. Mech.* 7:55–129

Barenblatt, G. I., Entov, V. M., Salganik, R. L. 1970. Some problems of the kinetics of crack propagation. In *Inelastic Behavior of Solids, Battelle Institute Materials Science Colloquia, 4th, Columbus, Ohio, 1969*, ed. M. L. Kanninen et al., pp. 559–84. New York: McGraw-Hill. 743 pp.

Barnett, D. M., Freund, L. B. 1975. An estimate of strike-slip fault friction stress and fault depth from surface displacement data. *Bull. Seismol. Soc. Am.* 65:1259–66

Barton, N. 1971. A relationship between joint roughness and joint shear strength. *Proc. Int. Symp. Rock Mech., Nancy*, Pap. I-8

Barton, N. 1973. Review of a new strength criterion for rock joints. *Eng. Geol.* 7: 287–332

Bilby, B. A., Cottrell, A. H., Swindon, K. H. 1963. The spread of plastic yield from a notch. *Proc. R. Soc. London Ser. A* 272:304–14

Bilby, B. A., Eshelby, J. D. 1968. Dislocations and the theory of fracture. See Liebowitz 1968, I:99–182

Biot, M. A. 1941. General theory of three-dimensional consolidation. *J. Appl. Phys.* 12:155–64

Brace, W. F. 1964. Brittle fracture of rocks. In *State of Stress in the Earth's Crust*, ed. W. R. Judd, pp. 111–74. New York: Elsevier

Brace, W. F., Bombolakis, E. G. 1963. A note on brittle crack growth in compression. *J. Geophys. Res.* 68:3709–13

Broek, D. 1974. *Elementary Engineering Fracture Mechanics.* Leyden: Noordhoff International. 408 pp.

Brown, W. F., Srawley, J. E. 1966. *Plane strain crack toughness testing of high strength metallic materials, ASTM Spec. Tech. Publ. No. 410.* Philadelphia: Am. Soc. Test. Mater. 129 pp.

Budiansky, B., Rice, J. R. 1973. Conservation laws and energy-release rates. *J. Appl. Mech.* 40:201–3

Budiansky, B., Rice, J. R. 1979. An integral equation for dynamic elastic response of an isolated 3-D crack. *Wave Motion* 1: 187–92

Cherepanov, G. P. 1968. Cracks in solids. *Int. J. Solids Struct.* 4:811–31

Cleary, M. P. 1976. Continuously distributed dislocation model for shear bands in geological materials. *Int. J. Num. Methods Eng.* 10:679–702

Cleary, M. P., Rudnicki, J. W. 1976. The initiation and propagation of dilatant rupture zones in geological materials. In *The Effect of Voids on Material Deformation*, ed. S. C. Cowin, pp. 13–30. New York: Am. Soc. Mech. Eng. Appl. Mech. Div. Vol. 16

Clifton, R. J., Simonsen, E. R., Jones, A. H., Green, S. J. 1976. Determination of critical stress intensity factor K_{IC} from internally pressurized thick walled vessels. *Exp. Mech.* 16:233–38

Cotterell, B., Rice, J. R. 1979. Slightly curved or kinked cracks. *Int. J. Fract.* In press

Coulson, J. H. 1972. Shear strength of flat surfaces in rock. In *Proc. US Symp. Rock Mech., 13th, Urbana, Ill. 1971*, ed. E. J. Cording, pp. 77–105

Das, S., Aki, K. 1977a. A numerical study of two-dimensional spontaneous rupture propagation. *Geophys. J. R. Astron. Soc.* 50:643–68

Das, S., Aki, K. 1977b. Fault plane with barriers: a versatile earthquake model. *J. Geophys. Res.* 82:5658–70

Dieterich, J. H. 1972. Time-dependent friction in rocks. *J. Geophys. Res.* 77: 3690–97

Dieterich, J. H. 1978. Time-dependent friction and the mechanics of stick-slip, *J. Geophys. Res.* 116:790–806

Dieterich, J. H. 1979. Modeling of rock friction, part I: experimental results and constitutive equations. *J. Geophys. Res.* 84:2161–68

Dieterich, J. H., Conrad, G. 1978. Mechanism of unstable slip in rock friction experiments. *EOS, Trans. Am. Geophys. Union* 59:1206

Dmowska, R. 1973. Crack fault model and surface deformation associated with dip-slip faulting. *Publ. Inst. Geophys. Polish Acad. Sci.* 62:124–39

Dugdale, D. S. 1960. Yielding of steel sheets containing slits. *J. Mech. Phys. Solids* 8:100–4

Edmunds, T. M., Willis, J. R. 1976. Matched asymptotic expansions in non-linear fracture mechanics—I. Longitudinal shear of an elastic perfectly-plastic specimen. II. Longitudinal shear of an elastic work hardening plastic specimen. *J. Mech. Phys. Solids* 24:205–38

Eshelby, J. D. 1956. The continuum theory of lattice defects. In *Progress in Solid State Physics*, ed. F. Seitz, D. Turnbull. 3:79–144

Eshelby, J. D. 1970. Energy relations and the energy-momentum tensor in continuum mechanics. See Barenblatt et al 1970, pp. 77–115

Evernden, J. F., ed. 1977. *Proc. Conf. II, Experimental Studies of Rock Friction with Application to Earthquake Prediction*, convened under auspices of National Earthquake Hazards Reduction Program, 28–30 April 1977.

Freund, L. B. 1976a. Dynamic crack propagation. In *The Mechanics of Fracture*, ed. F. Erdogan, pp. 105–34. New York: Am. Soc. Mech. Eng. Appl. Mech. Div.

Freund, L. B. 1976b. The analysis of elastodynamic crack-tip stress fields. In *Mechanics Today*, ed. S. Nemat-Nasser, 3:55–91. Elmsford, NY: Pergamon

Freund, L. B. 1979. The mechanics of dynamic shear crack propagation. *J. Geophys. Res.* 84:2199–209

Freund, L. B., Barnett, D. M. 1976. A two-dimensional analysis of surface deformation due to dip slip faulting. *Bull. Seismol. Soc. Am.* 66:667–75

Fukao, Y. 1979. Tsunami earthquakes and subduction processes near deep-sea trenches. *J. Geophys. Res.* 84:2303–14

Fung, Y. C. 1965. *Foundations of Solid Mechanics.* Englewood Clifts, NJ: Prentice-Hall. 525 pp.

Goulty, N. R., Burford, R. O., Allen, C. R., Gilman, R., Johnson, C. E., Keller, R. P. 1978. Large creep events on the Imperial fault, California. *Bull. Seismol. Soc. Am.* 68:517–21

Griffith, A. A. 1920. The phenomena of rupture and flow in solids. *Philos. Trans. R. Soc. London Ser. A* 221:163–98

Handin, J. 1966. Strength and ductility. In *Handbook of Physical Constants*, ed. S. P. Clark Jr., *Geol. Soc. Am. Mem.* 97:238–89

Heaton, T. H., Helmberger, D. V. 1978. Predictability of strong ground motion in the Imperial Valley: modeling the M4.9 November 4, 1976 Brawley earthquake. *Bull. Seismol. Soc. Am.* 68:31–48

Husseini, M. I., Jovanovich, D. B., Randall, M. J., Freund, L. B. 1975. The fracture energy of earthquakes. *Geophys. J.* 43: 367–85

Ida, Y. 1972. Cohesive force across the tip of a longitudinal shear crack and Giffith's specific surface energy. *J. Geophys. Res.* 77:3796–805

Ida, Y. 1973a. Stress concentration and unsteady propagation of longitudinal shear cracks. *J. Geophys. Res.* 78:3418–29

Ida, Y. 1973b. The maximum acceleration of seismic ground motion. *Bull. Seismol. Soc. Am.* 63:959–68

Ingraffea, A. R., Heuze, F. E., Ko, H. Y., Gerstle, K. 1977. An analysis of discrete fracture propagation in rock loaded in compression. See Abou-Sayed 1977, pp. 2A4-1–2A4-7

Ishida, M., Kanamori, H. 1977. The spatio-temporal variation of seismicity before the 1971 San Fernando earthquake, California. *Geophys. Res. Lett.* 4:345–46

Jaeger, J. C., Cook, N. G. W. 1976. *Fundamentals of Rock Mechanics.* New York: Halsted. 585 pp. 2nd ed.

Kanamori, H., Anderson, D. L. 1975. Theoretical basis of some empirical relations in seismology. *Bull. Seismol. Soc. Am.* 65:1073–95

Kanamori, H., Cipar, J. J. 1974. Focal processes of the great Chilean earthquake, May 33, 1960. *Phys. Earth Planet. Inter.* 9:128–36

Kassir, M. K., Sih, G. C. 1966. Three-dimensional stress distribution around an elliptical crack under arbitrary loadings. *J. Appl. Mech.* 33:601–11

King, C.-Y., Nason, R. D., Tocher, D. 1973. Kinematics of fault creep. *Philos. Trans. R. Soc. London Ser. A* 274:355–60

Knauss, W. G. 1974. On the steady propagation of a crack in a viscoelastic sheet. In *Deformation and Fracture of High Polymers. Battelle Inst. Mater. Sci. Colloq., 7th, Kronberg im Taunus, 1972*, ed. H. H. Kausch, J. A. C. Hassell, R. I. Jaffee. New York: Plenum. 644 pp.

Knott, J. F. 1973. *Fundamentals of Fracture Mechanics.* New York: Halsted. 273 pp.

Knowles, J. K., Sternberg, E. 1972. On a class of conservation laws in linearized and finite elastostatics. *Arch. Rational Mech. Anal.* 44:187–211

Kobayashi, T., Fourney, W. L. 1978. Experimental characterization of the development of the microcrack process zone at a crack-tip in rock under load. In *Proc. US Symp. Rock Mech., 19th, Stateline, Nev. 1978*, ed. Y. S. Kim, pp. 243–46

Kostrov, B. V., Nikitin, L. V. 1970. Some general problems of mechanics of brittle fracture. *Arch. Mech. Stosowanej* 22:749–76

Kovach, R. L., Nur, A., Wesson, R. L., Robinson, R. 1975. Water-level fluctuations and earthquakes on the San Andreas fault zone. *Geology* 3:437–40

Kranz, R. L. 1979a. Crack growth and development during creep of Barre granite. *Int. J. Rock Mech. Min. Sci. Geomech. Abstr.* 16:23–35

Kranz, R. L. 1979b. Crack-crack and crack-pore interactions in stressed granite. *Int. J. Rock Mech. Min. Sci. Geomech. Abstr.* 16:37–47

Lawn, B. R., Wilshaw, T. R. 1975. *Fracture of Brittle Solids.* New York: Cambridge Univ. Press. 204 pp.

Liebowitz, H., ed. 1968. *Fracture: An Advanced Treatise.* Volumes I-VII. New York: Academic

Lindh, A., Fuis, G., Mantis, C. 1978. Seismic amplitude measurements suggest foreshocks have different focal mechanisms than aftershocks. *Science* 201: 56–59

Martin, R. J. 1972. Time-dependent crack growth in quartz and its application to creep in rocks. *J. Geophys. Res.* 77: 1406–19

Martin, R. J., Durham, W. B. 1975. Mechanisms of crack growth in quartz. *J. Geophys. Res.* 80:4837–44

McGarr, A., Spottiswoode, S. M., Gay, N. C., Ortlepp, W. D. 1979. Observations relevant to seismic driving stress, stress drop, and efficiency. *J. Geophys. Res.* 84: 2251–63

Melosh, H. J. 1976. Non-linear stress propagation in the earth's upper mantle. *J. Geophys. Res.* 81:5621–32

Melosh, H. J. 1978. Reply to Savage & Prescott, 1978. *J. Geophys. Res.* 83:5009–10

Minster, J. B., Jordan, T. H. 1978. Present-day plate motions. *J. Geophys. Res.* 83: 5331–54

Mueller, H. K., Knauss, W. G. 1971. Crack propagation in a viscoelastic sheet. *J. Appl. Mech.* 38:483–88

Palmer, A. C., Rice, J. R. 1973. The growth of slip surfaces in the progressive failure of overconsolidated clay. *Proc. R. Soc. London Ser. A* 332:527–48

Paris, P. C., Sih, G. C. 1965. Stress analysis of cracks. In *Fracture Toughness Testing and Its Applications, ASTM Spec. Tech. Publ. 381*, pp. 30–76. Philadelphia: Am. Soc. Test. Mater. 409 pp.

Rice, J. R. 1966. An examination of the fracture mechanics energy balance from the point of view of continuum mechanics. In *Proc. Int. Conf. Fract., 1st*, ed. T. Yokobori, 1:283–308. Tokyo: Japanese Soc. Strength and Fracture

Rice, J. R. 1968a. Mathematical analysis in the mechanics of fracture. See Liebowitz 1968 II:191–311

Rice, J. R. 1968b. A path independent integral and the approximate analysis of strain concentration by notches and cracks. *J. Appl. Mech.* 35:379–86

Rice, J. R. 1973. The initiation and growth of shear bands. In *Plasticity and Soil Mechanics*, ed. A. C. Palmer, pp. 263–74. Cambridge, England: Cambridge Univ. Eng. Dept.

Rice, J. R. 1976. The localization of plastic deformation. In *Proc. Int. Cong. Theor. Appl. Mech., 14th*, ed. W. T. Koiter, 1: 207–20. Amsterdam: North-Holland

Rice, J. R. 1978a. Thermodynamics of the quasi-static growth of Griffith cracks. *J. Mech. Phys. Solids* 26:61–78

Rice, J. R. 1978b. Models for the inception of earthquake rupture. *EOS, Trans. Am. Geophys. Union* 59:1204

Rice, J. R. 1979a. The mechanics of quasi-static crack growth. In *Proc. US Natl. Congr. Appl. Mech., 8th, UCLA., June*

1978, ed. R. E. Kelly, pp. 191–216. North Hollywood, Calif: Western Periodicals
Rice, J. R. 1979b. Theory of precursory processes in the inception of earthquake rupture. *Gerländs Beitr. Geophys.* 88: 91–127
Rice, J. R., Cleary, M. P. 1976. Some basic stress diffusion solutions for fluid-saturated elastic porous media with compressible constituents. *Rev. Geophys. Space Phys.* 14:227–41
Rice, J. R., Rudnicki, J. W. 1979. Earthquake precursory effects due to pore fluid stabilization of a weakening fault zone. *J. Geophys. Res.* 84:2177–94
Rice, J. R., Rudnicki, J. W., Simons, D. A. 1978. Deformation of spherical cavities and inclusions in fluid-infiltrated elastic materials. *Int. J. Solids Struct.* 14:289–303
Rice, J. R., Simons, D. A. 1976. The stabilization of spreading shear faults by coupled deformation-diffusion effects in fluid-infiltrated porous materials. *J. Geophys. Res.* 81:5322–34
Richter, C. F. 1958. *Elementary Seismology.* San Francisco: Freeman. 768 pp.
Rudnicki, J. W. 1977a. The inception of faulting in a rock mass with a weakened zone. *J. Geophys. Res.* 82:844–54
Rudnicki, J. W. 1977b. *Localization of Deformation, Brittle Rock Failure and a Model for the Inception of Earth Faulting.* PhD thesis. Brown Univ., Providence, R.I.
Rudnicki, J. W. 1977c. The effect of stress induced anisotropy on a model of brittle rock failure as localization of deformation. See Abou-Sayed 1977, pp. 3B4-1–3B4-8
Rudnicki, J. W. 1979. The stabilization of slip on a narrow weakening fault zone by coupled deformation–pore fluid diffusion. *Bull. Seismol. Soc. Am.* 69:1011–26
Rudnicki, J. W., Rice, J. R. 1975. Conditions for the localization of deformation in pressure-sensitive dilatant materials. *J. Mech. Phys. Solids* 23:371–94
Ruina, A. 1978. Influence of coupled deformation-diffusion effects on retardation of hydraulic fracture. See Kobayashi & Fourney 1978, pp. 274–82
Rutter, E. H., Mainprice, D. H. 1978. The effect of water on stress relaxation of faulted and unfaulted sandstone. *Pure Appl. Geophys.* 116:634–54
Sacks, I. S., Suyehiro, S., Linde, A. T., Snoke, J. A. 1977. The existence of slow earthquakes and the redistribution of stress in seismically active regions. *EOS, Trans. Am. Geophys. Union* 58:437
Savage, J. C. 1971. A theory of creep waves propagating along a transform fault. *J. Geophys. Res.* 76:1954–66
Savage, J. C., Burford, R. O. 1973. Geodetic determination of relative plate motion in central California. *J. Geophys. Res.* 78: 832–45
Savage, J. C., Prescott, W. H. 1978. Comment on Melosh, 1976. *J. Geophys. Res.* 83:5005–7
Schapery, R. A. 1975. A theory of crack initiation and growth in viscoelastic media: I. theoretical development, *Int. J. Fract.* 11:141–59
Schmidt, R. A. 1976. Fracture-toughness testing of limestone. *Exp. Mech.* 16: 161–67
Schmidt, R. A. 1977. Fracture mechanics of oil shale–unconfined fracture toughness, stress corrosion cracking and tension test results. See Abou-Sayed 1977, pp. 2A2-1–2A2-6
Schmidt, R. A., Huddle, C. W. 1977. Effect of confining pressure on fracture toughness of Indiana limestone. *Int. J. Rock Mech. Min. Sci. Geomech. Abstr.* 14: 289–93
Schmidt, R. A., Lutz, T. J. 1979. *K_{IC} and J_{IC} of Westerly granite—effects of thickness and in-plane dimensions.* In *Fracture Mechanics Applied to Brittle Materials*, ed. S. W. Freiman, *ASTM Spec. Tech. Publ. 678*, pp. 166–82. Philadelphia: Am. Soc. Test. Mater.
Scholz, C. H. 1972. Static fatigue of quartz. *J. Geophys. Res.* 77:2104–14
Sieh, K. 1978a. Prehistoric large earthquakes produced by slip on the San Andreas fault at Pallett Creek, California. *J. Geophys. Res.* 83:3907–39
Sieh, K. 1978b. Central California foreshocks of the great 1857 earthquake. *Bull. Seismol. Soc. Am.* 68:1731–49
Simons, D. A. 1977. Boundary layer analysis of propagating mode II cracks in porous elastic media. *J. Mech. Phys. Solids* 25: 99–116
Steketee, J. A. 1958. Some geophysical applications of the elasticity theory of dislocations. *Can. J. Phys.* 36:1168–98
Stuart, W. D. 1979. Quasi-static earthquake mechanics. *Rev. Geophys. Space Phys.* 17: 1115–20
Stuart. W. D., Mavko, G. M. 1979. Earthquake instability on a strike-slip fault. *J. Geophys. Res.* 84:2153–60
Swanson, P. L., Spetzler, H. 1979. Stress corrosion of single cracks in flat plates of rock. *EOS, Trans. Am. Geophys. Union* 60:380
Swolfs, H. S. 1972. Chemical effects of pore fluids on rock properties. In *Underground Waste Management and Environmental Implications*, ed. T. D. Cook, pp. 224–34. *Am. Assoc. Petrol. Geol. Mem. 18*
Tada, H., Paris, P. C., Irwin, G. R. 1973.

The Stress Analysis of Cracks Handbook. Hellertown. Pa: Del Research Corp.

Tapponnier, P., Brace, W. F. 1976. Development of stress-induced microcracks in Westerly granite. *Int. J. Rock Mech. Min. Sci. Geomech. Abstr.* 13:103–12

Thatcher, W. 1979. Systematic inversion of geodetic data in Central California. *J. Geophys. Res.* 84:2283–95

Wachtman, J. B. 1974. Highlights of progress in the science of fracture of ceramics and glass. *J. Am. Ceram. Soc.* 57:509–19

Wawersik, W. R., Brace, W. F. 1971. Post failure behavior of a granite and a diabase. *Rock Mech.* 3:61–85

Weaver, J. 1977. Three-dimensional crack analysis. *Int. J. Solids Struct.* 13:321–30

Willis, J. R. 1967. A comparison of the fracture criteria of Griffith and Barenblatt, *J. Mech. Phys. Solids* 15:151–62

Wilkening, W. W. 1978. J-integral measurement in geological materials. See Kobayashi & Fourney 1978, pp. 254–58

Ann. Rev. Earth Planet. Sci. 1980. 8:527–58
Copyright © 1980 by Annual Reviews Inc. All rights reserved

THE MOONS OF MARS

*10140

Joseph Veverka and Joseph A. Burns
Laboratory for Planetary Studies, Cornell University, Ithaca, New York 14853

Introduction

Asaph Hall (1878) discovered the two moons of Mars, Phobos and Deimos, just over one hundred years ago, in 1877 (Pollack 1977, Veverka 1978). During the same opposition, several other moons were reported, most notably by W. S. Holden (Gingerich 1970); none of these have been seen since, even by spacecraft, and so this review will confine itself to Phobos and Deimos.

Most of the observations of Phobos and Deimos made before the advent of spacecraft measurements in 1969 were necessarily restricted to positional determinations. The distinction of attempting the first physical studies of the satellites goes to Edward C. Pickering under whose direction the first series of visual photometric measurements were made at Harvard in the period between 1877 and 1882 (Pickering 1882).

While several images of Phobos were obtained during the Mariner 7 mission in 1969 (see section on "Dimensions" below), the first systematic attempt to study the satellites by spacecraft was made during the Mariner 9 mission in 1971–1972 (Pollack et al 1972, 1973). The best images obtained had a surface resolution of 100–200 m (Veverka et al 1974). Very extensive observations of both satellites were carried out during the Viking mission, which began in 1976 and is still continuing. The highest resolution images of the satellites (10 m for Phobos, 3 m for Deimos) were obtained during a series of close encounters with the Viking Orbiters (Veverka 1978). Viking Orbiter 1 flew within 100 km of Phobos in February 1977, and within 300 km in May of that year. Viking Orbiter 2 flew within 30 km of Deimos in October 1977.

Physical Properties

ORBITS Because of their proximity to Mars, their rapid motion, and their faintness, the small dark martian satellites are difficult to pinpoint precisely through a telescope (Pascu 1978). Nevertheless, ground-based observations of Phobos and Deimos have been used to ascertain accurate

0084-6597/80/0515-0527$01.00

Table 1 Orbital elements of martian satellites[a]

Orbital element	Phobos	Deimos
Semimajor axis, a	9378 km (2.76 $R_{♂}$)	23459 km (6.90 $R_{♂}$)
Eccentricity, e	0.015	0.00052
Inclination, I (with respect to Laplace plane)	1°.02	1°.82

[a] Data from Born & Duxbury (1975).

orbital elements that compare favorably with those based on Mariner 9 positions (Born & Duxbury 1975). As the numbers listed in Table 1 show, the martian moons travel along nearly circular paths of low inclination. The orbital inclinations are properly referenced, as here, relative to the Laplace plane (Burns 1972, Sinclair 1972), i.e. the mean plane about which orbits precess when the perturbations of both the Sun and the planetary oblateness are considered; at distances further from Mars, this plane tilts more and more away from the equator and toward the planet's orbital plane so that, while it is only 0.009° relative to the martian equator at Phobos' distance, at Deimos' position it is tilted by 0.92°. Hence, the martian satellites lie essentially in Mars' equatorial plane. For Mars, the synchronous orbit, along which the orbital period of a satellite is equal to the planetary spin period, is at semimajor axis $a_s = 6.03\, R_{♂}$. Thus, Deimos lies just beyond the synchronous position, whereas Phobos orbits well within it, revolving more than three times as fast as Mars spins.

The orbits of the martian satellites we see today are little modified by perturbations. The masses of the satellites are small enough that they essentially do not mutually interact. Planetary perturbations also have negligible effects. Mars' oblateness and, to a much lesser extent, the solar presence cause the node of the orbit plane to regress secularly about the pole of the Laplace plane (for Phobos, the period of this motion is 2.25 yr while for Deimos, it is 54.5 yr); pericenter precesses within the orbit plane at essentially the same rate (Sinclair 1978). However, because both orbits are almost circles lying in the martian equatorial plane, their external appearance is little changed by these precessions. Nevertheless, these motions permitted the martian oblateness J_2, as well as the location of the martian rotational pole, to be accurately determined (Burns 1972) even before the spacecraft era. Periodic perturbations, due to solar gravity and to the higher order terms in Mars' gravity field, are essentially negligible unless precise instantaneous positions, such as required for close flybys of the satellites, are desired. Although the rotation axis of Mars sweeps out a cone of half-angle 23°59′ every 173,000 yr under the action of the solar torque (Ward et al 1979), the satellite orbits are able to maintain an essentially constant inclination with respect to the Laplace pole (Goldreich 1965) even as the planet itself precesses.

SECULAR ACCELERATION AND TIDES A third of a century ago, Phobos was noted to be accelerating along its orbit (Sharpless 1945): this secular acceleration has important consequences since it implies that energy is being *lost* from the satellite orbit (Burns 1972, 1977). Sinclair (1972) evaluated the acceleration (given in units of 10^{-3} deg/yr^2) to be 0.96 ± 0.16, while more recently with a different solution technique, he found 0.663 ± 0.059; Shor (1975), using a broadened data base, instead determined 1.43 ± 0.15; the quoted values, for reasons of historical definition, are one-half the actual secular acceleration. In his calculation, Shor (1975) had improved Sinclair's data base by including early Russian observations and more recent Soviet measurements. Later Sinclair (1978) fixed the orbital elements of the satellites at those deduced from Mariner 9 observations and then applied Earth-based measurements to evaluate the secular accelerations; he, however, continued to ignore the critical early positions included by the Soviet author. Shor's result is perhaps to be preferred because it better fits subsequent observations made by Viking (G. H. Born, private communication, 1977). Results for the motion of Deimos are far less conclusive, but all authors measure a deceleration. Innumerable suggestions, summarized by Shklovskii & Sagan (1966) and Burns (1972), have been put forth to explain Phobos' acceleration; most celestial mechanicians (Burns 1972, Pollack 1977, Lambeck 1979) now accept tidal forces as the most probable cause. Earlier, tidal drag had been thought ineffective because of incorrect notions as to the mass of the satellites and the efficacy of terrestrial solid-body tides in evolving our Moon's orbit (Burns 1978).

Tidal forces produce an acceleration of Phobos because the tidal response of any nonperfect planet is delayed in time and, since Phobos is within the synchronous orbit, this means that the tidal bulge on Mars lags the inner satellite's position. The tidal bulge thereby exerts a retarding force on the satellite, causing it to fall nearer the planet and so move faster. In the case of Deimos, the tidal force (see figures in Burns 1977, 1978) pulls it forward, and therefore outward, but at a much reduced rate, owing to the a^{-6} dependence of the tidal torque. The measured acceleration of Phobos then allows one to compute the major tidal parameter for Mars: according to the various measurements of the acceleration, Q/k_2, the anelasticity factor divided by the second order Love number, is 575 (Shor 1975), 850 (Sinclair 1972), or 1250 (Sinclair 1978). Simple theoretical models of Mars (Burns 1978, Lambeck 1979) give $k_2 \sim 0.1$ and so the martian Q ranges from about 50 to 150 (Smith & Born 1976, Pollack 1977, Lambeck 1979), very reasonable values that are compatible with results for the solid Earth tides (Burns 1977, Pollack 1977, Lambeck 1979). Accepting tides as the cause of the secular acceleration of Phobos, one can integrate Phobos' orbit forward in time (see section on "Origin") to find that, if the object retains its integrity, its orbit should collapse onto the planet's surface in $t^* = (3/13)n_o/\dot{n}_o$, where n_o is the observed mean motion and the dot signifies a time derivative. With Shor's acceleration, the orbital collapse time is 3×10^7 yr hence (Burns 1978).

DIMENSIONS The first direct estimate of the size of Phobos was obtained by Smith (1970) from a low resolution Mariner 7 image. He estimated the cross section of Phobos visible in that frame as 22 km by 18 km, in agreement with later values derived by Duxbury (1974) from systematic observations by Mariner 9 in 1971–72. Duxbury fitted triaxial ellipsoids to the irregular shapes of the satellites, obtaining the following dimensions: 27 ± 2 km by 21.6 ± 1.4 km by 18.6 ± 1.4 km for Phobos and 15^{+6}_{-2} km by 12.2 ± 1 km by 11 ± 2 km for Deimos. The values quoted are the diameters for a model ellipsoid whose semidiameters are A, B, C. The long axis —$2A$— of both satellites always points to the center of Mars (ignoring librations, see next section); the shortest axis —$2C$— points along the spin axis; the intermediate axis —$2B$— points in the direction of orbital motion. The convention is to use spherical, planetocentric coordinates on the satellites with longitude measured west from the sub-Mars point, while north and south is as on Mars.

The Mariner 9 values for the dimensions of Phobos and Deimos have been continuously improved through ongoing observations by the Viking Orbiters; the best current values according to T. C. Duxbury (personal communication, 1979) are 27 km by 21 km by 19 km for Phobos, and 15 km by 12 km by 10 km for Deimos. The ratio of the longest to the shortest axis is approximately equal to 1.5 in each case.

It is important to realize that Phobos and Deimos are only approximately ellipsoidal. Both satellites, having obviously experienced a long history of cratering, are in fact irregular objects; departures of 2–3 km from the ellipsoidal fits do occur.

The shape of Phobos is much closer to that of a triaxial ellipsoid than is that of Deimos, the major departure of Phobos being caused by the largest crater, Stickney. On the other hand, Deimos is only crudely triaxial. In fact, its surface consists of a series of facets separated by prominent ridges; a conspicuous saddlelike depression, which does not have the morphology of a crater, occurs near the south pole (Veverka & Thomas 1979). It appears unlikely that the overall shape of either satellite is determined by martian tides, as had been suggested for Phobos by Soter & Harris (1976).

The fact that both Phobos and Deimos are fundamentally irregular must be kept in mind when estimating surface areas and volumes. The uncertainties involved are difficult to assess rigorously, but P. C. Thomas (personal communication, 1979) has attempted to estimate realistic values based on the Duxbury measurement of the mean axes but taking into account the actual topographic detail on the satellites. The resulting volumes are 5000 km^3 for Phobos and 1000 km^3 for Deimos, with uncertainties of probably $\pm 15\%$. The respective surface areas are about 1000 km^2 for Phobos and 400 km^2 for Deimos.

ROTATIONS Mariner 9 observed both Phobos and Deimos to be in the rotational state expected for tidally evolved moons (Burns 1972): this is the minimum energy configuration and has a satellite's long axis (of length $2A$) approximately aligned with the planet-satellite line, while its shortest axis ($2C$) is roughly normal to the orbit plane. The time of damping to this synchronous state is $\sim 10^5$–10^6 yr for Phobos and $\sim 10^8$–10^9 yr for Deimos (Peale 1977, cf Burns 1972), assuming a satellite Q of 10^2–10^3. The long axis of Phobos has been found, as predicted (Burns 1972), to librate by approximately $\pm 5°$ about the satellite-planet line (Duxbury 1977). The libration is due to a near resonance between the in-plane free libration period of Phobos and the orbital period of the satellite; the latter is the period of a torque, which drives librational motion, since the orbital angular velocity varies along the orbit, thereby allowing misalignment to occur between the satellite long axis and the Mars-Phobos line of centers.

MASS AND MEAN DENSITY The mass of Phobos has been determined from perturbations on the Viking Orbiter 1 spacecraft during a series of close encounters in February and May, 1977. This value 9.6 (± 2.0) $\times$ 10^{18} gm (Christensen et al 1977, Tolson et al 1978, T. C. Duxbury, personal communication, 1979) combined with the volume estimate (5000 km^3) given above yields a mean density of 2.0 (± 0.5) gm/cm^3 (Table 2).

The mass of Deimos was determined from perturbations on Viking Orbiter 2 during its series of close flybys (Duxbury & Veverka 1978). While these data are still being analyzed, the best current estimate for

Table 2 Bulk properties of martian satellites

Property	Phobos	Deimos
Axes (radii)		
a (km)	13.5	7.5
b	10.5	6.0
c	9.0	5.0
Mass (10^{18} gm)	9.6 ± 2.0	2.0 ± 0.7
Volume (km^3)		
Duxbury[a]	5200 ± 600	1300 ± 300
Thomas[b]	5000 ± 500	1000 ± 150
Density (gm/cm^3)		
Duxbury	1.9 ± 0.5	1.5 ± 0.6
Thomas	2.0 ± 0.5	1.9 ± 0.7

[a] Volume of model ellipsoid (T. C. Duxbury, personal communication, 1979).
[b] Volume allowing for departures from model ellipsoid (P. Thomas, personal communication, 1979).

the mass of Deimos is $2.0(\pm 0.7) \times 10^{18}$ gm, which corresponds to a formal mean density of 1.9 (± 0.7) gm/cm^3 (Table 2). The implication of these mean densities for the compositions of the satellites is discussed below.

GEOMETRIC ALBEDO Presently, the best values of the geometric albedos are derived by combining spacecraft data on the absolute sizes of Phobos and Deimos with apparent magnitudes (reduced to mean opposition) obtained at the telescope (Veverka 1977a). In the case of Deimos, an independent indirect confirmation of the albedo exists from telescopic polarization measurements (Zellner 1972; Section 9 below).

Telescopic observations of the brightnesses of Phobos and Deimos are reviewed by Veverka (1977a). Perhaps the best available values for the mean opposition magnitudes are those of Zellner & Capen (1974) who find $V = 11.4 \pm 0.2$ for Phobos, and 12.45 ± 0.05 for Deimos. Combining these with the most recent estimates of the dimensions (Table 2), we find $p = 0.05 \pm 0.01$ and 0.06 ± 0.01 for the geometric albedos of Phobos and Deimos, respectively. These values refer to the *V* filter of the *UBV* system (Harris 1961) for which the effective wavelength is about 0.56 μm. In this calculation, we have assumed that the telescopic observations were made near elongation so that the $A \times C$ cross section of the satellites was being observed.

Relying on absolute photometry from Viking Orbiter frames Klaasen et al (1979) found albedos of 0.066 for Phobos and of 0.069 for Deimos near 0.56 μm. However, absolute photometry of the Viking Orbiter cameras is somewhat uncertain and these values should be treated with caution. Nevertheless, the results are consistent with 1.15 ± 0.10, the estimate of the relative albedos between Phobos and Deimos made by Noland & Veverka (1976) on the basis of Mariner 9 photometry. We note that both spacecraft studies were based on images in which the two satellites are resolved and therefore involve no assumptions concerning the dimensions.

Pollack et al (1978) have evaluated the geometric albedo of Phobos to be 0.05 ± 0.01, based on observations of the satellite from the surface of Mars using the Viking Lander cameras. Pang et al (1980) have obtained an independent estimate of 0.05 determined from brightness observations using the Canopus tracker on Mariner 9. Both of these studies require independent knowledge of the dimensions. Finally, telescopic observations of the polarization of Deimos (Zellner 1972) can be used to obtain an indirect value of the geometric albedo. This estimate depends on a calibration of the "polarization-albedo" relation (Bowell & Zellner 1974, Zellner et al 1977, Veverka 1977b) and comes out to be about 0.06 using the latest calibration of the relationship.

All of these measurements have been critically reviewed by French (1980) who concludes that the best estimates of the geometric albedos are 0.05 ± 0.01 and 0.06 ± 0.01 for Phobos and Deimos, respectively, where the error bars reflect uncertainties in both the brightnesses and the dimensions.

High resolution Mariner 9 and Viking Orbiter images show little or no albedo variation on Phobos. However, on Deimos there are conspicuous patches of relatively brighter material which cover about 10% of the surface area (Thomas & Veverka 1980). This material is about 30% brighter than the average background (Noland & Veverka 1977b, French 1980). Certain images of Phobos obtained at large phase angles show relatively dark markings on the floors of some craters. Goguen et al (1978) have shown that this appearance is due to a difference in photometric functions (and therefore in texture) and not of albedos—the markings are inconspicuous at small phase angles.

COLORS AND SPECTRAL REFLECTANCE CURVES Phobos and Deimos are both grey throughout the visible portion of the spectrum. Viking Orbiter images taken through various filters between 0.4 and 0.6 μm (Figure 1) show little, if any, color variation on the surfaces (to accuracies of $\pm 5\%$). Within the accuracy of the Viking photometry ($\pm 5\%$) the relatively brighter material on Deimos has the same color as the background between 0.4 and 0.6 μm. Telescopic measurements of the B–V colors exist for both satellites and a value of U–B is available for Deimos

(Zellner & Capen 1974). For Phobos, the *B–V* measurement of +0.65 as well as the low albedo are consistent with the characteristics of C-objects in the asteroid classification of Bowell et al (1978). Even though Deimos' *U–B* value is somewhat low, Deimos can be marginally considered a C-object (Veverka & Thomas 1979).

The Mariner 9 ultraviolet spectrometer made a number of observations of Phobos and Deimos between 0.20 and 0.35 μm. These spectra have been reduced and combined with other observations by Pang and his co-workers to arrive at approximate spectral reflectance curves for the satellite between about 0.2 and 0.6 μm. For Phobos they found a curve which is generally similar to those of Type I and II carbonaceous chondrites (Pang et al 1978, 1980) and which, in the range of overlap (0.35–0.60 μm), matches that of the asteroid Ceres. Supported partly by arguments based on near-infrared spectra which suggest that the surface material on Ceres may be similar to that in Type II carbonaceous chondrites (Lebofsky 1978), Pang et al (1980) interpret their spectral data to mean that Phobos' surface has a carbonaceous chondrite composition. The same suggestion was put forth by Pollack et al (1978) on the basis of color measurements of Phobos between 0.4 and 0.9 μm made by the Viking I Lander cameras.

No Viking Lander color measurements of Deimos exist. However, Pang et al (1980) have combined the Mariner 9 UVS (ultraviolet

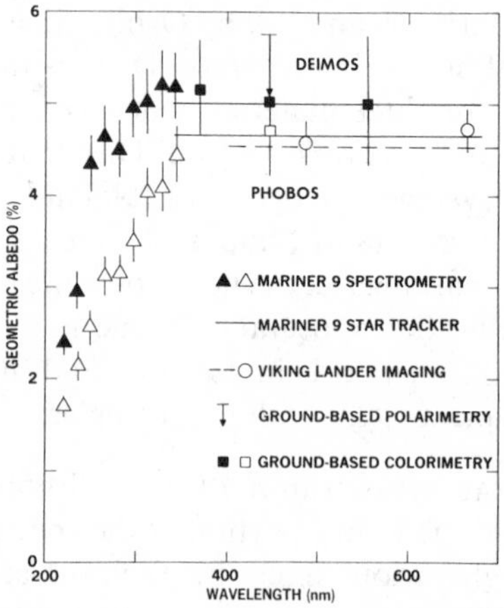

Figure 1 Spectral reflectance curves of Phobos and Deimos according to Pang et al (1980). While there is general agreement about the shapes of these curves, the absolute values of the geometric albedos quoted by other authors tend to be somewhat higher. Values of 0.05±0.1 for Phobos, and 0.06±0.01 for Deimos near 0.56 μm, seem appropriate.

spectrometer) data between 0.25 and 0.35 μm with some broadband measurements extending out to about 0.6 μm (telescopic *U*, *B*, *V*, Mariner 9 Star Tracker data, etc) to construct a spectral reflectance curve that is generally similar to that of Phobos (Figure 1).

They suggest that Deimos has a composition like that of Phobos—specifically that it consists of a material similar to those in Type I or Type II carbonaceous chondrites.

PHOTOMETRIC PROPERTIES AND PHASE CURVES From a preliminary reduction of Mariner 9 data, Pollack et al (1972) concluded that the phase curve of Phobos was generally comparable to that of the Moon, as would be expected for a body covered by a low albedo regolith having an intricate texture like that of the outer layer of our Moon.

A thorough analysis of the Mariner 9 imaging data by Noland & Veverka (1976, 1977a,b) expanded on this conclusion and extended it to Deimos. These spacecraft observations covered a range of phase angles between 20° and 90° and thus did not include the region near zero phase at which the opposition effect would be expected. Noland & Veverka demonstrated that a lunar type photometric law was a suitable first order function for both satellites—again an indication that both surfaces consist of dark complex regoliths; they also specifically looked for, but did not find, enhanced brightness in the near specular direction which might be expected if significant exposures of uncomminuted rock were present. Noland & Veverka computed values for the phase coefficients of the surface materials to be 0.020 and 0.017 mag/deg, respectively; the phase coefficients for the whole objects (including shape effects) were 0.033 and 0.030 mag/deg. These values are in reasonable agreement with some telescopic measurements (e.g. Zellner & Capen 1974); see Veverka (1977a) for a review. The intrinsic phase coefficients of the two satellites bracket those of the Moon, a fact that can be interpreted to mean that Deimos is a little smoother than the Moon, but that Phobos is rougher. (See also above section on albedo.) Morphological reasons for this difference are discussed by Veverka & Thomas (1979). Pang et al (1980) have reduced some Mariner 9 star tracker observations of Phobos to derive a photometric function between phase angles of 72° and 98°. Their results are generally consistent with those of Noland & Veverka (1976).

The most complete recent data on the photometric behavior of the surface materials of Phobos and Deimos are those of Klaasen et al (1979), who analyzed Viking observations between phase angles of 2° and 120°. They find that both photometric functions are lunarlike and essentially independent of wavelength between 0.4 and 0.6 μm. In the range of overlap with the data of Noland & Veverka (1977a,b), the agreement is satisfactory.

The best values for the intrinsic phase coefficients of the surface materials between 20° and 80° are 0.024 mag/deg for Phobos and 0.020 mag/deg for Deimos. At smaller phase angles both satellites show a near-opposition surge in brightness. The phase coefficients between 0° and 20° are 0.039 mag/deg and 0.036 mag/deg. Thus near $\alpha = 0°$, Phobos is some 0.3 mag brighter than would be predicted by extrapolating the (20°–80°) phase coefficient to zero phase; for Deimos, the increment is similar.

Values of the phase integral q were first derived by Noland & Veverka (1976) from Mariner 9 observations; they found $q \sim 0.5$, like that of the Moon. More extensive coverage (Klaasen et al 1979), at both smaller and larger phase angles, makes it clear that the phase curves are actually steeper than suspected by Noland & Veverka. The best current values of q are 0.27 and 0.32, extremely low values! An implication of these very low values is that the Bond albedos ($A_B = p \cdot q$) are of the order of only 1–2% for the satellites; for comparison the lunar value is about 7%.

POLARIZATION High quality telescopic measurements of the linear polarization of light scattered from Deimos were performed by Zellner (1972). The polarization curve between phase angles of 5° and 30° shows a well-developed negative branch (Dollfus 1961, Veverka 1977b) whose characteristics are consistent with those expected of a dark surface having the intricate texture of a regolith. As mentioned above, an estimate of the geometric albedo (0.06) can be made from the slope of the positive branch of the polarization curve.

Telescopic polarization measurements of Phobos are difficult due to its proximity to the planet. Several sequences of polarization measurements using filters and a vidicon detector were made of Phobos during the Mariner 9 mission (Noland et al 1973) with a precision considerably inferior to telescopic measurements. Nevertheless, the conclusion that, between 70° and 90°, the polarization of Phobos is higher than that extrapolated for Deimos from Zellner's measurements is probably valid. These polarization results are in agreement with the tendency for the surface of Phobos to appear both darker and rougher than that of Deimos.

THERMOPHYSICAL PROPERTIES Radiometric observations of Phobos at 10 and 20 μm by the Mariner 9 infrared spectrometer (Gatley et al 1974) showed that the surface is covered by a material whose thermal conductivity is extremely low. Model calculations show that the thermal diffusivity of this layer is about 10^{-6} cal cm^{-1} s^{-1} K^{-1} and the layer is at least a few millimeters thick. These properties are explained best in terms of a low density particulate layer whose texture is similar to that of the outer layers of the lunar regolith.

The extensive infrared observations of Phobos and Deimos made by the Viking Orbiter Infrared Thermal Mapper are currently being analyzed.

Surface Morphology

INFERENCES ABOUT SURFACES The physical properties described above suggest that both satellites are covered by regoliths. In this context, the term regolith is used to describe a particulate, probably fragmented surface layer, possibly, but not necessarily, derived by impact comminution of surface materials. While all of the physical measurements demonstrate the presence of a particulate surface layer, none can place a useful constraint on its depth.

From the inferred cratering history of Phobos and Deimos, one can show (Thomas & Veverka 1980)—if all ejecta products are retained—that at least 100 m of regolith have been produced on Phobos by impacts. The corresponding lower limit for Deimos is about 20 m. On Phobos, the regolith seems to be laterally homogeneous over the satellite on scales larger than 200 m. On Deimos, prominent lateral inhomogeneities associated with brighter patches and streamers are observed in Figure 2 (Noland & Veverka 1976, Thomas & Veverka 1980; see below).

There is some evidence that the regoliths on both satellites are layered (Veverka & Thomas 1979). Regolith depth on Phobos has been estimated from the morphology of the grooves that cover the surface of the satellite (see below) to average some 100–200 m (Thomas et al 1979a), an estimate consistent with that based on the morphology of craters on the satellite (Veverka & Thomas 1979). Many craters on Deimos are filled in with 10–20 m of sediment (Thomas & Veverka 1980), most likely derived from ejecta, providing an estimate of the minimum average depth of regolith on the outer satellite. There is strong evidence that, at least on Deimos, the regolith is thinnest near local gravitational highs (ridges and crater rims) and is thickest in local lows (Thomas & Veverka 1980). This trend can be understood from the Hill curve studies of Dobrovolskis et al (1979) described below.

The appreciable thicknesses of the regoliths on both satellites suggest that an important fraction of crater ejecta are retained or re-accreted by the satellites.

GENERAL APPEARANCE OF THE SURFACES The present irregular shapes of the satellites and the heavily cratered appearance of the surfaces no doubt result from a long history of impact cratering, the most severe segment of which probably occurred during the first 500–600 million years of the solar system's existence. The three largest craters on Phobos are

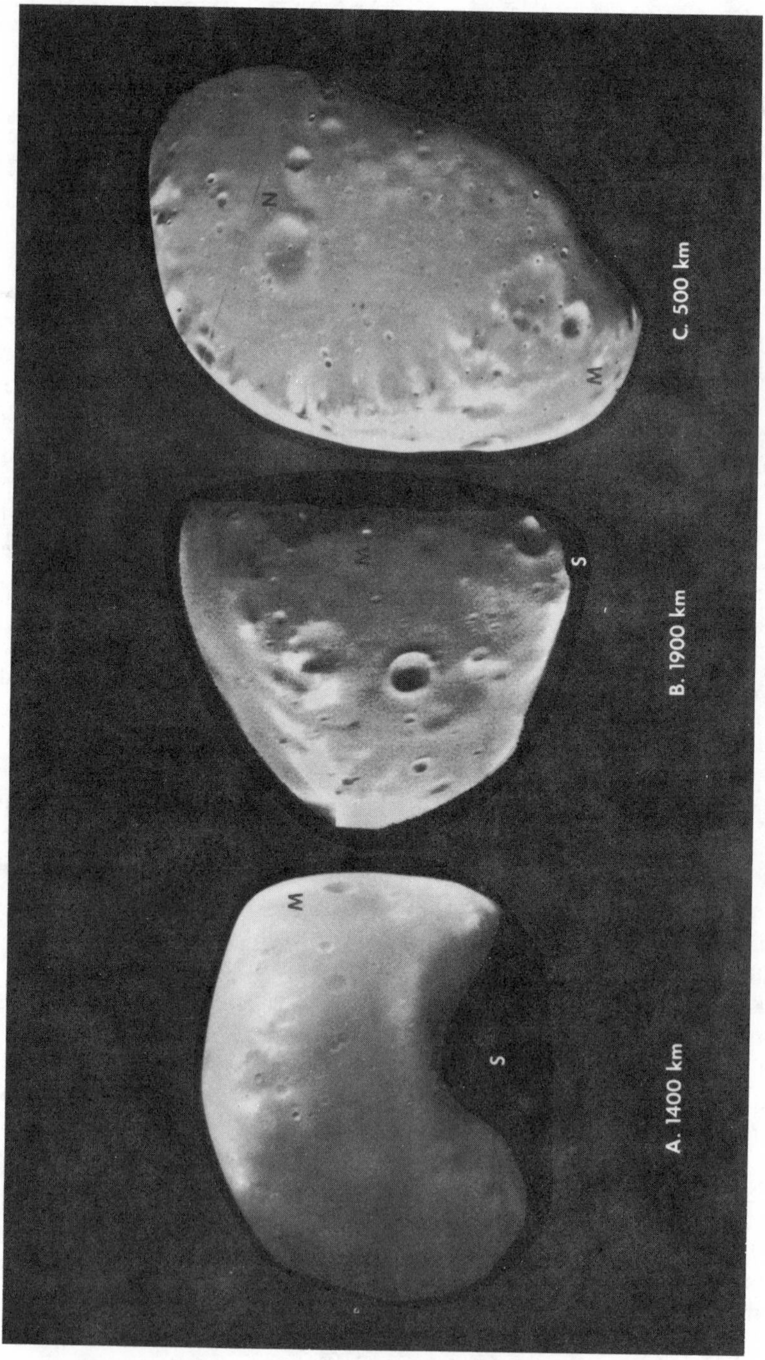

Figure 2 Three views of Deimos obtained by Viking Orbiter 2 (from Duxbury & Veverka 1978).

Stickney (diameter = 10 km), Hall (5 km), and Roche (5 km). The largest distinct crater on Deimos is less than 3 km across.

While both satellites are heavily cratered (see next section), Deimos has a distinctly smoother appearance. It is now known that this difference is real and is due to the fact that most craters on Deimos, unlike those on Phobos, are filled with sediment.

The freshest craters on Phobos are bowl-shaped and have morphologies similar to those of small lunar craters, with depths some 20% of the diameter, and with distinct raised rims (Thomas 1978). The satellites are too small to show any craters with central peaks or rings (Hartmann 1972, 1973, Veverka & Thomas 1979). The craters that are seen on the martian moons do not display any systematic change in morphology with size that could be interpreted as evidence of a harder substrate (Oberbeck & Quaide 1967).

No secondary craters have been identified positively on either satellite, although decameter blocks, probably emplaced as ejecta, are seen on both surfaces (Thomas 1978, Veverka & Thomas 1979).

CRATER DENSITIES Crater densities on Phobos and Deimos are similar, and are comparable to those in the lunar highlands (extrapolated to small

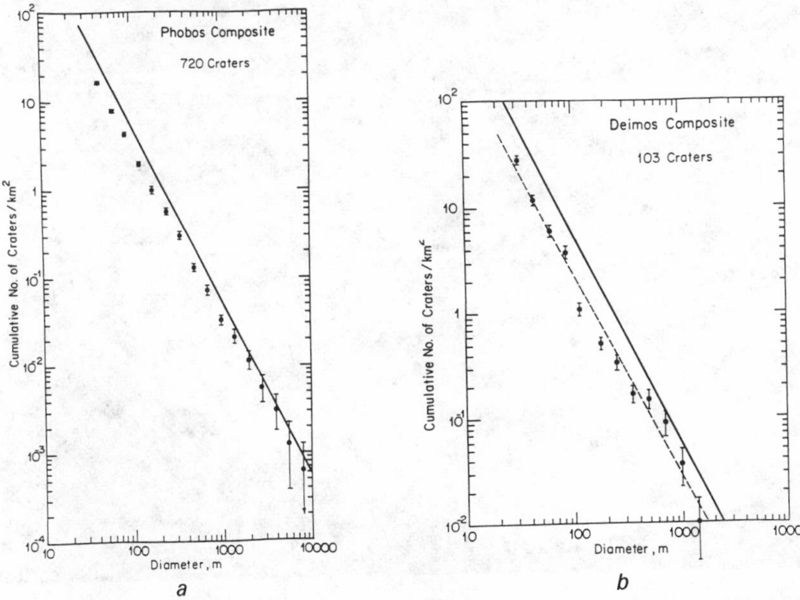

Figure 3 (*a*) Cumulative crater density on Phobos (from Thomas et al 1979b). The solid line is an extrapolation of the lunar highland crater densities from Hartmann (1973). (*b*) Cumulative crater density on Deimos. The dashed line is a linear fit to the data points; the solid line is the lunar curve as in 3*a*.

crater sizes from the data of Hartmann 1973; see Figure 3). Thus, both surfaces are very heavily cratered. There are no large areas on either satellite for which crater densities differ by more than a factor of three (Thomas 1978, Thomas et al 1979b). This evidence and the fact that large craters are uniformly distributed on both satellites suggest that no large scale spallation events have occurred in the recent cratering history. The derivation of absolute ages is not possible in any rigorous manner at present (Thomas et al 1979b) because we do not know the history of the impacting flux at Mars, or the degree to which some of the smaller craters might be caused by re-accreting ejecta from large craters, etc. The best guess is that the observed surfaces are at least a few billion years old, and may well date back to the heavy bombardment that ended some 4 billion years ago (Thomas et al 1979b).

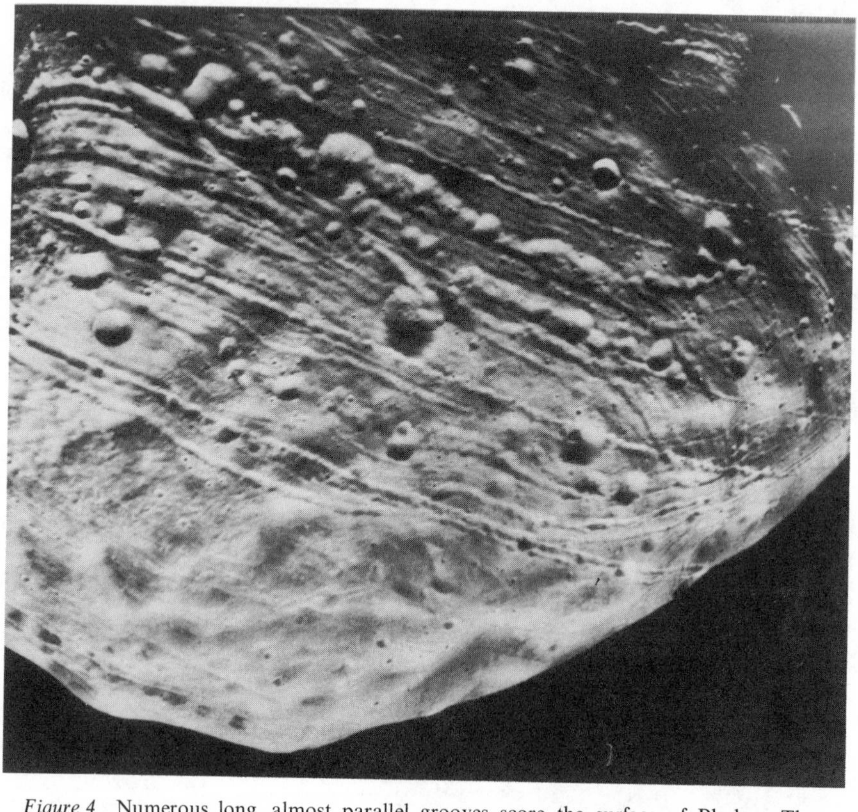

Figure 4 Numerous long, almost parallel grooves score the surface of Phobos. They are typically 100 to 200 m wide and 20 or so m deep. Viking 1 image about 12 km from left to right. Stickney is out of the picture at upper left. From Thomas et al (1979a).

Although absolute ages remain uncertain, crater densities can be used to establish the relative ages of various surface features. As already mentioned, there are no large areas on either satellite that are substantially younger or older than the average surfaces. Specifically, since one sees a significant number of small impact craters even on the floors of the grooves (Thomas et al 1978), these features are quite old. Thomas et al (1979a) estimate that the grooves are as old as the crater Stickney and date back to the first half of the cratering history recorded on Phobos. No convincing case can be made that the various sets of grooves that occur on the surface of the inner satellite differ significantly in age.

GROOVES ON PHOBOS A major surprise of the Viking mission was the discovery of long, linear depressions (Figure 4) on Phobos, initially called "striations" (Duxbury & Veverka 1977, Veverka & Duxbury 1977), but later termed "grooves" by Thomas (1978). These linear depressions cover the surface of Phobos (Figure 5) and are partially composed of coalesced pits in a regolith at least 100–200 m deep. They are typically 100–200 m wide and 10–20 m deep. The longest can be traced about halfway around Phobos, i.e. for about 30 km across its surface. The paths of the grooves cut the surface along sets of parallel planes (Figure 6*a*); about half a dozen prominent families of subparallel grooves can be identified (Thomas et al 1979a). The best developed families tend to have plane normals that lie in the *A–C* plane of Phobos.

The morphology of the grooves is strongly correlated with distance from the crater Stickney (Figure 6*b*). The deepest and widest grooves occur near the crater; they diminish in prominence with increasing distance from Stickney and become absent on the trailing side of Phobos, roughly antipodal to Stickney (see Figure 5; Thomas et al 1978, 1979a, Thomas 1979). Based on relative crater counts and superposition arguments, Thomas et al in these two papers argue that the formation of grooves occurred only during a very restricted time interval in the history of Phobos, a period closely following the formation of Stickney. The close association of the distribution of the grooves, and of their morphology, with the location of Stickney, suggested to Thomas et al that the impact that formed Stickney probably opened fractures along pre-existing planes of weakness. These openings subsequently developed into grooves either by the drainage of regolith into the fractures, or by the ejection of regolith overlying the fractures, or by a combination of both processes. However, no theory has been developed to account for the actual pattern of fractures supposedly responsible for the grooves. The modes of fracturing of a triaxial ellipsoid under a random severe impact remain to be investigated even in the simplest approximation of a homo-

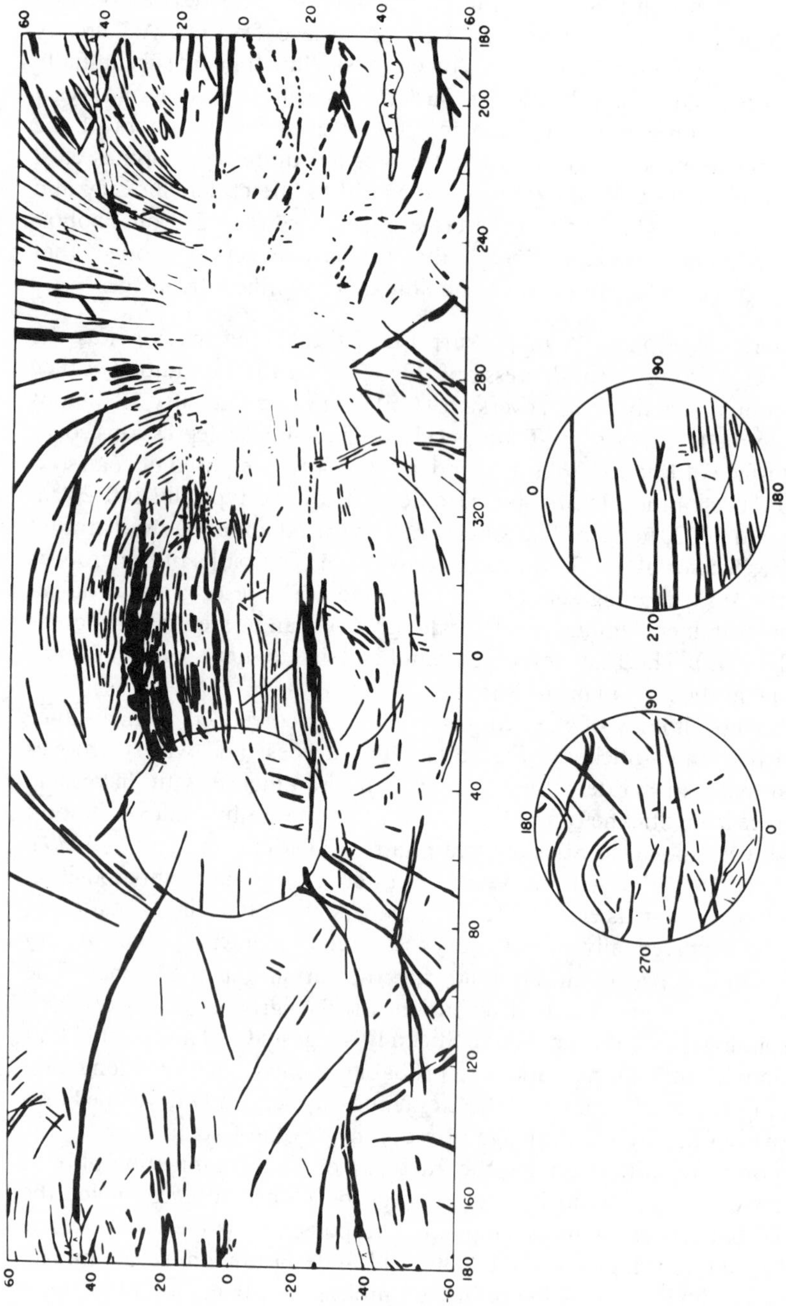

Figure 5 A sketch map of prominent grooves on Phobos. The outline of the crater Stickney is shown near (0°, 50°W). From Thomas et al (1979a).

geneous body. The more realistic case of a possibly inhomogeneous body with pre-existing planes of weakness is probably intractable. Thomas et al (1979a) argue that the fundamentally pitted nature of the Phobos grooves and the likelihood that some may have raised rims require at least a modification of the original groove morphology by the movement or ejection of regolith by outgassing. Internal heating due to the Stickney

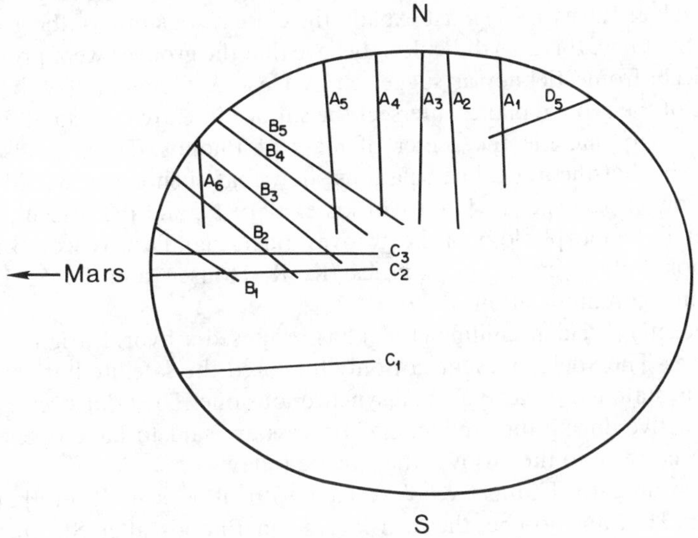

a

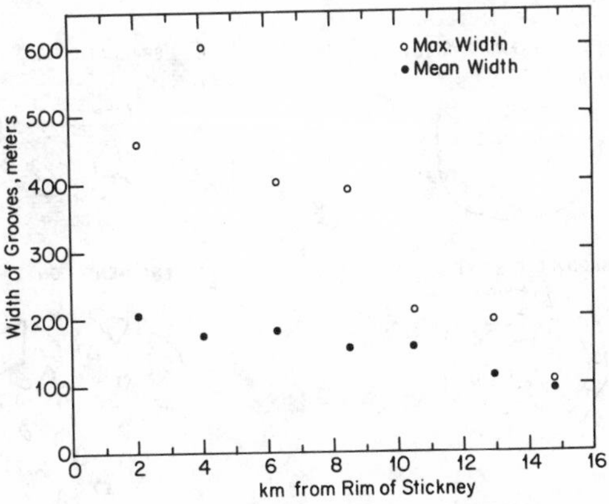

b

Figure 6 (*a*) Traces of some ot the major families of grooves on Phobos. From Thomas et al (1979a). (*b*) Dependence of groove characteristics on distance from Stickney. From Thomas et al (1979a).

impact itself may have been strong enough to produce steam from the satellite's supposed carbonaceous chondrite material; this vapor would have been released preferentially along the fractures.

Other proposed schemes of groove formation face more immediate difficulties (Thomas et al 1979a): the tidal rupturing mechanism of Soter & Harris (1977) cannot account for the relatively old age of the grooves, while any mechanism invoking fragmentation during capture (Pollack & Burns 1977) must explain the close association of the grooves with Stickney. It is also difficult to believe that the grooves were produced by ejecta from Stickney as suggested by Head & Cintala (1979) because some of the groove planes intersect the surface in extremely small circles that do not include the center of mass of Phobos. There is also the problem that theoretical calculations show that such ejecta would have very low impact speeds (Dobrovolskis et al 1979), and the difficulty that in detail the morphology of the grooves differs significantly from known examples of chains of secondaries (K. R. Housen and D. R. Davis, personal communication, 1979).

Recently, Weidenschilling (1979) has proposed a hybrid origin for the grooves. The Stickney event not only fractured the satellite but changed its spin state temporarily from a synchronous one. The tidal stresses that were active during the "re-locking" process are said to have opened up the fractures into the grooves that we see today.

According to Thomas & Veverka (1979), it is significant that the craters Hall and Roche, the next largest on Phobos after Stickney, do not have associated grooves. They review available experimental data on

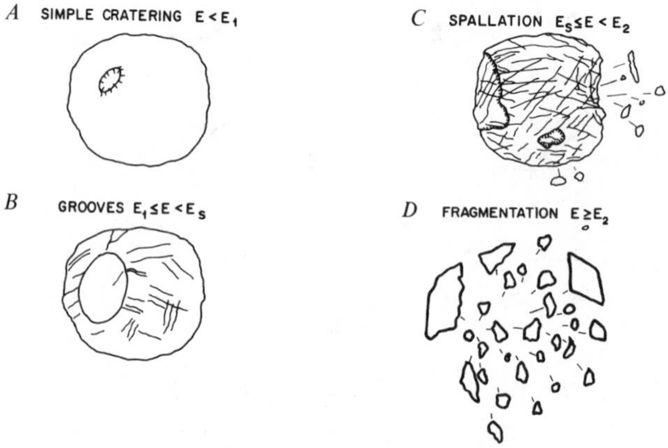

Figure 7 Schematic diagram showing the results of four kinds of impacts involving progressively higher energies. (*A*) Simple cratering. (*B*) Cratering plus groove formation. (*C*) Spallation. (*D*) Fragmentation. From Thomas & Veverka (1979).

the shattering of targets by impacts and propose a qualitative sequence of damage with increasing energy density (ergs/gm) of impact, shown in Figure 7: the formation of an isolated crater, followed by the formation of a crater and associated fracturing, followed by more severe fracturing accompanied by spallation from the antipodal point, and culminating in massive spallation (i.e. destruction) of the target. The important point is that the difference in energy density between the first and last step is less than a factor of ten for a given material. Thomas & Veverka estimate, on the basis of the data that they review, that the Stickney event involved an energy close to that needed for the destruction of Phobos. They estimate that the energy involved in the formation of Stickney was about 10^7 (ergs/gm of Phobos), and that the maximum that Phobos can withstand is about 3 times this value. The impacts that produced Hall and Roche are calculated to have involved energies of only 10^6 ergs/gm, not sufficient to enter the fracturing regime.

In the view of Thomas & Veverka (1979), the absence of grooves on Deimos is explained most readily by the fact that the largest crater on the present surface of the satellite is less than 3 km across. Thomas & Veverka estimate that a crater 5 km across would be required to produce conspicuous fracturing of the satellite. If these ideas of Thomas & Veverka have merit, grooves would occur on some asteroids. If tidal stresses play an important role in opening up fractures to form grooves, as proposed by Weidenschilling (1979), grooves could occur on other close satellites such as Amalthea, but not on asteroids.

Crater Ejecta

BRIGHTER STREAMERS AND CRATER FILL ON DEIMOS Phobos is fairly homogeneous in albedo, but on Deimos one sees brighter patches (Figure 8) which, although of roughly the same grey color, are about 30% lighter than the surroundings (Noland & Veverka 1977b). Thomas & Veverka (1980) have demonstrated that this material is moving downhill—in spite of the very low gravity of only $\sim 1\ \mathrm{cm/s^2}$ (or 10^{-3} g)—and is probably derived from crater rims and ridges near topographic highs (Figure 9). The observed brightness difference could be explained if the brighter material had a finer texture (smaller mean particle size) than the rest of the Deimos regolith (French 1980). Thermal creep and shaking due to impacts appear to be sufficient to explain the downhill movement (Thomas & Veverka 1980), but it remains unclear why no similar phenomena have been noticed on the surface of Phobos.

Small craters on Deimos tend to have easily identifiable fill, whereas those on Phobos do not (Thomas 1978). Since some of this fill occurs in craters with prominent rims whose interiors would seem to be immune from the accumulation of external sediments moving downslope, Thomas & Veverka (1980) have suggested that this fill could represent ballistically emplaced ejecta. If this view is correct, the observed amounts of fill would indicate that about one-half of the ejecta produced during the recorded cratering history of Deimos were deposited widely over its surface, either directly or after recapture from Mars orbit (see below).

Several other lines of evidence suggest that some ejecta are retained directly on both satellites. The strongest of these is the occurrence of dark halo craters (Veverka & Thomas 1979). However, the fraction of ejecta retained directly, or after recapture, must differ strongly for the two satellites, if we are to understand the radically different appearance of the two surfaces. One possible explanation is that most ejecta on Phobos did not escape and were closely localized near the parent crater, whereas, due to the smaller gravity of Deimos, ejecta on the outer satellite, while mostly retained, are not so localized and therefore form more widespread and uniform deposits. The possibility that most ejecta generated during cratering events are retained on both satellites agrees with the existence of thick regoliths on both bodies (Thomas & Veverka 1980, Veverka & Thomas 1979). This possibility is also consistent with

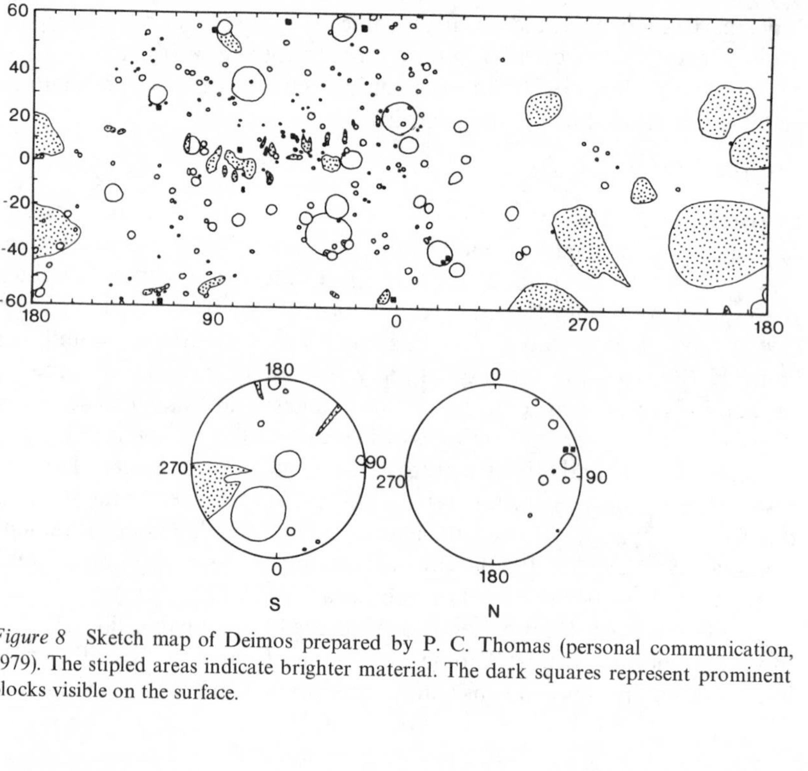

Figure 8 Sketch map of Deimos prepared by P. C. Thomas (personal communication, 1979). The stipled areas indicate brighter material. The dark squares represent prominent blocks visible on the surface.

some aspects of theoretical calculations that attempt to trace the history of ejecta following an impact on one of the satellites.

CRATERING MECHANICS ON PHOBOS AND DEIMOS The pockmarked surfaces of the martian satellites give abundant evidence that meteoroids have collided with Phobos and Deimos. What happens to the material generated by these impacts? Once the particle's ejection speed and direction is specified, the motion of a particle relative to the satellite's surface is governed by the gravitational attraction of the satellite and of Mars, as well as by the fact that the satellite lies in a noninertial reference system due to its orbital motion. Goguen & Burns (1978), along with Housen & Davis (1978), demonstrated that the nonsphericity of the satellites is important, both because of the change in the satellite's gravity

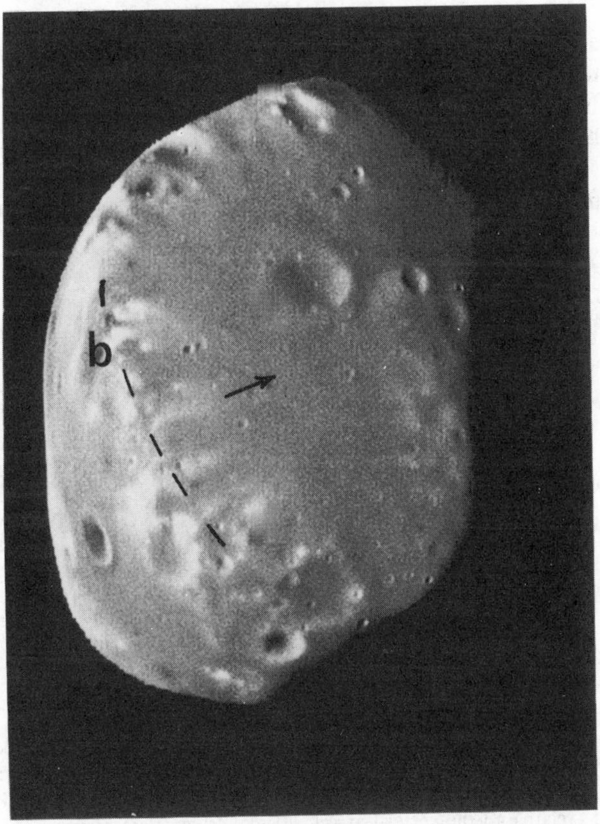

Figure 9 Downslope movement of brighter material on Deimos. The outline of a ridge is shown as is the direction of local movement. From Veverka & Thomas (1979).

and in the launch and impact points for ejecta. All investigations on the dynamics of satellite ejecta up to present at best model the satellites as triaxial ellipsoids, with axes like those given by Duxbury (1977), and use the Viking determination of satellite masses (Tolson et al 1978); deviations of the satellite surfaces from the model ellipsoidal shape are likely to influence only the local motion of debris, but such variations are difficult to handle analytically and so they have only been considered qualitatively. Housen & Davis (1978), and Dobrovolskis et al (1979) find that the ejecta behavior for velocities less than 10 m/s depends strongly on the longitude of impact, as well as on the speed and direction of ejection. Debris ejected in the retrograde sense (relative to satellite motion) generally travels farther from its source than prograde material (see Figure 10), because ejecta are retained at higher velocities when they move retrograde. However, for a *given* velocity, prograde particles travel farther. This happens especially for objects thrown farther than a few degrees of longitude because then the particle is in flight for a substantial fraction of the satellite's orbital period: hence the Coriolis effect comes into play. This asymmetry between prograde and retrograde ejecta is especially noticeable at Phobos' present position, but is less for distant Deimos, as it would have been in the past for Phobos, when tidal evolution had not yet moved it inward and when most craters were produced. The

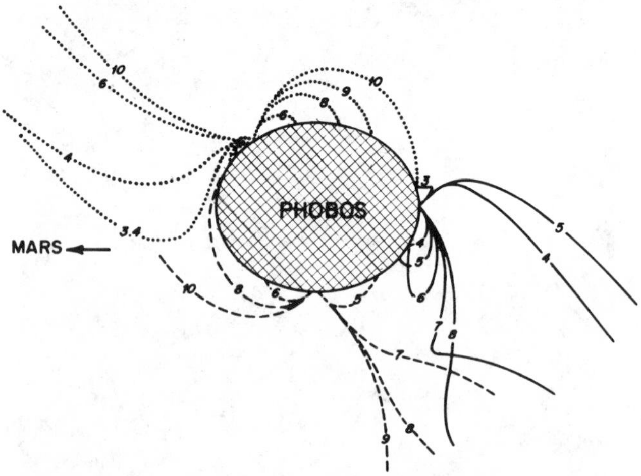

Figure 10 The motion of ejecta in the neighborhood of Phobos at its current position. The gravitational attraction of a spherical Mars and of the triaxial satellite are accounted for, as is the rotation of the coordinate system due to the satellite's orbital motion. Particles are ejected at 45° relative to the vertical at velocities of 10 and less meters per second. The distortion shown in this figure would be less for Deimos, or Phobos in its past positions. After Dobrovolskis et al (1979).

abrupt termination of the ejecta sheet on the prograde side of a crater (see Figure 10) is perhaps pertinent to the case of Stickney whose ejecta blanket seems narrowly confined on its sub-Mars side (Thomas 1978). There are few differences between debris thrown out at the sub-Mars point from that at the anti-Mars point, but these positions look considerably different from those at $\pm 90°$ in longitude from them (see Figure 10). Dobrovolskis et al (1979) compute that, in general, prograde ejecta require lower speeds to escape. They find that, for example, prograde debris from the sub-Mars point of Phobos, propelled at 45° from vertical, escapes at velocities greater than about 3 m/s while retrograde debris leaves at velocities greater than about 6 m/s; from the leading hemisphere, these velocities are 5 m/s and 8 m/s. For Deimos, the distinction is slight: for instance, at the sub-Mars point, the two speeds are about 4 m/s versus 5 m/s. These very low escape velocities make it unlikely that secondary craters will be gouged out. Many pathological orbits have been found in these numerical explorations; they include loops, cusps, "folded" ejecta blankets and even a temporary satellite of Deimos.

The problem of ejecta loss from the martian satellites can also be considered by looking at the (Hill) curves of constant pseudoenergy in a coordinate system that rotates with the satellite (Goguen & Burns 1978, Dobrovolskis, Burns & Goguen 1979). Deimos is entirely enclosed by its Roche lobe (that Hill curve which opens at the inner Lagrange point to circumplanetary space). Curiously, Deimos' mean surface lies nearly along an equipotential, which suggested to Soter & Harris (1976) that the satellites might have their shapes determined by tidal distortion. In contrast, the surface of Phobos currently overflows its Roche lobe except for regions within a few km of its sub- and anti-Mars points. This does not mean that material will necessarily leave the inner moon's surface, but merely that escape is permitted energetically; surface material is usually retained, however, because the local gravity has an inward component everywhere. Nevertheless, as material is jostled about by collisions on either satellite, it should preferentially slump so as to fill an equipotential surface; this "flow" of material is only visible on Deimos (Thomas & Veverka 1980). The loss of surface debris from satellites through the Roche lobe has recently been invoked by Dermott et al (1979), who claim the particles in the rings of Uranus come from small satellites residing within each ring.

The differences seen in the motions of ejecta near Phobos versus those near Deimos are of considerable interest. First, we expect crater ejecta blankets to be more asymmetric on Phobos. Second, particles residing on Phobos may more readily leak away. It may be this property that cleans its surface relative to Deimos. However, mitigating against this possibility is the fact that easy escape from Phobos is a recent phenomenon since the satellite's putative tidal evolution has only just brought it inside the martian Roche limit.

DUST BELTS Most ejecta that escape the satellites should end up in orbits near those of the two moons. On this basis Soter (1971) has suggested that dust belts have surrounded, or perhaps even *do* surround, Mars. The predicted spatial densities of these belts depend upon the recent cratering histories of the satellites and upon the efficiency of loss mechanisms. At present, the belts should be quite rarified. Prior to the arrival of Mariner 9 and Viking, Soter (1971) estimated that space vehicles could orbit Mars for several years before being struck by anything larger than a sand grain.

The dynamical basis of satellite dust belts is easily demonstrated: as we have noted, escape velocities from the satellites are less than 8 m/s, which is to be compared against the circular velocities of the satellites about Mars of 2.1 and 1.4 km/s for Phobos and Deimos, respectively. Escape velocities from the satellite orbits are thus more than two orders of magnitude larger than those from the satellite surfaces. Thus most particles that depart the satellites with ease have orbits about Mars not much different than those of the satellites themselves, i.e. they reside within a tube narrowly centered on the source satellite. Much of this debris should ultimately re-impact the originating moon (Soter 1971), other ejecta may strike the partner satellite, while the remainder will leave the system altogether, either impacting Mars or escaping to interplanetary space. Particles larger than a cm are influenced principally by the martian gravity field once they leave the vicinity of the satellite. Soter (private communication, 1977) has computed the mean re-encounter time for these particles, including orbital precession, as $\tau \simeq P_o v/S^2$, where P_o is the orbital period of the satellite, v is the velocity the particle has with respect to the satellite after leaving its immediate vicinity, given in units of satellite orbit velocity v_o, and $S = R/a$. For reasonable v, $\tau \sim (a^4/R^2)v_{ej}$ with v_{ej} the ejection velocity, and so the life expectancies of particles ejected from Deimos, are more than 100 times larger than those of particles leaving Phobos at the same velocity. For ejection velocities of order 10 m/s, the re-collision times are O (10^2 yr) for Phobos-generated detritus.

Radiation pressure forces can produce important perturbations for small circumplanetary particles. Burns et al (1979; see Burns & Soter 1979) derive a necessary condition for this force to increase a particle's orbital eccentricity to a large enough value so that the particle either strikes the planet or escapes the system. For these outcomes to happen, they find that β, the ratio of radiation pressure to solar gravity, must be at least 1/3 $(v/v_{\odot})$, where $v_{\odot}$ is the planet's orbital velocity. For particles in orbit about Mars, this yields $\beta \gtrsim 0.02$. Mie calculations (Burns et al 1979) show that spherical particles composed of any of eight cosmochemically significant materials having particles sizes between 0.02 and 5 μm will be ejected during a single martian year. Thus, if the grains making up the regoliths of the satellites are small and do not clump in the crater explosion, they will be quickly lost from martian orbit and not re-impact the satellites. Poynting-Robertson drag will cause larger particles orbiting Mars to collapse with a characteristic time scale of $2 \times 10^7 \rho s/Q_{pr}$ yr (Burns et al 1979, equation 55) where ρ and s are particle density and radium in cgs units, whereas Q_{pr} is the radiation pressure efficiency factor, which is of order unity for particles larger than a few microns.

Origin

Since Phobos' orbit is known to be evolving at present, the question as to what the primordial orbits of Phobos and Deimos were like should be asked before attempting to distinguish possible origins for the satellites. For circular orbits, simple integrations back in time (Burns 1972, 1977, Pollack 1977), assuming the current rates of orbital evolution are due to frequency-independent planetary tides, find that Phobos would have been much closer ($5.96\,R_{\male}$ for Shor's acceleration, but $5.52\,R_{\male}$ and $5.24\,R_{\male}$ for the slower Sinclair accelerations) to the synchronous orbit location; Deimos, on the other hand, because it is little affected by tides, would only be a bit nearer a_s. The inclusion of frequency-dependent tides, in which the lag angle is a function of tidal frequency, allows a more significant eccentricity to develop in the past for both satellites (contrast curves *a* and *b* in Figure 11*A*). The resulting elongate orbits suggested to Singer (1968, 1971) that the martian satellites might be captured asteroids; however, the time to evolve inward from highly eccentric orbits can be prohibitive, especially for Deimos. Smith & Tolson (1977; see Burns 1978) find nevertheless that the orbits of the two satellites can cross in less than the age of the solar system if $Q_{\male} < 100$; they believe that intersecting orbits require an intersatellite collision on a time scale of 10^3–10^4 years.

Lambeck (1979) generalizes the problem, pointing out that, if the satellites are carbonaceous chondritic in composition, their rigidity and Q are likely to be much less than those of Mars: hence tides induced in the satellites by Mars can have significant consequences on the orbital evolution of the satellites (compare the various panels of Figure 11). Such tides (Goldreich 1963) in general cause satellite orbits to circularize as time increases. Choosing very favorable Q's (i.e. rapid evolution time scales), Lambeck (1979) finds that Phobos could easily evolve in 4.5×10^9 years from a parabolic orbit to a nearly circular one having today's secular acceleration at its present distance. Deimos, however, remains intransigent: it is too far from the planet to be moved measurably by tides. The only way Lambeck (1979) can see to explain its present orbit, if Deimos is a captured body, would be for the satellite to have been much more massive originally in order that its orbital evolution would have proceeded more swiftly.

The inclinations of both satellites, now quite low, are very little affected by tides; as emphasized by Burns (1972, 1978), Pollack (1977), and Lambeck (1979), this places severe constraints on a direct capture from heliocentric orbit since there is little unique about the martian equatorial plane for an incoming particle on a heliocentric orbit (cf Pollack et al

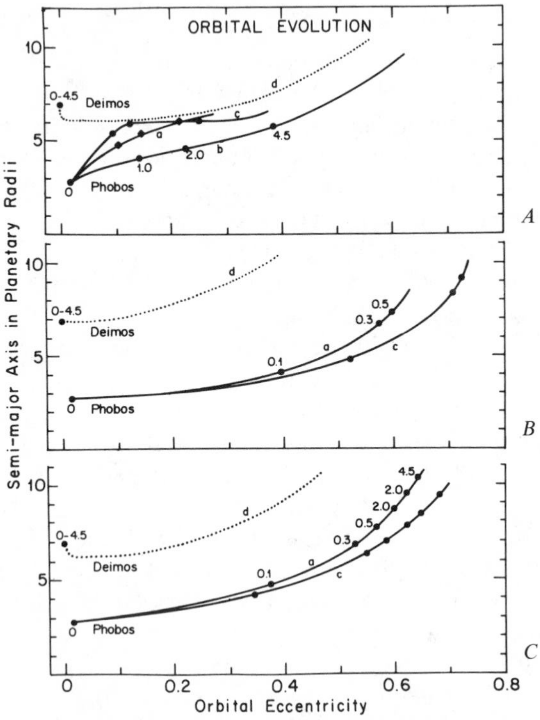

Figure 11 The orbital evolution of Phobos and Deimos due to tides. Phobos' path is represented for constant Q (curve *a*), for Q proportional to frequency ω (curve *b*), and for Q proportional to $\omega^{-1/3}$ (curve *c*); Deimos' (curve *d*) is given only for constant Q. The numbers alongside the solid circles give the time in billion years ago at which Phobos reached that point; these times can be scaled linearly for other values of Q. Deimos does not evolve noticeably from its position in 4.5 billion years. (*A*) Only martian tides ($Q = 50$) considered. (*B*) Only satellite tides ($Q = 10$) included. (*C*) Both satellite and planetary tides act. From Lambeck (1979).

1979). Thus the low inclinations seen today suggest that the moons either formed in situ or, if captured, had their orbits modified later by forces other than tides.

For completeness, we mention that the meteoroid collision that produced Stickney, if properly placed, could change Phobos' e by about 0.01 without much modifying i; if this had happened, the initial condition ($e = 0.015$) upon which Figure 11 is based may mislead us into believing that Phobos' orbit was quite elongate at one time. In addition, we point out that the significance for orbital evolution of 3:1 and 2:1 resonances between the lumpy martian gravity field, mainly due to the Tharsis bulge, and the orbital period of Phobos remains unexplored (see Burns 1978). And, lastly, we remind the reader that, for lack of a better assumption, all orbital evolution calculations take Q constant even though there is good reason to believe that there have been appreciable changes in the martian interior temperature and therefore in Q.

The evidence cited above that the surfaces of Phobos and Deimos might be like carbonaceous chondrites, material presumably formed far out in the solar system, has persuaded several investigators to seek ways the satellites could have been captured. As we have seen, direct capture into an ultimately low inclination orbit is implausible. Furthermore, the orbital evolution time scale seems difficult to satisfy, in particular for Deimos.

To capture any object from a heliocentric orbit into a planetocentric orbit requires that energy be lost in order to turn the hyperbolic path seen from the planet into an elliptic one. Indeed, this energy loss is another difficulty faced by enthusiasts of tidal capture. Even though Figure 11 suggests that Phobos' original orbit may have been highly eccentric (perhaps parabolic), if capture were to take place through tides alone it would require that Phobos' approach velocity to Mars be essentially zero since tides dissipate little energy during a single pass. Thus Phobos' heliocentric orbit would have had to match closely Mars': this is just one more low probability event.

More believable schemes to explain capture of the martian satellites invoke gas drag in order to decelerate hyperbolic planetoids as they approach Mars. The gas drag mechanism has been applied to a variety of satellite systems by Pollack et al (1977, 1979). In particular, they propose that the outer satellites of the jovian planets were captured when planetoids on heliocentric paths collided with the gaseous nebulae out of which the giant planets themselves supposedly grew. They demonstrate that, to a good approximation, the fractional decrease in particle velocity is the same as the amount (measured in units of planetoid mass) of nebular mass traversed. Computing the velocity loss for a particle that passes through a model primordial jovian nebula

(Bodenheimer 1977), Pollack et al (1979) find that solar-orbiting planetoids that intersect the gas cloud at less than several hundred planetary radii and that have radii $R \lesssim 100$ km would be acquired by Jupiter. The initial orbits are typically quite elongated and inclined relative to Jupiter but they regularize quickly upon repeated passages through the nebula; at the same time the orbits collapse at a rate $\propto R^{-1}$. The appreciable inclinations and eccentricities of the orbits of today's irregular jovian satellites, as well as their clustering into two groups, implies that the nebula must have abruptly disappeared within a few orbital periods after capture. This final stipulation can be satisfied since, at the onset of the dissociation of molecular hydrogen in the planet's interior, the jovian nebula is thought to have undergone hydrodynamic collapse from a size of hundreds of planetary radii down to tens in a few years.

Under what circumstances could a gas drag theory account for the capture of the martian satellites? First, to make capture probable, Mars should have to be enveloped by a gaseous cloud at least of mass $0.01\,M_{\male}$. Second, to slow the orbital evolution (and to have this occur preferentially near the synchronous orbit where the satellites seem to have originated) requires either (*a*) low densities at a_s or (*b*) v_n, the nebular velocity, at a_s be very near v_o, the orbital velocity of the satellite, so that the interaction is reduced. This last requirement implicates either viscous or electromagnetic forces (cf Burns 1978, Pollack et al 1979). Numerical integrations of the orbital evolution equations (Burns 1976) by Kilgore et al (1978) for particles moving through models of the martian nebula, which is supported only by gravity and pressure forces, demonstrate that the eccentricities and inclinations of satellite orbits evolve more rapidly than do their semimajor axes. However, the evolution of the latter is still rapid (days to weeks) on cosmic time scales unless the nebular density is very low, making capture itself improbable, or unless $v_o = v_n$ somewhere. The gas drag scheme in the first case also has problems in halting the evolution at distances close to today's synchronous orbit. Another approach to gas drag capture is by Hunten (1979) who sought to capture the martian satellites by collision into a protoatmosphere of Mars. Due to the relatively rarified atmospheric density, capture is feasible only for very gentle approach velocities. In order to fortuitously halt the orbital evolution before total orbital collapse, Hunten (1979) calls upon a T-Tauri wind to remove the solar nebula; this releases the external back-pressure that contained the extended protoatmosphere. To summarize both of these studies: capture is possible, but not easy.

The more traditional view of the origin of Phobos and Deimos sees the satellites forming through accretion (Goldreich 1965, Burns 1972, 1978,

Pollack 1977). They accumulate out of the disk that is thought to have surrounded each planet during the last stages of its growth. The disk is made up of particles attracted by the planet's gravity, but unable to collapse to its surface because of excess angular momentum (Safronov & Ruskol 1977); these particles flatten into a thin sheet through inelastic collisions, which damp out-of-plane motions as well as noncircular ones. The satellites then coalesce out of this disk, particles at first coagulating due to surface forces and then being drawn together through gravitational effects. This process is believed to produce satellites on nearly circular, low inclination orbits with fairly regular spacing. The accretion scenario, notwithstanding—or perhaps because of—its own indefinitiveness, seems to account well for the satellites of the giant planets and in addition, based on the regularity of the martian satellite orbits, may explain the existence of Phobos and Deimos.

The characteristics of the martian satellites are not quite what would be predicted by the purest of accretion theories but slight modifications can be made to overcome inconsistencies. The martian satellites are considerably smaller than the regular satellites of the giant planets; however, Mars itself is tiny compared to the jovian planets. The moons of Mars seem to have been formed much closer together than most regular satellites. It is possible, however, that a single large satellite of Mars originated (Hartmann et al 1975) as a single entity only to be shattered by an impact into Phobos, Deimos, and a myriad of small pieces, the latter being eliminated subsequently by radiation forces. Even though Phobos and Deimos are now on different sides of the synchronous orbit position, the current position of a_s may be deceiving since the synchronous orbit distance moved inward about 10% when Mars differentiated (Pollack & Burns 1977). Phobos' position within the synchronous orbit distance is damaging to the idea of accretion just as it was a bit of an embarrassment for the Laplace nebula theory. Questions raised about an accretion origin in view of the Viking results of low density and possible carbonaceous chondritic composition can be addressed. Harris (1977) suggests that the carbonaceous material could be merely an outer coating accreted during the later stages of formation. Moreover, he attempts to argue that the low densities of Phobos and Deimos may reflect their accumulation in a low gravity environment and not an intrinsically low density of the material in the satellites.

Conclusion

In the century since their existence became known, Phobos and Deimos have challenged our understanding of how the solar system works. As the smallest of the extraterrestrial denizens to be viewed closely, they have been invaluable in providing a testing place for theories on the relative importance of various interplanetary processes. In the next century, as planetary exploration seeks new limits, the martian moons will continue to play this role.

ACKNOWLEDGMENT

We are grateful to P. C. Thomas for providing maps of Phobos and Deimos, and for numerous discussions. This work was supported by the NASA Mass Data Analysis Program under Grants NSG7593 and NSG7547, and by the NASA Planetary Geology Program under Grant NSG7156.

Literature Cited

Bodenheimer, P. 1977. Calculations of the effect of angular momentum on the early evolution of Jupiter. *Icarus* 31:356–68

Born, G. H., Duxbury, T. C. 1975. Phobos and Deimos ephemerides from Mariner 9 TV data. *Celest. Mech.* 12:77–88

Bowell, E., Chapman, C. R., Gradie, J. C., Morrison, D., Zellner, B. 1978. Taxonomy of asteroids. *Icarus* 35:313–35

Bowell, E., Zellner, B. 1974. Polarizations of asteroids and satellites. In *Planets, Stars and Nebulae*, ed. T. Gehrels, pp. 381–404. Tucson: Univ. Arizona Press

Burns, J. A. 1972. Dynamical characteristics of Phobos and Deimos. *Rev. Geophys. Space Phys.* 10:463–82

Burns, J. A. 1976. An elementary derivation of the perturbation equations of celestial mechanics. *Am. J. Phys.* 44:944–49. See also *Am. J. Phys.* 45:1230

Burns, J. A. 1977. Orbital evolution. In *Planetary Satellites*, ed. J. A. Burns, pp. 113–56 Tucson: Univ. Arizona Press

Burns, J. A. 1978. The dynamical evolution and origin of the martian moons. *Vistas Astron.* 22:193–210

Burns, J. A., Soter, S. L. 1979. On the loss of particles from planetary rings. *Bull. Am. Astron. Soc.* 11:557–58

Burns, J. A., Lamy, P. L., Soter, S. L. 1979. Radiation forces on small particles in the solar system. *Icarus* 40:1–48

Christensen, E. J., Born, G. H., Hildebrand, C. E., Williams, B. G. 1977. The mass of Phobos from Viking flybys. *Geophys. Res. Lett.* 11:555–57

Dermott, S. F., Gold, T., Sinclair, A. T. 1979. The rings of Uranus: Nature and origin. *Astron. J.* 84:1225–34

Dobrovolskis, A. R., Burns, J. A., Goguen, J. 1979. Life near the Roche limit: Unusual behavior of ejecta from Phobos, Deimos and Amalthea. *Bull. Am. Astron. Soc.* 11:595

Dollfus, A. 1961. Polarization studies of planets. In *Planets and Satellites*, ed. G. P. Kuiper, B. M. Middlehurst, pp. 343–99. Chicago: Univ. Chicago Press

Duxbury, T. C. 1974. Phobos: Control network analysis. *Icarus* 23:290–99

Duxbury, T. C. 1977. Phobos and Deimos: Geodesy. In *Planetary Satellites*, ed. J. A. Burns, pp. 346–62. Tucson: Univ. Arizona Press.

Duxbury, T. C., Veverka, J. 1977. Viking imaging of Phobos and Deimos: An overview of the primary mission. *J. Geophys. Res.* 82:4203–11

Duxbury, T. C., Veverka, J. 1978. Deimos encounter by Viking: Preliminary imaging results. *Science* 201:812–14

French, L. 1980. *Photometric Properties of Carbonaceous Chondrites and Related Material.* PhD thesis. Cornell University, Ithaca, NY

Gatley, I., Kieffer, H., Miner, E., Neugebauer, G. 1974. Infrared observations of Phobos from Mariner 9. *Astrophys. J.* 190:497–503

Gingerich, O. 1970. The satellites of Mars: Prediction and discovery. *J. Hist. Astron.* 1:109–115

Goguen, J., Burns, J. A. 1978. Escape from the martian satellites. *Bull. Am. Astron. Soc.* 10:593

Goguen, J., Veverka, J., Thomas, P., Duxbury, T. C. 1978. Phobos: Photometry and origin of dark markings on crater floors. *Geophys. Res. Lett.* 5: 981–84

Goldreich, P. 1963. On the eccentricity of satellite orbits in the solar system. *Mon. Not. R. Astron. Soc.* 126:257–68

Goldreich, P. 1965. Inclination of satellite

orbits about an oblate precessing planet. *Astron. J.* 70:5–9

Hall, A. 1878. *Observations and orbits of the satellites of Mars with data for ephemerides in 1879.* Washington DC: US Gov't. Print. Off.

Harris, A. W. 1977. On the origin of the satellites of Mars. *Bull. Am. Astron. Soc.* 9:519

Harris, D. L. 1961. Photometry and colorimetry of planets and satellites. In *Planets and Satellites,* ed. G. P. Kuiper, B. M. Middlehurst, pp. 272–342. Chicago: Univ. Chicago Press

Hartmann, W. K. 1972. Interplanet variations in scale of crater morphology—Earth, Mars, Moon. *Icarus* 17:707–13

Hartmann, W. K. 1973. Martian cratering, 4. Mariner 9 initial analysis of cratering chronology. *J. Geophys. Res.* 78:4096–4116

Hartmann, W. K., Davis, D. R., Chapman, C. R., Soter, S., Greenberg, R. 1975. Mars: Satellite origin and angular momentum. *Icarus* 25:588–94

Head, J., Cintala, M. 1979. PGPI Abstract. *NASA TM 80339,* p. 19

Housen, K., Davis, D. R. 1978. Ejecta distribution patterns on Phobos and Deimos. *Bull. Am. Astron. Soc.* 10:593–94

Hunten, D. M. 1979. Capture of Phobos and Deimos by protoatmospheric drag. *Icarus* 37:113–23

Kilgore, T. R., Burns, J. A., Pollack, J. B. 1978. Orbital evolution of "Phobos" following its "capture." *Bull. Am. Astron. Soc.* 10:593

Klaasen, K., Duxbury, T. C., Veverka, J. 1979. Viking photometry of Phobos and Deimos. *J. Geophys. Res.* In press

Lambeck, K. 1979. On the orbital evolution of the martian satellites. *J. Geophys. Res.* 84:5651–58

Lebofsky, L. 1978. Asteroid 1 Ceres: Evidence for water of hydration. *Mon. Not. R. Astron. Soc.* 182:17P–21P

Noland, M., Veverka, J. 1976. Photometric functions of Phobos and Deimos. I. Disc-integrated photometry. *Icarus* 28:405–14

Noland, M., Veverka, J. 1977a. Photometric functions of Phobos and Deimos. II. Surface photometry of Phobos. *Icarus* 30:200–11

Noland, M., Veverka, J. 1977b. Photometric functions of Phobos and Deimos. III. Surface photometry of Deimos. *Icarus* 30:212–23

Noland, M., Veverka, J., Pollack, J. B. 1973. Mariner 9 polarimetry of Phobos and Deimos. *Icarus* 20:490–502

Oberbeck, V. R., Quaide, W. L. 1967. Thickness of lunar surface layer. *J. Geophys. Res.* 72:4697–4704

Pang, K. D., Pollack, J. B., Veverka, J., Lane, A. L., Ajello, J. M. 1978. The composition of Phobos: Evidence of a carbonaceous chondrite surface from spectrum analysis. *Science* 199:64–66

Pang, K. D., Rhodes, J. W., Lane, A. L., Ajello, J. M. 1980. Deimos: Spectral evidence for a carbonaceous chondrite composition. *Nature* 283:277–78

Pascu, D. 1978. A history of the discovery and positional observation of the martian satellites: 1877–1977. *Vistas Astron.* 22:139–48

Peale, S. J. 1977. Rotation histories of the natural satellites. In *Planetary Satellites,* ed. J. A. Burns, pp. 87–112. Tucson: Univ. Arizona Press

Pickering, E. C. 1882. Satellites of Mars in 1881–82. *Ann. Harvard Coll. Obs.* 33:159–61

Pollack, J. B. 1977. Phobos and Deimos. In *Planetary Satellites,* ed. J. A. Burns, pp. 319–45. Tucson: Univ. Arizona Press

Pollack, J. B., Burns, J. A. 1977. An origin by capture of the martian satellites? *Bull. Am. Astron. Soc.* 9:518–19

Pollack, J. B., Veverka, J., Noland, M., Sagan, C., Hartmann, W. K., Duxbury, T. C., Born, G. H., Milton, D. J., Smith, B. A. 1972. Mariner 9 television observations of Phobos and Deimos. *Icarus* 17:394–407

Pollack, J. B., Veverka, J., Noland, M., Sagan, C., Duxbury, T. C., Acton, C. H. Jr., Born, G. H., Hartmann, W. K., Smith, B. A. 1973. Mariner 9 television observations of Phobos and Deimos. 2. *J. Geophys. Res.* 78:4313–26

Pollack, J. B., Burns, J. A., Tauber, M. E. 1977. Concerning the capture of the outer satellites. *Bull. Am. Astron. Soc.* 9:455

Pollack, J. B., Veverka, J., Pang, K., Colburn, D., Lane, A. L., Ajello, J. M. 1978. Multicolor observations of Phobos with the Viking Lander cameras: Evidence for a carbonaceous chondrite composition. *Science* 199:66–69

Pollack, J. B., Burns, J. A., Tauber, M. E. 1979. Gas drag in primordial circumplanetary envelopes: A mechanism for satellite capture. *Icarus* 37:587–611

Safronov, V. S., Ruskol, E. L. 1977. The accumulation of satellites. In *Planetary Satellites,* ed. J. A. Burns, pp. 501–12. Tucson: Univ. Arizona Press

Sharpless, B. P. 1945. Secular accelerations in the longitudes of the satellites of Mars. *Astron. J.* 51:185–86

Shklovskii, I. S., Sagan, C. 1966. *Intelligent Life in the Universe.* San Francisco: Holden-Day

Shor, V. A. 1975. The motions of the martian satellites. *Celest. Mech.* 12:61–75

Sinclair, A. T. 1972. The motions of the satellites of Mars. *Mon. Not. R. Astron. Soc.* 155:249–74

Sinclair, A. T. 1978. The orbits of the satellites of Mars. *Vistas Astron.* 22:133–38

Singer, S. F. 1968. The origin of the Moon and its geophysical consequences. *Geophys. J. R. Astron. Soc.* 15:205–26

Singer, S. F. 1971. The martian satellites. In *Physical Studies of Minor Planets (NASA SP-267)*, ed. T. Gehrels, pp. 399–405. Washington DC: US Gov't. Print. Off.

Smith, B. A. 1970. Phobos: Preliminary results from Mariner 7. *Science* 168:828–30

Smith, H. R., Tolson, R. H. 1977. The Q of Mars and the early orbit of Phobos. *EOS, Trans. Am. Geophys. Union* 58:1181

Smith, J. C., Born, G. H. 1976. Secular acceleration of Phobos and Q of Mars. *Icarus* 27:51–54

Soter, S. L. 1971. The Dust Belts of Mars. *Cornell Ctr. Radiophys. Space Res. Rep. 462*

Soter, S. L., Harris, A. 1976. The equilibrium shape of Phobos and other small bodies. *Icarus* 30:192–99

Soter, S. L., Harris, A. 1977. Are striations on Phobos evidence of tidal stress? *Nature* 268:421–22

Thomas, P. 1978. *The Morphology of Phobos and Deimos*. PhD thesis. Cornell University, Ithaca, NY

Thomas, P. 1979. Surface features of Phobos and Deimos. *Icarus* 40:223–43

Thomas, P., Veverka, J. 1979. Grooves on asteroids: A prediction. *Icarus* 40:395–406

Thomas, P., Veverka, J. 1980. Down-slope movement of material on Deimos. *Icarus*. In press

Thomas, P., Veverka, J., Duxbury, T. C. 1978. Origin of the grooves on Phobos. *Nature* 273:202–4

Thomas, P., Veverka, J., Bloom, A., Duxbury, T. 1979a. Grooves on Phobos: Their distribution, morphology and possible origin. *J. Geophys. Res.* In press

Thomas, P., Veverka, J., Chapman, C. R. 1979b. Crater densities on the satellites of Mars. *Icarus*. In press

Tolson, R. H., Duxbury, T. C., Born, G. H., Christensen, E. J., Diehl, R. E., Farless, D., Hildebrand, C. E., Mitchell, R. T., Molko, P. M., Morabito, L. A., Palluconi, F. D., Reichert, R. J., Taraji, H., Veverka, J., Neugebauer, G., Findlay, J. T. 1978. Preliminary results of the first Viking close encounter with Phobos. *Science* 199: 61–64

Veverka, J. 1977a. Photometry of satellite surfaces. In *Planetary Satellites*, ed. J. Burns, pp. 171–209. Tucson: Univ. Arizona Press

Veverka, J. 1977b. Polarimetry of satellite surfaces. In *Planetary Satellites*, ed. J. Burns, pp. 210–31. Tucson: Univ. Arizona Press

Veverka, J. 1978. The surfaces of Phobos and Deimos. *Vistas Astron.* 22:163–92

Veverka, J., Duxbury, T. C. 1977. Viking observations of Phobos and Deimos: Preliminary results. *J. Geophys. Res.* 82: 4213–23

Veverka, J., Thomas, P. C. 1979. Phobos and Deimos: A preview of what asteroids are like. In *Asteroids*, ed. T. Gehrels, pp. 628–50. Tucson: Univ. Arizona Press

Veverka, J., Noland, M., Sagan, C., Pollack, J. B., Quam, L., Tucker, R. B., Eross, B., Duxbury, T. C., Green, W. 1974. A Mariner 9 atlas of the moons of Mars. *Icarus* 23:206–9

Ward, W. R., Burns, J. A., Toon, O. B. 1979. Past obliquity oscillations of Mars: The role of the Tharsis uplift. *J. Geophys. Res.* 84:243–59

Weidenschilling, S. J. 1979. A possible origin for the grooves of Phobos. *Nature* 282: 697–98

Zellner, B. H. 1972. Minor planets and related objects. VIII. Deimos. *Astron. J.* 77:103–85

Zellner, B. H., Capen, R. C. 1974. Photometric properties of martian satellites. *Icarus* 23:437–44

Zellner, B. H., Leake, M., Lebertre, T., Duseaux, M., Dollfus, A. 1977. The asteroid albedo scale. I. Laboratory polarimetry of meteorites. *Proc. 8th Lunar Sci. Conf.*, pp. 1091–1110

Ann. Rev. Earth Planet. Sci. 1980. 8:559–608
Copyright © 1980 by Annual Reviews Inc. All rights reserved

REFRACTORY INCLUSIONS IN THE ALLENDE METEORITE

×10141

Lawrence Grossman[1]

Department of the Geophysical Sciences, The University of Chicago, 5734 South Ellis Avenue, Chicago, Illinois 60637

INTRODUCTION

The fall of the Allende Type 3 carbonaceous chondrite in Chihuahua, Mexico, on February 8, 1969 (King et al 1969, Clarke et al 1970) marked the beginning of a decade of remarkable discoveries which caused an unprecedented, explosive increase in our knowledge of physical and chemical processes that occurred at the birth of the solar system. Prior to its fall, studies of carbonaceous chondrites were severely limited by the rarity of these types of meteorite, their small sizes, and the relatively small sample sizes of their chips and powders obtainable from museum curators who were cognizant of the scientific value of these meteorites and their short supply. At a time when many of the world's geochemical laboratories were in close communication with one another and were perfecting their analytical techniques to derive the maximum information from a minimum of sample in preparation for the return of the first lunar rocks only six months thence, there were suddenly several tons of a single carbonaceous chondrite available for study. Large amounts of material were collected within a few days of the fall, before terrestrial weathering could cause significant degradation of it (Levy et al 1970), and were distributed widely within the scientific community shortly thereafter (Rancitelli et al 1969). So much material was then available that Morgan et al (1969) proposed that a large amount of homogenized powder be prepared and disseminated as a sorely needed standard for interlaboratory comparison of meteorite analyses. This suggestion was promptly taken up by the Smithsonian Institution (Mason 1975) and it is hoped that the compilation of papers (Jarosewich & Clarke 1980) presenting analytical results on 4 kg of such material will soon be published.

Individual stones weighing several kg each were cut into thin slices

[1] Also Enrico Fermi Institute, University of Chicago.

0084-6597/80/0515-0559$01.00

(Clarke et al 1970, Grossman & Ganapathy 1975), thereby exposing to view for the first time hundreds of cm^2 of slab surfaces of a carbonaceous chondrite. Such a view is shown in Figure 1, from which Allende is seen to consist of a heterogeneous mixture of different types of inclusion, distinguishable from one another here by their different sizes, shapes, and colors, embedded in a fine-grained, black matrix. Studies of individual inclusions demonstrate that these different inclusion types have different mineralogical and chemical compositions and, therefore, their own independent histories. Except for its size, Allende does not appear to be unusual among other Type 3 carbonaceous chondrites. The heterogeneity of these meteorites and the fact that each inclusion type has its own origin are important reasons why many earlier studies of carbonaceous chondrites based on analyses of small, bulk samples had such a relatively low information return. In the case of Allende, major advances were made because the large amounts of material available permitted access to and sampling and

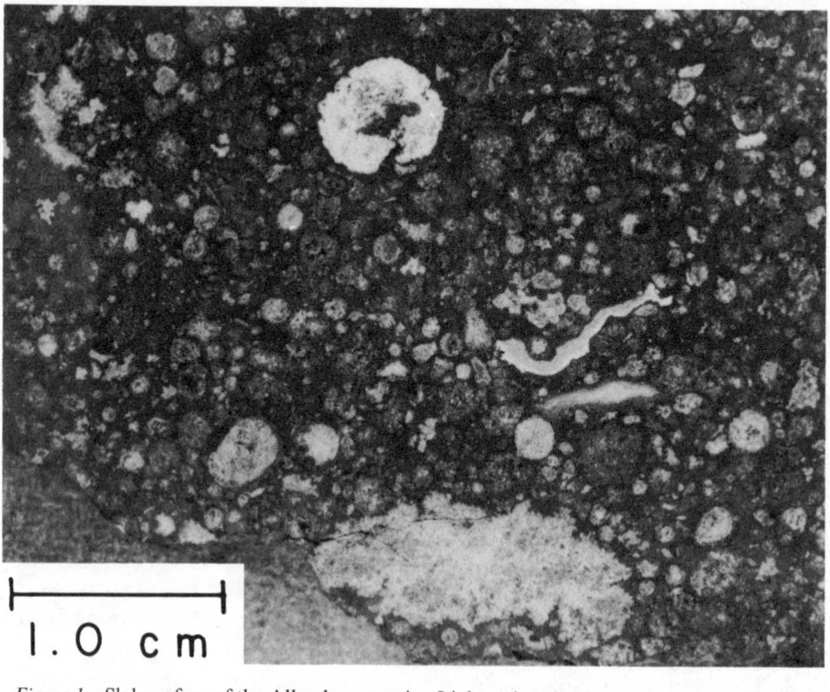

Figure 1 Slab surface of the Allende meteorite. Light-colored object at top center is one of the coarse-grained inclusions which are the subject of this paper. The two narrow, elongated objects at middle right are fine-grained inclusions (Grossman et al 1975, Grossman & Ganapathy 1976a) and the large object at bottom center is an amoeboid olivine aggregate (Grossman & Steele 1976, Grossman et al 1979a). A dark clast (Fruland et al 1978) is barely visible at middle left. Most of the other light-colored objects are chondrules.

Table 1 Chemical formulae of minerals mentioned in the text

Type of compound	Mineral	Chemical formula
Oxides	Corundum	Al_2O_3
	Hibonite	$CaO \cdot 6Al_2O_3$
	Perovskite	$CaTiO_3$
	Spinel	$MgAl_2O_4$
	Thorianite	ThO_2
	Baddeleyite	ZrO_2
	Magnetite	Fe_3O_4
Silicates	Wollastonite	$CaSiO_3$
	Rhönite	$Ca_4(Mg, Al, Ti)_{12}(Si, Al)_{12}O_{40}$
Melilite Series	Gehlenite (Ge)	$Ca_2Al_2SiO_7$
	Åkermanite (Åk)	$Ca_2MgSi_2O_7$
Plagioclase Series	Anorthite (An)	$CaAl_2Si_2O_8$.
	Albite (Ab)	$NaAlSi_3O_8$
Clinopyroxenes	Diopside	$CaMgSi_2O_6$
	Hedenbergite	$CaFeSi_2O_6$
	Fassaite	$Ca(Mg, Al, Ti)(Si, Al)_2O_6$
Olivine Series	Forsterite	Mg_2SiO_4
	Fayalite	Fe_2SiO_4
Garnets	Grossular	$Ca_3Al_2Si_3O_{12}$
	Andradite	$Ca_3Fe_2Si_3O_{12}$
Feldspathoids	Nepheline	$NaAlSiO_4$
	Sodalite	$3NaAlSiO_4 \cdot NaCl$

analysis of individual, mm- to cm-sized inclusions. Often, individual inclusions were studied by a combination of different techniques in one or more laboratories. This approach required new techniques for handling and analysing samples in the μg to mg size range, many developed in the lunar sample program, because of the number of different experiments being performed on splits of small amounts of materials.

This paper reviews the literature on only those objects in Allende commonly referred to as "coarse-grained inclusions." Although these inclusions comprise little more than 5 per cent of the mass of Allende, they have been the subject of more than 85 per cent of the papers published about this meteorite. A brief review of some of the other types of Allende inclusion is given by Wark (1979). Table 1 lists the chemical formulae of minerals discussed in this review.

PETROGRAPHY AND MAJOR MINERALS

The coarse-grained inclusions are usually the largest, most prominent, white inclusions seen on slab surfaces (Figure 1). They range from <1 mm to 2 cm in size and usually contain crystals at least 0.5 mm and often up to 2–3 mm in longest dimension. They can be completely spherical, a fact which led some workers, e.g. Clarke et al (1970), Gray et al (1973) and Martin & Mason (1974), to call them chondrules, but are often highly irregular and convoluted in shape (Figure 2). Some are fragments of once-larger inclusions. Some have natural cavities containing wollastonite needles (Fuchs 1971, Grossman 1975a, Allen et al 1978). Others are doughnut-shaped, their central holes filled with minerals typical of the matrix of Allende. Some early work (Arrhenius & Alfvén 1971, Grossman 1973, Seitz & Kushiro 1974) suffered from failure to distinguish these inclusions from other Ca-rich bodies in this meteorite (Grossman & Ganapathy 1975) and more recent work (R. N. Clayton & Mayeda 1977, Wasserburg et al 1977, Lorin et al 1978, MacPherson & Grossman 1979) indicates that important information has been overlooked in considering all members of even this restricted group of inclusions to have a common origin.

In a petrographic and electron microprobe survey of a large number of coarse-grained inclusions, Grossman (1974, 1975a) found that there were basically two kinds: those with minor ($<5\%$) clinopyroxene and those with major ($>35\%$) clinopyroxene. He pointed out that pyroxene in the first kind, Type A, is relatively pure diopside ($Al_2O_3 < 9\%$, $Ti < 0.7\%$) and is restricted to thin (<50 μm) rims around the inclusions, while that in the second kind, Type B, occurs as coarse crystals in inclusion interiors and contains 15–22% Al_2O_3 and 1.8–10.8% Ti. Clarke et al (1970) were the first to call these Ti-, Al-rich clinopyroxenes "fassaite." One inclusion, called "intermediate" by Grossman (1975a), has major amounts of pyroxene intermediate in composition between these two (9–18% Al_2O_3, 0.9–2.4% Ti). Later work showed, however, that fassaite is also present in Type A inclusions, both as a minor phase in their interiors (Allen et al 1978) and in their rims (Wark & Lovering 1977a). The latter workers also found diopside in rims on Type B inclusions. Because rims are clearly distinguishable petrographically from inclusion interiors, account for such a small volume fraction of each inclusion, and probably have little genetic relation to the interiors, it now seems unwise to use compositions of rim phases as part of a classification scheme. Thus, ignoring rim phases, all primary pyroxenes in the interiors of Type A, Type B, and intermediate inclusions contain $>9\%$ Al_2O_3 and $>0.9\%$ Ti and are all classifiable as fassaites (Deer et al 1978). The major difference between Type A and Type B inclusions is still the abundance of pyroxene and, if this is the sole criterion for distinguishing inclusions, the intermediate one would now belong to Type B. Because fassaite is the only dark-colored, coarsely crystalline phase in the inclusions, the presence of abundant dark material is often sufficient to distinguish a Type B from a Type A on a slab surface.

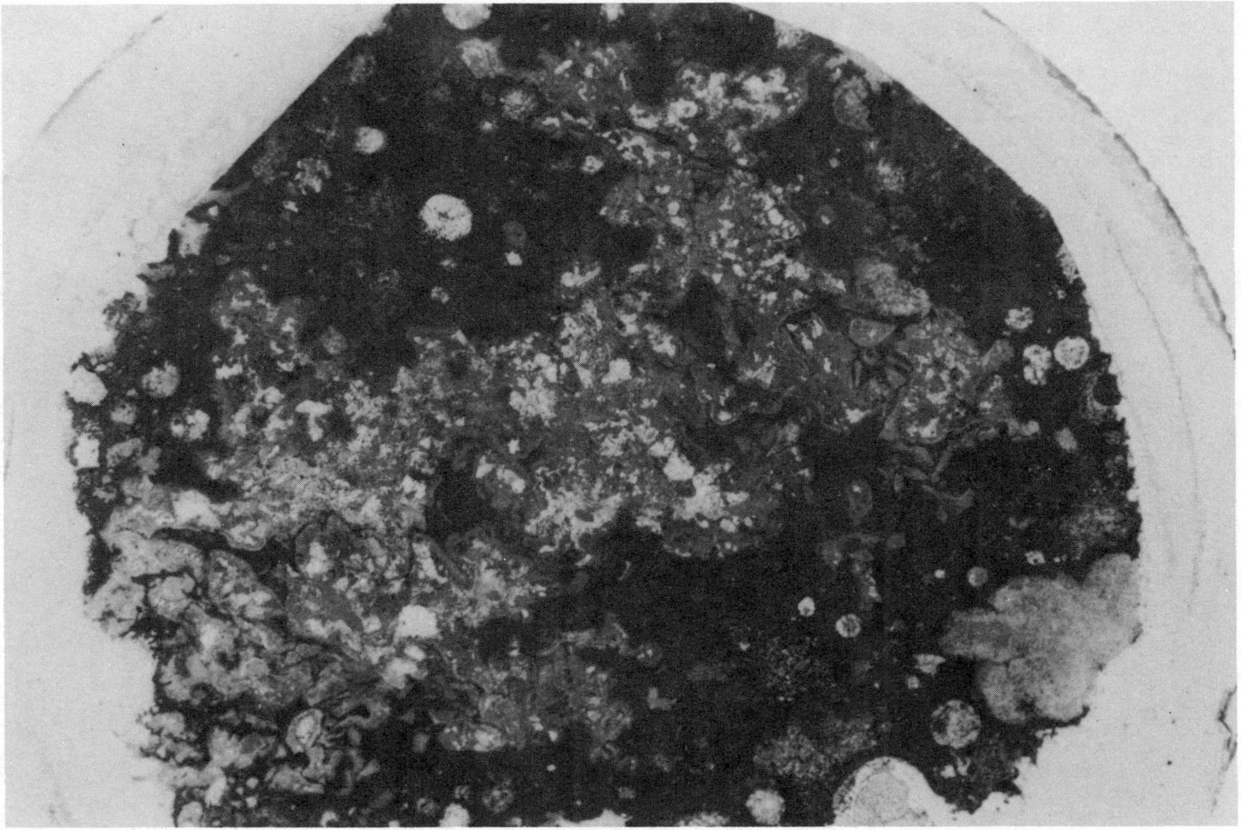

Figure 2 Fluffy Type A inclusion, TS24F1 of Grossman (1975a), consisting of melilite (white), hibonite and spinel (not visible), and fine-grained alteration products (grey). In places, highly contorted external surface of inclusion is marked by a light-colored rim, which separates the interior of the inclusion from the matrix of Allende (black). Transmitted light; width of field 2.5 cm.

Type A Inclusions

MacPherson & Grossman (1979) pointed out that two subtypes exist. Those in the most common subtype are extremely contorted in shape, are very heavily altered and contain abundant cavities (Figure 2). These we will call "fluffy." The others are more nearly spherical, less heavily altered and are more compact. Of the primary phases in both subtypes, melilite is by far the most abundant, $>75\%$ by volume. In fluffy Type A's, melilite ranges in composition from $Åk_0$ to $Åk_{33}$, but in compact ones, from $Åk_8$ to $Åk_{80}$. (As seen in Table 1, melilite refers to a solid solution series between pure gehlenite and pure åkermanite. The notation $Åk_x$ indicates a phase intermediate between these two end-members which is composed of x mole % åkermanite and $[100-x]$ mole % gehlenite.) Spinel, nearly pure $MgAl_2O_4$ in composition (Grossman 1975a), is usually the next most common phase, 5–20% by volume of the primary phase assemblage. Fluffy Type A's sometimes contain rare, orange-red, V-rich spinels (up to 5% V_2O_3, 20% FeO). Hibonite (Keil & Fuchs 1971, Haggerty & Merkel 1976, Allen et al 1978) is usually the next most abundant phase, $\sim 5\%$, followed by perovskite, ~ 1–3%. Occasionally, rhönite occurs as a minor phase in Type A's (Fuchs 1971, El Goresy et al 1977b).

From published petrographic descriptions, inclusions A and B of Fuchs (1971), 6/16 of Blander & Fuchs (1975), TS27F1 of Grossman (1975a), the inclusion in Figure 5 of Kurat (1975), and CG-11 of Allen et al (1978) are fluffy Type A's. Textural relations among primary phases are often obscured or obliterated by veins and cavities containing secondary alteration products. Where visible, most of the primary assemblage consists of interlocking masses of blocky or lath-shaped melilite crystals which can be up to 200 μm in longest dimension and often contain deformation lamellae. A wide range of objects are poikilitically enclosed by melilite (i.e. totally enclosed within single melilite crystals): rounded prisms of fassaite ($\leqq 30$ μm), spinel octahedra ($\leqq 10$ μm), rounded, elongated hibonite grains ($\leqq 10$ μm), rhönite, euhedral perovskite crystals ($\leqq 5$ μm), rounded fassaite grains ($\leqq 30$ μm) enclosing perovskite, and spinel containing and/or rimmed by perovskite ($\leqq 5$ μm). Euhedral fassaite ($\leqq 110$ μm) and ragged, elongated hibonite grains are sometimes found between melilites. Euhedral hibonite crystals, up to 250 μm long, are sometimes intergrown with and embayed by melilite and contain perovskite ($\leqq 10$ μm). A peculiar feature of CG-11 is the reversed zoning of its melilites whose compositions vary from $Åk_{17}$ in cores to $Åk_3$ in rims (Allen et al 1978). In TS27F1, a severely deformed, 500 μm melilite grain is contiguous to a polycrystalline mosaic of interlocking melilite grains (10–30 μm) and ragged spinels, suggesting shock-induced recrystallization.

Allen et al (1978) studied grain-to-grain differences in trace and minor element contents of spinels and perovskites in a fluffy Type A and found that concentration variations of a factor of six were common within each of these phases.

Grossman (1975a) described and illustrated a compact Type A, TS2F1. This type of inclusion is usually composed almost entirely of several 2–3 mm blocky crystals of melilite, poikilitically enclosing spinel, perovskite, fassaite, and rhönite. Spinel crystals are euhedral, commonly reach more than 50 μm in size, and may form chains up to 50 crystals long. Spinel, 1–6 μm perovskite cubes, purple anhedral perovskites ($\sim$70 μm), and 1–10 μm prisms of fassaite occur as isolated grains within melilite. The long axes of the fassaite prisms are aligned in three directions within single melilite crystals and one of these directions is parallel to a set of edges on the perovskite cubes. Also within melilite, perovskite grains ($\leqq$5 μm) are found at spinel-melilite contacts and wedged between spinel crystals and intergrowths of vermiform perovskite with fassaite occur, sometimes with rhönite (El Goresy et al 1977b) and/or spinel. The only mode of occurrence of hibonite in compact Type A's is as acicular crystals within 100 μm of their innermost rim layer.

The literature contains much more petrographic, as well as chemical and isotopic, information about Type B inclusions than Type A's, a fact that may lead the casual reader to the conclusion that Type B's are much more abundant. Grossman (1975a) pointed out, however, that a significant fraction of the Type A's are only a few tenths of a mm in size, are perhaps fragments of once larger inclusions, and are thus not particularly prominent on slab surfaces. Therefore, fewer thin sections are made of these inclusions and, due to their relatively small sizes, they are discriminated against in chemical and isotopic studies. The two inclusion types are probably comparable in abundance, though not in mass.

Type B Inclusions

Petrographic details and illustrations of Type B inclusions are found in Clarke et al (1970), Blander & Fuchs (1975) (inclusions 6/1 and 10/2), Kurat et al (1975) (an inclusion in Bali), R. N. Clayton et al (1977) (sample A13S4), Fuchs (1978), and El Goresy et al (1979b). Grossman (1975a) surveyed the petrography and mineral chemistry of a large number of Type B's, two of which, among others, were further studied by petrographic and electron and ion microprobe techniques by Hutcheon et al (1978a, b), Steele et al (1978), and Steele & Hutcheon (1979).

Figure 3 shows a typical Type B inclusion. Such inclusions contain 35–60% fassaite, 15–30% spinel, 5–25% plagioclase, and 5–20% melilite. Spinel, nearly pure $MgAl_2O_4$, occurs as grains from <1 μm to >100 μm in size, poikilitically enclosed by 0.1–5 mm long laths of plagioclase, An_{100}, and melilite, $Åk_{15-80}$, and by anhedral, sub-equant fassaite grains up to 3–4 mm in size. Grains of fassaite, anorthite, and shocked melilite fit snugly together, with fassaite grains often filling angular interstices between anorthite and melilite laths. Anorthite laths sometimes pierce fassaite grains. Occasionally, corrosion textures are observed which suggest that melilite has replaced anorthite and fassaite in some cases and that melilite and fassaite have replaced anorthite in others. Characteristic of many Type B's is a thick outer mantle of melilite with minor spinel which surrounds a core of fassaite, anorthite, and more Åk-rich melilite (Kurat et al 1975, R. N. Clayton et al 1977). The presence or absence of such a mantle is the basis of Wark & Lovering's (1977a) subdivision of Type B inclusions into B1's and B2's, respectively. In some B1's the outer reaches of this mantle have a zone of 1–25 μm long hibonite needles which project inward toward the center of the inclusion (Blander & Fuchs 1975,

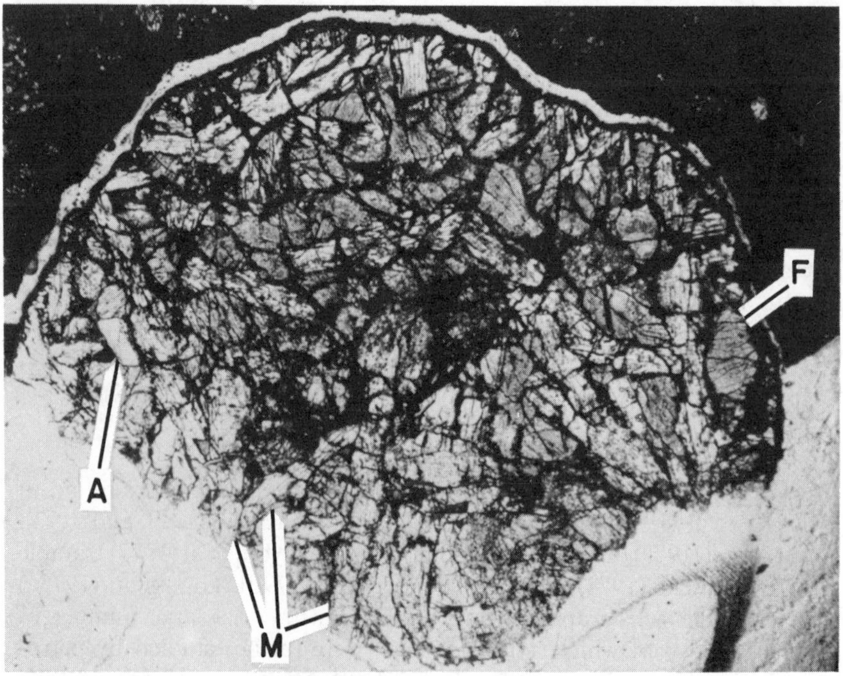

Figure 3 Portion of Type B inclusion, containing long, narrow melilite crystals (white, *M*), anhedral, sub-equant fassaites (grey, *F*), sparse, stubby laths of anorthite (white, *A*), and spinel as inclusions in these phases, especially fassaite. Thin white rim is epoxy where the inclusion has separated from the dark Allende matrix during preparation. Transmitted light; inclusion is 1.1 cm in diameter.

Wark & Lovering 1978a). MacPherson & Grossman (1979) noted that melilite laths in these mantles become progressively more Åk-rich along their lengths toward the center of the inclusion and that stubby, radially oriented laths appear to have had their growth interfered with by longer, neighboring laths growing at angles to them. Grossman (1975a) described fassaite crystals containing very dense concentrations of spinel at one end and virtually none at the other. El Goresy et al (1979b) described spinel "framboids" inside pyroxene and melilite crystals. These are spheroidal shells, up to several hundred μm in diameter, of spinel crystals which enclose fassaite, melilite, anorthite, and spinel. Wark et al (1979) also noted these, but considered "ribbons," continuous chains of spinel on Types B1 and B2 inclusion rims, to be larger versions of them and "palisades," long, curving chains of linked individual spinel crystals in Types A and B2 interiors, to be segments of larger ones.

Compared to those in fluffy Type A inclusions, grain-to-grain variations in trace and minor element contents of spinels within the same Type B inclusion are small, only about a factor of two (Steele & Hutcheon 1979), although huge variations in Na and Mg contents of anorthites were reported in another Type B by Hutcheon et al (1978b).

Other Types

Although use of the term "intermediate" to describe inclusions with fassaites of particular compositions was found superfluous above, that for which Grossman (1975a) coined the term differs from Types A and B also in having an anorthite-rich (50%), melilite-free mode. Wark & Lovering (1978a) also use the term "intermediate" or Type I for several other inclusions that contain $\sim$50% anorthite, but which are melilite-bearing. The latter differ also from Types A and B by having "chilled" textures, a different outer rim sequence, and different trace phases.

Other melilite-free inclusions from Allende and Leoville were described by Lorin et al (1978). They consist of ophitic fassaite-anorthite intergrowths, contain spinel framboids, but have no rims. They may be related to Grossman's (1975a) "intermediate" inclusion, a re-examination of which in this laboratory revealed an ophitic texture and spinel framboids.

Blander & Fuchs (1975) illustrated an inclusion, their sample 9/16, in which euhedral forsterite and spinel crystals, some in framboids, are poikilitically enclosed by fassaite and in which melilite is $Åk_{67-88}$. It is similar to one studied by Dominik et al (1978), except that in the latter, spinels are either coated by perovskite and enclosed by forsterite or fill interstices between forsterites. Furthermore, this entire assemblage has a thick mantle of fassaite which encloses spinel ribbons, melilite ($Åk_{88}$), and anorthite. Forsterite contains 1.2–1.8% CaO. It is very similar to inclusion

A16S3 (R. N. Clayton et al 1977). Another olivine-bearing inclusion (Haggerty 1978a) differs from these in containing no melilite.

Valence of Titanium

The presence of Ti^{3+} in Allende fassaites is indicated by their dark color, pleochroism, and features in their polarized absorption spectra (Dowty & Clark 1973a,b, Burns & Huggins 1973), as Ti is the only transition metal present in more than trace amounts. Dowty & Clark (1973a) also found that about 70% of the Ti in their pyroxene had to be calculated as Ti^{3+} to obtain a chemical formula with 4.00 cations per six oxygen atoms. Mason (1974) and Grossman (1975a) further noted that the cation deficit calculated assuming that all Ti is Ti^{4+} increases with Ti content of the fassaite. Rhönite was first discovered by Fuchs (1971) as a minor phase in the inclusions and was later found to contain large amounts of Ti^{3+} (Haggerty 1976, 1977, Fuchs 1978).

Haggerty (1976, 1977, 1978b) proposed the use of Ti^{3+}/Ti^{4+} ratios in phases in the inclusions as indicators of condensation temperatures, based largely on his measurements of such ratios in phases exhibiting critical textural relationships with one another which imply a crystallization sequence. The validity of this cosmothermometer is in doubt, however, as one critical texture is the commonly observed symplectic intergrowth of spinel, perovskite, fassaite, melilite, and hibonite, which Haggerty interprets as decomposition products of rhönite, an inference disputed by El Goresy et al (1977b) and by Fuchs (1978). Furthermore, some of the critical textures are in Type B inclusions which, as shown later, are probably melt products whose textures no longer reflect the condensation sequence.

Proposed Modes of Origin

SOLID CONDENSATES Much of the interest in coarse-grained inclusions stems from their mineralogical similarity to phase assemblages predicted from thermodynamic calculations to be high-temperature condensates from a gas of solar composition. Early work on condensation calculations and their application to these inclusions was reviewed by Grossman & Larimer (1974). In her study of a Type A inclusion in the Vigarano meteorite, Christophe Michel-Lévy (1968) was the first to suggest that the inclusions are samples of high-temperature condensates, in this case, those predicted in the calculations of Larimer (1967). Marvin et al (1970) and Larimer & Anders (1970) noted the similarity between the minerals in such inclusions from Allende and Ca-, Al-, Ti-rich compounds which Lord (1965) had calculated were near saturation at high temperature in a gas of solar composition and they ascribed a high-temperature condensation origin to them. The work of Grossman (1972) was the first theoretical investigation of the sequence of condensation of phases from a cooling gas of solar composition at constant pressure where full equilibrium calculations were employed. Those calculations for the abundant condensable elements were extended to include more solid solutions and were updated in response to newer thermodynamic and solar system abundance data in a series of papers (Grossman & Clark 1973, Grossman & Olsen 1974, Grossman 1975b, 1977a,b, Lattimer & Grossman 1978, Lattimer et al 1978, Grossman et al 1979b). Grossman (1972) showed that Ca, Al, and Ti would condense totally as corundum, melilite, perovskite, and spinel at temperatures above the point where significant fractions of the most abundant condensable elements, Mg, Si, and Fe, would begin to condense. He showed further that the assemblage melilite-spinel-perovskite is in complete chemical equilibrium with a gas of solar composition. Using results from Lattimer & Grossman (1978), the temperature range indicated is 1533–1438 K at a reference total pressure of 10^{-3} atm and about 75° lower for each tenfold decrease in pressure considered. This is precisely the mineral assemblage of Type A inclusions, except for hibonite and very minor amounts of rhönite and fassaite. Even the åkermanite content of melilite in these inclusions is in accord with predictions for high-temperature condensates (Grossman 1972, 1975a). Using many arguments first introduced by Grossman & Clark (1973), Allen et al (1978) searched for, but could not find, any textural or mineralogical evidence for the crystallization sequence expected if the fluffy Type A inclusion they were studying had originally crystallized from a melt. MacPherson & Grossman (1979) felt that the highly irregular shapes and intense alteration of fluffy Type A's implied that they formed by accretion of grains or grain clusters into highly porous aggregates and suggested that those inclusions are composed of direct gas-to-solid nebular condensates that have not been melted or metamorphosed since they accreted. It follows that all phases present in these inclusions have to be considered as candidates for high-temperature condensates, even though lack of thermodynamic data for some of them (hibonite, fassaite, and rhönite) prevents predictions of their stabilities in a solar gas (Allen et al 1978). Hence, hibonite is generally considered the first major element-bearing phase that condensed, instead of corundum which plays this role in the calculated condensation sequence (Grossman 1972, Blander & Fuchs 1975). Carrying the analogy further, hibonite must be a relict phase in these melilite-, spinel-rich inclusions since the latter phases are calculated to coexist only after corundum has reacted completely with the gas to form them. There are also good theoretical reasons to believe that fassaite could be a high-temperature condensate. Diopside is such a phase (Grossman

1972) and Grossman & Clark (1973) estimated that large amounts of $CaTiAl_2O_6$ and $CaAl_2SiO_6$ might dissolve in it in the temperature range of interest. The phase Ti_3O_5 reaches unit activity in a solar gas at temperatures slightly below the condensation temperature of forsterite (Lattimer & Grossman 1978) and it is thus certainly conceivable that significant amounts of trivalent Ti could have condensed in solid solution in pyroxene at higher temperature. It also follows that textural and chemical properties of phases in these inclusions should be directly interpretable in terms of physico-chemical conditions in the solar nebular gas. An example of this approach is the suggestion by Allen & Grossman (1978) that melilite grains in a fluffy Type A whose cores are richer in the low-temperature åkermanite component than their rims may have condensed during a period of falling pressure. This approach is still largely unexplored, however, because of the relatively little work done on fluffy Type A's, because of the hesitation of many workers to believe that any coarse-grained inclusions escaped melting, and because of the extensive obscuration of primary features in these inclusions by secondary alteration products. The major mineral phases of compact Type A's are also identical to those expected from high-temperature condensation, but, in these inclusions, textural evidence for an origin by aggregation of grains is less compelling than in fluffy Type A's. They are much less altered, implying that they were not as porous. Their shapes are near-spherical, suggesting that they may have crystallized from melt droplets. Hibonite, supposedly the first condensate, occurs only at their margins, rather than in their centers. In an SEM (scanning electron microscope) study of a compact Type A, Hutcheon (1977) observed epitaxial growth of perovskite on surfaces of spinel crystals which, in turn, are poikilitically enclosed by melilite. He pointed out that the implied crystallization sequence, spinel, perovskite, and then melilite, is not predicted by the equilibrium condensation model. Again, however, relatively little work has been done on Type A's.

LIQUIDS Far more petrographic work has been done on Type B's and it is upon studies of these that most arguments for a liquid origin for coarse-grained inclusions are based. Ever since their discovery, they were referred to as chondrules (Clarke et al 1970) because their near-spherical shape and the large sizes of their constituent crystals suggested that they were solidified melt droplets (Mason 1975). The term "igneous texture" was used to describe these tightly interlocking assemblages of coarse crystals (Mason & Martin 1974). Furthermore, the crystallization sequence inferred from textural relations in these inclusions was found to be the same as that expected for solidification of melts of similar composition (Clarke et al 1970, Mason 1975). The validity of the latter argument is questionable, however, since the synthetic system with which the inclusions were compared contained no Ti whose presence in two valence states in the inclusions may have an important effect on the stability of fassaite, one of the major phases. Blander & Fuchs (1975) made detailed petrographic observations of Type B inclusions, pointed out how these textures were identical to those seen in terrestrial igneous rocks of different composition, and concluded that the inclusions must have crystallized from melts. *But it has never been shown that the same textures cannot be produced by condensation of silicates from slowly cooled vapors of moderate pressure.* Blander & Fuchs (1975) argued against a condensation origin for the same inclusions by showing also that the crystallization sequence recorded in their textures is different from that predicted by the equilibrium condensation model. Similarly, Hutcheon (1977) observed epitaxial growth of perovskite on spinel crystals that are poikilitically enclosed by fassaite and pointed out that the implied crystallization sequence, spinel, perovskite, and then fassaite, does not conform to the predictions of this model. Against this, it must be argued that we simply may not know the exact condensation sequence because of the unknown effect of fassaite upon it, that departures from equilibrium may have occurred during condensation, and that phases may not have accreted in their order of condensation. Among the most compelling of Blander & Fuchs' arguments that Type B inclusions are not direct solid condensates is that they contain melilite and spinel, which condense above forsterite, and anorthite, which condenses below forsterite (Lattimer & Grossman 1978), but they do not contain forsterite itself. Because of the fassaite problem, however, even this argument may not be definitive, particularly since condensation calculations predict that anorthite forms by a reaction between spinel and diopside, the most likely phase that would be replaced by fassaite in the condensation sequence.

A revealing example of the ambiguity of the textural evidence in deciding between a liquid or solid origin lies in the different interpretations of spinel framboids. El Goresy et al (1979b) suggest that they formed by condensation of spinel around preexisting solid condensates and were themselves overgrown by later solid condensates. Wark et al (1979), on the other hand, favor an origin by incorporation of small droplets with solid spinel rims by larger droplets. Framboids have never been observed in fluffy Type A's.

Perhaps the strongest case for a once-molten Type B is the one studied by MacPherson & Grossman (1979). The fact that melilite crystals have had their growth interfered with by one another implies that the inclusion formed by in situ growth of melilite, rather than by random accretion of condensate melilite grains that had been suspended in the gas. Furthermore, because the low-temperature åkermanite component increases in concentration along the length of these crystals toward the center of the inclusion, they grew from the edge of the inclusion inward as temperature fell. It is difficult to see why condensation of solids from a vapor should lead to formation of a spherical shell that was filled in by later condensates. It is much easier to see how the observed textures could have been produced by crystallization of a suspended melt droplet, which would have cooled and solidified from the outside in by radiating heat away from its surface. This inclusion must have been molten, but whether this is true for other Type B's that do not show these chemical and textural properties is less clear.

How liquids of these compositions could have formed is also not known. Grossman & Clark (1973) showed that equilibrium condensation at pressures slightly above 10^{-3} atm would occur at temperatures slightly above minimum solidus temperatures for these compositions, producing partial melts. Complete melting would require condensation at much higher pressures than are normally considered for the inner part of the solar nebula (Cameron & Pine 1973), but which may have prevailed in giant gaseous protoplanets, which have been proposed as a source for these inclusions by Podolak & Cameron (1974) and Consolmagno & Cameron (1980). Blander & Fuchs (1975) argued that even low pressure condensation would produce melt droplets of these compositions because metastable, subcooled liquids would condense instead of the equilibrium solids. It is also possible that such melt droplets are secondary in origin, formed by melting of preexisting solid condensates in energetic nebular processes, such as lightning discharges (Whipple 1966, Cameron 1966, Sonnett 1979) or impacts between small, high velocity solid objects (Whipple 1972, Cameron 1973, Kieffer 1975).

In studies of coarse-grained inclusions in Lancé (Kurat 1970, 1975) and Bali (Kurat et al 1975), an origin by crystallization from liquids was inferred from textures. These workers were convinced that the bulk chemical compositions were established in gas-condensed phase reactions but, noting the thermodynamic equivalence of partial condensation and partial evaporation, they suggested that these liquids formed as evaporation residues by repeated impact events on parent bodies of chondritic composition. Chou et al (1976) also argued that the inclusions are volatilization residues, but of interstellar matter. The latter view was shared by Notsu et al (1978) and Hashimoto et al (1979) who produced Ca-, Al-rich residues in high-temperature vaporization experiments on a bulk sample of Allende and on the fine-grained fraction of the Murchison C2 chondrite, respectively. We will return to some of these ideas later.

Phase relations Knowledge of the sequence of crystallization of minerals expected from melts having the compositions of coarse-grained inclusions would be an invaluable aid to interpretation of their textures. Very little is known, however, about the low-pressure, liquid-crystal phase relations in the system applicable to these objects, $CaO–Al_2O_3–MgO–TiO_2–Ti_2O_3–SiO_2$. Although no compositions studied by Yagi & Onuma (1967) along the join $CaMgSi_2O_6–CaTiAl_2O_6$ are applicable to the inclusions, two studied by Yang (1976) and two by K. Onuma & Kimura (1978) in the plane $CaMgSi_2O_6–CaAl_2SiO_6–CaTiAl_2O_6$ appear to be close to inclusion bulk compositions. In none of these studies, however, were oxygen fugacities buffered so that Ti^{3+} could coexist with Ti^{4+}. This is also true in Seitz & Kushiro (1974), who determined the crystallization sequence in a melted fine-grained inclusion. Although that particular inclusion is close in composition to the coarse-grained ones discussed herein, it contains 2.3% FeO and 1.1% Na_2O which are not present in the primary phase assemblage of coarse-grained inclusions. Using petrographic and electron microprobe techniques, Butler (1977) studied a slowly cooled blast furnace slag that was remarkably similar in composition to coarse-grained inclusions, except for a total of 4% Fe, S, K_2O, Na_2O, and MnO. He found that melilite crystallized before Ti-, Al-pyroxene, in agreement with the crystallization sequence inferred by MacPherson & Grossman (1979) for their Type B inclusion. This reviewer considers Butler's (1977) results to be particularly significant because the slag crystallized under reducing conditions, as indicated by the presence of metallic Fe, and the pyroxene stoichiometry suggests that Ti may be present both as Ti^{3+} and Ti^{4+}.

MICRO-MINERALOGY AND MICRO-TEXTURES

Trace Phases

Palme & Wlotzka (1976) discovered an Fe-Ni particle containing minor to major amounts of Mo, Ru, Rh, W, Os, Ir, and Pt in a Type B inclusion. Grossman (1973) had calculated that Mo, Ru, W, Os, and Ir could have condensed totally as pure metals at temperatures above or within the range of condensation temperatures of the major phases in the inclusions and had predicted their presence therein. Palme & Wlotzka (1976) considered the condensation of those elements, together with Re, Pt, Rh, Fe, Ni, and Co, into a common alloy using an ideal solution model. Because the bulk composition of their metal particle has roughly chondritic proportions of refractory siderophiles, they inferred from their calculations that it last equilibrated with a solar gas between 1460 and 1500 K at 10^{-3} atm, in accord with temperatures inferred above from the silicate and oxide assemblage. Within the particle, however, are regions in which some refractories are concentrated relative to others and regions of sulfides. These were attributed to low-temperature exsolution from a primary condensate alloy and to a secondary alteration process involving introduction of volatile S. At about the same time, the tremendous potential of the SEM/X-ray analyser for characterizing such phases was demonstrated by Lovering et al (1976) who reported four trace phases, $<2\ \mu m$ in size, rich in rare earth elements (REE), Th, Nb, U, and Zr from Type B inclusions and by Wark & Lovering (1976) who analysed ten refractory siderophile-rich alloys, 0.5–3 μm in diameter, from a Type A. For more details of the compositions of the refractory lithophile-rich phases and others, see Lovering et al (1979). On the basis of previous condensation calculations, which showed that all these elements are refractory in a gas of solar composition (Grossman & Larimer 1974), these phases were all interpreted as condensates.

In a series of contributions (El Goresy et al 1977a, 1978a,b, 1979a), it was shown that the trace phases have two major modes of occurrence, either individual metal nuggets, ~ 1–2 μm in size, or complex assemblages, "Fremdlinge," of metal alloys, sulfides, phosphates, oxides, and silicates. These aggregates are usually 3–15 μm in size and are composed of smaller grains, some in the submicron range. Although Wark & Lovering (1978a, b) and Lovering et al (1979) find Fremdlinge only in Type B inclusions, El Goresy sees them in both A's and B's, but notes that individual beads and Fremdlinge are not usually found together in the same inclusion.

Both El Goresy et al (1978b) and Wark & Lovering (1978b) agree that the individual metal nuggets contain Mo, Ru, Rh, W, Re, Os, Ir, and Pt in roughly chondritic proportions, although these elements may be segregated from one another into separate regions within each nugget, as seen by the former workers and inferred by the latter workers. Usually, these nuggets contain substantial amounts of Fe ($\sim 10\%$) and Ni ($\sim 1\%$), but some in Type B's have much lower concentrations. Wark & Lovering (1978b) made the important observation that nuggets within a single host melilite crystal are closer in composition to one another than to nuggets within another crystal in the same Type A inclusion, implying that the inclusion was never molten. This reviewer agrees with their assessment that it is of utmost importance to confirm this observation with more statistics, but suggests that results be obtained separately for fluffy and compact Type A's. El Goresy et al (1978b) also made an important observation whose confirmation would have a profound effect on views of the origin of the inclusions. Inside some metal grains, they found several percent Zr, with no evidence for its presence as a separate oxide phase. This is important because Zr condensation calculations show that even in a gas as reducing as one of solar composition, ZrO_2 is 10^7 times more stable than metallic Zr at its condensation temperature. This factor is so great that even if tremendously nonideal solutions of Zr in platinum metal alloys are postulated, Zr should still condense in oxidized form. Furthermore, it is doubtful that, after condensing in this way, Zr could have been reduced to the metal during later, lower-temperature secondary reactions. If the observation is correct, it would be difficult to deny that the grains in question formed under much more reducing conditions than those of a solar gas.

Blander et al (1980) discovered Pt-metal nuggets in a Type B in which refractory siderophiles were not in cosmic proportion to one another. They proposed isolation of grains from gas at a temperature above that where all these elements would be totally condensed, later oxidation by other elements in the inclusion, uncertainties in condensation models due to assumptions of ideal solution, and a source region of nonsolar composition as possible explanations. As in previous work (Fuchs & Blander 1977), they stressed the possible effects of supersaturation, perhaps due to coating over of early formed nuclei by other condensates, on deviation of the subsequent course of condensation from the equilibrium path.

El Goresy et al (1978b) classify Fremdlinge into three types. Type 1 are compact and contain 50–70% by volume of Fe-Ni metal, sulfides and refractory siderophile alloys, and the remainder phosphates, oxides, and silicates. Re, Os, Ir, Pt, Rh, and Ru are not alloyed with Fe-Ni, but are present as submicron grains, either individually or in predominantly binary, ternary, or quaternary alloys. Situations are common in which a grain of pure Os is found only a few μm from one of pure Ir (Wark & Lovering 1978b). Zn, Ga, Ge, Sn, and As are sometimes alloyed with Ni, Fe, and/or Pt metals. In the Fe-Ni alloys, a wide range of Ni/Fe ratios can be found within the same inclusion. Mo and W form sulfides and alloys with Pt metals. V and Nb occur in oxides and sulfides. Silicates are clinopyroxene, melilite, anorthite, wollastonite, nepheline, sodalite, and K-, Na-, Ca-phases. Type 2 contain 70–90% Fe-Ni. Mo and W sulfides and Pt metal alloys occur inside silicates and Fe-Ni. Type 3 contain 0–70% Fe-Ni which, with sulfides, usually occupies the core of the Fremdling, around which is a spongy rim of oxides, phosphates, and silicates. Pt metal alloys are found throughout. Other phases are ZrO_2, pyroxene, spinel, anorthite, nepheline, and sodalite. Individual Fremdlinge are sometimes enriched in a single element, such as Mo, V, or Sc, relative to all the other rare elements normally encountered.

El Goresy et al (1978b) and Wark & Lovering (1978b) consider the individual metal nuggets to be solar nebular condensates, largely because they contain solar proportions of refractory siderophiles. The former workers ascribe the origin of Fremdlinge to accretion of components that condensed in a large number of events from a large number of chemically distinct reservoirs, many of which were nonsolar in composition and possibly pre–solar nebular in origin. Many constituents of the Fremdlinge, Ca-phosphates, nepheline, sodalite, Fe-Ni sulfides, and alloys of volatile transition metals such as Ga, Ge, Zn, and Sn, are simply not stable in a gas of solar composition at the high temperatures at which their enclosing host phases condense. The first major problem for El Goresy's model is thus preservation of these phases during the later high-temperature event that produced the major phases in the inclusions, a fact that caused El Goresy et al (1978b) to assert that not even the major phases condensed from a gas of solar composition.

The second major problem for this model is that it assumes that all minerals in the Fremdlinge and textural relations between them were produced during condensation and have not been modified during later melting events. Liquid-crystal phase equilibrium data are lacking for multicomponent Pt metal systems. From the high melting points of the end-members, however, it is conceivable that some alloy compositions would remain solid even at the high temperatures required for total melting of the major phase assemblage of the inclusions, as the siderophiles would not dissolve in molten silicates. This would not be the case for the lithophiles, however. Even though melting points of some phases present, such as thorianite and baddeleyite, are $>2500°C$, they may dissolve totally in silicate melts at much lower temperatures. Surely, the Fe-Ni sulfide fraction and nepheline and sodalite would have melted if the major primary phases did. Thus, many mineralogical and textural features which El Goresy attributes to condensation would not have survived such a melting event. As discussed above, fluffy Type A inclusions may have never been molten and Type B1 inclusions like that studied by MacPherson & Grossman (1979) probably were. Which other inclusions were molten is still not clear. Although El Goresy has not yet combined his data on Fremdlinge with complete petrographic descriptions of their host inclusions, such a synthesis would be extremely valuable.

Wark & Lovering (1978b) challenged El Goresy's basic assumption that all features in Fremdlinge predate the phases that enclose them, arguing that open-system secondary alteration processes are responsible for many of their characteristics. Their observation that corroded Fe-Ni grains are surrounded by magnetite and intersect cracks filled with it is certainly conclusive evidence that this has occurred in some cases. Magnetite is only stable in a solar gas below 400 K. Furthermore, the silicate assemblage of Fremdlinge includes anorthite, wollastonite, nepheline, and sodalite, phases characteristic of secondary alteration products that form at the expense of primary silicate and oxide phases of the inclusions (Allen et al 1978) and therefore must postdate any melting event. If the silicates and magnetite are indeed secondary, the possibility exists that all of the enigmatic volatiles entered Fremdlinge in such reactions. El Goresy et al (1978b) insist, however, that the overwhelming majority of Fremdlinge show no connection to magnetite-filled cracks. What fraction of the Fremdlinge have been so affected is still unclear. Nor is it clear what effects, other than introduction of volatiles, may be due to secondary alteration. Neither this process nor exsolution (Wark & Lovering 1978b), however, seems capable of producing such features as virtually pure Os and Ir grains only a few μm apart, entire Fremdlinge composed predominantly of a single refractory element, inclusions of MoS_2 in Fe-Ni that has not reacted to form a sulfide (El Goresy et al 1978b), or Ca-phosphates containing 3% RuO_2 (El Goresy et al 1979a). These and other observations suggest that different constituents of Fremdlinge, some of which are now intimately associated with one another, were formed in different physico-chemical environments. In addition, Ru-bearing phosphate must have formed in a system far more oxidizing than a solar gas.

Alteration Products

Many papers dealing with petrography of coarse-grained inclusions mention dark, opaque, fine-grained regions in thin sections of them, but it was the SEM work of Allen et al (1978) that provided the most information about this material in fluffy Type A's. They found that such regions consist of cavities and veins whose walls are lined with euhedral crystals of grossular and which contain mats of acicular wollastonite crystals, euhedral nepheline crystals projecting inward, and anorthite. Some wollastonite needles are bent through nearly 90°, but are not broken, indicating great strength such as is found for vapor-deposited whiskers. Their presence and euhedral crystals partially filling void spaces are clear evidence for their origin by condensation from a vapor. The fact that this material corrodes and veins major mineral phases means that minerals in the cavities are secondary alteration products of those primary phases. Thus, after formation of hibonite, melilite, and spinel in fluffy Type A's, these minerals reacted partially with a gas phase to form the alteration products. Allen et al (1978) found that alteration products occupy more than 75% of the inclusion they studied. The intense alteration suffered by these inclusions implies that the reacting vapor had easy access to the primary phases, one of the major reasons why MacPherson & Grossman (1979) believe these inclusions were loose aggregates at that stage.

Large amounts of similar alteration products are found in Type B's where they are not as abundant as in A's. The small grain sizes of the alteration phases make the altered regions too opaque to study by normal optical techniques. Hutcheon et al (1978b) studied several Type B's by cathodoluminescence. This technique has the advantageous capability of portraying different phases in these regions in characteristic and vivid colors. This study revealed that chemical and/or structural changes have occurred in anorthite, melilite, and fassaite next to fractures and embayments containing alteration products. Since the delicate textures of the alteration products would have been destroyed during melting, their presence in once-molten inclusions implies that vapor phase reactions occurred after melting and crystallization.

Interpretation of the alteration products as condensates from a vapor presents important difficulties. Neither wollastonite nor nepheline nor sodalite can condense from a gas of solar composition at equilibrium. Either the alteration process was not an equilibrium one or the reacting gas was not solar in chemical composition. Arrhenius & Alfvén (1971) and Arrhenius (1972, 1978) took the coexistence of volatile Na in nepheline and sodalite with refractories to mean that they condensed simultaneously due to ionization-controlled sublimation from a low-pressure, high-temperature plasma in which grains were not in thermal equilibrium with the gas. From the textural relations, however, it is now clear that Na was deposited in the inclusions in an event that postdated formation of the primary refractory phases. The presence of nepheline, anorthite, and wollastonite in Fremdlinge led to the suggestion, above, that some of them may have been affected by the same alteration process as the major phases. Although the mechanism for introduction of Na is not understood, other volatiles such as are found in some Fremdlinge may have accompanied it.

Both Clarke et al (1970) and Fuchs (1974) noted that grossular decomposes to wollastonite, melilite, and anorthite above 1073 K. Unless grossular nucleated metastably above this temperature, this is an upper limit to the condensation temperature of this phase in the alteration products. Fuchs (1974) interpreted such grossular in a Type B inclusion as a devitrification product of a glass below 1073 K, supporting the contention of Blander & Fuchs (1975) that the inclusions crystallized from a melt. The textural observations of Allen et al (1978) suggest that this interpretation is incorrect.

Rims

Even before the fall of Allende, Christophe Michel-Lévy (1968) had noted a spinel-olivine rim around a Type A inclusion in the Vigarano meteorite. Frost & Symes (1970) saw pyroxene rims on a perovskite-bearing inclusion and Kurat (1970) described nepheline-, perovskite-, and diopside-rich rims on Type A's in Lancé. In Vigarano, Reid et al (1974) illustrated rims on Type A's. In Allende, Blander & Fuchs (1975) described diopside rims around cavities inside a Type A inclusion and perovskite-spinel and Na-rich rims on Type B's. Diopside rims around Type A's and lining cavities in their interiors, spinel rims on Type B's, and olivine rims around inclusions of both types were reported by Grossman (1975a).

These reports were greatly augmented and clarified by SEM observations of polished sections and the data were synthesized into the following picture by Wark & Lovering (1977a,b, 1978a). The outermost region of every coarse-grained inclusion consists of a narrow (usually $\lesssim 60\ \mu m$) sequence of mineralogically distinct zones. Furthermore, there are mineralogical differences between the rims on Type A, on Type B, and on intermediate inclusions. From inside to outside, the sequence in Type A rims is Fe-bearing spinel + perovskite, nepheline + sodalite + melilite + forsterite + anorthite, fassaite which gradually changes its composition outward to diopside, and, finally, hedenbergite + andradite. Allen et al (1978) discovered in a fluffy Type A a $300\ \mu m$ thick layer consisting largely of porous masses of clinopyroxene prisms on the exterior surface of the outermost zone seen by Wark & Lovering. Type B rims consist of Fe-bearing spinel + perovskite, forsterite + nepheline, and hedenbergite + diopside.

Textures within each zone are not clastic. There is usually little void space and crystals are snugly intergrown with one another. The resulting massive appearance of the rims suggests that their constituents crystallized in situ and are not accreta that originally crystallized elsewhere. Because of the paucity of FeO in the minerals in inclusion interiors and its presence in rim phases, Christophe Michel-Lévy (1968) thought that rims are high-temperature reaction products between inclusion interiors and the FeO-rich matrix of the meteorite, a suggestion also made by Reid et al (1974). Because the diopside rims observed by Kurat (1970) and Grossman (1975a) are FeO-free, these workers stated that this phase could not be a reaction product between the inclusions and the meteorite matrices but, instead, between them and a gaseous, pre-agglomeration medium. Perhaps the best evidence that reaction with the matrix was not involved is the observation of both Blander & Fuchs (1975) and Wark & Lovering (1977a) that rims are absent from jagged surfaces of inclusion fragments, but present on the smooth, rounded surfaces of the same fragments, even though all surfaces are in contact with the matrix. This implies the following sequence: inclusion solidification, rimming, fragmentation, and incorporation into the meteorite.

The rim sequence could have been produced by reaction with a single medium, perhaps even at fixed pressure and temperature, such as in the case of alteration zones formed in terrestrial contact metasomatic processes. In this event, rim mineralogy would be strongly dependent on the composition of the inclusion interior. Thus, simply because they have different rim sequences from one another does not necessarily lead to Wark & Lovering's (1977a) conclusion that Type A and Type B1 inclusions were rimmed in different physico-chemical regimes, because the chemical compositions of these different inclusion types are very probably different. On the other hand, each layer could have formed by precipitation from, or reaction with, a separate physico-chemical environment, in which case the mineralogy would not necessarily depend on the composition of the inclusion interior. Only if the layers formed in this way does the Wark & Lovering suggestion follow. It is the thick melilite mantles of Type B1 inclusions that differentiate them from B2's. Thus, during rim formation, each of these inclusion types had different mineral assemblages in chemical communication with the external medium and, if the first of the above alternatives is correct, a different rim sequence should have formed on each. If the second alternative is correct and all Type B's formed in the same place, they should all have the same rim layers. Which is correct is not yet known, as Type B2 rims have not been studied as thoroughly as B1 rims (Wark & Lovering 1977a).

Observations that rims overlie alteration products imply that the alteration event predated rimming (Allen et al 1978). Because rims formed before the inclusions were incorporated into the meteorite and because alteration products never breach the rim, the source of vapor for the alteration process could not have been the Allende matrix. Perhaps the alteration phases were the first result of the same process that later formed rims.

Serious problems arise if rims are interpreted as condensates. Nepheline, sodalite, and andradite are not known to condense from a gas of solar composition at equilibrium. Either they condensed in a nonequilibrium process or they formed from a gas whose composition was not solar. This is also suggested by the occurrence of diopside exterior to nepheline and sodalite in Type A rims, indicating condensation of Na-bearing phases prior to diopside. New approaches will be required to learn more about the origin of alteration products and rims. One promising possibility is the inversion of condensation calculations. In this approach, any mineral assemblage of interest could be assumed to be an equilibrium condensate assemblage and the composition of the gas phase with which it is in equilibrium could be calculated. If the compositions so calculated have any astrophysical significance, new insights about the inclusions would be suggested.

BULK CHEMICAL COMPOSITION

Major Elements

Complete major element compositions of individual coarse-grained inclusions have been published for the following samples: one, labeled inclusion 1, in Gray & Compston (1974), three in Graham (1975), five, labeled Group I, in Conard (1976), nine, labeled Group I, in Mason & Martin (1977), four, Groups I, III, V, VI, in Taylor & Mason (1978), and another, CG-11, in Davis et al (1978b). The latter is an extremely heavily altered fluffy Type A whose composition is 29.4% CaO, 37.6% Al_2O_3, 4.3% MgO, 1.0% TiO_2, 25.1% SiO_2, 1.7% FeO, and 0.8% Na_2O. From its mineralogical description, the one analysed by Gray & Compston (1974) is probably also a fluffy Type A. Its composition is almost identical to that studied by Davis et al (1978b). Although the types of inclusions studied by Taylor & Mason (1978) are unknown, those analysed by Graham (1975), Conard (1976), and Mason & Martin (1977) are probably Type B's. Their compositions are in the range 21–37% CaO, 21–33% Al_2O_3, 5–15% MgO, 1.0–1.7% TiO_2, 24–36% SiO_2, 0.3–3.1% FeO, and 0.06–1.7% Na_2O. As Grossman & Larimer (1974) and Grossman (1975a,b, 1977a,b) pointed out, these bulk compositions are close to those of high-temperature condensate assemblages calculated to form before condensation of significant fractions of Mg, Si, and Fe, except for the presence of relatively small amounts of FeO and Na_2O which must be ascribed to introduction of volatiles during secondary alteration and rimming processes (Wänke et al 1974, Grossman 1975a, Grossman & Ganapathy 1975).

Do Types A and B inclusions have the same bulk chemical composition? Do they plot precisely along the predicted trajectory of condensates in composition space? These are important, first-order questions that have not been answered by the above data for the following reasons. First, because crystal sizes of the major phases are often large compared to the inclusions in which they are found and because inclusions like the B1's have thick, monomineralic mantles, the size of an unbiassed sample for major element analysis should be comparable to that of the entire inclusion. Samples smaller than this will only be representative if they are splits of homogenized powders made from nearly entire inclusions and this was definitely not the case in Conard (1976), whose samples were obtained and described by Gray et al (1973), in Davis et al (1978b), or in Taylor & Mason (1978). There is no indication that such precautions were taken by Gray & Compston (1974), Graham (1975), or Mason & Martin (1977). Second, because of the presence of relatively large amounts of secondary alteration products, bulk analysis will not in general yield the composition of the primary mineral assemblage alone (Davis et al 1978a), even when great care is taken to ensure representativeness. Davis et al (1978a) concluded that Type A's are lower in MgO than B's, based on data for CG-11 and unpublished data from this laboratory for three other A's and several B's. The significance of this conclusion, however, was obscured by the above problems. Wark & Lovering (1978a) reached a similar conclusion about MgO and found other differences as well, but their broad-beam electron microprobe analyses of polished sections apparently also suffer from contamination by alteration products and may have the matrix correction problems often encountered when this technique is applied to coarse-grained samples. These criticisms also apply to the data used by McSween (1977). Better estimates of major element compositions of primary phase assemblages are those of Grossman (1975a). Those were computed from means of large numbers of spot microprobe analyses of the major phases and ranges of modal analyses, excluding alteration products, observed in many inclusions. From these, MgO and SiO_2 are lower and CaO higher in Type A's than B's. This technique was applied individually to two Type A and three Type B inclusions by Grossman & Ganapathy (1976b). Although the above results were confirmed for MgO and CaO, the SiO_2 data overlap in this study. It is clear that the latter approach must be applied to many more inclusions before the questions posed above can be answered with precision.

Trace Elements

PREDICTIONS OF CONDENSATION MODELS Results of calculations of equilibrium condensation temperatures of many trace elements from a gas of solar composition are given by Grossman (1973, 1977b), Grossman & Larimer (1974), Boynton (1975a, 1978a), Grossman & Ganapathy (1976b), Palme & Wlotzka (1976), Ganapathy & Grossman (1976), Grossman et al (1977), Davis & Grossman (1979), and Blander et al (1980). They show that Os, Re, Ir, Ru, W, and Mo could have condensed completely as pure metals and that Zr, Hf, Y, Sc, and some REE could have condensed as pure oxides at temperatures above the accretion temperature of the inclusions (1438 K at 10^{-3} atm, for example). None of the pure crystalline phases of Pt, Rh, V, Ta, Th, U, Pu, and the remaining REE for which thermodynamic data exist are capable of condensing these elements totally above this temperature. Calculations show, however, that these elements can do so if they form ideal solid solutions either with other siderophiles in the case of Pt and Rh (Palme & Wlotzka 1976) or in condensed oxides and silicates of more abundant elements in the case of Th, U, and Pu (Ganapathy & Grossman 1976), REE except Eu (Grossman

& Ganapathy 1976b), Ta, and V. It is thus not certain that these elements can condense above this temperature because the complete absence of activity coefficients for them in relevant phases dictates use of simple ideal solution models, which overestimate condensation temperatures for those components that actually exhibit positive deviations from ideality (Boynton 1975a, 1978a). On the other hand, perhaps only a few of the elements discussed here condensed in the forms calculated, but rather as more stable pure compounds, such as some of the phases found in Fremdlinge, for which there are no thermodynamic data. Substantial concentrations of those trace elements that can condense above the accretion temperature of the inclusions should be found in the inclusions, because those elements may have condensed in solid solution in major condensate phases, they may have nucleated upon them, or they may have acted as condensation nuclei for them (Grossman 1973).

GENERAL OBSERVATIONS The Allende literature contains many trace element or isotopic analyses of objects for which there are either no accompanying mineralogical or petrographic descriptions or descriptions that are not detailed enough to tell if the inclusions in question are coarse-grained ones. The following are sources of trace element data for bulk inclusions that are either definitely or probably coarse-grained: one in Gast et al (1970), eight in Gray et al (1973; labeled Ca-Al chondrules), 15 in Grossman (1973; omitting sample 8), two in Tanaka & Masuda (1973; inclusion O and Ca-Al rich chondrule), one in Wetherill et al (1973; low Na), one in Osborn et al (1974; Ca-Al chondrule), one in Wänke et al (1974), 10 in a series of papers by Grossman & Ganapathy (1975, 1976b) and Grossman et al (1977), three in Chen & Tilton (1976; ChL-1, ChL-2, and WA), three in Chou et al (1976), five in Conard (1976; Group I), three in Tatsumoto et al (1976; N17, N18, N19), one in Palme & Wlotzka (1976), one in Drozd et al (1977; 3666-I1), 10 in Mason & Martin (1977; Group I), three in Nagasawa et al (1977; 5, 6, 7), one in Davis et al (1978b), and four in Taylor & Mason (1978; Groups I, III, V, VI).

The conclusion reached from all these data is that the average inclusion is enriched in each of the above refractory elements, except V, Rh, and Mo, plus Eu, Ba, Sr, and Nb, by a factor of 15–20 relative to Type 1 carbonaceous (C1) chondrites. These elements span a tremendous range of chemical properties and geochemical behavior. The only thing common to them all is that their condensation temperatures from a gas of solar composition are above or within the range of condensation temperatures of the major mineral phases in the inclusions. Furthermore, excluding samples that were obviously grossly contaminated by Allende matrix material during their extraction from the meteorite, all the more volatile elements, with the occasional exception of Na, Cl, Br, and Au, are depleted in coarse-grained inclusions relative to C1 chondrites by amounts ranging from 10% to a factor of 320. *The trace element results are thus very powerful evidence that coarse-grained inclusions formed in a gas-condensed phase fractionation process.* It is concluded that Eu, Ba, Sr, and Nb condense in the same temperature range as the other refractories, even though calculations have not yet shown how. Although V (Conard 1976), Rh (Taylor & Mason 1978), and Mo (Mason & Martin 1977) are also enriched relative to C1's, they are not as enriched as other refractories. It should be noted, however, that calculations show that V is one of the last refractories to condense as a pure oxide (Grossman & Larimer 1974) and, unless it formed a solid solution instead, it would not have condensed totally above the accretion temperature of the inclusions. Similarly, Rh is the last refractory siderophile to condense totally into an alloy (Palme & Wlotzka 1976). Mo is a mystery, however, as it is one of the first elements to condense as a pure phase. Perhaps its low enrichment is due to an error in its Cl abundance (Palme & Wlotzka 1976, Blander et al 1980).

Although trace element data cannot be used to chose decisively between a condensation or evaporation origin for the inclusions, they do constrain the type of partial evaporation process that would be acceptable. The key observation is the equal enrichment of refractory siderophiles and refractory lithophiles. In the model advocated by Kurat et al (1975), the inclusions result from repeated cycles of volatilization and condensation during meteorite bombardment of chondritic bodies. These processes should have been accompanied by melting events on the parent body, during which lithophiles would have fractionated from siderophiles due to the tendency of the latter to form melts that are immiscible with and denser than lithophile-rich melts. Thus, although partial evaporation of dust aggregates is a possibility (Chou et al 1976), the trace element composition of the inclusions seems to rule out parent-body volatilization processes.

Condensation calculations predict even lower concentrations of volatiles than are found in the inclusions, except for such elements as Fe, Ni, Co, and Cr (Grossman et al 1979b), which may have begun to condense just above the accretion temperature of the inclusions. Na and Cl are known to be major constituents of phases in alteration and rim assemblages. Wänke et al (1974) found volatiles to be enriched in the outer part relative to the inner part of an inclusion, indicating their association in this and other inclusions with secondary alteration products, rims, and adhering particles of Allende matrix.

Grossman & Ganapathy (1976b) and Grossman et al (1977) determined concentrations of 21 refractory elements in nine inclusions and found highly variable ratios of some elements to others in individual inclusions, implying that refractories fractionated from one another during condensation and entered inclusions in separate components. When the mean concentration of each refractory element was divided by its C1 abundance, however, almost the same number arose in each case, 17.5 ± 0.4 (see Figure 4). Thus, as a group, coarse-grained inclusions did not fractionate refractories from one another relative to C1 chondrites, suggesting total condensation of each element from a gas of solar composition and indiscriminate incorporation of all condensate components into coarse-grained inclusions as a whole. The composition of C1 chondrites is generally regarded as that of the total condensable matter of the solar system. An enrichment factor of 17.5 relative to C1 chondrites will result if all of a particular refractory element in C1 chondrites is concentrated into 100/17.5 = 5.7 wt % of the total condensable matter. As Grossman et al (1977) argue, this value is in excellent agreement with independent estimates from condensation calculations of the fraction of the total condensable matter that should have condensed above the accretion temperature of the inclusions. This is true, even allowing for the presence of

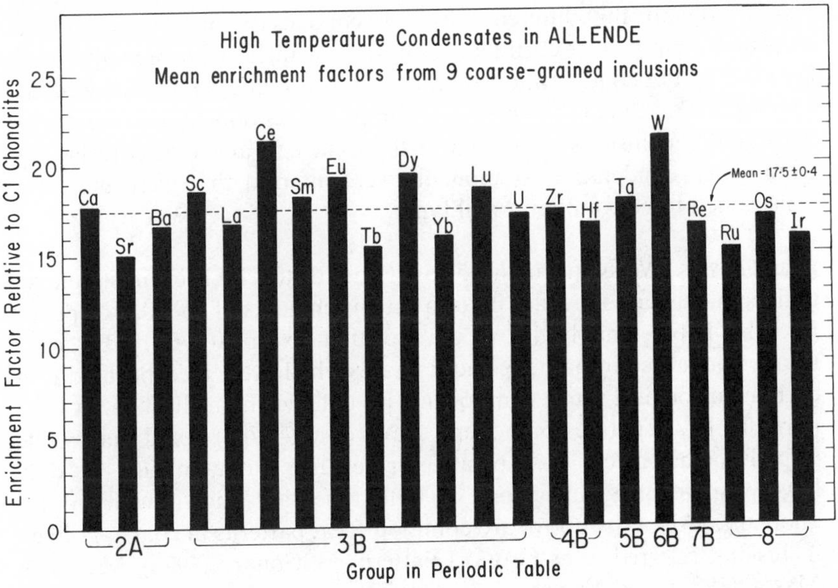

Figure 4 Refractory elements are uniformly enriched relative to C1 chondrites in this collection of coarse-grained inclusions. This indicates total condensation and indiscriminate incorporation of each of these elements into the inclusions, regardless of differences in their chemical properties. From Grossman et al (1977).

substantial amounts of alteration products, provided they introduced no additional refractories.

Applying linear regressions and factor analysis to their chemical data for bulk inclusions, Grossman et al (1977) concluded that Os, Ir, Ru, Re, and most of the W were carried into the inclusions either in a single refractory siderophile element alloy or in a group of such alloys that behaved like a single component, i.e. they did not separate from one another during their incorporation into the inclusions. Similarly, Sc, Zr, Hf, Ta, U, and REE except Eu entered in pyroxene or in one or more trace phases that did not separate from each other. Sr and some Eu were brought in with melilite, either in its crystal structure or in inclusions, and Ba was carried in alone in a separate phase.

Of all the inclusions for which bulk trace element analyses are reported, only two in Grossman & Ganapathy (1975, 1976b) and Grossman et al (1977) and one in Davis et al (1978b) are known to be Type A's. These are simply too few data points upon which to make comparisons between A's and B's. They were used, however, together with unpublished data from this laboratory for three additional Type A's to conclude that Sc-REE correlations are different in the two inclusion types (Davis et al 1978a). If this is verified by further studies of more Type A's, the conclusion would be that different refractory condensate components entered each type, implying that different physico-chemical environments were involved in the condensation of each. This would also seriously affect conclusions reached in the previous paragraph, as pooled data for Types A, B, and I inclusions were used in that study to arrive at correlations from which condensate components were inferred that were therefore constrained to be common to all types.

RARE EARTHS When normalized to C1 chondrites, REE in most coarse-grained inclusions show flat patterns at enrichments of 15–20, except for Eu. This is presumably due to the fact that, except for Eu, REE were totally condensed into one or more phases which were incorporated into each inclusion as a single component. Eu entered the inclusions, at least partially, in a separate component. Davis et al (1978a) noted that, while Type B inclusions can have positive or negative Eu anomalies, A's only have positive ones. Inclusions of both types are known with no Eu anomalies. All of the above are common REE patterns in coarse-grained inclusions, referred to as Group I patterns by Conard (1976), Mason & Martin (1977), and Nagasawa et al (1977).

A radically different REE pattern, referred to as Group II, was first found in Allende fine-grained inclusions by Tanaka & Masuda (1973). [Space does not permit detailed discussion of these, but petrographic descriptions are in Blander & Fuchs (1975; inclusion 7/9), Grossman et al (1975), and Grossman & Ganapathy (1975) and chemical data in Conard (1976), Grossman & Ganapathy (1976a), Mason & Martin (1977), and Nagasawa et al (1977).] These are characterized by being relatively uniformly enriched in light REE by factors of 20–60, having deep negative Eu anomalies, steadily decreasing enrichments from Gd to Er, sharp positive Tm anomalies, and a steep drop through Yb to Lu. Boynton (1975a) used thermodynamic calculations to show that a pattern with these general characteristics would result in fine-grained inclusions if they condensed with the REE remaining in the gas after prior removal of the most refractory REE in an earlier condensate. This is very strong evidence that a REE-bearing component in these inclusions formed in gas-condensed phase fractionation processes. In particular, this component cannot be a vaporization residue, or else it would have the inverse pattern. This is supported by volatilization experiments on REE-doped, Ca-, Al-rich glasses which yield residues that are enriched in heavy REE relative to light ones (Nagasawa & Onuma 1979). Arrhenius (1978) argued that Group II patterns formed by plasma condensation and that thermodynamic calculations neglecting charged species are an oversimplification of the real situation, regardless of their success at modeling the patterns. Boynton (1975b) showed, however, that REE patterns unlike any so far seen in Allende would result if REE condensation were controlled by ionization potentials.

Boynton (1975a) assumed that discrepancies between the pattern observed by Tanaka & Masuda (1973) and the one he calculated by assuming ideal solution of REE in the removed phase are due to nonideal effects and computed "activity coefficients" that would make the two patterns match as closely as possible. Davis & Grossman (1979) felt that relative activity coefficients required by this model were too high for the temperature being considered, 1650 K, and found that when such a model was applied to 20 Group II patterns in the literature, relative activity coefficients required by the model had to vary by unreasonably large amounts from inclusion to inclusion over a narrow temperature range. Instead, Davis & Grossman (1979) showed that most discrepancies could easily be explained by incorporation of a variable, but usually small, amount of a second REE component with a flat REE pattern into each inclusion.

Boynton (1978a) derived relative activity coefficients for Th, U, Pu, and Cm based on those in Boynton (1975a) and found that nonideal effects of the calculated magnitude enhanced considerably the solar nebular volatilities of Th, U, and Pu relative to the ideal solution calculations of Ganapathy & Grossman (1976). He further asserted that these workers were incorrect in concluding that Th, U, and Pu were each so refractory that they should be present in unfractionated proportion to one another in large collections of coarse-grained inclusions, claiming that, due to his newly calculated volatility for U, this element should be depleted relative to other refractories. If, as Davis & Grossman (1979) suggest, REE activity coefficients derived in Boynton (1975a) are vastly in error, then Boynton's (1978a) conclusions about Th, U, and Pu derived therefrom are also in error. This seems to be confirmed by Grossman et al's (1977) finding that U has the same enrichment factor as all other refractories in coarse-grained inclusions.

Group II REE patterns are sometimes found in coarse-grained inclusions: CG-5 (Type B) of Grossman & Ganapathy (1976b) and 4691 (Type B) of Mason & Martin (1977). The conclusions reached by Boynton (1975a) and Davis & Grossman (1979) are thus also applicable to these samples. Two REE components are present, one that was isolated from its gaseous reservoir after prior loss of a higher-temperature condensate and one from a source that suffered no prior loss. All solar nebular condensation calculations consistently underestimate the Tm abundance in inclusions with Group II REE patterns. Davis & Grossman (1979) suggested therefore that the first of the above components condensed under slightly more reducing conditions than those in a gas of solar composition, in which case Tm would have been more volatile. If these inclusions were molten after condensation, the above two components may no longer be physically separable. No petrographic differences are apparent between these and normal coarse-grained inclusions.

Eu and Yb are the most volatile REE according to calculations and, in fine-grained inclusions, sometimes show large negative anomalies superimposed on otherwise flat REE patterns. Such "Group III" patterns (Conard 1976, Mason & Martin 1977) are apparently also found occasionally in coarse-grained inclusions, as are so-called "Group VI" patterns, which are flat except for positive Eu and Yb anomalies (Taylor & Mason 1978). Group III inclusions must have incorporated a REE component that stopped equilibrating with the gas at a temperature high enough to prevent complete condensation of Eu and Yb, but low enough that all the more refractory REE had condensed totally (Boynton 1978b). Not only does the REE component in Group VI inclusions appear to have equilibrated at a low enough temperature that all REE condensed totally into it, but it also contains excess Eu and Yb, presumably due to prior condensation of all the more refractory REE in some other nebular reservoir. Finally, although most Group II inclusions have negative Eu and Yb anomalies, indicating incomplete condensation of Eu and Yb into them, some, including coarse-grained inclusions A-2 (Conard 1976) and

CG-6 (Grossman & Ganapathy 1976b) have positive Eu and Yb anomalies superimposed on an otherwise normal Group II pattern. It must contain the same two REE components as other Group II inclusions, but at least one of them, or perhaps a third, contains excess Eu and Yb that probably originated in the same way as the excesses in Group VI inclusions. No petrographic descriptions of coarse-grained inclusions with Group III or Group VI patterns have been published. No exceptional mineralogical characteristics were noted by Gray et al (1973) for A-2 or by Grossman (1975a) for CG-6, indicating that there are no other obvious signs of the unusual history recorded by its REE pattern.

MINERAL ANALYSES The ion microprobe has been used to study trace elements in individual mineral phases. Although Hutcheon et al (1977) reported <50 ppb Li in major phases of Types A and B inclusions, Phinney & Whitehead (1978) found ~ 1 ppm in all major phases in Type B's. Thus, Li does not appear to be enriched in the inclusions relative to C1's and is probably not refractory. The element Be, however, shows substantial enrichments in melilite and anorthite in Phinney & Whitehead's (1978) inclusion and may thus be refractory.

Trace element analyses of mineral separates from Type B inclusions, particularly for REE, are the subject of several papers. Mason & Martin (1974) found that melilite from one inclusion has a positive Eu anomaly and steadily decreasing enrichments from La to Sm while fassaite from the same inclusion has a negative Eu anomaly and steadily increasing enrichments toward heavy REE. These complementary patterns were rationalized on crystallochemical grounds. They pointed out, as did Grossman & Ganapathy (1976b), that this is difficult to reconcile with the fact that bulk trace element contents seem to be independent of inclusion mineralogy and argued for a two-stage process—incorporation of REE into the inclusions during accretion of condensate components and redistribution of REE during crystallization after a later melting event. N. Onuma et al (1974) made the same observation and reached the same conclusion but proposed an alternative, that melilite and pyroxene partitioned REE between them during condensation in a closed system. This remains a viable hypothesis, as the cases of independence of trace elements on mineralogy documented by Grossman & Ganapathy (1976b) could be explained by nebular heterogeneity. Nagasawa et al (1976) determined melilite/liquid REE partition coefficients, combined them with data from other experiments on pyroxene/liquid partitioning to predict the equilibrium REE distribution between melilite and pyroxene, and found that the latter is in good agreement with data for mineral separates studied by Mason & Martin (1974), N. Onuma et al

(1974), and Nagasawa et al (1977). This verifies the conclusion of Mason & Martin (1974) and N. Onuma et al (1974) that REE are in the melilite and pyroxene crystal structures. The fact that this could be verified for mineral separates which undoubtedly contain Fremdlinge implies that the bulk of the REE do not reside in trace phases. Nagasawa et al (1976) concluded, as did Mason & Martin (1974), that melilite and pyroxene co-crystallized under closed-system conditions within the inclusion.

A hibonite-rich sample from a fluffy Type A was found to be rich in REE, with positive Eu and Lu anomalies (Davis et al 1978b). Impure mineral separates from compact Type A's (samples 1 and 2) were analyzed by Nagasawa et al (1977).

ISOTOPIC COMPOSITION

Three reviews of the isotopic compositions of Allende inclusions have appeared recently (R. N. Clayton 1978, Podosek 1978, and Lee 1979), so there is no need here to re-explore all ramifications of the data. Only the bearing of isotopic data on the origin of coarse-grained inclusions will be reviewed.

Normal Inclusions

OXYGEN R. N. Clayton et al (1973) discovered large oxygen isotopic anomalies in all types of Allende inclusions, including coarse-grained ones. On a diagram such as Figure 5, all samples produced from an isotopically homogeneous reservoir by mass-fractionation accompanying normal chemical processes should lie along a line whose slope is 0.5, such as the dashed line for terrestrial materials. Data points for inclusions, however, plot along the line *AD* whose slope is close to unity. The oxygen isotopic compositions of the inclusions are thus controlled predominantly by nuclear, rather than chemical, processes. R. N. Clayton et al (1973) suggested that *AD* is a mixing line between normal solar system oxygen at higher $\delta^{17}O$ and $\delta^{18}O$ than *D* and another oxygen component whose isotopic composition is vastly enriched in ^{16}O at lower $\delta^{17}O$ and $\delta^{18}O$ than *A*. Noting that astrophysical environments exist where ^{16}O can be nucleosynthesized in the absence of ^{17}O and ^{18}O, such as the explosive carbon-burning zone of a supernova, they further proposed that grains must have condensed in a region so close to the site of ^{16}O nucleosynthesis that matter produced in this process had not yet mixed significantly with oxygen from other nucleosynthetic processes. They suggested that such ^{16}O-rich grains must have found their way into the interstellar gas cloud that underwent gravitational collapse and fragmentation to form the solar nebula and that they survived this process, presumably because temperatures were not high enough or were not high for long enough to completely evaporate them. The solar nebula was thus pictured as being composed of gas containing a homogenized mixture of oxygen isotopes produced in a number of nucleosynthetic environments (normal solar system oxygen) and previously formed dust grains whose oxygen was vastly enriched in ^{16}O relative to it. It was proposed that the major mineral phases of the inclusions condensed from this gas with nearly solar oxygen isotopic composition, but that when the inclusions accreted, they also incorporated a few per cent of their oxygen in the form of preexisting ^{16}O-rich grains. That different inclusions plot in different places along *AD* was explained by different proportions of new and old condensates in each. Later work on mineral separates (R. N. Clayton et al 1977), however, revealed inconsistencies in this model. Fassaite and spinel from Type B inclusions were found to plot close to *A*, with nearly a 5% excess of ^{16}O relative to terrestrial samples. This meant that, if the exotic oxygen

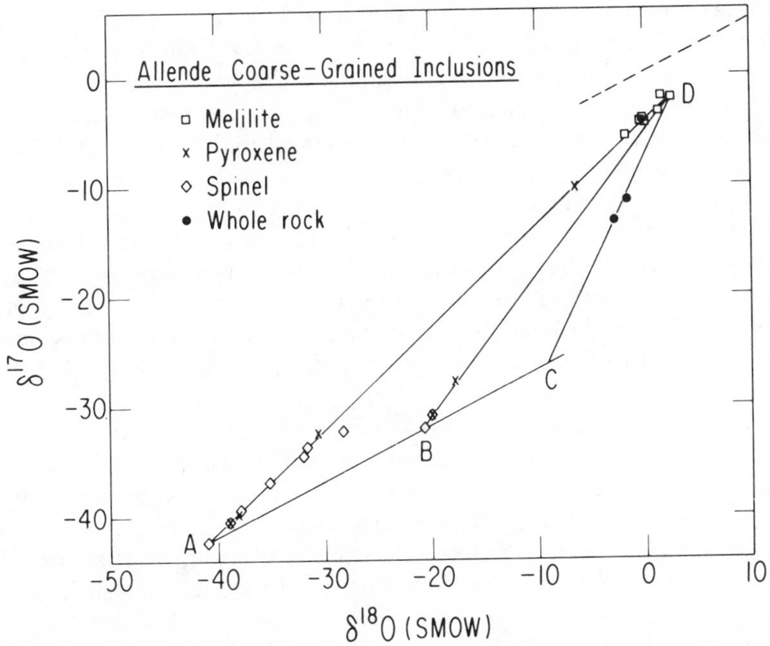

Figure 5 Variations in $^{17}O/^{16}O$ and $^{18}O/^{16}O$ ratios in Allende inclusions relative to the SMOW standard. Normal inclusions lie along *AD*, but FUN inclusions EK 1-4-1 and C1 plot along *BD* and *CD*, respectively. Melilite from normal inclusions and EK 1-4-1 always plots near *D* and fassaite and spinel near *A* and *B*. Dashed line for terrestrial samples and line *ABC* are mass-fractionation lines, having slopes of 0.5. From R. N. Clayton & Mayeda (1977).

component were pure ^{16}O, there would have to be about 5% interstellar grains in those mineral separates and even more if the exotic component had ^{17}O and ^{18}O in it. There simply were no candidate phases present at that concentration in the mineral separates, unless they were fractions of the fassaite and spinel themselves. This alone would not have been sufficient to dismiss the interstellar grain hypothesis, as Lattimer et al (1978) showed that most phases that condense from a gas of solar composition also condense from gases having even the extremes of composition that might be expected in many astrophysical environments. The real difficulty is that in every Type B inclusion investigated, the fassaites and spinels plot near *A* and the melilites near *D*. The interstellar grain hypothesis then requires the same ratios of exotic to nebular melilite, exotic to nebular fassaite, and exotic to nebular spinel in every inclusion and there seems to be no way to make the mixing process so precise.

The possibility remains that all crystals of fassaite and spinel in every inclusion have the same oxygen isotopic composition near *A* and all melilites have the same composition near *D*. It is extremely difficult, but not impossible, to reconcile this with an origin for Type B inclusions by crystallization from a melt. One possibility is that condensation occurred in a gas whose isotopic composition was near *A*, that melting took place during or after condensation, and that spinel and fassaite crystallized first upon cooling. The partially molten inclusions then found themselves immersed in a gas whose composition was close to *D*. The rate of oxygen exchange between gas and remaining liquid was rapid, so that melilite which later crystallized from the melt had more nearly normal oxygen isotopic composition. If fassaite and spinel exchanged their oxygen much more slowly with the melt, their much more ^{16}O-rich compositions could have been preserved. The required crystallization sequence is that observed by Seitz & Kushiro (1974) but, as already mentioned, those experiments may not be relevant to many Type B inclusions. Although spinel was undoubtedly the first-crystallizing phase in many of them, one of the inclusions analysed by R. N. Clayton et al (1977) was one in which MacPherson & Grossman (1979) suggested that melilite was first; yet, the oxygen isotopic distribution is the same in all of them. If all Type B inclusions were molten, it would thus have to be argued that the isotopic variations were produced after complete solidification (Blander & Fuchs 1975, Chou et al 1976, MacPherson & Grossman 1979). In such a model, all phases crystallized near *A* and attempted to exchange their oxygen with a gas near *D*. R. N. Clayton et al (1977) pointed out that, if oxygen diffusion is much more rapid in melilite than in fassaite and spinel, melilite would have approached *D* in composition, but fassaite and spinel would have been little changed from their initial composition. The presence of alteration products and rims indicates that the inclusions did react with a gas phase after solidification of primary phases and before incorporation into the Allende parent body. Although it is not known whether alteration products or rims or both were analysed, the discovery by R. N. Clayton et al (1977) that a nepheline-, sodalite-, grossular-rich sample from near the rim of an inclusion plots right at *D* suggests very strongly that oxygen exchange and formation of alteration products and/or rims are related.

The alteration process would also work if the inclusions were never molten, but here there is an alternative, that fassaite and spinel condensed from a gas of isotopic composition *A* and melilite from another whose composition was close to *D*. In this model, condensation must have taken place more rapidly than homogenization of ^{16}O-rich gas with normal gas, which is also a requirement in the exchange model.

Melilite and spinel from compact Type A's have isotopic compositions similar to their counterparts in Type B's, but mineral separates from fluffy Type A's have not yet been analysed.

MAGNESIUM Whether excess ^{16}O entered the solar nebula in presolar grains or as gas, other elements should also show isotopic anomalies in Allende inclusions. Early work showed relatively small Mg isotopic variations (Gray & Compston 1974, Lee & Papanastassiou 1974) that were not due to mass-fractionation but that could not be unambiguously interpreted in terms of enrichment or depletion of specific isotopes, although Gray & Compston (1974) favored excess ^{26}Mg. Lee et al (1976) found clear evidence for excess ^{26}Mg in the coarse-grained inclusion studied by Gray & Compston (1974) and in other types of Allende inclusions. A study of mineral separates, each containing many crystals, showed that ^{26}Mg excesses are linearly correlated with ^{27}Al/^{24}Mg ratios of the major mineral phases in WA, a Type B inclusion (Lee et al 1977a). Spinel and fassaite, with very low ^{27}Al/^{24}Mg ratios, show virtually no excess ^{26}Mg. Melilite, with moderate ^{27}Al/^{24}Mg, has a 4‰ excess and anorthite, whose ^{27}Al/^{24}Mg ratio is over 200, has a 9% excess. The existence of a positive correlation between ^{26}Mg/^{24}Mg and ^{27}Al/^{24}Mg ratios fulfills a necessary condition for isochronism and is taken as clear evidence for in situ decay of ^{26}Al. One interpretation of the data is that the constituent phases crystallized from a homogeneous reservoir, either gaseous or liquid, of Mg and Al isotopes in which the initial ^{26}Al/^{27}Al ratio was $\sim 5 \times 10^{-5}$, the latter value being obtained from the slope of the correlation line. These results were confirmed by mass spectrometric measurements on directly loaded individual crystals of anorthite, melilite, and spinel (Lee et al 1977b) and by an ion microprobe study of anorthite (Bradley et al 1978) from the same inclusion. ^{26}Mg/^{24}Mg ratios of individual crystals in polished sections of other Type B inclusions were measured by ion microprobe (Lorin et al 1977, Hutcheon et al 1978a,b). These data are in good agreement with those for WA in that they are correlated with $^{27}Al/^{24}Mg$ ratios along the same line, even within individual anorthite crystals (Bradley et al 1978, Hutcheon et al 1978b).

Taken together, the above studies show that ^{26}Al was present in the crystal structures of the major mineral phases of the inclusions and that it decayed in situ. As Esat et al (1979) cautioned, however, melilite from WA lies $\sim 1‰$ off the best fit line through the other WA data points. D. D. Clayton (1977a) used this observation as an argument in favor of his proposal that Allende inclusions are actually sintered mixtures of micron-sized presolar grains in which ^{26}Al had long since decayed away completely. He argued that the gross correlation between ^{26}Mg excesses and $^{27}Al/^{24}Mg$ ratios exists because the ^{26}Mg excesses were originally present in Al-rich interstellar phases and ^{26}Mg was kinetically hindered from diffusing far from the ^{27}Al originally associated with it during recrystallization into the mm-sized phases now observed. In this view, the correlation is merely a mixing line between phases that formed with large ratios of $^{27}Al/^{24}Mg$ and, as a result, of $^{26}Mg/^{24}Mg$ as well and phases with low ratios of $^{27}Al/^{24}Mg$ and $^{26}Mg/^{24}Mg$. The lack of collinearity of WA data points is taken as evidence for only slight spatial segregation of Mg from Al in this model. One of the major problems with this explanation, however, is that it ignores the fact that, if the presolar grains model is correct, major spatial segregation of Mg from Al *has* occurred in the process of making mm-sized grains of chemically distinct phases from the material required by D. D. Clayton (1977a) to be homogeneous in composition on a $10\,\mu m$ scale originally. An alternative, and far more probable, explanation for the lack of *precise* isochronism suggested by Esat et al (1979) is late disturbance of the Al-Mg system by an isotopic exchange process similar to that which affected oxygen in melilite.

Because of the short half-life of ^{26}Al (7.2×10^5 yr) and the fact that this isotope was most likely produced in the carbon-burning zone of a supernova, the previous existence of ^{26}Al in the Allende inclusions implies that such an explosion occurred at most a few million years prior to their formation and in the vicinity of the birthplace of the solar system. This led to the suggestion by Cameron & Truran (1977) that a nearby supernova explosion triggered the collapse of the interstellar cloud to form the solar nebula. An alternative suggestion, that ^{26}Al was produced locally in reactions involving energetic protons emitted from an early active sun, has been investigated by Heymann & Dziczkaniec (1976), D. D. Clayton et al (1977), and Lee (1978). Such models do not appear plausible, however, because isotope anomalies that have not been observed should have also been produced in other elements and because a mechanism is unknown for production of the required high fluences of protons which, collectively, must have carried off a substantial fraction of the sun's gravitational binding energy.

R. N. Clayton et al (1973) originally proposed that interstellar grains were the carriers of ^{16}O because of the relative ease with which isotopically different gases could mix and erase spatial variations in isotopic composition. If, however, the supernova trigger is correct and if the ^{16}O comes from the same supernova, the time scale for condensation could easily have been shorter than that for gas mixing. Thus, it is not clear that gas is ruled out as the ^{16}O carrier, although Lattimer et al (1978) suggested reasons why grains are more likely to be responsible and computed condensation sequences for zones of different composition in expanding supernova ejecta. Even if grains were the original carriers, they may have evaporated locally in the contracting nebula, creating isotopically anomalous gas pockets in which rapid condensation may have occurred.

Ion probe studies of inclusions other than WA also show deviations from the normal Al-Mg isochron. Lorin & Christophe Michel-Lévy (1978a) found hibonite grains in what appear to be Type A inclusions from both Leoville and Allende that plot well above the normal isochron upon which fall data points for other minerals in the same inclusions. Similarly, hibonite in the fluffy Type A studied by Steele & Hutcheon (1979) plots above an isochron through data for melilite, spinel, and other hibonites which has a lower slope than usual. Such results were attributed to early formation of hibonite and other phases and exchange of Mg isotopes in melilite and anorthite during later reactions in which oxygen isotopes were also exchanged (Lorin & Christophe Michel-Lévy 1978b). This appears to be in conflict, however, with the observations of Hutcheon et al (1979) who found petrographic evidence that hibonite was more altered than melilite in another fluffy Type A, some of whose hibonite plots above the straight line through primary melilite and other hibonite points. Other explanations are possible. If Lorin & Christophe Michel-Lévy's (1978a,b) inclusions are fluffy Type A's, they may be mechanical mixtures of phases whose Al and Mg isotopic compositions were never homogenized with one another. Inferred differences in condensation age would be $>10^6$ years, too great to interpret the inferred differences in initial $^{26}Al/^{27}Al$ in this way. Instead, perhaps they record Al isotopic heterogeneities in the gas cloud from which these phases condensed.

No traces of radiogenic ^{26}Mg were found by Lorin et al (1978) in their study of the petrographically unusual ophitic inclusions. Anorthite in the core of one was very enriched in heavy Mg isotopes by a mass-fractionation process, while spinel and fassaite in the core and anorthite near the edge were not. It is difficult to reconcile these internal Mg isotopic variations with a melting event, which is the favored explanation for the textures, unless isotopic compositions were modified later, by preferential exchange. Oxygen isotopic measurements of mineral separates from such inclusions might be revealing. Why the Mg isotopes are different in these inclusions is unknown, although their bulk compositions must be different from those of normal inclusions, another indication of a different history.

OTHER ELEMENTS Epstein & Yeh (1977) and Yeh & Epstein (1978) found Si isotopic variations in Allende inclusions that appear to be due to mass-fractionation. R. N. Clayton et al (1978) showed that this is true for both Type A and Type B inclusions and that the magnitude and sign of the variations are what would be expected for fractionation during high-temperature condensation of silicates from a gas whose $\delta^{30}Si = 0$. Patchett (1979) showed that mass-fractionation processes have enriched both coarse- and fine-grained inclusions in the light isotopes of Sr. These could be chemical processes specific to the condensation of Sr or characteristic of additional trace elements that entered the inclusions with Sr, such as Eu in the coarse-grained ones (Grossman et al 1977). Tatsumoto & Shimamura (1979) reported variations in the $^{238}U/^{235}U$ ratio in Type B inclusions which, because of their size and the mass numbers involved, are probably due to nuclear effects. Heydegger et al (1979) reported very small nuclear effects in the $^{50}Ti/^{49}Ti$ ratio in fassaites from Type B inclusions.

If incompletely homogenized supernova debris is the source of the ^{16}O excesses, then nuclear isotope effects should be present in many other elements in these samples. A major problem thus arises in that, aside from the aforementioned possibilities of nuclear effects in U and Ti, diligent searches for such effects in Ba (McCulloch & Wasserburg 1978a), Ca (Lee et al 1978), K (Begemann & Stegmann 1976, Birck et al 1977, Stegmann & Begemann 1979), and Si and Mg (above papers) in normal inclusions have been unsuccessful. This is a particularly chronic problem for models in which the inclusions are pictured as sintered aggregates of interstellar grains (D. D. Clayton 1977a) or volatilization residues of such material (Chou et al 1976, Notsu et al 1978, Hashimoto et al 1979).

FUN Inclusions

The story is different, however, in three other inclusions. Samples from two Type B's, C1 and EK 1-4-1, are the only ones, of a very large number of Allende inclusions whose oxygen isotopic compositions have been measured, which do not plot along *AD* in Figure 5 (R. N. Clayton et al

1977, R. N. Clayton & Mayeda 1977), but rather along *CD* and *BD*, respectively. Although pure mineral separates were not analysed in the case of C1, the familiar pattern in which melilite plots near *D* and spinel and fassaite at much lower $\delta^{17}O$ values, in this case near *B*, is observed for EK 1-4-1. R. N. Clayton & Mayeda (1977) suggested that the original bulk isotopic compositions of EK 1-4-1 and C1 were at *B* and *C*, respectively, lying along a mass-fractionation line *ABC* with the original bulk composition of usual Allende inclusions. They called upon a kinetic effect in the gas phase prior to condensation of these two inclusions to enrich them in the heavy oxygen isotopes relative to other inclusions and upon the same exchange process that modified the isotopic composition of melilite in the latter to do so in these inclusions also. It is also possible to replace the mass-fractionation step with a mixing process in which a second isotopically anomalous nucleosynthetic component lying beyond *C* is added to the inclusions (Lee et al 1978). This is considered unlikely, however, as the mixing line so generated would fortuitously have to have the slope of a mass-fractionation line, 0.5.

C1 (Lee et al 1976) and EK 1-4-1 (Wasserburg et al 1977) have also undergone large mass-dependent enrichments in their heavy Mg isotopes. The ratio of the Mg effect in C1 to that in EK 1-4-1 is the same as the inferred ratio of their original oxygen effects, but the latter are about one-half as great. As pointed out by R. N. Clayton & Mayeda (1977), coexistence of mass-fractionated oxygen and Mg is expected from their model, though the relative magnitudes may be a problem. Within each inclusion, all phases have the same Mg isotopic composition (Wasserburg et al 1977), except for C1 anorthite (Esat et al 1978). Evidently, in contrast to oxygen, later isotopic exchange with an external reservoir introduced insignificant Mg relative to that originally present, except for anorthite. Si is much more severely mass-fractionated toward the heavy isotopes in EK 1-4-1 (Yeh & Epstein 1978) and C1 (R. N. Clayton et al 1978) than in normal inclusions, C1 again being more fractionated than EK 1-4-1. In contrast, very small Ca mass-fractionations, opposite in sign to those for Mg, O, and Si, were found in these inclusions (Lee et al 1978).

These so-called "FUN" inclusions are remarkable in another way: they are the only ones to show clear-cut nuclear anomalies in elements other than oxygen: Ca (Lee et al 1978), Nd in EK 1-4-1 only and Ba (McCulloch & Wasserburg 1978a), Sm (Lugmair et al 1978, McCulloch & Wasserburg 1978b), Sr (Papanastassiou & Wasserburg 1978), Si (R. N. Clayton et al 1978, Yeh & Epstein 1978), and possibly Mg (Wasserburg et al 1977). Furthermore, in surprising contrast to oxygen, for each of these inclusions, the isotopic compositions of these elements are the same within analytical uncertainty in all mineral separates and splits analysed, except for Mg in C1 anorthite.

Throughout this section, an alternative working hypothesis has been entertained that the inclusions were never molten and different phases within them obtained their different isotopic compositions by condensing in isotopically distinct locales. In the case of FUN inclusions, this model would have fassaite and spinel condense from a gas phase related to that which spawned their counterparts in other inclusions by a mass-fractionation process. The melilites in FUN inclusions would have to condense from a separate reservoir of more nearly normal oxygen isotopes, just as the model requires for other inclusions. These cannot be the same melilites as are in the usual inclusions, however, because they have the same nuclear anomalies in other elements as the fassaites and spinels in FUN inclusions. The hypothesis thus requires two isotopically distinct oxygen reservoirs within each of three reservoirs (one for usual inclusions, one for EK 1-4-1, and one for C1), which are isotopically different in other elements. This logical extension of the alternative hypothesis seems very improbable and implies that the hypothesis may have to be dropped as a possible explanation for even the usual Allende inclusions.

A third isotopically unusual inclusion is known, HAL. So far, only its Ca and Mg isotopic compositions have been reported (Lee et al 1979). Mass-dependent heavy Ca isotope enrichment has occurred and small nuclear effects are also present. If ^{26}Al was once present in HAL, its abundance was, as in C1 (Esat et al 1978), far less than in usual Allende inclusions. Esat et al (1979) reported that another inclusion, Egg-3, may be of the FUN-type, exhibiting mass-fractionated Mg, but also ^{26}Mg excesses.

One of the most fascinating things about C1 is that its gross mineralogical composition and textural characteristics are just like those of isotopically normal Type B inclusions (Gray et al 1973), indicating that the same physico-chemical processes leading to inclusion formation occurred in the isotopically distinct regions in which C1 and normal inclusions originated. The same can probably be said for EK 1-4-1, although here the state of the remaining sample makes detailed petrographic comparison difficult. HAL, however, is not only isotopically different from C1 and EK 1-4-1, but it is also mineralogically exotic (Allen et al 1980), classifiable neither as Type A nor Type B. HAL's mineralogical and isotopic differences stem either from alteration after condensation from an isotopically normal reservoir (Lee et al 1979) or from differences in physico-chemical conditions during condensation from an isotopically distinct region (Allen et al 1980). REE patterns of C1 (Conard 1976) and HAL (Tanaka et al 1979) have pronounced negative Ce anomalies. These can only be produced during condensation if the gas phase is extremely oxygen-rich, such as in supernova ejecta from which H has previously been burned out (Boynton 1978c, Tanaka et al 1979). Calculations show that Ce becomes very volatile under these conditions, a prediction apparently confirmed by volatilization experiments in oxidizing gases (Nagasawa & Onuma 1979). REE in these two inclusions condensed in a chemically distinct environment from those in all other inclusions.

Lee (1979) showed that the simplest interpretation of nuclear effects in Ba, Nd, and Sm in FUN inclusions is that C1 incorporated an excess of nuclides synthesized in the *p*-process and EK 1-4-1 has excesses of both *p*- and *r*-process nuclides. Although it is possible that both nucleosynthetic processes can take place in different zones of the same supernova, the distribution of nuclear effects in C1 and EK 1-4-1 shows that matter from each of these processes was carried in a separate component. It is not known whether the carriers of *p*- and *r*-process–enriched heavy elements were solid or gaseous prior to mixing with mass-fractionated low-Z elements. Nor is it understood why nuclear anomalies in heavy elements occur only in inclusions with mass-fractionated low-Z elements. One attempt at explaining this (D. D. Clayton 1980) reinterprets the line *ABC* in Figure 5 as being due to mixing of isotopically anomalous interstellar dust with light-isotope enriched gas sputtered from it, but fails to explain many details of the observations, such as the very tight adherence of many inclusions to the line *AD*.

AGE INFORMATION

Information about the age of the inclusions has been obtained from the U-Pb, Rb-Sr, and K-Ar systems and from decay products of extinct nuclides.

Rb-Sr System

Gray et al (1973), Nyquist et al (1973), Wetherill et al (1973), Nagasawa & Jahn (1976), and Tatsumoto et al (1976) observed extremely low $^{87}Sr/^{86}Sr$ ratios in inclusions with very low Rb/Sr ratios. The lowest $^{87}Sr/^{86}Sr$ ratio measured by Gray et al (1973), 0.69877, is lower than all other solar system materials and identifies the inclusions as the oldest known solids in the solar system. Data from a variety of materials in Allende (Gray et al 1973, Tatsumoto et al 1976) and from mineral separates from individual inclusions (Nagasawa & Jahn 1976) fail to define isochrons, leading to the conclusion that element redistribution has occurred, perhaps recently. The conclusions of these workers are unaffected by the findings of Patchett (1979), since Sr isotopic compositions were corrected for mass-fractionation in the Rb-Sr studies.

U-Th-Pb System

Pb isotopic compositions of a variety of materials separated from Allende define straight lines on $^{207}Pb/^{204}Pb$–$^{206}Pb/^{204}Pb$ diagrams, yielding model ages of 4.57×10^9 yr (Chen & Tilton 1976) and 4.55×10^9 yr (Tatsumoto et al 1976). Furthermore, the data define chords on concordia diagrams, suggesting initial ages of 4.57×10^9 and 4.55×10^9 yr and more recent disturbance at 0.28×10^9 yr and 0.11×10^9 yr, respectively. Although these results could be affected by U isotopic variations, the $^{238}U/^{235}U$ ratio is near normal in one of Tatsumoto et al's (1976) inclusions that departs significantly from their $^{207}Pb/^{206}Pb$ isochron (Tatsumoto & Shimamura 1979). In an ion microprobe study of Th-, U-rich micro-phases in Type B inclusions, Lovering et al (1976) detected no ^{204}Pb and used observed $^{207}Pb/^{206}Pb$ ratios to obtain an average age of 4.60 and a range of 4.53–4.65×10^9 yr for six different grains, presumably by assuming the terrestrial $^{238}U/^{235}U$ ratio. On some of the same grains and others, Wark & Sewell (1979) calculated a mean age of 4.56 and a range of 4.32–4.84×10^9 yr simply from the ratio of U to Th to Pb, assuming no common Pb and the terrestrial U isotopic composition. Perhaps some of the scatter in these ages is due to the U isotopic variations known to be present in bulk Type B inclusions (Tatsumoto & Shimamura 1979).

K-Ar System

Although one olivine-bearing coarse-grained inclusion is known to have an $^{40}Ar/^{39}Ar$ age equal to the canonical age of the solar system (Dominik et al 1978), three others have apparent ages greater than 4.6×10^9 yr, one studied by Jessberger et al (1977), 4.9×10^9 yr, and two by Jessberger & Dominik (1979), 4.90 and 4.98×10^9 yr. As the technique assumes the samples have normal $^{39}K/^{40}K$ ratios, one possible explanation for high ages is that the inclusions contain isotopically anomalous K. Stegmann & Begemann (1979) found, however, that the K isotopic composition in one of Jessberger & Dominik's (1979) inclusions with a high age was indistinguishable from that of terrestrial K. The reason for the high ages is unknown. One possibility, suggested by D. D. Clayton (1977b), is that the inclusions contain ^{40}Ar produced by decay of interstellar ^{40}K prior to formation of the solar system. Why presolar ^{40}Ar was not degassed during sintering and fusion of the interstellar grains to form the inclusions is, however, a severe difficulty for this model. D. D. Clayton (1975) also proposed that ^{41}K excesses, from decay of now-extinct ^{41}Ca, should be found in materials derived from presolar grains and suggested that his model of forming inclusions by partial fusion of presolar condensates would be in great difficulty if ^{41}K excesses are not observed in Allende inclusions (D. D. Clayton 1977b). The latter appears to be the case, given the results of Begemann & Stegmann (1976), Birck et al (1977), and Stegmann & Begemann (1979).

Extinct Nuclides

In addition to ^{26}Al, other extinct nuclides have left their decay products in the inclusions. ^{136}Xe from fission of ^{244}Pu was measured in a Type B inclusion by Drozd et al (1977), in a melilite separate by Marti et al (1977), and in EK 1-4-1 by Papanastassiou et al (1978). The ratio of ^{244}Pu to other refractories is similar to that in other meteorites. The presence of fission tracks (Shirck 1974, Drozd et al 1977, Podosek et al 1977) shows that ^{244}Pu decayed after the major mineral phases formed. Scheinin et al (1976) and Lugmair et al (1978) claimed to have found a small excess of ^{142}Nd in fassaite from a Type B and attributed it to decay of extinct ^{146}Sm. Excess ^{129}Xe from ^{129}I decay was found by Scheinin et al (1976) in the same inclusion. Papanastassiou et al (1978) reported excess ^{129}Xe in EK 1-4-1 and noted that the inferred initial $^{129}I/^{127}I$ ratio is approximately that of other chondrites. Because I is probably in Cl-bearing sodalite, an alteration product, the excess ^{129}Xe implies that the alteration process occurred very shortly after inclusion formation.

CONCLUSION

Before the fall of Allende, the prevailing point of view was that the solar nebula was a well-stirred, homogeneous, quiescent gas cloud, hot enough in its center that preexisting interstellar dust grains were completely evaporated, with condensation occurring later, during a period of slow, monotonic cooling. A mere decade of research on Allende has changed this picture so dramatically that it will never be the same. In coarse-grained inclusions alone, only 5% of the meteorite, we have uncovered remarkably well-preserved evidence for a supernova explosion that occurred just before condensation, for incompletely homogenized material from several nucleosynthetic sources, and for solar nebular regions of different chemical and isotopic composition. It is doubtful that we would have come this far in such a short time had it not been for the large size of Allende, for major technological advances in analytical techniques and instrumentation, and for the excitement about and intense interest in the problem of solar system origin shared by an international group of scientists with diverse approaches to the subject.

ACKNOWLEDGMENTS

In preparing this review, I have benefitted from discussions with R. N. Clayton, A. M. Davis, G. J. MacPherson, and T. Tanaka. Special thanks go to M. Bar Matthews and G. J. MacPherson for bibliographic and editorial assistance and to M. Bowie for typing the manuscript. This work was supported by the National Aeronautics and Space Administration through Grant NGR 14-001-249 and by the Alfred P. Sloan Research Foundation.

Literature Cited

Allen, J. M., Grossman, L. 1978. Solar nebula condensation: Implications from Allende inclusion mineralogy. *Meteoritics* 13:383–84 (Abstr.)

Allen, J. M., Grossman, L., Davis, A. M., Hutcheon, I. D. 1978. Mineralogy, textures and mode of formation of a hibonite-bearing Allende inclusion. In *Proc. Lunar Planet. Sci. Conf., 9th*, pp. 1209–33. Pergamon. 3973 pp.

Allen, J. M., Grossman, L., Lee, T., Wasserburg, G. J. 1980. Mineralogy and petrography of HAL, an isotopically unusual Allende inclusion. *Geochim. Cosmochim. Acta*. In press

Arrhenius, G. 1972. Chemical effects in plasma condensation. In *Proc. Nobel Symp. 21*, pp. 117–29. Stockholm: Almqvist & Wiksell. 389 pp.

Arrhenius, G. 1978. Chemical aspects of the formation of the solar system. In *The Origin of the Solar System*, ed. S. F. Dermott, pp. 521–81. New York: Wiley. 668 pp.

Arrhenius, G., Alfvén, H. 1971. Fractionation and condensation in space. *Earth Planet. Sci. Lett.* 10:253–67

Begemann, F., Stegmann, W. 1976. Implications from the absence of a ^{41}K anomaly in an Allende inclusion. *Nature* 259: 549–50

Birck, J.-L., Lorin, J.-C., Allègre, C. 1977. Potassium isotopic determination in some meteoritic and lunar samples: Evidence for irradiation effects. *Meteoritics* 12:179–80 (Abstr.)

Blander, M., Fuchs, L. H. 1975. Calcium-aluminum-rich inclusions in the Allende meteorite: Evidence for a liquid origin. *Geochim. Cosmochim. Acta* 39:1605–19

Blander, M., Fuchs, L. H., Horowitz, C., Land, R. 1980. Primordial refractory metal particles in the Allende meteorite. *Geochim. Cosmochim. Acta* 44:217–24

Boynton, W. V. 1975a. Fractionation in the solar nebula: Condensation of yttrium and the rare earth elements. *Geochim. Cosmochim. Acta* 39:569–84

Boynton, W. V. 1975b. The application of rare-earth elements to infer gas-grain temperature differentials in the solar nebula. *Meteoritics* 10:369–71 (Abstr.)

Boynton, W. V. 1978a. Fractionation in the solar nebula, II. Condensation of Th, U, Pu and Cm. *Earth Planet. Sci. Lett.* 40: 63–70

Boynton, W. V. 1978b. The chaotic solar nebula: Evidence for episodic condensation in several distinct zones. In *Protostars and Planets*, ed. T. Gehrels, pp. 427–38. Tucson: Univ. Ariz. Press. 756 pp.

Boynton, W. V. 1978c. Rare-earth elements as indicators of supernova condensation. In *Lunar Planet. Sci. IX*, pp. 120–22. Houston: Lunar Planet. Inst. 1333 pp.

Bradley, J. G., Huneke, J. C., Wasserburg, G. J. 1978. Ion microprobe evidence for the presence of excess ^{26}Mg in an Allende anorthite crystal. *J. Geophys. Res.* 83: 244–54

Burns, R. G., Huggins, F. E. 1973. Visible-region absorption spectra of a Ti^{3+} fassaite from the Allende meteorite. *Am. Mineral.* 58:955–61

Butler, B. C. M. 1977. Al-rich pyroxene and melilite in a blast-furnace slag and a comparison with the Allende meteorite. *Mineral. Mag.* 41:493–99

Cameron, A. G. W. 1966. The accumulation of chondritic material. *Earth Planet. Sci. Lett.* 1:93–96

Cameron, A. G. W. 1973. Accumulation processes in the primitive solar nebula. *Icarus* 18:407–50

Cameron, A. G. W., Pine, M. R. 1973. Numerical models of the primitive solar nebula. *Icarus* 18:377–406

Cameron, A. G. W., Truran, J. W. 1977. The supernova trigger for formation of the solar system. *Icarus* 30:447–61

Chen, J. H., Tilton, G. R. 1976. Isotopic lead investigations on the Allende carbonaceous chondrite. *Geochim. Cosmochim. Acta* 40:635–43

Chou, C.-L., Baedecker, P. A., Wasson, J. T. 1976. Allende inclusions: Volatile-element distribution and evidence for incomplete volatilization of presolar solids. *Geochim. Cosmochim. Acta* 40:85–94

Christophe Michel-Lévy, M. 1968. Un chondre exceptionnel dans la météorite de Vigarano. *Bull. Soc. fr. Minéral. Cristallogr.* 91:212–14

Clarke, R. S. Jr., Jarosewich, E., Mason, B., Nelen, J., Gómez, M., Hyde, J. R. 1970. The Allende Mexico meteorite shower. *Smithsonian Contrib. Earth Sci.* 5:1–53

Clayton, D. D. 1975. ^{22}Na, Ne-E, extinct radioactive anomalies and unsupported ^{40}Ar. *Nature* 257:36–37

Clayton, D. D. 1977a. Solar system isotopic anomalies: Supernova neighbor or presolar carriers? *Icarus* 32:255–69

Clayton, D. D. 1977b. Interstellar potassium and argon. *Earth Planet. Sci. Lett.* 36:381–90

Clayton, D. D. 1980. Origin of Ca-Al-rich inclusions: II Sputtering and collisions in the three phase interstellar medium. Preprint. Submitted to *Astrophys. J.*

Clayton, D. D., Dwek, E., Woosley, S. E. 1977. Isotopic anomalies and proton irradiation in the early solar system. *Astrophys. J.* 214:300–15

Clayton, R. N. 1978. Isotopic anomalies in the early solar system. *Ann. Rev. Nucl. Part. Sci.* 28:501–22

Clayton, R. N., Mayeda, T. K. 1977. Correlated oxygen and magnesium isotopic anomalies in Allende inclusions: I. Oxygen. *Geophys. Res. Lett.* 4:295–98

Clayton, R. N., Grossman, L., Mayeda, T. K. 1973. A component of primitive nuclear composition in carbonaceous meteorites. *Science* 182:485–88

Clayton, R. N., Onuma, N., Grossman, L., Mayeda, T. K. 1977. Distribution of the pre-solar component in Allende and other carbonaceous chondrites. *Earth Planet. Sci. Lett.* 34:209–24

Clayton, R. N., Mayeda, T. K., Epstein, S. 1978. Isotopic fractionation of silicon in Allende inclusions. In *Proc. Lunar Planet. Sci. Conf., 9th,* pp. 1267–78. Pergamon. 3973 pp.

Conard, R. 1976. *A study of the chemical composition of Ca-Al-rich inclusions from the Allende meteorite.* M.S. thesis. Oregon State Univ., Corvallis. 129 pp.

Consolmagno, G. J., Cameron, A. G. W. 1980. The origin of the FUN anomalies and the high temperature inclusions in the Allende meteorite. *The Moon and the Planets.* In press

Davis, A. M., Grossman, L. 1979. Condensation and fractionation of rare earths in the solar nebula. *Geochim. Cosmochim. Acta* 43:1611–32

Davis, A. M., Allen, J. M., Grossman, L. 1978a. Major and trace element characteristics of coarse-grained inclusions in Allende. *EOS, Trans. Am. Geophys. Union* 59:314 (Abstr.)

Davis, A. M., Grossman, L., Allen, J. M. 1978b. Major and trace element chemistry of separated fragments from a hibonite-bearing Allende inclusion. In *Proc. Lunar Planet. Sci. Conf., 9th,* pp. 1235–47. Pergamon. 3973 pp.

Deer, W. A., Howie, R. A., Zussman, J. 1978. *Rock Forming Minerals,* Vol. 2A. New York: Wiley. 668 pp. 2nd ed.

Dominik, B., Jessberger, E. K., Staudacher, Th., Nagel, K., El Goresy, A. 1978. A new type of white inclusion in Allende: Petrography, mineral chemistry, ^{40}Ar-^{39}Ar ages, and genetic implications. In *Proc. Lunar Planet. Sci. Conf., 9th,* pp. 1249–66. Pergamon. 3973 pp.

Dowty, E., Clark, J. R. 1973a. Crystal structure refinement and optical properties of a Ti^{3+} fassaite from the Allende meteorite. *Am. Mineral.* 58:230–42

Dowty, E., Clark, J. R. 1973b. Crystal structure refinement and optical properties of a Ti^{3+} fassaite from the Allende meteorite: Reply. *Am. Mineral.* 58:962–64

Drozd, R. J., Morgan, C. J., Podosek, F. A., Poupeau, G., Shirck, J. R., Taylor, G. J. 1977. ^{244}Pu in the early solar system? *Astrophys. J.* 212:567–80

El Goresy, A., Nagel, K., Dominik, B., Ramdohr, P. 1977a. Fremdlinge: Potential presolar material in Ca-Al-rich inclusions of Allende. *Meteoritics* 12:215 (Abstr.)

El Goresy, A., Nagel, K., Ramdohr, P. 1977b. Type A Ca-, Al-rich inclusions in Allende meteorite: Origin of the perovskite-fassaite symplectite around rhönite and chemistry and assemblages of the refractory metals (Mo, W) and platinum metals (Ru, Os, Ir, Re, Rh, Pt). *Meteoritics* 12:216 (Abstr.)

El Goresy, A., Nagel, K., Ramdohr, P. 1978a. The Allende meteorite: Fremdlinge and their noble relatives. In *Lunar Planet. Sci. IX,* pp. 282–84. Houston: Lunar Planet. Inst. 1333 pp.

El Goresy, A., Nagel, K., Ramdohr, P. 1978b.

Fremdlinge and their noble relatives. In *Proc. Lunar Planet. Sci. Conf., 9th*, pp. 1279–1303. Pergamon. 3973 pp.

El Goresy, A., Nagel, K., Dominik, B., Ramdohr, P., Mason, B. 1979a. Ru-bearing phosphate-molybdates and other oxidized phases in Fremdlinge in Allende inclusions. *Meteoritics* 14:390–91

El Goresy, A., Nagel, K., Ramdohr, P. 1979b. Spinel framboids in Allende inclusions: A possible sequential marker in the early history of the solar system. In *Proc. Lunar Planet. Sci. Conf., 10th*. In press

Epstein, S., Yeh, H.-W. 1977. The $\delta^{18}O$, $\delta^{17}O$, $\delta^{30}Si$ and $\delta^{29}Si$ of oxygen and silicon in stony meteorites and Allende inclusions. In *Lunar Planet. Sci. VIII*, pp. 287–89. Houston: Lunar Sci. Inst. 1082 pp.

Esat, T. M., Lee, T., Papanastassiou, D. A., Wasserburg, G. J. 1978. Search for ^{26}Al effects in the Allende FUN inclusion C1. *Geophys. Res. Lett.* 5:807–10

Esat, T. M., Papanastassiou, D. A., Wasserburg, G. J. 1979. Trials and tribulations of ^{26}Al: Evidence for disturbed systems. In *Lunar Planet. Sci. X*, pp. 361–63. Houston: Lunar Planet. Inst. 1426 pp.

Frost, M. J., Symes, R. F. 1970. A zoned perovskite-bearing chondrule from the Lancé meteorite. *Mineral. Mag.* 37:724–26

Fruland, R. M., King, E. A., McKay, D. S. 1978. Allende dark inclusions. In *Proc. Lunar Planet. Sci. Conf., 9th*, pp. 1305–29. Pergamon. 3973 pp.

Fuchs, L. H. 1971. Occurrence of wollastonite, rhönite, and andradite in the Allende meteorite. *Am. Mineral.* 56:2053–68

Fuchs, L. H. 1974. Grossular in the Allende (Type III carbonaceous) meteorite. *Meteoritics* 9:11–18

Fuchs, L. H. 1978. The mineralogy of a rhönite-bearing calcium aluminum rich inclusion in the Allende meteorite. *Meteoritics* 13:73–88

Fuchs, L. H., Blander, M. 1977. Molybdenite in calcium-aluminum-rich inclusions in the Allende meteorite. *Geochim. Cosmochim. Acta* 41:1170–74

Ganapathy, R., Grossman, L. 1976. The case for an unfractionated $^{244}Pu/^{238}U$ ratio in high-temperature condensates. *Earth Planet. Sci. Lett.* 31:386–92

Gast, P. W., Hubbard, N. J., Wiesmann, H. 1970. Chemical composition and petrogenesis of basalts from Tranquillity Base. In *Proc. Apollo 11 Lunar Sci. Conf.*, pp. 1143–63. Pergamon. 1936 pp.

Graham, A. L. 1975. The Allende meteorite: Chondrule composition and the early history of the solar system. *Smithsonian Contrib. Earth Sci.* 14:35–40

Gray, C. M., Compston, W. 1974. Excess ^{26}Mg in the Allende meteorite. *Nature* 251:495–97

Gray, C. M., Papanastassiou, D. A., Wasserburg, G. J. 1973. The identification of early condensates from the solar nebula. *Icarus* 20:213–39

Grossman, L. 1972. Condensation in the primitive solar nebula. *Geochim. Cosmochim. Acta* 36:597–619

Grossman, L. 1973. Refractory trace elements in Ca-Al-rich inclusions in the Allende meteorite. *Geochim. Cosmochim. Acta* 37:1119–40

Grossman, L. 1974. Mineral chemistry of Ca-Al-rich inclusions in the Allende meteorite. *EOS, Trans. Am. Geophys. Union* 55:332 (Abstr.)

Grossman, L. 1975a. Petrography and mineral chemistry of Ca-rich inclusions in the Allende meteorite. *Geochim. Cosmochim. Acta* 39:433–54

Grossman, L. 1975b. Dust in the solar nebula. In *The Dusty Universe*, ed. G. B. Field, A. G. W. Cameron, pp. 268–92. New York: Neale Watson. 323 pp.

Grossman, L. 1977a. Chemical fractionation in the solar nebula. In *The Soviet-American Conference on Cosmochemistry of the Moon and Planets*, ed. J. H. Pomeroy, N. J. Hubbard. *NASA SP-370*, pp. 787–96. Washington, DC: US Gov't. Print. Off. 929 pp.

Grossman, L. 1977b. High-temperature condensates in carbonaceous chondrites. In *Comets, Asteroids, Meteorites—Interrelations, Evolution and Origins*, ed. A. H. Delsemme, pp. 507–16. Toledo, Ohio: Univ. Toledo. 587 pp.

Grossman, L., Clark, S. P. Jr. 1973. High-temperature condensates in chondrites and the environment in which they formed. *Geochim. Cosmochim. Acta* 37: 635–49

Grossman, L., Ganapathy, R. 1975. Volatile elements in Allende inclusions. In *Proc. Lunar Sci. Conf., 6th*, pp. 1729–36. Pergamon. 3637 pp.

Grossman, L., Ganapathy, R. 1976a. Trace elements in the Allende meteorite—II. Fine-grained, Ca-rich inclusions. *Geochim. Cosmochim. Acta* 40:967–77

Grossman, L., Ganapathy, R. 1976b. Trace elements in the Allende meteorite—I. Coarse-grained, Ca-rich inclusions. *Geochim. Cosmochim. Acta* 40:331–44

Grossman, L., Larimer, J. W. 1974. Early chemical history of the solar system. *Rev. Geophys. Space Phys.* 12:71–101

Grossman, L., Olsen, E. 1974. Origin of the high-temperature fraction of C2 chondrites. *Geochim. Cosmochim. Acta* 38:173–87

Grossman, L., Steele, I. M. 1976. Amoeboid olivine aggregates in the Allende meteorite. *Geochim. Cosmochim. Acta* 40: 149–55

Grossman, L., Fruland, R. M., McKay, D. S. 1975. Scanning electron microscopy of a pink inclusion from the Allende meteorite. *Geophys. Res. Lett.* 2: 37–40

Grossman, L., Ganapathy, R., Davis, A. M. 1977. Trace elements in the Allende meteorite—III. Coarse-grained inclusions revisited. *Geochim. Cosmochim. Acta* 41: 1647–64

Grossman, L., Ganapathy, R., Methot, R. L., Davis, A. M. 1979a. Trace elements in the Allende meteorite—IV. Amoeboid olivine aggregates. *Geochim. Cosmochim. Acta* 43: 817–29

Grossman, L., Olsen, E., Lattimer, J. M. 1979b. Silicon in carbonaceous chondrite metal: Relic of high-temperature condensation. *Science* 206: 449–51

Haggerty, S. E. 1976. Titanium cosmothermometry and rhonite decomposition in Allende. *Meteoritics* 11: 294 (Abstr.)

Haggerty, S. E. 1977. Refinement of the Ti-cosmothermometer in the Allende meteorite and the significance of a new mineral, $R^{2+}Ti_3O_7$, in association with armalcolite. In *Lunar Sci. VIII*, pp. 389–91. Houston: Lunar Sci. Inst. 1082 pp.

Haggerty, S. E. 1978a. The Allende meteorite: Solid solution characteristics and the significance of a new titanate mineral series in association with armalcolite. In *Proc. Lunar Planet. Sci. Conf., 9th*, pp. 1331–44. Pergamon. 3973 pp.

Haggerty, S. E. 1978b. The Allende meteorite: Evidence for a new cosmothermometer based on Ti^{3+}/Ti^{4+}. *Nature* 276: 221–25

Haggerty, S. E., Merkel, G. A. 1976. Primordial condensation in the early solar system: Allende mineral chemistry. In *Lunar Sci. VII*, pp. 342–44. Houston: Lunar Sci. Inst. 1008 pp.

Hashimoto, A., Kumazawa, M., Onuma, N. 1979. Evaporation metamorphism of primitive dust material in the early solar nebula. *Earth Planet. Sci. Lett.* 43: 13–21

Heydegger, H. R., Foster, J. J., Compston, W. 1979. Evidence of a new isotopic anomaly from titanium isotopic ratios in meteoric materials. *Nature* 278: 704–7

Heymann, D., Dziczkaniec, M. 1976. Early irradiation of matter in the solar system: Mg (p, n) scheme. *Science* 191: 79–81

Hutcheon, I. D. 1977. Micro-mineralogy of calcium-aluminum-rich inclusions from Allende. In *Lunar Sci. VIII*, pp. 487–89. Houston: Lunar Sci. Inst. 1082 pp.

Hutcheon, I. D., Steele, I. M., Solberg, T. N., Clayton, R. N., Smith, J. V. 1977. Ion microprobe studies of lithium in Allende inclusions. *Meteoritics* 12: 261 (Abstr.)

Hutcheon, I. D., Steele, I. M., Clayton, R. N., Smith, J. V. 1978a. An ion microprobe study of Mg isotopes in two Allende inclusions. *Meteoritics* 13: 498–99 (Abstr.)

Hutcheon, I. D., Steele, I. M., Smith, J. V., Clayton, R. N. 1978b. Ion microprobe, electron microprobe and cathodoluminescence data for Allende inclusions with emphasis on plagioclase chemistry. In *Proc. Lunar Planet. Sci. Conf., 9th*, pp. 1345–68. Pergamon. 3973 pp.

Hutcheon, I. D., MacPherson, G. J., Steele, I. M., Grossman, L. 1979. A petrographic and ion probe isotopic study of Type A coarse-grained inclusions. *Meteoritics* 14: 427

Jarosewich, E., Clarke, R. S. Jr. 1980. The Allende meteorite reference sample. *Smithsonian Contrib. Earth Sci.* In press

Jessberger, E. K., Dominik, B. 1979. Gerontology of the Allende meteorite. *Nature* 277: 554–56

Jessberger, E. K., Staudacher, Th., Dominik, B., Herzog, G. F. 1977. ^{40}Ar-^{39}Ar dating of the Pueblito de Allende meteorite. *Meteoritics* 12: 266–69 (Abstr.)

Keil, K., Fuchs, L. H. 1971. Hibonite $[Ca_2(Al, Ti)_{24}O_{38}]$ from the Leoville and Allende chondritic meteorites. *Earth Planet. Sci. Lett.* 12: 184–90

Kieffer, S. W. 1975. Droplet chondrules. *Science* 189: 333–40

King, E. A. Jr., Schonfeld, E., Richardson, K. A., Eldridge, J. S. 1969. Meteorite fall at Pueblito de Allende, Chihuahua, Mexico: Preliminary information. *Science* 163: 928–29

Kurat, G. 1970. Zur Genese der Ca-Al-reichen Einschlüsse im Chondriten von Lancé. *Earth Planet. Sci. Lett.* 9: 225–31

Kurat, G. 1975. Der kohlige Chondrit Lancé: Eine petrologische Analyse der komplexen Genese eines Chondriten. *Tschermaks Min. Petr. Mitt.* 22: 38–78

Kurat, G., Hoinkes, G., Fredriksson, K. 1975. Zoned Ca-Al-rich chondrule in Bali: New evidence against the primordial condensation model. *Earth Planet. Sci. Lett.* 26: 140–44

Larimer, J. W. 1967. Chemical fractionations in meteorites—I. Condensation of the elements. *Geochim. Cosmochim. Acta* 31: 1215–38

Larimer, J. W., Anders, E. 1970. Chemical fractionations in meteorites—III. Major element fractionations in chondrites. *Geochim. Cosmochim. Acta* 34: 367–87

Lattimer, J. M., Grossman, L. 1978. Chemical condensation sequences in supernova ejecta. *The Moon and the Planets* 19:169–84

Lattimer, J. M., Schramm, D. N., Grossman, L. 1978. Condensation in supernova ejecta and isotopic anomalies in meteorites. *Astrophys. J.* 219:230–49

Lee, T. 1978. A local proton irradiation model for isotopic anomalies in the solar system. *Astrophys. J.* 224:217–26

Lee, T. 1979. New isotopic clues to solar system formation. *Rev. Geophys. Space Phys.* 17:1591–1611

Lee, T., Papanastassiou, D. A. 1974. Mg isotopic anomalies in the Allende meteorite and correlation with O and Sr effects. *Geophys. Res. Lett.* 1:225–28

Lee, T., Papanastassiou, D. A., Wasserburg, G. J. 1976. Demonstration of ^{26}Mg excess in Allende and evidence for ^{26}Al. *Geophys. Res. Lett.* 2:109–12

Lee, T., Papanastassiou, D. A., Wasserburg, G. J. 1977a. Aluminum-26 in the early solar system: Fossil or fuel? *Astrophys. J. Lett.* 211:L107–10

Lee, T., Papanastassiou, D. A., Wasserburg, G. J. 1977b. Mg and Ca isotopic study of individual microscopic crystals from the Allende meteorite by the direct loading technique. *Geochim. Cosmochim. Acta* 41:1473–85

Lee, T., Papanastassiou, D. A., Wasserburg, G. J. 1978. Ca isotopic anomalies in the Allende meteorite. *Astrophys. J. Lett.* 220: L21–25

Lee, T., Russell, W. A., Wasserburg, G. J. 1979. Calcium isotopic anomalies and the lack of aluminum-26 in an unusual Allende inclusion. *Astrophys. J. Lett.* 228:L93-98

Levy, R. L., Wolf, C. J., Grayson, M. A., Gilbert, J., Gelpi, E., Updegrove, W. S., Zlatkis, A., Oro', J. 1970. Organic analysis of the Pueblito de Allende meteorite. *Nature* 227:148–50

Lord, H. C. III. 1965. Molecular equilibria and condensation in a solar nebula and cool stellar atmospheres. *Icarus* 4:279–88

Lorin, J.-C., Christophe Michel-Lévy, M. 1978a. Radiogenic ^{26}Mg fine-scale distribution in Ca-Al inclusions of the Allende and Leoville meteorites. In *Fourth Int. Conf. Geochron. Cosmochron. Isotope Geol., U.S.G.S. Open-File Rep. 78-701*, pp. 257–59. Washington, DC: US Gov't. Print. Off. 476 pp.

Lorin, J.-C., Christophe Michel-Lévy, M. 1978b. Heavy rare metals and magnesium isotopic anomalies studies in Allende and Leoville calcium-aluminum rich inclusions. In *Lunar Planet. Sci. IX*, pp. 660–62. Houston: Lunar Planet. Inst. 1333 pp.

Lorin, J.-C., Shimizu, N., Christophe Michel-Lévy, M., Allègre, C. J. 1977. The Mg isotope anomaly in carbonaceous chondrites: An ion-probe study. *Meteoritics* 12:299–300 (Abstr.)

Lorin, J.-C., Christophe Michel-Lévy, M., Desnoyers, C. 1978. Ophitic Ca-Al inclusions in the Allende and Leoville meteorites: A petrographic and ion-microprobe study. *Meteoritics* 13:537–40 (Abstr.)

Lovering, J. F., Hinthorne, J. R., Conrad, R. L. 1976. Direct ^{207}Pb/^{206}Pb dating by ion microprobe of uranium-thorium-rich phases in Allende calcium-aluminium-rich clasts (CARC's). In *Lunar Planet. Sci. VII*, pp. 504–6. Houston: Lunar Sci. Inst. 1008 pp.

Lovering, J. F., Wark, D. A., Sewell, D. K. B. 1979. Refractory oxide, titanate, niobate and silicate accessory mineralogy of some Type B Ca-Al-rich inclusions in the Allende meteorite. In *Lunar Planet. Sci. X*, pp. 745–47. Houston: Lunar Planet. Inst. 1426 pp.

Lugmair, G. W., Marti, K., Scheinin, N. B. 1978. Incomplete mixing of products from *r*-, *p*-, and *s*-process nucleosynthesis: Sm-Nd systematics in Allende inclusion EK 1-04-1. In *Lunar Planet. Sci. IX*, pp. 672–74. Houston: Lunar Planet. Inst. 1333 pp.

MacPherson, G. J., Grossman, L. 1979. Melted and non-melted coarse-grained Ca-, Al-rich inclusions in Allende. *Meteoritics* 14:479–80

Marti, K., Lugmair, G. W., Scheinin, N. B. 1977. Sm-Nd-Pu systematics in the early solar system. In *Lunar Sci. VIII*, pp. 619–21. Houston: Lunar Sci. Inst. 1082 pp.

Martin, P. M., Mason, B. 1974. Major and trace elements in the Allende meteorite. *Nature* 249:333–34

Marvin, U. B., Wood, J. A., Dickey, J. S., Jr. 1970. Ca-Al rich phases in the Allende meteorite. *Earth Planet. Sci. Lett.* 7:346–50

Mason, B. 1974. Aluminum-titanium-rich pyroxenes, with special reference to the Allende meteorite. *Am. Mineral.* 59:1198–1202

Mason, B. 1975. The Allende meteorite: Cosmochemistry's Rosetta Stone? *Acc. Chem. Res.* 8:217–24

Mason, B., Martin, P. M. 1974. Minor and trace element distribution in melilite and pyroxene from the Allende meteorite. *Earth Planet. Sci. Lett.* 22:141–44

Mason, B., Martin, P. M. 1977. Geochemical differences among components of the Allende meteorite. *Smithsonian Contrib. Earth Sci.* 19:84–95

McCulloch, M. T., Wasserburg, G. J. 1978a. Barium and neodymium isotopic anomalies in the Allende meteorite. *Astrophys. J. Lett.* 220: L15–19

McCulloch, M. T., Wasserburg, G. J. 1978b. More anomalies from the Allende meteorite: Samarium. *Geophys. Res. Lett.* 5: 599–602

McSween, H. Y. Jr. 1977. Chemical and petrographic constraints on the origin of chondrules and inclusions in carbonaceous chondrites. *Geochim. Cosmochim. Acta* 41: 1843–60

Morgan, J. W., Rebagay, T. V., Showalter, D. L., Nadkarni, R. A., Gillum, D. E., McKown, D. M., Ehmann, W. D. 1969. Allende meteorite: Some major and trace element abundances by neutron activation analysis. *Nature* 224: 789–91

Nagasawa, H., Jahn, B.-M. 1976. REE distribution and Rb-Sr systematics of Allende inclusions. In *Lunar Sci. VII*, pp. 591–93. Houston: Lunar Sci. Inst. 1008 pp.

Nagasawa, H., Onuma, N. 1979. High temperature heating of the Allende meteorite II. Fractionation of the rare earth elements. In *Lunar Planet. Sci. X*, pp. 884–86. Houston: Lunar Planet. Inst. 1426 pp.

Nagasawa, H., Schreiber, H. D., Blanchard, D. P. 1976. Partition coefficients of REE and Sc in perovskite, melilite, and spinel and their implications for Allende inclusions. In *Lunar Sci. VII*, pp. 588–90. Houston: Lunar Sci. Inst. 1008 pp.

Nagasawa, H., Blanchard, D. P., Jacobs, J. W., Brannon, J. C., Philpotts, J. A., Onuma, N. 1977. Trace element distribution in mineral separates of the Allende inclusions and their genetic implications. *Geochim. Cosmochim. Acta* 41: 1587–1600

Notsu, K., Onuma, N., Nishada, N., Nagasawa, H. 1978. High temperature heating of the Allende meteorite. *Geochim. Cosmochim. Acta* 42: 903–7

Nyquist, L. E., Hubbard, N. J., Gast, P. W., Bansal, B. M., Wiesmann, H., Jahn, B. 1973. Rb-Sr systematics for chemically defined Apollo 15 and 16 materials. In *Proc. Lunar Sci. Conf., 4th*, pp. 1823–46. Pergamon. 3290 pp.

Onuma, K., Kimura, M. 1978. Study of the system $CaMgSi_2O_6$-$CaFe^{3+}AlSiO_6$-$CaAl_2SiO_6$-$CaTiAl_2O_6$: II. The join $CaMgSi_2O_6$-$CaAl_2SiO_6$-$CaTiAl_2O_6$ and its bearing on Ca-Al-rich inclusions in carbonaceous chondrites. *J. Fac. Sci., Hokkaido Univ.*, ser. IV, 18: 215–36

Onuma, N., Tanaka, T., Masuda, A. 1974. Rare-earth abundances in two mineral separates with distinct oxygen isotopic composition from an Allende inclusion. *Meteoritics* 9: 387–88 (Abstr.)

Osborn, T. W., Warren, R. G., Smith, R. H., Wakita, H., Zellmer, D. L., Schmitt, R. A. 1974. Elemental composition of individual chondrules from carbonaceous chondrites, including Allende. *Geochim. Cosmochim. Acta* 38: 1359–78

Palme, H., Wlotzka, F. 1976. A metal particle from a Ca, Al-rich inclusion from the meteorite Allende, and condensation of refractory siderophile elements. *Earth Planet. Sci. Lett.* 33: 45–60

Papanastassiou, D. A., Wasserburg, G. J. 1978. Strontium isotopic anomalies in the Allende meteorite. *Geophys. Res. Lett.* 5: 595–98

Papanastassiou, D. A., Huneke, J. C., Esat, T. M., Wasserburg, G. J. 1978. Pandora's box of the nuclides. In *Lunar Planet. Sci. IX*, pp. 859–61. Houston: Lunar Planet. Inst. 1333 pp.

Patchett, P. J. 1979. Sr isotopic fractionation in Allende and other solar-system materials. *Meteoritics* 14: 513

Phinney, D., Whitehead, B. 1978. Light elements in minerals of an Allende inclusion. In *Lunar Planet. Sci. IX*, pp. 893–94. Houston: Lunar Planet. Inst. 1333 pp.

Podolak, M., Cameron, A. G. W. 1974. Possible formation of meteoritic chondrules and inclusions in the precollapse Jovian protoplanetary atmosphere. *Icarus* 23: 326–33

Podosek, F. A. 1978. Isotopic structures in solar system materials. *Ann. Rev. Astron. Astrophys.* 16: 293–334

Podosek, F. A., Shirck, J. R., Taylor, G. J. 1977. ^{244}Pu geochemistry and geochronology. In *Lunar Sci. VIII*, pp. 781–83. Houston: Lunar Sci. Inst. 1082 pp.

Rancitelli, L. A., Perkins, R. W., Cooper, J. A., Kaye, J. H., Wogman, N. A. 1969. Radionuclide composition of the Allende meteorite from nondestructive gamma-ray spectrometric analysis. *Science* 166: 1269–71

Reid, A. M., Williams, R. J., Gibson, E. K. Jr., Fredriksson, K. 1974. A refractory glass chondrule in the Vigarano chondrite. *Meteoritics* 9: 35–46

Scheinin, N. B., Lugmair, G. W., Marti, K. 1976. Sm-Nd systematics and evidence for extinct ^{146}Sm in an Allende inclusion. *Meteoritics* 11: 357–58 (Abstr.)

Seitz, M. G., Kushiro, I. 1974. Melting relations of the Allende meteorite. *Science* 183: 954–57

Shirck, J. 1974. Fission tracks in a white inclusion of the Allende chondrite—evidence for ^{244}Pu. *Earth Planet. Sci. Lett.* 23: 308–12

Sonnett, C. P. 1979. On the origin of chondrules. *Geophys. Res. Lett.* 6: 677–80

Steele, I. M., Hutcheon, I. D. 1979. Anatomy of Allende inclusions: Mineralogy and

Mg isotopes in two Ca-Al-rich inclusions. In *Lunar Planet. Sci. X*, pp. 1166–68. Houston: Lunar Planet. Inst. 1426 pp.
Steele, I. M., Smith, J. V., Hutcheon, I. D., Clayton, R. N. 1978. Allende inclusions: cathodoluminescence petrography: anorthite and spinel chemistry: Mg isotopes. In *Lunar Planet. Sci. IX*, pp. 1104–6. Houston: Lunar Planet. Inst. 1333 pp.
Stegmann, W., Begemann, F. 1979. Allende meteorite—old age but normal isotopic composition of potassium. *Nature* 282: 290–91
Tanaka, T., Masuda, A. 1973. Rare-earth elements in matrix, inclusions, and chondrules of the Allende meteorite. *Icarus* 19: 523–30
Tanaka, T., Davis, A. M., Grossman, L., Lattimer, J. M., Allen, J. M., Lee, T., Wasserburg, G. J. 1979. Chemical study of an isotopically-unusual Allende inclusion. In *Lunar Planet. Sci. X*, pp. 1203–5. Houston: Lunar Planet. Inst. 1426 pp.
Tatsumoto, M., Shimamura, T. 1979. Isotopic composition of uranium in Allende and other chondrites. *Meteoritics* 14: 543–44
Tatsumoto, M., Unruh, D. M., Desborough, G. A. 1976. U-Th-Pb and Rb-Sr systematics of Allende and U-Th-Pb systematics of Orgueil. *Geochim. Cosmochim. Acta* 40: 617–34
Taylor, S. R., Mason, B. H. 1978. Chemical characteristics of Ca-Al inclusions in the Allende meteorite. In *Lunar Planet. Sci. IX*, pp. 1158–60. Houston: Lunar Planet. Inst. 1333 pp.
Wänke, H., Baddenhausen, H., Palme, H., Spettel, B. 1974. On the chemistry of the Allende inclusions and their origin as high temperature condensates. *Earth Planet. Sci. Lett.* 23: 1–7
Wark, D. A. 1979. Birth of the presolar nebula: The sequence of condensation revealed in the Allende meteorite. *Astrophys. Space Sci.* 60: 59–75
Wark, D. A., Lovering, J. F. 1976. Refractory/platinum metal grains in Allende calcium-aluminium-rich clasts (CARC's): Possible exotic presolar material? In *Lunar Sci. VII*, pp. 912–14. Houston: Lunar Sci. Inst. 1008 pp.
Wark, D. A., Lovering, J. F. 1977a. Marker events in the early evolution of the solar system: Evidence from rims on Ca-Al-rich inclusions in carbonaceous chondrites. In *Proc. Lunar Sci. Conf., 8th*, pp. 95–112. Pergamon. 3965 pp.
Wark, D. A., Lovering, J. F. 1977b. Marker events in the early evolution of the solar system: Evidence from rims on calcium-aluminium-rich inclusions in carbonaceous chondrites. In *Lunar Sci. VIII*, pp. 976–78. Houston: Lunar Sci. Inst. 1082 pp.
Wark, D. A., Lovering, J. F. 1978a. Classification of Allende coarse-grained Ca-Al-rich inclusions. In *Lunar Planet. Sci. IX*, pp. 1211–13. Houston: Lunar Planet. Inst. 1333 pp.
Wark, D. A., Lovering, J. F. 1978b. Refractory/platinum metals and other opaque phases in Allende Ca-Al-rich inclusions (CAI's). In *Lunar Planet. Sci. IX*, pp. 1214–16. Houston: Lunar Planet. Inst. 1333 pp.
Wark, D. A., Sewell, D. K. B. 1979. Total U-Th-Pb mineral ages of three Ca-Al-rich inclusions in the Allende meteorite. In *Lunar Planet. Sci. X*, pp. 1289–91. Houston: Lunar Planet. Inst. 1426 pp.
Wark, D. A., Wasserburg, G. J., Lovering, J. F. 1979. Structural features of some Allende coarse-grained Ca-Al-rich inclusions: Chondrules within chondrules. In *Lunar Planet. Sci. X*, pp. 1292–94. Houston: Lunar Planet. Inst. 1426 pp.
Wasserburg, G. J., Lee, T., Papanastassiou, D. A. 1977. Correlated O and Mg isotopic anomalies in Allende inclusions: II. Magnesium. *Geophys. Res. Lett.* 4: 299–302
Wetherill, G. W., Mark, R., Lee-Hu, C. 1973. Chondrites: Initial strontium-87/strontium-86 ratios and the early history of the solar system. *Science* 182: 281–83
Whipple, F. L. 1966. Chondrules: Suggestion concerning the origin. *Science* 153: 54–56
Whipple, F. L. 1972. On certain aerodynamic processes for asteroids and comets. In *Proc. Nobel Symp. 21*, ed. A. Elvius, pp. 211–32. Stockholm: Almqvist & Wiksell. 389 pp.
Yagi, K., Onuma, K. 1967. The join $CaMgSi_2O_6$-$CaTiAl_2O_6$ and its bearing on the titanaugites. *J. Fac. Sci., Hokkaido Univ.*, ser. IV, 13: 463–83
Yang, H.-Y. 1976. The join $CaMgSi_2O_6$-$CaAl_2SiO_6$-$CaTiAl_2O_6$ and its bearings on the origin of the Ca- and Al-rich inclusions in the meteorites. *Proc. Geol. Soc. China* 19: 107–26
Yeh, H.-W., Epstein, S. 1978. $^{29}Si/^{28}Si$ and $^{30}Si/^{28}Si$ of meteorites and Allende inclusions. In *Lunar Planet. Sci. IX*, pp. 1289–91. Houston: Lunar Planet. Inst. 1333 pp.

AUTHOR INDEX

(Names appearing in capital letters indicate authors of chapters in this volume.)

C

D

N

CUMULATIVE INDEXES

CONTRIBUTING AUTHORS VOLUMES 4–8

CHAPTER TITLES VOLUMES 4–8

OCEANOGRAPHY, ATMOSPHERIC SCIENCE AND PALEOCLIMATOLOGY

ORDER FORM ANNUAL REVIEWS INC.

Please list on the order blank on the reverse side the volumes you wish to order and whether you wish a standing order (the latest volume sent to you automatically upon publication each year). Volumes not yet published will be shipped in month and year indicated. Prices subject to change without notice. Out of print volumes subject to special order.

NEW TITLES FOR 1980

ANNUAL REVIEW OF PUBLIC HEALTH ISSN 0163-7525
Vol. 1 (avail. May 1980): $17.00 (USA), $17.50 (elsewhere) per copy

ANNUAL REVIEWS REPRINTS: IMMUNOLOGY, 1977–1979 ISBN 0-8243-2502-8
A collection of articles reprinted from recent *Annual Review* series
Avail. Mar. 1980 Soft cover: $12.00 (USA), $12.50 (elsewhere) per copy

SPECIAL PUBLICATIONS

ANNUAL REVIEWS REPRINTS: CELL MEMBRANES, 1975–1977 ISBN 0-8243-2501-X
A collection of articles reprinted from recent *Annual Review* series
Published 1978 Soft cover: $12.00 (USA), $12.50 (elsewhere) per copy

THE EXCITEMENT AND FASCINATION OF SCIENCE, VOLUME 1 ISBN 0-8243-1602-9
A collection of autobiographical and philosophical articles by leading scientists
Published 1965 Clothbound: $6.50 (USA), $7.00 (elsewhere) per copy

THE EXCITEMENT AND FASCINATION OF SCIENCE, VOLUME 2: Reflections by Eminent Scientists
Published 1978 Hard cover: $12.00 (USA), $12.50 (elsewhere) per copy ISBN 0-8243-2601-6
Soft cover: $10.00 (USA), $10.50 (elsewhere) per copy ISBN 0-8243-2602-4

THE HISTORY OF ENTOMOLOGY ISBN 0-8243-2101-7
A special supplement to the *Annual Review of Entomology* series
Published 1973 Clothbound: $10.00 (USA), $10.50 (elsewhere) per copy

ANNUAL REVIEW SERIES

Annual Review of ANTHROPOLOGY ISSN 0084-6570
Vols. 1–8 (1972–79): $17.00 (USA), $17.50 (elsewhere) per copy
Vol. 9 (avail. Oct. 1980): $20.00 (USA), $21.00 (elsewhere) per copy

Annual Review of ASTRONOMY AND ASTROPHYSICS ISSN 0066-4146
Vols. 1–17 (1963–79): $17.00 (USA), $17.50 (elsewhere) per copy
Vol. 18 (avail. Sept. 1980): $20.00 (USA), $21.00 (elsewhere) per copy

Annual Review of BIOCHEMISTRY ISSN 0066-4154
Vols. 28–48 (1959–79): $18.00 (USA), $18.50 (elsewhere) per copy
Vol. 49 (avail. July 1980): $21.00 (USA), $22.00 (elsewhere) per copy

Annual Review of BIOPHYSICS AND BIOENGINEERING* ISSN 0084-6589
Vols. 1–8 (1972–79): $17.00 (USA), $17.50 (elsewhere) per copy
Vol. 9 (avail. June 1980): $17.00 (USA), $17.50 (elsewhere) per copy

Annual Review of EARTH AND PLANETARY SCIENCES* ISSN 0084-6597
Vols. 1–7 (1973–79): $17.00 (USA), $17.50 (elsewhere) per copy
Vol. 8 (avail. May 1980): $17.00 (USA), $17.50 (elsewhere) per copy

Annual Review of ECOLOGY AND SYSTEMATICS ISSN 0066-4162
Vols. 1–10 (1970–79): $17.00 (USA), $17.50 (elsewhere) per copy
Vol. 11 (avail. Nov. 1980): $20.00 (USA), $21.00 (elsewhere) per copy

Annual Review of ENERGY ISSN 0362-1626
Vols. 1–4 (1976–79): $17.00 (USA), $17.50 (elsewhere) per copy
Vol. 5 (avail. Oct. 1980): $20.00 (USA), $21.00 (elsewhere) per copy

Annual Review of ENTOMOLOGY* ISSN 0066-4170
Vols. 7–24 (1962–79): $17.00 (USA), $17.50 (elsewhere) per copy
Vol. 25 (avail. Jan. 1980): $17.00 (USA), $17.50 (elsewhere) per copy

Annual Review of FLUID MECHANICS* ISSN 0066-4189
Vols. 1–11 (1969–79): $17.00 (USA), $17.50 (elsewhere) per copy
Vol. 12 (avail. Jan. 1980): $17.00 (USA), $17.50 (elsewhere) per copy

Annual Review of GENETICS ISSN 0066-4197
Vols. 1–13 (1967–79): $17.00 (USA), $17.50 (elsewhere) per copy
Vol. 14 (avail. Dec. 1980): $20.00 (USA), $21.00 (elsewhere) per copy

Annual Review of MATERIALS SCIENCE ISSN 0084-6600
Vol. 1–9 (1971–79): $17.00 (USA), $17.50 (elsewhere) per copy
Vol. 10 (avail. Aug. 1980): $20.00 (USA), $21.00 (elsewhere) per copy

(continued on reverse)

*Price will be increased to $20.00 (USA), $21.00 (elsewhere) per copy effective with the 1981 volume.

Annual Review of MEDICINE: Selected Topics in the Clinical Sciences* ISSN 0066-4219
Vols. 1–3, 5–15, 17–30 (1950–52, 1954–64, 1966–79): $17.00 (USA), $17.50 (elsewhere) per copy
Vol. 31 (avail. Apr. 1980): $17.00 (USA), $17.50 (elsewhere) per copy

Annual Review of MICROBIOLOGY ISSN 0066-4227
Vols. 15–33 (1961–79): $17.00 (USA), $17.50 (elsewhere) per copy
Vol. 34 (avail. Oct. 1980): $20.00 (USA), $21.00 (elsewhere) per copy

Annual Review of NEUROSCIENCE* ISSN 0147-006X
Vols. 1–2 (1978–79): $17.00 (USA), $17.50 (elsewhere) per copy
Vol. 3 (avail. Mar. 1980): $17.00 (USA), $17.50 (elsewhere) per copy

Annual Review of NUCLEAR AND PARTICLE SCIENCE ISSN 0066-4243
Vols. 10–29 (1960–79): $19.50 (USA), $20.00 (elsewhere) per copy
Vol. 30 (avail. Dec. 1980): $22.50 (USA), $23.50 (elsewhere) per copy

Annual Review of PHARMACOLOGY AND TOXICOLOGY* ISSN 0362-1642
Vols. 1–3, 5–19 (1961–63, 1965–79): $17.00 (USA), $17.50 (elsewhere) per copy
Vol. 20 (avail. Apr. 1980): $17.00 (USA), $17.50 (elsewhere) per copy

Annual Review of PHYSICAL CHEMISTRY ISSN 0066-426X
Vols. 10–21, 23–30 (1959–70, 1972–79): $17.00 (USA), $17.50 (elsewhere) per copy
Vol. 31 (avail. Nov. 1980): $20.00 (USA), $21.00 (elsewhere) per copy

Annual Review of PHYSIOLOGY* ISSN 0066-4278
Vols. 18–41 (1956–79): $17.00 (USA), $17.50 (elsewhere) per copy
Vol. 42 (avail. Mar. 1980): $17.00 (USA), $17.50 (elsewhere) per copy

Annual Review of PHYTOPATHOLOGY ISSN 0066-4286
Vols. 1–17 (1963–79): $17.00 (USA), $17.50 (elsewhere) per copy
Vol. 18 (avail. Sept. 1980): $20.00 (USA), $21.00 (elsewhere) per copy

Annual Review of PLANT PHYSIOLOGY* ISSN 0066-4294
Vols. 10–30 (1959–79): $17.00 (USA), $17.50 (elsewhere) per copy
Vol. 31 (avail. June 1980): $17.00 (USA), $17.50 (elsewhere) per copy

Annual Review of PSYCHOLOGY* ISSN 0066-4308
Vols. 4, 5, 8, 10–30 (1953, 1954, 1957, 1959–79): $17.00 (USA), $17.50 (elsewhere) per copy
Vol. 31 (avail. Feb. 1980): $17.00 (USA), $17.50 (elsewhere) per copy

Annual Review of SOCIOLOGY ISSN 0360-0572
Vols. 1–5 (1975–79): $17.00 (USA), $17.50 (elsewhere) per copy
Vol. 6 (avail. Aug. 1980): $20.00 (USA), $21.00 (elsewhere) per copy

*Price will be increased to $20.00 (USA), $21.00 (elsewhere) per copy effective with the 1981 volume.

To ANNUAL REVIEWS INC., 4139 El Camino Way, Palo Alto, CA 94306 USA
(Tel. 415-493-4400)

Please enter my order for the following publications:
(Standing orders: indicate which volume you wish order to begin with)

______________________, Vol(s). ____ Standing order ____

______________________, Vol(s). ____ Standing order ____

______________________, Vol(s). ____ Standing order ____

______________________, Vol(s). ____ Standing order ____

Amount of remittance enclosed $________ California residents please add applicable sales tax.
Please bill me ☐ Prices subject to change without notice.

SHIP TO (include institutional purchase order if billing address is different)

Name ______________________

Address ______________________

____

______________________ Zip Code ____________

Signed ______________________ Date ____________

☐ Please add my name to your mailing list to receive a free copy of the current Prospectus each year.
☐ Send free brochure listing contents of recent back volumes for *Annual Review(s)* of

____

NEW TITLES

ANNUAL REVIEWS REPRINTS: IMMUNOLOGY, 1977–1979 (to be published March 1980)

Topics in membrane biology, antibodies, cellular immunology, and clinical immunology selected from recent *Annual Review* series by Dr. Irving Weissman of Stanford University

---

ANNUAL REVIEW OF PUBLIC HEALTH (Volume 1 to be published May 1980)

Health and Disease in the United States, *L. A. Fingerhut, R. W. Wilson, and J. J. Feldman*
Quality Assessment and Quality Assurance in Medical Care, *P. J. Sanazaro*
Public Health and Individual Liberty, *D. E. Beauchamp*
Planning and Regulation of Medical Services, *T. Bice*
Public Health Nursing: The Nurse's Role in Community-Based Practice, *R. de Tornyay*
Long-Term Care: Can Our Society Meet the Needs of Its Elderly?, *R. Kane and R. Kane*
Economic Evaluation of Public Health Programs, *L. B. Lave*
The Future of Health Departments: The Governmental Presence, *G. E. Pickett*
To Advance Epidemiology, *R. A. Stallones*
Epidemiology As a Guide to Health Policy, *M. Terris*
The Need for Assessing the Outcome of Common Medical Practices, *J. E. Wennberg*
Biostatistical Implications of Design, Sampling, and Measurement to Health Science Data Analysis, *G. G. Koch, D. B. Gillings, and M. E. Stokes*
Strategies for Reimbursement of Short-Term Hospitals, *C. A. Watts, W. Dowling, and W. Richardson*
Scientific Bases for Identifying Potential Carcinogens and Estimating their Risks, *Interagency Regulatory Liaison Group, Work Group on Risk Assessment*

FROM

NAME____________________

ADDRESS____________________

____

PLACE
STAMP
HERE

ANNUAL REVIEWS INC.

4139 EL CAMINO WAY

PALO ALTO, CALIFORNIA 94306, USA